The Brains and Lives of Cephalopods

The Brains and Lives of Cephalopods

Marion Nixon
and
the late J. Z. Young

OXFORD
UNIVERSITY PRESS

This book has been printed digitally and produced in a standard specification in order to ensure its continuing availability

OXFORD
UNIVERSITY PRESS

Great Clarendon Street, Oxford OX2 6DP

Oxford University Press is a department of the University of Oxford.
It furthers the University's objective of excellence in research, scholarship,
and education by publishing worldwide in

Oxford New York

Auckland Cape Town Dar es Salaam Hong Kong Karachi
Kuala Lumpur Madrid Melbourne Mexico City Nairobi
New Delhi Shanghai Taipei Toronto

With offices in

Argentina Austria Brazil Chile Czech Republic France Greece
Guatemala Hungary Italy Japan South Korea Poland Portugal
Singapore Switzerland Thailand Turkey Ukraine Vietnam

Oxford is a registered trade mark of Oxford University Press
in the UK and in certain other countries

Published in the United States
by Oxford University Press Inc., New York

Reprinted 2011

ISBN 978-0-19-852761-9

Printed and bound in Great Britain by CPI Antony Rowe,
Chippenham and Eastbourne

'But if anything is mis-stated in this book, one should prefer whatever proves to be more accurate.'
From the Prologue to *Íslendingabók* by Ari Thorgilsson the Learned (1067–1148).

Preface

Cephalopods are an ancient group of exclusively marine molluscs. They inhabit the open waters of the sea as well as the sea floor from the shore to the abyssal depths, in deep trenches, and even such seemingly hostile environments as hydrothermal vents. They appear able to occupy most of the many habitats offered by the sea and its substratum. Their life histories and reproductive habits show considerable diversity and many are short lived. In size they range from the tiny *Idiosepius*, only 20 mm in length, to *Architeuthis*, the giant squid with an overall length of some 9 m including the very long tentacles. Cephalopods are the fastest of all marine invertebrates, most moving by jet propulsion. The fossil record shows how abundant they were in the past, and even today there are huge populations of squids in the Southern Ocean, indicating a very successful group of animals. One of the reasons for this success is their elaborate behaviour, and underpinning this, their well developed sense organs and large, complex brain.

John Young, who unfortunately did not live to see the completion of this book, was one of the outstanding zoologists of the twentieth century. Throughout his very long career his major interest was the nervous system and its functional organisation at all levels (Boycott 1998), and which he continued to study until well into the 1990s. After graduation this interest led him to study cephalopods at the Naples Zoological Station and it was there that he made what is for many his most famous discovery: the squid giant axon. Studying the brain made John realise the great potential of a comparative approach to learning, and this he began to investigate in cephalopods, first in *Sepia* and later in *Octopus*, both found along the coasts of the Mediterranean. His original aim was to discover the physical basis of learning and memory and together with his many collaborators, notably Brian Boycott, Stuart Sutherland and Martin and Joyce Wells, and he pursued this with a remarkable series of experiments over many years (Wells, M. J. 1966); meanwhile he published some 150 research papers on cephalopods alone (Nixon and Messenger 1977, Boycott 1998), as well as the monograph entitled *The anatomy of the nervous system* of *Octopus vulgaris* in 1971. Since then the brain of this cephalopod has proved of considerable value in the study of neurobiology, behaviour, sensory capacities and learning (Hanlon and Messenger 1996, Budelmann *et al.* 1997).

Alongside the learning experiments there were anatomical studies based mainly on serial sections of the brains of *Octopus*, *Sepia* and *Loligo*, stained with Cajal's silver method (Stephens 1971). These beautiful preparations can yield a great deal of information in the hands of a gifted scientist like John who soon noticed considerable variation in the size, appearance and organisation of the numerous lobes of the brains of different cephalopods, whose habits, habitats, and life styles differed markedly. For example, there are lobes concerned with touch learning in *Octopus*, which uses its long arms and innumerable suckers to explore the world, but not in *Sepia* or *Loligo*, which do not use touch as a distance receptor (Young 1964). By studying the brain alone then one can infer something about the life of the animal, and as John wrote (1977 a) 'it is clear..........that the sense organs and nervous system can tell us a great deal about the way of life of various cephalopods. We badly need further information about habits from data obtained at sea. These are of course very difficult to obtain. We are gradually learning how to piece the information together and to correlate it with studies of the structure and organization of the brain. Such knowledge, gained inductively, can be used, with caution, for deductions about the probable habits of species not yet investigated at sea'. Since then the development of submersibles and remote operated vehicles has made it possible to observe living cephalopods in the sea and to test proposed hypotheses concerning their functional morphology; for example *Mastigoteuthis* (Roper and Vecchione 1997).

At the time of his retirement from the Chair of Anatomy at University College London, in 1974, John had already amassed a collection of serial sections of brains from numerous cephalopod species, some dating back to his first visit to Naples in 1929. This led him to suggest writing a book describing the brains and lives of these molluscs. And so we obtained further specimens for sectioning histologically and began to assemble information from the literature, and from friends and colleagues, about the distribution, ecology, behaviour and life histories of these animals. In the following account we have combined this information with descriptions of the anatomy and functional organisation of the brains and sense organs at the level of the light microscope in an attempt to give the reader some idea of the lives of many members of forty four of the families of the class, only three the Sepiadariidae, Psychroteuthidae and Walvisteuthidae being omitted.

It is a great sadness that John Young did not live to see the publication of this book. At least his descriptions, often accompanied by his own meticulous line drawings of the nervous system, many of which are included here (unfortunately not all with his discretely placed initials), thus ensuring that they will not be lost for future generations. It is my hope, as it would have been John's, that this account, whatever its limitations, will stimulate young scientists to go out and discover much more about the lives and brains of these beautiful and fascinating animals.

We are grateful for the help of many of our friends and colleagues during the preparation of this book. In particular it gives great pleasure to thank John Messenger who read a large proportion of the book and provided valuable help and constructive criticism during its preparation, and Ian Gleadall for much useful advice and helpful criticism of several chapters. We owe thanks to Desmond Donovan, Katharina Mangold, John Temple and Martin Wells who all gave many helpful comments and constructive criticism on various chapters of the book, and to James Armstrong, Jane Astafiev, Noel Finney, Tessy Hogan, Keith P. Ryan, E. Maconnachie and Pamela Stephens for photographic, technical and secretarial help. Many thanks are due to John Temple, Robert Ireland, Alan Griffiths and Michael M. Wilcock who carefully scrutinised and corrected the etymology and derivation of generic names.

Raye (Raymonde Parsons) Young gave great happiness, support and encouragement to John Young as well as typing many of his manuscripts, and contributed much help to his productive years after retirement. Raye also made the line drawing of a diver holding his capture of a mass of eggs belonging to the squid *Thysanoteuthis*.

We are grateful to The Royal Society of London, The Marine Biological Association U.K., Plymouth and The Wellcome Trust, London, for grants to us both; to The Wellcome Institute for the History of Medicine, London, the Department of Psychology, University of Oxford, and the Geology Department, Birkbeck College, University of London, for the provision of workspace.

We thank The Natural History Museum, London, the Institute of Oceanographic Sciences, Southampton, and the Marine Biological Association, Plymouth, in the United Kingdom; the Stazione Zoologica 'Anton Dohrn', Napoli, Italy; the Observatoire Océanologique de Banyuls, France; the National Museum of Natural History, Washington, and the Rosentiel School of Marine and Atmospheric Science, University of Miami, Florida, in the U.S.A., together with many of our colleagues for their help in lending or giving us specimens for this study.

Marion Nixon

Contents

1 Introduction

1.1 The cephalopods

Cephalopoda form one of the three largest classes of the phylum Mollusca, and are among the most attractive of all the invertebrates living in the sea, with their astonishing capacity to change colour apparently instantaneously. We know they have received the attention of humans since at least the sixteenth century BCE, as octopuses, squids, and argonauts adorn pottery and other objects surviving from that time. Later Aristotle (384–322 BCE; see Thompson 1910) made careful observations and recorded many details of the cephalopods living along the Mediterranean coasts, including aspects of their anatomical structure, natural history, and behaviour. Aristotle and other classical writers gave methods for capturing cephalopods so their value as a food source was presumably well known, and recipes for squids and octopuses were included in a compilation of Roman and Greek cookery of the fourth century CE, ascribed to Apicius (Edwards 1984). The exploitation of cephalopods has continued until today, and currently the annual catch is some 2 million tons.

The name Cephalopoda is derived from the Greek for head and foot; the latter being the term for the ring of appendages, the arms and tentacles, attached to the head. Most cephalopods have eight flexible, muscular arms with one or more rows of suckers, and the squids and cuttlefishes also possess a pair of long tentacles bearing suckers, confined in the majority to the expanded tip, the club. Behind the appendages and the head is the mantle, a sac enclosing the gills and viscera. Many cephalopods have a pair of fins, one on either side of the mantle. *Nautilus* is the only living cephalopod with an external shell. It has no fins, and the so-called 'foot' comprises around 90 slender appendages, which are attached to the head and are devoid of suckers.

Cephalopod molluscs are exclusively marine and abundant, and have been so over a span of 500 million years. During this period they have shown considerable diversity. The ancestral forms possessed an external, chambered, shell; best known in fossil nautiloids and ammonoids. The shell became chambered as a result of structural innovation, but among living cephalopods it is found in only three families: the Nautilidae, the Spirulidae and the Sepiidae. In all living cephalopods, with the exception of *Nautilus*, the shell has become internalized (covered by the mantle integument) and reduced in size. The oldest undisputed fossil cephalopods are from strata of the late Cambrian period. Since then cephalopods have passed through several severe faunal crises when conditions were such as to endanger the continuity of species or groups; the record indicates their appearances, radiations, and extinctions (Fig. 1.1). In the recent past a new diversity emerges, as is evident from the numerous families of oceanic squids living today.

Cephalopods are found around the coasts and from the surface to the depths of the sea; some live mostly on the sea bed and others in the water column. Coastal waters are largely the habitat of cuttlefishes, sepiolids, loliginid squids, and one subfamily of octopodids. Some species live between the tides: one octopus is known to emerge to forage in rock pools, an idiosepiid and a few myopsin squids enter estuaries, but no cephalopod has invaded fresh water. The oceans and seas are dominated by squids, and several octobrachian families. Finned octopods and vampire squid inhabit the deep waters of the oceans and seas; some cephalopods have been observed in deep trenches, and a few to visit hydrothermal vents (Voss 1988a, Tunnicliffe *et al.* 1998, González *et al.* 1998). Cephalopods are apparently able to exploit most of the habitats available in the sea.

The largest living marine invertebrate is the giant squid, *Architeuthis*, the mantle of which can reach a length of more than 2.5 m. Fossil forms too attained a considerable size: the largest known nautiloid has a straight shell some 4 m in length, and the shell of the largest ammonite is almost 2 m in diameter (Lehmann 1981, Stevens 1988). There is a considerable difference in the size of *Architeuthis* and the smallest cephalopod, *Idiosepius*, which—with a mantle less than 20 mm in length—is more than two orders of magnitude smaller.

Cephalopods are the fastest of all marine invertebrates. The majority move by jet propulsion. This is achieved by rapid contractions of the muscles of the mantle wall to expel water from the mantle cavity as a jet, through a muscular and highly mobile funnel. The funnel can be directed to provide effective steering. Many of the slender, streamlined, muscular squids move swiftly, in short bursts of high speed, and some travel over long distances. Some squids develop sufficient propulsion to break through the surface of the sea and 'fly'. Many squids and cuttlefishes move by undulation of the fins as well as by jet propulsion. Some cephalopods have evolved buoyancy mechanisms: in a number the tissue fluids are lighter than seawater so that they are nearly neutrally buoyant, and thus can drift in currents. The nautiluses, cuttlefishes, and spirulids achieve neutral buoyancy by the presence of liquid and gas in the chambers of the shells. The liquid and gas move by means of the active epithelium of the siphuncle. The structure and function of the chambered shells gives lift to these cephalopods so that they are close to neutral buoyancy. The mode of locomotion is often reflected in the shape and form encountered among living cephalopods: they range from firm, muscular, and streamlined forms to others which are soft, gelatinous, and almost shapeless, with many variations between these extremes.

The cephalopods are unique in the animal kingdom for their ability to change colour within milliseconds due to the presence of thousands of pigment-filled sacs in the skin, the chromatophores, which

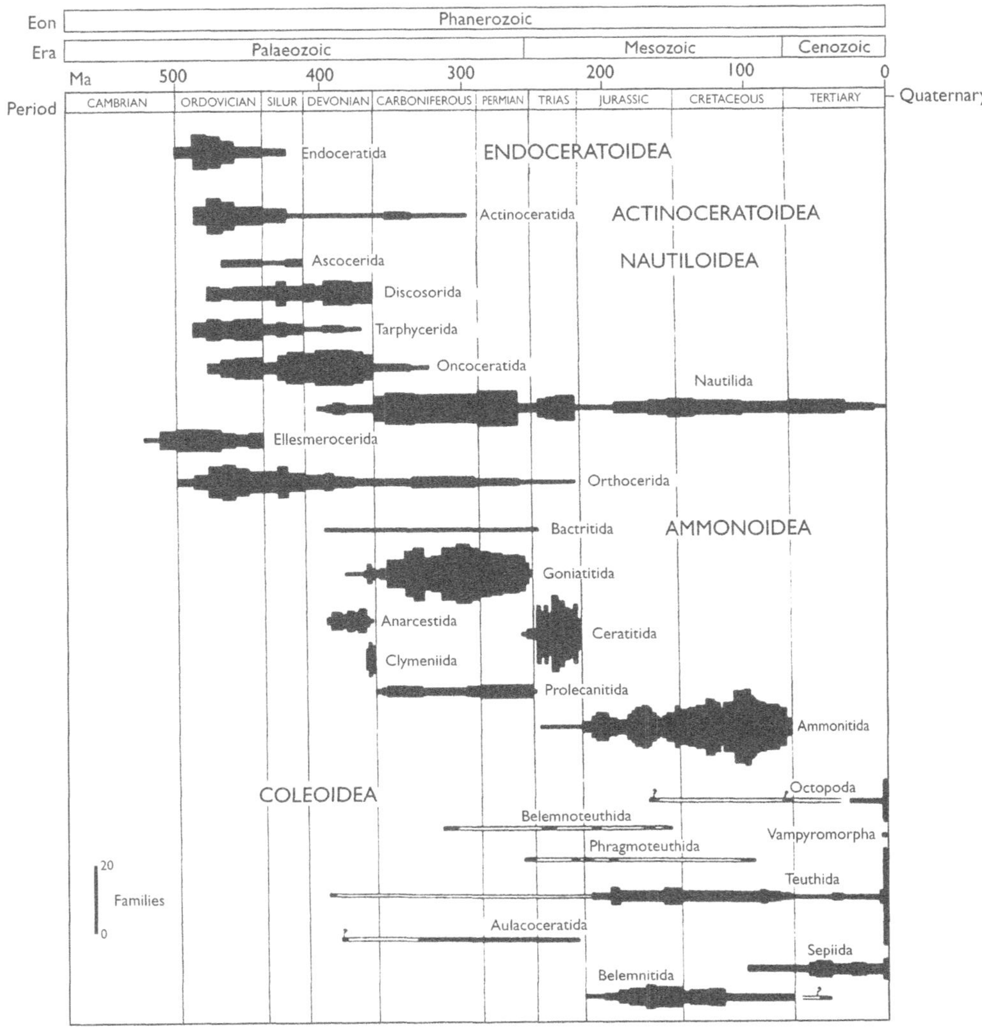

Fig. 1.1. Outline history of the Cephalopoda in time. Their diversity is indicated by the number of families known (House 1988).

are under the direct control of the central nervous system. These animals display a wide variety of body patterns which have the greatest expression amongst those forms living in coastal waters. They can match their background, give courtship displays, startle predators, and communicate with each other by changes in colour and body shape and form. Squids living in open waters have less varied patterns. These are largely for counter-shading but may have other functions as yet not observed. Many midwater squids possess light organs, which usually act to obscure the animal's shadow when it is visible to predators lurking below them. Bioluminescence serves in some species for sexual signalling, and in others may act to startle, or to lure. Most cephalopods possess an ink sac and will eject ink to startle a pursuing predator. The resulting dark cloud, or pseudomorph, hangs suspended in the water, distracting the predator and making escape possible.

Cephalopods are highly developed invertebrates and have a complex nervous system with a central brain, far larger than that of other molluscs, and approaching that of vertebrates such as fish and reptiles (see Fig. 1.5) (Packard 1972). Laboratory studies have demonstrated that cephalopods have the capacity to learn quickly and discriminate well. The sense organs are elaborate and complex: the eyes, for example, are large and well developed, and the size of those of the giant squid provide another record within the animal kingdom.

1.2 Living Cephalopoda: an overview

The history of this class is long (Fig. 1.1). Cephalopods are generally considered to have arisen from an ancient, primitive, mollusc closely allied to the class Monoplacophora, possessing most of the characteristics considered to have been present in an ancestral mollusc (House 1988). During this long period of time—some 500 million years—the cephalopods have undergone structural innovations, the most important of which was the appearance of a chambered shell, and its subsequent internalization and reduction or loss. The main features of evolution in this class are described by Teichert (1988). The evidence from both fossil and living cephalopods indicates their considerable taxonomic diversity. They comprise five subclasses, and all living cephalopods belong to just two of these subclasses: the Nautiloidea and the Coleoidea (Table 1.1).

Various systems of classification have been devised for the cephalopods since the end of the eighteenth century when Cuvier (1795) divided his Classe Mollusques into three orders: Céphalopodes, Gastéropodes, and Acéphales. Céphalopodes were defined as molluscs with a free head supporting an arm crown. Céphalopodes, as originally conceived, comprised only the subdivision now generally referred to as Coleoidea (Bather 1888) but was later extended to include forms with an external shell. The formal term Cephalopoda dates from 1802 in the English translation of Cuvier's *Lectures on comparative anatomy* (Cuvier 1802, Donovan 1996).

Table 1.1. The classification of Recent Cephalopoda and the etymology or derivation of their names (for references cited see pages 369–371)

Taxon	Etymology or derivation
Class Cephalopoda Cuvier, 1795	Head–foot
Subclass Nautiloidea Agassiz, 1848	
Order Nautilida Agassiz, 1848	
Family **Nautilidae** Blainville, 1825	
Nautilus Linnaeus, 1758	Sailor[a]
Subclass Coleoidea Bather, 1888	Sheath, deposited by the mantle
Superorder Decabrachia Haeckel, 1866	Ten arms
Order Spirulida Haeckel, 1896	
Family **Spirulidae** Rafinesque, 1815	
Spirula Lamarck, 1799	A small twist or coil
Order Sepiida Gray, 1849	
Family **Sepiidae** Leach, 1817	
Sepia Linnaeus, 1758	Greek name
Sepiella Gray, 1849	Little *Sepia*
Metasepia Hoyle, 1885b	Changed
Family **Sepiadariidae** P. Fischer, 1882	
Sepioloidea Orbigny, 1845[b]	*Sepia*-shaped
Sepiadarium Steenstrup, 1881	Little, tiny, a diminutive of Greek name for meerschaum cuttlefish
Order Sepiolida Grimpe, 1921	
Family **Sepiolidae** Leach, 1817	
Subfamily Rossiinae Appellöf, 1898	
Rossia Owen, 1835	After James Clark Ross
Semirossia Steenstrup, 1887b	Half
Neorossia Boletzky, 1971	New
Subfamily Heteroteuthinae Appellöf, 1898	
Heteroteuthis Gray, 1849	Different squid (see Teuthida)
Stoloteuthis Verrill, 1881	Stump
Nectoteuthis Verrill, 1883	Swimming
Iridoteuthis Naef, 1912a	Rainbow
Sepiolina Naef, 1912a	Little *Sepiola*
Subfamily Sepiolinae Appellöf, 1898	
Sepiola Leach, 1817	Little *Sepia*
Inioteuthis Verrill, 1881	Back of head
Euprymna Steenstrup, 1887a	With a good stern (referring to plump, thick rear end of body)
Sepietta Naef, 1912a	Little *Sepia*
Rondeletiola Naef, 1921	After Guillaume Rondelet
Order Idiosepiida Grimpe, 1921	
Family **Idiosepiidae** Appellöf, 1898	
Idiosepius Steenstrup, 1881	Distinct
Order Teuthida Naef, 1916	Teuthis, Greek for squid
Suborder Myopsina Orbigny, 1841[b]	Closed eye
Family **Pickfordiateuthidae** Voss, 1953	
Pickfordiateuthis Voss, 1953	After Grace Evelyn Pickford
Family **Loliginidae** Lesueur, 1821	
Loligo Lamarck, 1798	Latin for squid
Sepioteuthis Blainville, 1824	*Sepia* squid

Table 1.1. *Continued*

Taxon	Etymology or derivation
Loliolus Steenstrup, 1856	Little *Lolius*
Lolliguncula Steenstrup, 1881	Little *Loligo*
Uroteuthis Rehder, 1945	Tail
Suborder Oegopsina Orbigny, 1845	Open eye
Family **Chtenopterygidae** Grimpe, 1922	
Chtenopteryx Appellöf, 1890	Comb-wing (Appellöf is assumed to have mistakenly added 'h')
Family **Enoploteuthidae** Pfeffer, 1900	
Enoploteuthis Orbigny, 1844 (in Rüppell)	Armed
Abralia Gray, 1849	Graceful (or wandering)
Abraliopsis Joubin, 1896	Little *Abralia*
Watasenia Ishikawa, 1913	After Shozaburo Watase
Family **Ancistrocheiridae** Pfeffer, 1912	
Ancistrocheirus Gray, 1849	Fish-hook hand
Family **Pyroteuthidae** Pfeffer, 1912	
Pterygioteuthis H. Fischer, 1896	Wing
Pyroteuthis Hoyle, 1904	Fire
Family **Lycoteuthidae** Pfeffer, 1908a	
Subfamily Lycoteuthinae Pfeffer, 1908a	
Lycoteuthis Pfeffer, 1900	Wolf
Nematolampas Berry, 1913	Thread-lamp
Selenoteuthis Voss, 1959	Moon
Subfamily Lampadioteuthinae Berry, 1916	
Lampadioteuthis Berry, 1916	Lamp
Family **Histioteuthidae** Verrill, 1881a	
Histioteuthis Orbigny, 1841[b]	Sail
Family **Bathyteuthidae** Pfeffer, 1900	
Bathyteuthis Hoyle, 1885c	Deep
Family **Psychroteuthidae** Thiele, 1920	
Psychroteuthis Thiele, 1920	Cold
Family **Architeuthidae** Pfeffer, 1900	
Architeuthis Steenstrup, 1857b	Ruling
Family **Neoteuthidae** Naef, 1921	
Neoteuthis Naef, 1921	New
Alluroteuthis Odhner, 1923	Different-tailed
Nototeuthis Nesis and Nikitina, 1986a	Southern
Family **Ommastrephidae**, Steenstrup, 1857c	
Subfamily Todarodinae, Adam, 1960	
Todarodes Steenstrup, 1880	Similar to '*Todarus*'
Martialia Rochebrune and Mabille, 1889	Of Mars
Nototodarus Pfeffer, 1912	Southern *Todarus*
Subfamily Illicinae Posselt, 1891	
Illex Steenstrup, 1880	Seductive
Todaropsis Girard, 1890	Like *Todarus*
Subfamily Ommastrephinae Posselt, 1891	
Ommastrephes Orbigny, 1834 (1834–1847)	Eye-rolling
Hyaloteuthis Gray, 1849	Glassy
Dosidicus Steenstrup, 1857a	Subject to jurisdiction
Sthenoteuthis Verrill, 1880	Strong
Eucleoteuthis Berry, 1916	Famous
Ornithoteuthis Okada, 1927	Bird
Family **Thysanoteuthidae** Keferstein, 1866	
Thysanoteuthis Troschel, 1857	Fringe
Family **Onychoteuthidae** Gray, 1849	
Onychoteuthis Lichtenstein, 1818	Claw
Onykia Lesueur, 1821	Claw
Ancistroteuthis Gray, 1849	Fish-hook
Moroteuthis Verrill, 1881a	Stupid
Kondakovia Filippova, 1972	After N.N. Kondakov

Table 1.1. *Continued*

Taxon	Etymology or derivation
Family **Gonatidae** Hoyle, 1886	
Gonatus Gray, 1849	Knee
Gonatopsis Sasaki, 1920	Like a joint
Berryteuthis Naef, 1921	After Samuel Stillman Berry
Eogonatus Nesis, 1972	East
Family **Brachioteuthidae** Pfeffer, 1908b	
Brachioteuthis Verrill, 1881	Arm
Family **Cycloteuthidae** Naef, 1923	
Cycloteuthis Joubin, 1919	Circle
Discoteuthis R. E. Young and Roper, 1969	Disc
Family **Octopoteuthidae** Berry, 1912	
Octopoteuthis Rüppell, 1844	*Octopus* squid (the adult lacks tentacles)
Taningia Joubin, 1931	After Vedel Täning
Family **Lepidoteuthidae** Pfeffer, 1912	
Lepidoteuthis Joubin, 1895	Scale
Family **Pholidoteuthidae** Clarke, 1980	
Tetronychoteuthis Pfeffer, 1900	Four-claw
Pholidoteuthis Adam, 1950	Scale
Family **Walvisteuthidae** Nesis and Nikitina, 1986b	
Walvisteuthis Nesis and Nikitina, 1986b	After Walvis Ridge
Family **Chiroteuthidae** Gray, 1849	
Chiroteuthis Orbigny, 1841[b]	Hand
Grimalditeuthis Joubin, 1898a	For the Crown Prince of Monaco, of the House of Grimaldi
Planctoteuthis Pfeffer, 1912	Wandering
Asperoteuthis Nesis, 1980	Rough
Family **Mastigoteuthidae** Verrill, 1881b	
Mastigoteuthis Verrill, 1881b	Whip
Family **Joubiniteuthbidae** Naef, 1922	
Joubiniteuthis Berry, 1920	After Louis Joubin
Family **Promachoteuthidae** Naef, 1912b	
Promachoteuthis Hoyle, 1885c	Challenger, after H. M. S. *Challenger*
Family **Batoteuthidae** R. E. Young and Roper, 1968	
Batoteuthis R. E. Young and Roper, 1968	Bramble
Family **Magnapinnidae** Vecchione and Young, 1998	
Magnapinna Vecchione and Young, 1998	Large fin
Family **Cranchiidae** Prosch, 1847	
Subfamily Cranchiinae Pfeffer, 1912	
Cranchia Leach, 1817	After John Cranch
Leachia Lesueur, 1821	After William Elford Leach
Liocranchia Pfeffer, 1884	Smooth
Subfamily Taoniinae Pfeffer, 1912	
Taonius Steenstrup, 1861	Of a peacock
Megalocranchia Pfeffer, 1884	Great
Galiteuthis Joubin, 1898b	After Monsieur Gal
Bathothauma Chun, 1906	Deep wonder
Sandalops Chun, 1906	Sandal eye
Helicocranchia Massy, 1907	Spiral
Liguriella Issel, 1908	Little *Liguria* (the name of the research ship)
Teuthowenia Chun, 1910	After Richard Owen
Mesonychoteuthis Robson, 1925	Middle
Egea Joubin, 1933	After Monsieur Ege
Superorder Octobrachia Haeckel, 1866	Eight arms
Order Vampyromorphida Robson, 1929	
Family **Vampyroteuthidae** Thiele, 1915	
Vampyroteuthis Chun, 1903	Vampire
Order Cirroctopodida J.Z. Young, 1989	Curl octopod
Family **Cirroteuthidae** Keferstein, 1866	

Table 1.1. *Continued*

Taxon	Etymology or derivation
Cirroteuthis Eschricht, 1836	Curl
Froekenia Hoyle, 1904	after Clara Fröken
Cirrothauma Chun, 1911	Curl wonder
Family **Stauroteuthidae** Grimpe, 1916	
Stauroteuthis Verrill, 1879	Cross or stake
Family **Opisthoteuthidae** Verrill, 1896	
Opisthoteuthis Verrill, 1883	Backwards or behind
Grimpoteuthis Robson, 1932	After Georg Grimpe
Order Octopodida Leach, 1817	
Family **Amphitretidae** Hoyle, 1886	
Amphitretus Hoyle, 1885a	A perforation on both sides
Family **Bolitaenidae** Chun, 1911	
Bolitaena Steenstrup, 1859	Greek for small cuttlefish
Eledonella Verrill, 1884	Little *Eledone* (see below)
Japetella Hoyle, 1885a	Little *Japetus*, after Japetus Steenstrup
Family **Vitreledonellidae** Robson, 1932	
Vitreledonella Joubin, 1918	Glassy little
Family **Octopodidae** Orbigny, 1840[b]	
Subfamily Octopodinae Grimpe, 1921	
Octopus Cuvier, 1797	Eight foot
Cistopus Gray, 1849	Pouch foot, for the small pockets on web between each arm
Scaeurgus Troschel, 1857	On the left hand (position of hectocotylus)
Amphioctopus P. Fischer, 1882	A 'fin' (artefact) on both sides
Pteroctopus P. Fischer, 1882	Wing
Enteroctopus Rochebrune and Mabille, 1889	For long, intestine-like arms
Hapalochlaena Robson, 1929	Soft cloak
Danoctopus Joubin, 1933	Named after research ship *Dana*
Robsonella Adam, 1938	Little Robson, after Guy Colbourn Robson
Euaxoctopus Voss, 1971	Easily broken, referring to arms
Ameloctopus Norman, 1992	Without ink
Subfamily Eledoninae Grimpe, 1921	
Eledone Leach, 1817	Greek name for a kind of *Octopus*
Velodona Chun, 1915	Veil gift, for large web
Pareledone Robson, 1932	Near *Eledone*
Tetracheledone Voss, 1955	In four parts (the pads of the funnel organ)
Vosseledone Palacio, 1978	After Gilbert L. and Nancy A. Voss
Subfamily Graneledoninae Voss, 1988	
Graneledone Joubin, 1918	Grain
Thaumeledone Robson, 1930	Wonder
Bentheledone Robson, 1932	Deep
Megaleledone Taki, 1961	Large
Subfamily Bathypolypodinae Robson, 1928	
Bathypolypus Grimpe, 1921	Deep, many-footed
Benthoctopus Grimpe, 1921	Deep
Grimpella Robson, 1928	After Georg Grimpe
Teretoctopus Robson, 1929	Smooth
Vulcanoctopus González and Guerra, 1998	After Vulcanus, God of fire
Family **Tremoctopodidae** Tryon, 1879	
Tremoctopus Chiaie, 1830	Hole, for cephalic water pores
Family **Argonautidae** Cantraine, 1841	
Argonauta Linnaeus, 1758	Sailor on the *Argo* of Jason[a]
Family **Alloposidae** Verrill, 1881	
Haliphron, Steenstrup, 1859	Possibly Greek, for scatterbrained or useless
Family **Ocythoidae** Gray, 1849	
Ocythoe Rafinesque, 1814	Swift-running

[a] The Greek name was given by Aristotle to the paper nautilus *Argonauta*, not to *Nautilus* which occurs only in the Pacific Ocean.

[b] This refers to Orbigny in Férussac and d'Orbigny (1834–1848).

1.3 Classification of living cephalopods

The system of classification used in this book is given in Table 1.1. From the superordinal to the suprafamilial level it follows Doyle *et al.* (1994) and Donovan ((a) in press); below this level it mostly follows Voss (1977) and incorporates the findings from subsequent studies. Sweeney and Roper (1998) included most of the changes subsequent to 1977 in their classification system. After examining the status of phylogeny and classification of coleoid cephalopods R.E. Young *et al.* (1998b) commented: 'Once a stable phylogeny based on cladistic analyses using morphological, molecular and fossil data is achieved, a new classification may be needed', and we fully concur with this view.

1.3.1 Class Cephalopoda

Cephalopoda are bilaterally symmetrical and possess an external calcareous shell, which in most living forms is present as an internal supporting structure. Their appendages, the arms and/or tentacles, encircle the mouth. The mantle cavity serves to provide a respiratory flow of water over the gills and in the majority it also provides the jet propulsive force for locomotion. The nervous system is centralized, the eyes are well developed, and statocysts and other sensory systems are present. The buccal mass includes an upper and lower beak; most have a radula. The sexes are separate. The eggs are rich in yolk and postembryonic development is direct, with or without a planktonic phase. An ink sac is usually present.

Subclass Nautiloidea

The subclass Nautiloidea is today represented by only one living genus (Table 1.1). Its members possess an external shell. The animal itself inhabits the most recently formed part of the shell, the body or living chamber, which is open to the external environment. Behind the living chamber is the phragmocone, which is divided into numerous chambers by walls, or septa. The first-formed chamber is termed the protoconch. All of the chambers, from the protoconch to the living chamber, are traversed by a membranous tube, the siphuncle (Stenzel 1964).

Order Nautilida

The order includes about 10 living species of the genus *Nautilus*, of the family Nautilidae, and its ancestry is described by Teichert and Matsumoto (1987). The classification of the variety of shell forms found among the nautiloids and the main features of evolution of the subclass are given by Holland (1987), Teichert (1988), and House (1988).

Living *Nautilus* has an external shell, which can reach a diameter of about 220 mm (Saunders 1987), and a siphuncle. The animal possesses nearly a hundred small, delicate tentacles arranged in two circles around the mouth. These tentacles have no suckers or hooks, and so are quite different from the tentacles encountered in members of the other subclass, the Coleoidea. A fleshy hood covers the head and bases of the tentacles. The beaks have thick calcareous tips, and the wide radula is distinctive as it has nine teeth and four marginal plates. There are two pairs of gills. The funnel, or hyponome, is not fused ventrally (Stenzel 1964).

Subclass Coleoidea

The term Coleoidea comes from *koleos*, Greek for sheath. The coleoids were characterized by Bather (1888) as cephalopods 'in which the protoconch is typically preserved by an external sheath deposited by the mantle: the shell comes to be enveloped by the mantle, and may partly, even wholly, disappear.' This subclass includes all living cephalopods with the exception of *Nautilus*. It comprises 2 superorders, 6 orders, 46 families, 140 genera, and about 700 species (Table 1.1). The history of the classification of the coleoids has been reviewed by Donovan (in press b).

The subclass Coleoidea is characterized by the following features: the first shell chamber, or protoconch, is typically represented by the external sheath; the shell is internal, when present, and may be calcareous or chitinous. Ten appendages are usually present, one pair of which may be modified or lost. The beaks are of chitin. The majority of the coleoids possess a radula, but it may be reduced or absent; seven teeth and two marginal plates are usually present in each transverse row. Only one pair of gills is present. The majority possess an ink sac. The skin usually has pigment-containing sacs, the chromatophores, the expansion of which is under nervous control.

Superorder Decabrachia

The decabrachians possess 10 arms, the fourth pair being modified as tentacles. Four orders belong to this superorder: Spirulida, Sepiida, Sepiolida, and Teuthida.

Order Spirulida

The order includes only the family Spirulidae, and today is represented by a single genus and species, *Spirula spirula*. A spirally coiled internal shell with a chambered part, the phragmocone, is present in the posterior part of the body. The shell is of aragonite, and lacks a nacreous layer. A large luminous organ lies between the fins. The eye has a circular lid. The arms are short with a broad web. Both ventral arms are hectocotylized in the male. The radula is absent.

Order Sepiida

The order Sepiida includes the family Sepiidae, the second most speciose family of the entire class. One other family, the Sepiadariidae, is usually included in the order but its position remains uncertain (Boletzky 1995).

Sepiida possess an internal calcareous shell, the cuttlebone or sepion, which has a phragmocone. The shell wall lacks a nacreous layer. The fins are narrow and extend along the whole length of the mantle; they are not connected posteriorly. The arms and tentacles bear only suckers, those on the tentacles being borne on a club. The most ventral pair of arms are considerably broader than the others and have a swimming membrane—an elongate muscular vane along the aboral surface of the arms that provides support during swimming. The left ventral arm is hectocotylized in the male. When the animal is not attacking prey the tentacles are retracted into special pockets at their bases, below the eyes. Only one oviduct, the left, develops. Accessory nidamental glands are present. The radula is present. The ink sac is well developed.

Order Sepiolida

The order comprises two families: the Sepiolidae with thirteen genera and the Idiosepiidae with one.

Sepiolida are small as adults. The mantle is short and posteriorly rounded. The internal shell, the gladius, is chitinous, and may be rudimentary. Anteriorly the dorsal mantle may be free or fused with the head. The anterior edge of the ventral mantle may cover the base of the funnel. The head is as wide as the mantle, or nearly so. The eyes have a closed anterior chamber, like those of Myopsina (see below). The fins are rounded and well developed anteriorly, and are large in relation to the mantle. The arms are short and bear suckers, which in the male are usually larger than those of the female and of different sizes. The tentacles are retractile. The radula is present.

Order Idiosepiida

The family Idiosepiidae was placed in a new order by Grimpe (1921). A fully developed gladius is present (Boletzky 1995).

Order Teuthida

This is the largest order of the superorder Decabrachia, and traditionally has included two suborders, the Myopsida and the Oegopsida. These terms were introduced by Alcide d'Orbigny in 1845. In the former the eyes are without immediate contact with the external environment, whereas in the latter the eyes are in direct contact with the sea. The suffix 'ida' (used at the ordinal level) has been changed recently to the more appropriate 'ina', thus Myopsina and Oegopsina (Sweeney and Roper 1998). The phylogenetic position of the Myopsina is problematical as the derivation is uncertain; the ancestry of Oegopsina is not known and their phylogeny is a special challenge because of the wide range in morphology (R.E. Young *et al.* 1998b).

Teuthida possess a shell represented by an internal support, the gladius, formed of chitin. It is thin, horny, and resembles a pen (the name 'calamary' for the common squid *Loligo*, comes from calamarius, pertaining to a calamus or pen), or is sagittate in shape. The fins are located posteriorly or along the length of the mantle, and usually merge along the midline or near the posterior end. The tentaclular clubs and arms have suckers and may have hooks. The radula is present.

Myopsina includes the family Loliginidae, with 6 genera and 42 species, and the Pickfordiateuthidae with 1 genus and species. The position of this last family is uncertain. The eye has a closed anterior chamber. The suckers on the arms are in two longitudinal rows. The tentacles can contract, and their clubs have four longitudinal rows of suckers and no hooks. The left or both ventral arms are hectocotylized. Only the left oviduct develops. Nidamental glands are present.

Oegopsina comprises 27 families, with 73 genera and more than 200 species. The anterior eye chamber is open. The arms and tentacles of many species have hooks as well as suckers. The tentacles can contract. Both oviducts are usually developed. Accessory nidamental glands are absent. Hectocotylization in the male may be present or absent. Many species have photophores on the external surface of body, head, appendages, and/or the ventral surface of the eyeball.

Superorder Octobrachia

This superorder comprises three orders: Vampyromorphida, Cirroctopodida, and Octopodida, in which the second pair of the ancestral ten appendages are modified or lost (Doyle *et al.* 1994). All the arms (except the filamentous, modified arms of the Vampyromorphida) bear suckers. Cirroctopodida have fins and bear cirri on the arms, whereas Octopodida have neither fins nor cirri, and these two groups were earlier placed in the suborders Cirrata and Incirrata by Grimpe (1916). The suborder Cirrata was raised to ordinal level, Cirroctopodida, by Young (1989). This change allows all other octopods to remain in the order Octopoda, which was defined as 'Pedes 8. Corpus apterum.' by Leach in 1817. This definition precedes by nearly two decades the capture of the first finned octopod and its description by Eschricht in 1836.

Order Vampyromorphida

The order includes only one family, with a single genus and species *Vampyroteuthis infernalis*. The characteristic features include eight arms, bearing a single row of suckers and paired cirri, and a pair of slender, extensible filaments, located between the two most dorsal pairs of arms. The filaments are the modified second pair of arms as their nerves contribute to the circumoral commissure before joining the brain, just as the arm nerves do in all other coleoid cephalopods (Young 1977a). A broad gladius extends the whole length of the mantle (Pickford 1949a). The ink sac is absent. One pair of fins is present at hatching; a second pair develops to form the adult fins while the former pair is resorbed. Numerous small photophores and a single pair of large photophores are present. The radula is present.

Order Cirroctopodida

The order includes 3 families, 6 genera and 32 species. The characteristics of the order include a pair of large blade- or paddle-shaped fins, supported by a strong U-, V- or saddle-shaped fin cartilage, the gladius. Paired cirri along the arms accompany uniserial suckers. The mantle is globose and is fused with the head. The web is deep, and may reach to the ends of the arms; a secondary or intermediate web is present or absent. The ink sac is absent. The radula is reduced, homodont or absent. There is no hectocotylus. The left oviduct only is developed.

Order Octopodida

The order incudes 8 families, over 30 genera and more than 200 species. The family Octopodidae, the only one whose members are mainly benthic in habit, includes *Octopus* which is notable for being the most speciose genus of the entire class. All of the remaining families are pelagic in habit. The order is characterized by eight arms and a body without fins (Leach 1817). The gladius may be represented by a pair of small cartilaginous rods, or is absent. Both oviducts are developed. The right or left third arm of the male is hectocotylized. The arms bear only suckers, and these are in one or two longitudinal rows. The web does not usually extend to the tips of the arms. The ink sac is present in shallow-water octopods but is lost in deep-water forms. The radula is present.

1.4 Marine environments

The waters of the oceans and seas, and of the sea floor from the coasts to the deepest reaches, provide the environments and habitats for the cephalopods, which are almost exclusively marine. Oceanic waters comprise those from the continental slopes across the deep ocean floor. Coastal waters are the 'neritic' province, which lies over the continental shelf and comprises the littoral and sublittoral zones. The littoral zone lies between the highest and lowest levels of spring tides; the sublittoral zone between the lowest limit of the littoral zone and the edge of the continental shelf.

The continental shelf, the submerged perimeter of a continent, cants gently seawards, with a gradient of about 1 : 1000 and depths ranging from 20 to 550 m (Marshall 1979). The shelf gives way to the continental slope where the gradients are steeper, about 1 : 40, with depths to about 2000 m. The continental slope then merges with the continental rise where the gradients are gentler, 1 : 100 to 1 : 700, and depths are 5000 m or more. At the foot of the continental rises lie the abyssal plains, the most level areas of the sea floor. The deepest parts of the oceans are the deep sea trenches; these are chasms in the abyssal plains and form the hadal regions with depths descending from 6000 m to 11 500 m (Fig. 1.2). The mid-ocean ridge system is a vast and complex mountain chain that winds round the world for a distance of 5000 km, and forms a third of the sea floor. The mountains, or ridges, rise from the abyssal plain to within 2000 m of the surface, and occasionally even break through the surface in the form of mid-oceanic islands. Besides these islands there are submarine volcanoes, or seamounts, with steep slopes; these are abundant in the Pacific. Seamounts with a flat-topped profile occur in the deep oceans and are termed guyots.

The marine environment is complex, with many contributing factors (Table 1.2). The most obvious feature is perhaps the movement of water: apart from wave action near the surface, the whole oceanic water mass is in continual flow as part of a greater system of circulation in the oceans, dominated by latitude-related temperature differences and the rotation of the earth (Marshall 1979). The temperature varies widely, ranging from the icy waters of the Arctic and Antarctic to the tropical waters around the equator. The surface layer is separated from cooler underlying waters by a layer in which the temperature decreases rapidly with depth: the seasonal thermocline. There is a second rapid decrease in temperature, the permanent thermocline, below which is the more or less uniformly cold (5 °C) deep sea, comprising most of the earth's seawater. Energy and illumination provided by the sun is mostly absorbed in the top 1 m of water, with no more than 5% penetrating to 50 m depth. At a depth of 10 m most of the energy is in the blue-green part of the spectrum, though there is still a trace of visible red light. In the clearest ocean water the greatest reach of blue-green light is probably close to 1200 m. Factors such as salinity and density also contribute to the variety of conditions in the marine environment.

1.5 Distribution of cephalopods

1.5.1 General view

Cephalopods occur in all oceans and seas from the Arctic to the Antarctic, with the exception of the Baltic and the Black Sea. A few species enter estuaries but none has invaded fresh water. They have adapted to most depths: they are found on the sea floor in coastal and open waters, in the water column from the surface to abyssal depths, in deep trenches, on seamounts and guyots (Nesis 1993a), and even as visitors to hydrothermal vents (Tunnicliffe *et al.* 1998).

The marine environment can be divided by depth, although the depth ranges are very approximate and the separation is not rigid. A number of terms are used to describe the habits of animals living in various depth ranges (Table 1.3). Organisms that drift includes animals of all sizes that are adapted for life at all levels between the air–sea interface and the sea–substrate interface; they are termed

Table 1.2. The characteristics of the major ecological zones of the ocean

Ecological zone	Upper limit (m)	Lower limit (m)	Temperature in temperate zones (°C)	Area (% of total ocean bed)	Light
Littoral	Shoreline	100–300	25–5	8	Well lit
Bathyal	100–300	1000–4000	15–5	16	Almost dark
Abyssal	1000–4000	6000–7000	<4	75	Dark
Hadal	6000–7000	11 500	3.6–1.2	1	Dark

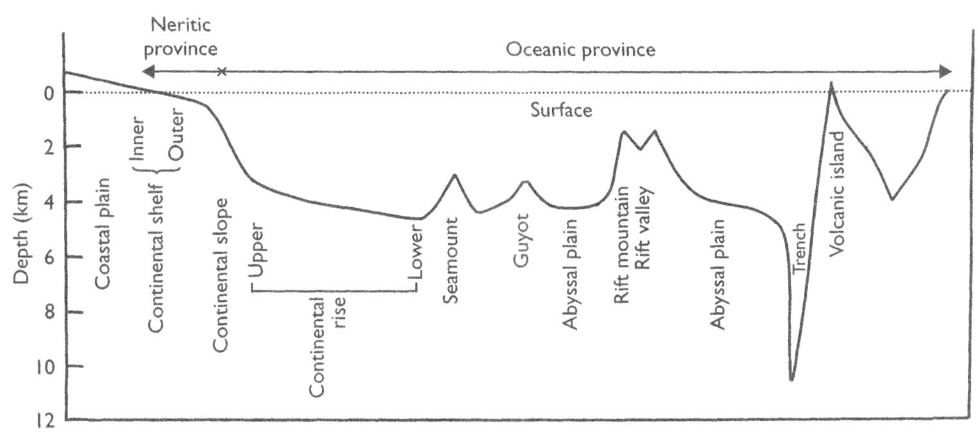

Fig. 1.2. Major features of the sea, from the edge of the coast to the greatest depths of the ocean.

Table 1.3. Life in the sea. A descriptive classification of the marine environment, showing the major ecological realms and zones. The major divisions of coastal and oceanic waters cannot be rigidly separated from one another, and the figures given must be regarded as very approximate

Life in the sea		Province	Zone	Depth (m)
Pelagic life	Neuston Life in the top few centimetres Plankton Passive floaters Nekton Active swimmers	Neritic		
		Oceanic	Epipelagic	0–100
			Mesopelagic	100–1000
			Bathypelagic	1000–4000
			Abyssopelagic	4000–6000
			Hadopelagic	6000–11 500
Benthic life	Littoral system On continental shelf			
	Deep sea system		Bathybenthic	200–4000
			Abyssobenthic	4000–6000
			Hadobenthic	6000–11 500

pelagic. Animals that tend to drift or wander are termed *planktonic*, and those with powers of locomotion are *nektonic* (swimming). Small organisms living in the first few centimetres of water below the surface form the *neuston*. The *epipelagic* zone, the uppermost waters of the open ocean, overlies the *mesopelagic*, or twilight, zone. Beyond is the *abyssopelagic* zone of perpetual darkness, and deepest of all is the *hadal* region. The animals of the sea floor form the *benthos*, and they include many species of cephalopods that, although predominantly bottom-living, can also enter the water column.

Movements of cephalopods can include vertical, horizontal, diel, and seasonal migratory patterns, and some species may exhibit more than one pattern of migration. The first record of vertical movement by these animals was by d'Orbigny (1841), who noted that they occur on the surface at night but not during the day. Vertical movement is now known in many species and several patterns of migration have been found (Clarke and Lu 1974, 1975, Lu and Clarke 1975a, 1975b, Roper and Young 1975). Among the pelagic cephalopods some migrate into the upper 200 m at night but live at moderate depths during the day, whereas others which live deeper move up to depths of 200–300 m. At night some species spread beyond their daytime depths. Others are non-migrators and remain in the same stratum of the water column throughout life.

Many cephalopods make in- or offshore journeys seasonally; others migrate during growth and development, *ontogenetic movements* (Boyle 1983b, Mangold 1987, Vecchione 1987). The hatchlings of some oceanic species are associated with seamounts, islands, and coastal areas, whereas the young of some of the bottom-living coastal species are found in open waters and far from the coast. Several oegopsin squids move over long distances, most notably in the western Atlantic and off the coasts of Japan (O'Dor and Coelho 1993, Hanlon and Messenger 1996); one of the more remarkable journeys is that of ommastrephid squids travelling over 2000 km, at an average speed of 20 km/day (O'Dor 1983).

The distribution of the major groups of cephalopods differs between coastal areas and the oceans and seas. In coastal waters members of the mostly bottom-living Sepiida, Sepiolida, and Octopodinae predominate, along with the nektonic Myopsina squids. The oceans and seas are dominated by the families of teuthin squids; other ocean-living coleoids belong to the Spirulida, Vampyromorpha, Cirroctopoda, and some to the Octopoda. There are some interesting asymmetries in distribution; for example, there are no sepiids native to the American continent. Furthermore, the number of species living along the eastern margins of the oceans is lower than on the western margins: the significance of this is not known (Nixon (b) in press). In the oceans it is estimated that 53% of all species are exclusive to the Indo-Pacific, and 29% are confined to the Atlantic Ocean (Mangold and Boletzky 1989). The richest cephalopod fauna is found along the complex coastlines of the western Pacific (Norman 1997). There are differences in the distribution of species of oegopsin squids around the world, and this is illustrated by the families represented in the diet of sperm whales (Fig. 1.4) (Clarke 1983, 1996b). On passing northwards in the Atlantic Ocean, from 11 °N to the colder waters of 60 °N, all at longitude 20°W, the number of species captured diminishes (Fig. 1.3). Some species of cephalopods are currently considered to have a worldwide or circumtropical distribution, whereas others are restricted to certain areas, for example to waters with a low temperature.

1.5.2 Major groups

Nautilida

The exact geographic range of the genus *Nautilus* is uncertain but it is found in the Indo-Pacific, between 30 °N and 30 °S and 90 °E to

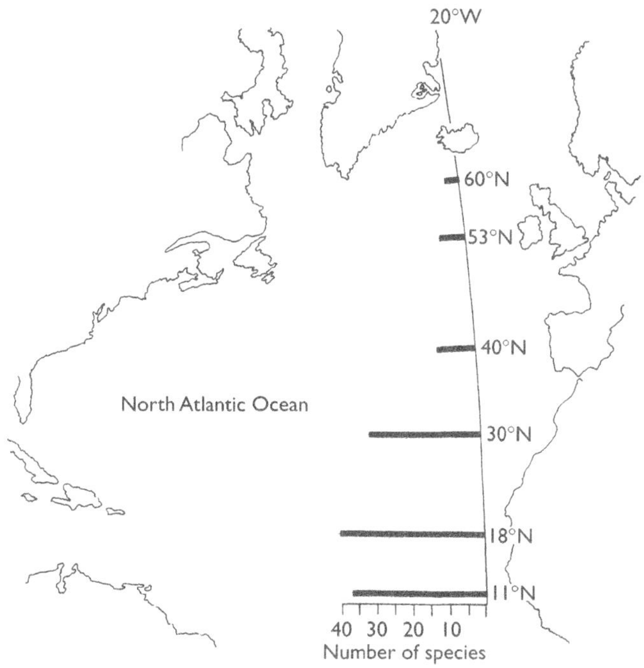

Fig. 1.3. The number of species captured (with opening/closing nets) at the same longitude but at different latitudes in the North Atlantic Ocean. Note the decrease in species on passing northwards (based on data from Clarke and Lu 1974, 1975; Lu and Clarke 1975a, 1975b; Nixon in press b).

185 °E (Saunders 1987). It is associated with islands and oceanic reefs, and appears adapted to vertical, fore-reef habitats. It has a vertical range from 90 to 650 m but is usually found between 150 m and 300 m. The shell is sufficiently strong to withstand the pressure of the sea down to the depths at which it lives, but there is a margin of safety before the shell implodes (Denton 1974). The upper depth limit is mainly dependent on temperature. *Nautilus* migrates vertically each day, moving from deep water to 100–150 m around sunset, and descending in the morning to 250–350 m (Saunders and Ward 1987).

Coleoidea

Decabrachia

Spirulida occur around the world from 40 °N to 40 °S, and from the Caribbean Sea eastwards around the continent of Africa and across the Indian Ocean to near New Caledonia in the Pacific, and is associated with land masses (Bruun 1955). It is mesopelagic, in depths of between 10 m and 1000 m, and migrates vertically each day, being taken at less than 300 m in darkness but below 500 m in daylight (Clarke 1969).

Sepiida occur between 60 °N to about 40 °S but are entirely absent from the Americas. They live along coasts in tropical and temperate waters, from the surface to depths of about 600 m. The depth to which they can descend is reflected in the strength of the cuttlebone. The daylight hours are spent mostly on the sea floor, often buried in the substrate, but in darkness they migrate vertically.

Sepiolida are small and coastal-living. Most are benthic in habit, and often bury themselves in mud or sand, but many have been captured some distance above the bottom. Rossiinae inhabit cold waters on the fringes of most of the world's oceans, at depths of nearly 2000 m. Sepiolinae can be found from 65 °N to 40 °S, living mainly in warm and relatively shallow waters, although some penetrate to depths of nearly 1000 m. Most are benthic in habit. Heteroteuthinae are found widely in the tropical and subtropical waters of the Pacific and Atlantic Oceans and in the Mediterranean Sea. They are pelagic and bathypelagic, and may be found to depths of 2000 m or more; they also have frequent contact with the bottom and may deposit their eggs there.

Idiosepiida are the smallest known decabrachians and are found in coastal waters; at least one species inhabits estuaries and mangrove swamps. Their geographic distribution is poorly known but they do occur around the coasts of southern Africa, the western edges of the Pacific Ocean, the South China Sea, and Australia. An adhesive organ on the dorsal surface of the mantle enables them to attach themselves to seaweed.

The Teuthida includes two suborders, Myopsina and Oegopsina. The former are found worldwide between 50 °N and 60 °S, but the number of species is greater along the western edges of the oceans and seas (Nixon in press (b)). Most are neritic and semipelagic, and are usually found in depths of 20–250 m, but they can range from the surface to depths of 500 m. They are associated with the sea floor when spawning. The Oegopsina have a worldwide distribution in all oceans from the Arctic to the Antarctic. They are the overwhelmingly dominant cephalopods of the oceanic pelagic realm where they are found from depths of about 3000 m to the surface; some species even 'fly' above the surface. There is little knowledge of their habitat requirements but some genera include species that associate with the bottom.

Octobrachia

Vampyromorphida includes only one genus and species. It has a worldwide distribution between about 40 °N and 35 °S, at depths of 700 m to 1500 m, and occurs perhaps as deep as 3000 m (Pickford 1952, Clarke and Lu 1974, 1975).

Cirroctopodida are found worldwide from the Arctic to the Antarctic. The majority live at considerable depths but they are known to occur from 100 m to below 7000 m (Voss 1988a). They are mostly pelagic, and some, although entirely pelagic, live near to but always above the seafloor.

Octopodida are found on the fringes of coasts as well as in open waters from the Arctic to the Antarctic, and from the surface to depths of 4000 m. Only one of the eight families is benthic in habit, the Octopodidae, which includes more than 90% of all octopod species. The other families are epi- to meso- and bathypelagic in habit.

1.6 Life histories and reproduction

The life histories of cephalopods show much variety. *Nautilus* (Nautilida) takes some 15 years to reach maturity and then continues to live for several more years (Saunders 1983, 1984, Ward 1987). This longevity contrasts with that of coleoids, such as cuttlefish, squids, and octopuses, which have much shorter life spans. Most of the coleoids have life spans of less than 1 year although a few may live

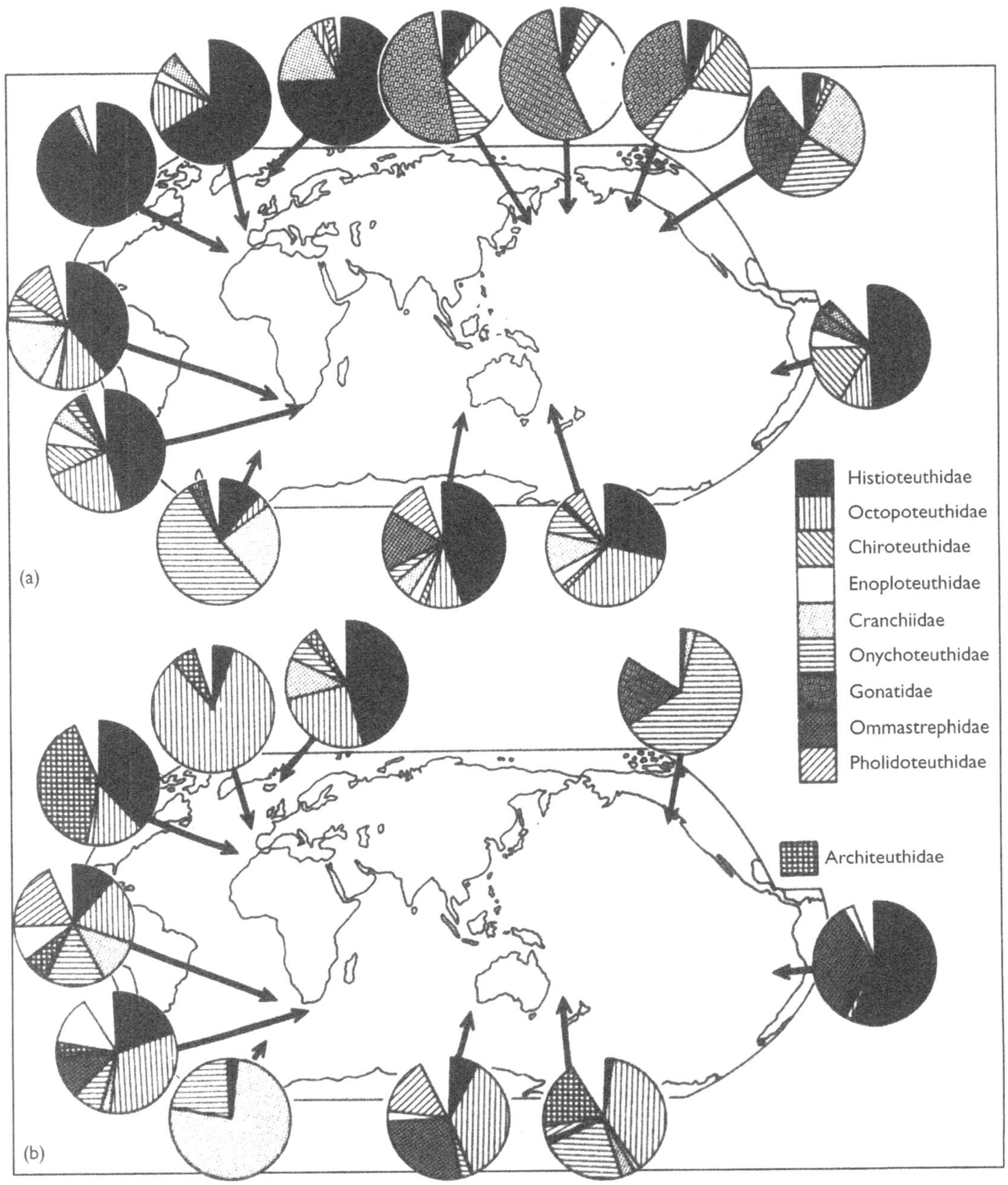

Fig. 1.4. Oceanic squid families identified from the contents of the stomachs of sperm whales captured in various parts of the world: (a) The composition by family, based on the number of lower beaks of cephalopods eaten by sperm whales. (b) Biomass of squid consumed, estimated from the numbers and sizes of the lower beaks and the corresponding mean body weight (Clarke 1983).

for about 5 years. Small, warm water, species can complete their life spans in a few months (Table 1.4). During these periods of time cephalopods become mature, and some species can reach a considerable size. This indicates fast growth, and evidence for this comes from laboratory and field studies. A rapid growth rate was recognized in laboratory studies and the uptake of food ingested found to be high (Nixon 1968, Boyle 1983b, 1987). Growth can be linked to age only when the actual age of the animal can be determined, and this comes from the evidence of the growth of the statolith (Jackson 1989, Arkhipkin 1996). The short life spans of squids and their fast growth is due to their efficient digestion, protein-based metabolism, and ability to sustain continued growth (Jackson and O'Dor 2001).

Reproductive habits are best known for the coastal-living coleiods, and for a few species of oegopsin squids of the family Ommastrephidae (Boyle 1983b). The sexes are separate (gonochoristic) in all cephalopods, and external dimorphism is apparent in many species (Mangold 1987). Dimorphism takes several forms. It may be restricted to a specialized arm in males, the *hectocotylus*, which transfers the spermatophores containing the spermatozoa to the females. Adults can differ in size, the male being larger than the female in some species, and *vice versa*. Among pelagic octopods (e.g. *Tremoctopus*) sexual dimorphism is extreme as the males are dwarfs. There may also be differences in body patterning between the sexes (Hanlon and Messenger 1996).

Table 1.4. The life span of cephalopods estimated from incremental growth, laboratory culture studies, or tag and recapture of animals in the sea. Based on data from Boyle (1983a,b, 1987) and Saunders (1984). F, female; M, male.

Genus and species		Life span (months)
Nautiloidea		
	Nautilida	
	Nautilus	240
Coleoidea		
	Sepiida	
	Sepia officinalis	18–24
	Sepiolida	
	Sepiola robusta	6–9
	Sepietta oweniana	7–8
Teuthida		
	Myopsina	
	Loligo pealei	13–22
	Loligo opalescens	14–36
	Loligo vulgaris	18–24 (F), 36–42 (M)
	Oegopsina	
	Illex illecebrosus	12–18
	Gonatus fabricii	24–36
	Dosidicus gigas	12–36 (F), 9–12 (M)
Octopodida		
	Octopus briareus	8–17
	Octopus cyanea	12–15
	Octopus joubini	6.5–11.5
	Octopus maya	10
	Octopus vulgaris	12–20
	Eledone cirrhosa	15–20
	Eledone moschata	15–24
	Enteroctopus (Octopus) dofleini	36 (F) 48–60 (M)
	Bathypolypus arcticus	48

The males of some species copulate with immature females, and the spermatozoa can remain viable for a long period of time. Other coleoids mate only with mature females, when spawning usually follows shortly after mating, and death soon afterwards. The mode of transfer of spermatophores by the male to the female varies: in most Octopoda the male introduces the hectocotylus into the mantle of the female and deposits a spermatophore. The males of epipelagic octopods have a complex hectocotylus; this breaks, or autotomizes, and is left in the mantle of the female where it continues to flex and move slowly for some time after separation from the male (Voss and Williamson 1971).

The number and size of eggs varies widely. Octobrachians produce from 10 to 500 000 eggs (Hochberg *et al.* 1992), teuthins from several thousand to several hundred thousand, or perhaps even millions, and sepiids 1000 or more eggs (Mangold 1987). The eggs range in size from less than 1 mm to some 35 mm in length. In the ovary the eggs may be markedly different in size, all of the same size, or in batches. The eggs may be spawned on a single occasion followed by mass mortality, or released singly or in small numbers at intervals over a period of time. Evidence has been accumulating that spawning is intermittent in species in which the ovarian eggs display marked disparity in size. In *Nautilus* the presence of more than one size class of eggs in the ovary was described and illustrated by Owen (1832); it is now known that spawning is intermittent and a few eggs at a time are laid over a long period of time (Mangold 1987, Mangold *et al.* 1993). Ovoviviparity, in which the fertilized eggs develop in the oviduct, is found in one of the octopods, *Ocythoe*.

In some species, for example the common cuttlefish *Sepia officinalis* (Nixon and Mangold 1998), the newly hatched young resembles the adult in external appearance, adopts a similar lifestyle, and remains mostly in the same habitat throughout its life. In other species, for example the common octopus, *Octopus vulgaris* (Nixon and Mangold 1996), the young differ markedly from the adult in shape and form and may move from one habitat to another during ontogenesis; these young are termed *paralarvae* (R.E. Young and Harman 1988). Assemblages of cephalopod paralarvae, collected near the Hawaiian Islands, were found to form 'island-associated' and 'oceanic' groups. The island-associated group included octopus species with benthic adults, six species with mesopelagic adults, and three species with epipelagic adults. The paralarvae of only one species was collected exclusively offshore (Bower *et al.* 1999). The wide variation in the shape and form of hatchlings is illustrated in a manual for their identification (Sweeney *et al.* 1992). The morphology of juveniles was the subject of a review by Okutani (1987).

1.7 Chromatophores, body patterns, and photophores

A characteristic feature of coleoids, especially of shallow-water species, is their capacity to change the colour of their skin within milliseconds, to the observer the change seems instantaneous. The colour change is due to the presence in the skin of tens of thousands of pigment-containing sacs, the chromatophores, which have radial muscles attached to them and which are under neural control. Contraction of the muscles expand the sac and so allows the pigment to spread; relaxation of the muscles enables the elastic sac to retract, reducing the area of pigment that is visible. When the pigment sacs are retracted then light can reach the underlying reflecting cells that produce interference colours, especially blues and greens (Packard and Hochberg 1977, Hanlon and Messenger 1988, 1996, Messenger 2001). The external appearance of coleoid cephalopods is the result of a combination of chromatic, textural, postural, and locomotor components, this is termed the *body pattern*. The ability to combine these components in various ways allows coleoids to create different body patterns with great speed. Even the shape and form of the animal can change, and the erection of skin papillae may further alter the body pattern. The body patterns and the behaviour exhibited by living coleoids of coastal waters can be an aid to their identification (Hanlon 1988).

Body patterns are involved in many aspects of the life of coleoids including courtship, communication, camouflage (cryptic coloration or crypsis), and in relation to predators and to prey. The body patterning may be correlated with habitat. Those species that are active in daylight and live in complex environments have a rich repertoire of patterns, especially of cryptic coloration (Hanlon and Messenger 1988). One octopod genus, *Hapalochlaena*, possesses iridescent blue rings or lines, which presumably act as a warning device since at least one of its species secretes toxic saliva which is lethal to humans (S.K. Sutherland 1983). Coleoids inhabiting murky or deep water, or of nocturnal habits, have rather limited body patterns. The deep-living vampire squid, *Vampyroteuthis*, is uniformly black and as its chromatophores lack muscles it is not likely to undergo colour changes (Young 1977a). Coleoids of open waters do not show the varied chromatic patterns found in the coastal-water forms.

Besides changing their appearance abruptly, it is a characteristic of most living coleoids that they can emit ink from the ink gland complex, comprising the ink sac, the ink gland and ink sac duct (Budelmann *et al.* 1997). Aristotle observed that the ink sac is largest and the ink most abundant in the cuttlefish, and that the animal discharges the ink when 'frightened.' The ink, which is a suspension of almost pure melanin (Flood *et al.* 1990), can be ejected with mucus as a dense 'smoke screen' or as a small mucus-bound blob that hangs suspended like a cloud in the water: such a *pseudomorph* distracts the predator while the cephalopod escapes. The ink may also serve as an alarm substance in loliginids (Lucero *et al.* 1994).

Light organs, or *photophores*, occur in many coleoids. One squid, *Lycoteuthis*, remained alive after capture during the *Valdivia* expedition and was observed to produce an almost magical light; its 'body appears to be covered with a diadem of precious stones.' (Chun 1915). It was the brilliance of this animal that gave Carl Chun the desire to investigate the structure of these organs and to embark on his study of cephalopods, which resulted in the monograph published in 1910 and 1915. Bioluminescence is widespread in oegopsin squids living in oceanic waters between 400 m and 1200 m in depth, and is also found in a few coastal-living species of myopsin squids. Light organs are present in some species belonging to the Spirulida, Sepiolida, Vampyromorphida, and Octopodida (Herring 1977, 1983, R.E. Young 1983), but as yet none has been found in Sepiida or Cirroctopodida.

The distribution of light organs on the body can be indicative of their function. In many oegopsin squids most of the photophores are on the ventral surface, and are often numerous below the eyes. This type of distribution usually serves for ventral counter-illumination, to conceal the opaque structures of the animal when viewed from below by potential predators (R.E. Young 1977, Denton 1990). In the almost complete absence of observations of living animals in the sea the value of light organs in the life of these animals is poorly understood, but in some species they may provide sexual signals (R.E. Young 1983, Hanlon and Messenger 1996). In one species arm-tip flashes appear to startle or deter predators (Roper and Vecchione 1993).

1.8 Central nervous system and receptors

All modern cephalopods have a well-developed nervous system and central brain, which is relatively enormous when compared with that of other molluscs, other invertebrates, and even some fishes and reptiles (Fig. 1.5) (Packard 1972). The cephalopod brain is complex and has numerous lobes to which different functions can be attributed largely as a result of experiments carried out on *Octopus*, *Sepia*, and *Loligo* (Young 1977a).

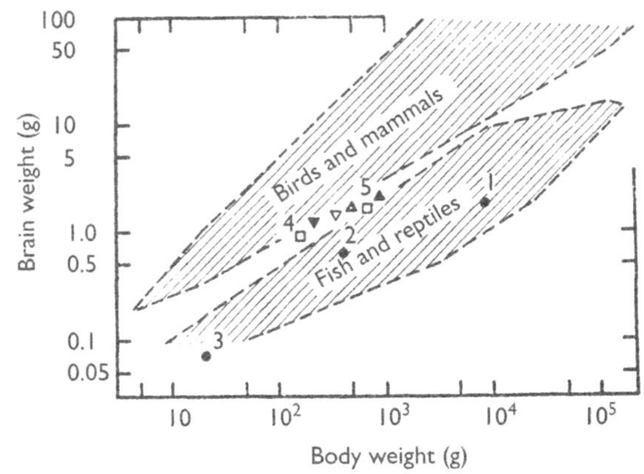

Fig. 1.5. Brain and body weight relationships of cephalopods and vertebrates. Log–log plots of brain and body weights of adult cephalopods are compared with those of higher and lower vertebrates (birds and mammals, and fishes and reptiles, respectively). 1, *Octopus vulgaris*, 2, *O. salutii*, and 3, *O. defilippi* (Octopoda); 4, *Illex*, and 5, *Todarodes* (Oegopsina);∇, *Loligo* (Myopsina); Δ, *Sepia* (Sepiida) (Packard 1972; with permission of Cambridge University Press).

The cerebral ganglia of modern *Nautilus* may be more elaborate than those of fossil Nautiloidea or of fossil Coleoidea. However, the organization is sufficiently similar to the brains of living coleoids for us to be fairly sure that the latter have descended from an animal somewhat similar to *Nautilus*. Furthermore, the similarities between the brains of extant coleoids makes it plausible to suggest that they all have a common ancestor. In nearly all of the characters of its brain *Vampyroteuthis infernalis* (Vampyromorphida) is considered to represent a stage long after the separation of the coleoid and nautiloid stocks (Young 1977a, 1988a).

Cephalopods possess several types of receptor organs. The eyes are typically large and well developed. The giant squid, *Architeuthis*, possesses the largest eyes found in the animal kingdom, with a recorded diameter of 200 mm. Like many other molluscs, cephalopods also possess extraocular photoreceptor organs (Mauro 1977, R.E. Young 1978, N.A. Voss 1988b). These are present in decabrachians and octobrachians, but their function is not well understood: in some oegopsin squids they serve to monitor downwelling light to regulate the light emitted from their own photophores (R.E. Young *et al.* 1979a). Cephalopods have paired organs of balance, the *statocysts*, which have features correlated with locomotor behaviour (Young 1988b). Octopuses and perhaps many other cephalopods are sensitive to chemical and tactile stimuli: receptors sensitive to these stimuli are found all over the body (Hanlon and Messenger 1996). A system similar to the lateral line of fish is also present, at least in the early stages of life, in some sepiids, myopsins, and octopods (Budelmann 1994, Budelmann *et al.* 1997).

1.9 Behaviour and learning

The agile, swift-moving cephalopods, with their ability to escape by jet propulsion, are difficult to observe. In the sea these fast-moving invertebrate animals compete with fish and studies of a few species have revealed complex behaviour and a remarkable capacity for learning. An excellent and comprehensive account of the behaviour of cephalopods is given by Hanlon and Messenger (1996).

Investigation of the behaviour and habits of oceanic cephalopods poses many problems. A few animals survive capture and can be kept in aquaria on board ship; it was in this way that Schmidt (1922) observed the movements of *Spirula spirula*. As techniques and facilities for holding live animals on ships improved, so experimental studies became feasible: for example, the demonstration of ventral counter-illumination in two oceanic enoploteuthid squids, *Abralia* and *Abraliopsis* (R.E. Young *et al.* 1979a); and the recording of responses of the photophores and measurement of the luminescence of several oegopsins to investigate the function of these organs (R.E. Young *et al.* 1982, Herring *et al.* 1985).

The development of deep-sea submersibles, remotely operated vehicles, underwater cameras, and video-recording devices now allows glimpses into the life of cephalopods inhabiting the depths of the sea. One example is of rare cirroctopods swimming either by means of their fins or hanging suspended with the web open, like an umbrella (Roper and Brundage 1972). Another is of characteristic positions adopted, such as the 'cockatoo' posture of the cranchiid squid *Teuthowenia*, and the 'J' posture of the ommastrephid squid *Ornithoteuthis* (Vecchione and Roper 1991). *In situ* observations and video-recordings from manned submersibles, combined with anatomical studies of the specimens observed, have revealed relationships between the structure and function of locomotor and feeding apparatus in some cirroctopods (Vecchione and Young 1997).

In shallow coastal waters cephalopods are readily visible, thus making it somewhat easier to investigate their habits. The common cuttlefish, squid, and octopus were observed in antiquity in the Mediterranean as Aristotle recorded some of their colour patterns, spawning habits, development, diet, and modes of locomotion. Towards the end of the nineteenth century the establishment of seawater aquaria and marine biological stations made possible the close observation of some coastal cephalopods (Lee 1875, Lo Bianco 1908). Direct observation of cephalopods in shallow water has been greatly aided by the development of self-contained underwater breathing apparatus (scuba), as short- and long-term diving, saturation diving, and even experiments in the field have become feasible. One long-term study followed the life cycle of a shallow-water reef squid, on the Caribbean coast of Panama, and recorded changes in habitat, migratory patterns, feeding habits, predatory and social interactions, courtship displays, and skin patterning (Moynihan and Rodaniche 1982). Communication with each other by means of changing patterns has been described in several coastal cephalopods (Moynihan 1985). An intensive, short-term study gave considerable insight into the behaviour and mode of life of several species inhabiting a bay in the Virgin Islands, in the Caribbean (Hochberg and Couch 1971). During observations of cephalopods living around a continental island off the coast of Queensland, Australia, a previously unsuspected mode of ambling locomotion with arms IV was reported in one sepiid; furthermore, the hatchlings and young of several species of oceanic squids were captured along the shores (Roper and Hochberg 1988). Below the surface of a bay off the coast of South Africa, echo-trace records suggested that shoals of squids were present below the boat: this was confirmed by scuba divers who found a population of loliginids spawning on the sea floor and documented their behaviour (Sauer *et al.* 1992). Investigations of intraspecific relationships, social interactions, and other forms of behaviour have become possible under laboratory conditions with the introduction of large tanks with windows (O'Dor *et al.* 1977, Hanlon *et al.* 1983).

The modern study of cephalopod neurobiology dates back to the late 1940s when, at the Stazione Zoologica in Naples, J.Z. Young initiated the comparative study of learning, behaviour, and neuroanatomy in cephalopods. *Octopus vulgaris* proved to be an excellent laboratory animal and adaptable to long-term studies. Young took it as a general model for the cellular mechanisms of memory, typically through studying the effects of small brain lesions on the abilities of this animal during visual and tactile discrimination. The octopus is well suited as a model since it learns readily and its central nervous system comprises two sets of anatomically distinct, paired centres, involved in the learning process (reviews by: Boycott and Young 1950, Young 1961, 1964, 1965c, 1983, Wells and Wells 1957, Wells 1966, 1978, Wells and Young 1965, Sanders 1975). In parallel with the discrimination and behavioural studies, the anatomy of the brain and the receptors of *O. vulgaris* were described in detail by Young in 1971. The great interest in the neurobiology of cephalopods in recent years has stemmed largely from these earlier studies of *O. vulgaris* (Robertson and his colleagues 1990, 1994, 1995, 1996, Budelmann 1994, 1995, Abbott *et al.* 1995, Messenger 1996, 2001, Budelmann *et al.* 1997).

As we have seen the cephalopods are enormously varied in shape and form and can occupy every habitat of the ocean from the surface to its depths, and from the coast down the continental shelf and continental slope to the sea floor: 'Each habitat requires different behaviour... By examining the details of the organization of the brain appropriate to each habitat we can draw conclusions about the significance of the patterns of connectivity.' (Young 1977a). With the greatly enhanced capabilities to observe cephalopods in the sea it becomes increasingly possible to determine the relationship between behaviour and learning, and the anatomy and morphology of the animal and its brain.

1.10 Identification, orientation, terminology and measurements of cephalopods and their brains

Several illustrated keys for the identification of cephalopods are available and include Nesis (1987) for cephalopods of the world, Okutani (1995) for cuttlefish and squids of the world, and Roper *et al.* (1984, 1985) for species of interest to fisheries. In addition Sweeney and his colleagues (1992) have provided keys for the 'larval' and juvenile stages. More restricted guides are available, including those for the cephalopods of Hong Kong (Voss and Williamson 1971, Gleadall 1997, Norman and Hochberg 1994), the Philippines (Voss 1963, Norman and Sweeney 1997), Japanese waters (Okutani *et al.* 1987), and Iberian waters (Guerra 1992). There are keys for the order Teuthida (Roper *et al.* 1969), the sepiolids living in the

Mediterranean (Bello 1995), and the genus *Euprymna* (Norman and Lu 1997a).

Terms used to describe cephalopods are provided in glossaries by Roper and Voss (1983), and Roper *et al.* (1984). Modifications and additions to these terms are given by Sweeney and his colleagues (1992), Norman and Lu (1997a), and Norman and Sweeney (1997).

Cephalopods are bilaterally symmetrical, and their orientation in life is shown in Fig. 1.6. The nautilid, *Nautilus*, lives in the last-formed chamber of the shell from which its head emerges and its orientation is shown in Fig. 1.6a. The coleoid cephalopods vary greatly in shape and form and this led Hoyle (1886) to adopt the position of the animal during horizontal swimming to describe its orientation; thus the head is anterior, the apex of the visceral mass posterior, the funnel ventral, and the opposite side dorsal (Fig. 1.6b, c); this also applies to *Nautilus*.

Nautilus has more than 90 delicate circumoral tentacles. These have transverse ridges but are devoid of suckers, and are quite unlike the pair of tentacles found in decabrachians. Coleoidea have up to 10 circumoral appendages (arm crown) and these are termed right (R) or left (L). In decabrachians four pairs of arms are numbered I–IV from the dorsal to the ventral side of the head, thus the dorsal-most pair are arms I and the ventral-most arms IV. They have in addition a pair of appendages between arms III and IV, and these are modified to form tentacles (Fig. 1.6b). The length of the arms, from dorsal to ventral, can be expressed by a formula: thus IV > III > II > I shows that the fourth pair of arms is the longest and the first the shortest. The order differs depending on species, so that arms I may be the shortest and arms III the longest. The inner surface of the arms is oral and the outer surface is aboral. The appendages of decabrachians bear suckers and/or hooks. Vampyromorpha possess eight arms with suckers, and a pair of modified appendages between arms I and II; these are retractile filaments and lack suckers. The remaining octobrachians possess just eight arms with suckers.

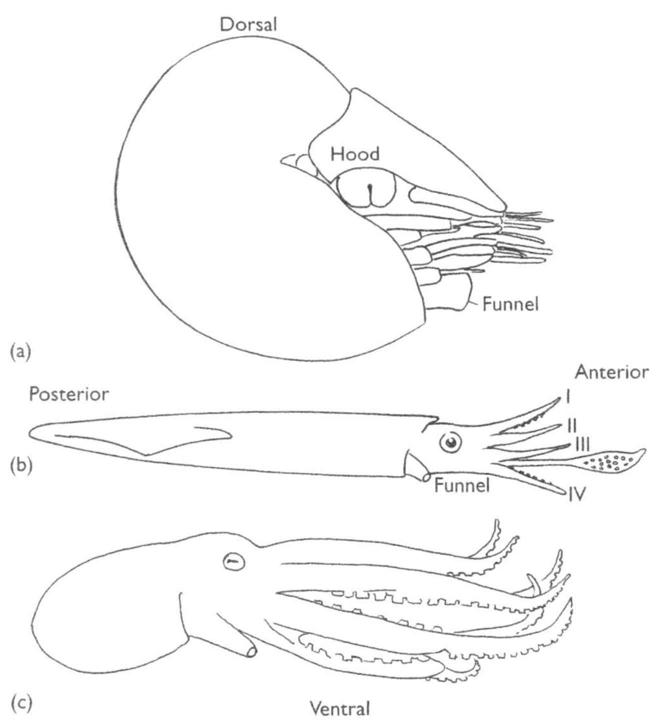

Fig. 1.6. The orientation of cephalopods is based on the position adopted by the animal during normal, horizontal locomotion: (a) Nautilida, (b) Decabrachia, and (c) Octobrachia.

The size of coleoid cephalopods is usually given as the dorsal mantle length (ML). This is measured between the anterior edge and the posterior apex of the mantle in decabrachians (Fig. 1.7a, b), and from an imaginary line drawn through the centre of both eyes to the posterior end of the mantle in octobrachians (Fig. 1.7c) (Mangold-Wirz 1963, Roper and Voss 1983). The total length (TL) is difficult to measure since the arms and tentacles may contract or relax, depending on the condition of the animal, and usually changes after preservation. The arms can alter in shape and form during anaesthesia and preservation; shrinkage in small octopuses can reach 30% (Nixon 1971). Total length, however, does better represent the size of a coleoid as the arms usually form a major part of the animal. Furthermore, changes in proportion often take place during ontogeny: at hatching the arms may be very short compared with the mantle length but later increase in length until many times that of the mantle. Weight is a more sensitive indicator and provides a much better measure of the changes with the growth of these soft-bodied animals (Nixon 1971, Forsythe and Hanlon 1989). Ideally the weight, the dorsal mantle length, and the total length should all be determined. Sometimes the length of the gladius (GL) of teuthins, or of the cuttlebone of sepiids, is measured.

The volumes of various lobes of the brain of a number of specimens were estimated from areas measured on suitable samples of serial histological sections (1 in 10 or 1 in 20, or for the larger optic lobes 1 in 30). Each lobe was measured on one side of the brain, and using a microscope with a drawing tube the outlines of the lobes were traced directly on to a digitizing pad. This information with the aid of a computer allowed the volumes of the lobes to be estimated (Maddock and Young 1987).

The nomenclature for the various lobes of the central nervous system mainly follows that of Dietl (1878), whose names are accurate and have priority. These names, with subdivisions and additions, have been used to describe the brain of *Sepia officinalis* (Boycott 1961), *Loligo* (Young 1974, 1976, 1977b, 1979, Messenger 1979a), and *Octopus vulgaris* (Young 1971).

1.11 Specimens, illustrations, abbreviations, and citations to authors

The specimens available for this study were provided with the help of institutions from around the world and the kindness of many colleagues. Whole specimens, or specific tissues, were preserved in neutral formalin, before being subjected to histological methods. The brains were mostly stained with Cajal's reduced silver method and then sectioned serially. Staining techniques for other tissues include haematoxylin and eosin, Bodian's, and Holmes' (Stephens 1971); cuticular tissue was stained with Spirit blue and counterstained with Masson's trichrome (Elder and Owen 1967). For scanning electron

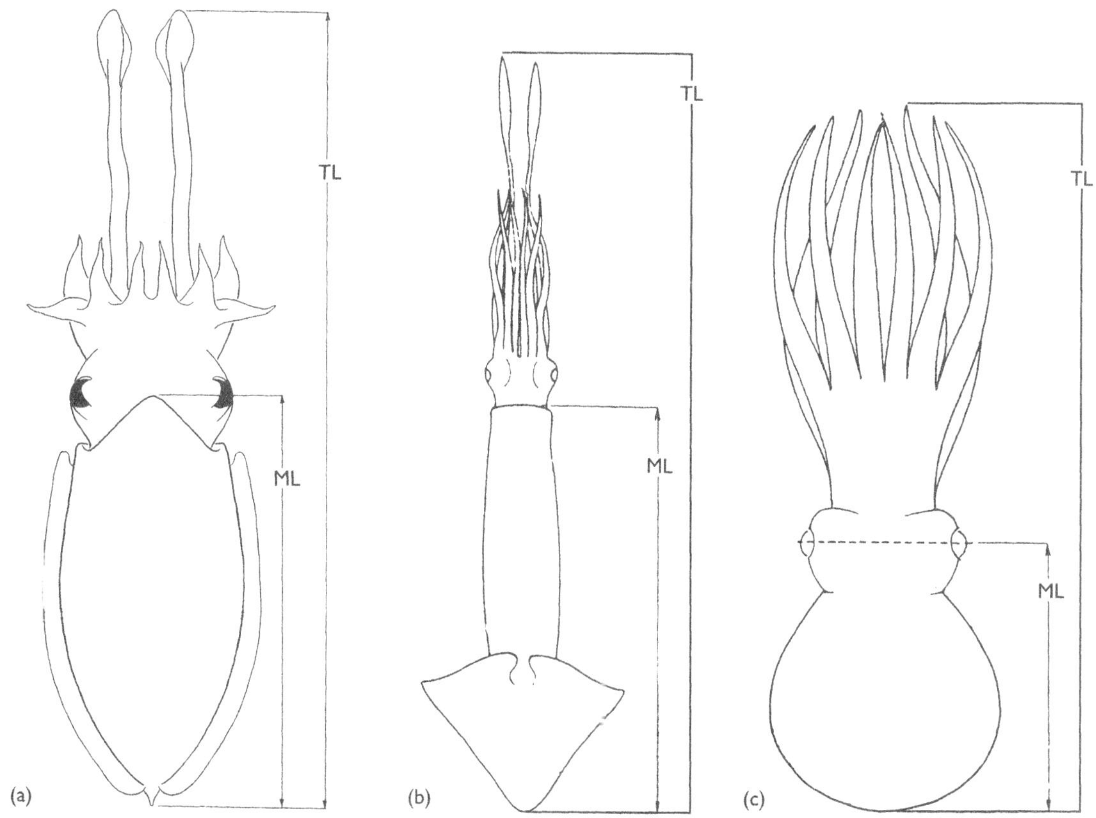

Fig. 1.7. Points between which the mantle length (ML) and total length (TL) are measured in Decabrachia (a, Sepiida and b, Teuthida), and in the Octobrachia (c, Octopoda) (Mangold-Wirz 1963, Roper and Voss 1983).

microscopy the tissues were treated conventionally (Boyde 1974, Bone and Ryan 1976). The histological sections described in this book are now held in The J.Z. Young Collection in The National Museum of Natural History, Smithsonian Institution, Washington, DC 20560, USA.

The majority of the photomicrographs of histological sections are original. Illustrations from other sources are cited in the legend and given in the list of references. The line drawings of the central nervous system are reconstructions by J.Z. Young from serial histological sections. Abbreviations are listed on pp. 372–379.

Where there are references to different authors with the same name then the initials of the most cited author are dropped. Thus Clarke refers to M.R. Clarke, Sanders to G.D. Sanders, Sutherland to N.S. Sutherland, Voss to G.L. Voss, Ward to P.D. Ward, and Young to J.Z. Young.

2 Central nervous system and sensory systems: an outline

2.1 Introduction

In early molluscs the central nervous system probably consisted of a network of nerve cell bodies spread out along cords. This can be seen in extant molluscs of the class Amphineura. Reduction is apparent in *Chiton* (Polyplacophora), with only two pairs of longitudinal cords united by transverse cords, lateral and ventral to the oesophagus, and a cerebral commissure above. The system of ganglia and connectives found in lamellibranch and gastropod molluscs has probably developed by a reduction of perioesophageal rings and aggregations of nerve cells to form ganglia in the longitudinal cords (Bullock 1965). Owen (1832) recognized that the central nervous system of *Nautilus* (the only living representative of the Nautilida) was similar to that in other molluscs such as *Aplysia* (Gastropoda), and in all the remaining extant cephalopods (Coleoidea), although in the latter the primitive pattern is largely obscured by the brain's greater concentration.

In all living cephalopods the nervous tissue forms a compact circumoesophageal brain (Fig. 2.1) and is more highly developed than in any of the other molluscan groups. The greatest centralization (cerebralization) among cephalopods is found in the octobrachians and is achieved by shortening of the connectives between the superior buccal and brachial lobes (Fig. 2.2). The brain is less compact in

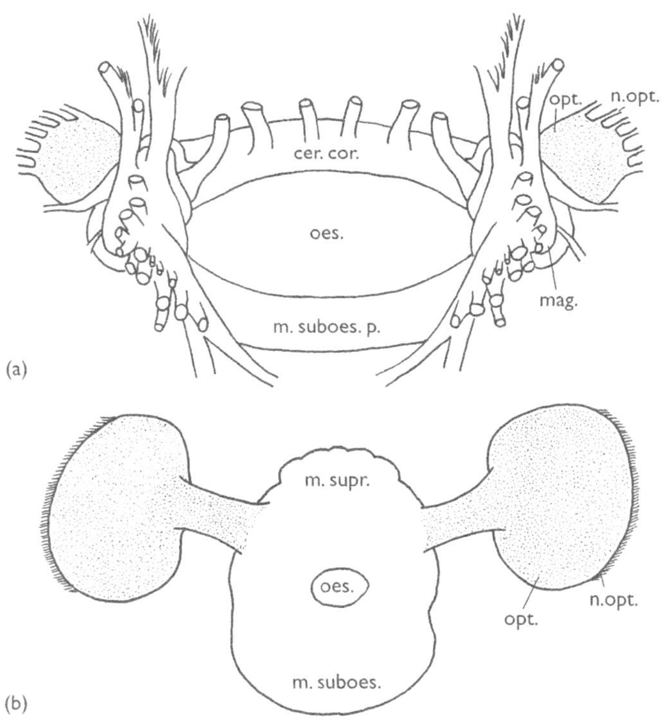

Fig. 2.1. Diagram showing the organization of the central nervous system around the oesophagus of (a) *Nautilus* (Nautilida), and (b) *Octopus* (Octopodida), both seen from in front (modified from Young 1965a).

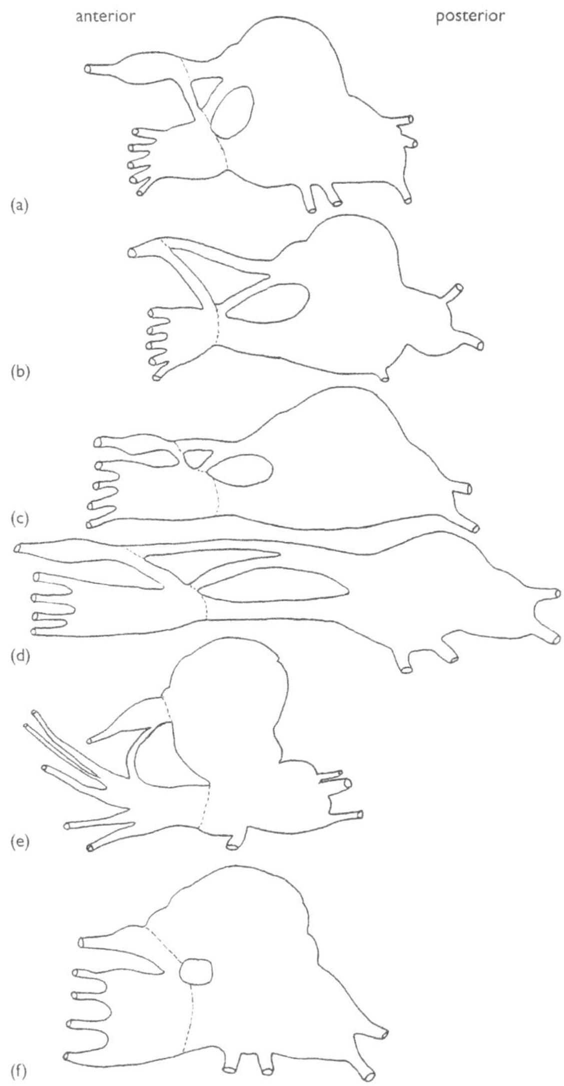

Fig. 2.2. The central nervous system, seen from the side, and shown in outline (but not to the same scale). The superior buccal and the brachial lobes lie on the left and anterior to the broken line: (a) Sepiida, *Sepia*; (b) Sepiolida, *Sepiola*; (c) Teuthida, Myopsina, *Loligo*; (d) Teuthida, Oegopsina; *Ommastrephes*; (e) Vampyropmorphida, *Vampyroteuthis*; (f) Octopodida, *Octopus*.

decabrachians and in the nautilids. The central nervous system of cephalopods comprises many lobes and includes the paired cerebral, buccal, labial, pedal, palliovisceral, brachial, optic, olfactory, and peduncle ganglia found in molluscs of other classes. The left and right ganglia of each pair are joined by a commissure, and connectives (with pathways or tracts) in the anterior–posterior direction between the ganglia on one side.

Nautilus has a simpler brain, in the literal sense of having fewer parts, than is found in the Coleoidea. However, the nervous system of *Nautilus*, with 13 discernible lobes, is vastly more complex than that of any non-cephalopod mollusc. The brain of *Nautilus* is a system of three ganglionated nerve cords (Fig. 2.1(a), Fig. 2.3). These comprise a supraoesophageal cord, a cerebral cord, and a suboesophageal cord with two parts, namely the anterior (brachiopedal) and posterior (palliovisceral). The cerebral cord shows no internal division into lobes. It receives visual and tactile inputs. The anterior suboesophageal cord lies laterally and is continuous on each side with the cerebral and posterior suboesophageal cords. The anterior part of the anterior suboesophageal cord gives rise to the nerves of the tentacles and hood and the posterior part is the origin of the hyponome (funnel) nerves. The posterior suboesophageal cord gives rise to a series of pallial and visceral nerves. The lateral region of the brain, where all three nerve cords join, contains the largest cells of the brain. It receives some static nerve fibres and is the origin of the nerves of the ocular tentacles.

The brains of Coleoidea are usually more centralized, and more tightly packed around the oesophagus than that of *Nautilus*. There are also clear internal and external features allowing the division of the brain into 25 major lobes, some of which are further subdivided, making 37 or 38 lobes in all (Figs 2.4, 2.5). The central nervous system comprises the supra- and suboesophageal masses, the two parts being connected by the

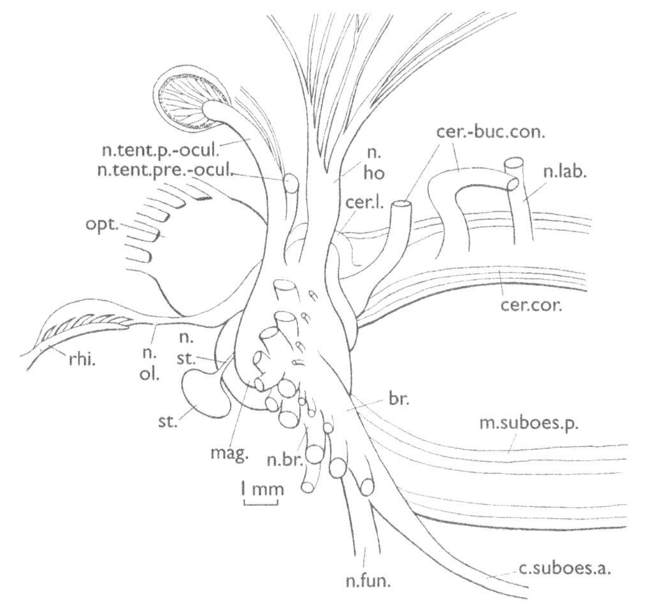

Fig. 2.3. Nautilida, *Nautilus*. Drawing of reconstruction of the central nervous system from serial transverse sections, seen from in front (Young 1965a).

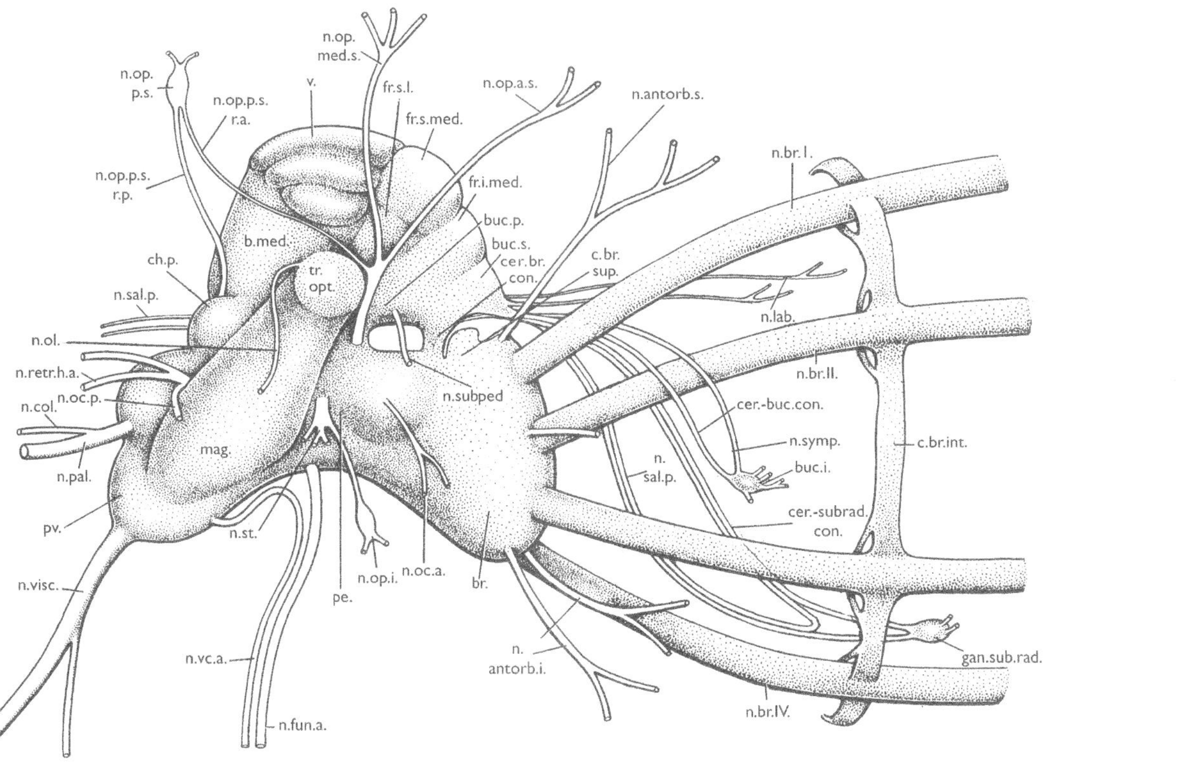

Fig. 2.4. Octopodida, *Octopus vulgaris*. Drawing of reconstruction of the central nervous system, drawn from serial sections, seen from the right side (Young 1971).

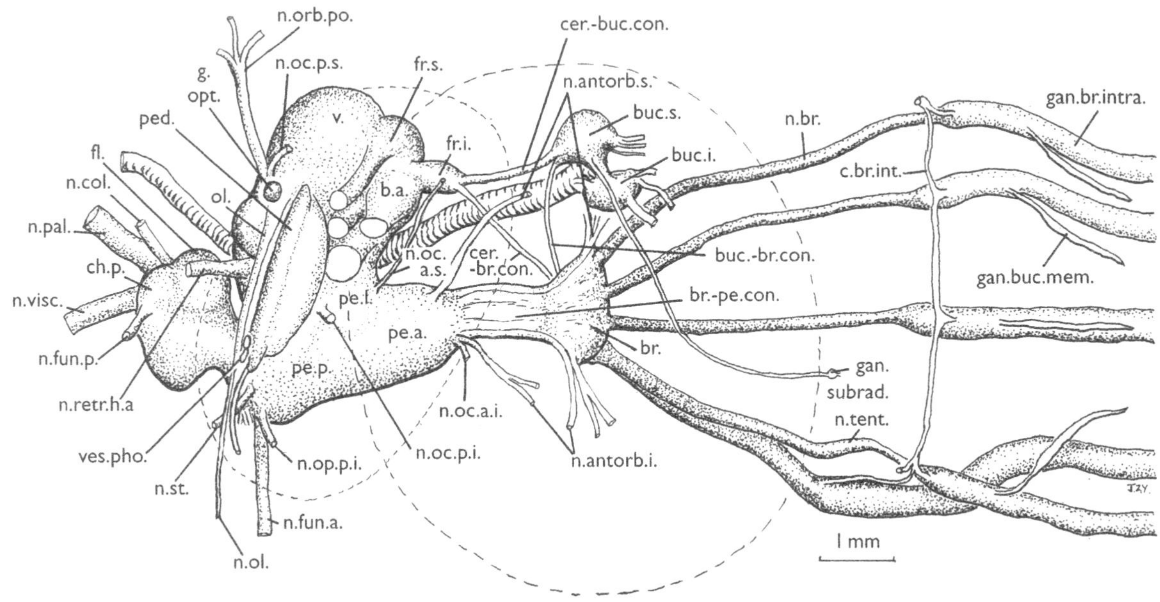

Fig. 2.5. Teuthida, *Loligo vulgaris*. Drawing of reconstruction from tracings of transverse sections of the brain of a specimen (50 mm mantle length), seen from the side. The outlines of the eye and optic lobe are shown by broken lines (Young 1976).

perioesophageal, or magnocellular, lobes. Of these parts, the supraoesophageal mass is structurally the most complex. The suboesophageal mass can be divided into anterior, middle, and posterior parts.

Cephalopods are predatory, highly mobile and most have relatively short life spans, although *Nautilus* is an exception as it lives for 16 years or more. In addition to a well-developed nervous system, cephalopods are equipped with all the major sense organs some of which rival those found among the vertebrates. They have elaborate and highly sophisticated eyes and statocysts as well as other sense organs including extraocular photoreceptors, mechanoreceptors, and chemoreceptors.

2.2 Central nervous system

2.2.1 *Suboesophageal mass and magnocellular lobes*

The anterior suboesophageal cord has two distinct parts in *Nautilus* (Fig. 2.3). The anterior part innervates the numerous delicate tentacles, and is directly continuous with the supraoesophageal cord, and so may be regarded as essentially 'cerebral' (Kerr 1895). The posterior part gives rise to the nerve to the hyponome, or funnel, and is continuous dorsally with the lateral (magnocellular) region. The posterior suboesophageal mass has visceral nerves arising in its posteroventral region, and pallial nerves in the posterodorsal margin, and is termed palliovisceral. The lateral (perioesophageal) region receives afferent fibres from the eyes, tentacles, lips, statocysts, and olfactory (or rhinophore) organs (Young 1965a).

In the Coleoidea, the suboesophageal mass is divided into three parts: anterior, middle, and posterior (Fig. 31.28). The anterior suboesophageal (brachial lobe) part is connected to the middle and posterior parts by the brachiopedal connective. The middle suboesophageal part comprises several lobes including chromatophore, pedal, and funnel lobes. The posterior suboesophageal part includes palliovisceral, funnel, chromatophore, and vasomotor lobes. Decabrachians possess in addition a fin lobe.

The posterior suboesophageal mass is separated from the rest by a transverse connective tissue septum attached to the walls of the cranium. All the nerve fibres pass through numerous foramina in this septum. The anterior portion of the suboesophageal mass, in front of the cephalic arteries and septum, is separable into pedal and brachial lobes. The whole region in front of the septum is often compared with the pedal ganglia of other molluscs, and this term is certainly applicable to the part of the pedal lobe that innervates the funnel. The nature of the rest of the pedal lobe is more debatable since the lateral pedal lobe innervates eye muscles, and so would seem to be part of the head (Young 1971). The question whether the anterior suboesophageal mass should be regarded as 'cerebral' rather than 'pedal' remains to be resolved. Its two sides are connected by a commissure above the oesophagus and this may suggest that it is 'cerebral'. However, this commissure is present only in octopods, being absent from Decabrachia and *Vampyroteuthis* (Vampyromorphida).

The magnocellular lobes run round the side of the brain, partly above and partly below the oesophagus; and also partly in both the middle and posterior suboesophageal masses (Figs 2.4, 5.19). The suboesophageal mass and the magnocellular lobes contain a series of lower motor centres, from which the axons of motoneurons proceed to directly innervate muscles or glands. Included are motoneurons serving the head retractor muscles, funnel, fins, and chromatophores,. The eye muscles are innervated by the oculomotor centres of the anterior lateral pedal lobes. Decabrachia possess a giant-fibre system which consists of a small number of conspicuous cells and may provide a model for the organization of these suboesophageal centres (Young

1939). The two first-order giant cells lie in the ventral magnocellular lobe where jet propulsion is initiated under the influence of fibres from the optic lobes, the statocysts, and probably from other parts of the brain. The axons of these two first-order giant cells pass to the palliovisceral lobe. Here they meet and either fuse (*Loligo*) or exchange synapses (*Sepia*), thus ensuring simultaneous action on both sides of the body. The axons then terminate on second-order cells, which here also come under the influence of fibres from the higher centres of the basal and precommissural lobes. The second-order axons proceed to the various muscle systems involved in movement by jet propulsion. Some are motoneurons ending at the retractor muscles of the head and funnel. Two large axons proceed to the stellate ganglia, where they synapse with third-order axons that pass to circular muscles in the mantle.

Locomotion by jet propulsion also involves particular postures of the arms, and for this there may be nerve fibres proceeding forwards from the region of the first-order giant cell. Little is known in detail about this except in *Pterygioteuthis* (Pyroteuthidae), where there is a conspicuous anterior giant nerve fibre system, linked synaptically with the main one (Young 1977a). The final destination of the anterior fibres is not known, but there are further cells and synapses in the brachial lobes and/or ganglia in the arms. These giant cell chains provide an example of the multisynaptic organization that is characteristic of the lower motor centres of the suboesophageal lobes. Such chains provide for the patterns of movement of all those parts of the body, activated separately or in combination with others, that are involved in any given action. For example, the paired fins operate separately in gentle swimming but there is a commissure that allows the paired fins to function together in rapid movement; and the arms may be used alone, coordinated with the fins and mantle during jet propulsion.

The pathways for many motor activities originate from the region of the junction of the pedal, palliovisceral, and magnocellular lobes. This region receives fibres from the supraoesophageal lobes (the basal, precommissural, and subvertical lobes) and from other parts of the suboesophageal lobes. There are neurons here whose axons proceed directly to the arms, and the neuropil receives fibres from the arms, supraoesophageal lobes, and statocysts. In *Loligo*, for example, control of the fins is from the posterior pedal lobes. From here there are two pathways to the fin lobe, one direct and the other to the anterior palliovisceral lobe, from which a tract runs to the fin lobe. The posterior pedal lobe also sends fibres to the retractor muscles of the head and funnel, and their actions are thus coordinated with those of the fins. Control of positioning of the arms for locomotion is from the magnocellular region in both Octobrachia and Decabrachia. Delicate control of the arms is by the neurons of the anterior pedal and brachial regions, but little is known of how they interact. Electrical stimulation of the anterior pedal lobe of the decabrachian *Sepia officinalis* (Sepiidae) elicited movements that were recognizable as components of normal behaviour such as raising the first and second pairs of arms (Boycott 1961). Tracts proceed from the anterior pedal lobe directly to the arms, bypassing the brachial and prebrachial lobes. Stimulation of these tracts produces movement and colour change in individual arms.

There are great differences in the suboesophageal lobes among the cephalopods. For example the magnocellular lobes are greatly developed in some squids, such as *Mastigoteuthis* (Mastigioteuthidae) and several of the Cranchiidae, and the elaborate organization is rather like that of a higher centre (Young 1977a).

2.2.2 *Supraoesophageal mass*

In *Nautilus* the supraoesophageal mass is a greatly developed form of the cerebral cord found in non-cephalopod molluscs. The cerebral cord is a strip of tissue in communication laterally with the brachial, palliovisceral, magnocellular, optic, and olfactory lobes (Fig. 2.3). Above the union of all these parts the cerebral cord carries a slight swelling, the lateral cerebral lobe. The two pairs of cerebrobuccal connectives arise from the anterolateral edge of the supraoesophageal cord. At the origin of the more medial of these there are slight swellings that may be called posterior buccal lobes (their relations are similar to those of the lobes with the same name in Coleoidea). From the rest of the length of the anterodorsal face of the supraoesophageal cord arise a series of labial nerves (Fig. 2.6). The supraoesophageal cord does not show internal division into the clearly marked lobes so characteristic of coleoid cephalopods. It is not, however, a homogeneous structure and regions with characteristic differences can be discerned, although there are hardly any external signs of division. The regions are (1) a posterior buccal region, (2) a plexiform zone, (3) a laminated zone within this, (4) central and commissural zones, and (5) lateral cerebral lobes (Young 1965a).

In Coleoidea the supraoesophageal mass can be divided into numerous lobes, some of which have relatively large cells. The buccal ganglia, which are higher motor centres for the control of feeding, lie far forward in some coleoids but in others are united with the rest of the supraoesophageal mass (Fig. 2.2). Other higher centres include the basal lobes, which are concerned with the control of motor activities such as locomotion. The more dorsal part of the supraoesophageal mass consists of lobes receiving inputs from the eyes and the arms. These lobes have distinctive patterns of organization, with small cells and bundles of interweaving fibres, and they form centres for processing chemotactile and visual memories. Each centre consists of four lobes: median inferior frontal, subfrontal, lateral inferior frontal, and posterior buccal (chemotactile); and median superior frontal, vertical, lateral superior frontal and subvertical (visual) (Fig. 2.7). Together with the optic lobes, these two centres allow for integration presumably to ensure responses to complex sets of signals from receptors. These groups of supraoesophageal lobes may be termed the 'highest' centres of the brain (Young 1963b, 1965c, 1983, 1991b, 1995a, b).

2.2.3 *Buccal lobes and the inferior frontal lobe system*

In *Nautilus* and Decabrachia the superior buccal lobes are separated from the rest of the brain and lie close to the rear of the buccal mass (Figs 2.5, 2.6). In *Octopus* (Fig. 2.4), *Cirrothauma* (Fig. 27.18), and *Vampyroteuthis* (Fig. 26.9) the superior buccal lobes are part of the central brain. In *Octopus* the superior buccal lobe is a motor centre,

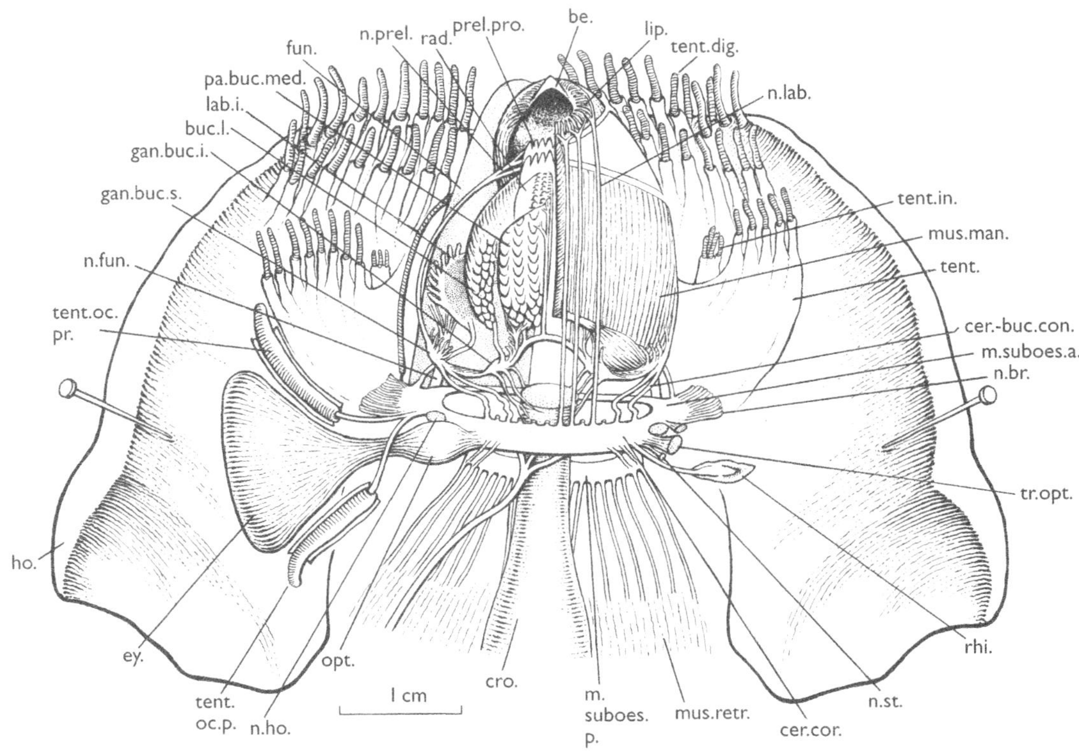

Fig. 2.6. Nautilida, *Nautilus*. Drawing of dissection of the nervous system and buccal mass. The dissection has been carried deeper on the left than the right. The tentacles are shown somewhat diagrammatically (Young 1965a).

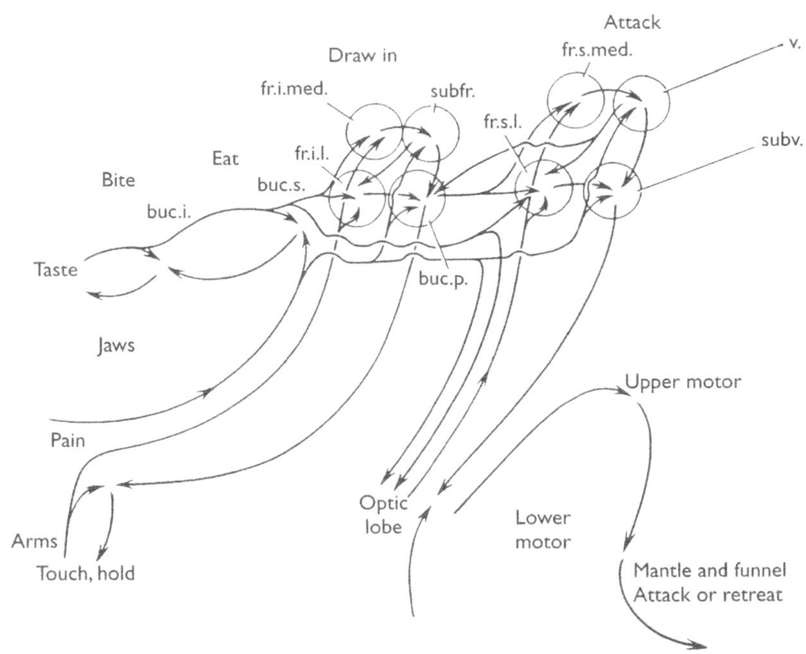

Fig. 2.7. Octopodida, *Octopus vulgaris*. Diagram of the two pairs of centres, arranged in parallel, in the visual and tactile memory systems (modified from Young 1964).

regulating the operations of eating, including the actions of the posterior salivary glands, the salivary papilla, and the buccal mass. The inferior buccal ganglia give rise to the 'sympathetic nerves' which contribute to an elaborate plexus in the walls of the oesophagus and crop (Young 1967).

In Decabrachia, the most anterior part of the central brain is a small inferior frontal lobe (Fig. 2.5). In Octobrachia (Fig. 2.4) the whole inferior frontal system is part of the central brain and is especially well developed, presumably because of the greater use of the arms for handling prey. In *Octopus vulgaris* the inferior frontal lobe system receives fibres direct from the arms and some of its cells send axons to them (Budelmann and Young 1985). The superior buccal lobe is continuous posteriorly with a posterior buccal lobe, which probably represents the inferior frontal lobe of Decabrachia. In *Octopus vulgaris* large tracts of fibres enter the superior buccal lobe direct from the arms and pass through the lateral inferior frontal lobes to a median inferior frontal lobe where chemotactile inputs from the arms and the lips are integrated. Numerous bundles of fibres proceed from the median inferior frontal lobe to the underlying subfrontal lobe, which contains numerous small amacrine cells and a few larger ones going to the posterior buccal lobe, from which efferent axons pass direct to the arms. The circuit within the lateral inferior frontal to median inferior frontal to subfrontal to posterior buccal lobes has been shown to be the system for the storage of chemotactile memory in *Octopus vulgaris* (Fig. 2.7) (Wells 1978, Young 1983, Robertson 1994). It constitutes a special development of the much simpler system found in Decabrachia. The numerous small cells of the subfrontal lobes are continuous with the inner layer of small cells present in the posterior buccal and superior buccal lobes. These small cells are characteristic of those regions of cephalopod ganglia that are concerned with the analysis of sensory input, as investigated for instance in the stellate ganglion (Young 1972). The small cells of the subfrontal lobes, together with the large cells of the posterior buccal lobes, serve to restrain the tendency of the arms to draw in objects previously associated with pain, or the absence of food (Wells and Wells 1957, Wells and Young 1965): after surgical ablation of these lobes the arms grasp all objects encountered, and are subsequently unable to relinquish them (Young 1983).

2.2.4 *Vertical lobe system*

In *Nautilus* the cerebral cord probably constitutes a 'higher' centre allowing for the interaction of signals coming from receptors, but no experimental evidence is yet available.

In Coleoidea, the vertical lobe system consists of a set of lobes which, in experimental studies of *Octopus vulgaris*, have been found to be largely concerned with the establishment of visual and tactile memories (Young 1963b, 1965c, 1983, 1991b, Wells 1978) (Fig. 2.7). This part of the brain may be considered as a modulator of the actions of the systems that control visual and tactile responses, especially the restraining influence of nocifensor (pain) signals received from other parts of the body. The whole system can be regarded as a backward development of the centres at the front of the supraoesophageal lobes that are concerned with obtaining food. In octopods, the masses of small cells of the vertical lobe system are part of a continuous series extending forward in the subfrontal, posterior buccal and the superior buccal lobes. A tactile memory system is not found in decabrachians and the subfrontal and inferior frontal lobes are absent. The basic plan and connectives of the vertical lobe system, with its four parts, is otherwise remarkably similar in Decabrachia, Octobrachia, and Vampyromorphida, and presumably developed in their common ancestor.

Within the major circuit of the vertical lobe system are upper and lower circuits (Fig. 2.7). Visual, tactile, and gustatory impulses enter through the lateral superior frontal lobe. The latter is organized as a matrix of interacting axes, among which associations are found between the signals of input events and their consequences: an arrangement somewhat similar to that in mammals, for instance in the hippocampus (Young 1991b, 1995a). In the lateral superior frontal lobe there is interaction and signals pass to the median superior frontal lobe (which has an interweaving plexus) where there is a matrix of cells (Fig. 2.8). The axons of these cells of the median superior frontal lobe pass through a plexus to the vertical lobe. Here they meet large numbers of minute amacrine cells and a few larger, much branched, neurons whose axons provide a third matrix. Signals pass either to the subvertical and precommissural lobes or back to the lateral superior and inferior frontal lobes, thus constituting an upper circuit. The vertical lobe also receives numerous fibres from below, which are presumed to bring nocifensor signals. The interaction of these signals with the numerous small amacrine cells and the impulses entering from the superior frontal lobe is an essential factor in the establishment of visual and tactile memories (Fig. 2.9). The output of the vertical lobe system is through the subvertical lobe to at least three destinations: (1) back to the optic lobes, (2) back to the inferior frontal lobe system, and (3) down to the suboesophageal lobes via the large cells of the precommissural lobe. There are also connections to the basal lobes, including the lateral subvertical to anterior basal lobe tracts.

The vertical lobe system of Decabrachia differs from that of Octobrachia mainly in the form of the vertical lobe itself. This lobe is dome-shaped in Decabrachia with a relatively thin outer layer of cells and extensive central mass of large cells and neuropil. In Octobrachia the vertical lobe shows variation. In Octopoda it is divided into five lobules, with thick cell walls and relatively scant neuropil, in Cirroctopodida the system consists of islands of cells separated by neuropil with no clear distinction between vertical,

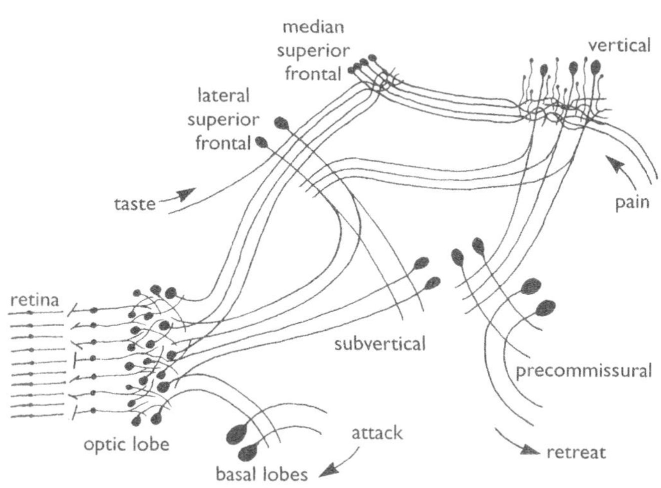

Fig. 2.8. Octopodida, *Octopus vulgaris*. Diagram of the matrices of the visual memory system (Young 1995a; with permission of Oxford University Press).

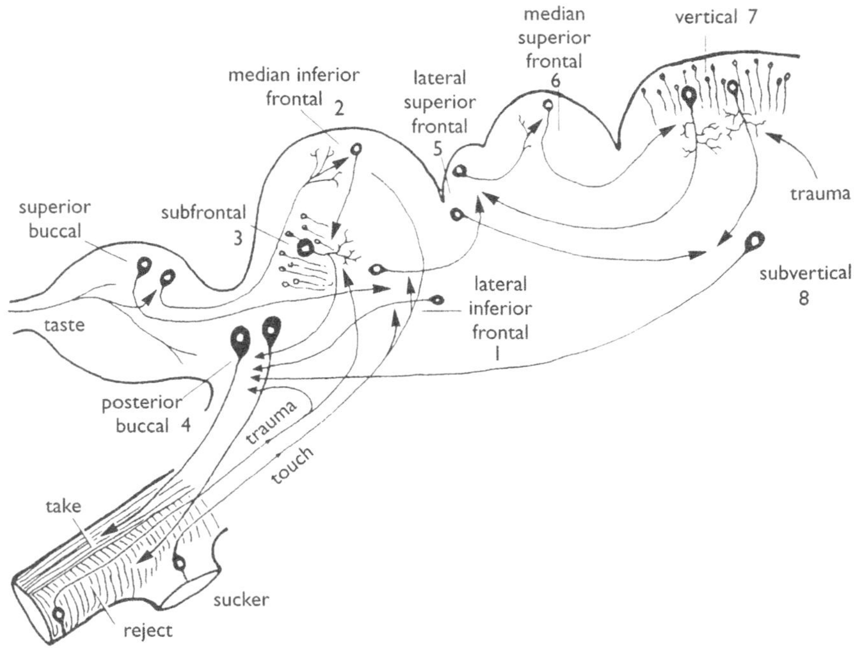

Fig. 2.9. Octopodida, *Octopus vulgaris*. Diagram of the matrices of the tactile memory system. The successive matrices are shown, of which the first four are purely tactile (Young 1995a; with permission of Oxford University Press).

subvertical and dorsal basal lobes, and in Vampyromorphida it is represented by islands of neuropil surrounded by masses of small cells.

2.2.5 *Basal lobes*

In *Nautilus* the central and ventral (commissural) regions of the cerebral cord contain larger cells than the dorsal parts, and appear to provide the output pathways to other parts of the central nervous system (Young 1965a). In this respect they resemble the comparable basal regions of the coleoid brain.

In Coleoidea the basal lobes form a system of higher motor centres regulating the direction and speed of movements (Young 1971). This system includes the peduncle lobe and the anterior basal and median basal lobes; all have a rather similar organization, though the median basal lobe differs somewhat. The common features include the pattern of connections of the lobes and the presence of numerous very fine parallel fibres, similar to those of the vertebrate cerebellum (Messenger 1967b, Hobbs and Young 1973, Woodhams 1977). The basal lobes seem to be largely concerned with regulating the performance of actions initiated elsewhere in the brain. Electrical stimulation shows that these lobes are concerned with the production of coordinated actions rather than movement of individual parts (Boycott 1961).

The main inputs to the basal lobes are from the optic lobes, the statocysts, the other basal lobes, the ventral magnocellular lobes, the lateral pedal lobes, the subvertical lobe, and the precommissural lobe (Young 1971). The visual input comes from all parts of the optic lobes, maintaining topographical organization, at least for the peduncle lobes (Messenger 1983), and probably the anterior basal lobes.

An important part of the output of the basal lobes passes to the oculomotor centre of the anterior lateral pedal lobe. The anterior and median basal lobes, the anterior lateral pedal lobes, the peduncle lobes, and the magnocellular lobes are all involved in the control system for compensatory eye movements, the statocyst–oculomotor reflex (Budelmann and Young 1984). Among invertebrates this system attains the highest level of complexity in the cephalopods (Budelmann and Tu 1997).

The anterior basal lobe has numerous connections with the optic lobes. In *Loligo* this lobe consists of two parts, anterior and posterior, each with a distinct set of numerous very fine parallel fibres (Young 1977b).

The median basal lobe lies at the rear of the supraoesophageal mass and is concerned especially with the control of movements of the mantle, collar, and funnel in locomotion, and in respiration and feeding.

The output from each higher motor centre passes to the lower and intermediate motor centres (in the suboesophageal lobes), to the optic lobes, and to the other basal lobes. Each higher motor centre has a region of large neurons and a 'spine' region, with small cells and a neuropil containing very fine fibres. In the peduncle lobe, and perhaps others, the optic input may excite the output cells in an orderly sequence, the timing of which may be regulated by the fine fibres acting as delay channels since speed of conduction is directly related to diameter in cephalopod neurons (Pumphrey and Young 1938).

An interbasal lobe is present in Decabrachia on each side of the brain between the anterior basal lobe and the median basal lobe (Young 1977b). It has been little investigated but in *Sepia* electrical stimulation here causes tentacle ejection (Boycott 1961). The lobe in a comparable position in octopods is small, and as yet of unknown function.

The lateral basal lobes can just be discerned externally as slight swellings, on the lower posterior face of the supraoesophageal mass,

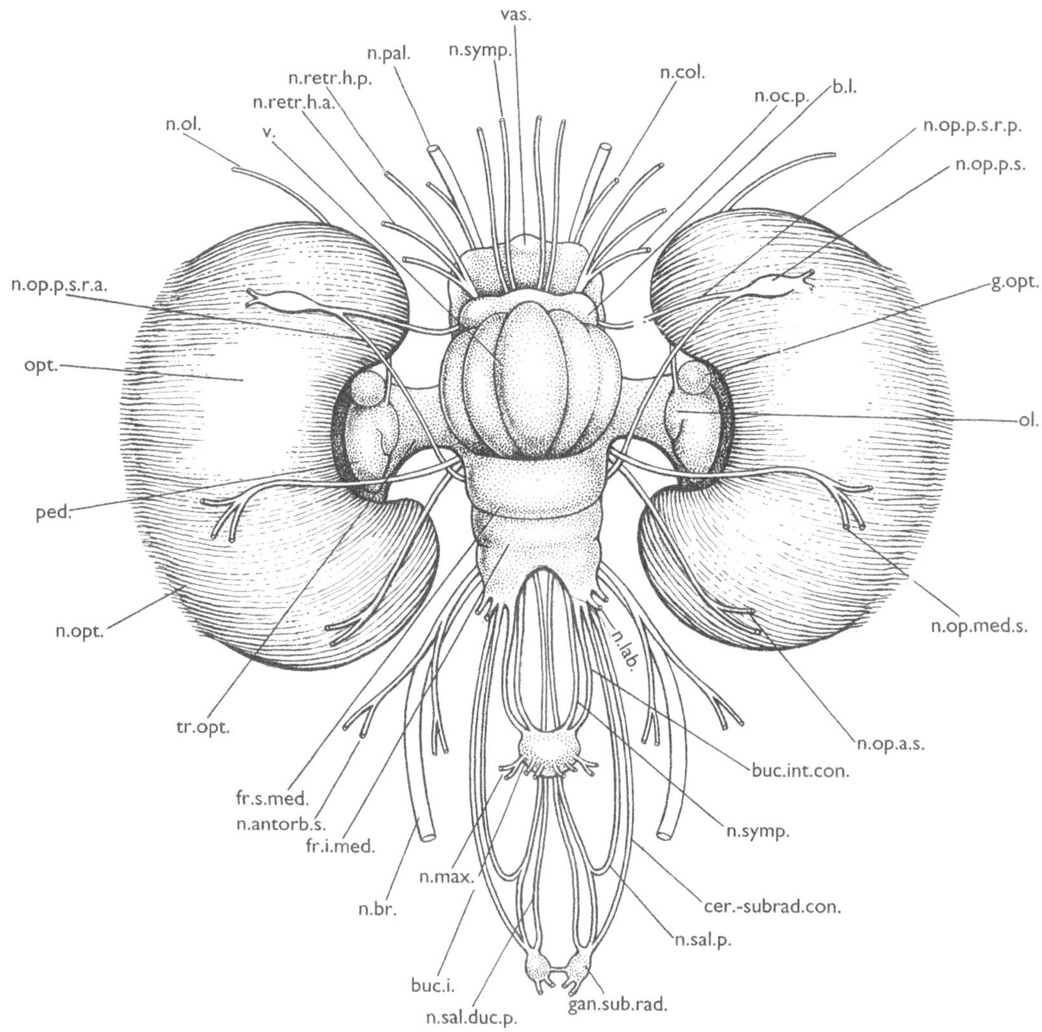

Fig. 2.10. Octopodida, *Octopus vulgaris*. Drawing of the central nervous system, showing the position of the optic and peduncle lobes; seen from above (Young 1971).

on either side of the median basal lobe (Young 1971, 1977b). They are concerned with the control of the chromatophores and muscles of the skin. Functionally they lie between the optic lobes, which initiate the body patterns displayed, and the final motoneurons of the chromatophore lobes (Boycott 1961).

The dorsal basal lobe lies above the median basal lobe and it consists of groups of cells, often isolated as islands, receiving largely optic tract fibres (Young 1977b). Fibres leave the posterior region of the dorsal basal lobe to run to the optic gland. In octopods there arises here a 'subpedunculate nerve', which runs down and along the floor of the orbit to the pharyngoophthalmic vein (Young 1971, Wells 1978).

2.2.6 *Optic lobes*

The optic lobes can be considered as special developments of the molluscan cerebral cord and are always the largest lobes of the brain (Figs 2.1, 2.10). They lie on either side of the central brain, each in a venous sinus immediately behind the eye.

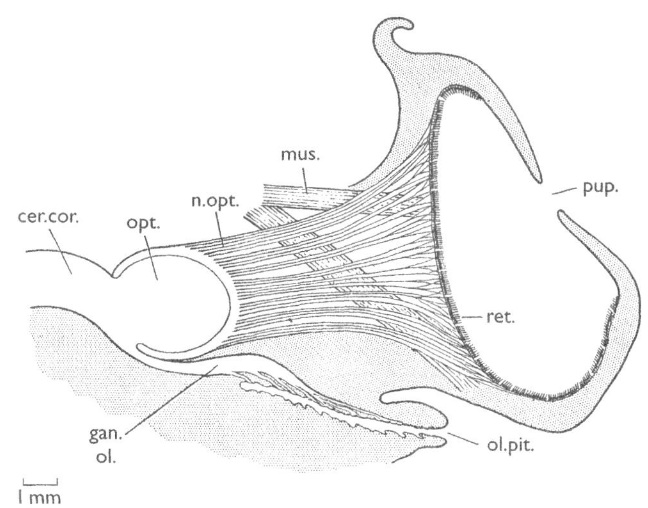

Fig. 2.11. Nautilida, *Nautilus*. Drawing of a section through the eye and optic lobe showing the pinhole eye, retina, and optic nerves without the notable chiasma seen in coleoids (Young 1965a).

Each optic lobe is ovoid in form in *Nautilus* (Fig. 2.6) and optic nerves from the retina enter over its lateral surface (Fig. 2.11) (Young 1965a), and although a chiasma is apparently lacking there are some signs of a dorsoventral one (Muntz and Raj 1984). Each optic lobe is formed of layers, continuous with the comparable regions of the hind part of the cerebral cord and are lateral extensions of the cord.

In many coleoids the optic lobes are kidney- or bean-shaped (Fig. 2.10), but there is considerable variation in their shape and form from species to species The organization of the optic lobes is elaborate and essentially the same among the cephalopods, although in Cirroctopodida that of the outer, cortical layers is simpler. There is an outer cortex and a central medulla. The cortex has three layers: an

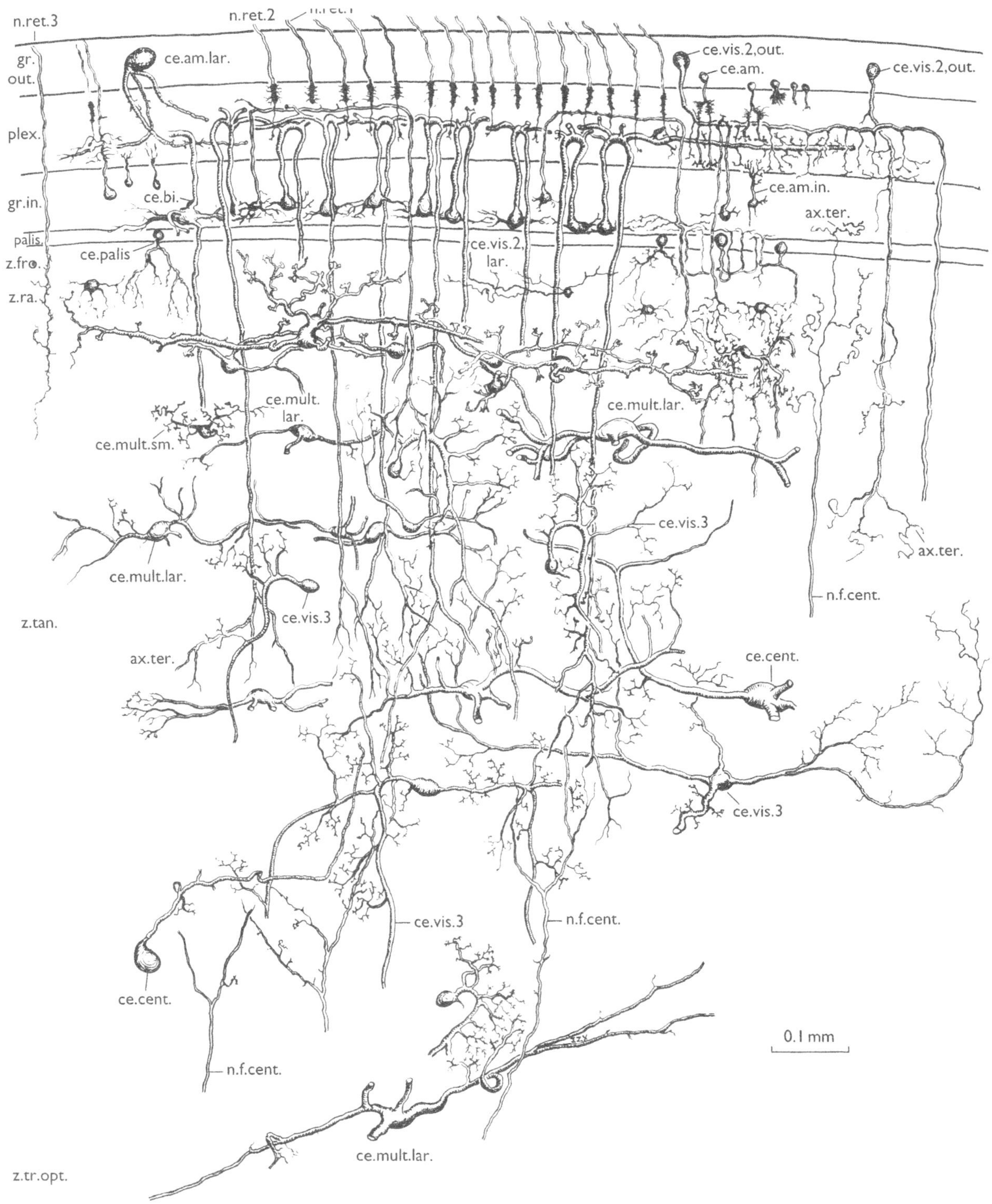

Fig. 2.12. Teuthida, *Loligo*. Cells of the optic lobe drawn from sagittal section showing especially the axons of the largest second-order visual cells, the large multipolar cells of the medulla, and the third-order cells (Young 1974).

outer and an inner granule cell layer with a plexiform zone between (Fig. 2.12). In Octopoda each optic lobe receives optic nerves after they have passed through a dorsoventral chiasm, making a precise re-inverted projection on to the cortex of the optic lobe (Fig. 2.13) (Young 1962a, b). However, there is no chiasm in *Cirrothauma murrayi*, the eye of which lacks a lens (Chun 1910a, Aldred *et al.* 1983). The preservation of the topographical arrangement is probably a fundamental feature of the whole organization of the visual system of the majority of coleoids. The optic nerve fibres mostly end in the plexiform zone, a system of tangential layers in which the optic fibres make contact with the dendrites of the second-order visual neurons (Young 1971, Lam *et al.* 1974). These cells have an unusual form, with a trunk running up to the plexiform zone, giving off tangential dendritic branches and then turning down again to the medulla. The dendrites have a highly organized planar orientation most commonly in the transverse or longitudinal planes according to species (Young 1962a). They are of various lengths, some very long, so their receptive fields presumably cover a large fraction of the total visual field. On either side of the plexiform zone the inner and outer granule cell layers include amacrine cells of various sizes, and cells which give rise to dopaminergic efferent fibres to the retina (Messenger 1996). The cortex of the optic lobe thus carries a system (the 'deep retina' of Cajal 1917), which shows a certain likeness to the deeper layers of the vertebrate retina; the neurons may have functional similarity, although in cephalopods these layers are separated from the eye by the long optic nerve fibres.

The axons of the second-order neurons proceed inwards to the centre of the optic lobe in parallel bundles, surrounded by columns of nerve cell bodies. Many of these are small amacrine cells, others are multipolar horizontal cells with widely spreading dendrites, remarkably like their counterparts in vertebrates. The columns of cells join progressively towards the centre of the lobe and give rise to islands of cells whose axons proceed to the optic tract. Fibres running outwards from the brain and the opposite optic lobe reach to all parts of the medulla of the optic lobe but probably not to the plexiform zone.

The pattern of connections within the optic lobes seems remarkably constant throughout the class, and is already present in *Nautilus*, although the granule layers are not distinct. In the cirroctopod *Cirrothauma murrayi*, which has no lens, the granule cell layers are present but irregular, suggesting that in most cephalopods the small granule cells may be concerned with features such as edge detection. The optic lobes are involved in analysing the visual input and regulate visual behaviour and learning, and in addition are the site for the visual memory store and the expression of the chromatic components of the skin (Young 1962a, 1964).

2.2.7 *Peduncle lobes*

The peduncle lobes lie on the optic tracts (Fig. 2.10) and each has a conspicuous set of fine parallel fibres in its 'spine' (Messenger 1967b, 1979a). They are concerned with the regulation and control of locomotion, primarily on the basis of visual cues. In Decabrachia the peduncle lobes also receive information from extraocular photoreceptor organs, or photosensitive vesicles, and may serve in the regulation of vertical movements, such as diel, ontogenetic, or reproductive migrations (R.E. Young 1978), or of the intensity of light emitted by

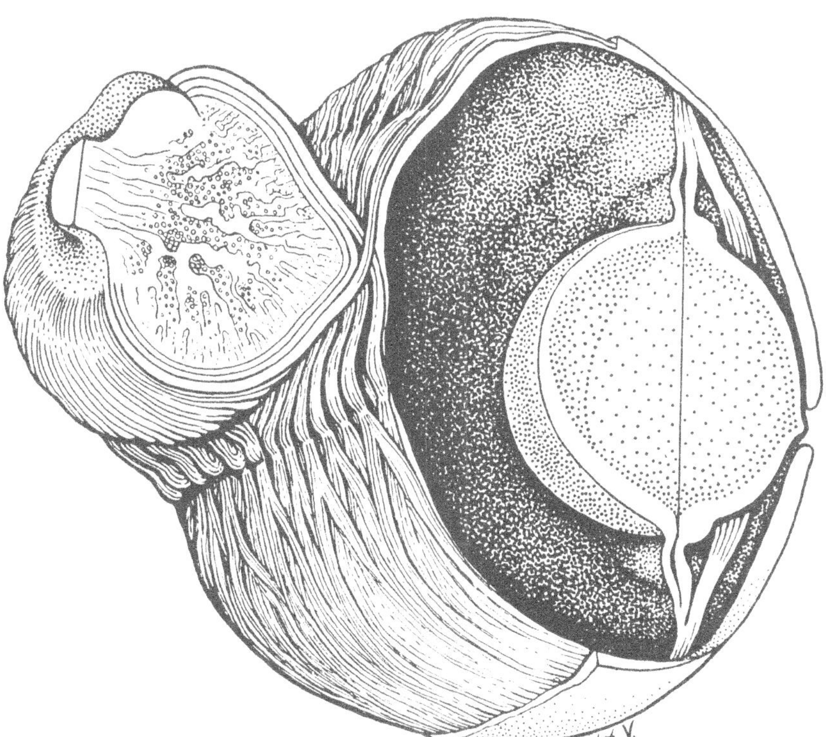

Fig. 2.13. Octopodida, *Octopus vulgaris*. Drawing of a dissection of the eye and optic lobe to show the arrangement of the optic nerves and their chiasma (Young 1962a).

photophores involved in countershading (R.E. Young and Roper 1976). With these added functions the peduncle lobes are larger in Decabrachia than in Octobrachia. The peduncle lobes are involved in other functions, including the control of chromatophores and erection of papillae in the skin.

2.2.8 *Olfactory lobes*

The olfactory lobes of *Nautilus* are essentially lateral extensions of the cerebral cord, arising behind the optic stalk, and passing ventrally and backwards. Each lobe receives the olfactory nerve from the rhinophores (Figs 2.3, 2.6). The layers characteristic of the cerebral cord can be recognized in this lobe (Young 1965a).

In coleoids the olfactory lobes lie medial to the peduncle lobes and are closely associated with the dorsal basal area (Fig. 2.10). The neuropil of the olfactory lobes becomes confluent with that of the basal zone of the peduncle lobes. The lobes, with the more medial dorsolateral lobes, receive fibres from the olfactory organs (Messenger 1979a).

2.3 Functional organization of the central nervous system

The main functions of the lobes of the central brain are given in Table 2.1. The efferent pathways are often interrupted by ganglia between the brain and the effector organs, for instance the brachial ganglia, the stellate ganglia, and the buccal ganglia. Many reflex actions are thus performed locally and are not under detailed central control. This is especially so in octopods where the axial nerve cords of the arms of *Octopus vulgaris* contain many more nerve cells than are present in the brain (Table 2.2) (Young 1971). The intricate operations of the suckers and movements of the arms are therefore locally controlled. This principle of decentralization leaves the central ganglia of the system free to organize the behaviour of the whole animal. This is achieved by a hierarchical arrangement of the neurons in which the lower centres control smaller groups of muscles and the higher centres regulate major actions, for example those of the fins or the arms. The highest centres initiate responses of the whole animal, partly in accordance with past experience. These properties have been demonstrated using electrical stimulation and surgical lesion techniques (F.K. Sanders and Young 1940, Boycott and Young 1950, Boycott 1960, 1961, Chichery and Chanelet 1976, Chichery and Chichery 1991a, 1991b).

For the purposes of description it is convenient to recognize the following types of 'centre', although this is at best only a crude categorization (Boycott 1961):

- The *lower motor centres* are strictly only those that contain final motoneurons. Electrical stimulation of these centres therefore produces movements of isolated groups of muscle fibres. One example is the expansion of patches of chromatophores on the mantle as a result of stimulating the posterior chromatophore lobe.
- The *intermediate motor centres* send fibres to the peripheral ganglia (e.g. stellate) where the final motoneurons lie, and thus produce movements of whole parts, such as the arms or the mantle, but not of individual groups of muscles.
- *The higher motor centres* when stimulated produce combined movements of different groups of effectors. For instance stimulation of any part of the anterior basal lobe gives turning movements of the head, eyes, funnel, and fins in the manner shown by a normal animal.

Table 2.1. A summary of the main functions ascribed to the major lobes of the brain, derived from studies of *Octopus vulgaris* (Young 1971)

Lobes	Function(s) of the lobes
Suboesophageal mass	
Anterior suboesophageal mass	Actions of arms and suckers
Middle suboesophageal mass	Intermediate and lower motor centres involved in most actions of animal
Posterior regions suboesophageal mass	Actions of mantle and some of those of the viscera
Supraoesophageal mass	
Anterior basal lobe	Control of posture and movement of head and eyes
Dorsal basal lobe	Probably controls actions of defence and avoidance
Median basal lobe	Concerned in institution and control of movements of funnel and mantle
Lateral basal lobes	Control of chromatophores and muscles of skin
Peduncle lobes	Part of motor control system, coordinates motor activity, colour changes and ink-ejection
Superior buccal lobe	Motor control of feeding
Vertical lobe, median superior frontal lobe, lateral superior frontal lobes and subvertical lobe	Regulation of exploratory and learning behaviour and memory
Median inferior frontal lobe lateral inferior frontal lobes, subfrontal lobes and posterior buccal lobe	Closely inter-related lobes which together form a functional system concerned with chemotactile information from arms
Perioesophageal lobes	
Magnocellular lobes	Intermediate motor centre between supraoesophageal higher and suboesophageal lower motor centres; involved in defence and fast escape reactions. In Decabrachia the giant fibre system originates here

Table 2.2. The numbers of cells in the lobes of the central nervous system of *Octopus vulgaris* (Young 1963b)

	Lobes		Number of nerve cells
	Optic		128 940 000
Supraoesophageal mass	Vertical		25 066 000
	Subvertical		810 000
	Frontal	superior	1 854 000
		inferior	1 085 000
	Subfrontal		5 308 000
	Subpedunculate		144 000
	Basal	anterior	380 000
		dorsal	1 796 000
		inter	57 000
		lateral	127 000
		median	245 000
	Buccal	superior	150 000
	Peduncle		142 000
	Olfactory		136 000
Suboesophageal mass	Palliovisceral		108 000
	Chromatophore		526 000
	Vasomotor		1 307 000
	Brachial		341 000
	Pedal		241 000
Perioesophageal lobes	Magnocellular		581 000

- The *silent or correlation areas* are so called as they cannot be effectively activated, and are in the highest regions of the brain. Lesion experiments have shown that these areas are concerned with learning and probably other aspects of behaviour such as motivation, searching, and hunting.
- The *primary sensory areas* receive information from receptors for correlation in such a way as to provide responses appropriate to changes in the environment. The optic lobes are concerned with the analysis of complex incoming signals. The peripheral layers of the optic lobes are silent areas, but electrical stimulation of the centre of these lobes elicits a variety of reponses.

The functions of the lobes of the brain are described in detail in Chapters 5, 8, and 31, and summarized in Table 2.1.

New techniques have become available and appear to be promising tools for the study of both the physiology and morphology of the cephalopod brain. They include the tracing of labelled pathways with three-dimensional magnetic resonance microscopy, brain-slice recordings, implanted electrodes to record from intact animals, and mapping of metabolic brain activity (Bullock and Budelmann 1991, Williamson and Budelmann 1991, Novicki *et al.* 1992, Budelmann *et al.* 1995).

2.4 The brain and its lobes

Cephalopods have active lifestyles, especially when compared with other molluscs, and must routinely process much information. The visual information from the eyes is considerable and the brain has large visual areas; in the two optic lobes there are almost 100 million cells in a specimen of *Octopus vulgaris* weighing nearly 2 kg (Young 1963c). There are other sensory inputs, most notably from the arms and suckers (Graziadei 1971).

Cephalopods have the largest and most complex nervous system of all invertebrates. The brain-to-body ratio exceeds that of most fishes and reptiles (Packard 1972). The brain of *Octopus vulgaris* grows extensively during the lifetime of the animal. The cell numbers in the brain increase from 2×10^5 at hatching to 2×10^8 in the adult (Young 1963c, 1971, Packard and Albergoni 1970, Giuditta *et al.* 1971). The numbers of cells in the various lobes of the brain of *Octopus vulgaris* are shown in Table 2.2. Growth of the brain has been demonstrated in 10 species of oceanic squids (Zuev 1975); presumably similar increases take place in both the number and the size of cells in the brains of other cephalopods.

Twenty-five major lobes have been characterized in the brain of coleoid cephalopods, and further subdivisions have been recognized, making nearly 40 lobes in all (Young 1971, Budelmann 1995). The volumes of numerous lobes of the brains of 23 species of Decabrachia and 11 Octobrachia, all from the Mediterranean Sea, were measured from histological sections in a classical study by Wirz (1959). Using the same method Frösch (1971) measured the brains of newly hatched young of 10 of the species investigated by Wirz (1959) (Table 2.3). Maddock and Young (1987), using similar methods, determined the volumes of the whole central nervous system and the optic lobes (Table 2.4), and the volumes of the 30 distinct lobes of the brain of 38 species of Decabrachia and 24 Octobrachia, and also the 2 major parts of the brain of *Nautilus* (Table 2.6). In coleoids the lobes were organized into 12 functional sets (Table 2.5) and their volumes expressed as percentages of the total brain volume (Table 2.6). Principal components analysis (Blackith and Reyment 1971) was used to investigate the relationships between the different species, as shown by the relative sizes of the lobes of the brain.

Comparison of the functional sets of lobes reveals that the inferior frontal system is relatively larger in octobrachians than in decabrachians, and the optic lobes are generally smaller in the former group (Maddock and Young 1987) (Table 2.6). *Vampyroteuthis* lies somewhat between these two groups. Even the division into orders is not apparent from the dimensions of the parts of their brains. There is, however, striking similarity between the vertical lobe and the giant-fibre system of *Sepia* and *Loligo* (Young 1977a). Among Octobrachia there is a sharp division between the Cirroctopodida and the Octopodida due to the presence in the former of a fin lobe and of relatively large brachial lobes and a small vertical lobe system. Among Octopodida benthic forms differ from pelagic forms as the brachial lobes and inferior frontal lobe system in the former are relatively larger than in the latter. Among Decabrachia the giant squid *Architeuthis* is odd as it has a relatively large brachial lobe and quite small optic lobes (Table 2.6). *Vampyroteuthis* has a supraoesophageal mass which forms a larger portion of the total volume of the brain than it does in all of the other coleoids measured, as does its vertical lobe system The brain of *Nautilus* differs from that of coleoids, but some comparisons can be made. The supraoesophageal lobes are

Table 2.3. The volumes of the lobes of the brains of newly-hatched specimens (italicized) (Frösch 1971) and adults (Wirz 1959) of several coleoid species captured in the Mediterranean Sea. The same methods were used in both sets of measurements

	Octopus vulgaris	*Eledone cirrhosa*	*Argonauta argo*	*Sepia officinalis*	*Rossia* sp.	*Sepiola* sp.	*Sepietta oweniana*	*Sepietta obscura*	*Loligo vulgaris*	*Loligo (Alloteuthis) media*
Superior buccal lobe	5.12	4.03	4.48	1.87	2.75	2.35	2.45	2.60	1.97	2.57
	3.31	*3.56*	*2.75*							
Inferior frontal lobe	2.55	3.36	3.46	1.11	1.73	1.31	1.32	1.57	1.33	1.62
	2.96	*3.90*	*1.23*	*0.93*	*1.37*		*1.14*		*1.32*	*1.00*
Superior frontal lobe	3.65	3.90	4.23	4.37	3.84	3.83	4.31	3.54	3.51	3.40
	3.12	*3.81*	*2.95*	*3.38*	*2.70*		*2.74*		*3.08*	*3.69*
Subvertical lobe	0.78	1.47		2.16	1.89	1.61	1.83	1.52	1.45	1.54
	0.62	*0.69*		*1.30*	*0.61*		*0.32*			*0.87*
Subfrontal lobe	0.18	0.45	0.19							
	0.75	*1.29*	*0.38*							
Vertical lobe	1.95	3.26	2.08	6.94	6.01	4.52	5.20	4.35	3.43	4.02
	4.91	*5.78*	*4.32*	*11.86*	*8.28*		*5.70*		*9.85*	*13.37*
Anterior basal lobe	8.78	7.19	6.14	8.49	9.64	9.73	9.31	9.05	10.25	11.76
	3.25	*3.81*	*3.20*	*5.61*	*4.24*		*5.42*		*5.20*	*4.94*
Posterior basal lobe	26.70	24.80	23.44	30.60	28.91	28.75	29.23	28.96	32.31	30.95
	15.15	*15.65*	*23.01*	*20.36*	*18.14*		*27.25*		*21.40*	*23.59*
Brachial lobe	8.31	11.39	7.50	4.65	8.87	6.34	5.78	6.65	4.41	4.55
	17.51	*16.81*	*12.49*	*7.81*	*12.87*		*9.42*		*9.41*	*7.40*
Pedal lobe	16.65	14.97	17.86	23.90	25.71	22.01	23.88	24.66	26.12	26.79
	18.02	*17.05*	*20.99*	*18.14*	*19.99*		*22.08*		*21.05*	*19.06*
Palliovisceral lobe	25.86	26.51	31.97	19.39	15.28	23.51	20.74	21.23	18.63	16.91
	24.08	*21.49*	*28.68*	*22.11*	*22.94*		*25.93*		*19.43*	*19.32*
Inferior buccal ganglion	2.31	2.07	2.94	1.17	1.79	1.67	1.58	1.50	1.16	1.16
Stellate ganglion	2.18	2.77	2.57	2.73	1.85	2.86	2.28	2.25	1.55	1.52
Gastric ganglion	0.88	0.98	1.45	0.48	0.64	0.97	0.51	0.51	0.46	0.49
Subradular ganglion	0.40	0.35		0.023					0.0056	
Optic lobe	57.40	72.40	59.17	120.80	108.30	96.50	109.00	106.80	93.49	79.01
	61	*82*	*104*	*108*	*104*		*136*		*134*	*180*

small but even so are larger than in some coleoids (e.g. *Sandalops*, a cranchiid squid, and Cirroctopodida). The optic lobes of *Nautilus* form a smaller portion of the brain than in any of the coleoids measured apart from *Cirrothauma murrayi* (Maddock and Young 1987): in both the eye lacks a lens (Griffin 1900, Chun 1913, Young 1965a, Aldred *et al.* 1983).

Changes in the proportions of the lobes of the brain take place between hatching and the juvenile or adult stage of life, those in some lobes being quite marked (Table 2.3) (Wirz 1959, Frösch 1971). Several of the changes have been correlated with morphological developments and changes in behaviour, habitat, and habits between hatching and the juvenile or adult stage of life in both *Sepia officinalis* (Nixon and Mangold 1998) and *Octopus vulgaris* (Nixon and Mangold 1996). In the latter, the buccal lobes form a larger proportion of the brain at hatching than at settlement on the sea floor; in contrast, the brachial lobes increase considerably at this time, coincident with the rapid growth of the arms and suckers.

In *Sepia officinalis*, the basal lobes form a much larger proportion of the brain at hatching than in the juvenile (Table 2.3). These lobes are involved in various activities including swimming, jet propulsion, and movements of the head, arms, tentacles, and fins. The relatively large volume of these lobes in the hatchling is in accord with the time spent actively in the water column (Nixon and Mangold 1998). The vertical lobe system forms 7.0% of the total brain volume at hatching and 11.8% in juveniles (Wirz 1954, 1959, Frösch 1971) and is concerned with the learning of visual tasks. Improvements in learning during the first 4 months of life in *Sepia officinalis* mirror the development of its vertical lobe system (Messenger 1973). As yet there are no similar studies in other newly hatched species of cephalopod. Their ability to learn is likely to differ, but these highly successful, short-lived, animals are all likely to have a capacity to learn.

2.5 Sensory systems

Cephalopods have numerous sense organs, and some rival those found in vertebrates in their complexity (Budelmann 1996, Budelmann and Tu 1997).

Table 2.4. The volumes of the central nervous system and of the optic lobes for each specimen. These were mostly identified to species level and are listed alphabetically. The abbreviation for each is given on the left, but these are not in strict alphabetical order (*Ocythoe tuberculata* weighed 700 g but its mantle length was not measured; ROSS refers to *Neorossia caroli*) (Maddock and Young 1987)

Abbreviation	Specimens	Mantle length (mm)	CNS volume (mm^3)	Optic lobe volume (mm^3)
ABR	*Abraliopsis morrisi*	32	9.8	33.0
ALLO	*Alloposus mollis*	—	33.8	126.4
AMPH	*Amphitretus pelagicus*	32	3.8	5.6
ARCH	*Architeuthis sp.*	1570	846.8	919.8
ARGO	*Argonauta argo*	—	10.6	24.0
BAT	*Bathothauma lyromma*	165	10.0	36.4
BATP	*Bathypolypus sponsalis*	25	26.4	11.6
BAY	*Bathyteuthis abyssicola*	30	3.0	4.0
BENO	*Benthoctopus piscatorum*	89	129.4	63.6
CHIRO	*Chiroteuthis veranyi*	68	10.8	39.4
CTE	*Chtenopteryx sicula*	36	5.2	19.0
CIR	*Cirroteuthis sp.*	75	8.6	3.4
CIRR	*Cirrothauma murrayi*	155	7.2	1.0
CRAN	*Cranchia scabra*	125	19.6	119.2
DISCO	*Discoteuthis lacinosa*	51	21.8	75.8
EGE	*Egea inermis*	198	15.8	37.4
ELED	*Eledone moschata*	—	43.8	15.4
ELLA	*Eledonella pygmaea*	75	8.8	20.0
GAL	*Galiteuthis glacialis*	297	14.8	81.6
GON	*Gonatus fabricii*	110	20.6	80.8
GRIM	*Grimalditeuthis bonplandi*	67	5.0	22.8
GRIMP	*Grimpoteuthis sp.*	160	47.2	22.8
HEL	*Helicocranchia papillata*	—	1.8	5.4
HET	*Heteroteuthis dispar*	11	3.4	9.2
HIST	*Histioteuthis miranda*	60	33.4	98.2
ILL	*Ilex illecebrosus*	162	49.0	149.4
JAP	*Japetella diaphana*	25	1.8	3.4
JOUB	*Joubiniteuthis portieri*	36	3.0	4.0
LEA	*Leachia dislocata*	132	4.6	13.2
LOL	*Loligo forbesi*	50	65.0	218.6
LOLL	*Lolliguncula brevis*	75	11.6	19.2
LYCO	*Lycoteuthis diadema*	80	36.0	153.8
MAST	*Mastigoteuthis schmidti*	60	11.6	17.0
MEG	*Megalocranchia sp.*	335	74.6	123.8
NAUT	*Nautilus pompilius*	120	109.1	29.8
ROSS	*Neorossia caroli*	60	6.6	14.8
NEO	*Neoteuthis sp.*	23	3.8	14.2
OCTEU	*Octopoteuthis danae*	70	38.2	101.4
OBIM	*Octopus bimaculatus*	—	49.8	57.0
ODEF	*Octopus defilippi*	—	12.6	23.8
ODOF	*Octopus dofleini*	—	926.2	412.6
OMAC	*Octopus macropus*	—	66.6	49.8
OSAL	*Octopus salutii*	—	91.6	93.0
OVUL	*Octopus vulgaris*	80	92.6	79.0

Table 2.4. *Continued*

Abbreviation	Specimens	Mantle length (mm)	CNS volume (mm^3)	Optic lobe volume (mm^3)
OCY	*Ocythoe tuberculata*	—	98.4	165.6
ONY	*Onychoteuthis banksi*	68	4.8	11.0
OPIST	*Opisthoteuthis sp.*	37	22.8	35.6
PICK	*Pickfordiateuthis pulchella*	21	2.4	6.8
PTERO	*Pteroctopus tetracirrhus*	—	8.6	13.0
PTE	*Pterygioteuthis giardi*	23	1.4	4.2
PYRO	*Pyroteuthis margaritifera*	24	4.4	15.0
SAN	*Sandalops sp.*	112	10.8	23.4
SGAE	*Scaeurgus unicirrhus*	—	50.8	87.8
SEP	*Sepia officinalis*	80	163.8	232.4
SEPI	*Sepiola rondeleti*	23	5.6	12.6
SEPT	*Sepioteuthis sepioidea*	150	55.4	205.8
SPIR	*Spirula spirula*	21	7.0	15.6
TAO	*Taonius pavo*	540	35.8	192.6
TEU	*Teuthowenia megalops*	265	51.4	209.4
TREM	*Tremoctopus violaceus*	—	44.0	114.4
VAMP	*Vampyroteuthis infernalis*	45	10.6	36.4
VITR	*Vitreledonella richardi*	66	4.8	11.8

Table 2.5. The abbreviations, in alphabetical order, used for the combinations of lobes included in each of the functional sets; their volumes are given in Table 2.6, as are those for the supra- (SUPRA) and suboesophageal (SUB) masses

Abbreviations	Lobes
BRAC	Brachial
CHROM	Chromatophore
FINL	Fin
INFF	Inferior frontal Subfrontal Superior buccal Median inferior frontal
LATB	Lateral basal
MEDB	Median basal Dorsal basal Interbasal Dorsolateral Subpedunculate
OPTIC	Optic
PALL	Pallial Magnocellular
PARA	Anterior basal Peduncle Olfactory
PEDAL	Pedal
PSV	Photosensitive vesicles
VERT	Superior frontal Vertical Subvertical Precommissural

Table 2.6 The volumes of the functional sets of lobes, measured on one side and then expressed for the two sides together, are given as percentages of the *whole* central nervous system, excluding the optic lobes; the figures for the latter represent both together. The total volume of the brain, the brachial, pallial, and optic lobes of *Nautilus* were included to allow comparison with the Coleoidea (Maddock and Young 1987). A few corrections to the table have been made since its publication. The specimens measured are listed in Table 2.4 and the abbreviations for the functional sets of lobes in Table 2.5

Species	INFF	VERT	PARA	MEDB	LATB	BRAC	PEDAL	PALL	CHROM	FINL	OPTIC	PSV	SUPRA	SUB
Vampyromorphida														
Vampyroteuthis	5.6	27.6	13.2	10.7	0.0	8.5	15.6	16.0	0.0	2.8	378.4	0.2	57.1	42.9
Decabrachia														
Spirulida														
Spirula	3.1	14.2	15.3	15.6	0.0	11.2	17.1	20.2	0.0	2.7	223.8	1.1	48.7	51.3
Sepiida														
Sepia	1.4	23.5	7.0	12.6	1.3	6.1	16.3	26.0	1.8	3.5	141.9	0.0	46.4	53.6
Sepiolida														
Neorossia	3.3	15.0	9.4	11.8	0.9	8.7	20.5	26.6	1.3	2.4	227.5	0.0	40.5	59.5
Heteroteuthis	3.4	17.8	13.1	13.2	0.8	9.5	17.2	21.8	0.3	2.9	293.7	0.2	48.2	51.8
Sepiola	4.6	10.8	11.2	9.9	1.1	13.1	18.9	24.7	2.0	3.7	232.8	0.0	37.6	62.4
Teuthida														
Loligo	1.1	20.6	10.9	12.1	0.7	12.7	18.8	17.2	1.9	3.8	336.0	0.2	45.7	54.3
Lolliguncula	3.7	17.7	8.2	13.1	1.4	11.0	18.3	21.0	2.0	3.4	168.6	0.0	44.4	55.6
Sepioteuthis	0.5	23.1	13.6	10.8	1.7	5.1	18.0	20.5	2.0	4.6	371.4	0.0	49.8	50.2
Alloteuthis	5.5	22.2	12.6	11.4	1.1	5.3	17.3	19.3	0.7	4.2	281.9	0.1	53.1	46.9
Pickfordiateuthis	3.2	27.2	8.8	13.9	1.1	5.6	16.3	20.4	1.1	2.4	278.5	0.0	54.2	45.8
Illex	2.0	14.2	11.9	12.9	0.5	10.4	21.2	21.8	0.6	3.0	287.1	1.0	42.9	57.1
Onychoteuthis	0.5	10.5	10.7	13.7	0.0	11.4	18.4	27.1	3.5	4.1	233.3	10.1	35.4	64.6
Abraliopsis	0.9	11.0	13.4	14.7	0.6	10.0	16.2	30.3	0.9	2.0	329.8	13.1	40.5	59.5
Pyroteuthis	0.8	11.9	10.4	14.6	1.2	10.3	21.9	22.3	2.2	4.2	342.4	17.7	39.0	61.0
Pterygioteuthis	1.3	12.4	11.2	14.4	2.1	9.3	18.1	25.2	0.8	4.9	320.0	14.8	41.7	58.3
Neoteuthis	2.6	12.1	10.8	16.4	1.5	6.0	21.9	24.6	0.0	4.2	373.1	0.8	43.4	56.6
Chtenopteryx	0.7	14.5	13.2	13.3	2.6	6.7	20.7	24.6	1.0	2.1	363.5	4.0	44.8	55.2
Gonatus	1.5	13.2	7.7	11.1	0.9	13.9	22.4	25.3	1.0	2.9	392.1	1.3	34.5	65.5
Lycoteuthis	2.9	16.6	13.9	12.7	0.6	12.7	19.3	17.5	0.5	3.0	398.0	3.7	46.9	53.1
Bathyteuthis	1.4	20.8	6.3	8.9	0.0	8.1	25.3	26.3	0.0	2.7	126.9	30.3	37.6	62.4
Chiroteuthis	1.8	13.8	12.0	14.6	0.6	12.7	21.5	20.3	0.2	2.2	366.6	0.8	43.0	57.0
Mastigoteuthis	2.6	6.4	8.7	13.0	0.5	10.5	13.3	40.1	0.0	4.6	145.7	1.3	31.5	68.5
Grimalditeuthis	1.8	15.6	14.2	9.2	1.2	10.6	20.9	18.8	1.5	5.3	712.1	0.4	42.9	57.1
Joubiniteuthis	4.0	7.7	13.1	13.7	0.4	17.1	17.5	22.5	0.0	3.6	133.8	3.3	39.2	60.8
Discoteuthis	1.0	12.1	11.3	9.6	0.0	12.6	20.9	19.2	0.0	13.1	348.5	8.0	34.2	65.8
Architeuthis	1.8	14.1	12.6	7.0	0.0	29.4	14.1	18.4	0.0	2.6	108.6	1.0	35.5	64.5
Octopoteuthis	2.7	15.4	10.2	8.3	0.7	15.2	17.3	19.3	1.5	8.8	287.4	3.7	37.9	62.1
Histioteuthis	2.1	15.4	11.9	8.4	0.0	13.7	22.7	19.7	0.0	4.9	300.2	0.2	39.0	61.0
Cranchia	2.1	15.7	20.3	13.4	0.3	13.5	13.5	16.5	0.5	3.5	609.6	7.2	52.4	47.6
Leachia	1.1	12.4	12.9	19.8	1.5	14.2	17.6	20.4	0.0	0.0	284.9	7.5	47.8	52.2
Bathothauma	1.6	16.4	9.8	11.9	0.3	12.4	26.1	19.2	0.0	1.8	367.6	6.1	40.6	59.4
Sandalops	1.3	8.3	7.9	9.6	1.1	10.9	18.6	38.9	0.0	3.2	218.6	54.1	28.3	71.7
Helicocranchia	1.0	12.6	11.5	16.4	1.1	6.1	14.5	34.8	0.5	1.4	288.4	0.5	42.7	57.3
Taonius	2.4	13.9	14.2	10.4	1.1	18.4	13.0	24.0	0.0	2.5	448.7	47.8	42.0	58.0
Galiteuthis	1.4	11.9	12.0	10.0	0.8	13.2	21.1	28.7	0.0	0.9	467.2	12.2	36.1	63.9
Egea	1.5	16.1	13.2	13.5	1.2	9.7	15.9	25.5	0.4	2.7	236.7	44.2	45.8	54.2
Megalocranchia	4.8	10.5	11.3	6.5	2.1	15.7	18.9	25.0	0.9	3.8	166.1	63.8	35.7	64.3
Teuthowenia	2.1	10.9	10.9	12.0	2.4	6.3	14.9	33.4	0.6	5.1	406.8	42.3	39.7	60.3
Octobrachia														
Cirroctopodida														
Cirroteuthis	6.3	4.0	7.1	7.1	0.0	17.6	21.6	28.1	0.0	8.3	38.6	0.0	24.4	75.6
Cirrothauma	10.0	2.9	7.7	4.5	0.0	20.3	23.4	24.0	0.0	6.4	13.5	0.0	25.7	74.3
Grimpoteuthis	9.3	3.4	5.4	5.2	0.0	12.0	19.5	40.4	0.0	4.9	48.3	0.0	23.2	76.8
Opisthoteuthis	9.2	3.8	9.2	15.8	0.0	13.7	20.1	28.2	0.0	0.0	156.9	0.0	38.0	62.0

Table 2.6 *Continued*

Species	INFF	VERT	PARA	MEDB	LATB	BRAC	PEDAL	PALL	CHROM	FINL	OPTIC	PSV	SUPRA	SUB
Octopodida														
Japatella	9.2	11.3	9.3	13.9	0.0	12.0	19.0	23.2	0.0	0.0	188.8	0.0	45.8	54.2
Eledonella	5.4	9.3	7.7	9.6	0.0	10.5	18.8	19.4	2.3	0.0	226.7	0.0	49.1	50.9
Vitreledonella	4.0	11.9	10.5	16.2	1.3	11.3	19.8	22.4	0.6	0.0	250.9	0.0	46.0	54.0
Amphitretus	6.2	10.9	12.9	10.0	0.8	14.0	21.8	19.2	3.8	0.0	145.8	0.0	41.3	58.7
Octopus vulgaris	8.5	12.6	5.4	4.7	1.7	15.3	27.8	18.8	5.1	0.0	85.0	0.0	33.0	67.0
O. bimaculatus	9.6	17.7	5.6	5.1	1.0	21.9	15.2	21.1	2.1	0.0	114.5	0.0	39.7	60.3
O. defilippi	8.1	14.6	7.1	9.8	0.0	21.1	22.5	15.0	1.3	0.0	188.0	0.0	40.1	59.9
O. dofleini	8.6	8.4	4.4	5.8	0.0	17.6	19.3	30.8	1.8	0.0	44.6	0.0	30.5	69.5
O. macropus	8.0	10.9	7.1	8.6	1.3	20.4	20.9	19.3	1.9	0.0	74.6	0.0	37.5	62.5
O. salutii	7.0	16.3	6.1	8.3	1.2	16.1	24.0	17.6	2.3	0.0	101.5	0.0	40.0	60.0
Scaeurgus	6.4	15.2	6.3	12.6	0.0	8.0	23.8	23.1	1.4	0.0	172.4	0.0	43.6	56.4
Pteroctopus	8.5	10.4	5.9	12.8	0.7	13.8	22.1	24.2	1.5	0.0	151.8	0.0	38.3	61.7
Eledone	9.3	11.4	5.2	8.7	1.1	20.9	17.0	23.3	2.1	0.0	35.3	0.0	36.7	63.3
Bathypolypus	9.6	11.2	7.5	7.0	0.6	13.7	26.5	21.2	2.0	0.0	44.2	0.0	36.6	63.4
Benthoctopus	10.7	7.6	6.7	7.3	0.0	18.0	21.2	24.5	1.7	0.0	49.2	0.0	34.6	65.4
Argonauta	3.7	12.0	10.4	11.1	2.2	7.0	26.4	25.5	0.4	0.0	227.2	0.0	40.7	59.3
Tremoctopus	2.2	13.9	8.5	14.2	1.5	8.7	24.6	24.9	0.9	0.0	259.4	0.0	40.9	59.1
Ocythoë	3.7	7.8	5.1	10.1	1.1	8.1	22.0	39.3	2.8	0.0	172.4	0.0	27.7	72.3
Alloposus	4.2	11.6	13.8	13.6	1.3	10.0	18.6	23.3	1.7	0.0	373.2	0.0	46.4	53.6
Nautilida														
Nautilus	0.0	0.0	0.0	0.0	0.0	28.6	0.0	29.7	0.0	0.0	27.3	0.0	41.8	58.2

2.5.1 Eyes and extraocular photoreceptors

All cephalopods possess eyes although differences are found in their morphology. *Nautilus* has a 'pinhole' eye (Fig. 2.11), which lacks a lens or any other dioptric apparatus but is capable of forming images (Mugglin 1939, Young 1965a). The eye has a cup-like structure and remains open via a pupil. The retina has 4 million photoreceptor cells (Messenger 1981). The visual acuity of the animal is low and its sensitivity is increased at the expense of resolution (Muntz 1994b).

Of the coleoids, *Cirrothauma murrayi* (Cirroctopodida) is remarkable as its eyes are simple open cups with no trace of a lens, iris, or ciliary body (Chun 1913). A retina is present but the rhabdoms are somewhat irregular (Aldred *et al.* 1978, 1983); presumably the eyes of this deep-sea animal act mainly as 'photon detectors' and are probably unable to form recognizable images. Coleoids usually possess a lens; cornea, pupil, iris and ciliary body are all present (Fig. 2.13). Most have large eyes and those of the giant squid, *Architeuthis* (Architeuthidae), have been recorded to reach 200 mm in diameter (Table 15.1), the largest in the animal kingdom. The eyes of most coleoids are spherical, but in *Amphitretus* (Amphitretidae) they are telescopic, in *Bathothauma* (Cranchiidae) stalked, and in *Histioteuthis* (Histioteuthidae) of unequal size. A lens produces an inverted image on the retina and the ciliary muscles allow for accommodation (Messenger 1981). The retina has receptor cells, which are primary sensory cells. There are an estimated 20 million receptor cells in the retina of *Octopus vulgaris* (Octopodidae), and Packard (1972) has postulated that in the large eyes of some squids there could be as many as 10^9 cells. In this animal, unlike vertebrates, the outer segments of the receptor cells point towards the lens, and so towards the light. The eye produces a perfectly formed image and *O. vulgaris* is capable of discrimination of form; not just the area or length of outline of an object (Messenger 1981). Behavioural studies have demonstrated that the eye is highly efficient in supplying a wide range of information to a coleoid cephalopod about its surroundings (Messenger 1991, Hanlon and Messenger 1996).

Extraocular photoreceptors, or photosensitive vesicles, are not known in *Nautilus*, but are present in coleoids. They show variation in size and form (R.E. Young 1978), structurally resemble photoreceptors, and respond to light (Mauro 1977). The significance of the extraocular photoreceptors is little understood, although in some species they are involved in the regulation of counter-illumination (R.E. Young 1977).

2.5.2 Statocysts and other mechanoreceptors

The statocysts of *Nautilus* are a pair of simple ovoid sacs, half-filled with numerous small, free-moving statoconia and half with endolymph. The inner surface of each statocyst is lined with two types of primary sensory hair cells. In addition to their function as gravity receptor organs they are able to detect rotary movements (angular accelerations) although they lack the specialized receptor systems found in coleoids (Neumeister and Budelmann 1997).

Among coleoids, two basic designs of statocyst are found. The decabrachian type of statocyst has an irregular shape and shows greater differentiation than that of the octobrachians (Fig. 2.14). It is the most complex of all invertebrate equilibrium receptor systems and has a level of sophistication that is equivalent to that of the vestibular system of vertebrates (Young 1960b, Stephens and Young

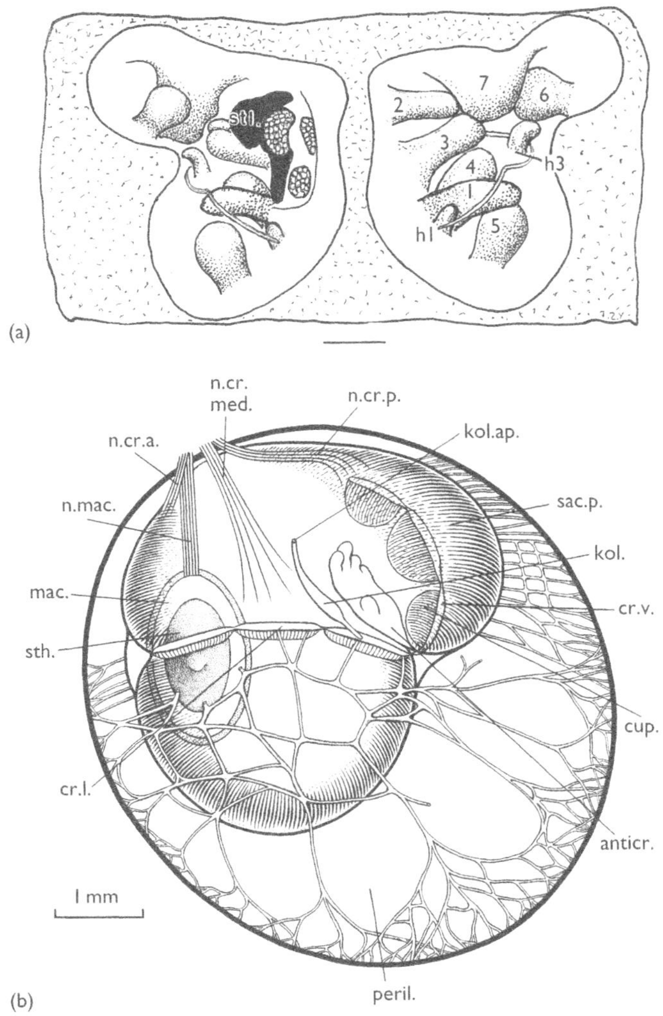

Fig. 2.14. The statocysts: (a) Drawing of the paired statocysts of the decabrachian *Illex coindetii* seen from in front (scale bar 1 mm) and (b) of one of the paired statocysts of *Octopus vulgaris* (Young 1960b, 1989).

1982). There are receptor organs for the detection of gravity (the macula/statolith (statoconia) system), three in octobrachians and one in decabrachians, and there is an angular acceleration receptor system (the crista/cupula system) which is divided into segments arranged in the three dimensions of space (Budelmann 1977, 1994). Function of the statocyst/statolith has been investigated in several oceanic squids and morphological features distinguish the statoliths of those living near the bottom and pelagic forms (Arkhipkin and Bizikov 2000). The control system for the statocyst–oculomotor reflex is complex (Budelmann and Tu 1997).

Other mechanoreceptors include a lateral line analogue which, in decabrachians and in octopod hatchlings, comprises lines of epidermal hair cells on the head and arms. It is sensitive to local water movements such that cuttlefish and squids are able to detect a moving fish, 1 m in length, from a distance of 30 m (Budelmann 1994, 1996). The innervation of this system is as yet unclear.

Epidermal sensory hair cells have been found in two groups and located on either side of the nuchal crest in a loliginid. These hair cells are polarized precisely parallel to the animal's transverse and horizontal axes. They may be involved in the control of head-to-body movements and are termed neck proprioceptors (Preuss and Budelmann 1995a, b).

2.5.3 Chemoreceptors

Chemoreceptors are found in all cephalopods, but clear characterization as 'olfactory', 'gustatory' or 'general chemical sense' organs remains to be demonstrated. *Nautilus* has paired organs, called rhinophores (Fig. 2.3): short, hollow tentacles, one below each eye, lined with a folded epithelium. Behavioural observations indicate that *Nautilus* also has receptor cells in the preocular tentacles (Bidder 1962).

In coleoids the olfactory organs, or pits, are small structures sited on either side of the head, posteroventral to each eye at the junction of the mantle and 'neck'; they vary in size and form (Watkinson 1909). The sensory epithelium comprises several morphological types of receptor cell (Budelmann *et al.* 1997). The olfactory nerve connects the organ with the olfactory lobe and contains about 1000 fibres (Young 1963c, Messenger 1979a). There is now good evidence in *Loligo* (Loliginidae) that the organs are chemoreceptors as they respond when squid ink, or its precursors, is applied to them (Lucero and Gilly 1995). However, their exact function, and appropriateness of the term 'olfactory', is unresolved.

Contact chemoreceptors occur in the distal region of the long digital tentacles of *Nautilus*, and in the suckers and lips of cuttlefish, squid, and octopuses. In coleoids the lips possess putative chemoreceptors (Emery 1975). The suckers are especially well-endowed since some tens of thousands of chemoreceptor cells are present in a single sucker of *Octopus vulgaris*, whereas there are a few thousand in the sucker of a cuttlefish (Graziadei 1965).

3 Nautilidae

3.1 *Nautilus*

Nautilus (Figs 3.1, 3.2) is the only extant representative of the order and is characterized by its large, external shell. It is found in the Indo-Pacific Ocean, between 54°E and eastwards to 172°W, and from 35°N to 35°S. Empty shells drift and have been recorded from 17°E to 158°W, and from 38°N to 41°S; their distribution is greater than that of the living animals (House 1987, Saunders 1987). *Nautilus* inhabits the fore-reef slope in depths of 80–600 m or sometimes deeper, with a preferred range of 150–400 m (Saunders 1984, Saunders and Ward 1987, Ward 1987).

Bidder (1962) visited New Caledonia to investigate feeding and locomotion in *Nautilus*, and she was followed by Denton and Gilpin-Brown (1966) who wanted to continue their researches into buoyancy in cephalopods. During the 1970s and 1980s this animal was the subject of numerous research projects which resulted in many publications, including books devoted to it, and these encompass much that is known of this genus (Hamada *et al.* 1980, Saunders and Landman 1987, Davis 1987, Ward 1987, 1988).

3.1.1 *The shell and buoyancy*

The shell has bilateral symmetry. It is formed of heavily calcified chambers, and is concerned with buoyancy, and it also provides protection for the soft body of the animal. The shell is of calcium carbonate in the form of aragonite, with an outer prismatic layer, a nacreous layer, and an inner prismatic layer (Saunders 1987). The chambers of the shell contain gas and liquid, each chamber being sealed off by calcareous walls. The most recently formed part of the shell is the living chamber with an inner surface of pearl. The chambers extend back to the first-formed, or protoconch, and all are penetrated by the siphuncle, a membranous tube (Fig. 3.3).

Fig. 3.1. *Nautilus macromphalus* in an aquarium in New Caledonia (photograph kindly provided by A.M. Bidder).

Fig. 3.2. *Nautilus*, a posthatchling with a shell diameter of 48 mm (photograph kindly provided by J.M. Arnold).

Fig. 3.3. *Nautilus macromphalus*. The photograph, taken immediately after the shell had been sawn open, shows a section through part of three septa and the siphuncle. The pellicle and chalky tube are moist, and the pellicle is thrown into folds around the septal neck (Denton and Gilpin-Brown 1966).

The early growth stages of the shell can be distinguished:

- the initial plate of organic matrix with cicatrix
- the formation of the shell composed of the initial organic plate and outer prismatic layer
- the formation of a three-layered shell-wall around the initial calcified shell
- the formation of a three-layered protoseptum on the inner surface of the initial calcified shell-wall (Mutvei and Doguzhaeva 1997).

The life of *Nautilus* is largely dominated by the production of its shell, as new material is added continuously to the aperture at approximately 0.15–0.60 mm/day, thus enlarging the living chamber (Landman and Cochran 1987). During growth the animal must move forward in the living chamber, away from the last-formed septum, and simultaneously secrete fluid to fill the vacated space (Greenwald and Ward 1987). This process of forward movement is not understood but does involve the repositioning of the muscles that attach the animal to its shell. From the first appearance of a new septum the chamber space remains full of cameral fluid for a period of several days to several weeks. This allows the new septum to thicken, so enabling it to withstand any differences in pressure and the virtual vacuum that is created in the chamber as liquid is emptied from it. Radiography at regular intervals revealed that the rate of chamber formation is not constant but increases as chambers formed become larger, the process taking from 2 to 3 weeks to as long as 13 weeks (Ward 1987). As sexual maturity is reached the growth of the shell is almost complete. The maximum size of the shell of an adult is 220 mm diameter, that of the male usually being larger than that of the female (Saunders 1987). Individual nautiluses have been found to live for as long as 4 years after reaching sexual maturity, and have a life span of 15–20 years (Saunders 1983, 1984).

A *Nautilus* weighing 1 kg or more in air weighs only a few grams in water (Denton and Gilpin-Brown 1966). This difference in weight results from the chambers of the shell containing a quantity of liquid that diminishes from the newest-formed chamber to the older chambers; most of the chambers contain very little liquid. The passage of liquid into the chambers is through the siphuncular region, with its active epithelium, which can pump the liquid out of the chambers. Gas enters the chambers by diffusion from tissue within the siphuncle; it contains nitrogen, oxygen, and argon. The pressure within the chambers of the shell is subatmospheric. The rate of emptying is less than 1 ml per chamber per day at a depth equivalent to 3.5 times atmospheric pressure (Ward and Martin 1978). The movement of fluid into, or out of, the chambers can be rapid under conditions of stress, such as sudden loss of a piece of shell, or weight increase by the addition of cameral liquid. Compensatory buoyancy changes occur most rapidly during the first 10 h after experimental interference in animals of 500–1000 g (Greenwald and Ward 1987), indicating that the siphuncle remains functional throughout the life of the animal.

3.1.2 Locomotion and activity

Nautilus moves entirely by jet propulsion; all the locomotor and respiratory work results from pressurizing the seawater in the mantle cavity by the actions of two sets of muscles that draw in and expel the water (O'Dor *et al.* 1990). These muscles are of approximately equal mass (Packard *et al.* 1980) and both consist of at least two types of muscle fibres (Hochachka *et al.* 1978). One set of muscles belong to the funnel, or hyponome, and its flaps; the other set attach the animal to the wall of the shell near the back of the living chamber. Both sets are attached to the cephalic cartilage. *Nautilus* usually moves backwards with its head and some of its tentacles trailing behind. It is capable of manoeuvring and of swimming forwards, but is perhaps least efficient in this last direction. When actively moving the animal exhibits gentle rocking and pitching motions (Bidder 1962, Denton and Gilpin-Brown 1966). After expulsion of water both sets of muscles relax allowing water to enter the mantle cavity and it is this alternation during retraction and extension that is responsible for the rocking and pitching movement. At the same time the eyes move so that the conspicuous iris grooves remain approximately vertical (Packard *et al.* 1980), and stabilization of the eyes also involves the statocysts (Hartline *et al.* 1979).

Nautilus can move slowly when the funnel contracts, with little input from the shell muscle. In rapid movements both sets of muscles are involved. In the sea, in a depth below the effect of wave surge, *Nautilus pompilius* has a maximum velocity of 0.25 m/s (Ward *et al.* 1977) and a similar rate was recorded for *N. macromphalus* (Packard *et al.* 1980). O'Dor *et al.* (1990) found that acceleration in *N. pompilius* was 0.37 m/s with an average speed of 0.14 m/s: these authors demonstrated the cost of transport to be low and the animal very efficient in its low-speed, low-energy environment.

The funnel is flexible and can curve to either side, as well as backwards and to some extent upwards. It is most efficient when pointing downwards, so that the animal moves upwards as water is expelled. The funnel lies below and in front of the centres of gravity and buoyancy (Denton and Gilpin-Brown 1966) and accounts for the manoeuvrability of the animal. In the sea divers watched as 'the nautilus jetted off the bottom in flight, carrying the lobster with it', and moving 'vertically through the depths quite easily' (Ward 1988).

Diurnal movements were followed by tracking *N. belauensis* as it entered shallower water at night and moved deeper during the day (Ward *et al.* 1984, Ward 1987). One ascended to 150–100 m beginning between 1700 and 1830 h (sunset was at 1815 h) and between 0430 and 0500 h descended to 250–350 m, where it remained during the hours of daylight. The maximum rate of vertical ascent was about 2 m/min and of descent 3 m/min. It is not known whether all species undergo such vertical migrations. Another study of a freely moving *Nautilus* in the sea confirmed vertical movement to be mostly crepuscular, but not exclusively, up at dusk and down at dawn (O'Dor *et al.* 1993). Divers observed nautiluses feeding on cliff walls in darkness (Ward 1988).

Nautilus pompilius can endure very low oxygen tension for long periods (Wells *et al.* 1992) and under such conditions its activity is intermittent, with brief periods of jet propulsion separated by long periods (15–20 min) of rest. During inactive periods ventilatory movements cease and the heartbeat slows. The onset of activity is heralded by an increase in frequency and amplitude of the heartbeat and a resumption of visible respiratory movements. Recovery, even after a prolonged period in near anoxic conditions, is rapid when the animal enters well-oxygenated waters. Such a capacity allows the animal to survive attacks from potential predators by withdrawal into its shell, and this is its only strategy of defence.

3.1.3 Life history

Willey (1902) was the first to attempt to follow the reproductive processes and embryonic development of this animal in the field. In captivity *N. pompilius* displays no courtship behaviour; instead, the male grasps the female with several tentacles so that the apertures of the shell are in approximation and the hoods are pulled together. It is not known how spermatophores are transferred and only one is found in Valenciennes' organ (Haven 1977a, b). The sperm is markedly different from that of the coleoid cephalopods. There is a laterally compressed acrosome, a laterally compressed, rod-shaped nucleus, and two longitudinal furrows in the nucleus, each accommodating an elongate mitochondrion (Arnold 1987, Healy 1995).

Owen (1832) illustrated the single ovary of a mature female *N. pompilius* with oocytes of several size classes present, as did Griffin (1900). The number of eggs in each size class varies from 6 to 10, and the largest are ovate (25 × 16 mm). The female manipulates an egg capsule into a small hole under coral with the inner crown of tentacles; this takes about an hour, in which time the egg is firmly set on the substratum (Hamada *et al.* 1987). Eggs are usually laid in batches of 5–8 in aquaria (Arnold 1987), each batch being separated by an interval of time. *Nautilus* can be described as iteroparous, as egg-laying is separated by long periods of time (Mangold *et al.* 1993).

Each egg has an inner and outer capsule. These are initially flexible but harden after a few hours in seawater. The outer capsule is sac-like and its base adheres to and follows the contours of the substratum. The oocyte in the inner capsule is ovate and is about 27 mm in length. An extraembryonic circulatory system is present, and this is unique among the living cephalopods in which embryogenesis has been described. The discovery and investigation of embryos of *Nautilus* provided an invaluable insight into their development. A striking feature is the large yolk mass, indicating a long developmental period. In the embryo the eyes lie on either side of the head, and the tentacles arise in two bilateral groups above the yolk sac and below the eyes and funnel-retractor muscle complex (Arnold 1987). An advanced embryo has a tiny cap-like shell (about 5 mm in diameter) to which it is firmly attached by the quite dense cephalic retractor muscles, with a distinct median depression, the cicatrix (two lines of melanin) (Mutvei *et al.* 1993). A well-developed eye complex, a cluster of tentacular buds, and a long hyponome are all present (Fig. 3.4).

Living embryos show four distinct movements:

- a rhythmic contraction of the retractor and associated muscles (a respiratory action)
- a rapid contraction which pulls the embryo against the yolk sac
- a sudden circumferential contraction of a band of large muscle cells which form a deep indentation
- a slow rotation of the embryo itself inside the capsule (Arnold and Carlson 1991).

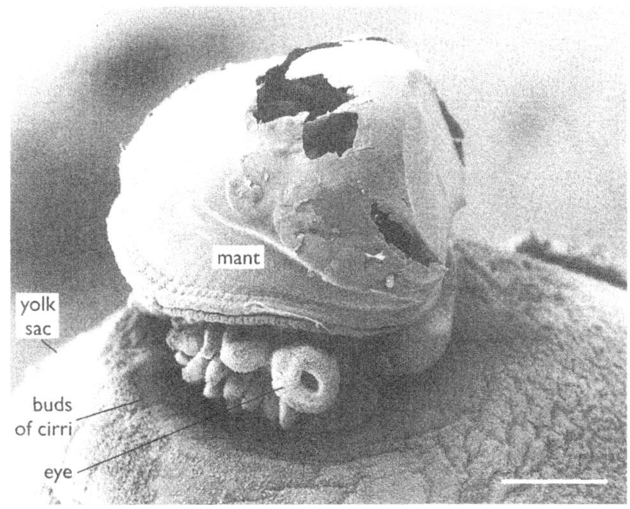

Fig. 3.4. *Nautilus belauensis.* Prechambered stage in lateral view. The shell is present, the eye and funnel are prominent features, and the cirral buds are developing (scale bar 1 mm) (photograph kindly provided by J.M. Arnold).

Several adults (*N. belauensis*) were maintained in tanks and eggs were laid at intervals over a long period, with an average rate of 1.3 eggs per female per month, and after 13 months the first successful hatching took place in an aquarium. Hatching took place at night, and the young nautilus would promptly began to feed on pieces of offered shrimp. It resembled the adult and the shell was 20–25 mm in height. Several eggs hatched and survived until shell diameters of up to 38 mm had been reached (Carlson 1991). In another study a single female of the same species laid 43 eggs during a period of 15 months and 3 (kept at 24.4 °C) hatched after about 12 months (Okubo *et al.* 1995).

At low tide in a lagoon (water depth 0.25–1.0 m) behind a reef in Fiji, between January and March, three small, live *Nautilus* were captured in daylight. These animals were found in holes in submerged rocks, with an opening of about 150 mm. Their shells measured 23.2–26.1 mm in diameter and eight or nine septa were present in two of the shells (Davis and Mohorter 1973). A small animal is shown in Fig. 3.2.

3.1.4 Tentacles

The tentacles are small, numerous, and delicate (Fig. 3.1). They are retractile, and can be divided into pre- and postocular, digital, and buccal (labial) tentacles. These groups of tentacles have different functions; the digital tentacles alone form three functional sets (Bidder 1962).

Each of the pre- and postocular tentacles has a cirrus which is retractile into a sheath, and has a ridge-like processes normal to its longitudinal axis. The epithelium of the postocular tentacle has ciliated cells bearing macrocilia, which project between the folds of the ridges to the surface, and also non-ciliated cells (Barber and Wright 1969).

There are 19 pairs of digital tentacles (Owen 1832). Each cirrus can be extended for 90–110% of its retracted length, taking 5–10 s to do so (Kier 1987). Once extended the cirrus is capable of bending and twisting. The tentacle has a relatively large axial nerve cord, subcentrally located on the oral side. The nerve cord is surrounded by radial muscle fibres which spread outwards to the periphery as sheets, and so divide the longitudinal muscles into bundles. The oral surface of each cirrus has a series of ridges and grooves by means of which *Nautilus* adheres to, and retains its hold on, a foreign surface.

The pre- and postocular tentacles, together with the four most lateral digital tentacles, spread outwards in a 'cone of search' for food (Bidder 1962). Bait can be located in less than 1 min, at a distance of 600 mm, indicating an initial chemosensory response; this is followed by a tactile response. Food encountered by the inner ring of lateral and dorsal digital tentacles is grasped and drawn in below the hood (Haven 1972, Ward 1987). The natural diet of *Nautilus* must include crustacean exoskeletons and moults, as fragments occur in their crops (Ward and Wicksten 1980, Magnier and Laboute 1978, Ward 1987). The faeces contain abundant sand grains, polychaete setae, and the remains of crustaceans and of bivalve molluscs (M.J. Wells, pers. comm.).

3.1.5 Buccal mass and digestive tract

The buccal mass is huge when compared with the animal itself (Fig. 3.5) (Morton and Nixon 1987, Nixon 1988). The beaks are chitinous and almost black in colour and have white, calcareous tips; this last feature separates *Nautilus* from all other extant cephalopods, although a few fossil forms had beaks with associated calcareous concretions (Nixon 1996). The tip of the upper beak takes the form of an arrowhead, whereas that of the lower beak is jaw-like, and in some species bears teeth on the oral surface. The calcified tips of the beak form a powerful shearing mechanism (Nixon 1988).

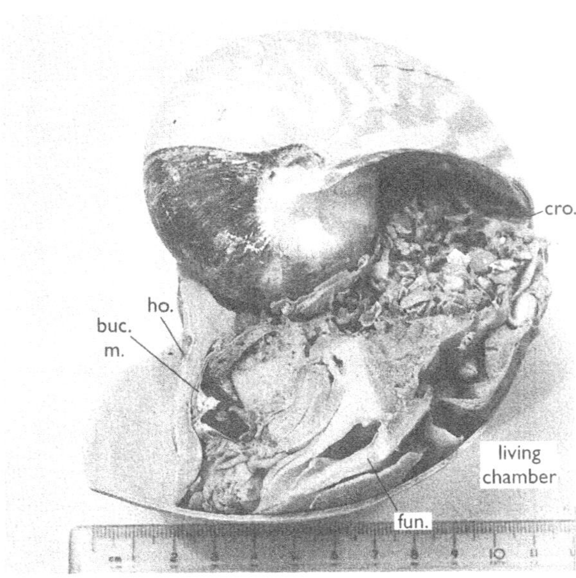

Fig. 3.5. *Nautilus pompilius.* The shell has been partially cut away and the hemisection shows the buccal mass in relation to the animal and the living chamber of the shell (maximum diameter of shell 127 mm, ML 84 mm, and anteroposterior length of buccal mass 35.5 mm).

The radula is unique among cephalopods as it has 13 elements across each transverse row. These comprise a central rhachidian tooth with, on either side, two lateral teeth, two marginal teeth, and two marginal plates (Nixon 1995, 1998). The radula arises in a blind-ended sac from which it runs downwards, anteriorly along the floor of the buccal complex, and then over the bending plane, before terminating (Fig. 3.6).

On either side of the radula is the radular support system. A study by Messenger and Young (1999) shows the presence of a 'bolster' within each radular support. The bolster is massive and extends along the radula and into the hyaline shield. Each bolster consists of a muscular, dorsal part and a ventral rod. Of the muscular part the dorsal fibres are longitudinal, otherwise most fibres are transverse. The two bolsters are joined anteriorly by a thin band of transverse muscle fibres, which form a sling below the radula. The rod forms more than half of the cross-section of the bolster and also reaches into the hyaline shield. A thin membrane encloses the rod so that it is quite separate from the surrounding bolster. Together the rod and bolster apparently form a muscular hydrostat support system (Kier 1988). The bolsters are enclosed by a membrane and the volume must remain constant, so that contractions of the transverse muscles elongate the bolsters. The exact function of the rod is not yet understood. The system causes the radula to move forwards and upwards, as in *Octopus* and *Sepia*, and the teeth to be elevated and splayed to make an effective 'rake'.

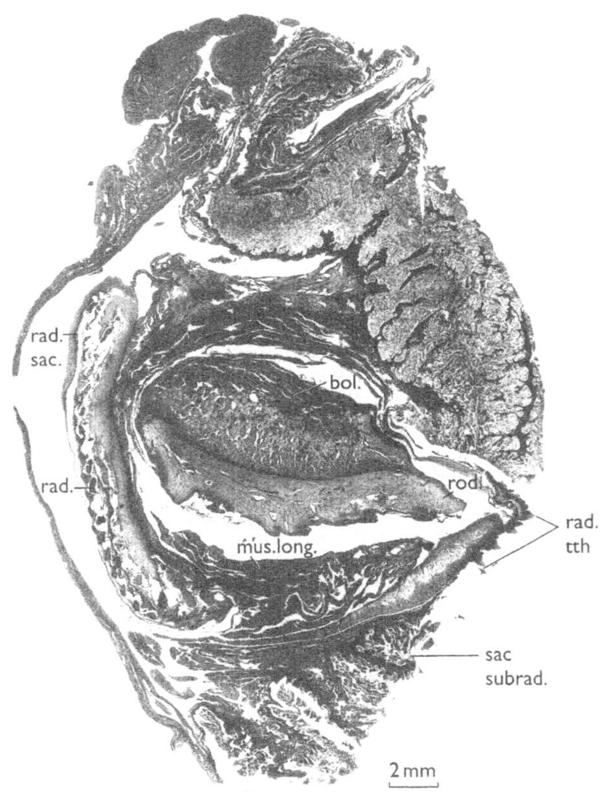

Fig. 3.6. *Nautilus pompilius.* A medial sagittal section of the buccal mass, showing the long radula, the radular support, the bolster and rod (Masson's trichrome) (Messenger and Young 1999).

Below the bending plane of the radula there is a series of transversely organized invaginations, or sacs (Fig. 3.6). The uppermost may be concerned with the dissolution of the radula, its teeth and ribbon. The second invagination has infolded walls and its epithelium contains mucus cells. The ventral-most sac has a lining of tall columnar cells. The function(s) of these invaginations is not known.

A palp, bearing tall papillae, lies on each side of the radula and there is in addition a median structure, also with a papillate surface, described as a 'salivary process' and a 'tongue' respectively by Griffin (1900). Histologically the three parts constitute a single functional structure, but the lateral parts include some secretory tissue. The nature of the radular apparatus was considered 'problematical' by this author, and subsequently Messenger and Young (1999) came to a similar conclusion.

The oesophagus leaves the buccal mass to pass through the central nervous system before entering the crop. It is deeply folded and has no lining of chitin such as is found in other living cephalopods. Mucus presumably aids the passage of the often tough, sharp and large pieces of material (up to 5 mm^2) along the oesophagus (Haven 1972). The oesophagus has folds which presumably provide sufficient extensibility for such fragments to pass along readily and yet prevent excessive expansion where it is surrounded by the central nervous system. Owen (1832) commented that 'the capability of propelling such rude and angular pieces through a narrow canal into the gizzard without rupturing the thin lining... is not one of the least extraordinary examples of the powers of living matter'.

The crop can become distended with food and its wall was described as 'thin but tough with a smooth surface' by Owen (1832) (Fig. 3.5). The food retained in the crop can form as much as one-fifth of the total weight of the animal (Ward 1988), and can be stored in the crop for as long as 2 weeks before digestion (K. Mangold, pers. comm.). It has been estimated that 50 g of food can last a *Nautilus* for more than 2 months at the reef bottom (Wells 1987, O'Dor *et al.* 1993); this would minimize the need to search for food. The stomach is 'lined with a thick cuticular membrane, delicately furrowed and adapted to numerous fine ridges' (Owen 1832); here ingested particles are reduced in size (Ward and Wicksten 1980).

3.1.6 Olfactory organs

Below each eye there is a sac, this is the olfactory organ, or rhinophore (Figs 3.7, 3.8). Each olfactory organ is a short, hollow tentacle with expanded upper and lower lips, opening to the exterior via a pore which is patent even in the adult. The sac is lined with a folded and highly vascular epithelium with three types of cell. There are flask-shaped, ciliated cells, which may be receptors, ciliated cells with striated roots, and mucous cells, all resting on a basal lamina (Young 1965a, Barber and Wright 1969). The structure of this organ is similar to that in *Octopus* (Woodhams and Messenger 1974) and in *Lolliguncula* (Emery 1975).

No ganglion cells have been seen below the epithelium of the olfactory organ, but they occur along the course of the nerves leading away from it (Young 1965a). The nerve fibres collect to form a set of bundles and further nerve cells mix with the fibres; the whole may be called the olfactory ganglion (Fig. 3.9). The cells of this ganglion are scattered and many are rather large. Medial to the end of the canal the cells become fewer and leave the olfactory nerve as a strand lying below the optic lobe.

The nerve leads to the olfactory lobe, which is essentially a lateral extension of the cerebral cord arising behind the optic stalk and then passing ventrally and backwards to receive the olfactory nerve (Fig. 3.8). The layers characteristic of the cerebral cord can be recognized in the olfactory lobe. The dorsal and lateral parts of the olfactory

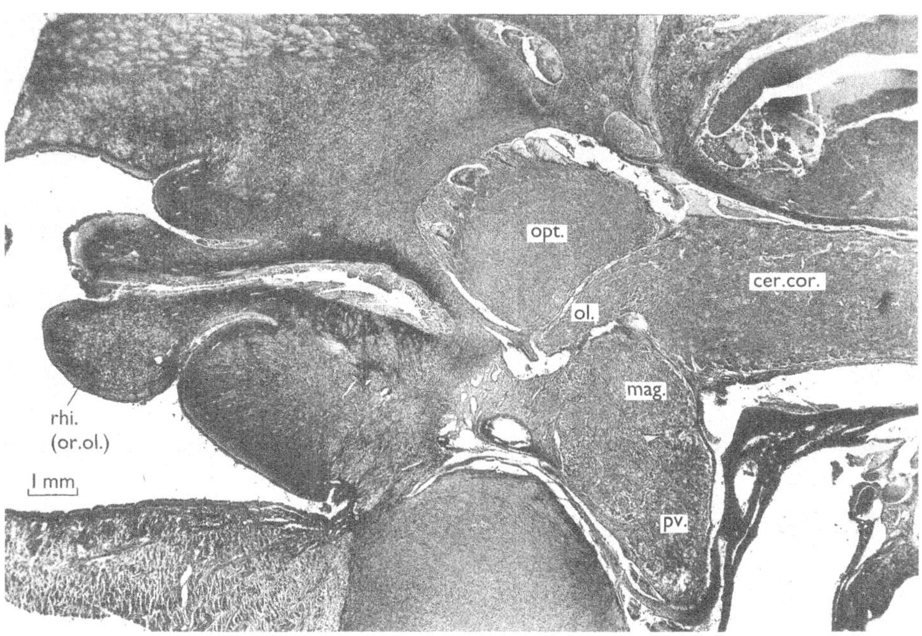

Fig. 3.7. *Nautilus.* Transverse section through the head showing the rhinophore, or olfactory organ, the olfactory lobe, and the central nervous system.

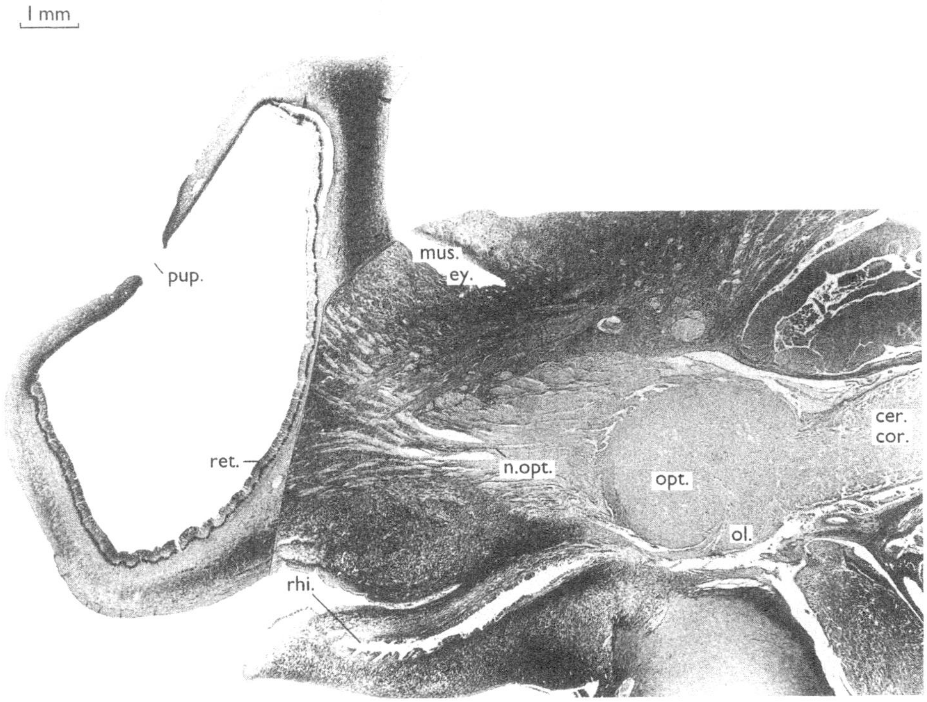

Fig. 3.8. *Nautilus.* Transverse section through the head to show the pinhole eye, the retina, the optic nerves and the optic lobe, and the olfactory lobe of the brain.

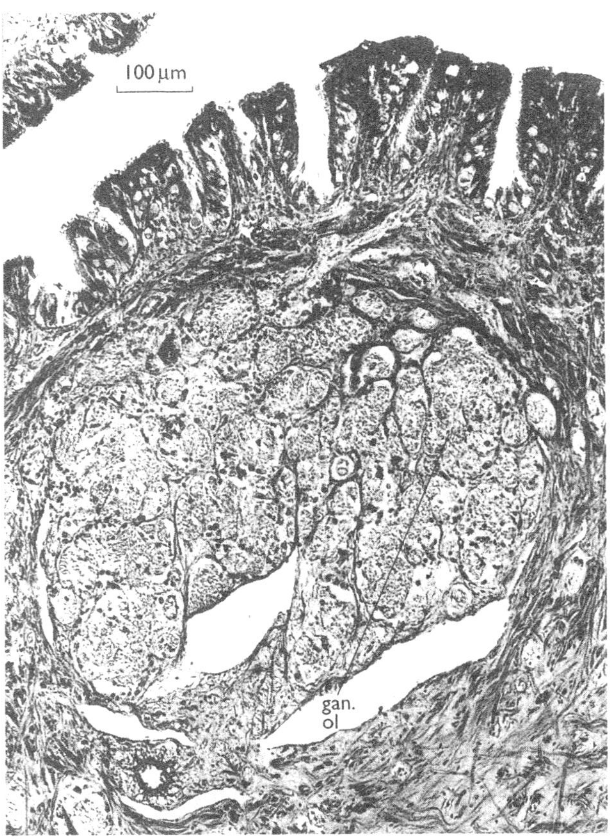

Fig. 3.9. *Nautilus.* Section through the olfactory ganglion and the epithelial lining of the olfactory organ.

lobe consist of cords of neuropil, fibres and small cells running transversely and continuous with the plexiform layer of the cerebral cord. Below this is a laminated zone, with thin bands of fibres which are continuous with the central and ventral parts of the cerebral, or supraoesophageal, cord. Connections have been traced to the optic, cerebral, lateral cerebral, magnocellular, funnel, and palliovisceral regions (Young 1965a).

3.1.7 Eyes

The eye of *Nautilus* has no lens and no cornea, and appears to operate on the principle of a pinhole camera. It has a small aperture forming the pupil, and is open to seawater in the absence of a lens (Figs 2.11, 3.8). Such an eye is found only in *Nautilus* and is quite distinct from those of all other living cephalopods, the coleoids. The anatomy of this eye has been described at light and ultrastructural level (Barber and Wright 1969), and visual behaviour, visual sensitivity, and retinal responses demonstrated experimentally (Hurley *et al.* 1978, Muntz and Raj 1984, Muntz and Wentworth 1987, Muntz 1994b).

The eye is a prominent feature (15 × 10 mm) with the long axis horizontal (Muntz and Raj 1984). The eyes lie on either side of the head, immediately below the hood; they are not contained in orbits (Fig. 3.8). Each eye has mobility by virtue of attached muscles (Owen 1832, Willey 1902) and can move around the optic stalk by four extraocular eye muscles (Mugglin 1939).

The retina covers most of the cavity of the eye (Fig. 3.8) (Muntz and Raj 1984). It is composed of supporting cells and photoreceptor cells (Barber and Wright 1969). Each eye contains 4 million receptors (Messenger 1981). In each receptor outer segment there are several rhabdomeres occurring almost equally at all angles, with substantial or complete overlap between the microvilli of neighbouring receptors. There is, however, no apparent regular grouping of the rhabdomeres from adjacent receptors to form a rhabdom (Muntz and Wentworth 1987), such as is found in coleoids (Messenger 1981). The supporting cells of the retina possess both microvilli and small cilia: in the presence of cilia the cells of *Nautilus* differ from those of coleoids. Both types of cell contain pigment granules, but only the visual cells contain a complex, lamellated, membranous structure, the myeloid body, above and below the nucleus. The function of these bodies is uncertain although in *Octopus* they are implicated in the metabolism of visual pigment (Breneman *et al.* 1986). These bodies are structurally more complicated in *Nautilus* than in coleoids (Muntz and Wentworth 1987).

A superficial, pigmented channel runs ventrally from the aperture, the iris groove. It runs from the open pupil on the outside of the iris and has a lining of ciliated epithelium. The external surface of the iris consists of columnar epithelial cells with microvilli interspersed with mucous cells (Muntz and Raj 1984). Foreign particles pipetted on to the outer surface of the iris were trapped in mucus and transported ventrally (at about 0.9 mm/min) down the iris groove and away from the pupil: in this way the groove may provide some protection for the eye (Muntz 1987a).

Willey (1902) watched the pupil 'contract from a comparatively wide orifice, 2.5 mm in diameter, to a mere slit with apposed margins' and noted that the aperture of the pupil was 'capable of slight accommodation to the intensity of light'. The pupil constricts or dilates with changes in illumination, closing (in 90 s) in response to a sudden increase in light intensity, and opening (in 50 s) when the light decreases. Furthermore, stimulation of the left retina by a small spot of light caused pupillary constriction in the right eye (Hurley *et al.* 1978).

Visual acuity of *Nautilus* has been determined from the optometer response; the minimum separable angle was between 5.5° and 11.25° (Muntz and Raj 1984). Visual sensitivity was found to be 468 nm, the wavelength of maximal absorption of rhodopsin, and typical of deep-sea animals (Muntz 1987b). It is likely that under optimal conditions with a clear sea some reception of downwelling light is possible even to depths of 770 m, and bioluminescence can almost certainly be detected. Illuminated traps caught significantly more *Nautilus*, indicating positive phototaxis (Muntz 1994a).

The phototactic behaviour to stimuli of two different sizes was tested in a Y-maze, the larger stimulus being four times the area of the smaller one. When the stimuli were of equal intensity the larger one was usually chosen. As the intensity was reduced for the larger compared with the smaller stimulus, the preference decreased and eventually reversed. When the intensity of the larger stimulus was about 0.5 log units less than that of the smaller one, at which point the radiant flux from each stimulus was approximately the same, the two stimuli were equally preferred. This suggests that summation must occur within the visual system over visual angles of 13° or more (Muntz 1994b). These findings agree with previous experimental data suggesting that sensitivity is increased at the expense of resolution (Muntz and Raj 1984).

An ocular compensation reflex stabilizes the eye about the pitch axis against rocking motions of the animal that result from locomotory propulsive movements. Under stable conditions the eye has 1–2 spontaneous rotations. After ablation of one or both statocysts the compensatory movements of each eye are mediated largely by the statocyst of the same side (Hartline *et al.* 1979).

The eye of *Nautilus* seems a poor device, and yet has apparently persisted for some 400 million years. Its ability to discriminate form appears limited, but it can detect and orientate towards light (Muntz 1987b, 1994a). The visual behaviour, well-developed retina, and statocyst-mediated stabilization all suggest that vision is important in the life of this animal, even in the absence of dioptric apparatus.

3.1.8 Optic lobes

The optic lobes are lateral extensions of the supraoesophageal cord (Figs 2.6, 2.11, 3.8) and form 27.3% of the total brain volume (Maddock and Young 1987). The lobes are separated from the cerebral cord by slight depressions and are continuous ventrally with the magnocellular lobe (Young 1965a).

The organization of the optic lobe is simpler than in coleoids. In *Nautilus* the lobe is divided into layers. The plexiform layer, up to 400 μm thick, receives the optic nerve fibres. There is no outer granular cell layer but the whole plexiform layer contains scattered cells of 5–14 μm diameter. Within the plexiform zone is a well-defined tangential layer of cells, inside of which the laminated zone is of rather less regular tangential layers of cells. The lamination is formed by a series of layers of cells and neuropil. These layers become progressively less regular proceeding inwards, so that the inner members form islands of cells, rather similar to those present in the optic lobes of coleoid cephalopods.

The optic nerve fibres enter all over the lateral surface of the optic lobe. The longest optic nerve fibres are some 10 mm or more in length. The optic nerve fibres appear to pass from the retina to the optic lobe without forming a dorsoventral chiasma (Fig. 3.8), but more recently Muntz and Raj (1984) have detected signs of such a chiasm.

3.1.9 Central nervous system

The central nervous system of *Nautilus* was first described by Owen in 1832. Young (1965a, 1987) gave a comprehensive description and found it to differ somewhat from other extant cephalopods, the coleoids. The system is based on simple ganglionated nerve cords forming a ring which widely encircles the oesophagus and lies just behind the huge buccal mass (Fig. 2.6). Closer examination reveals that there are various differentiated lobes, and when seen from the side or medially there is a striking resemblance to the nerve centres concentrated around the oesophagus of living coleoids. The plan of the central nervous system of *Nautilus* presents a 'primitive' condition, yet constitutes a general indication of conditions found in coleoids.

The nervous system of *Nautilus* is made up of one supraoesophageal and two suboesophageal cords, namely the supraoesophageal, the pedal, and the pallioviscéral nerve cords. The three cords meet laterally in the perioesophageal region, which corresponds to the position of the magnocellular lobe of coleoids. This lateral region contains some of the most important motor centres of *Nautilus*, and is the 'central' part of the whole system. At this level the three cords converge and afferent fibres from the eyes, tentacles, lips, statocysts, olfactory organs, and hood all enter. This region thus influences the actions of the buccal mass, tentacles, funnel, and head retractor muscles, but it is not yet understood how the central coordination of action is produced.

Anterior suboesophageal cord

There are two distinct parts to the anterior suboesophageal cord; one is directly continuous with the supraoesophageal cord and the other, more posterior part, gives rise to the nerve to the funnel and is continuous dorsally with the magnocellular region (Fig. 3.10). It may be argued that the more anterior part is truly 'brachial' and the posterior part is 'pedal'. The former part is essentially cerebral and innervates the hood as well as the tentacles. A pair of strap-like structures, the main part of the anterior suboesophageal cord, lie at the sides, and narrow ventromedially into a thin strand, the anterior suboesophageal commissure. The nerves to the tentacles arise from the anterolateral borders of the main lateral structures. This more anterior, brachial part gives rise to the nerves that run to the arms, hood, and ocular tentacles. It seems likely that they are of cerebral origin, rather than part of the foot as is usually considered (Young 1987). These direct pathways from the tentacles to the supraoesophageal centres confirm the 'cephalic' nature of the tentacles (Young 1965a).

Behind the region which gives rise to the brachial (tentacular) nerves is that from which the funnel nerves arise, and although these parts are continuous and exchange fibres, the funnel part is continuous with the magnocellular region. The boundary between the anterior and posterior part of the anterior suboesophageal cord is distinct although it is not marked by any fibrous septum. The funnel and magnocellular parts contain some cells of up to 30 μm diameter, and give the impression of a motor region, while the brachial portion contains numerous small cells and neuropil with interweaving bundles of fibres. The nerve cord in each tentacle is ganglionated, with nerve cells at the periphery.

Posterior suboesophageal cord

The posterior suboesophageal (pallioviscéral) cord is circular in cross-section and has a thick peripheral layer of large, pear-shaped cells, 20–40 μm in diameter, with smaller cells, about 10 μm in diameter, in the deeper layers. The centre is mainly of neuropil with islands of medium and small cells. Visceral nerve bundles leave partly above and partly below commissural bundles in the posteroventral region of the cord, as in coleoids. Pallial nerves arise from the posterodorsal margin of the cord and run mainly to the large head retractor muscles, by which the animal withdraws into the living chamber of the shell. Some bundles of nerves run through the muscles to the wall of the mantle which has few muscles.

No stellate ganglia are present in *Nautilus*.

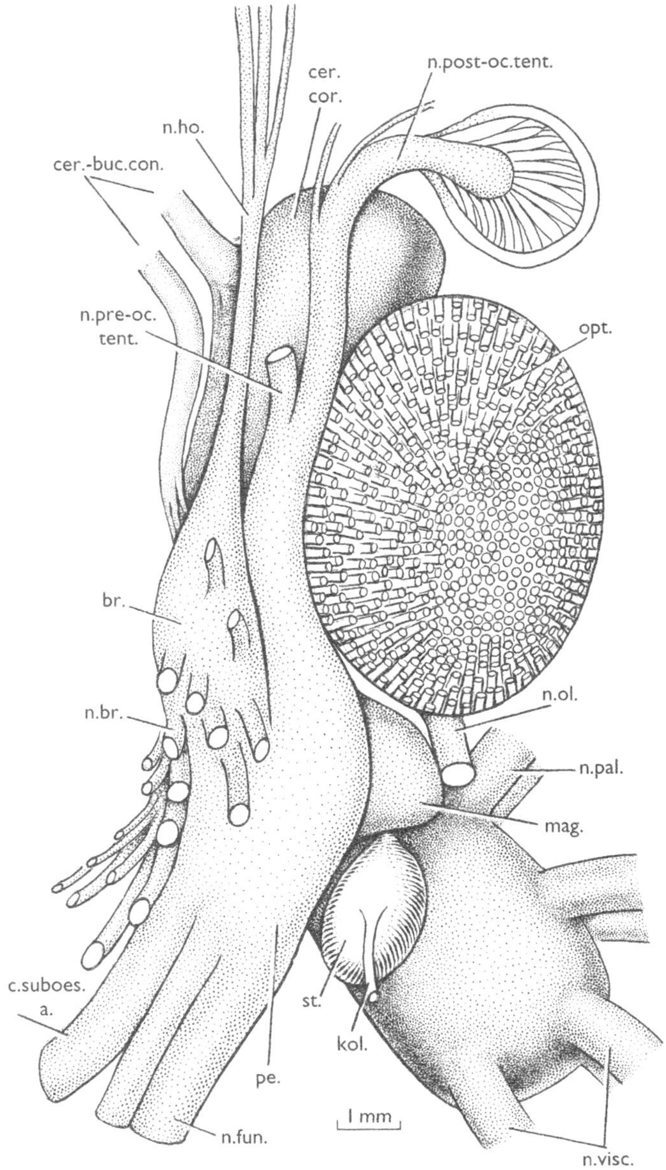

Fig. 3.10. *Nautilus*. A reconstruction of the outlines of the main parts of the central nervous system made from the tracings of serial sagittal sections as seen from the side (Young 1965a).

Magnocellular lobe

This lobe is immediately lateral to the union of the three main cords when seen in sagittal section (Fig. 3.10). It lies anterior to the statocyst and static nerve and receives the upper end of the funnel portion of the anterior suboesophageal cord. The lobe contains an irregular row of large, round cells, 50 μm in diameter, lying medially in front of the statocyst (Figs 3.11, 3.12). The rest of the lobe consists of bundles of fibres and islands of medium and small cells, much as in the upper part of the anterior and posterior suboesophageal cords.

Anteriorly the magnocellular lobe is not sharply marked off from the brachial portion of the anterior part of the anterior suboesophageal cord, yet the internal composition is distinct. The brachial part consists of islands of medium and small cells and the magnocellular portion mainly of neuropil with an outer posterior wall of larger

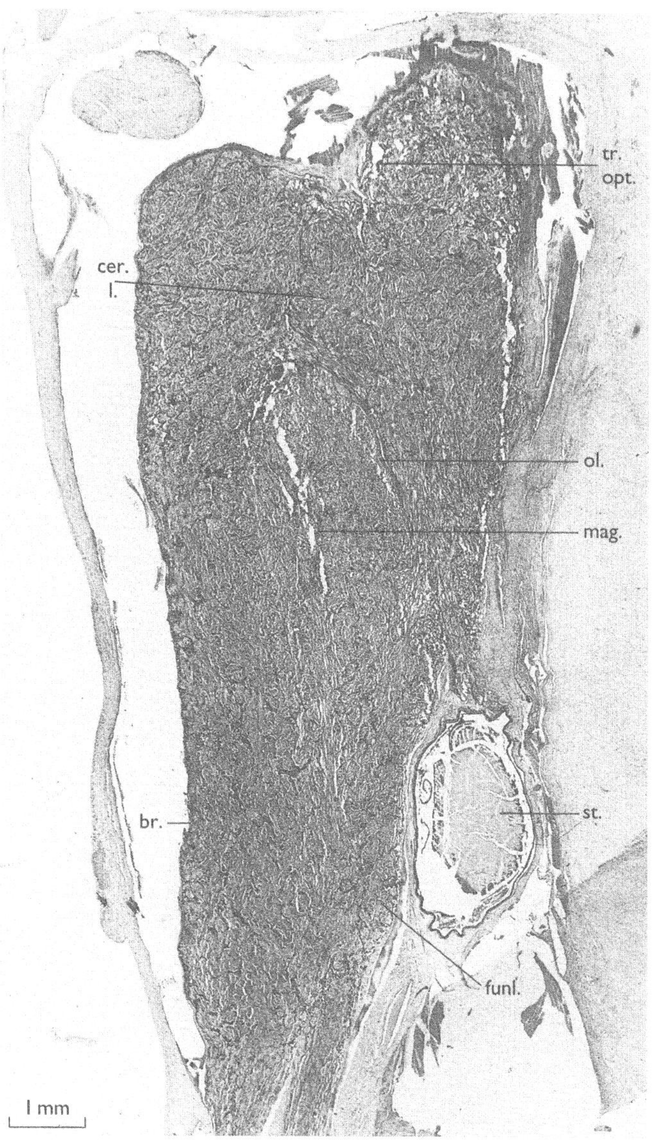

Fig. 3.11. *Nautilus*. Sagittal section to show the brachial, funnel, and magnocellular lobes.

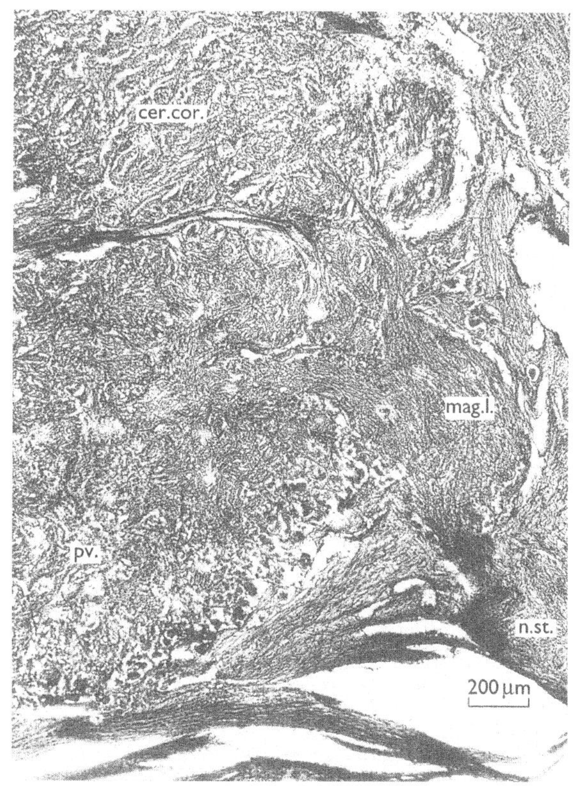

Fig. 3.12. *Nautilus*. Transverse section showing the entry of the static nerve and some of its fibres running to the posterior cord.

cells. In the narrowest part the magnocellular lobe expands laterally, forming a small extension to receive nerves from the statocyst.

It was previously said that the magnocellular lobe gives rise to the nerves to the two ocular tentacles (Young 1965a), but these in fact come from the upper end of the brachial lobe, which runs up in front of the magnocellular lobe.

Supraoesophageal cord

The supraoesophageal (cerebral) cord shows little sign of division externally (Figs 3.10, 3.13) although there is a slight swelling above the entry of the brachial cord, indicating the lateral cerebral lobe (Fig. 3.14). The supraoesophageal cord, although it lacks the clearly marked internal division so characteristic of the supraoesophageal mass of coleoids, is by no means a homogeneous structure. The divisions are not sharp, but five regions with characteristic differences can be recognized. These are the posterior buccal region, a plexiform zone, a laminated zone within the plexiform zone, a central zone in the middle of the cord, and a commissural zone (Fig. 3.13). This condition foreshadows the differentiation to be found in the brain of coleoids.

The posterior buccal region contains large cells, up to 30 μm in diameter, and gives rise to nerves which run to the lips and to the superior buccal ganglia. These nerves probably contain afferent and efferent fibres and the large cells of this region presumably give rise to motor fibres, controlling movements of the oral papillae and the buccal mass.

The plexiform zone occupies the region immediately behind the buccal region and around the dorsal and posterior surfaces of the cord. The outer part consists of bundles of fibres running into the cord from other centres, deep to which is an inner plexiform zone of islands of neuropil surrounded by layers of mostly small cells.

The laminated zone lies within the inner margin of the plexiform zone and is marked by a conspicuous sheet of closely packed cells, running tangentially. The cells are mostly small and often grouped around blood vessels, which are abundant. Deep to the tangential layer is a zone of neuropil, the laminated zone, about 500 μm in width, with few cells, this is a clearly differentiated region of the cord. Bundles from the plexiform layer converge on the laminated zone from in front, above and behind. Its output presumably passes ventrally to the central part of the cord. The laminated zone ends moderately sharply below with the beginning of the cells of the region called the central zone.

The central zone has a greater proportion of larger neurons, up to 25 μm in diameter, than of smaller ones. The region consists mostly of laterally running fibres. A characteristic of the central zone is the presence of large vascular spaces and much connective tissue. This zone is probably the output region of the cord, in addition to the

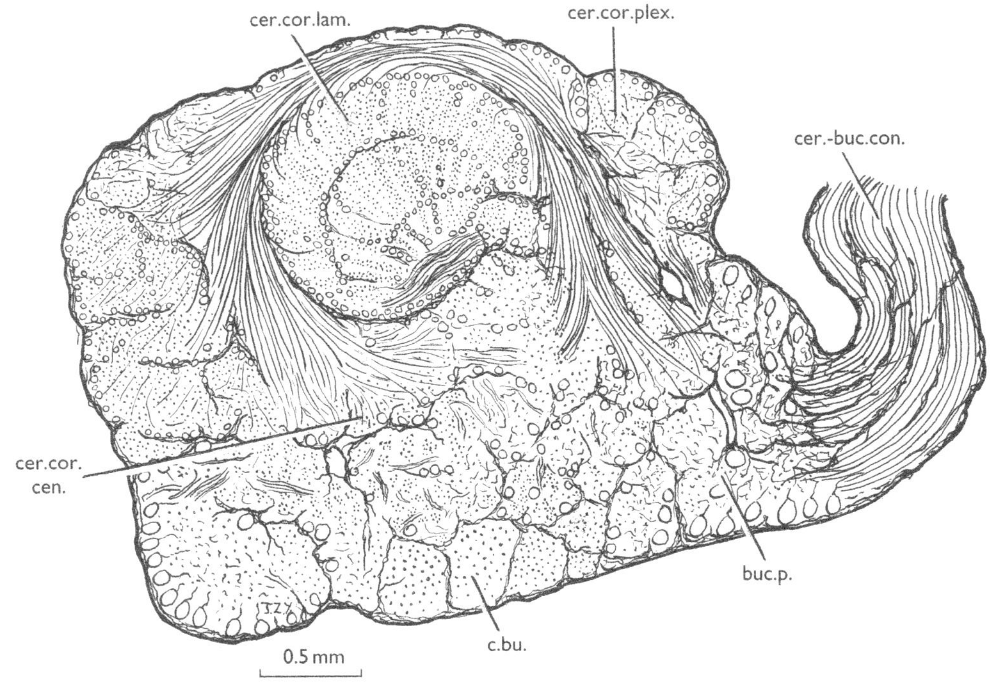

Fig. 3.13. *Nautilus.* Drawing of a sagittal section of the supraoesophageal cord. The cells are shown larger than they really are (Young 1965a).

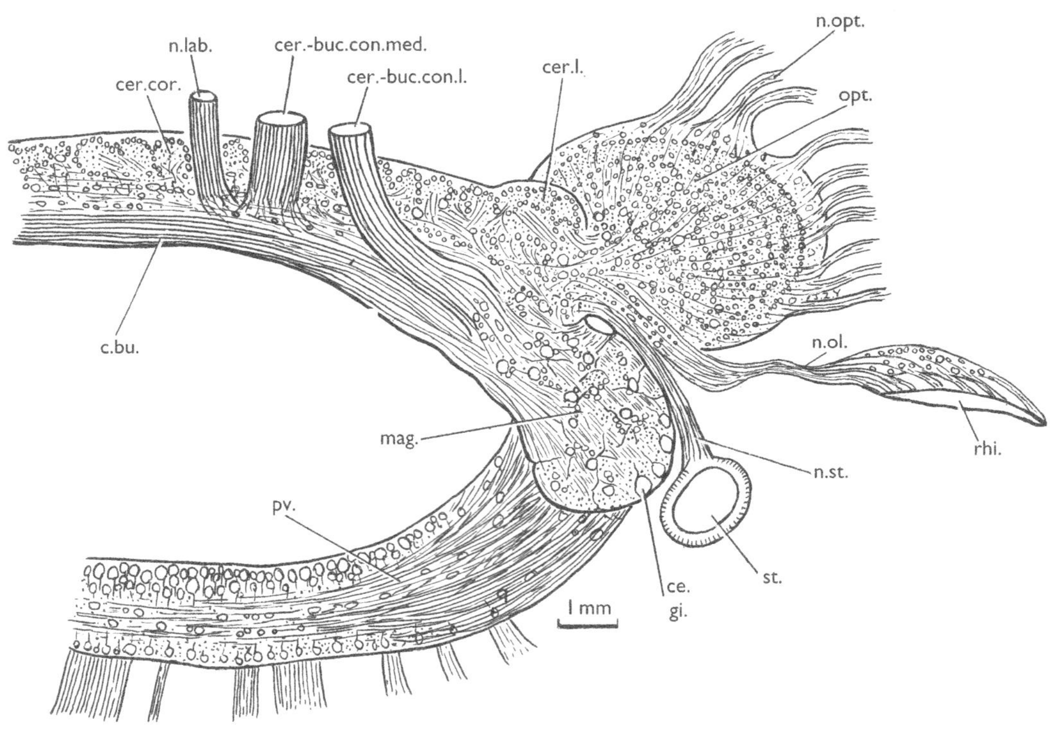

Fig. 3.14. *Nautilus.* Drawing from transverse sections of the central part of the nervous system, reconstructed from photographs. The sizes of the nerve cells are shown in correct proportion but are larger than in reality (Young 1965a).

buccal region, which may be considered as an anterior extension of the central zone. The central zone is continuous laterally with the magnocellular lobe and posterior suboesophageal cord and with the centre of the optic lobe. More anteriorly it is continuous with the brachial part of the anterior suboesophageal cord.

In the median part of the cord the cells become fewer and fewer proceeding ventrally, and the central region passes imperceptibly into the commissural region. This is occupied by bundles of commissural fibres.

The lateral cerebral lobes (Fig. 3.14) are especially developed regions of the plexiform zone, and lie at the sides above the brachial portion of the anterior suboesophageal cord. The lobes exchange bundles of fibres with probably all the lobes in the neighbourhood, many to the plexiform zone, and large bundles to the brachial lobes, as well as other connections.

3.1.10 Statocysts

The two statocysts are located ventrolaterally on either side of the brain (Fig. 3.14). They are embedded in fibrous tissue where the three main cords of the central nervous system meet. Each statocyst is ovoid in shape, with a single undivided cavity (Fig. 3.11). The lower half of the lumen of the statocyst is filled with numerous, small, free-moving, calcareous statoconia and the upper half with endolymph (Young 1965a). The statoconia are crystals of aragonite and amorphous calcium phosphate, with some strontium and magnesium (Lowenstam *et al.* 1984). Most of the statoconia are ovoid in form but some are spherical (Neumeister and Budelmann 1997). The statocyst is oriented with its long axis transversal, and approximately 45° to the horizontal plane of the animal, and the lateral end is higher than the medial end. It communicates with the exterior by Kölliker's canal (Fig. 3.10).

Neumeister and Budelmann (1997) made a detailed morphological and behavioural study of the statocysts. They found the lining of the statocyst sac to be formed of a single layer of columnar epithelial cells, comprising four types of cell: namely type A and type B receptor hair cells, supporting cells, and ciliated cells at the entrance of Kölliker's canal. Below the epithelium a nerve plexus separates the epithelial cells from the statocyst wall, which consists of a thick layer of connective tissue. The inner surface of the sac is lined with regularly distributed primary sensory hair cells (130 000–150 000) excluding the area around the opening of Kölliker's canal. The hair cells of type A have 10–15 kinocilia arranged in a single ciliary row, and are present in the ventral half of the statocyst. Type B hair cells carry 8–10 irregularly arranged kinocilia and are present on the dorsal half of the statocyst. Both of these types of cells are morphologically polarized. There is a distinct borderline at the equator of the statocyst where the two hair cell types meet.

Behavioural studies were undertaken to determine whether the statocyst could sense angular accelerations and the statocyst-dependent funnel movements were measured during sinusoidal oscillations of restrained *Nautilus* around a vertical body axis. The funnel movements were either in the same direction or in the opposite direction from that of the applied rotation. Compensatory funnel movements were observed during optokinetic stimulation and during combined optokinetic and statocyst stimulation. The experiments demonstrated that the statocysts can detect rotary movements, or angular acceleration, as well as function as gravity receptor organs.

Kölliker's canal proceeds laterally from the statocyst towards the surface by a small pore at the base of the olfactory tentacle, or rhinophore, below the eye. The inner surface of the canal is ciliated. Young (1965a) found that the canal remains open in the adult. The area around the orifice of the canal into the statocyst and an area ventral to the opening is densely covered with ciliated cells, each bearing 30–50 irregularly arranged kinocilia.

The gross morphology of the statocyst of *Nautilus* differs markedly from that of the coleoid cephalopods but is similar to that of other molluscs. Morphological and behavioural findings indicate that the statocysts of *N. pompilius* function as gravity receptor organs and can detect rotatory movements (angular acceleration) in the absence of the specialized receptor systems (crista/cupula systems) present in the statocysts of coleoid cephalopods. This indicates that both statocysts and visual inputs control compensatory funnel movements. Although the eyes and visual system are less complex in *Nautilus* than in coleoids, nevertheless visual and mechanoreceptive information converge for the control of locomotion in space (Neumeister and Budelmann 1997).

4 Spirulidae

4.1 *Spirula*

The only species of the family, *Spirula spirula* (Fig. 4.1), is characterized by an internal, coiled, chambered shell which makes it unique among extant cephalopods. It has a worldwide distribution in tropical and subtropical waters between latitudes 40°N and 40°S, but is absent from the Mediterranean, the Red Sea, and the Sulu Sea (Bruun 1943). It migrates vertically, being found mostly below 500 m in daylight and above 300 m in darkness. The movements, both up and down, take several hours, or may be continuous for some of the population. Adults and young occur at similar depths (Clarke 1969).

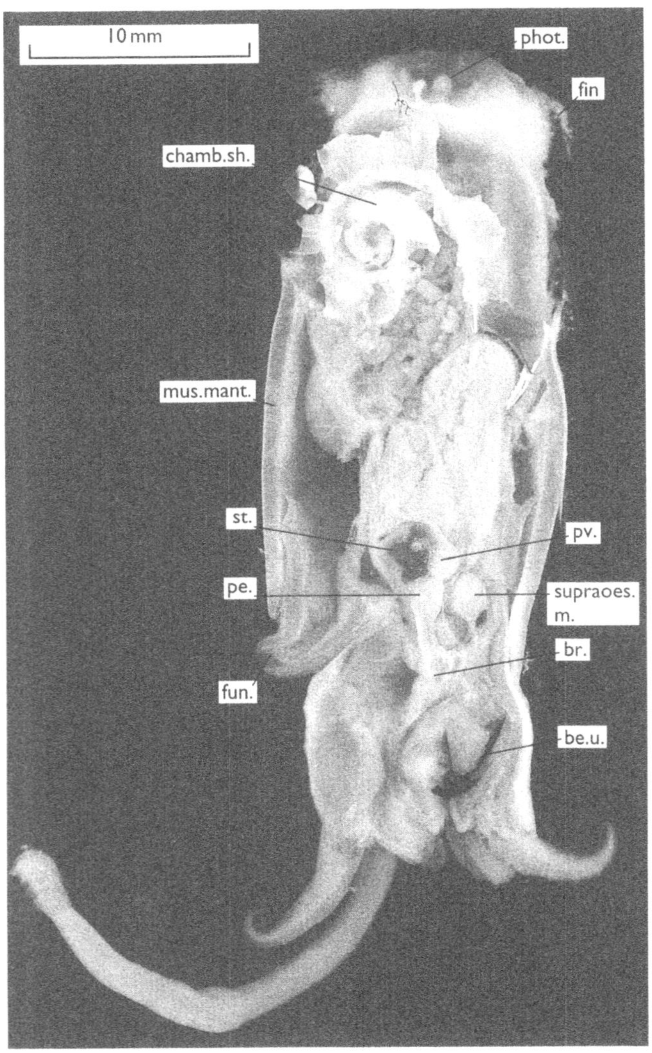

Fig. 4.1. *Spirula spirula*. Hemisection to show the main features of the animal (30 mm ML).

4.1.1 The shell, buoyancy, and locomotion

Empty shells of *Spirula* are found in their thousands on tropical beaches, yet little was known of the animal itself before the beginning of the twentieth century. Franc Péron discovered *Spirula* during his voyage around the world between 1800 and 1804 (Chun 1910a), and only 13 specimens were known until the *Dana* expedition of 1920–1922. During this voyage further captures were made, including some living animals which were observed on board ship (Schmidt 1922).

The shell, first described by Appellöf (1893), is of aragonite (Bøggild 1930). It has a lamellar structure with an external spherulitic-prismatic layer and an inner prismatic layer (Mutvei 1964, 1970). The shell lies in the posterior half of the mantle (Fig. 4.2) and is already present at hatching. The shell wall has inner and outer plates, and is as thin as the septa; it is characterized by the absence of a nacreous layer (Doguzhaeva 1996).

A living *Spirula* is close to neutral buoyancy (Denton *et al.* 1967). The newly formed chambers of the shell are filled with liquid, which is pumped out by osmotic activity when the walls of the chamber become sufficiently strong to withstand the pressure. Gas enters gradually from the tissues, the pathway being via a porous region at the upper corner of each chamber, between the cavity and the siphuncle, through which liquid can pass in or out (Fig. 4.2). The gas

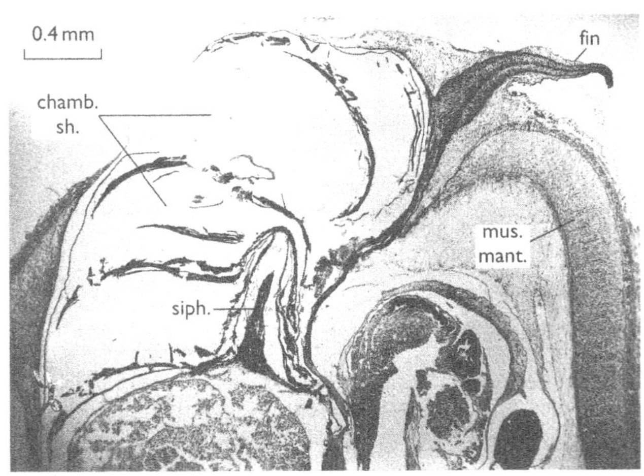

Fig. 4.2. *Spirula spirula* (32 mm ML). Longitudinal section through the mantle and shell to show chambers of shell, the siphuncle and fin (haematoxylin and eosin).

within the chambers of the shell is always less than atmospheric pressure. While still inside the animal the shell can withstand implosion pressure equivalent to depths of 1360–2020 m (Denton and Gilpin-Brown 1971). The findings confirm the conclusion from captures of *Spirula* that it is mesopelagic and does not penetrate much below 1000 m (Clarke 1969).

When newly hatched the animal possesses a shell which already has two to three chambers. At at a mantle length of 3 mm there are three chambers; a mature female (30 mm ML) has 30 (Denton and Gilpin-Brown 1971), and a mature male (43 mm ML) has 39 chambers in the shell. The position of the shell in the body and its function as a buoyancy device determine that the animal lives head-down. This posture has been observed in an aquarium where the animal remained so suspended for hours. *Spirula* is also capable of swift and jerky movements, upwards, downwards, or laterally. During such movements the fins are in continuous motion; after severance of one fin the animal could no longer swim downwards or remain underwater (Schmidt 1922).

The mantle and funnel are both employed during movement. The muscle of the mantle wall is thick, with mostly circular muscles, an outer layer of longitudinal fibres and radial bundles (Fig. 4.3). Radial connective tissue fibres run obliquely in different directions. The funnel is muscular and relatively large (Fig. 4.1). A complete giant nerve fibre system is also present, indicating the capacity of the animal to make rapid movements.

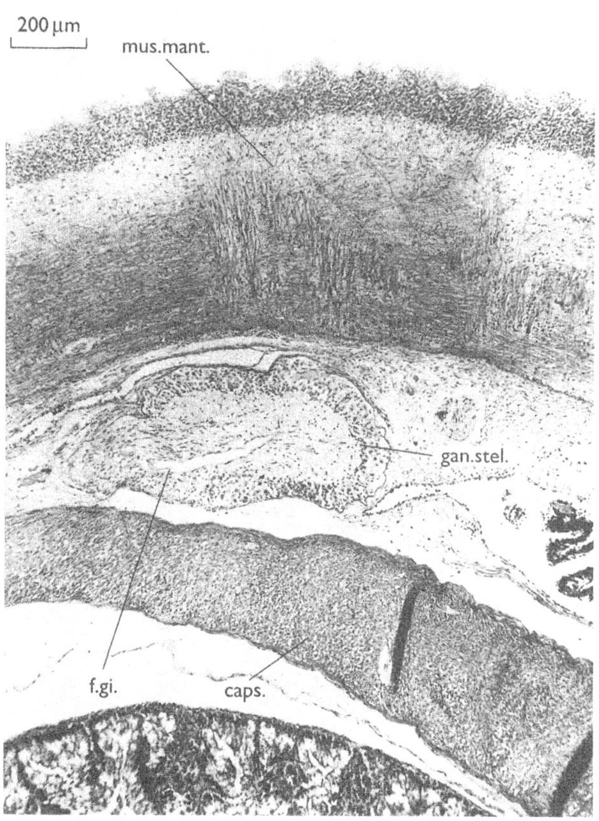

Fig. 4.3. *Spirula spirula*. Transverse section showing the thick musculature of the mantle and the stellate ganglion with a giant fibre (decalcification, haematoxylin and eosin).

4.1.2 Life history

Males are mature at 38 mm ML; the maximum size recorded is 46 mm ML (Clarke 1970). Ventral arms IV are hectocotylized and notable for their asymmetric modification; the hectocotylus of the right arm is larger and has a long deep groove, whereas that of the left arm is slender and ends with a series of 'fingers'. These arms are probably used together (Kerr 1931). Hectocotylization is first apparent in animals of 20 mm ML, and its development on the two arms has been followed (Bruun 1943). Morphology of the spermatophores has been described at the light microscope level (Kerr 1931) and of the spermatozoa at the ultrastructure level (Healy 1990a). Females begin to mature at about 24 mm ML and nidamental glands become enlarged at 30 mm ML. At this size spermatophore sacs were found on the buccal membranes (Clarke 1970). In the ovary eggs of varying sizes are found, and mature ones (1.7 mm long) are ovoid in shape (Chun 1910a, 1910b, 1915). The disparity in the size of the ovarian eggs indicates that spawning is intermittent (Mangold *et al.* 1993).

Young have been collected from mud deposits, suggesting that the eggs may be laid on the bottom (Bruun 1955). Small specimens of 2–4 mm ML have been caught from the surface to 1250 m depth, in daylight and darkness, off the Canary Islands (Denton and Gilpin-Brown 1971). At this size the mantle and funnel form a relatively large portion of the animal, suggesting that it moves actively in the water column. With growth there is a change in form, and maturity is reached in some 12–15 months; the egg stage is presumably protracted and the adults probably have a life span of 18–20 months (Clarke 1970).

4.1.3 Morphology

The arms and the tentacles have dense and complex muscles. The eight arms are short and tapering, and have four longitudinal rows of suckers, each attached by a short peduncle. The arm suckers are strengthened by an inner ring which has blunt teeth of almost uniform height and size. This is encircled by an infundibulum with polygonal processes and projecting pegs. The cuticular surface is covered with pores approximately 0.07 µm in diameter. The tentacular suckers are slightly smaller than those on the arms, but their features are similar (Nixon and Dilly 1977).

The beaks are darkly tanned, and the rostrum of the lower beak protrudes and is sharp. The radula is vestigial and all that remains is an invagination (300 µm in length) with a lining of columnar epithelium (Figs 4.4, 4.5). The buccal complex shows vestiges of a pair of bolsters, each within the radular supports. On either side of the buccal complex are the paired lateral buccal palps which bear stout teeth on their anteromedial surfaces (Fig. 4.6). The salivary papilla carries the single duct from the paired posterior salivary glands, which are a conspicuous part of the mantle cavity (Kerr 1931). The oesophagus has a thick cuticular lining and it opens into the gizzard, the cuticle of which is even thicker. The digestive gland has two distinct parts (as in *Sepia*) opening by separate ducts to the caecum. In our sections the gizzard contains some whole copepods and ostracods and others that are half-digested.

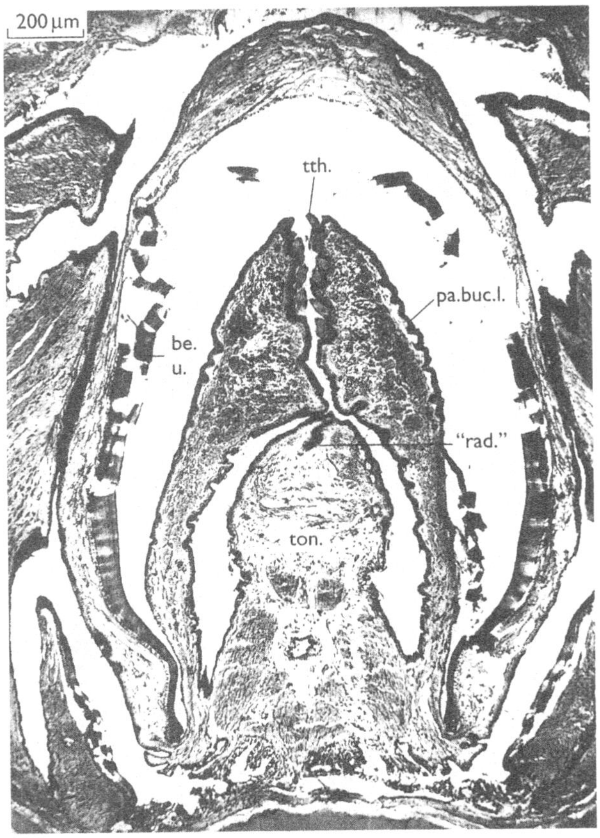

Fig. 4.4. *Spirula spirula*. Transverse section of the buccal mass showing the vestigial radula, the toothed lateral buccal palps, and the 'tongue'.

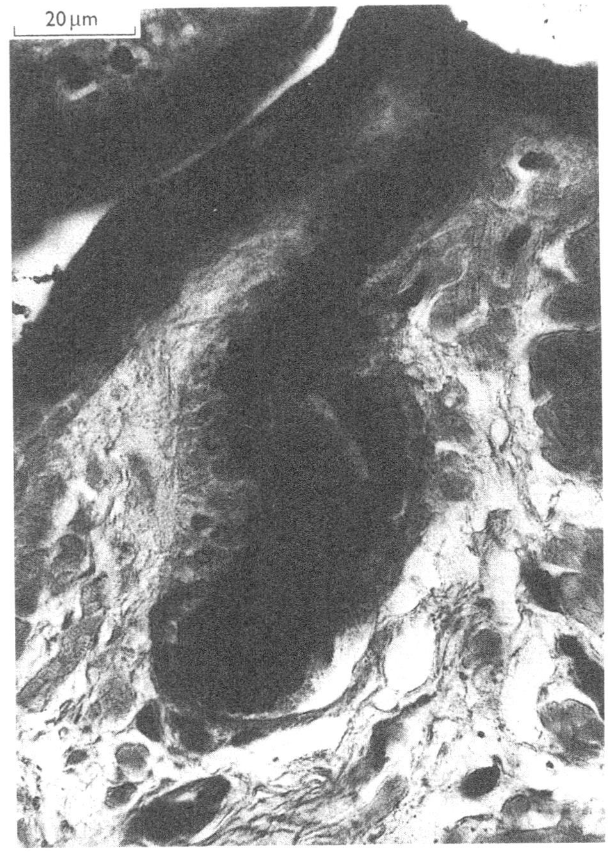

Fig. 4.5. *Spirula spirula*. The same section as in previous figure but at higher magnification to show the vestigial radula and amorphous rod.

4.1.4 Photophore

At the hind end of the body a photophore lies between the fins set in a depression, formed by an invagination with an opening 1–3 mm in diameter (Chun 1910a) (Figs 4.1, 4.7). The luminous organ comprises a basal reflector of collagen fibrils, a central photogenic region of lamellate tissue, and a distal spherical mass of tissue with a lamellar arrangement, or 'lens'. The lens is clearly delineated only in animals of more than 15 mm ML. The photophore emits a steady, continuous, and long-lasting pale yellow-green light (Schmidt 1922, Herring *et al.* 1981). It has an emission maximum of 471 nm, but its function is obscure; the light emitted can be modified as its intensity can increase 5–10-fold in 1–5 s; the subsequent decline is gradual and lasts for up to 1 min (Herring 1983).

4.1.5 Receptors and central nervous system

The eyes form a prominent part of the head. Two conjunctive extraocular eye muscles are present and they resemble those described in *Sepia* and *Sepiola* (Budelmann and Young 1993). The lens is transparent down to 310 nm (Messenger 1981). The rhabdoms of the retina are up to 200 μm in length at the centre, and are perhaps longer ventrally than dorsally (Fig. 4.8).

The optic lobes form 224% of the total brain volume (Table 2.6). The inner granular cell layer of the optic lobe is thick, and its cells are organized into columns. This columnar organization continues deep into the lobe, but in irregular form. The peduncle lobe is placed vertically and

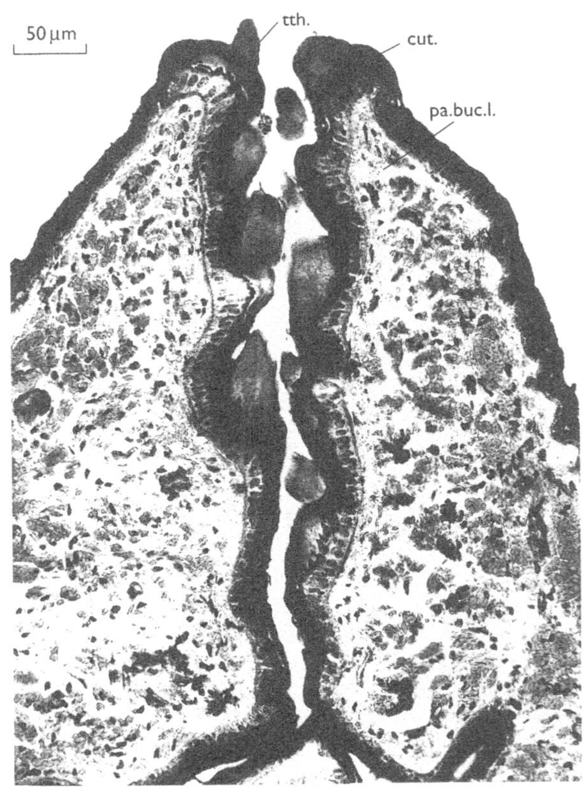

Fig. 4.6. *Spirula spirula*. The lateral buccal palps covered by a thick cuticle bearing stout teeth on their anteromedial surfaces.

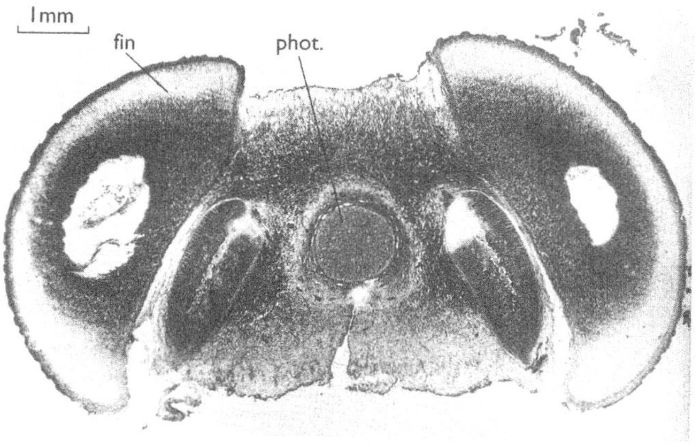

Fig. 4.7. *Spirula spirula*. Transverse section through the hind end of the body showing the fins and luminous organ with photophore cells.

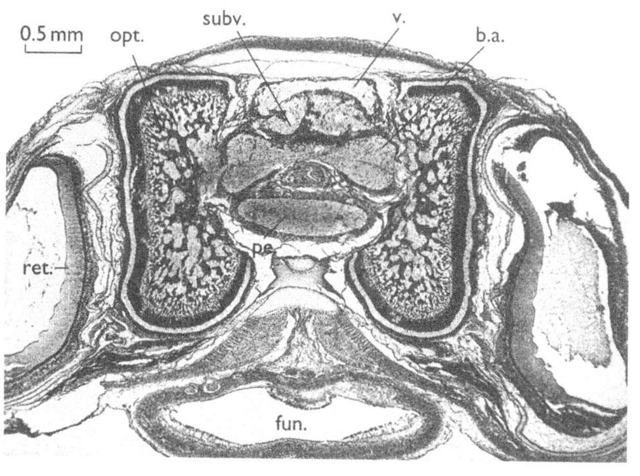

Fig. 4.8. *Spirula spirula*. Transverse section through the head and eyes.

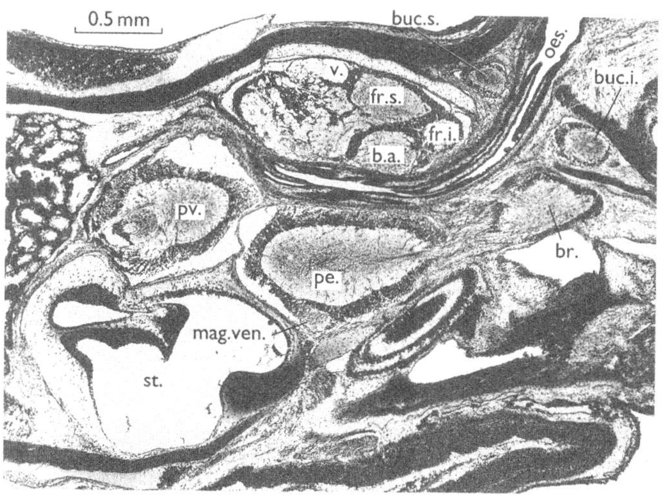

Fig. 4.9. *Spirula spirula*. Sagittal section through the head to show the arrangement of the central nervous system and the statocyst.

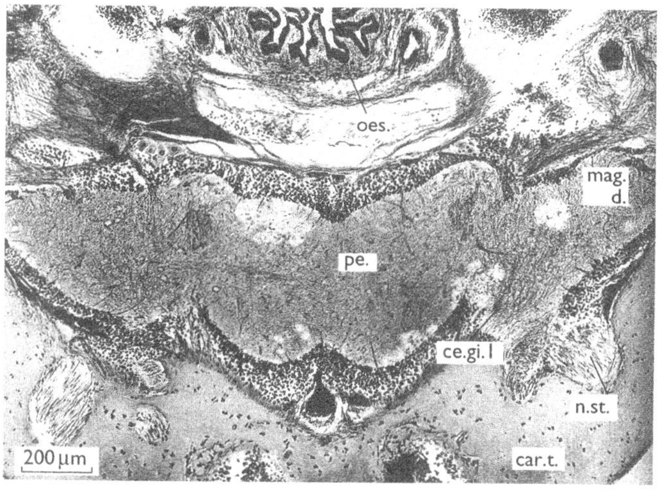

Fig. 4.10. *Spirula spirula*. Transverse section of suboesophageal mass and the first-order giant cells.

the optic gland dorsally on the optic tract. The paired extraocular photoreceptor organs each comprise 4–6 sacs of photosensitive vesicles, which lie on the medial surface of the optic lobe and beneath the optic tract. They form 1.1% of the total brain volume (Table 2.6).

The central nervous system follows the general plan found in decabrachians, but is without special similarity to that of either Sepiolidae or Sepiidae. The superior buccal ganglia lie far forward (Fig. 4.9). The inferior frontal lobe system forms 3.15% of the total brain volume (Table 2.6). The anterior suboesophageal mass is somewhat separated from the middle suboesophageal mass. There is no distinct prebrachial lobe. Many fibres run directly from the brachial nerves to the magnocellular lobe.

First-order giant nerve cells lie near the entry of the static nerves and they are elongated with large lateral dendrites (Figs 4.10, 4.11), and a nucleus about 25 µm in diameter in an animal of 13 mm ML. Their axons cross in the palliovisceral lobe, without fusion. There is one large fibre in the mantle connective, but those in the stellate ganglion and stellar nerves are of only moderate size (Fig. 4.3). In a mature male (33 mm ML) the most posterior nerve contains three fibres 35 µm, 21 µm, and 10 µm in diameter. The penultimate nerve contains one fibre 35 µm in diameter. The fin lobes receive prominent tracts from the anteroventral palliovisceral lobes. The fin lobes form 2.7% of the total brain volume (Table 2.6). There are no distinct chromatophore lobes.

The anterior basal, peduncle, and olfactory lobes occupy a considerable portion of the supraoesophageal mass. The inferior frontal lobe system forms 3.1% and the vertical lobe system 14.2% of the total brain volume (Table 2.6). The vertical lobe itself is relatively small, unlike the superior frontal lobe (Fig. 4.9).

The statocyst has thick walls of cartilage laterally and medially, but thinner anteriorly and posteriorly (Fig. 4.9). The macula princeps is almost completely divided into dorsal and ventral areas. The maculae neglectae are set obliquely. The cavity has small anteroventral and posterior sacs and large dorsal sacs. All five hamuli are present and conspicuous in size and in length (Fig. 4.12). The second hamulus stands below the crista, above which is a ridge. The third hamulus bends over the crista and the fourth is large with a broad base. There are only four anticristae, and these have somewhat different positions from those of other coleoid cephalopods. All lie far back and leave a spacious, unobstructed anterior chamber. In this the large mass of endolymph acts across the anterior transverse crista and longitudinal crista. On the medial wall there is a large protuberance, which may correspond to

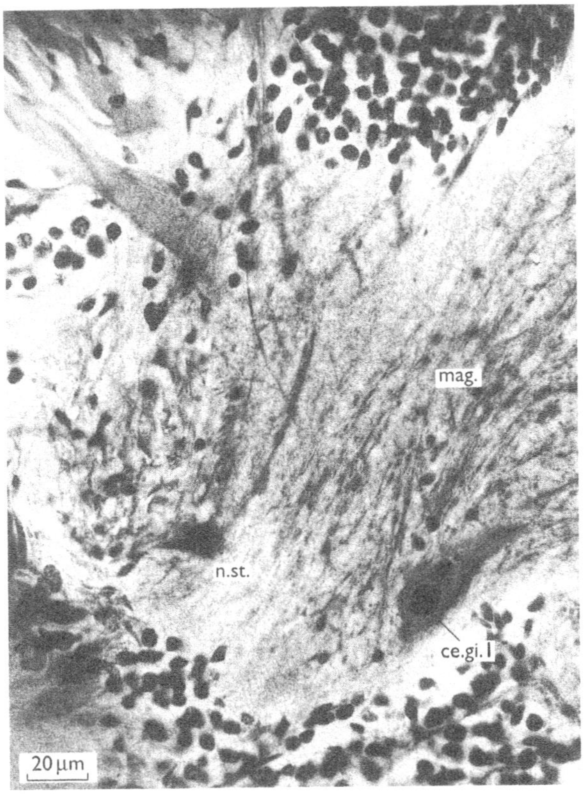

Fig. 4.11. *Spirula spirula.* First-order giant cell.

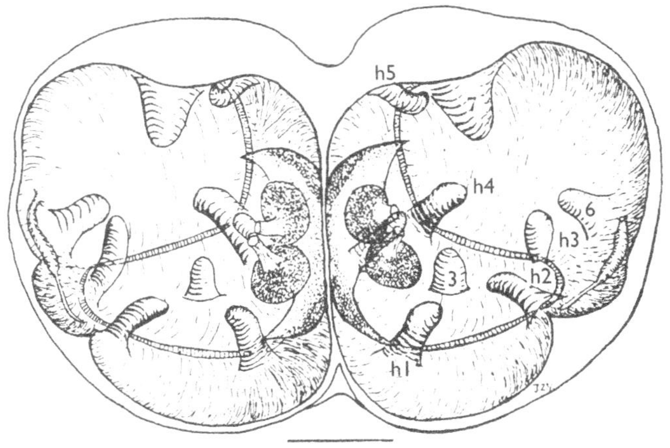

Fig. 4.12. *Spirula spirula.* Drawing of statocyst seen from in front (scale bar 1 mm) (Young 1989).

anticrista 2 or 3 of *Sepia*, and one below it standing on the floor of the chamber is in the position of anticrista 6, and the opening of Kölliker's canal is at its base. The largest anticrista is in the roof and can perhaps be compared with anticrista 7 in *Sepia*. Turning is likely to be slow owing to the inertial mass of endolymph in the more anterior region. Anticristae 6 and 7 provide only a partially closed channel for flow across the vertical crista. The large hamuli at each turn suggest rotation at the same speed in all directions (Young 1989). Discriminant analysis of several parameters of the statocyst place *Spirula* amongst the buoyant squids (Maddock and Young 1984).

5 Sepiidae

5.1 Introduction

Sepiidae live in tropical and warm temperate waters around oceanic islands and on the continental shelf and upper continental slope between 20°W and 180°E, from 40°N to 45°S (Khromov 1998, Khromov *et al.* 1998). The genus *Sepia* includes more than 90 species. It occurs from southern Norway southwards to southern Africa and in the Mediterranean, but is absent from Greenland, Iceland, the Baltic, and the Black Sea (Boletzky 1983a). Numerous species occur along the coasts of the Indo-Pacific Ocean and the waters of Australia (Adam 1979, Norman 1997, Lu 1998, Nixon in press b). Cuttlebones are washed on to the beaches of the Americas, but these have been identified as belonging to species from West Africa or Europe (Voss 1974). *Sepia* is typically an inhabitant of the continental shelf and upper slope waters, mostly at depths of 2–250 m, but some go to 500 m and a few penetrate depths of 1000 m (Adam and Rees 1966). One species may inhabit different depths in different areas, for example *Sepia elegans* at 30–220 m in the Mediterranean (Mangold-Wirz 1963) but 2–45 m off the north-west coast of Spain (Guerra 1985a). Sepiids are largely benthic or benthopelagic in habit, and mostly found where the bottom is of rock, sand, or mud, amongst seaweed grasslands or in coral reefs. There are two other genera in the family, *Metasepia* and *Sepiella*, of which little is known in comparison with *Sepia*.

5.2 *Sepia*

Sepia is the best-known genus of the family, and *S. officinalis* the most widely investigated member (Fig. 5.1). It is readily recognized from its dorsoventrally flattened form, and lateral fins extending along most of the length of the mantle. The mantle follows the contours of the internal shell, cuttlebone, or sepion, which is calcareous and forms an adjustable buoyancy device (Denton 1971) (Figs 5.2, 5.3).

5.2.1 *Life history*

Various features and habits of *S. officinalis* at hatching and during the early weeks of its life cycle are given in Table 5.1 (Nixon and Mangold 1998). Its life cycle and reproductive habits were reviewed by Boletzky (1983a).

Mature adults move inshore to spawning grounds where the water is some 30–40 m in depth (Mangold-Wirz 1963). Confirmation of shoreward migration for spawning has come from tagged *S. officinalis* (Boucaud-Camou and Boismery 1991). Courtship displays include

Fig. 5.1. *Sepia officinalis*. A mature animal displaying the deimatic body pattern, with a black spot on each side of the mantle (photograph kindly provided by B.B. Boycott).

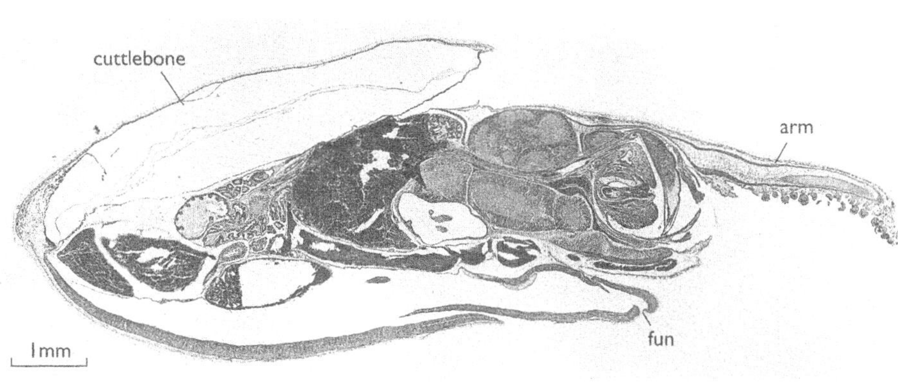

Fig. 5.2. *Sepia officinalis*. A medial sagittal section of hatchling (8.5 mm mantle length) to show its main features and the relatively large size of the cuttlebone at this stage (haematoxylin and eosin).

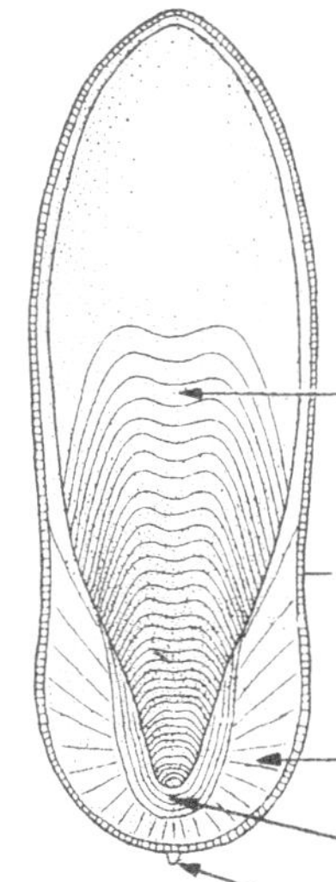

Fig. 5.3. *Sepia officinalis*. Cuttlebone, sepion, or shell (ventral view) to show its morphological features.

Table 5.1. A summary of various aspects of the early life, habits, and habitats of *Sepia officinalis* (Nixon and Mangold 1998)

Mature adults	Migrate to inshore shallow waters
Breeding period, Mediterranean	Winter, spring, and summer
Spawning site	Inshore spawning grounds
Depth of spawning site	5–60 m
Bottom habitat	Rocks, pebbles, sand, mud associated with sea grass
Number of eggs per individual	Up to 1000
Egg capsule length	25–30 mm
Eggs	Laid singly and attached to algae or other objects on the substrate
Egg-care	None
Embryo development	40–45 days at 20 °C
Mantle length at hatching	8–10 mm
Total length at hatching	13.6–17.0 mm
Weight	
at hatching	150–155 mg
at 40 days	1000 mg
Time of hatching	At transition from light to dark
Hatchling	Opaque, resembles adult, and moves away from egg to inhabit the same bottom habitat
Larva/paralarva – type	Non-dispersive
Denticles on beaks (oral surface)	Absent
Buccal mass length at hatching	30% of mantle length
Number of suckers on each arm	Numerous (>20)
Habitat	
in daylight	On or near bottom
in darkness	In water column
Response to light at hatching	Negative—buries itself or hides in daylight

Table 5.1. *Continued*

Chromatophores	
at hatching	350–500 mm^{-2} on dorsal surface
in juvenile	200–500 mm^{-2} on dorsal surface
Chromatic and body patterns	
at hatching	As variable as in adult
in juvenile	Highly variable
Locomotion	Jet propulsion and swimming
Buoyancy	Cuttlebone forms a neutral buoyancy mechanism
Ink sac	Present
Ink ejection	At or shortly after hatching
Prey—in the sea	Crustaceans
Feeding	Predatory – 'innate' response to shrimps Planktotrophic and benthic prey
Mode(s) of prey capture	
tentacular strike	Present
arms	Present
Paralysis of prey	Toxin(s) secreted by posterior salivary glands
Habits	Sand-covering, adheres to substrate by means of 'sucker' on anteroventral surface of mantle and arms, enters water column in darkness
Settlement on bottom	Benthopelagic at hatching
Receptors	
Eyes	Present
Extraocular light organs	Present
Statocysts	Present
Olfactory organs	Present
Epidermal ciliated receptor cells	Present
Lateral line analogue	Present
Life span	12–24 (36) months

elaborate body patterns and movements which determine sexual recognition. An acute zebra pattern (Hanlon and Messenger 1988) is displayed by both sexes, but is most conspicuous in the male (Tinbergen 1939). The male extends one ventral arm towards a potential partner and if accepted will grasp the female and manoeuvre her so that they are in a head to head position; at the same time he entwines her with his arms. Spermatophores are transferred by the male with his left, hectocotylized, arm IV from within his mantle into the bursa copulatrix, a special pouch below the buccal membrane of the female. Exact details of the mode of transfer of the spermatophores and the release of spermatozoa are not known.

Spawning can begin shortly after copulation, each egg being expelled by the female into her arms. The egg is flask-shaped, 25–30 mm in length, soft and gelatinous. The female *S. officinalis* stains the envelope of the egg with her own black ink. The arms draw out two processes from the gelatinous coat of the egg into a ring-shape to attach it to seaweed or some other object less than a centimetre in diameter. Eggs are laid at irregular intervals over many days, and a large female can continue for as long as 7 months (Boletzky 1987). The eggs are usually attached in clusters, the sight of which often stimulates other females to add their eggs (Bott 1938). The number of ova increases with the size of the female and the effective fecundity is as great as 5000 in a large individual. The size distribution of ovarian eggs is wide, and the function of the ovary is prolonged, allowing maturation of most of the potentially available ova. Thus *S. officinalis* is one of numerous coleoid species which are iteroparous, that is they spawn intermittently (Mangold *et al.* 1993). Development takes 40–45 days at 20 °C in the laboratory (Boletzky 1983a). There is a period of accelerated growth towards the end of development, between embryonic stages XVIII and XX (of Naef 1928, Fioroni 1991). The black capsule of the egg thins, because of an increase in the perivitelline fluid, and the embryo can apparently detect light changes as, in the laboratory, hatching consistently takes place just after transition from light to darkness (Paulij *et al.* 1991).

The ontogeny of *S. officinalis* during embryonic growth and early development has been reviewed by Fioroni (1991). At hatching the young bears a striking resemblance to the adult externally, and internally its tissues are highly differentiated (see Figs 5.2, 5.10) (Nixon and Mangold 1998). It can burrow by the action of the funnel as the jets of water expelled send sand swirling upwards, to descend on the upper surface of the mantle and head: this activity is counteracted by vigorous movements of the fins so the animal maintains its position (Boletzky 1974a). Recently hatched cuttlefish can adhere to a hard substrate by means of a ventral 'sucker' (Bather 1895). Such adhesive powers are retained at least until a mantle length of 30 mm is reached. The ventral area of adhesion is formed by the integumental muscle of the ventral mantle surface and the flattened underside of arms IV (Boletzky 1974a, Boletzky and Roeleveld 2000). When exposed to a strong current, hatchlings become slender and streamlined. Body patterns are already fully functional at hatching (Hanlon and Messenger 1988). Once the yolk is utilized the young swim in search of food, and will attack and capture mysids (Wells 1958). Young *S. officinalis* can capture freely swimming prey, even in complete darkness, apparently detecting small water movements generated by the prey (Budelmann and Bleckmann 1988, Bleckmann *et al.* 1991). The local movements are detected by lines of epidermal hair cells present on the head and arms (Naef 1928). These cells carry cilia (Budelmann *et al.* 1997) and constitute a mechanoreceptive organ analogous to the lateral lines found in fishes and amphibians (Budelmann and Bleckmann 1988, Budelmann *et al.* 1991, Bleckmann *et al.* 1991). The development of learning with age is apparent in young cuttlefish (Wells 1958, 1962), and improvement in performance has been correlated with the development of the vertical lobe system (see Fig. 5.12). The volumes of various lobes of the brain at hatching are shown in Fig. 5.18.

At hatching the mantle of *S. officinalis* is 6–9 mm in length. In laboratory-reared animals a mantle length of 150 mm is attained in some 200 days at 15 °C; the growth curve is typically sigmoid. The rate of growth is greater in males and they attain a larger size than females (Mangold-Wirz 1963). Males of 300 mm ML have been recorded; these are at least 2 years of age. Females of 250 mm ML are known, and an exceptionally large one of 300 mm ML was captured in the English Channel (Mattacola *et al.* 1984). Maturation occurs at different sizes: some males of 60–80 mm ML are mature but others are not, and females do not generally reach maturity until 110–250 mm ML. Thus males may have a longer period of reproductive activity than females. A cycle of shorter and longer generations occurs amongst females as large mature females migrate into shallow water in spring, whereas smaller mature ones move in later, in spring and early summer (Mangold 1966).

Sexual dimorphism in *S. officinalis* is apparent; females are smaller than males (Naef 1923) and the two sexes exhibit different body patterns during courtship (Tinbergen 1939). One of the largest members of the genus, *S. latimanus*, also shows different body patterns when courting. It was observed in the sea on the reef slope off the coast of Guam, during the annual mating season (Corner and Moore 1980). The males arrive first and take up positions on pinnacles of the reef to intercept and mate with the females as they arrive from deeper water. Males show specific precopulatory behaviour and body patterns, and these continue only if the female responds. In females the posture, movement, and body patterning is less elaborate, and she pales when ready to mate. On acceptance the female opens her arms and places them on the head of the male, at which time the copulatory arm is introduced into her mantle. Both cuttlefish then become blotched and skin papillae become visible. The female finds a suitable crevice in a coral branch to lay eggs. No ink is ejected over the egg, although a creamy white latex-like substance is seen on the arms and on the egg. The process of laying a single egg takes some 4–9 min. After depositing eggs females repeatedly copulate with other males.

5.2.2 The cuttlebone

The life of *Sepia* is dominated by the variable buoyancy system provided by the internal cuttlebone (Figs 5.3, 5.4). In *S. officinalis* this forms a very large portion of a hatchling (Fig. 5.2), and is some 9% of the volume of a juvenile (Denton 1961). It consists of a calcareous component, aragonite, a trace of strontium, and an organic component, a chitin–protein complex in which the chitin is in the β form (Hewitt 1975, Birchall and Thomas 1983).

The cuttlebone is formed of a dorsal shield, a ventral chamber zone, and a spine (Appellöf 1893). The dorsal shield has dorsal and inner calcified layers and a central, translucent layer of sclerotized chitin (Birchall and Thomas 1983). A specialized part of the inner layer constitutes the so-called 'fork', which may represent the ancestral phragmocone (Naef 1923). The ventral chamber is convex and consists of sealed cavities formed by lamellae, separated and supported by pillars (Fig. 5.4). The pillars are sigmoid in form and between them are suspended organic membranes (Bandel and Boletzky 1979, Birchall and Thomas 1983). The siphuncular zone lies on the ventral surface of the chamber zone and is an epithelial surface (Denton 1961); posteriorly this zone narrows and is partly surrounded by the fork. The chambers extend obliquely and each is sealed by the dorsal shield on one side and by the siphuncular epithelium on the ventral side (Denton 1971).

The chambers of the cuttlebone are thin and are laid down one below the other. The lamellar structure of the cuttlebone provides a record of

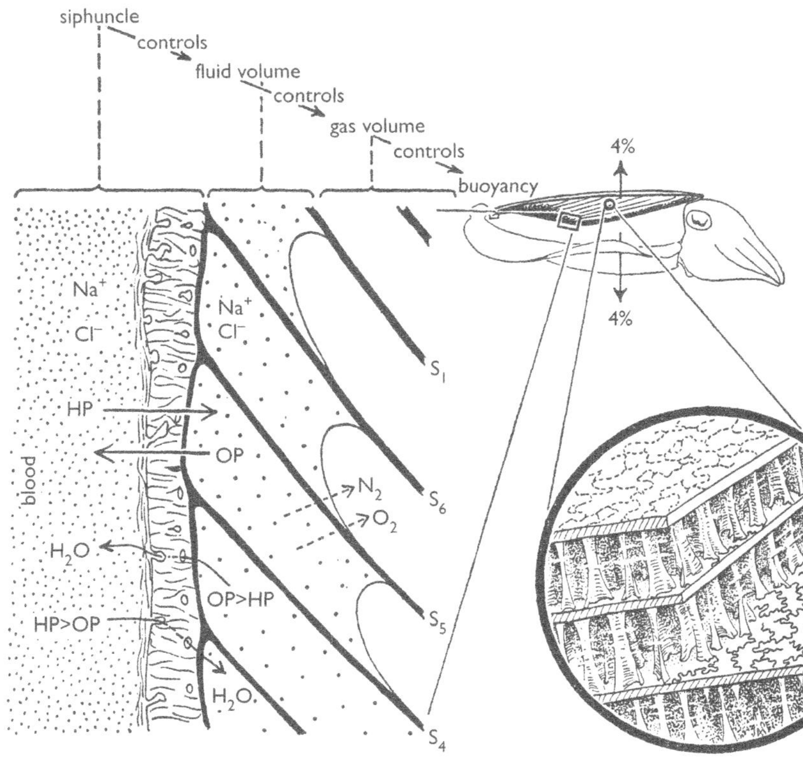

Fig. 5.4. *Sepia officinalis*. The hydrostatic mechanism of the cuttlebone. Withdrawal of salt ions from the fluid in contact with the siphuncular epithelium establishes an osmotic pressure (OP) that counteracts the hydrostatic pressure (HP) of the blood and the surrounding water. Excess of HP over OP pumps water into the cuttlebone, thus reducing the gas space; and excess of OP over HP pumps water out of the cuttlebone, so enlarging the gas space. The pressure in the gas space is determined by partial pressures of gases diffusing from the fluid. The pressure differential (between the aqueous phase and the gas phase) tending to crush the cuttlebone is distributed over successive septa (S) by the ribbon-like bridges of calcified chitin. Their attachment is twice as long on one edge compared with the other edge (Packard 1972, with permission of Cambridge University Press; data and figures from Denton and Gilpin-Brown 1961a, b, and Kälin 1967).

growth. The number of lamellae per unit length varies from 1 septum per day in some species (Choe 1963) to 1.7 per day in *S. officinalis* (Boletzky and Wiedmann 1978). The lamellae may become crowded under natural conditions in the English Channel and the North Sea (Denton and Taylor 1964) and the Bay of Biscay (Le Goff *et al.* 1998), but this was not seen in the cuttlebones of animals living in the Mediterranean (Boletzky 1983a). Under experimental conditions temperature affected the formation of chambers (Richard 1969). The cuttlebones of animals caught in the Bay of Biscay show a rhythm of deposition from a maximum of 18.7 lamellae/month in the summer, to 1.6 lamellae/month in winter (Le Goff *et al.* 1998). Changes in lamellar formation also occur in cuttlebones of *S. esculenta*, off southern Japan (Natsukari *et al.* 1991). Thus environmental conditions are involved in the formation of the cuttlebone. Septal crowding can be induced by starvation (Boletzky and Wiedmann 1978).

The buoyancy of the cuttlefish changes by means of its cuttlebone. The chambers of the cuttlebone are sealed laterally and anteriorly by the thick calcified chitin layer which forms the curved dorsal surface. They contain both gas and liquid; the oldest and most posterior chambers are almost full of liquid, and the newer ones with gas. When a new chamber is formed it is filled with liquid. Shortly afterwards this is extracted actively, so leaving a space which contains gas under a pressure lower than an atmosphere. The liquid is pumped in and out through the posterior ends of the chambers, where there is a specialized membrane; the changes in density are quite rapid. The volume of the total gas space can be varied by altering the balance between the hydrostatic pressure of the sea and the osmotic force operating between the liquid in the cuttlebone and the blood (Denton *et al.* 1961) (Fig. 5.4), thus allowing the animal to rise or sink passively with the changes in overall density that result (Denton and Gilpin-Brown 1961a, b).

When *S. officinalis* is kept in darkness, liquid is pumped out of the cuttlebone and the animal becomes close to neutral buoyancy; this is the period of active hunting in the water column. In daylight conditions the volume of liquid in the cuttlebone increases; the animal becomes denser and rests on the bottom, or even buries itself in the substrate. This system of buoyancy is depth limited but measurements show the design and strength of the cuttlebone is such that it can survive pressures greater than those encountered in life (Denton 1974). *S. officinalis* with cuttlebones 47–170 mm in length were lowered to depths of 200–500 m, and imploded between 150–200 m, whereas hatchlings with cuttlebones of 6–9 mm imploded at depths of 50–100 m (Ward and Boletzky 1984). The depths of implosion are thus close to those usually inhabited by *S. officinalis* at these stages of its life cycle; juveniles and adults are found down to 150 m and recently hatched young at 30–40 m (Boletzky 1983a).

5.2.3 Locomotion

The cuttlefish can move slowly by the actions of its fins and funnel, or swiftly by the rapid expulsion of jets of water through the funnel.

It can hover and move vertically in the water column by means of its fins and changes in buoyancy of the cuttlebone.

The elongated marginal fins lack any firm skeletal support or fluid-filled cavities for hydrostatic skeletal support. The fin musculature is arranged in three mutually perpendicular planes (Fig. 5.5) (Kier 1989, Johnsen and Kier 1993). Transverse muscle bundles extend parallel to the fin surface from the base of the fin to the fin margin. Dorsoventral muscle bundles extend from dorsal and ventral connective tissue fasciae to a median connective tissue fascia. A layer of longitudinal muscle bundles is situated adjacent to both the dorsal and the ventral surface of the median fascia. The muscle fibres are obliquely striated and include a core of mitochondria. A zone of fibres with a more extensive core of mitochondria is present in both the dorsal and ventral transverse muscle bundles (Kier 1989). These muscle masses have a different aerobic capacity (Kier *et al.* 1989). During vigorous swimming, dorsal bending of the fins probably involves simultaneous contractile activity of the dorsal and ventral muscle bundles. Ventral bending is the reverse, and both also involve contractions of the longitudinal bundles. Electromyography confirmed this analysis, but showed that during gentle swimming there was activity only in the dorsal portion during the up-phase of a wave, whereas rapid movement resulted in activity in both dorsal and ventral parts in both the up- and down-phases of the beat.

The fins undulate and so produce a metachronal wave. Waves moving backwards along both fins result in forward movement of the cuttlefish and waves running forwards move the animal backwards: these movements are aided by the expulsion of water through the funnel, which can be moved in all directions. The movements of the undulatory waves may be forwards along one fin and backwards along the other, so turning the cuttlefish about its vertical axis; at the same

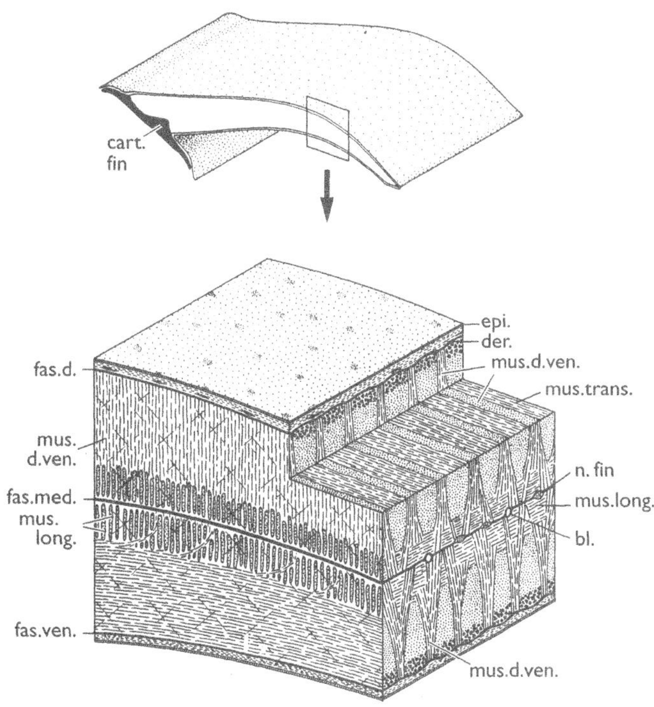

Fig. 5.5. *Sepia officinalis.* A schematic diagram of the fin to show the arrangement of the muscles in three planes, at right angles to one another (Johnsen and Kier 1993).

time the direction of the funnel is opposite to that of the animal's movement. The actions of the fins are mainly under the control of the anterior basal lobe of the brain (Boycott 1961). Numerous receptors, probably mechanoreceptors, are present in the fins, especially along their lateral margins, as mechanical stimuli evoke responses in the distal portion of the fin nerve: the receptors may use peripheral mechanical information during swimming to regulate the beat of the fins (Kier *et al.* 1985).

The mantle cavity is involved in respiration as well as locomotion. The mantle has circular, radial, and longitudinal muscles, and together they form 30% of the body weight of an animal of 250 g (Trueman and Packard 1968). Respiratory activity is mostly confined to the anterior region of the mantle; respiratory flow is produced by the alternate activity of the radial muscles of the mantle and the musculature of the collar flaps (Bone *et al.* 1994). Locomotor activity takes place mostly in the posterior region of the mantle (Packard and Trueman 1974). Contraction of the circular muscles of the mantle results in expulsion of water from the mantle cavity. The longitudinal fibres in the retractor muscles of the head and funnel serve to pull these parts of the body back towards the mantle cavity, and, together with mantle muscles, increase the force by which the water is expelled.

These groups of muscle fibres can be activated by the single pair of first-order giant nerve cells in the magnocellular lobe of the brain. These cells do not directly innervate the muscles, but branch and make synaptic contact with second-order giant fibres in the palliovisceral lobe. There are five of these nerve fibres on each side. Some run to the retractor muscles of the head and funnel, and others to the stellate ganglion on each side. In the stellate ganglion the second-order fibres branch and make synaptic contact with a number of third-order giant fibres, which pass to the muscles of the mantle. The cells of origin of these fibres are dispersed throughout the ganglion, and are not collected into a giant fibre lobe, as in *Loligo*.

The giant-fibre system is fast-conducting and acts to ensure that the whole mantle contracts together to allow rapid escape movements by the expulsion of jets of water from the mantle through the funnel. The effectiveness of the power stroke depends on the velocity and mass of water ejected, the mass of the animal, and the magnitude of the drag forces derived from rapid movement through water (Trueman 1980).

5.2.4 Arms and tentacles

The external features of the eight arms and pair of tentacles were described by Tompsett (1939) (Fig. 5.6). The histology of the musculature and other features was studied by Girod (1884), and the brachial nerves by Budelmann and Young (1987).

The arms have four longitudinal rows of suckers. Each sucker has a soft rim encircling the cuticular infundibulum, within which is a chitinous inner ring with blunt teeth. The infundibulum has numerous rings of polygonal processes carrying pegs, the earliest formed being largest and closest to the inner ring. The latest-formed and smallest pegs lie peripherally (Fig. 5.7). The outermost ring of polygonal processes are flat and devoid of pegs (Nixon and Dilly 1977).

The tentacles can be projected from the sac-like pockets below the eyes (Fig. 5.6) with great speed to capture prey with their clubs (Fig. 5.8). The club of *S. officinalis* has 5–8 longitudinal rows of

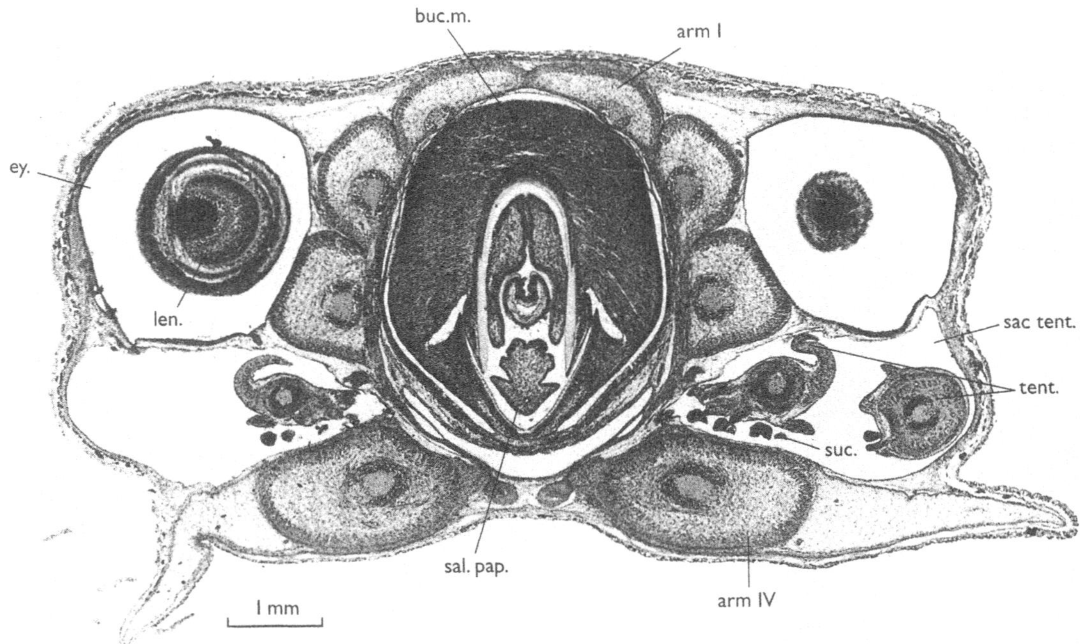

Fig. 5.6. *Sepia officinalis.* Transverse section of the head showing the buccal mass, the eyes, the arms, and the tentacles, each within its sac below the eye.

suckers. These are not of uniform size; there are about five notably enlarged ones on the manus. Each sucker has a rim surrounding the infundibulum. The latter has numerous rings of processes, each with a peg except the most peripheral processes, some of which have developing pegs (Fig. 5.9). The suckers of the tentacular club adhere to the firm exoskeleton of their crustacean prey, and the soft rim forms a seal with the substrate. The maintenance of adhesion depends on the integrity of the rim and infundibulum and can be maintained for about 30 min (Nixon and Dilly 1977).

The suckers of the arms and the tentacular clubs of a 12-day-old *S. officinalis* have a few small, isolated tufts of cilia (20 μm in length) protruding from pores on their outer surfaces (Nixon and Dilly 1977). These cilia are not likely to have a motile capacity since they are few in number and are sparsely scattered, but they may be chemoreceptors (Graziadei 1959, 1964, 1965).

5.2.5 Prey and prey capture

The food of *S. officinalis* includes crustaceans, bony fishes, molluscs and, less often, polychaetes and nemerteans (Nixon 1987). Off the coast of north west Spain the types of prey consumed changes with the growth: crustaceans form the main part of the diet of young cuttlefish, with some fishes and molluscs, but larger cuttlefish (ML > 100 mm) markedly increase their intake of fishes and molluscs although they continue to take crustaceans (Guerra 1985b). *S. officinalis* is buoyant and swims in mid-water in darkness and buries itself in gravel during the day (Denton and Gilpin-Brown 1961b). Furthermore, this cuttlefish takes prey from the plankton when in the water column and from the substrate when on the bottom (Guerra 1985b). There are occasions during the day when the cuttlefish, especially if hungry, will swim in search of prey. Once successful, it returns to the bottom. As the prey tends to be denser than seawater, its consumption results in an immediate increase in the density of the cuttlefish, so aiding its return to the bottom (Denton and Gilpin-Brown 1961a).

The hatchlings of *S. officinalis* attack mysids innately, only subsequently attacking other types of prey (Wells 1958, 1962, Nixon and Mangold 1998). In the sea the diet of hatchlings consists largely of small shrimps (Henry and Boucaud-Camou 1991). Throughout the juvenile stage crustaceans form the major part of their diet (Guerra 1985b). Small crustaceans are ingested whole, but larger ones are bitten into pieces small enough to pass along the oesophagus: the size of skeletal material ingested increases, as does the size of the prey, with the growth of the cuttlefish (Guerra *et al.* 1988a). Flesh is taken from the sides of the fish so that most of the bones remain in position, although some are ingested. When cephalopods are eaten the flesh is bitten into small pieces before ingestion.

The capture of a prawn by *S. officinalis* is spectacular. More than a century ago Richard Owen (1885) observed that the cuttlefish 'draws in the prey with a rapidity the eye can scarcely follow'. The first analysis of this action was made by F.K. Sanders and Young (1940), and subsequently by Messenger (1968), who divided the attack into three stages: attention, positioning, and seizure. Attention lasts for 1–10 s, during which time the eyes turn towards the prawn. The colour of the head and mantle changes while arms I, and sometimes arms II, are raised vertically and may move laterally. The head and body turn on a vertical axis until the head and eyes face the prey. Positioning lasts for 1–10 s, during which time the cuttlefish may approach, retreat, or remain still, in order to achieve the correct distance for attack. Arms III open and the pale tips of the tentacles emerge slowly and project for about one-third of their length. The time taken to seize prey with the tentacles is only 150–320 ms, while the cuttlefish hovers at the correct attack distance as the tentacles are projected on separate, but parallel, paths towards the prey (Fig. 5.8). The tentacular clubs take less than 32 ms to reach the prawn, and the force with which they strike carries

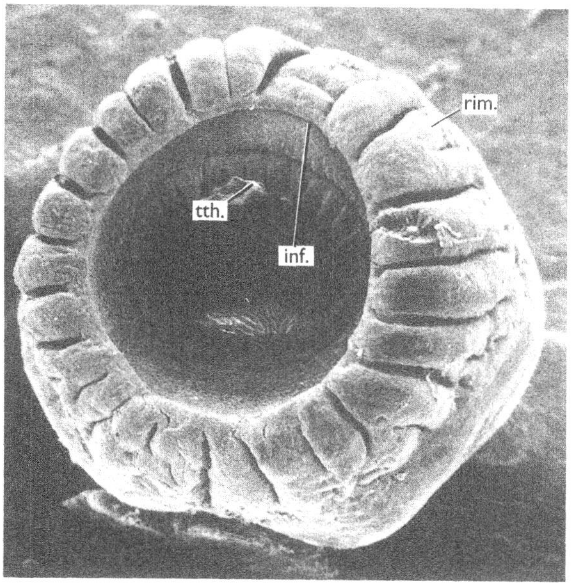

(a)

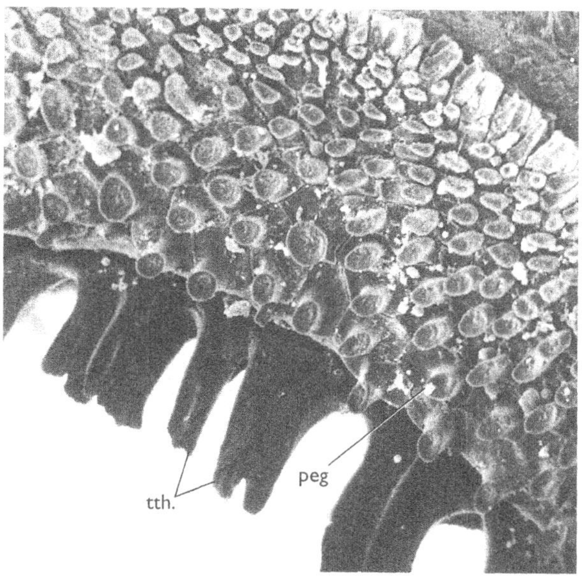

(b)

Fig. 5.7. *Sepia officinalis* (8.5 mm ML, 12 days old). Scanning electron micrographs of an arm sucker showing (a) the inner ring with conical-shaped teeth (field width 114 μm) and (b) the teeth of the inner ring and the pegs on the infundibulum, at higher magnification (field width 280 μm).

the prey forwards for 60–75 ms. The captured prawn is drawn back into the reflexed arms and bitten by the beak at the point of flexure (Wilson 1946). Five seconds later the prey exhibits convulsive movements (Messenger 1968) resulting from the paralysing action of the saliva, which contains cephalotoxin secreted by the posterior salivary glands (Ghiretti 1959, Koueta and Boucaud-Camou 1986).

Crabs are usually captured somewhat differently. After the first two stages of the attack sequence the cuttlefish will jump on and seize a crab with its arms, although the tentacles can be used (Messenger 1968). Investigation of the attack strategy revealed that the characteristics and size of the prey were influential in the mode of capture employed by the cuttlefish. The two strategies coexist, but the tentacular strike tends to be used when the cuttlefish is capturing crabs smaller than itself; it jumps on those that are larger (Duval *et al.* 1984,

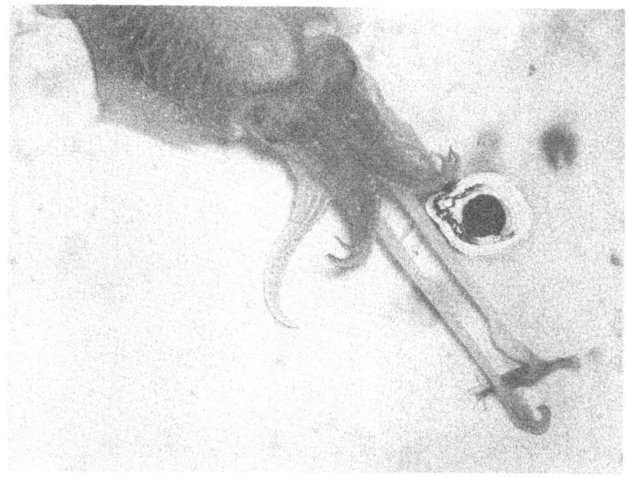

Fig. 5.8. *Sepia officinalis* seizing a shrimp with the clubs in a tentacular strike (photograph kindly provided by Dr J.B. Messenger).

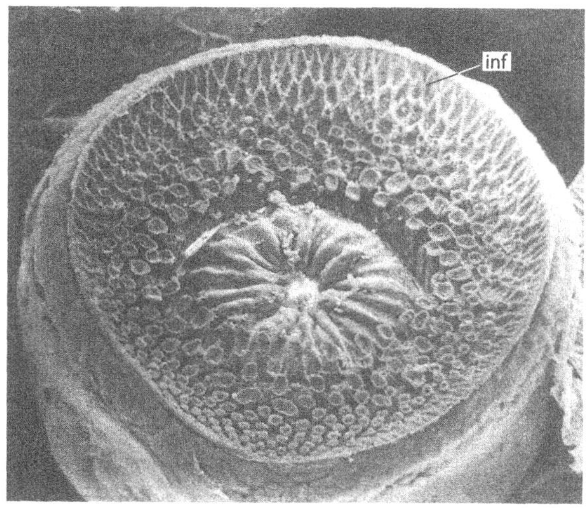

(a)

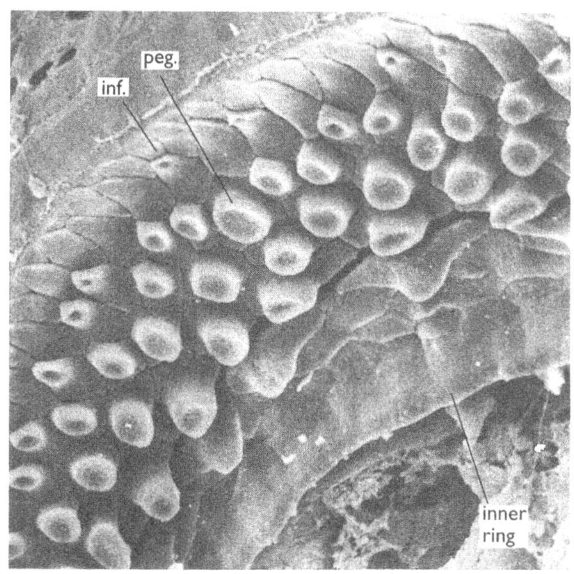

(b)

Fig. 5.9. *Sepia officinalis* (8.5 mm ML, 12 days old). Scanning electron micrographs of a tentacular sucker (field width 500 μm) (a) and at higher magnification (b) to show the blunt inner ring, and the pegs on the infundibulum. New pegs are developing at the periphery (field width 166 μm).

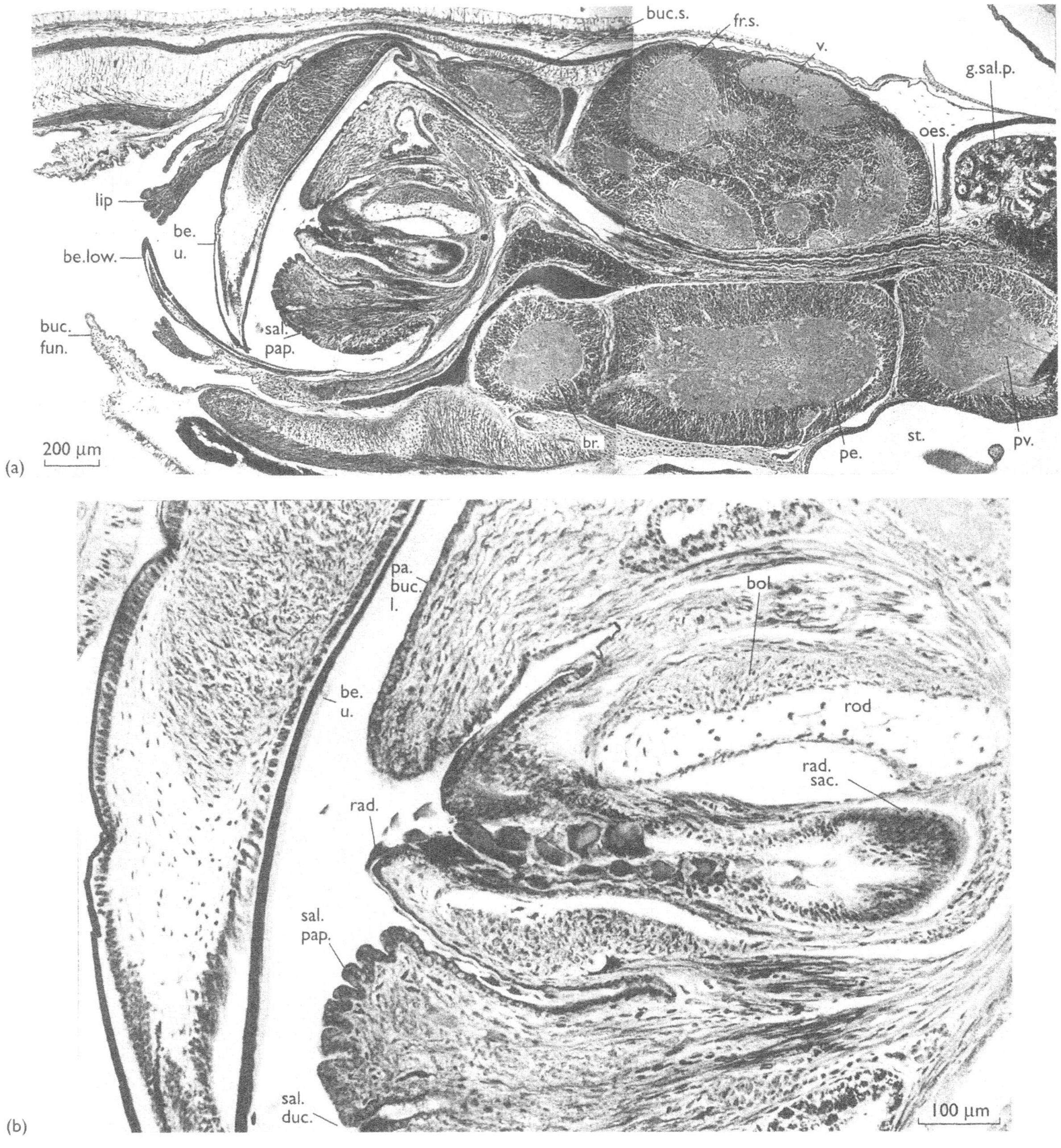

Fig. 5.10. *Sepia officinalis*. Sagittal section of a hatchling to show (a) the main features of the buccal mass and the central nervous system, and (b) the radula and its support system, at higher magnification (haematoxylin and eosin).

Chichery and Chichery 1991a). Once the captured crab is enclosed within the arms of the cuttlefish, it is rapidly manipulated so that the posterior margin is aligned with the buccal mass. In this position the introduction of the toxin-containing saliva from the posterior salivary glands is facilitated. Arms I and II are applied to the dorsal surface of the carapace, arms II encircle the crab laterally, and arms IV hold the underside. The whole process is equally successful in complete darkness. This last indicates tactile or chemosensory input for which the suckers are well endowed with receptors (Graziadei 1959, 1964).

In the wild, crustaceans form 75% of the diet of young and juvenile cuttlefish and they consume much of the exoskeleton (Guerra 1985b, Guerra *et al.* 1988a, Henry and Boucaud-Camou 1991). This implies that the calcium required for building the cuttlebone is of dietary origin. Confirmation comes from starved young cuttlefish, as the chambers of their cuttlebones have crowded septa (Boletzky and Wiedmann 1978).

5.2.6 *Feeding and the buccal mass*

Outer and inner lips surround the beaks (Fig. 5.10a). The outer lip is a simple fold of tissue around the much folded inner lip. The inner surface of the inner lip has a glandular patch of goblet cells, the labial gland (Boucaud-Camou 1968). The buccal mass is almost spherical and lies in a hollow, the peribuccal sinus, formed by the bases of the encircling arms (Fig. 5.6). It can rotate, and the beaks can protrude from the surrounding lips. Electrical stimulation of the median basal lobe of the brain elicits movements of the buccal mass, and stimulation of the inferior and superior buccal ganglia movements of the beak (Boycott 1961).

The beak is a chitin–protein complex (Hunt and Nixon 1981) and is secreted by underlying tall columnar epithelial cells (Dilly and Nixon 1976c). It is dark brown in colour except for the peripheral growing edges which are translucent. The upper and lower beaks of *S. officinalis* have sharply pointed rostra, which are thicker and stiffer than those of most squids (Clarke 1986). Together the beaks form a firm, almost spherical chamber within which is the radula, the radular support system, the lateral buccal palps, the salivary papilla of the posterior salivary glands, the anterior salivary glands, and the submandibular salivary gland (Fig. 5.10).

The radula runs forward from the radula sac, in which the teeth form, to the bending plane at the front of the buccal complex where it bends acutely before terminating (Fig. 5.10). The radula has seven teeth in each transverse row: a central rhachidian tooth, with two lateral teeth and a marginal tooth on either side (Nixon 1995). There is a support on either side of the radula. This is the bolster, which has a complex arrangement of mostly transverse muscles, and within it a cylindrical structure, the rod (Fig. 5.10b). The bolster lies in a sac and is attached posteriorly and laterally, but is free anteriorly to press against the wall of the hyaline shield of the radula. During direct electrical stimulation of the bolster its tip curves strongly upwards (Messenger and Young 1999). As the bolster is encased in a sheath of connective tissue its volume cannot change during muscular activity, so that it functions as a muscular hydrostat (Kier 1982, 1988, 1989). The rod, within the bolster, is readily seen in sections since its large cells stain poorly, apart from their nuclei (Fig. 5.10b). A viscid fluid can be expelled from the cut end of a rod taken from a freshly killed specimen. The rod runs the whole length of the bolster, below the transverse muscles, and has its own sheath (Fig. 5.11). The function of the rod is not fully understood, but the actions of the rod and bolster together push the anterior part of the hyaline shield forwards and upwards. This movement splays the lateral teeth of the radula laterally and raises the central rhachidian teeth vertically as the ribbon passes over the bending plane (Messenger and Young 1999).

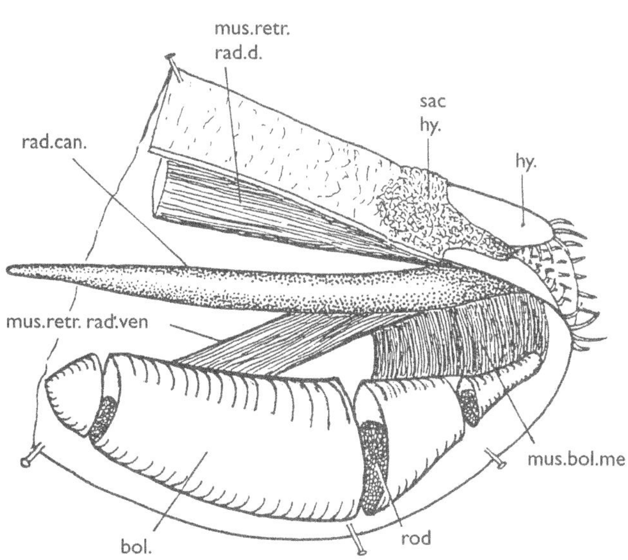

Fig. 5.11. *Sepia officinalis.* Drawing to show the arrangement of the rod within the bolster, the hyaline shield of the radula, and the radular teeth raised on passing over the bending plane (Messenger and Young 1999).

Lateral to the radular support system lie the paired lateral buccal palps. These have a cuticular covering and backward-pointing teeth on the anteromedial surface. The lateral buccal palps are mobile, and during cyclical feeding movements come forward to almost enclose the radula anteriorly. Posteriorly the palps unite to form part of the channel which leads into the oesophagus.

The salivary glands can be divided into 'distinct' glands and 'diffuse' glands (Boucaud-Camou 1968). The first type include the single submandibular gland below the radular complex, the paired anterior salivary glands within the lateral buccal palps, and the paired posterior salivary glands, which are free and lie dorsally in the mantle just behind the head. The posterior salivary glands have a common duct that runs forward to pass through the salivary papilla, which lies below the radular complex, and then opens on the anterior face of the papilla. Diffuse glands are present on the lips, the labial glands, the salivary papilla, the lingual glands, and in the buccal epithelium. Three types of cell have been recognized in the labial, submandibular, and anterior salivary glands, and include mucous cells, salivary cells, and enterochromaffin cells (Boucaud-Camou and Boucher-Rodoni 1983). The posterior salivary glands secrete cephalotoxin, which immobilizes the prey (Ghiretti 1959).

5.2.7 Learning

The capacity of *S. officinalis* to learn was first demonstrated by F.K. Sanders and Young in 1940, who also tested the effects on learning of ablation of various lobes of the central nervous system. The cuttlefish exhibits a series of reactions to a prawn entering its visual field and these were used to test the animal. When a prawn passes from sight the cuttlefish will follow, and this is repeated on a further disappearance of the prawn. After ablation of the vertical lobe, or the vertical and superior frontal lobes, no obvious effects could be observed on swimming, obstacle avoidance, or the ability to capture prawns. However, the operated cuttlefish would no longer pursue a prawn disappearing from its visual field. These lobes are electrically inexcitable (Boycott 1961).

The presentation of a prawn behind a glass screen arouses a series of reactions in a cuttlefish that finally result in the projection of the tentacles to capture the prey. The frequency of the response diminishes but a memory trace remains even after an interval of 18 h. Evidence of discrimination was obtained by presenting a prawn, or a prawn and a white circle behind glass: the cuttlefish learned not to project its tentacles towards the prawn and white circle but to continue to attack the prawn when it was shown alone (F.K. Sanders and Young 1940). The cuttlefish has also been shown to be able to discriminate between squares of different sizes (Messenger 1977).

During the first few weeks of life young *S. officinalis* will attack small crustaceans such as *Mysis*. If this prey is presented behind a glass screen then the young cuttlefish will continue to attack until almost exhausted, whereas adults soon cease to attack under the same conditions (Wells 1958, 1962). Wells noted that the vertical lobe was relatively small in the newly hatched cuttlefish when compared with the adult. Messenger (1973) pursued this observation and, using the same experimental procedure, measured learning at several stages of development. Cuttlefish at 7 days of age show little evidence of learning, at 1 month some learning was displayed, and at 2 months the animals were capable of both learning and retention. During this period considerable development had taken place in the vertical lobe (Fig. 5.12). Hunting behaviour was initially absent but had developed by an age of 4 months, and even then the cuttlefish did not spend long in continuing its pursuit of the prawn (Messenger 1973, 1977). Chichery and Chichery (1992a, b) tested cuttlefishes and found that at 11–22 months of age their capacity to learn remained similar to that in young animals; however, retention was poor, suggesting that long-term memory processes had deteriorated in older animals. A staining technique showed spontaneous terminal degeneration in parts of the central nervous system of the oldest animals.

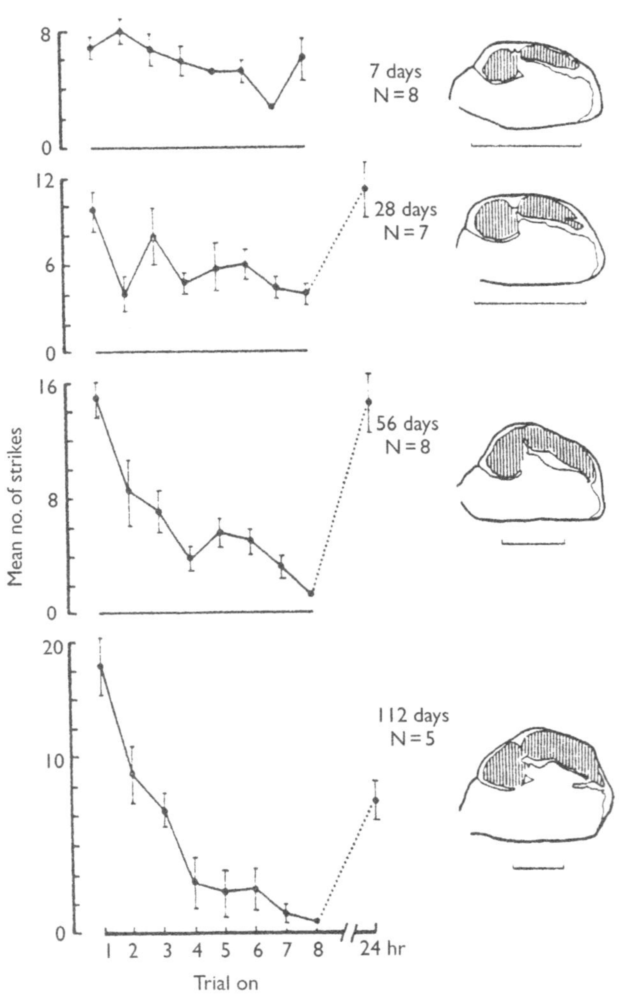

Fig. 5.12. *Sepia officinalis*. Changes with age in the ability of cuttlefish to learn not to strike at inaccessible prey. The graphs on the left show that the number of strikes made by the cuttlefish at a shrimp, *Mysis*, behind glass declines little in animals of up to 4 weeks of age, but at 8 weeks a steep learning curve is apparent, although the capacity for retention after 24 h was poor. At 16 weeks the animals display both learning and retention. Developmental changes in the vertical-superior frontal lobe system are shown on the right. At 112 days of age this attains proportions similar to those in the adult (Messenger 1973).

5.2.8 Body patterns: chromatic and textural components

Sepia can exhibit a dozen or so strikingly different body patterns and change from one to another with great rapidity. This is because the skin possesses millions of chromatophore organs, which are essentially pigment sacs, capable of a cycle of expansion and retraction lasting less than 1 s (Cloney and Brocco 1983). In addition there are reflective tissues and muscles that erect skin papillae. Together these give colour, tone, pattern, and texture to the skin so a wide variety of body patterns can be assembled for camouflage, threat, and courtship.

The pigment sacs, or chromatophores, are thin-walled cytoelastic sacs (Fig. 8.17). Each has muscle fibres radiating outwards from the central sac, lying in the same plane as the body surface (Florey 1969; see Chapter 8 for their structure and function). *S. officinalis* has brown, red, and yellow chromatophores (Naef 1923, Hanlon and Messenger 1988), their coloration being due to pigments which chemically are mostly ommochromes (Messenger 2001). The yellow ones are the most superficial, the brown the deepest in the skin. There is considerable overlap when the chromatophores expand.

Below the chromatophores lie the leucophores and iridophores (or iridocytes). These structures reflect the predominant wavelengths in the immediate environment. The function of the reflecting elements is in cryptic coloration and is almost certainly limited to intensity matching, which occurs only when the chromatophores are retracted, or 'switched out', as they are in bright illumination. Iridophores and leucophores are revealed when the chromatophores are retracted, and so reflect the incident light: the skin of the cuttlefish then takes on the colour and tone of its surroundings (Messenger 1974). In horizontally polarized light a pattern of pale stripes is revealed along the arms (Shashar *et al.* 1996).

Each iridophore is formed of individual oval or circular flexible plates of chitin arranged in multilayer stacks with fluid between (Denton and Land 1971). The iridophores produce interference colours of pink, green, or blue iridescence when viewed at the appropriate angle. They are numerous and are regularly organized on the ventral surface of the mantle where there are few chromatophores (Kawaguti and Ohgishi 1962), and on the eyelids which, with those of the ventral mantle, reflect mainly green light (Denton and Land 1971).

Leucophores have an irregular shape; they contain a creamy guanine-like substance and appear white in reflected light. They lie deepest of all in the skin and have been identified in the white spots along the fin. Leucophores are electron dense, elongated, much-branched structures with many ovoid, stalked clubs over their surface (Froesch and Messenger 1978).

The body patterns of newly hatched and young *S. officinalis* were analysed by Hanlon and Messenger (1988). The full range of adaptive coloration seen in the adult is exhibited by the hatchlings, but at this stage it is used mainly for concealment and the diversion of potential predators, rather than for signalling to conspecifics. The patterns of colour shown can be described by the system of hierarchical classification devised for *Octopus vulgaris* by Packard and Sanders (1971). In this system a 'body pattern' consists of several 'components' each including 'units' made up of diverse 'elements' comprising chromatophores, leucophores, iridophores, and skin muscles. The 'units' of the colour patterns

> are transient phenomena resulting from the activity of cells in the central nervous system. According to the motor program currently in operation, some of the chromatophores in the static array will be expanded at any one moment, whereas others will remain retracted. A few minutes later the program may change and with it the appearance of a particular patch of skin, resulting in a pattern of change of the whole animal (Hanlon and Messenger 1988).

The units involve the activation of certain definite sets of motoneurons of the chromatophore lobes of the brain and thus sets of fibres in the peripheral nerve. These can be stimulated artificially. The actions of the units are grouped into chromatic 'components' which are easily defined and occur repeatedly in the same relative position. Fifty-four of these components are recognized, some light, others dark, including lines, spots, triangles, squares, bars, and splotches on various parts of the body. Combinations of these components produce numerous body patterns: some classed as 'chronic' are maintained for hours at a time, and include those described as 'uniform light', 'stipple', or 'disruptive'. Others are classed as 'acute body patterns' and are displayed only for seconds; these include 'uniform blanching (or darkening)', 'flamboyant', 'deimatic', and 'intense zebra'.

Several discrete patterns are exhibited by juvenile *S. officinalis* (Holmes 1940, 1955, Hanlon and Messenger 1988). When swimming, a young cuttlefish displays broad stripes of brown and creamy-white down both sides of the mantle, lateral arms, and fins. The underside is white. When it is resting on a well-lit sandy substrate, a disruptive mottle pattern is displayed. A background of large dark and light areas provokes a large white square or stripe to develop across the back of the mantle, and this breaks up the outline of the body into large units. A deimatic (dymantic) or startle pattern is achieved by the sudden contraction of all chromatophores at the same time, with the exception of two pronounced black spots in the middle of the back, the brownish-black chromatophores around the eyes, and the red chromatophores on the edge of the fins (Fig. 5.1).

During courtship a mature male *S. officinalis* exhibits zebra-like black and white stripes on the broad side of the lateral arms. If the other animal is a male, flight ensues in response to this display, but if it is a female then she may accept the advances (Tinbergen 1939). Such intraspecific displays involve brightness differences (Hanlon and Messenger 1988).

Each body pattern of the chromatophores is accompanied by a particular textural pattern produced by the muscles of the skin, and described as 'smooth', 'coarse', and 'papillate'. There are also postural and locomotor patterns and these include 'waving arms' or 'flattened body'. Laboratory observations of patterning were confirmed by releasing newly hatched cuttlefish in the sea. Hanlon and Messenger (1988) saw that 'they were able to adapt to every substrate encountered' and gave 'direct evidence that their concealment patterns were effective against fish as well as ourselves, because motionless cuttlefish were repeatedly overlooked by passing fish' including predatory fish. Only when a cuttlefish moved was it detected by the carnivorous fish *Serranus cabrilla*, and when the predator 'moved directly towards it in an attack posture, the cuttlefish responded at a distance of up to 3 m, even when other fishes were nearer'. The cuttlefish usually oriented towards the attacker and showed a dark flamboyant pattern but 'upon the forward rush of the attacking fish, the young *Sepia* instantly changed to Uniform Blanching, inked and jetted 1 to 2 m backwards' (Hanlon and Messenger 1988). Attacks by fish were successful only when the cuttlefish was distracted by the observer.

There are many more chromatophores on the dorsal surface of the mantle than on the ventral surface. Those on the dorsal surface are mostly kept tonically expanded, making this side darker. Ventral chromatophores respond to rotation of the animal through 90° when those on the now uppermost half of the ventral and dorsal sides expand and so darken this surface (Fig. 5.13). Such darkening would maintain the counter-shading effect of a dark upper and a pale under surface when the animal becomes momentarily disorientated, as the response lasts only for a few seconds. The counter-shading reflex is abolished by ablation of the statocysts (Ferguson and Messenger 1991). Unilateral destruction of the gravity receptor system shows that the pitch counter-shading reflex is driven bilaterally, whereas the roll counter-shading reflex is driven unilaterally by input from the ipsilateral statocyst, and in part by light. Evidence from brain lesions shows the pathways for the counter-shading reflex to run through the lateral basal lobes of the brain (Ferguson *et al.* 1994).

The changes in colour are due to the neural control of the chromatophores: the muscles of the chromatophores are directly innervated by nerves whose cell bodies are concentrated in two anterior and two posterior chromatophore lobes of the brain (Boycott 1953). Electrical stimulation of the anterior chromatophore lobe produces expansion of the chromatophores of the front part of the head and the arms, that of the posterior chromatophore lobe darkening of the mantle, fin and visceral mass (Boycott 1961, Messenger 2001).

The activity of the nerves supplying a specific set of chromatophores has been successfully recorded by Messenger and Miyan (1986) using the deimatic response (Hanlon and Messenger 1988) of *S. officinalis* to threat. This pattern is dramatic and comprises all the features seen under natural conditions, including expanded pupils, dark rings round

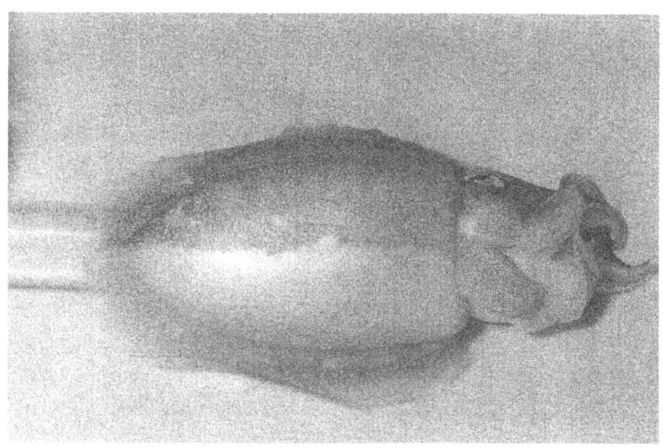

Fig. 5.13. *Sepia officinalis*. The counter-shading reflex after rotation of a lightly anaesthetized cuttlefish through 90° (Ferguson *et al.* 1994).

the eyes, and a pair of large black spots on a pale mantle (Fig. 5.1). Recordings from a nerve branch innervating the chromatophores of the black spot show a decrease in the firing rate in response to an approaching hand or 'threat'. An inhibitory relationship exists between the dark spot and its pale surround. This is almost certainly brought about centrally, as paling was not obtained by stimulation of the nerve alone.

Of the species of *Sepia* so far investigated, each apparently has its own characteristic patterns. Those for camouflage are extremely effective, so much so that they make it almost impossible for the human eye to detect motionless cuttlefish, whether seen against a uniform background of sand or gravel, or when among seaweed and rocks (Hanlon and Messenger 1988). Conversely, the patterns exhibited by a cuttlefish when about to attack will repel conspecifics, frighten enemies, and make it conspicuous (Adamo and Hanlon 1996).

5.2.9 Ink and its function

The ink sac is a characteristic feature of coleoids. Aristotle (see Thompson 1910) noted the ink as being most abundant in the cuttlefish and its discharge most copious. The ink is secreted by an ink gland, accumulates in an ink sac, and is retained by sphincters (Tompsett 1939, Budelmann *et al.* 1997). In *S. officinalis* the pigment consists of uniformly spherical granules of melanin (56–161 nm) (Flood *et al.* 1990), homogeneously dispersed in mucus (Schraermeyer 1994).

When the ink is discharged into the sea the shape formed may resemble the animal itself (a pseudomorph), or form a large cloud that conceals the cuttlefish and from behind which it can escape. Ejection of ink is elicited by electrical stimulation of the palliovisceral lobe of the brain (Boycott 1961). There is also evidence of the existence of receptors in some of the cells of the ink gland and in localized areas of the brain regions responsible for the control of the ejection of ink (Palumbo *et al.* 1999).

5.2.10 Eyes

The eyes are prominent and bulge slightly on either side of the broad head of *S. officinalis* (Fig. 5.1). Each eye, like that of *Loligo* (Fig. 8.21), comprises all of the major components found in a vertebrate eye, which it resembles superficially (Tompsett 1939), but the photoreceptors point towards the light and the retina is simpler. The cornea is not continuous with the sclera, but is a transparent window in the thin skin over the eye. This has a small duct anteriorly that is open to the sea. The pupil changes in size in response to different intensities of light. The iris is composed of an external epithelium, a chromatophore layer, an iridophore layer, a layer of muscle and collagen fibres, and a pigment epithelium. The lens, suspended by ciliary muscles, is spherical and consists of an anterior and a posterior portion separated by a complex septum. It is composed of crystalline proteins and the refractive index varies from 1.53 at the centre to 1.33 at the periphery (Bon *et al.* 1967).

The wall of the eyeball has a retina, a sclera, a choroid, and an iridescent argentea. The retina comprises receptor cells, supporting cells, and retinal glial cells. Electron microscopy of the retina (Shashar *et al.* 1996) shows the same orthogonal structure that is found in *Loligo* and *Octopus* (Messenger 1981). There is a sharply demarcated strip of longer rhabdoms lying above the equator, and the proximal layers of the retina, especially the nerve plexuses, are thicker in this strip. Within the strip the rhabdoms are shortest at the centre of the eye, longer anteriorly, and longest of all posteriorly (Young 1963a).

Accommodation is not well understood. Electrical stimulation can alter the shape of the isolated eye, by contraction of the muscles of the retina and sclera pushing the lens outwards and of the ciliary muscle drawing it inwards (Alexandrowicz 1927). However, the lens has a short focal length because of its optical properties, and the receptors are long (300 μm), so there may be little need for accommodation (Pumphrey 1961).

The shape of the pupil affects the sharpness of the image on different parts of the retina. The lid of the iris, when contracted, forms a W-shape, which allows the formation of two distinct pupils (Muntz 1977). This is correlated with two areas of specialization at the anterior and posterior poles of the retina where the receptors are longer. Each of these two areas receives light through one of the pupil's vertical segments only. The combination of a lens that obeys Matthiesson's ratio (focal length approximately 2.5 times the radius) and a constricted pupil presumably results in vision being more efficient for some directions than for others. The two areas of specialization can be correlated with the capture of prey in front and swimming backwards (Naef 1923, Messenger 1968). The pupils respond to light. If illumination is decreased by 3 log units, the pupil dilates slowly, taking about 30 s to reach maximum size. A similar increase in illumination brings about closure of the pupil in less than 5 s (Muntz 1977). The pupillary response is rapid when the cuttlefish is frightened or attacking prey, but details of the neural control of the iris are not known. It is interesting that the pupil can be dilated during signalling, and the pupil is wide open during the acute zebra pattern shown by sexually mature males (Hanlon and Messenger 1996).

Viewing cuttlefish through an imaging polarized light analyser has revealed a prominent polarization pattern of stripes on the arms, around the eyes, and on the dorso-anterior part of the head. Behavioural experiments followed the reactions of cuttlefish to their images reflected from a mirror, and their responses changed when polarization of the reflected images was distorted (Shashar *et al.* 1996).

The eye lies in an orbit formed anteriorly by a muscular wall and posteriorly by the orbital cartilage, and has elaborate extrinsic and intrinsic muscles (Alexandrowicz 1927). *S. officinalis* has a total of twelve extraocular muscles attached to each eye and an additional muscle which passes between the two trochlear cartilages (Fig. 5.14). Five muscles, all innervated by the anterior oculomotor nerve, are mainly concerned with turning the eye(s) medially forwards for binocular vision during fixation and prey capture. One muscle turns the eye backwards, one downwards, and two upwards, and the remaining three muscles act mainly to produce rotation of the eye. There is one anterior conjunctive muscle attached to the trochlear cartilage. The tendon of this muscle joins that of the corresponding muscle of the contralateral eye. The muscle is short but powerful and moves the two trochlear cartilages together, perhaps with some medial and forward movement of both eyes. The four eye muscle nerves originate in the ipsilateral anterior lateral pedal lobe of the middle suboesophageal mass (Budelmann and Young 1993).

The question of whether cephalopods have colour vision has been long debated, especially in view of the ability of these animals to change colour. Evidence has been accumulating since the 1970s that

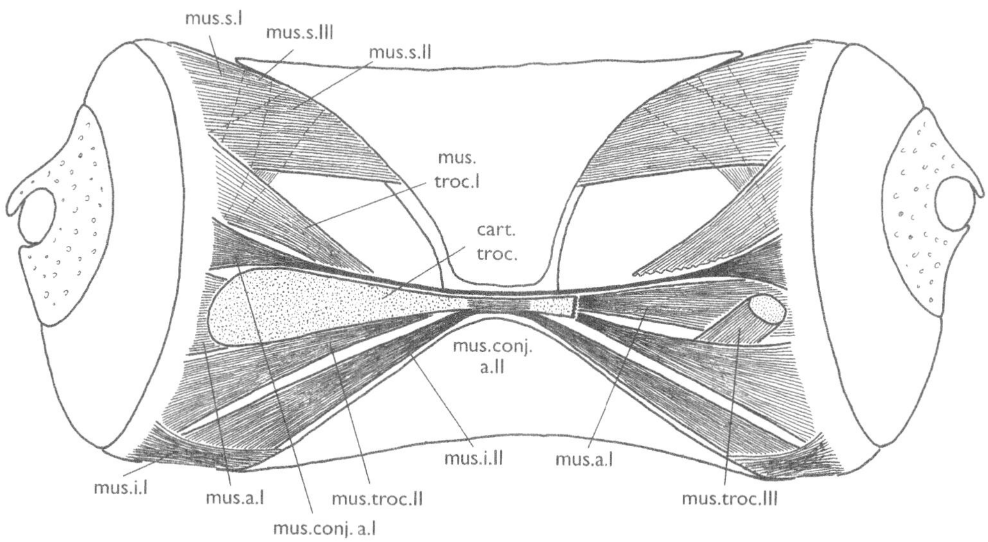

Fig. 5.14. *Sepia officinalis.* Drawing to show the extraocular muscles on the anterior surface of the eyeballs, and the conjoint anterior conjunctive muscles (Budelmann and Young 1993).

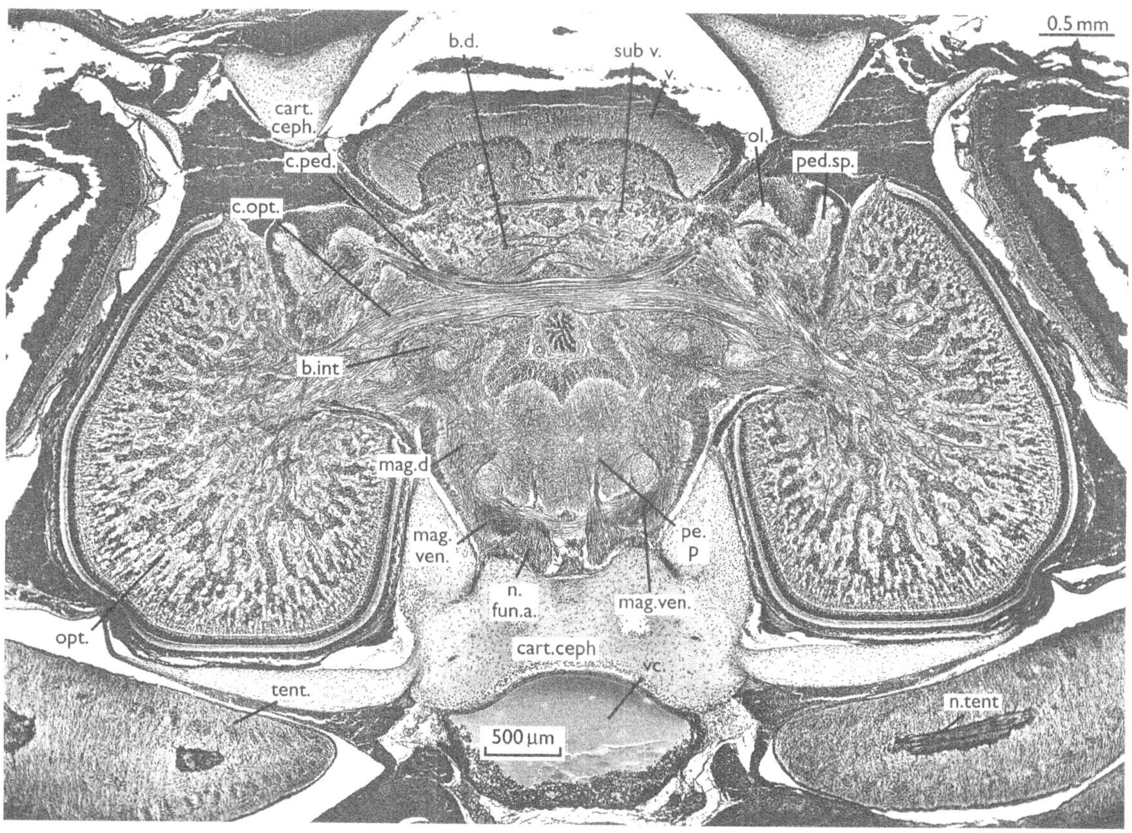

Fig. 5.15. *Sepia officinalis.* Transverse section through the brain of a young specimen, at the level of the anterior magnocellular lobe (Boycott 1961).

Octopus vulgaris is colour blind (Messenger 1981), and recent photographic evidence suggests that *S. officinalis* is too. When kept on specially designed backgrounds of stones of different hues, the latter does not respond to different wavelengths in the substrate as it regulates the chromatophores. However, when it perceives different intensities in the background it generates patterns in the skin: the higher the contrast in the background the bolder the pattern. It can produce patterns for crypsis (Marshall and Messenger 1996).

5.2.11 Optic lobes

Each optic lobe is almost as large as the rest of the central brain and is kidney-shaped, the outer side being convex and the inner side concave (Fig. 5.15). The bundles of optic nerve fibres, after passing through a dorsoventral chiasma, enter the optic lobe through the outer granular layer and cross each time that they divide. This presumably 'rectifies' the inverted retinal image as it is mapped on to the optic lobe, and brings the visual and gravity 'maps' into correspondence (Young 1962c).

The tangential fibres in the inner layers of the plexiform zone are long and include the dendrites of the second-order neurons of the visual pathway. In some histological sections, tangential to the surface, they are seen to be the spreading branches of a central trunk. The main branches of the trunk run straight for distances of up to 1 mm. Some of the dendrites run either parallel or at right angles to each other, but others run in various directions (Fig. 5.16a). The dendrites carry collateral twigs at intervals and divide at the ends. The axons of these second-order visual cells run through the inner granule layer and then penetrate deep into the centre of the lobe (Fig. 5.16b). Within the inner granule cell layer the radial fibres run in bundles separated by columns of cells; the latter appear as islands in transverse section. Bundles of fibres also run between the islands of cells, and, passing towards the centre of the lobe, the columns of cells join to make larger and larger islands as the columns of fibres spread outwards.

Electrical stimulation of the outer parts of the optic lobes revealed that they were inexcitable, but the central part did evoke responses from many of the effector systems of the body (F.K. Sanders and Young 1940). Boycott (1961) confirmed that a wide range of motor responses can be elicited by stimulating the centre of the optic lobe. He obtained expansion or relaxation of the chromatophores and also patterning; fin waves could also be elicited, and protraction or retraction of the head. Chichery and Chanelet (1976) chronically implanted bipolar electrodes in the medulla and induced two distinct and complex responses: an ipsilateral rotation or the deimatic body pattern (shown when alarmed). Stimulation of the right optic lobe medulla resulted in rotation of the eyes, the contralateral eye turning forwards and the ipsilateral one backwards. The head rotated towards the side stimulated, and the funnel in the opposite direction. The combined action of the funnel and fins swung the animal towards the stimulus. The whole animal darkened and the ipsilateral pupil dilated. Stimulus trains of long duration (1.5 s) were needed to induce the complete rotation of the animal. A 'normal' looking deimatic response was elicited; the main features were the upward movement of the eyes and the bilateral dilation of the pupils, the cessation of respiratory movements only to begin again, head retraction, dorsoventral flattening with spreading of the fins, expansion of the chromatophores around the eyes and edges of the fins, and the appearance of two black eyespots on the mantle (Fig. 5.1).

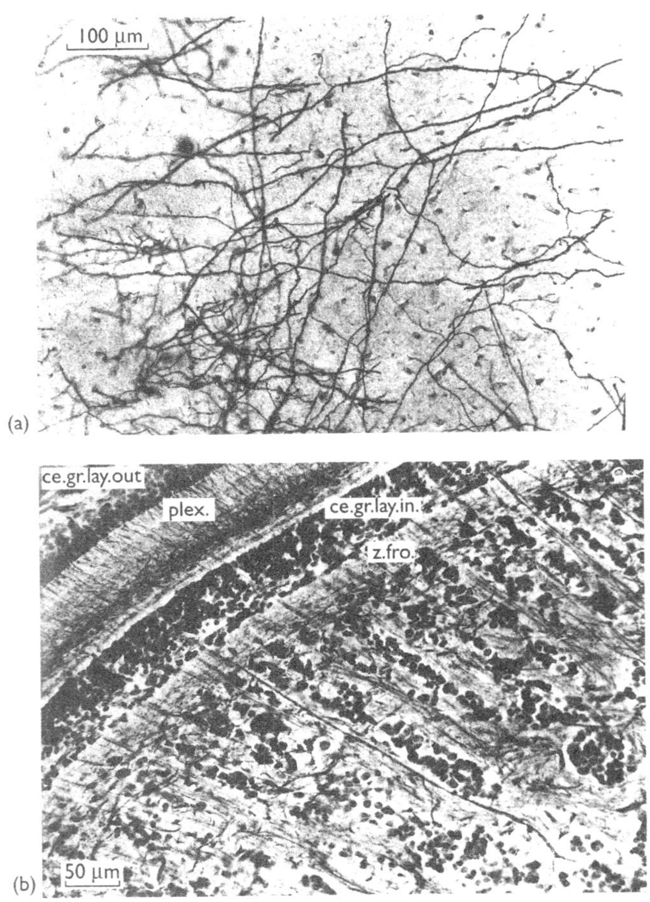

Fig. 5.16. *Sepia officinalis.* (a) Tangential section of the plexiform zone of the optic lobe to show the long dendrites of second-order neurons running largely at right angles to each other. (b) Transverse section of optic lobe showing the inner granular cell layer, and columns of cells and fibres in the outer part of the medulla.

5.2.12 *Central nervous system*

The brain of *S. officinalis* is enclosed in a cartilaginous cranium (Fig. 5.17) (Tompsett 1939) and lies between the eyes. The supra- and

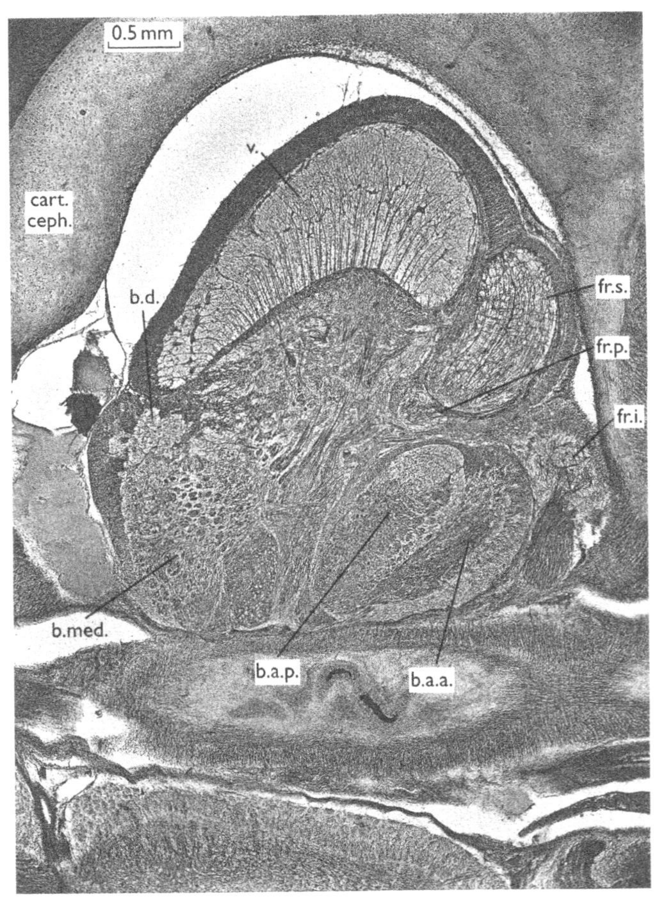

Fig. 5.17. *Sepia officinalis.* Sagittal section of the head showing the supra-oesophageal mass, near the midline (Boycott 1961).

suboesophageal masses are divided by the oesophagus, which passes through the brain, and two perioesophageal lobes, the magnocellular lobes, which run round the sides of the brain (Young 1939). On either side of the perioesophageal parts of the brain are the large optic lobes, joined to the supraoesophageal mass by short optic tracts, but not well defined. On each optic tract there are three structures: the peduncle lobe, the olfactory lobe, and the optic gland.

The supraoesophageal mass is subdivided into many lobes. Dorsally there are three lobes, the vertical, superior frontal, and inferior frontal lobes (Fig. 5.17). Below the vertical lobe is the subvertical lobe, beneath which is the precommissural lobe: the latter is so named because it lies in front of the ventral optic commissure. The remainder of the supraoesophageal mass consists of the basal lobes. The dorsal basal lobe forms the upper part of the posterior wall of the supraoesophageal mass, and it lies underneath the vertical lobe but behind the subvertical lobe. Immediately under the dorsal basal lobe is the median basal lobe, on either side of which are the lateral basal lobes. There are also paired interbasal lobes, which lie on either side, between the anterior basal lobe and the median and lateral basal lobes.

The suboesophageal mass is divided into the anterior, middle, and posterior suboesophageal masses (Fig. 5.17), and further subdivisions are recognized as a result of the experimental study of Boycott (1961). The anterior suboesophageal mass includes only the brachial lobe in decabrachians. The middle suboesophageal mass is divided into four parts: anterior pedal, posterior and lateral pedal lobes, with an anterior chromatophore lobe in the wall either side of the anterior pedal lobe. The posterior suboesophageal mass consists of a central palliovisceral lobe with paired fin and posterior chromatophore lobes on its dorsal side. At the back of the palliovisceral lobe lie the dorsal and ventral vasomotor lobes. On either side of the palliovisceral lobe and between the vasomotor lobes is the posterior part of the magnocellular lobe, in front of which lie the first-order giant fibres forming the ventral magnocellular lobe. Dorsally the magnocellular lobe continues round the oesophagus in front of the lateral basal lobes and joins the dorsal basal lobe (as in *Loligo*) (Young 1939).

Electrical stimulation of the basal lobes of late juveniles indicate that these lobes are involved in movements of the head, eyes, buccal mass, arms, tentacles, and fins, and in the control of respiration, locomotion, swimming, and chromatophores (Boycott and Young 1950, Boycott 1961, Chichery and Chanelet 1976, 1978). The relatively large proportion of the brain taken up by the basal lobes at hatching is compatible with the activities of the hatchlings in the water column, capturing prey and generating complex body patterns for concealment. The lateral basal and chromatophore lobes appear almost identical in hatchlings and adults, the posterior chromatophore lobe showing little change between 7 and 112 days after hatching: this is reflected in the large number of chromatic and body patterns displayed by hatchlings with the very numerous chromatophores they possess (Hanlon and Messenger 1988). The anterior basal lobe is concerned with the capture of prey and the accurate projection of the tentacles (Boycott 1961, Chichery and Chichery 1987).

The vertical lobe shows considerable change in size with growth, and its importance in learning is described earlier in this chapter. The inferior frontal lobe occupies a relatively larger proportion of the brain at hatching than it does in a juvenile (Table 2.3). Stimulation of this lobe in a juvenile results in some movements of the buccal mass (Boycott 1961) and progressive opening of the arms and grasping behaviour (Chichery and Chichery 1991a).

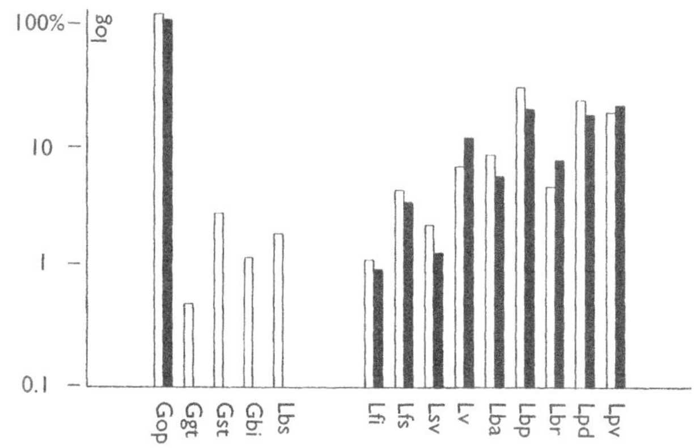

Fig. 5.18. *Sepia officinalis.* Various lobes of the brain shown as a percentage of the volume of the central nervous system, excluding the optic, plotted on a log scale (Wirz 1959, Frösch 1971). Open bars, newly hatched; filled bars, 'adult'. Gbi, inferior buccal ganglion; Ggt, gastric ganglion; Gop, optic ganglion; Gst, stellate ganglion; Lba, anterior basal lobe; Lbp, posterior basal lobe; Lbr, brachial lobe; Lbs, superior buccal lobe; Lfi, inferior frontal lobe; Lfs, superior frontal lobe; Lpd, pedal lobe; Lpv, palliovisceral lobe; Lsv, subvertical lobe; Lv, vertical lobe.

The central nervous system in cephalopods grows throughout life (Wirz 1959, Packard and Albergoni 1970). At hatching the lobes of the brain of *S. officinalis* can be distinguished (Fig. 5.10), and quantitative studies show some to be well developed at this stage but others less so when compared with those of the adult (Fig. 5.18) (Table 2.3) (Nixon and Mangold 1998). The relative volumes of some lobes change as the cuttlefish grows, and are notable in a few lobes. The changes are also a reflection of the habits of the animal at different stages of its life.

Anterior suboesophageal mass

This is a single lobe, the brachial lobe, associated with ten nerves, one from each of the eight arms and one from each tentacle (Fig. 5.19). The nerves contain afferent and efferent fibres. Many afferent fibres run from the arms and end in the brachial lobe (Fig. 5.20). A few run in the brachiopalliovisceral connectives to the magnocellular and palliovisceral lobes. Some afferent fibres pass up into the cerebrobrachial connective and reach to the inferior frontal and superior frontal lobes (Budelmann and Young 1987). The efferents derive mostly from the brachial lobe. Electrical stimulation of the dorsal part of the brachial lobe produced movements of all of the arms. These were unnatural in that they appeared as simultaneous, uncoordinated, contractions of the circular, longitudinal, and oblique muscles, and resulted in the arms being extended and held rigid. Movements of individual arms and tentacles were elicited by application of the electrode close to the pedal root of the brachial nerves, thus exciting fibres that arise further back in the anterior pedal lobe (Boycott 1961).

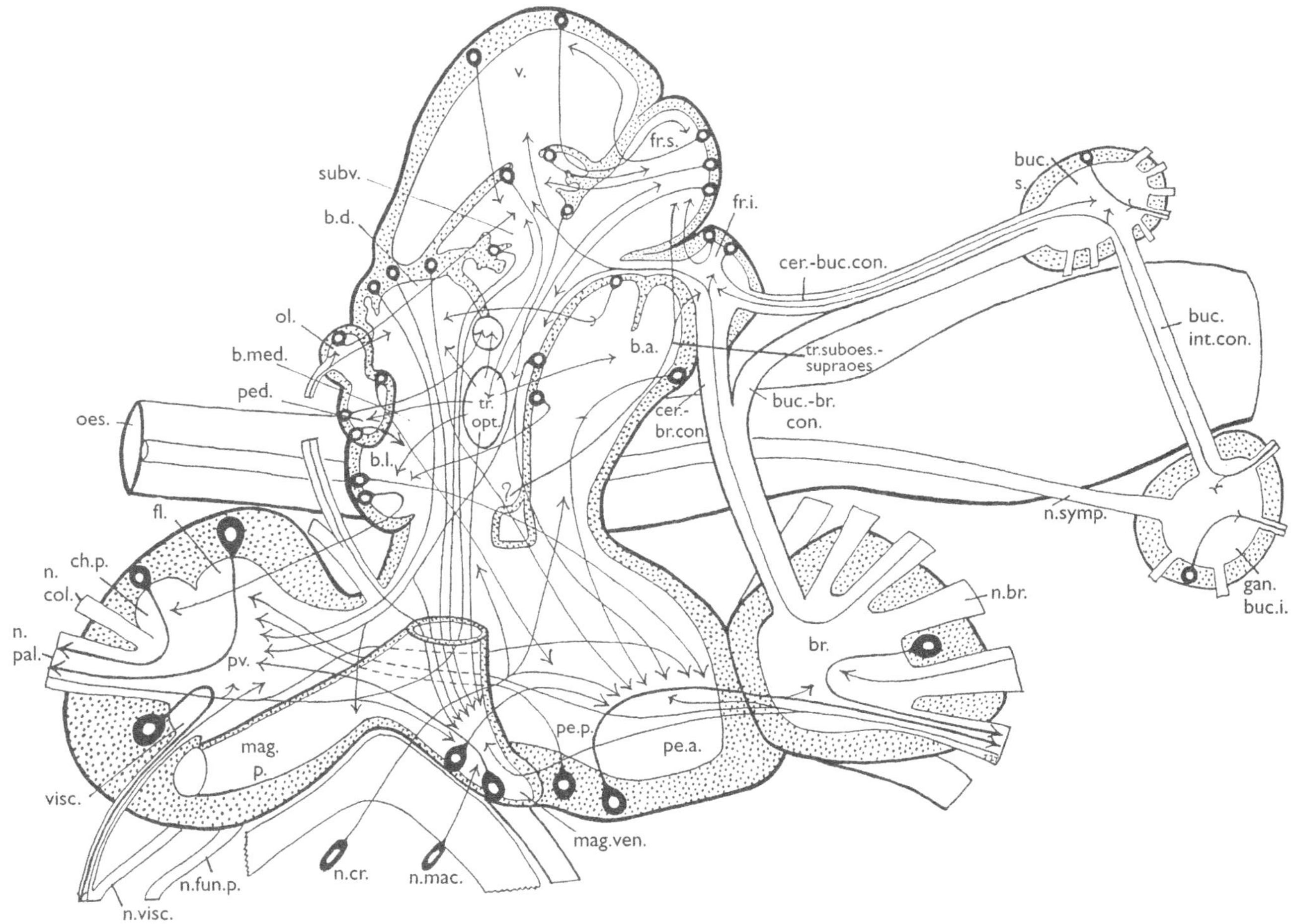

Fig. 5.19. *Sepia officinalis.* Drawing from sagittal sections of the central nervous system to show some pathways in the vertical lobe system and of the basal lobes.

Middle suboesophageal mass

This can be divided into the anterior, posterior, and lateral pedal lobes, and the anterior chromatophore lobe (Fig. 5.19). The middle pedal commissure marks the division between the anterior and posterior pedal lobes.

Electrical stimulation of the dorsal side of the anterior pedal lobe induced movements of the arms and tentacles, and paling of the head, mantle, and arms. Deeper penetration of the lobe with the electrode brought about extension of the tentacles. Stimulation approximately between the anterior chromatophore and the posterior pedal lobes often resulted in the first pair of arms being raised (Boycott 1961): an action seen during the 'attention' phase of an attack on a prawn (F.K. Sanders and Young 1940, Messenger 1968). This region is evidently an intermediate motor 'centre', whose cells control those of the lower centres of the arms. It also includes numerous large cells in the outer layers and smaller ones centrally.

The posterior pedal lobe gives rise to the nerves to the head retractor muscles and to the anterior funnel nerves. Stimulation of this lobe produces retraction of the head and movements of the tentacles and of the funnel (Boycott 1961). The posterior pedal lobe receives fibres from the anterior and median basal lobes, and from the precommissural and peduncle lobes. A large tract proceeds from the floor of the posterior pedal lobe to the fin lobe (Fig. 5.19); stimulation here produced movements of the whole fin, very similar to those following stimulation of the fin lobe itself. The posterior pedal lobe evidently performs some special function in integrating movements of the funnel, fins, head, and mantle.

The lateral pedal lobes comprise the anterior and posterior lateral pedal lobes. The anterior lateral pedal lobes contain the oculomotor centres and the six nerves for the eye muscles arise here from relatively large neurons in the outer cell layers. These are the final motoneurons running direct to the eye muscles (Budelmann and Young 1987), and stimulation of the lateral pedal lobes produced movements of the eyes (Boycott 1961). The inner layers of the smaller cells in the lobe include many efferent fibres to the statocyst (as in *Octopus*). The posterior lateral pedal lobe is directly continuous with the anterior lateral pedal lobe, but has a different function and output. Its large cells, of characteristic pear-shape, send their axons forward to the centre of the anterior pedal lobe. The posterior lateral pedal region is directly continuous with the magnocellular lobe and is an important centre involved in the initiation of many movements.

The anterior chromatophore lobes are groups of mainly large cells at the sides, dorsally and ventrally at the front of the anterior pedal lobe. Stimulation results in expansion of the chromatophores and erection of the papillae of the skin of the side of the head and arms. No combinations of expansion and contraction of the chromatophores

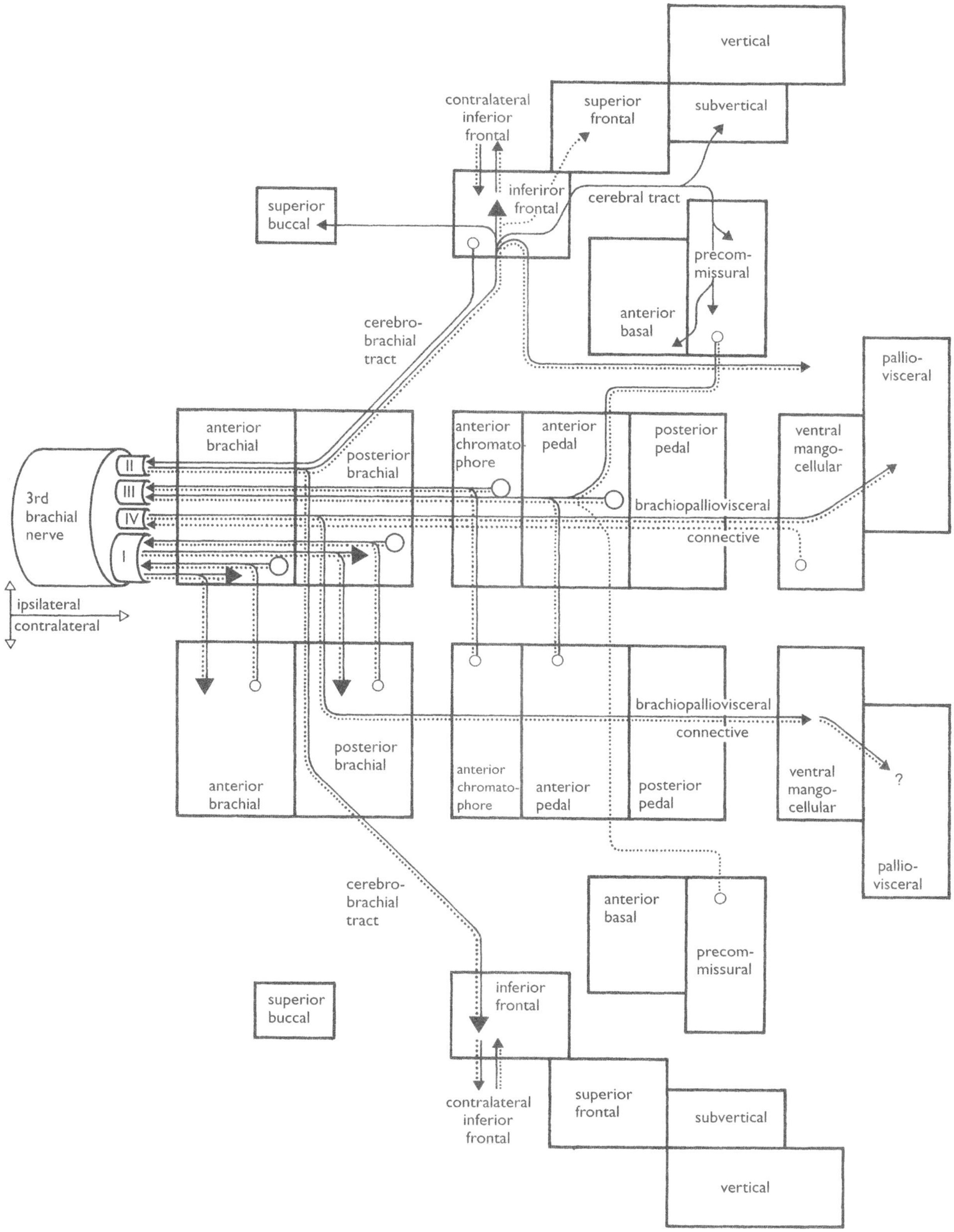

Fig. 5.20. *Sepia officinalis*. Schematic presentation of the afferent and efferent brain pathways of the third brachial nerve (dotted lines) as determined by centripetal cobalt fillings. Rectangles represent the ipsilateral and contralateral brain lobes as indicated, and open circles indicate the origin of the pathways and arrowheads their termination (the sizes of these symbols give an indication of the numbers of cells or fibres filled). When branched lines are shown they indicate the continuation of the pathway, not necessarily of individual fibres. I–IV represent fibre bundles. (The solid lines are of *Loligo vulgaris*.) (Budelmann and Young 1987.)

to provide patterns of colour were obtained. The fibres concerned probably run to the periphery without further synapse, so this area constitutes a lower motor centre of final motoneurons (Boycott 1961, Budelmann and Young 1987).

The middle pedal commissure is a conspicuous bundle marking the boundary between the anterior and posterior pedal lobes. Its large fibres come partly from the crista nerves, others from the posterior lateral pedal lobe and from the median basal lobe, and perhaps other sources.

Posterior suboesophageal mass

This region can be divided into a central palliovisceral lobe, with paired fin lobes and posterior chromatophore lobes on its dorsal side, and dorsal and ventral vasomotor lobes posteriorly, and the perioesophageal lobes.

The central palliovisceral lobe is a backward continuation of the posterior pedal lobe, separated from it by a transverse septum. From the palliovisceral lobe arise the posterior funnel, collar, posterior head retractor, and visceral nerves, and the mantle connective (or pallial nerve) leading to the stellate ganglion and fin. This region thus plays a part in respiration, locomotion, and various functions of the viscera. Stimulation of the middle of the dorsal side of the palliovisceral lobe produced a violent expiratory spasm, and just posteriorly retraction of the head and movements of the collar. Stimulation on the ventral side of the palliovisceral lobe induced funnel movements. The ink sac is innervated and after stimulating the palliovisceral lobe ink was sometimes ejected (Boycott 1961). Recently a possible excitatory glutamate–nitric oxide–cyclic guanosine monophosphate signalling pathway has been demonstrated in the fibres and neurons functionally related to the control of ink discharge (Palumbo *et al.* 1999).

The fin lobes are swellings on the dorsal side of the posterior suboesophageal mass (Fig. 5.19), and form 3.5% of the total brain volume (Table 2.6). Their cells are all large (Fig. 5.21) and the axons proceed directly to the muscles of the fins, bypassing the stellate ganglion. These are therefore the final motoneurons. Stimulation of one fin lobe always resulted in the whole ipsilateral fin being raised (Boycott 1961). There is no commissure between the fin lobes and no direct tract to them from the supraoesophageal lobes. They receive tracts from the posterior pedal lobe, and from the ventral side of the central palliovisceral lobe. These connections evidently mediate the complex interactions of fins, mantle, and funnel.

The posterior chromatophore lobes lie behind the fin lobes but their cells are much smaller (Fig. 5.21). The cells of the posterior chromatophore lobes are also final motoneurons with axons proceeding direct to the muscles of the chromatophores of the mantle, fins, and visceral mass, passing without interruption through the stellate ganglion. Stimulation of the posterior chromatophore lobes produced expansion of the chromatophores and erection of skin papillae of the ipsilateral, or both, side(s), depending on the strength of the stimulus, there being a commissure between the lobes (Boycott 1961). The lobes receive fibres from the lateral basal lobe.

The vasomotor lobes lie at the back of the palliovisceral lobe, and are composed of small cells.

Magnocellular lobes

The magnocellular lobes are perioesophageal, joining the optic and dorsal basal lobes above and the pedal and palliovisceral lobes below (Fig. 5.19). Each magnocellular lobe can be divided into three parts—dorsal, ventral, and posterior regions—although these are not sharply separated. Stimulation of the ventral part of the magnocellular lobe, which contains the first-order giant fibres, evoked strong expiratory spasms. No results were obtained from stimulation of the parts of the magnocellular lobe in the posterior suboesophageal mass or from the dorsal part around the oesophagus (Boycott 1961).

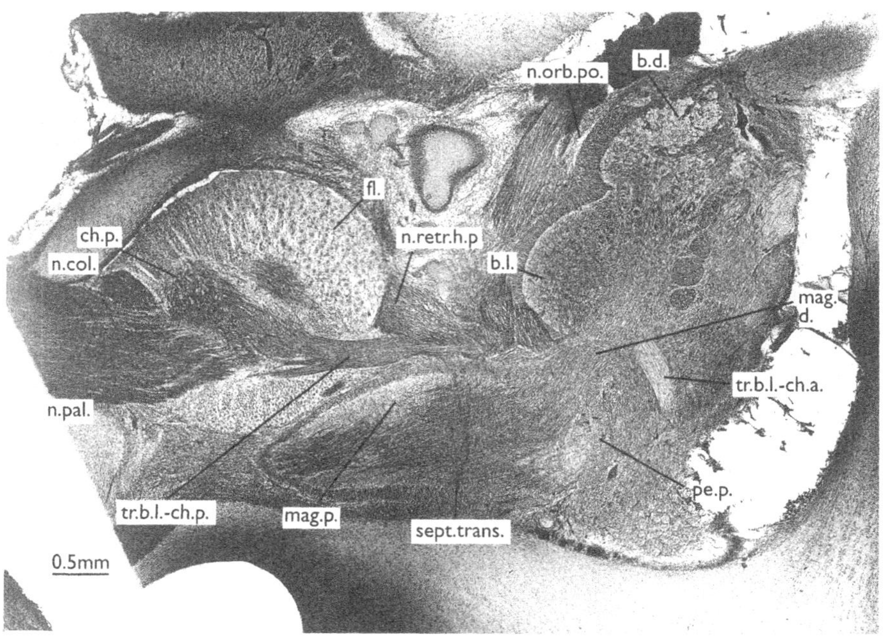

Fig. 5.21. *Sepia officinalis.* Sagittal section to show the relationships of the lobes at the sides of the brain. The fin lobe lies more medial than the posterior chromatophore lobe so that only its outer side is cut through (Boycott 1961).

The dorsal magnocellular lobe receives the large optic to magnocellular lobe tracts. These consist of two distinct bundles, the median tract of larger and the lateral tract of finer fibres (see *Loligo* and *Octopus*). Both parts consist entirely of descending fibres (as in *Octopus*) and both have some fibres crossing in the posterior magnocellular commissure. The fibres of the lateral tract end largely in the dorsal part of the lobe and may be concerned with visuo-static adjustments, since many of the cells of this and the posterior part of the lobe send efferent fibres to the statocyst (Budelmann and Young 1984). The larger optico-magnocellular fibres may operate giant fibre responses to visual stimuli.

The ventral magnocellular lobe and the first-order giant cells lie immediately above the macula princeps of the statocyst (Fig. 5.22). The lobe receives many fibres from the static nerves and from other parts of the brain, including the supraoesophageal lobes, optic lobes, and peripheral nerves. These inputs interweave throughout the lobe and with the dendrites of the first-order giant cells, as well as other large neurons of the lobe. The axons of these, like the giant axons, mostly pass backwards to the palliovisceral lobe to control the mantle, fins, and funnel, but some, like those of the nearby posterior lateral pedal lobe, run forwards and control the arms. Many fibres cross in the ventral magnocellular commissure. This area may be responsible for the execution of elaborate sequences of behaviour. Stimulation of the ventral magnocellular lobe resulted in strong expiratory actions (Boycott 1961), presumably produced by activating the giant fibres. Large fibres from the lateral part of the ventral magnocellular lobe pass backwards into the posterior magnocellular lobe. Here they interweave in bundles with fibres of the pallial nerve. The fibres leave the ventral side of the pallial nerve and spread out into the magnocellular lobe halfway along its course beside the palliovisceral lobe. The fibres divide into bundles that join the network of fibres along the lobe, some ending posteriorly, others reaching far forwards towards the ventral magnocellular lobe. Some are quite large (20 μm). They have not been followed to their ends but are probably afferent fibres. Similar bundles are seen in *Loligo*.

Giant-fibre system

The discovery of the giant nerve fibres in *Sepia* was reported by Young in 1934, followed in 1939 by a comprehensive description of the system in *Loligo*. He noted that this was a rediscovery, as Williams (1909) had observed the two first-order giant cells, the chiasm of their processes and the giant motor axons in *Loligo pealeii*. In a *Sepia officinalis* of 200 mm ML the first-order giant cell measures 50 μm in diameter at the level of the nucleus (20 μm). The cell is in fact no larger than many of the ordinary neurons of this region, but is nevertheless very easily distinguished by its curious position, partly in the neuropil. Its Nissl substance also differs markedly from that of the ordinary neurons. The shape of the cell has not been carefully studied, but it has a lateral dendrite and a large anterior dendrite, like that of *Loligo*, extending into the ventral magnocellular commissure. The two axons are not fused in *Sepia* (Fig. 5.23) (Martin 1977).

The second-order axons are more numerous in *Sepia* than in *Loligo*. In the mantle connective there are several fibres between 25 μm and 75 μm diameter (Young 1936), though it is not certain that all of them belong to the main system. In one specimen the posterior head retractor nerve contained 17 fibres, 15–50 μm diameter. These pass out partly in the dorsal and partly in the ventral root of the nerve. The visceral nerve contains five giant fibres, the largest being 50 μm in diameter. All of these probably run in the nerves to the funnel retractor muscle, though it is difficult to be certain of this. Finally, the posterior funnel nerve contains several large fibres; in one nerve there were eight (15–40 μm in diameter), but investigation of the exact relations of these fibres is very difficult.

The third-order giant fibres (Fig. 5.24) arising from cell bodies in the stellate ganglion are of moderate size, being about 270 μm in a specimen of 200 mm ML. Each stellar nerve contains several fibres of various sizes (Young 1936).

In the giant synapse of *Sepia officinalis* the second-order axons run parallel to the third-order axons and give off collaterals, which penetrate the enlarged sheaths of the third-order axons (Martin and Miledi 1986). Numerous small processes from the latter contact the

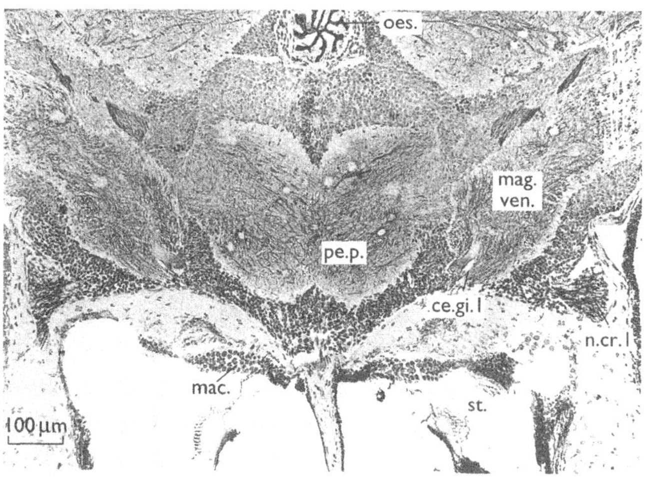

Fig. 5.22. *Sepia officinalis*. Transverse section to show the first-order giant cell and the macula of the statocyst.

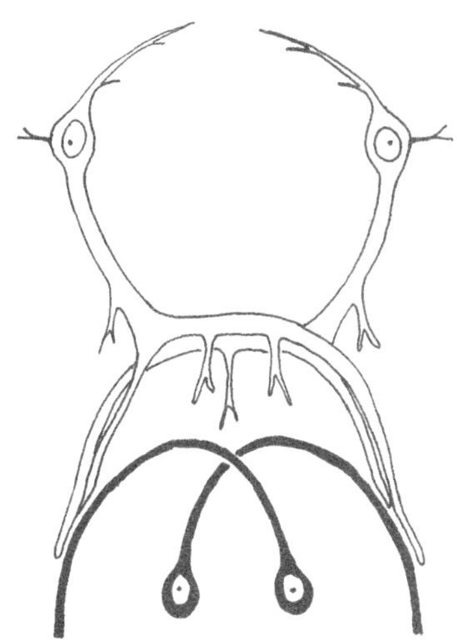

Fig. 5.23. *Sepia officinalis*. Schematic drawing of the course of the fibres in the chiasm of the first-order axons, and the second-order axons (black fibres) (Martin 1977).

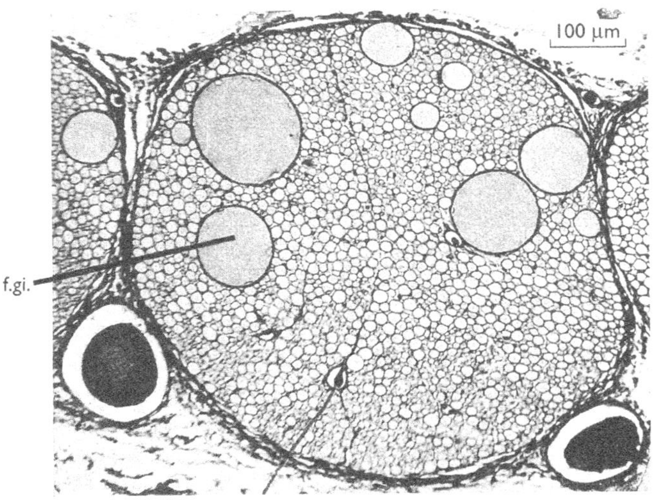

Fig. 5.24. *Sepia officinalis*. Transverse section of one of the stellar nerves showing the third-order giant fibres.

collaterals from different sides, forming synaptic junctions. Second-order axons enter the stellate ganglion and connect via large synapses to all third-order giant cells. The third-order giant axons are syncitial, arising by fusion of numerous axons from neurons in the dorsal wall of the stellate ganglion. There is no giant fibre lobe in the stellate ganglion of the cuttlefish.

Supraoesophageal mass

The supraoesophageal lobes are the 'highest' centres in the sense that they have few or no direct afferent or efferent connections with the periphery (except to the optic gland). They form two well-developed systems, so distinct that there are no direct connections between them: these are the higher motor centres and the 'memory' or 'association' centres. The higher motor centres consist of five lobes, the dorsal basal, median basal, lateral basal, peduncle, and interbasal lobes (Fig. 5.19). Stimulation of these produce elaborate parts of normal behaviour patterns (Boycott 1961). The association centres include the inferior frontal system (the inferior frontal and superior buccal lobes) and the vertical lobe system (the superior frontal, vertical, subvertical, and precommissural lobes) (Young 1965c). The vertical lobe system is a 'silent area' giving no obvious response to electrical stimulation (Boycott 1961), but experiments have shown it to be concerned with learning and memory.

Basal lobe system

The anterior and posterior parts of the anterior basal lobe, the median basal lobe, and the peduncle lobe will be considered here. These four parts have in common some signs of a 'cerebellar' type of structure in the form of sets of very fine parallel fibres (Young 1977b). This is less clear in the median basal lobe than in the other lobes. These lobes all receive inputs from the optic lobes, from the statocyst, and often from each other, but not from other sources. They all send fibres to the oculomotor centre of the anterior lateral pedal lobe and several of them also to other parts of the pedal, palliovisceral, and (perhaps) magnocellular lobes.

Anterior basal lobe

The anterior basal lobe has distinct anterior and posterior divisions concerned with the motor control of various actions (Figs 5.17, 5.25). The connections and cell patterns in *Sepia* are similar to those of *Loligo* for which they have been described in detail (Young 1977b), and of *Octopus* (Young 1971, Gleadall 1990).

Each division of the anterior basal lobe contains large efferent neurons with axons proceeding either to suboesophageal centres, to other basal lobes, or to the optic lobes (Fig. 5.26). There are also many small neurons whose axons form systems of parallel fibres. It has been suggested that these fine parallel fibres could act as a timing device as in the vertebrate cerebellum (Hobbs and Young 1973, Messenger 1979a), the larger neurons to initiate actions and the fine fibres to terminate actions by inhibition after appropriate delay.

Stimulation in the anterior anterior basal lobe initiates the opening of the arms like a rosette as if to receive prey (Boycott 1961, Chichery and Chichery 1987). With electrodes in the posterior anterior basal lobe stimulation resulted in the cuttlefish making movements similar to those when lifting itself out of sand: the ventral arms and funnel were turned downwards, so moving the animal upwards and backwards. Forward swimming movements with the fins then began with the arms pointing forwards as if attacking. Waves of contraction of both fins were also observed by Boycott (1961) during stimulation of the anterior basal lobe, and these proceeded in various directions; fin waves would begin at the front, as in forward swimming, or at the back for backward swimming. Sometimes the waves began anteriorly on the ipsilateral side and posteriorly on the contralateral side, as in turning. Lateral rotation of the arms, head, and eyes, either separately or together, was also seen. If the head and arms turned to the side which was stimulated then the ipsilateral eye turned backwards, the contralateral forwards: sometimes the opposite rotations occurred. The lobe thus functions as a 'higher' motor centre in that it controls and regulates several intermediate and lower motor centres (Boycott 1961, Messenger 1983). If the anterior basal lobe is removed from one side, the animal will spin continually towards that side (Boycott and Young 1950). This suggests that the lobe normally inhibits movement of the fins.

Lesions placed in the region of the fine parallel fibres of the lobe led to defects in the maintenance of ocular convergence during attack with the tentacles, and underestimates of the distance of the prey by as much as 20 mm (Chichery and Chichery 1987). This defect agrees with the conception that these fibres provide a timing device. Lesions at the centre of the lobe prevented the initial saccadic eye movements and subsequent rotation of the animal; the tentacles were projected, but without success. After lesions of the front of the anterior anterior basal lobe the cuttlefish showed rolling movements during acceleration. Lesions at the back of the posterior anterior basal lobe, also involved the precommissural lobe, and the animal failed to project its tentacles.

Median basal lobe

The median basal lobe contains large and small cells and parallel fibres like the anterior basal lobes, but their structure and function has been little investigated in *Sepia* or in any other cephalopod (Fig. 5.17). There are afferent and efferent connections with a very large range of

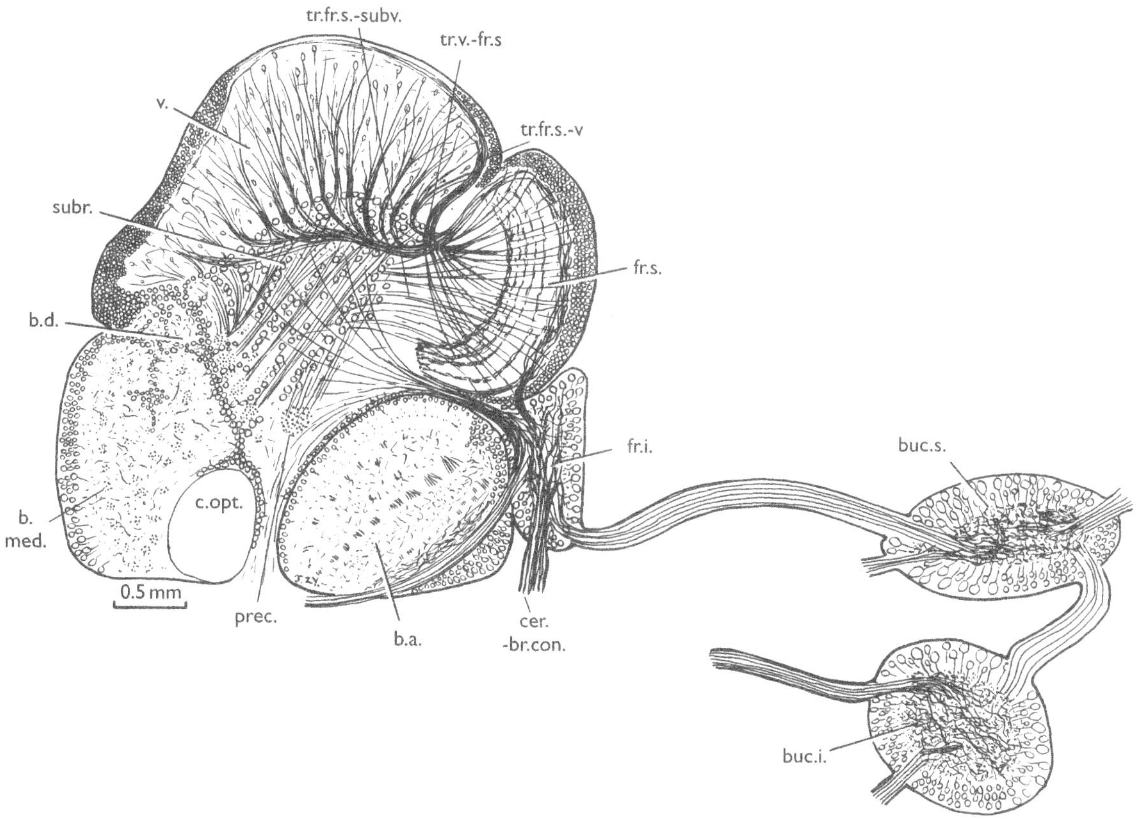

Fig. 5.25. *Sepia officinalis*. Drawing of a sagittal section of the suboesophageal mass, close to the midline, showing the pathways of the vertical lobe system (Young 1963b).

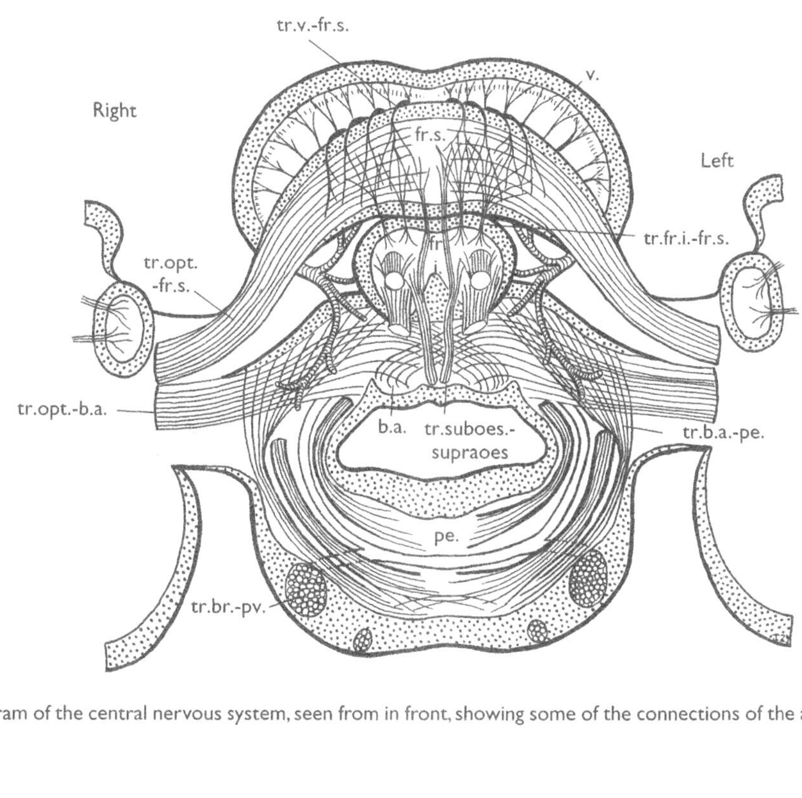

Fig. 5.26. *Sepia officinalis*. Diagram of the central nervous system, seen from in front, showing some of the connections of the anterior basal and superior frontal and vertical lobes.

suboesophageal centres, including the pedal, magnocellular, and palliovisceral lobes, and with the optic lobes (as in *Loligo*).

Electrical stimulaiton of the median basal lobe results in a variety of actions. It can produce gentle, coordinated, rhythmical respiratory movements. Sometimes the same stimulus can result in violent expiratory spasms, accompanied by folding the fins against the mantle, retraction of head and arms, and ejection of ink. These actions occur during an escape response via the fins and the giant-fibre system. An inspiratory spasm is occasionally seen, the mantle remaining inflated as long as the stimulus continued. This can occur when the animal is handled. Movements of both fins can also result from stimulation of this lobe, the waves proceeding anteriorly or posteriorly. The head can be protracted or retracted, but not rotated, and the first pair of arms raised as in attention. Movements of the buccal mass can also be elicited by stimulation. The median basal lobe is thus 'primarily responsible for the control of the suboesophageal centres concerned with the movements of the mantle, collar and funnel during breathing and swimming' (Boycott 1961). Thus it functions as a higher motor centre.

Lateral basal lobes

These lobes are anatomically distinct but are broadly continuous with the median basal lobes (Fig. 5.19). They have the well-defined function of controlling the chromatophores and muscles of the skin papillae (Boycott 1961). The cells of the lobes send axons to the anterior and posterior chromatophore lobes (Fig. 5.21).

The input to the lateral basal lobes is mainly from the optic lobes, and also from the median basal and peduncle lobes (as in *Loligo*). Electrical stimulation of one lateral basal lobe of *Sepia* produces uniform expansion of the chromatophores and erection of skin papillae of the head, arms, and mantle on one side (Boycott 1961). Higher stimulus strengths are more likely to elicit bilateral responses than weaker stimuli.

Interbasal lobe

This is a poorly defined, but important, region lying above and behind the median basal lobe (Fig. 5.27). The interbasal lobe receives nerve fibres from the optic lobes, and from areas surrounding the lobe. Together these influences are presumably concerned with prey capture: electrical stimulation of the interbasal lobe induces projection and retraction of the tentacles (Boycott 1961).

Dorsal basal lobe

The dorsal basal lobe is one of the least understood parts of the cephalopod brain (Figs 5.21, 5.25). Many of its cells are amacrine microneurons with no long axons. Others have long fine axons proceeding to the optic or median basal lobes. Most of the input also comes from these two sources. Electrical stimulation here does not elicit any movements (Boycott 1961).

Dorsolateral lobes

These are a pair of small lobules at the back of the supraoesophageal mass. They give rise to the fibres that innervate the optic gland, but the function of the dorsolateral lobe and the optic gland are not known in *Sepia*. However, cultivation of pieces of the optic gland with ovarian tissue was followed by maturation of the latter (Durchon and Richard 1967).

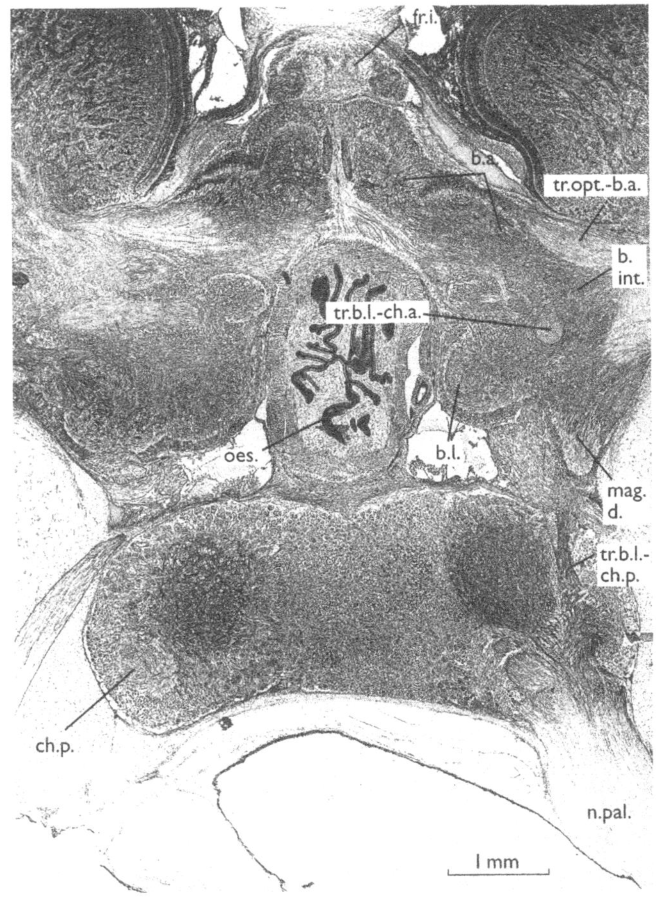

Fig. 5.27. *Sepia officinalis.* Horizontal section through the brain to show the interbasal lobe, and the lateral basal lobe to posterior chromatophore lobe tract.

Subpedunculate lobes

The subpedunculate lobes are a series of lobules above the median basal lobe. The axons of some of their cells proceed as a subpedunculate tract towards the optic commissure. This may be comparable with the subpedunculate nerve of *Octopus vulgaris*, which arises from the same region of the brain. Other cells of these lobules are very small and are probably neurosecretory.

Peduncle lobes

These are swellings on the two optic tracts (Figs 5.19, 5.28) and have attracted the attention of several researchers (Cajal 1917, Boycott and Young 1956, Boycott 1961). The peduncle lobes are best considered as part of the basal lobe system. The presence of a conspicuous set of parallel fibres first prompted the suggestion that the peduncle lobes may be analogous with the vertebrate cerebellum (Hobbs and Young 1973). The small cells and the parallel fibres are included in a distinct region of the peduncle lobe, the 'spine', which in *Sepia* runs in a horizontal plane along the upper surface of the lobe (as in *Octopus*). The input to the peduncle lobe comes from the optic lobe, the extraocular photoreceptor organs (the photosensitive vesicles), the statocysts, the other basal lobes, and from the contralateral peduncle lobe via a large peduncle commissure (Cajal 1917).

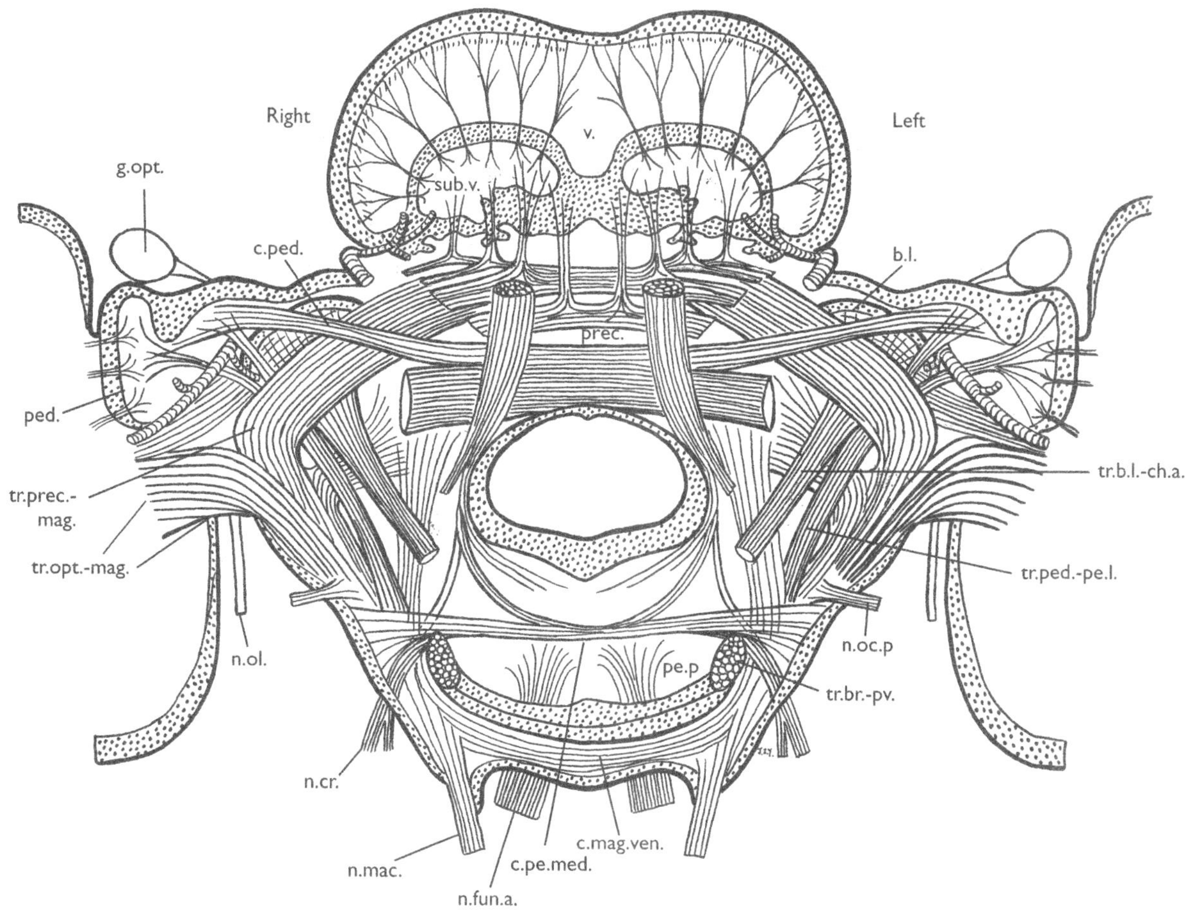

Fig. 5.28. *Sepia officinalis*. Drawing made from transverse sections showing some pathways in the brain at the level of the optic and peduncle lobe commissures, seen from in front.

Electrical stimulation of the peduncle lobe at threshold caused expansion of the chromatophores all over the body (Boycott 1961). A slightly stronger stimulus produced a wide variety of responses, such as were obtained after stimulation of the optic lobes, so that this may have been due to diffusion of the current to adjacent fibres of the optic tract. Experimental studies with *Octopus* show this lobe to be a higher motor centre, and this has been confirmed in *Sepia*. Using chronically implanted electrodes, similar responses were obtained from both the spine and basal zone of the peduncle lobe. At threshold there was an increase in the frequency of undulatory movement of the fin, darkening of the whole animal, and dilation of the ipsilateral pupil. With a stimulus above the threshold or of longer duration these responses intensified (Chichery and Chanelet 1978).

Olfactory lobes

Each olfactory lobe lies medial to the peduncle lobe, with which it is continuous (Figs 5.19, 5.29). It extends dorsally into several lobules and becomes continuous with the dorsolateral lobe. The olfactory lobe receives the olfactory nerve from below. This nerve is joined by a small bundle from the magnocellular lobe, whose significance is not understood.

Inferior frontal lobe

The inferior frontal lobe occupies the front part of the supraoesophageal mass and receives afferent fibres from the arms in the cerebrobrachial connective; there are possibly some efferent inferior frontal-brachial fibres in this connective going to the brachial lobe (Fig. 5.19). The inferior frontal lobe is also connected to the superior buccal ganglion through the cerebrobuccal connective (Hillig 1912) and there is a separate tract joining the inferior frontal lobe and the median basal lobe. Electrical stimulation of the inferior frontal lobe resulted in very slight movements of the buccal mass (Boycott 1961).

Superior frontal lobe

The superior frontal lobe has two parts, anterior and posterior (as in *Loligo*). The anterior portion has larger cells with axons proceeding to the subvertical lobe and perhaps beyond. Fibres passing to the precommissural lobe are seen in Fig. 5.25. This anterior part receives an input from the vertical lobe. In *Sepia* these fibres from the vertical lobe enter the superior frontal lobe not only from below (as in *Loligo*), but also from above, coming direct from the vertical lobe (Fig. 5.30). The posterior part of the superior frontal lobe has smaller cells, which send axons only to the vertical lobe: this part receives no fibres from the vertical lobe. Both parts receive many fibres from the optic, subvertical and inferior frontal lobes. Some fibres also come direct from the periphery both from in front (arms and lips?) and from behind, in the suboesophageal to supraoesophageal tract.

The superior frontal lobe thus allows for the bringing together of signals from the main afferent systems, except the statocyst. In the

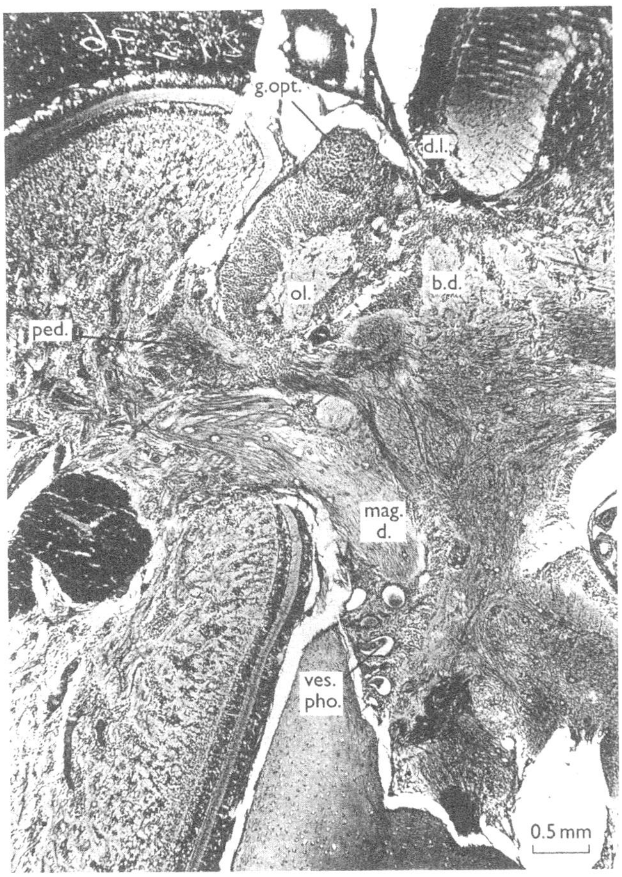

Fig. 5.29. *Sepia officinalis*. Transverse section of the brain to show the peduncle lobe, optic gland and photosensitive vesicles.

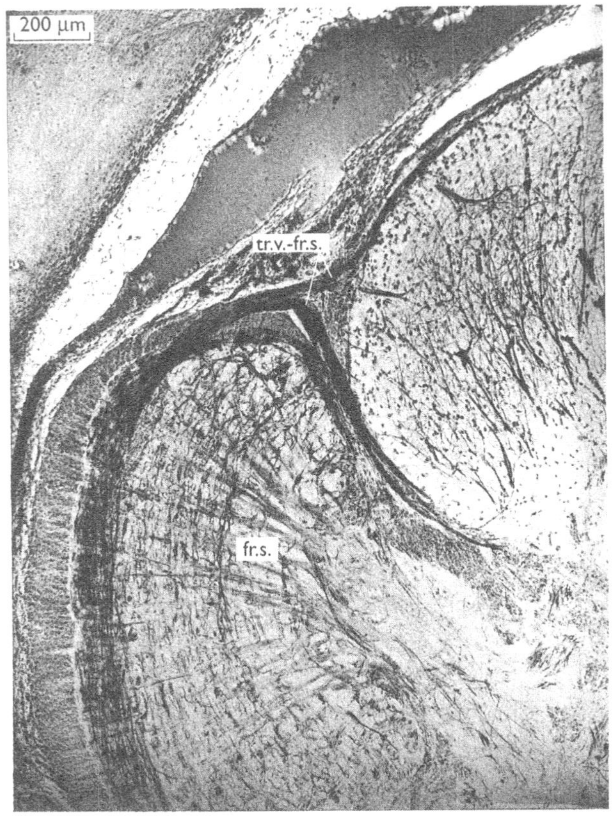

Fig. 5.30. *Sepia officinalis*. Sagittal section showing the two parts of the vertical to superior frontal lobe tract.

dense neuropil at the front of the superior frontal lobe the various inputs are also under the influence of recurrent fibres coming from the vertical lobe. The output from this part of the superior frontal lobe proceeds to the subvertical lobe where it again meets the influence of the descending fibres from the vertical lobe (Fig. 5.28). There is thus a complex of re-entrant paths.

A postfrontal lobe is present, as it is in *Loligo* where it was first described by Young in 1979. It comprises a set of cells around the hilum of the superior frontal lobe (Fig. 5.17) including large cells: their connections are not known in *Sepia*.

Vertical lobe

The vertical lobe is a prominent dome-shaped structure occupying the whole of the upper surface of the brain (Figs 5.17, 5.28). The structure and connections of this lobe have not been investigated in detail in *Sepia*, and, although there may be some points of difference, they are so like those of *Loligo* that no separate description is included in this chapter. The input comes from the superior frontal lobe in the form of interweaving bundles of fibres all over the upper surface. The only other input is through the subvertical lobe and this has not been investigated in *Sepia*.

The majority of the cells of the vertical lobe are microneurons (amacrines), either without axons or with short ones confined to the lobe. These form thick layers over the outer parts of the lobe (Fig. 5.31). The large output cells are scattered throughout the neuropil and seem to have the same forms as in *Loligo* (Young 1979). They run straight through the outer neuropil and join to form bundles. In the inner neuropil many have dendrites arising from curling trunks, some of which take indirect courses and thus sample a large volume of the neuropil. All the axons proceed either to the superior frontal lobe or to the subvertical and precommissural lobes.

Subvertical lobe

This lobe consists of large cells surrounding islands of neuropil. The fibres here run in irregular directions, very different from the organization of the neuropil of the vertical lobe (Fig. 5.31). The region serves as the main output of the vertical lobe, allowing combination with the other inputs to the system. The main outputs are to the optic lobes, the superior frontal lobe and to the precommissural lobe and suboesophageal centres (Fig. 5.28).

Precommissural lobe

The precommissural lobe is continuous with the subvertical lobe and provides the main output channel for the whole system (Figs 5.25, 5.28). Its input is mainly from the subvertical and the optic lobes, but also from other sources. The large cells of the precommissural lobe send axons directly to suboesophageal centres. These fibres (20 μm in diameter) arise far laterally, proceed medially, and send branches to the base of the median basal lobe (Fig. 5.32). Passing downwards and backwards they then send fine collaterals to the posterior pedal lobe, and to the magnocellular lobe, anterior to the transverse cerebral

Fig. 5.31. *Sepia officinalis*. Transverse section showing the numerous amacrine cells of the vertical lobe and the bundles of fibres of the superior frontal to vertical lobe tract.

septum. Passing through this septum they then divide into many further branches to the magnocellular and then to the central pallio-visceral lobe. The many collateral branches of these large fibres appear as a conspicuous series of fine trunks crossing the neuropil of the lobes in a dorsoventral direction, towards the branches of large fibres coming up from the magnocellular lobe. This is evidently an important final common path from the vertical lobe system.

5.2.13 Extraocular photoreceptor organs

The extraocular photoreceptor organs comprise a series of sacs, the photosensitive vesicles, that lie below and behind each optic tract and at the side of the magnocellular lobe (Fig. 5.29). They are pink in the living animal and lie within the cranium. The function of these vesicles has not been examined in any sepiid as yet.

5.2.14 Subpedunculate tissue: a neurosecretory system

This is a system of nerves that forms a neuropil under the epithelium of the vena cava. It is composed of neurons, the cell bodies of which are located in a layer of the palliovisceral lobe and in paired ganglionic trunks. These are lateral and medial trunks that emerge from the palliovisceral lobe with the posterior funnel and visceral nerves respectively. After passing some distance with these nerves the trunks take an independent course and give off tapering branches, which end

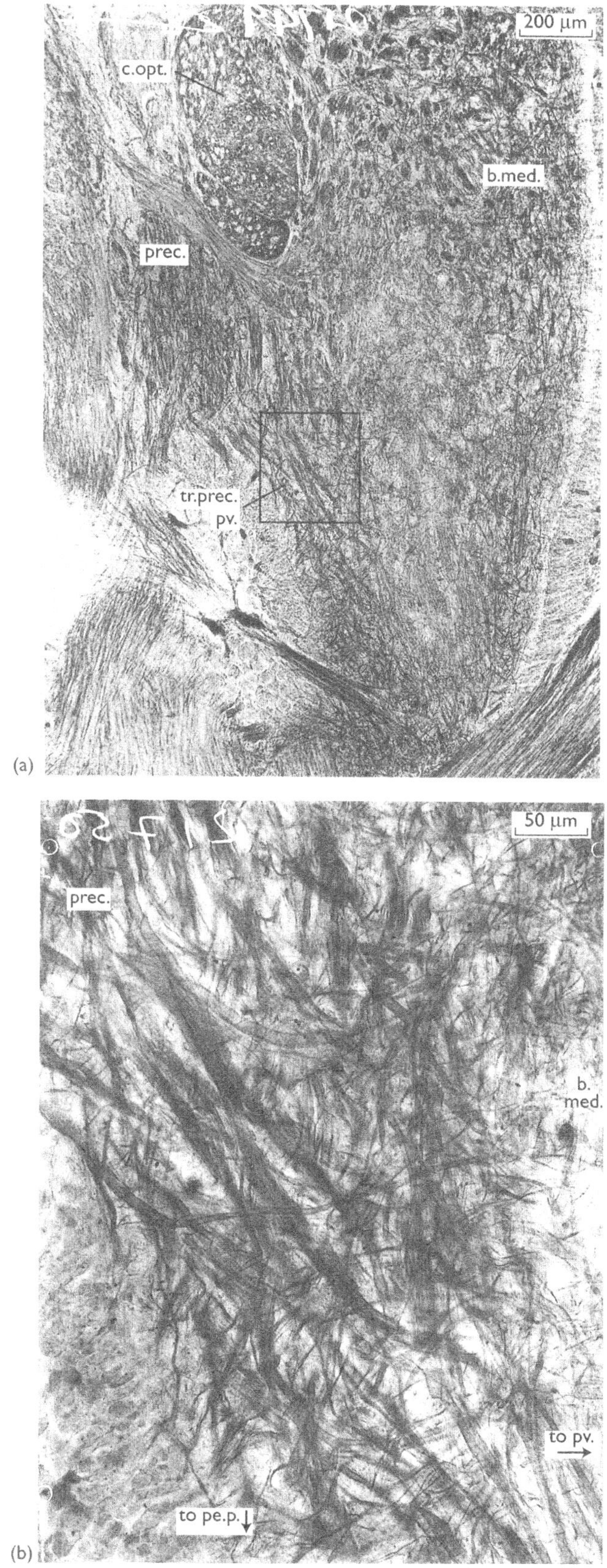

Fig. 5.32. *Sepia officinalis*. Sagittal section of the precommissural and median basal lobes showing the large fibres of the precommissural to palliovisceral lobe tract (a), and their branches to the posterior pedal lobe at higher magnification (b).

blindly in the loose connective tissue. The layer of the palliovisceral lobe and the lateral and medial trunks are made up of unipolar cells and bundles of fibres formed by their axons. These cells accumulate in cords and are especially abundant where the trunks give off nerves running to the vena cava. These nerves penetrate through the muscular coat of the vein to end as a dense network below its endothelium. This neuropil extends as a continuous layer in a large portion of the vena cava and into adjoining veins. Wherever the neuropil is present, the inside of the veins show longitudinal ridges which are well exposed to the blood, and this system presumably carries away secretory products. The neurosecretory system in *Sepia* is not arranged in such an orderly manner as in *Octopus* and *Eledone* (Alexandrowicz 1965) and the innervation is poorer.

5.2.15 Epidermal receptors: a 'lateral line' analogue

In young *S. officinalis* there are eight lines of epidermal hair cells on the head, four on each side, running on to the arms (Budelmann 1994). These are made visible as parallel lines of slightly expanded chromatophores, often with accompanying iridescent cells (Budelmann *et al.* 1991). On each side two lines run dorsally and two laterally (one above and one below the eye). The line below the eye is the shortest and does not continue on to an arm, unlike the other three lines. The lines are formed of sensory hair cells, with kinocilia up to 20 μm in length, and with axons that can be followed for a short distance beneath the epidermal line and towards the central nervous system (Sundermann 1983). The number of kinocilia varies from a few to more than 100 per cell; the number of cells varies per unit length of the lines, the greatest number being present on the lines above and below the eye (Budelmann *et al.* 1991). Each hair cell is morphologically polarized in one direction. Recordings of receptor potentials were made during stimulation of the head lines with water currents (Budelmann 1994). The sensitivity of these cells is such that it should allow the cuttlefish to detect a moving fish, 1 m body length, at a distance of 30 m (Budelmann *et al.* 1991). These authors demonstrated that this sensory system enables young cuttlefish to capture shrimps even in complete darkness, although with only 50% of the success rate obtained in light.

5.2.16 Olfactory organs

The olfactory organ itself lies just behind the eye and is covered by a brush border of microvilli and cilia. The pseudostratified epithelium of the organ consists of one type of epithelial cell and two types of sensory cells. The nuclei of all three cell types are in the basal part of the cells. The axons leading to the olfactory nerve lie above the basal lamina (Wildenburg 1991).

5.2.17 Statocysts

The paired statocysts lie below the brain and are embedded in the cartilage of the cranium (Fig. 5.2). The statocyst has an irregular shape and is structurally complex, with a level of sophistication equivalent to that of the vertebrate vestibular system (Fig. 5.33). Each

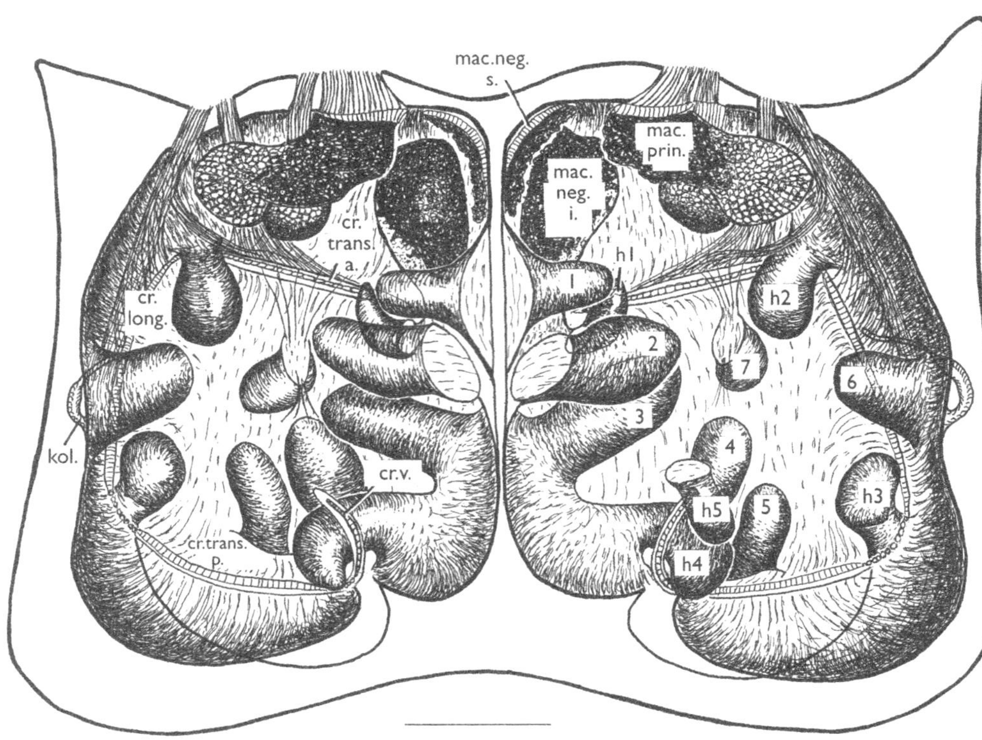

Fig. 5.33. *Sepia officinalis*. Drawing of the statocyst seen from above. The macula princeps carries a large statolith. The macula neglecta superior is in the vertical sagittal plane, the macula inferior in the horizontal plane (scale bar 1 mm) (Dilly *et al.* 1975).

statocyst has an irregularly shaped cavity containing receptor systems for the detection of gravity (the macula/statolith system) and angular acceleration (the crista/cupula system) (Budelmann 1977, 1994, Williamson 1991, 1995).

In each statocyst there are three maculae, the macula neglecta superior lying in the vertical sagittal plane, the macula neglecta inferior in the horizontal plane, and the macula princeps, which carries the statolith. The statolith is a small, hard, aragonitic structure that lies in the fluid-filled cavity. The statolith is less than 2 mm in length and has rounded protuberances in different planes: it is of value for estimating growth and age (Clarke 1998).

The crista ridge runs in three main planes along the statocyst wall. It is divided into four sections: three oriented horizontally, the anterior transverse crista, the longitudinal crista, and the transverse posterior crista, and one oriented vertically, the vertical crista. The beginning and end of each section of the crista is roofed over by a cartilaginous anticrista (Budelmann 1977). In each crista section there is a strip of sensory epithelium with hair cells arranged in rows. Intracellular recordings from the sensory hair cells show that those on the ventral side of the anterior transverse crista are secondary hair cells and physiologically polarized in the ventral direction. The hair cells of the dorsal side of the section are primary hair cells and physiologically polarized in the dorsal direction (Williamson 1991). A cupula is attached to each section of the crista. It is a sail-like structure, protruding freely into the fluid-filled statocyst, composed of a matrix of medium electron density and fibrous strands. It is movement of the cupula, caused by inertial movement of the fluid, that applies a shearing force to the hair cells and signals rotation.

The anticristae are rounded and differ in proportions and shapes. Anticrista 1 has a blunt pyramidal shape and contributes to the outer wall of the most anterior chamber. Anticristae 6 and 7 meet and touch in the adult, but have not been seen to fuse. Anticrista 7 is somewhat extended backwards and the channel has an inner wall for part of its inner circumference.

At hatching the statocysts form a large portion of the head. Their internal features are simpler and the cavity has fewer obstructions than in the juvenile (Bülow and Fioroni 1989). Their subsequent growth is at a slower rate than the rest of the animal. Hamulus 4 is well formed and the other hamuli develop rapidly in the days immediately after hatching (Maddock and Young 1984). The crista cupulae are already present at hatching, and so the angular acceleration receptor system is probably functional at this stage (Williamson 1995).

A discriminant analysis of numerous measurements of the statocyst of *Sepia* shows that it differs from oegopsin squids, both buoyant and non-buoyant, but is more like the latter (Maddock and Young 1984). The features of the statocyst suggest a need to monitor faster pitching and rolling movements, perhaps associated with the buoyancy system of the cuttlebone (Young 1989). *S. officinalis* can respond to low-frequency 'sound': it is not clear which sensory organ system detects vibration, but it is likely that it is the statocyst (Packard *et al.* 1990). The statocysts are also important in the counter-shading reflex patterns as after their ablation this reflex is abolished (Ferguson and Messenger 1991).

5.3 *Metasepia*

Metasepia is found in the tropical West Pacific and southern Japan, *M. tulbergi* from Japan and south-west to Taiwan and Hong Kong, and *M. pfefferi* in Australian waters from South Queensland to Western Australia (Voss and Williamson 1971, Okutani *et al.* 1987, Roper and Hochberg 1988). It inhabits shallow waters on the continental shelf, where the bottom is of silt or sand, at depths of 10–50 m.

Field observations by Roper and Hochberg (1988) of *M. pfefferi*, or Pfeffer's flamboyant cuttlefish, show it to possess extensive and elaborate body patterns. When undisturbed the animal is a uniform pale, grey-white to match its background, while the protective membranes of the arms enclose the suckers and have pale yellow stripes. A slight disturbance results in a highly contrasting black and white body pattern with a strong yellow fin line: the fins are translucent. Waves of colour occur and after repeated disturbance a series pass as a band-like wave moving through the dark field of the mantle; it is possible to see four such waves simultaneously, two moving anteriorly and two posteriorly (Fig. 5.34a, b). The skin generally has a highly complex sculpture with patterns of numerous spatulate papillae, including large primary and small secondary papillae. The primary papillae may be entirely white or yellow, or just tipped with white or yellow.

Metasepia displays a very unusual mode of locomotion, termed 'ambling'. In this it uses a pair of muscular ambulatory flaps which lie along the margin of the posterior third of the ventrolateral mantle. The movement is a slow and deliberate 'walk' in which the ventral arms act as legs and the ambulatory flaps slide or shuffle forwards (Fig. 5.34c). The animal can also move backwards, and if threatened on one side will move away sideways. This ambling gait appears to be the preferred mode of locomotion. These preliminary observations show this sepiid to have a rich repertoire of components of colour, texture, and posture, and Roper and Hochberg (1988) comment that 'in its habitat, *Metasepia pfefferi* looks like anything but a cuttlefish'.

5.4 *Sepiella*

Sepiella is found in tropical waters from West Africa to Australia. *S. ornata* is found along the coast of West Africa from 20 °N to 20 °S, *S. inermis* from the southern Red Sea along the northern coasts of the Indian Ocean to the Andaman Sea and the southern South China Sea. *S. japonica* is restricted to the coasts of China and Japan. Their vertical distribution is poorly known, but from such records as are available they have been caught between the surface and 200 m depth (Khromov 1998).

This genus is closely related to *Sepia* but distinguished from it by the presence of a round posterior or caudal gland. This opens posteriorly between and ventral to the fins and exudes a watery red fluid. The local name given to *Sepiella japonica* by fishermen of Hong Kong means 'blood faeces cuttlefish' and refers to this gland (Voss and Williamson 1971). The gland is associated with an iridescent layer, and on removing the skin 'flashes of golden-green and pink' were observed (Massy 1916): its function is not known. The cuttlebone has an elliptical outline and no posterior spine (Adam and Rees 1966, Roeleveld 1972, Khromov *et al.* 1998).

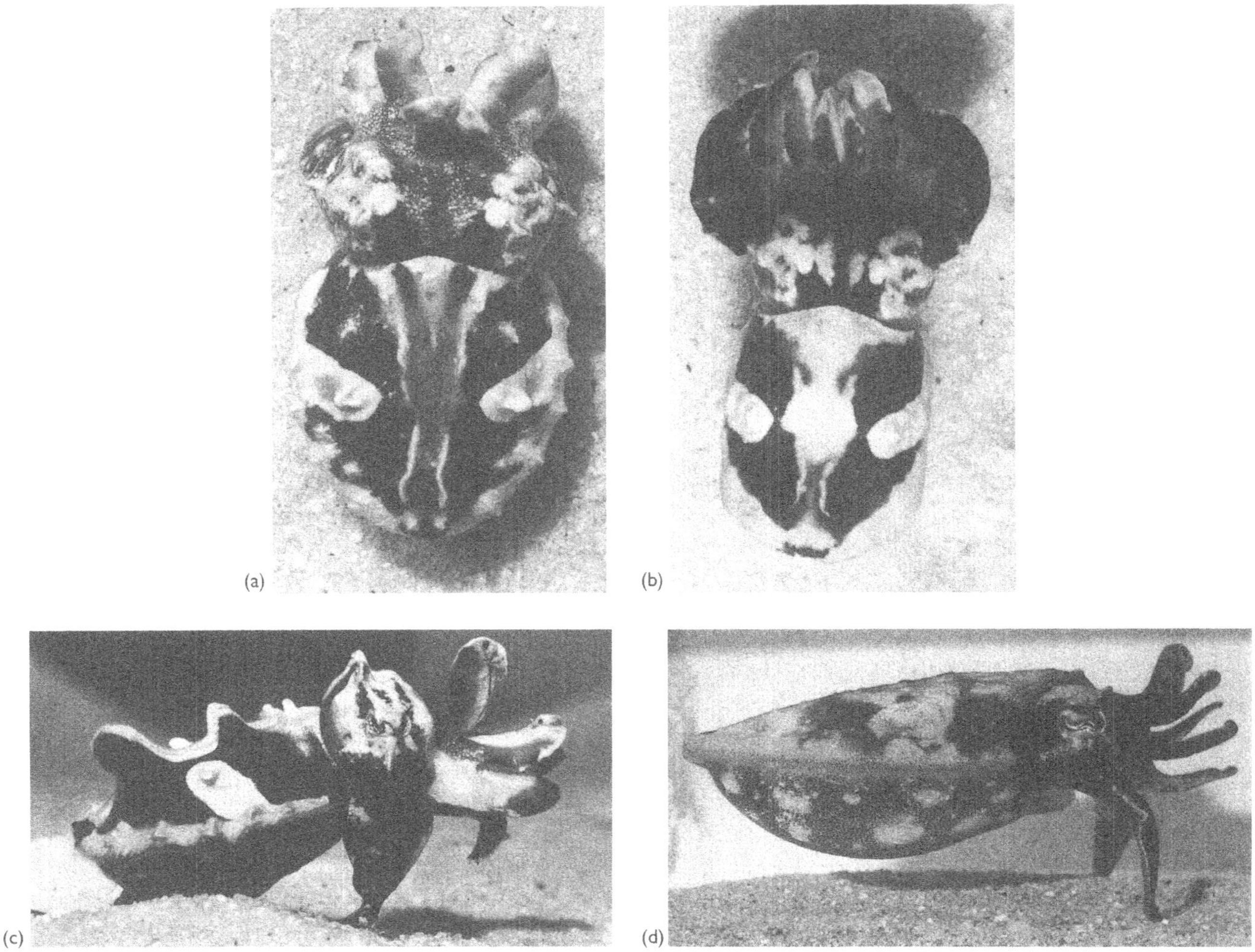

Fig. 5.34. *Metasepia pfefferi.* Photographs showing body patterns with wave passing over the dorsal mantle surface (a, b). The animal moving over the substratum in a slow walk (c), and 'ambling'. (Photographs kindly provided by C.F.E. Roper and F.G. Hochberg).

Sepiella japonica is very common along the coast of China, where it is found to depths of 50 m. It is fished commercially, and in some years it occurs in huge numbers in the Hong Kong area. Dorsally it is dark brown in colour, with brown chromatophores. A white line runs around the base of the fins (Voss and Williamson 1971). There are components of sexual body patterns; the male has large white spots and the female small ones (Yasumoto 1938). Spawning takes place along the coasts of Hong Kong in February and March (Voss and Williamson 1971). A female of 200 mm ML, the largest yet recorded, laid eggs (10.5×8.5 mm) that took 35–42 days to hatch (15.7–23.2 °C). The eggs have a coat of elastic, dark brown gelatinous substance and a smooth surface. The young have a mantle length of 3.4–5.0 mm at hatching (Choe 1966, Okutani 1978, Yamamoto 1982), 22 mm after 48 days, and 110 mm after 100 days.

The morphological development was followed from egg-laying until hatching. The cuttlebone is just discernible at 11 days and at the time of hatching has eight chambers. Dorsal chromatophores appear at 17 days, followed on the next day by a few ventral ones. The primordia of the eye first appears at 7.5 days; the retina and lens are visible at 14 days, and by 26 days it is black and strongly iridescent (Yamamoto 1982). The ultrastructure and the functional development of the retina was followed until hatching. Receptor potentials were elicited by a bright flash of light and were first recorded at 21 days; they increased gradually during subsequent development of the animal (Yamamoto 1985, Yamamoto *et al.* 1985).

Young *Sepiella* shoal and mostly remain swimming in the water column, except at night (Choe 1966). Small mysids are hunted and captured within 24 h of hatching, like *Sepia officinalis* (Wells 1958, 1962, Ohshima and Choe 1961). The stomach contents indicate that skeletal material from crabs is ingested, as such debris was found together with the bones from fish, although the heads of the latter were sometimes discarded. Small *Sepiella* show a preference for prawns, but fishes are eaten as the cuttlefish grows larger (Tang and Khoo 1974).

6 Sepiolidae

Sepiolids occur mostly in coastal waters over the continental shelf, from intertidal waters to depths of 1600 m. Rossiines and sepiolines are generally considered to live mostly on the bottom, and yet at night they are found in midwater (Bello and Biagi 1995). Heteroteuthines usually live in midwater but are now recognized to associate with the sea floor for at least part of the life cycle (Relini 1995, Villanueva 1995).

6.1 Rossiinae

Members of this subfamily live on the continental shelf and upper slope of the Arctic, Atlantic and Indo-Pacific Oceans and the Mediterranean Sea (Mangold-Wirz 1963, Okutani 1967, Hochberg and Fields 1980, Vecchione and Grant 1983, Nesis 1987, Reid 1991, Guerra 1992).

6.1.1 *Rossia*

The genus was first described by Owen (1835). *Rossia* (Fig. 6.1a) is found in the Arctic, the boreal Atlantic, and Pacific Oceans, in tropical Western Australia and in the Mediterranean Sea (Nesis 1987, Mangold and Boletzky 1988, Reid 1991). It is mostly benthic in habit from the sublittoral to the bathyal zones but enters the water column during the hours of darkness (Bello and Biagi 1995).

Life history

A female *R. palpebrosa* (65 mm ML) was caught during a voyage in the Canadian Arctic and illustrated with meticulous care by Owen (1835), and it had eggs of various sizes in its ovary. *R. australis* also has oocytes of various sizes up to 9.5 mm diameter in the ovary (Reid 1991). The range in size of ovarian eggs in both species suggests that spawning is intermittent over a period of time (Mangold *et al.* 1993). The female and male *R. pacifica* died 4 and 9 days respectively after the last eggs were laid (Summers 1985). *R. australis* reaches a maximum mantle length of 33.8 mm in males and 62.6 mm in females (Reid 1991). Female *R. pacifica* grows to 100 mm ML (Nesis 1993b).

Steenstrup (1881) recorded that *Rossia* eggs were deposited singly, one beside the other, and only attached to others by their sticky surfaces. One egg mass dredged at a depth of 350 m in Newfoundland waters contained 100–150 ovoid eggs (Aldrich and Lu 1968a). The eggs of *R. macrosoma* have been taken in bottom trawls (180–200 m

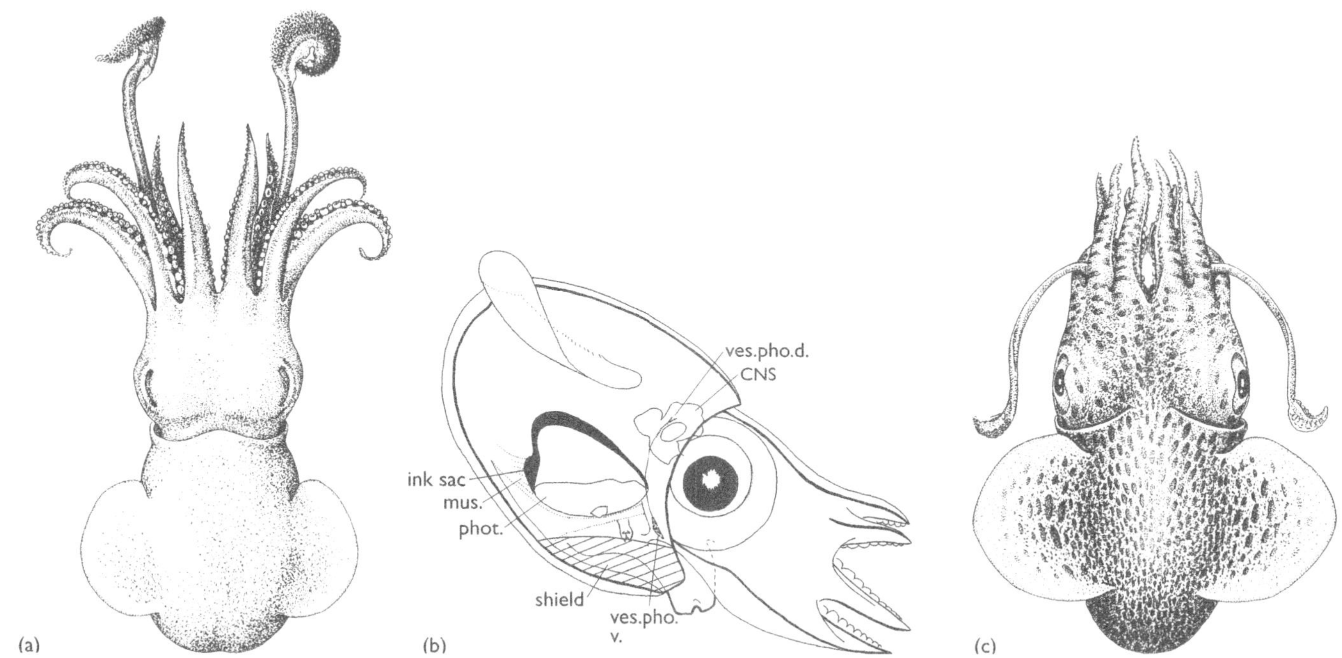

Fig. 6.1. Sepiolida. (a) Rossiinae, *Rossia macrosoma* (Guerra 1992); (b) Heteroteuthinae, *Heteroteuthis hawaiiensis* in lateral view to show the light organ, photosensitive vesicles and shield (R. E. Young 1977); (c) Sepiolinae, *Sepiola atlantica* (Guerra 1992).

depth) and found deposited in *Pinna* shells in the Mediterranean (Mangold-Wirz 1963, Boletzky and Boletzky 1973).

At hatching *R. macrosoma* (5 mm ML) swims actively using the fins, and jets from the funnel to 'hover' in a nearly horizontal position, and 'shoot backward... and rush forward... by means of vigorous jet propulsion,... sometimes taking a very slender shape' (Boletzky and Boletzky 1973). Some hatchlings bury themselves, some settle on sand in the open or under a shell, some adhere to the wall of the tank, and others crawl over the substratum with arms extended.

Juvenile rossiins crawl on their arms on the substratum, dig shallow depressions in which to rest with arms rolled under their heads, and swim (Hochberg and Fields 1980). *R. pacifica* exhibits an escape response and when startled will take off from the bottom, using fins and lateral arms. It then follows a repetitive cycle of inking, blanching, jetting and darkening, after which the animal drifts with arms spread and skin dark; finally it returns to the sea bed to bury itself with sand blown on to its back by the funnel (Shimek 1984).

Eyes and central nervous system

The eyes of *Rossia* occupy a large part of the head. Conjunctive muscles join in the midline, thus allowing binocular vision (Fig. 6.2) (Budelmann and Young 1993). The optic nerve fibres in the optic lobe have a regular arrangement and the incoming fibres cross as each bundle divides (as in *Loligo*; Young 1974). The individual fibres then finally pass through the outer granular layer in a highly regular pattern of rows, and this is maintained as they swell out in the plexiform layers (Fig. 6.3). Here they meet the tangential dendrites of the second-order visual cells, many of which are very long. Beyond the inner granular layer the axons of these cells form bundles separated by columns of small cells (as in *Loligo*).

The lobes of the brain of newly hatched and adult specimens were measured, using the same method, and can be compared in Fig. 6.4. The optic lobes are of almost the same proportion at both stages, and the motor centres are also well developed at hatching (Wirz 1959, Frösch 1971).

In the central nervous system the inferior frontal lobe receives large bundles of fibres from the cerebrobrachial connective, and from the magnocellular to supraoesophageal tract (Fig. 6.5). It sends fibres

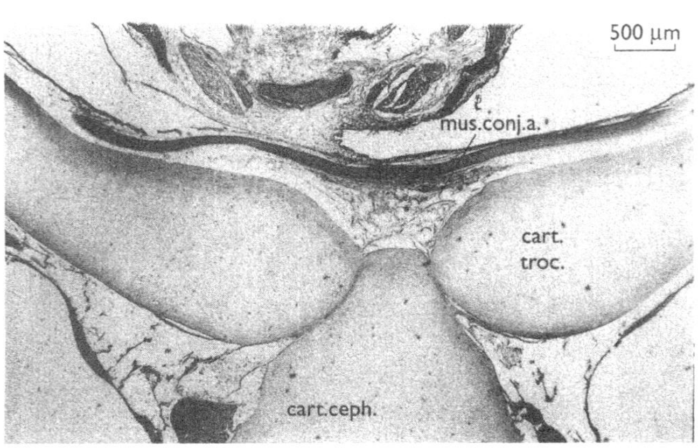

Fig. 6.2. *Rossia macrosoma* (65 mm ML). Horizontal section to show the trochlear cartilage and the conjunctive muscle.

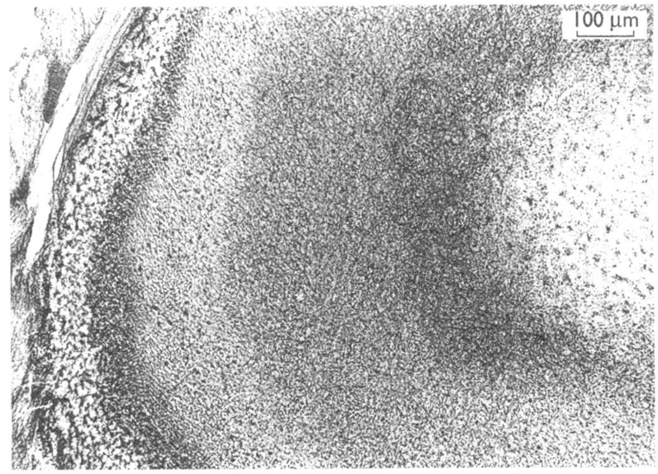

Fig. 6.3. *Rossia macrosoma* (65 mm ML). Horizontal section through the optic lobe.

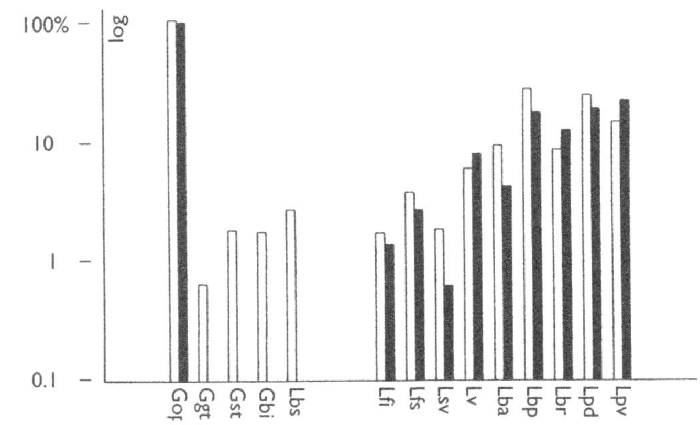

Fig. 6.4. *Rossia* sp. Histogram showing the volumes of each lobe as a percentage of the whole brain. Open bars, newly-hatched (Frösch 1971); filled bars, adults (Wirz 1959). Gbi, inferior buccal ganglion; Ggt, gastric ganglion; Gop, optic ganglion; Gst, stellate ganglion; Lba, anterior basal lobe; Lbp, posterior basal lobe; Lbr, brachial lobe; Lbs, superior buccal lobe; Lfi, inferior frontal lobe; Lfs, superior frontal lobe; Lpd, pedal lobe; Lpv, palliovisceral lobe; Lsv, subvertical lobe; Lv, vertical lobe.

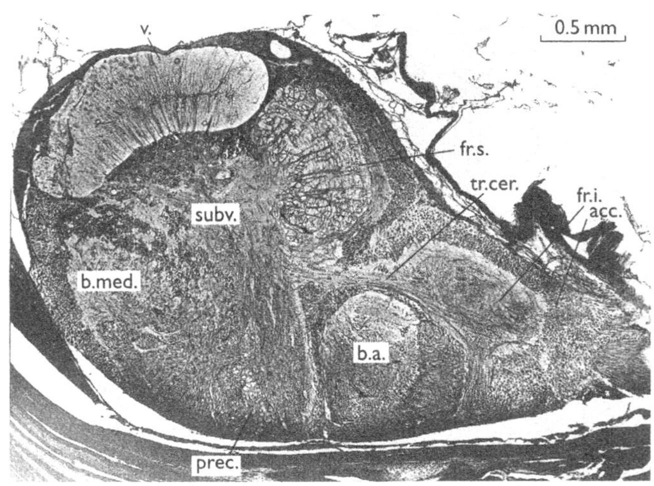

Fig. 6.5. *Rossia macrosoma*. Sagittal section through the central nervous system showing the accessory lobe of the inferior frontal lobe.

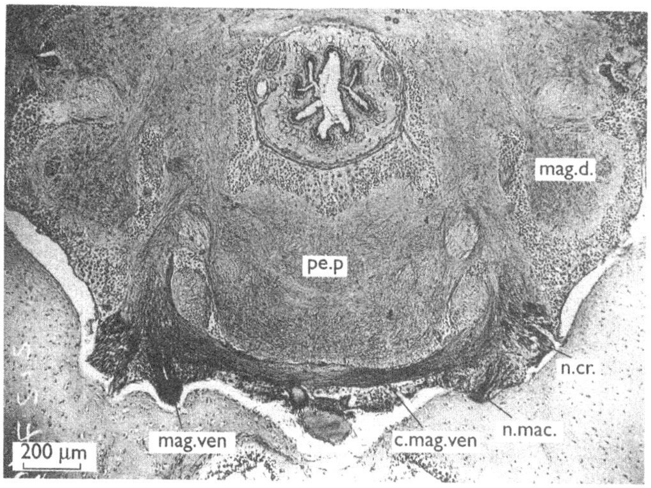

Fig. 6.6. *Rossia macrosoma*. Transverse section to show the magnocellular and pedal lobes.

to the interfrontal tract and to the cerebral tract. The wall of the inferior frontal lobe contains many small cells. The vertical lobe forms a prominent part of the central nervous system. An accessory lobe lies anteroventral to the fin lobe and it has many small and some large cells. The optic gland lies above the peduncle lobe. No first-order giant fibres could be found in the magnocellular lobe (Fig. 6.6). The stellar nerves were not examined. No photosensitive vesicles could be found in the sections; if they exist they must consist of only a few cells.

6.1.2 *Semirossia*

Members of this genus are known from the coasts of the western Atlantic, from Nova Scotia to Tierra del Fuego, and off southern Chile. They are found over muddy bottoms at depths of 85–450 m (Roper *et al.* 1984, Nesis 1987).

A bilobed light organ is present and is partly embedded in the ink sac. It consists of a central, voluminous chamber, and a reflector, with the free part covered by 'lens' tissue. The chamber of the photophore has a mass of very fine granules of varying sizes; these may be luminescent bacteria but no bacteriological tests have been carried out (Boletzky 1970).

6.1.3 *Neorossia*

Neorossia is known from the east and the north-west Atlantic, the Mediterranean Sea, and the Great Australian Bight (Mangold-Wirz 1963, Reid 1991, Bello 1995). It lives in waters 100–1740 m in depth (Villanueva 1992b, Würtz *et al.* 1995).

In *Neorossia caroli* the optic lobes form 227% of the total brain volume. The inferior frontal system forms 3.3% and the vertical lobe system forms 15% of the total brain volume (Table 2.6).

The statocyst is relatively short, broad, and shallow. The cavity is restricted by large anticristae 2 and 6 and hamuli. Anticristae 6 and 7 nearly meet to make a canal but do not fuse. Anticristae 3 and 4 are absent, and anticrista 5 lies far laterally and posteriorly (Young 1989). Measurements of the statocyst place *Neorossia* amongst non-buoyant cephalopods (Maddock and Young 1984).

6.2 Heteroteuthinae

Heteroteuthines are found widely in the tropical and subtropical waters of the Pacific and Atlantic Oceans and in the Mediterranean Sea (Nesis 1987). They live in depths of 50–160 m, and are mostly pelagic in habit, but associate with the bottom during part of the life cycle (Relini 1995, Sartor and Belcari 1995, Villanueva 1995).

6.2.1 *Heteroteuthis*

Heteroteuthis (Fig. 6.1b) is oceanic and inhabits the tropical and subtropical waters of the Atlantic, from Bermuda to La Plata, in the Caribbean Sea, from south-west Ireland to Guinea, off the islands of the Azores, Madeira, and Canaries and in the Mediterranean Sea, and in the Indo-Pacific (Nesis 1987). *H. hawaiiensis* has been caught from 150 to 700 m, and shows diel vertical migration, being found at 150–200 m in darkness (R.E. Young 1978).

Life history

Naef (1923) described a living specimen (about 1.5 mm ML) that was quite transparent, except for a few regularly arranged chromatophores; its tentacles were coiled as in the adult but its club had few suckers, and its luminous organ was already well developed. A large number of young *Heteroteuthis dispar* were captured near the Canary Islands, between 800 m and 960 m, among which was one of only 1 mm ML. These animals allowed Clarke (1969) to illustrate the changes with growth up to 22.7 mm ML. Specimens of up to 22 mm ML have been captured in the Mediterranean in bottom trawls between 319 and 1588 m (Villanueva 1992b, 1995). Relini (1995) has re-evaluated the distribution of *H. dispar* in relation to its life history, and the evidence available suggests that hatching takes place on the bottom in the slope zone (at >1000 m). The hatchlings then ascend to 800–400 m, where they live and grow for about a year while undergoing nocturnal movements towards the surface. In the second year the vertical movements broaden to include the deep slope waters and presumably contact with the bottom for at least the spawning females. One mature female was found with ovarian eggs in two developmental stages, so the eggs are presumably laid in batches as they mature. The maximum recorded size for *H. dispar* is 29 mm ML (Rees and Maul 1956).

Photophores

Heteroteuthis dispar gives a 'beautiful play of colours ... Despite its plump form ... is of very handsome appearance, especially because of its constant and vivid movements', and 'The ventral shield regulates the luminous effect' (Naef 1923). The body and arms have a silvery metallic sheen, the chromatophores are reddish-brown, and the fins translucent (Clarke and Herring 1971). The photophore of this species is large and sac-like and opens to the exterior by a pair of pores on its anterior face. It is surrounded by a thick layer of reflecting tissue made up of iridophores, whose structure differs considerably in the proximal, lateral and distal regions (Fig. 6.7). Two modes of luminescence were seen. In one a steady blue-green glow illuminated much of the ventral surface of the animal. In the other the luminous contents of the organ were ejected through the two anterior papillae,

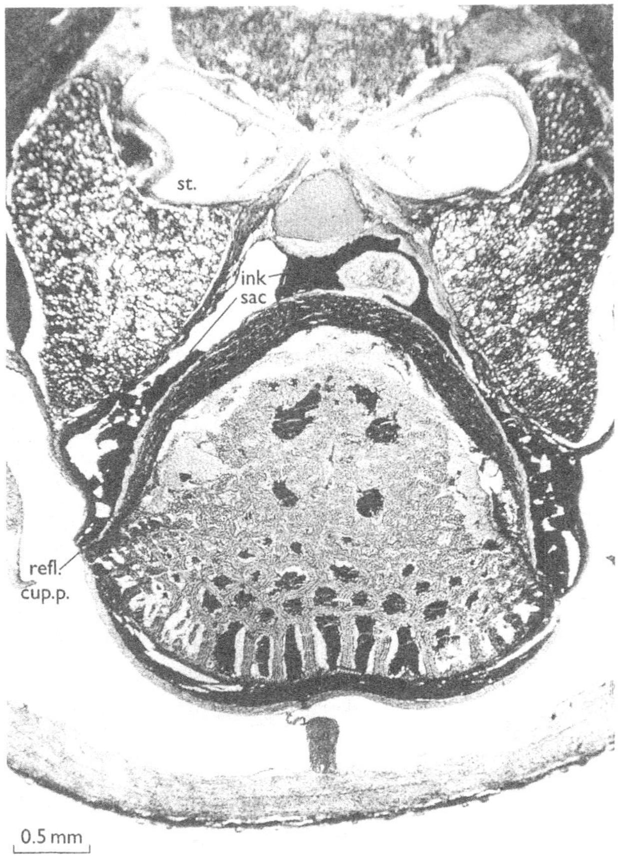

Fig. 6.7. *Heteroteuthis.* Transverse section of the light organ.

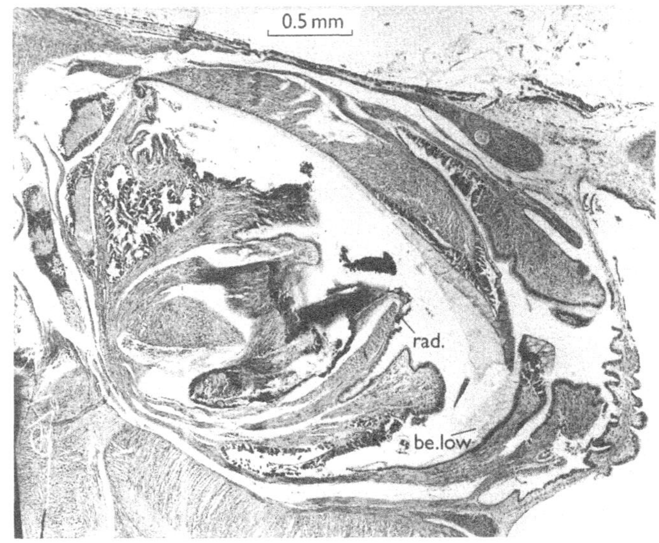

Fig. 6.8. *Heteroteuthis dispar* (11 mm ML). The buccal mass in sagittal section.

and appeared as a 'shower of "sparks" in the water'; the material was initially brightly luminous but its intensity decayed rapidly. On cutting open the light organ of a non-luminous animal the tissue remained dark for several seconds and as the material oozed out it began to glow. The cavity of the luminous gland contained a population of spherical vesicles, 0.3–10 mm, surrounded by dense interstitial material, but attempts to culture bacteria from the luminous material and the ink sac, were unsuccessful, although 'bacteria-like particles' were present (Dilly and Herring 1978). The emission maxima of the light organs lie between 485 and 490 nm (Herring 1983).

In *H. hawaiiensis* the photophore is partly embedded in the ink sac (Fig. 6.1b). The organ has a posterior, cup-shaped reflector surrounding a core, which contains branching glandular sacs embedded in a matrix of translucent tissue, and an anterior cap of iridophores. The glandular sacs empty via four ducts in two anterior papillae. The anterior cap is covered with a thick layer of muscle, and muscular bands pass to the photophore from anterior, lateral and posterior positions. The anteroventral surface of the mantle is highly modified to form the ventral shield. This has thick circular patches of numerous thin, disc-shaped iridophores, and over each patch there is a central chromatophore. Much light from the photophore passes through the ventral shield; effectively spreading it, as with frosted glass, over a surface approximately twice the width of the photophore. This heteroteuthine can luminesce continuously for 30 min, and intermittent observations indicate that the luminescence may last much longer (R.E. Young 1977). When handled the animal altered the light intensity widely from almost extinguishing it at one moment to turning it back on in the next, but the mechanism of control remains obscure. Extracts of the light organ of *H. hawaiiensis* tested positive for bacterial luciferase, and ultrastructural examination revealed bacteria-like bodies 2–3 mm in diameter (Leisman *et al.* 1980). The bioluminescence of this sepiolid may function for counter-shading during moonlit nights and ventral counter-shading at twilight; these are times when the animal is in the water column and so is exposed to potential predators rising from below (R.E. Young 1977). Experimental studies demonstrated that this species is capable of adjusting its luminescence to match overhead illumination (R.E. Young and Roper 1977). Some light from the photophore passes forward through the funnel to be reflected downwards by iridophores to conceal an obliquely oriented animal (R.E. Young 1977).

Buccal mass

The buccal mass is somewhat elongated. The beak is deep near its tip and not tanned in this specimen (Fig. 6.8). The muscles of the buccal mass are densely organized. The muscles of the bolster, or lateral support of the radula, contrast with the non-staining 'jelly like' material of the rod, with its scattered nuclei (Fig. 6.9). The radula has well-developed teeth and the ribbon extends some distance beyond the bending plane.

Receptors and central nervous system

The extraocular photoreceptor organs of *H. dispar* are each arranged in dorsal and ventral sets of photosensitive vesicles. They form 0.2% of the total brain volume (Maddock and Young 1987). The dorsal set consists of tiny sacs of vesicles below the peduncle lobe (Fig. 6.10), and the ventral set lies over the base of the funnel. A string of even smaller sacs joins the two sets of vesicles. The photosensitive vesicles are connected to the peduncle lobe by a bundle of fine nerve fibres. The position of the dorsal set of vesicles would allow them to receive light from above, and the ventral set could detect light emitted from the photophore.

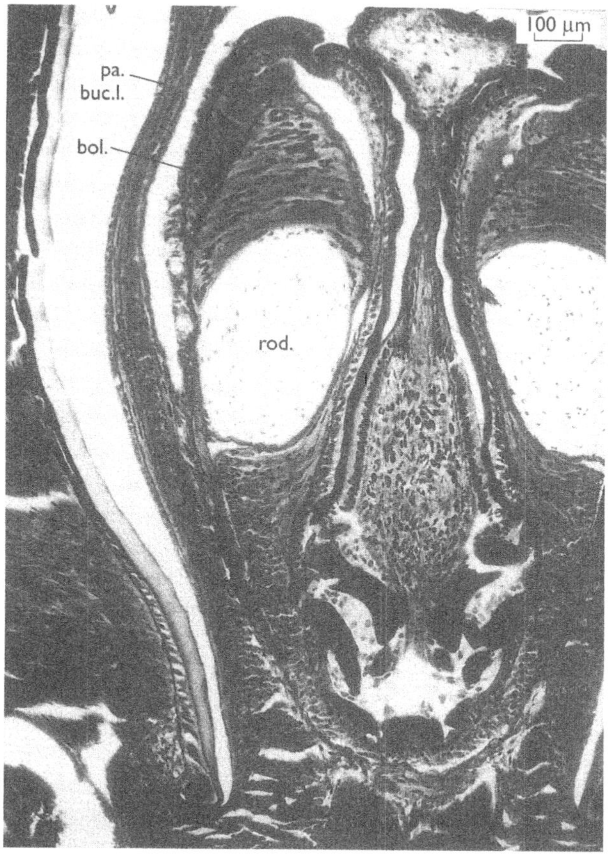

Fig. 6.9. *Heteroteuthis.* Transverse section to show the radula and its lateral supports, and bolsters, with the large poorly stained rods.

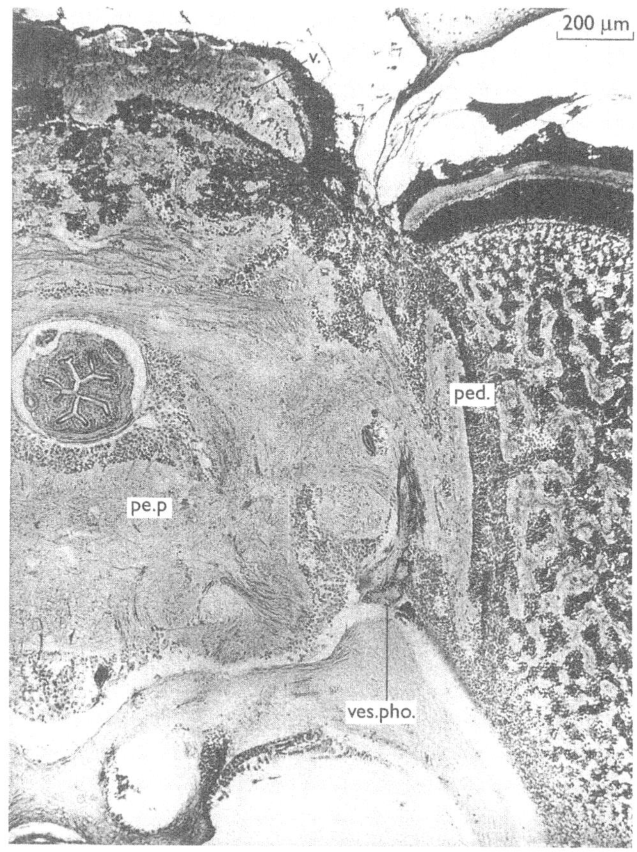

Fig. 6.10. *Heteroteuthis.* Transverse section of central nervous system showing the vertical lobe, part of the optic lobe and photosensitive vesicles.

The optic lobes form 293% of the total brain volume (Maddock and Young 1987). In the central nervous system the inferior frontal lobe system forms 3.4%, and the vertical lobe system, forms 17.8% of the total brain volume, largely due to the size of the superior frontal lobe (Table 2.6). No first-order giant cells or fibres could be seen in the sections, but there are isolated fibres in the mantle connective, some being 10 μm in diameter. In each stellate ganglion (both are flattened) there is a single fibre of 20 μm at the centre of the neuropil (Fig. 6.11). There are fibres of about 15 μm in the stellar nerves so it is possible that a giant-fibre system is present.

The statocyst is short and broad, and is little obstructed (Fig. 6.12). Hamuli 1, 3, 4, and 5 are present but only hamulus 4 is large (Young 1989).

6.2.2 *Nectoteuthis*

Nectoteuthis pourtalesi has been captured near the Antilles, in the West Indies, at a depth of 360 m (Verrill 1883).

Verrill (1883) described the animal after preservation as reddish brown on the dorsal and lateral surfaces, and the ventral shield dark purplish-brown from the innumerable, minute chromatophores. A light organ is present.

6.2.3 *Iridoteuthis*

Iridoteuthis iris was described by Berry (1909) from a juvenile collected near one of the main Hawaiian Islands. Forty-five specimens, including adults, were captured near the southern Emperor seamount, to the north-west of Hawaii, at 150–300 m depth (Harman and Seki 1990). They were apparently identical to the single juvenile described by Berry (1909). One other species, *I. maoria*, is found in the South Pacific (Dell 1959).

The features characteristic of the genus include a large mid-dorsal bulge on the mantle. The animal has a ventral shield about 80% of the mantle length, and this extends far forward to almost cover the large funnel. Its fins are broad, and their posterior edges surpass the posterior edge of the dorsal mantle. The head is broad and is dominated by the eyes, which can have a diameter equal to the mantle length. There is an internal photophore which is large with two prominent pores. The maximum sizes recorded are a male of 24.1 mm ML and a female of 28.4 mm ML (Harman and Seki 1990).

6.2.4 *Stoloteuthis*

Stoloteuthis has been captured on both sides of the Atlantic Ocean and in the Mediterranean Sea, in the Indo-Pacific, and around the Hawaiian Islands, Japan, and the Kerguelen Islands at depths of 170–700 m (Verrill 1878, 1881, Berry 1911, Haneda 1955, Nesis 1987, Relini and Massi 1991).

Stoloteuthis leucoptera is brown on the dorsal surface of the head and mantle, the ventral shield, arms IV, and the web between arms I

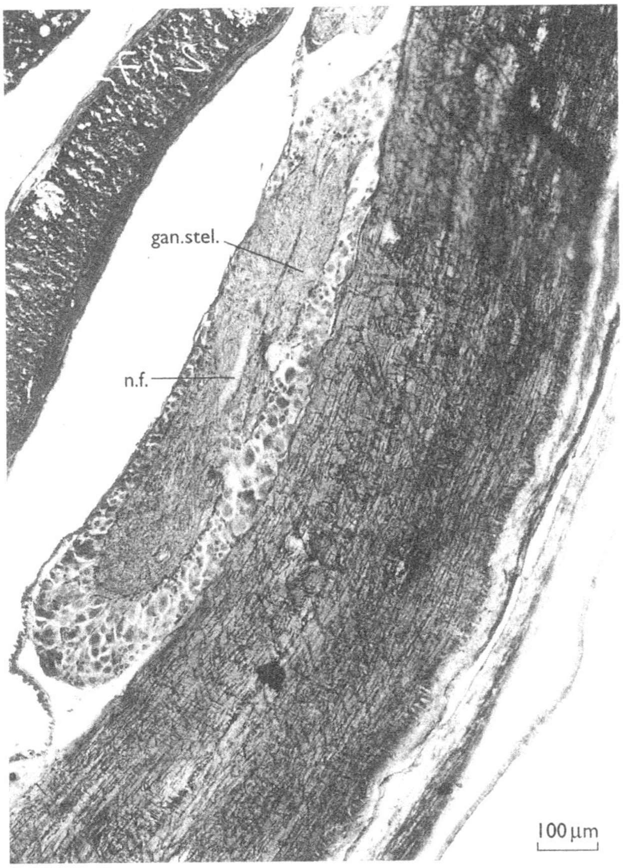

Fig. 6.11. *Heteroteuthis*. Transverse section through the stellate ganglion, showing large nerve fibre at the centre.

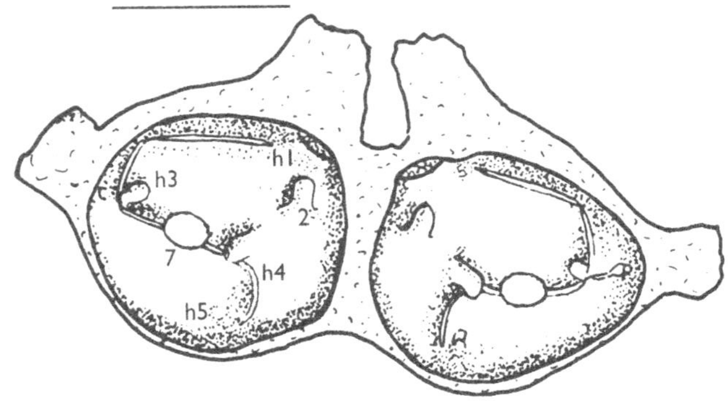

Fig. 6.12. *Heteroteuthis*. The statocyst drawn from above (scale bar 1 mm).

Table 6.1. The estimated density of sepiolids caught in the north-western Mediterranean in trawls between 147 and 1156 m depth (Villanueva 1995)

Subfamily/species	Estimates of number/km^2
Rossiinae	
Rossia macrosoma	60
Neorossia caroli	24
Heteroteuthinae	
Heteroteuthis dispar	98
Sepiolinae	
Sepiola neglecta	29
Rondeletiola minor	2109
Sepietta oweniana	2338

and III. The margin of the shield is blue, as is part of the head and dorsum, and ventrally there are blue spots below the eyes. The fins are translucent (Relini and Massi 1991). These authors comment that 'When alive in spite of their small size, they are outstanding for their silvery and blue hues'. Earlier Verrill (1881) described a living animal as 'very beautiful... owing to the elegance and brilliancy of its colours and the gracefulness of its movements. In swimming it moves its fin in a manner analogous to the motion of the wings of a butterfly'. Specimens of up to 17 mm ML have been caught, and a male of 11.5 mm ML was found to be mature (Relini and Massi 1991).

6.2.5 *Sepiolina*

Sepiolina nipponensis is found in warm and tropical waters and has been recorded from Japan and the Philippines. It is benthic and inhabits the lower continental shelf where the bottom is of mud or *Globigerina* ooze (Voss 1963, Okutani 1967). Large populations are occasionally trawled in a bay along the coast of Japan (Okutani 1995).

A small, round photophore lies on the ventral surface of the ink sac and it has two raised pores. This sepiolid has been seen to express brightly luminescent clouds into the sea; luciferase has been found in the mantle and coelenterazine in the digestive gland, suggesting chemical luminescence (Shimomura *et al.* 1980). Specimens of up to 25 mm ML have been found in the stomachs of lancet fish (Okutani and Kubota 1976).

6.3 Sepiolinae

The sepiolines inhabit coastal waters of the continental shelf and slope at depths of less than 5–600 m. They are nektobenthic and live where the bottom is of sand or mud, spending much time buried in the substrate, and yet they are also caught in midwater (Bello and Biagi 1995). There is some evidence of aggregation in the sepiolines since 5000 *Rondeletiola minor* were captured in a single trawl (Naef 1923), and 1250 *Sepietta oweniana* in one trawl lasting just 1 h (Belcari *et al.* 1989). Estimates from bottom trawls indicate that these two species occur in much greater densities than other sepiolids (Table 6.1) (Villanueva 1995).

6.3.1 *Sepiola*

Sepiola (Fig. 6.1c) lives in the coastal waters of the eastern Atlantic Ocean, off Iceland and south to Senegal, off Norway in the North Sea, in the central and western Mediterranean, and off Japan, the Philippines, and Korea in the Pacific Ocean (Okutani 1968, Knudsen 1981, Guerra 1982a, Relini and Bertuletti 1989). It is found in shallow waters mostly down to 35 m, but sometimes even to 450 m, where the bottom is of sand or mud; it undergoes vertical migrations, being also caught in midwater (Mangold-Wirz 1963, Bello and Biagi 1995).

Sepiola robusta is one of the best known species of the genus. It is dark in colour and closely matches the sandy mud. In this it burrows rapidly by blowing sand up from below the mantle with vigorous jets from the large funnel, the propulsive effects of the jet being counteracted by movements of the fins. These activities continue until sand almost wholly covers the animal, after which it remains motionless, except for respiratory movements, for much of the day. At night it enters the water column and can be captured in pelagic nets (Boletzky 1983b).

Life history

During mating the male takes the female by the neck and transfers spermatophores with its left, dorsal, hectocotylized arm to the bursa copulatrix. The spermatophores are released during spawning. One female (19 mm ML) laid some 50 eggs over a period of 30 days. The eggs, globular in shape (3.5–4.5 mm in diameter), are laid in water 10–60 m in depth, and are usually attached in a single layer to sea grass or ascidians, or in empty bivalve shells. Females die a few days after the end of spawning and males a week or so later. The eggs take 1 or 2 months to hatch at temperatures of 20 °C and 15 °C respectively. Its life cycle is of 6–9 months duration (Boletzky 1983b).

At hatching the young (about 2.2 mm ML) will bury itself and remain hidden in the sand during the day, like the adult. Chromatophores are present and the pattern and density is intermediate between entirely benthic and entirely planktonic forms (Fioroni 1981). At 1–2 days of age the hatchling will attack a mysid, first swimming towards it, hovering by means of the fins and funnel, while fixating the prey visually; attack follows, with a tentacular strike, and capture with the suckers of the club.

Photophores

The ventral light organ is already developed at hatching (Boletzky 1983b). It is a prominent feature in sagittal sections (Fioroni 1981). In *Sepiola robusta* and *S. atlantica* the light organs are paired, bean-shaped bodies, up to 30% of the mantle length, partly embedded in the ventral surface of the ink sac. Each organ has deeply folded epithelial pockets which open into the mantle cavity through a papilla halfway along the lateral margin, and enclose dense concentrations of luminous bacteria, *Photobacterium fischeri* (Herring *et al.* 1981). The whole structure is contained within a very thick reflective layer, formed of many sinuous, multiple layers of platelets. The middle of each organ is mostly obscured by an overlying portion of the ink sac in large animals. In living animals the organs are yellowish in colour, probably derived from the yellow pigment associated with the bacteria. The organs are overlain by a transparent layer, or 'lens', which appears to be composed entirely of muscle, with nerves and blood vessels present particularly along the inner margin. The light is directed ventrally and is not visible when the animal is viewed from the side or above. It is blue-green in colour to the dark-adapted eye. Luminescence in resting animals in constant darkness is steady but appears erratic when they swim. From below luminescence is seen as two horizontal bars, with brighter spots at their extremities. Brief illumination (5–10 s) of the animal increases the intensity of luminescence over the subsequent 10–15 s by up to 50%. This was followed by a decline over several minutes to the initial level of luminescence. Sepiolines have not been observed to emit luminescent particles from the mantle.

Receptors and central nervous system

The eyes are large and in *Sepiola rondeletii* they form as much as 50% of the body weight (Bullock 1965). The absorption spectrum of rhodopsin is 492 nm in *S. atlantica* (Muntz and Johnson 1978). Conjunctive eye muscles are present, and presumably allow binocular vision (Budelmann and Young 1993). In transverse section the eyes occupy two-thirds of the width of the head and the optic lobes the remaining third (Fig. 6.13) (Messenger 1981); the latter form 232% of the total brain volume in *S. rondeletii* (Maddock and Young 1987). The lens is formed of two distinct parts originating from the lentigenic body at different times, and separated by a septum (Willekens *et al.* 1984). The septum, formed by concentric zones of radially oriented elongated elements, is a distinct entity within the lens, and may form the barrier of high resistance observed between the two parts of the lens (Jacob and Duncan 1981, Messenger 1981).

The central nervous system of *Sepiola* was described by Dietl (1878), Schkaff (1914), and Thore (1939). It differs from that of *Sepia* in the wide separation of the brachial lobe from the pedal lobe, and in the absence of a system of giant nerve fibres.

The inferior frontal lobe system of *S. rondeletii* (Fig. 6.14) forms 4.6% of the total brain volume (Maddock and Young 1987). Large bundles of fibres run directly from the cerebrobrachial connectives to the interfrontal tract and cerebral tract. The vertical lobe system forms 10.8% of the total brain volume (Table 2.6). No first-order giant cell could be found in the several sets of serial sections available. The palliovisceral and magnocellular lobes form 24.7% of the total brain volume. The magnocellular lobe contains some large cells but no single outstanding one close to the entry of the macula nerve (Fig. 6.15). The commissure in the palliovisceral lobe contains several large fibres but none is especially large. Sections of the stellate ganglion and stellar nerves show no particularly large fibres, the largest being 9.8 μm in diameter. The mantle connective and fin nerves both

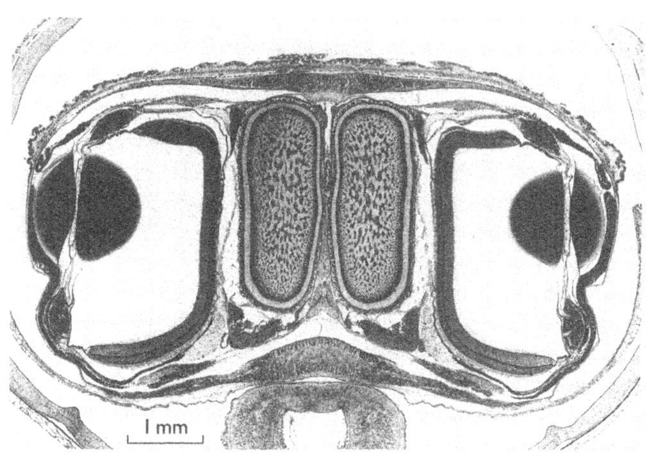

Fig. 6.13. *Sepiola.* The head in transverse section to show the eyes and optic lobes (photograph kindly provided by J.B. Messenger).

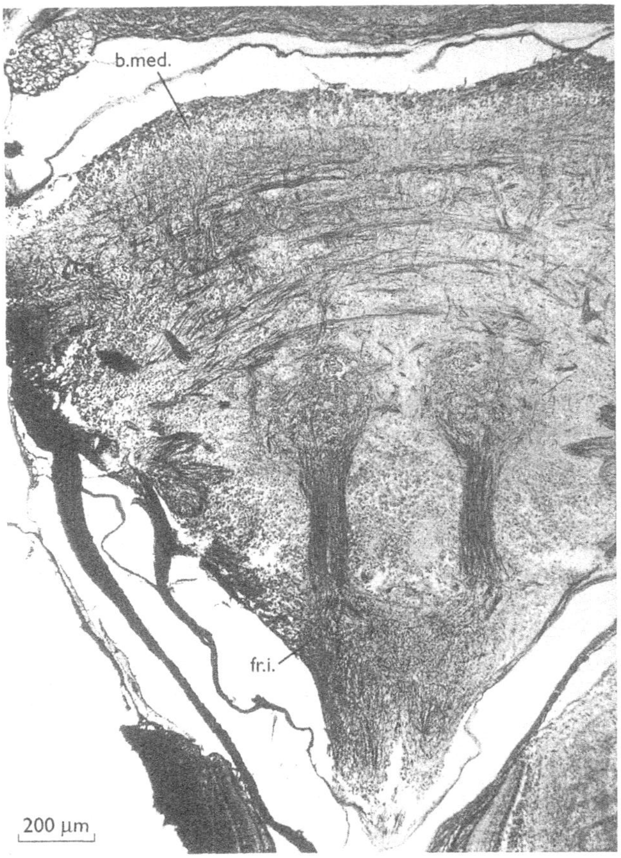

Fig. 6.14. *Sepiola rondeletii.* Horizontal section of the inferior frontal lobe.

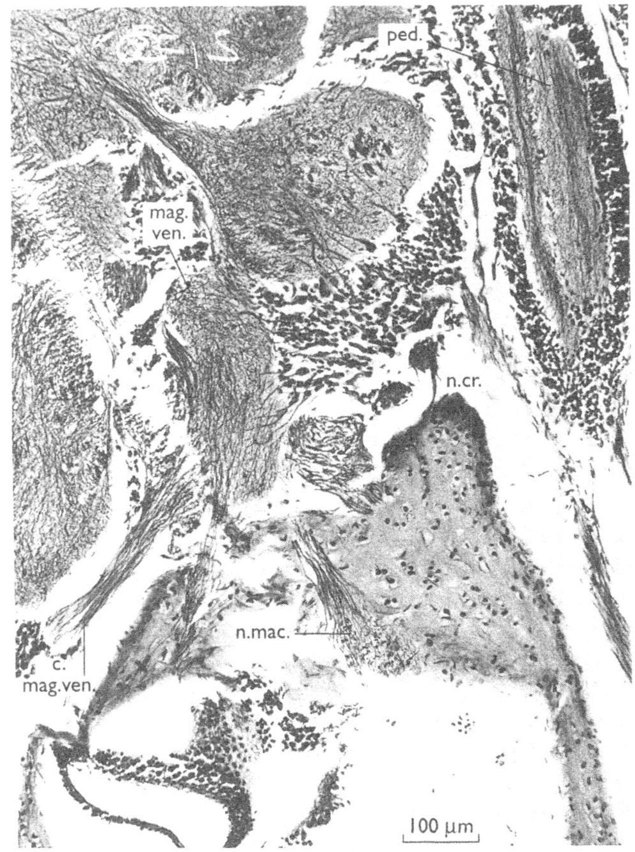

Fig. 6.15. *Sepiola rondeletii* (23 mm ML). Transverse section of the ventral magnocellular lobe.

contain fibres greater than 10 μm. *Sepiola atlantica* is capable of proceeding rapidly across a seawater tank by jet propulsion.

The extraocular receptor organ of *S. rondeletii* consists of about four minute sacs of photosensitive vesicles, lying between the peduncle and magnocellular lobes (Fig. 6.16).

The statocyst is broad, short and rather deep (Fig. 6.17) (Young 1989). Hamuli 1, 3, 4, and 5 are present but only two anticristae, one on the medial wall and the other, anticrista 7, in the roof. Measurements place *Sepiola* amongst the non-buoyant cephalopods (Maddock and Young 1984).

6.3.2 *Euprymna*

Members of this genus have an Indo-Pacific distribution mainly confined to the west Pacific between 10°S and 45°N. Here they are found mostly down to 60 m, although some go to depths of 200 m, where the bottom is of sand or mud (Voss and Williamson 1971). The distribution of the members of the genus have been mapped by Norman and Lu (1997a), who also give a generic diagnosis.

Life history

Euprymna scolopes is the best-known member of the genus and one for which we have some details of its life and habits. No precopulatory displays were seen when the animals were observed while mating in the sea or under laboratory conditions; instead the male swam rapidly to the female and grasped her from below (Singley 1982, 1983, Moynihan 1983b). Copulation lasted 35–80 min. Spermatophores were placed in the aperture of the oviduct via the hectocotylus. Fertilization presumably takes place deep in the oviduct since the female has no special storage organ. Eggs have been found attached to the underside of coral ledges adjacent to sandy areas of reef, in water less than 1 m deep. One female deposited 310 eggs over a period of 32 days. After hatching the young (1.6–1.9 mm ML) spend up to 30 days in the plankton, and at this age remain on the bottom during the day. The hatchlings prefer to take prey that is often large relative to their own size, using their arms to grasp the prey, until 25 days of age when the tentacles are evident. About 7 days after the end of spawning females usually die. Sexual maturity is reached in 2–6 months and the life cycle completed in less than 3 months, but it may be longer (Singley 1983, Hanlon *et al.* 1997).

Behaviour

Euprymna scolopes lives near the shore, spending most of the daylight hours buried in the sand and the hours of darkness foraging in the water column. It will back into a slope of sand and with its tentacles throw sand over any remaining exposed parts of the body or head. The sand adheres to the dorsal surface of the animal as it emerges to hunt and so forms a sort of carapace; when no longer needed the coat is shed as a unit, almost instantaneously (Singley 1982, Moynihan 1983b). In the sea escape responses include body pattern changes, jetting into midwater to hang motionless in different postures, and burial in the substrate (Anderson and Mather 1996). At all stages of life this sepiolid is active in the water column during the hours of darkness

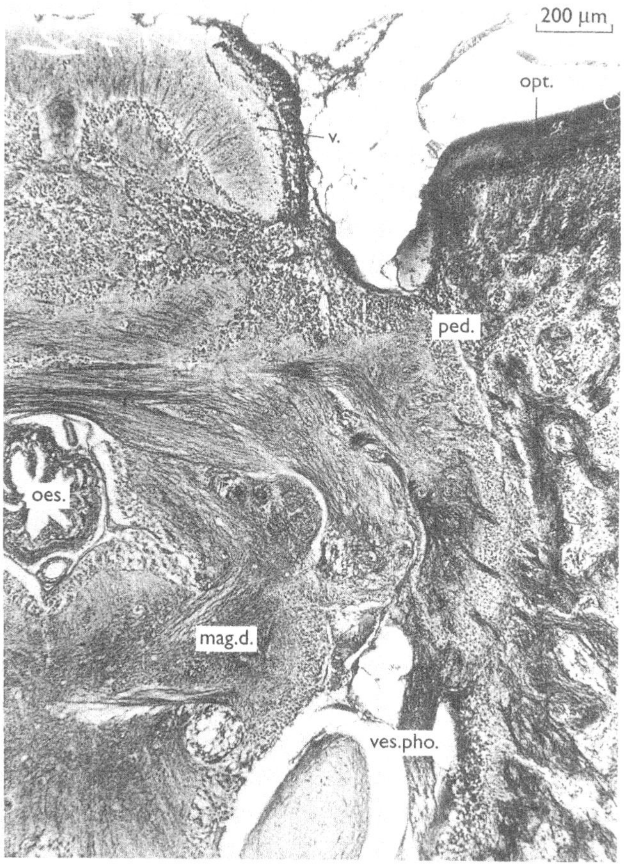

Fig. 6.16. *Sepiola rondeletii*. Transverse section of the brain to show the photosensitive vesicles.

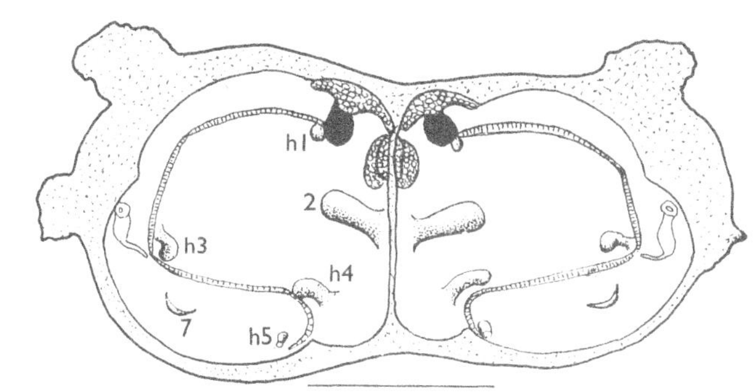

Fig. 6.17. *Sepiola rondeletii* (23 mm ML). Statocyst seen from above (bar 1 mm) (Young 1989).

(Hanlon *et al.* 1997). This pattern coincides with their predatory activity in the sea as 80% of those captured at night had fresh food, mostly errant polychaetes, in their stomachs, thus confirming their nocturnal habits (Shears 1988).

An Australian species, *Euprymna tasmanica*, shows similar behaviour (Norman and Lu 1997a). It is found from the surface to depths of 84 m, on sand and mud and often associated with sea grass beds. It is active at night but remains buried in the substrate during the day. When burying itself in the sand it uses arms III to rake grains of sand over the surface of the dorsal mantle and head, to which the sand adheres. This sand-coat remains attached if the sepiolid is flushed from the substrate in daylight, and would provide camouflage.

The ectoderm of *Euprymna* has ovate- and goblet-shaped secretory cells and polymorphic interstitial cells with a thick basement membrane below. The goblet cells are restricted to the dorsal epidermis, being especially abundant on the mantle and head with a few on the dorsal surface of the fins and arms, and they secrete an acid protein–polysaccharide. A complex system of nerve and glial cell processes lies between the secretory cells and the basement membrane, associated largely with the ovate cells, indicating that the latter have a rapid rate of secretion. The goblet cells secrete relatively slowly and Singley (1982), suggesting that, with their restricted distribution, they are important in sand adhesion, whereas the widely distributed ovate cells are concerned with the release of sand particles. De-adhesion occurs apparently instantaneously, but there is a refractory period lasting several minutes before sand again adheres to the dorsal surface and head of the animal. These observations support the hypothesis of the actions of the two types of secretory cell.

Euprymna scolopes is usually brownish in colour ranging from light to dark chocolate, and changing to assume the hues of its background (Moynihan 1983b). It displays disruptive, conspicuous, and occasionally intricate combinations of dark and light patterns to accompany its hunting and feeding activities. In captivity it will take shrimps by tentacular strikes and consume them tail first, the appendages and cephalothorax often being expelled.

Photophores

The visceral photophore of *E. scolopes* is associated with the ink sac and consists of a pair of lobular transparent lenses, a large bilobed reflector, and a saccular crypt housing symbiotic luminescent bacteria (Singley 1980, McFall-Ngai and Montgomery 1990). The organ ranges in size from 1.5 to 7.5 mm in length in specimens of 4–25 mm ML. All of the components of the photophore are present even in the smallest specimen examined. A light and ultrastructural study of the developmental stages of the light organ showed the increasing complexity of the crypt and its symbiont species. The bacteria cells are approximately 0.6 mm and are in narrow channels which consist of a single layer of epithelial tissue, surrounded by blood vessels and connective tissues; this forms the core tissue of the photophore (Ruby 1996). High densities of mitochondria are present in the cells adjacent to the bacteria (McFall-Ngai and Montgomery 1990). The emission of light is controlled by movements of both the reflector and the ink sac. The luminescent bacterial symbiont in this sepioline is *Vibrio fischeri* (Leisman in Hastings and Nealson 1981; Boettcher and Ruby 1990).

Wei and Young (1989) investigated the initiation of the symbiotic relationship in *E. scolopes* and their findings suggested that each generation acquires an infection from free-living bacteria rather than from the egg. The early events in the initiation and establishment of the *V. fischeri–E. scolopes* symbiosis indicate several steps in the process (Ruby and McFall-Ngai 1992). Access to initial inoculum of free-living luminous bacteria seems critical for the establishment of successful symbiosis. The symbiotic relationship is necessary for maximal bioluminescence, as after isolation the bacteria depress their luminescence (Boettcher and Ruby 1990). Development of the light organ requires the presence of the symbiotic luminous bacterium (Montgomery and McFall-Ngai 1994). The hatchlings initially lacked luminescence and no colony-forming units of *Vibrio fischeri* could be detected in the light organ. By day 5 the light organ contained 10^5 or more *V. fischeri* cells and the animals were luminous (Hanlon *et al.* 1997). A transient period of 12 h exposure of *E. scolopes* to symbiosis-competent bacteria is necessary to induce tissue regression in the light organ, over the next several days, and successful entrance into the light organ is necessary to induce morphogenesis (Doino and McFall-Ngai 1995). An overview of experimental studies of the cooperative association between *E. scolopes* and *V. fischeri* indicates that an elaborate dialogue occurs during the establishment of the symbiotic state (McFall-Ngai 1998). *E. morsei* is also reported to possess luminescent bacteria in its photophores (Haneda 1955).

Eyes

The eyes of *E. morsei* are sensitive to a vertical beam of polarized light and the animal showed preferential orientation in four directions relative to the electric vector (0°, 45°, 90°, 135°) (Jander *et al.* 1963). (Polarized light produces reflection from particles in the air or water that is most obvious at right angles to the electric vector, or *e*-vector.) When polarized light and light intensity patterns were presented simultaneously, *E. morsei* responded predominantly to the polarized light.

6.3.3 *Rondeletiola*

Rondeletiola minor is found in the eastern Atlantic from north-west Spain southwards to South Africa, and in the Mediterranean Sea, in depths of 30–700 m over sand or mud (Naef 1916, Knudsen 1981, Guerra 1982b, Relini and Bertuletti 1989). In most areas it appears to be rare and yet 5000 were caught in a single haul in the Mediterranean (Naef 1923).

There is a deeply embedded light organ on the ventral side of the ink sac (Bello 1995). A maximum size of 23 mm ML can be reached and both sexes are mature at 12 mm ML (Relini and Bertuletti 1989, Villanueva 1995).

6.3.4 *Sepietta*

Sepietta is found in the north-east Atlantic from Norway in the North Sea, southwards to Morocco and Madeira, and in the Mediterranean Sea (Naef 1916, Rees and Maul 1956, Mangold-Wirz 1963, Knudsen 1981, Guerra 1982b, Bergström and Summers 1983, Relini and Bertuletti 1989). It is an inhabitant of the upper slope of the continental shelf from 50 to 600 m in depth.

Sepietta oweniana is perhaps the best-known species of the genus. When alive it is a dense reddish-brown colour with iridescent green around the eyes. The juveniles swim rather awkwardly with mantle apex upwards and the arms trailing (Bergström 1985). Movement is by jet propulsion and synchronous beating of the fins. The funnel is relatively large and mobile. No light organs are found in this genus. Sizes are of up to 40 mm ML have been recorded (Bello 1995).

Life history

Near adults make inshore migrations. Mature animals are present throughout the year in the Mediterranean, and the minimum spawning size of *S. oweniana* is 18 mm ML in males and 20 mm in females (Relini and Bertuletti 1989). In the laboratory no courtship has been observed; instead the male grasps the head of the female, turns her upside down and holds her with his lateral arms (Bergström and Summers 1983). Spermatophores are deposited close to the opening of the left oviduct and spermatozoa are stored in the bursa copulatrix. The female spawns for about 2 days and dies within the next 24 h; the male dies after mating. Each female lays 130–200 eggs and attaches them to seaweed or to an ascidian, such as *Microcosmus sabatieri*. The eggs are spherical (2.7 × 2.3 mm) and grey-white in colour and are deposited to form masses in depths of 8–200 m. At 20°C hatching occurs in 30 days. The newly hatched young (2.5–2.8 mm ML) immediately takes shelter on the bottom. It will search day and night, swimming freely to capture prey, which is taken to the bottom for ingestion.

This behaviour changes with growth and development and at 10 weeks of age a diurnal pattern develops, and the young remains buried in the sand during daylight, only emerging at dusk to feed until dawn, from a position close to the bottom. A juvenile launches its attack from the substrate, using the fins to swim forward, while the funnel points downwards (Bergström 1985). Initially the tentacles are just discernible at the tip of the tight cone formed by the arms (Fig. 6.18). On approaching the prey the sepioline pales. On reaching the optimal distance the tentacles are ejected forward swiftly, the chromatophores expand, and colour returns. The tentacles are shot forward and brought back with the prey, to be enveloped in the arms and then moved to the buccal mass when the struggling ceases immediately. Stomach analyses show that crustaceans form the main part of its diet (Bergström 1985, Relini and Massi 1988).

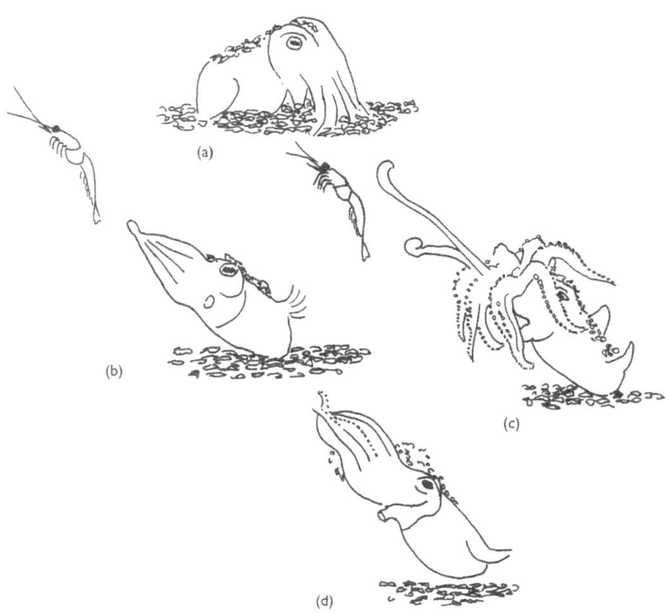

Fig. 6.18. *Sepietta oweniana*. Feeding behaviour. Clockwise from the bottom (a) resting, (b) positioning, (c) attack, and (d) returning to the bottom (Bergström 1985).

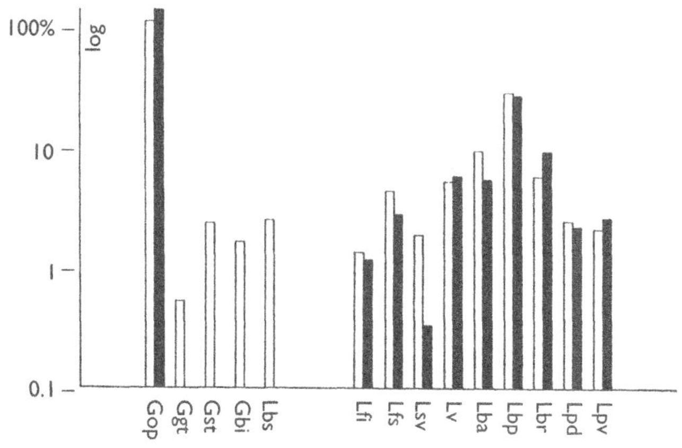

Fig. 6.19. *Sepietta oweniana.* Histograms showing the volumes of each lobe as a percentage of the whole brain. (See legend to Fig. 6.4 for abbreviations. Open bars, newly-hatched (Frösch 1971); filled bars, adults (Wirz 1959).

Central nervous system

The lobes of the central nervous system of *S. oweniana* were measured, using the same method, in newly hatched and adult specimens (Fig. 6.19) (Wirz 1959, Frösch 1971). Like *Rossia* the optic lobes are well developed in the former as is the higher motor centre.

6.3.5 *Inioteuthis*

This sepioline has been caught off the Andaman Islands, Japan, China, the Philippines, Indonesia, the Persian Gulf, and South Africa (Adam 1954, Voss 1963, Okutani 1967, Nesis 1987). It is littoral to sublittoral and is nektobenthic in habit.

There is no photophore. The dorsal surface of the head, arms and mantle has densely organized, purplish-brown chromatophores, minute to large in size. Specimens of up to 12.6 mm ML have been recorded (Verrill 1881).

7 Idiosepiidae

7.1 *Idiosepius*

Idiosepius (Fig. 7.1) is found in the coastal waters of Japan, the Philippines, Borneo, Sumatra, Maluki, Palau, and south of the Australian and African continents (Steenstrup 1881, Berry 1921, Voss 1962b, 1963, Roper and Hochberg 1987). Numerous specimens have been dipnetted along the mangrove mud banks in estuaries of the Townsville region of North Queensland, Australia (Jackson 1989). Adults have been found attached to seaweed in the littoral zone (Sasaki 1923, Moynihan 1983a), but young animals of up to 6 mm ML are caught in plankton nets (Okutani 1968).

The first species to be described of the only genus of the family was *Idiosepius pygmaeus*, and it was named by Steenstrup (1881) for its extraordinarily small size. It possesses several unusual and distinctive features including an adhesive organ on the dorsal surface of the mantle (Sasaki 1923), a toothed beak (Adam 1986), and stubs represent tentacles at hatching (Natsukari 1970). Steenstrup (1881) found the cuttlebone to be completely missing but noted that 'inside, below the mantle, there are solid strings situated in a rather peculiar way which might be said to function partly as a gladius, or parts of it'. Histological sections have subsequently revealed a fully developed gladius, reaching from the posterior to the anterior end of the mantle (Boletzky 1995). Females reach a mantle length of 22 mm and males 17 mm at maturity (Sasaki 1929, Jackson 1989).

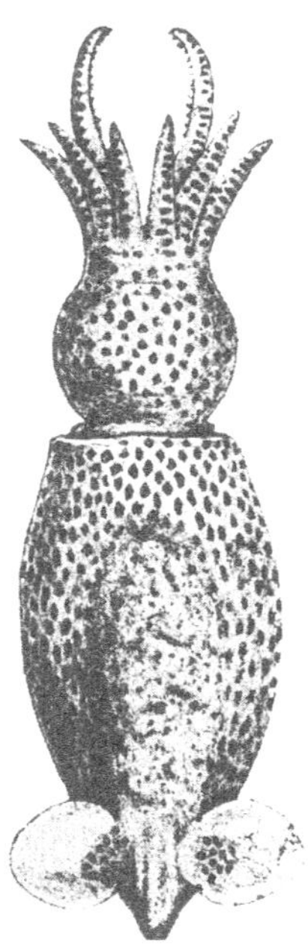

Fig. 7.1. *Idiosepius pygmaeus* (14 mm ML). Dorsal view of adult showing the adhesive organ (Sasaki 1923).

7.2 Behaviour

In the Palau Islands *Idiosepius* is found singly; it is active in daylight and floats horizontally at night (Moynihan 1983a). Movement is largely by jet propulsion, the small fins being used for fine control of movement. When undisturbed the animal is almost colourless, but it becomes dark on expansion of the many densely distributed chromatophores on the mantle. Various complex colour patterns are exhibited including stripes, bars, and false eyespots.

The distinctive adhesive organ is present on the dorsal surface of the mantle in both males and females, even in animals of only 2 mm ML (Okutani 1968). Animals brought from the sea at first swam freely in an aquarium, but as they approached seaweed each attached itself to the underside of a frond with its adhesive organ. All remained in this position for a long time, with arms held together above the head, while the head turned from left to right. When approached with small forceps they escaped swiftly. Attachment and detachment was effected rapidly, though the power of adhesion was considerable. The organ was not discernible in the living animal but could be seen after fixation as an area of longitudinal corrugations. The organ consists mostly of tall columnar cells, and these show cyclical activity and secretion of a sticky substance. Sensory cells may also be present (Sasaki 1929).

In captivity these animals made fast strikes with the tentacles, as do cuttlefish and many squids. The stimuli prompting such actions were not visible to the observer, but presumably resulted in the capture of prey. Another unusual mode of feeding was observed in which the wall of the tank was approached until the arms were in almost continuous contact with the surface. In this position the animals moved up and down and sideways while apparently 'nibbling' the surface. This activity increased after planktonic micro-organisms were released into the tank; at other times nibbling was intermittent (Moynihan 1983a). An unusual feature is the presence of small, irregular teeth along the rostral edge in both beaks. There is a relatively long crest in both upper and lower beak; the hood short in the former. The radula is unusual, too, as the central rhachidian tooth may have a simple or a bifid mesocone; seriation in these teeth is centripetal, and not centrifugal as in octopods (Adam 1986). The toothed beaks and bifid rhachidian teeth are perhaps involved in the so-called 'nibbling' behaviour.

7.3 Life history

Mature animals were observed laying eggs. The female attached herself to the tank and ejected a single egg from the funnel. The egg was received by the arms and stuck to the substratum by the tentacles. One egg was laid in about 30 s and the process was repeated. Each female laid one or two batches of eggs (1.6 × 1.4 mm), making a total of some 60 eggs. Each egg, enclosed in 8–10 layers of translucent membranes, takes 15–17 days to hatch at 18.5–22.6 °C. At hatching the young are up to 1.2 mm ML. The most notable feature of the hatchlings is the absence of tentacles: these form during postembryonic development (Natsukari 1970). In another laboratory study females laid 30–80 fertile eggs every 2–7 days, and the developmental stages from oviposition to hatching were followed (Yamamoto 1988). Lewis and Choat (1993) also observed that the female lays multiple batches of eggs and that fecundity is high in relation to the size of the animal. The newly hatched young of *Idiosepius* is planktonic and remains so until a mantle length of at least 6 mm is reached (Okutani 1968, Natsukari 1970).

Growth was followed in the thickness, structure, and organization of the mantle of juvenile females. The thickness increased and the production of muscle fibres found to be continuous; the muscle blocks also became larger as the animals grew in size. A predominance of small fibres and muscle blocks was found at the anterior end of the mantle, indicating this as the region of primary growth (Pecl and Moltschaniwskyj 1997).

An analysis of the rings of growth in the statoliths revealed that the youngest mature males were 22 days of age (Jackson 1993) and females were 13 mm ML and 59 days of age when mature. Thus *Idiosepius* is capable of passing through a number of generations in a year (Jackson 1989).

8 Pickfordiateuthidae and Loliginidae

The suborder is Myopsina is represented by Pickfordiateuthidae and Loliginidae, although the position of the first family is currently uncertain. These squids are found in coastal waters and are mostly neritic and semipelagic. The presence of a closed anterior chamber in the eyes of these squids separates them from the rest of the teuthids.

8.1 Pickfordiateuthidae

8.1.1 Pickfordiateuthis

Pickfordiateuthis pulchella, (Fig. 8.1a) is the only species and genus of the family, and it was first captured in the Florida Keys in less than 1 m of water (Voss 1953). It is quite common amongst sea grass as schools of 30–60 individuals occur in close formation, 1–3 m above the bottom (Hochberg and Couch 1971). It is known from the Florida Keys, the San Blas Islands, the U.S. Virgin Islands, the Netherlands Antilles, and further south in the coastal waters of Brazil (Voss 1953, 1956, Moynihan and Rodaniche 1982, Moolenbeek 1984, Haimovici *et al.* 1989).

The life history of this family is little known. One female (17 mm ML) had 27 spermatophores attached to the spermatophore receptacle on the buccal membrane, and the posterior end of its mantle cavity was filled with relatively large eggs (1.9 mm diameter), with one egg in the well-developed left oviduct (Voss 1953, 1956). Close inshore in Greater Lameshur Bay, St John (Virgin Islands), divers observed mate pairs in daylight close to the bottom, so egg-laying presumably takes place nearby (Hochberg and Couch 1971). This small myopsin squid reaches a mantle length of up to 22 mm.

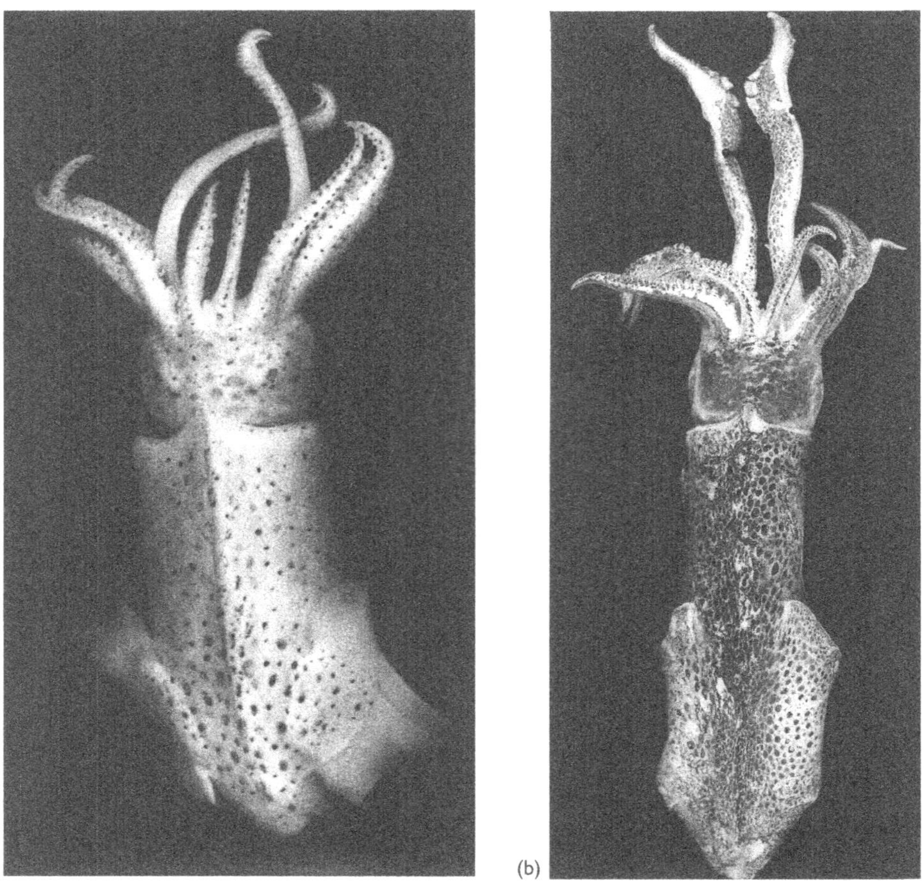

(a) (b)

Fig. 8.1. *Pickfordiateuthis pulchella* (21 mm ML) (a) and *Loligo vulgaris* (65 mm ML) (b).

The central nervous system closely resembles that of *Loligo*. The volume of the optic lobes is 278.5% of the total brain volume and the inferior frontal system 3.2% (Table 2.6). The vertical lobe system (27.2%) represents a larger proportion of the central nervous system than in most of the other cephalopods measured (Maddock and Young 1987). The peduncle lobe is small (3%), and the photosensitive vesicles, although present, were too small to measure. These features are clearly correlated with life in shallow water.

The first-order giant cells are typical and their axons fuse in the palliovisceral lobe, but retain some signs of their separate origins in our preparation. The stellate ganglion differs from that of *Loligo* (Young 1939) in that the giant-fibre lobe is less clearly demarcated, and contains mainly large cells whose axons fuse.

In the statocyst the hamuli and anticristae are all present, except at the first turn, but are mostly small. Anticrista 5 is placed far back (Young 1989).There is no complete horizontal canal. The statocyst is like that of a juvenile *Loligo*.

The central nervous system appears to be closely related to that of *Loligo*. The condition of the first-order giant fibres and of the statocyst are characteristic of young *Loligo*. This suggests that *Pickfordiateuthis* is probably a relatively recent offshoot from that genus, becoming sexually mature at a small size, appropriate for life in shallow water.

8.2 Loliginidae

Loliginids inhabit the continental margins of the shelf and upper slope areas of the world's oceans in tropical and temperate waters, where they occur from the surface to depths of 500 m with a horizontal distribution between latitudes 65°N and 60°S (Roper *et al.* 1984, Nixon in press a, in press b). They may be demersal, semipelagic, or nektobenthic in habit. Typically they exhibit diurnal, vertical movements aggregating near the bottom in daytime and dispersing in the water column in darkness. Several species school. Many migrate seasonally, retreating to overwinter in deeper waters and returning to shallower waters in spring. Most spawn in inshore waters. A few species enter estuaries where the salinity is 15–23‰.

There is in loliginids a thin transparent cornea covering the eye, fused with the orbit, and communicating anteriorly with the exterior through a minute opening, the lachrymal pore. This form of eye is characteristic of the members of the suborder Myopsina and separates them from other squids which belong to the suborder Oegopsina. Linnaeus (1758) included *Loligo* in his *Systema Naturae*, so the genus has a long history. The systematics of the family Loliginidae has for long been in disarray, but has recently been addressed (Sweeney and Vecchione 1998, Vecchione *et al.* 1998, F.E. Anderson 2000).

8.2.1 *Loligo*

Members of this genus (Fig. 8.1b) are found in coastal waters in the eastern and western Atlantic, the eastern and the north-west Pacific Oceans. The distribution is nearly continuous from southern Africa, around the north Atlantic, around South America and the west coast of the Americas to Japan (Vecchione *et al.* 1998). High-latitude barriers form gaps in the distribution of this temperate-tropical group of species. The coasts of the Indian and west Pacific Oceans are relatively rich in species (Nixon in press b). They mostly occur between the surface and 200 m but some species penetrate to depths of 500 m (Roper *et al.* 1984).

8.2.2 *Loliolus*

The two species of this genus occur along the coasts of the Indo-west Pacific (Vecchione *et al.* 1998). *Loliolus hardwickei* has a wide distribution and is found in estuaries where it can tolerate salinities of 24‰, and in shallow coastal waters to a maximum depth of 30 m where the bottom is sandy with patches of eel grass. The other species, *L. affinis*, is also widely distributed. The maximum sizes recorded are 33 mm ML for a male and 49 mm ML for a female *Loliolus hardwickei*, and 32 mm and 39 mm ML respectively for *L. affinis* (Lu *et al.* 1985).

8.2.3 *Lolliguncula*

The species of this genus are all found in warm, shallow inshore waters of North and South America with their low-latitude distributions centred at Panama, except *Lolliguncula mercatoria*, which occurs along the Atlantic coast of Africa (Vecchione *et al.* 1998). *L. brevis* occurs primarily in bays and in near shore waters to depths of 20 m, but it migrates to estuaries where the water is usually associated with low salinity and remains here during the warmer months of the year (Hixon *et al.* 1980). *Lolliguncula brevis* is found along the tropical western Atlantic coasts from Chesapeake Bay to La Plata, the Gulf of Mexico, and the Caribbean Sea (Nesis 1987). In the Apalachiola estuary the migration of *L. brevis* is influenced by temperature and salinity. Here in shallow waters of about 2 m depth these squids inhabit deep channels where the current velocity is high. Emigration takes place in the autumn and is associated with a fall in the temperature. Variation in population apparently follows the seasonal abundance of zooplankton and *L. brevis* may be the dominant invertebrate at certain periods (Laughlin and Livingston 1982).

The adult is small, with a maximum recorded mantle length of 90 mm (Hixon *et al.* 1980). At hatching the young has a mantle length of 1.8 mm and chromatophores are already present (McConathy *et al.* 1980) (the chromatophore patterns of this species are described later in this chapter). The development and morphology of the paralarva of *Lolliguncula brevis* has been described by Vecchione (1982). The paralarvae are planktonic and have been caught over the continental shelf in the coastal and estuarine waters of the Gulf of Mexico and are most abundant in waters with salinities of 26‰. These coastal waters have a range of dissolved oxygen concentrations. In some places there is less than 2 mL/L oxygen, and this squid inhabits such areas although the environment is considered hypoxic (Vecchione 1982, 1991). *L. brevis* is able to increase its capacity of oxygen extraction in hypoxic waters (Wells *et al.* 1988). The osmolality of the blood of *Lolliguncula brevis* is close to that of the surrounding seawater. Under experimental conditions it exhibits osmotic stress in salinities below 17‰ and dies after 48 h in water of 16.5‰ (Hendrix *et al.* 1981). This finding suggests that behavioural mechanisms are important in the regulation of the extracellular osmotic concentration, so that the animal avoids long periods in very low salinities. The haemocyanin of specimens from Chesapeake Bay is apparently insensitive to external changes in salinity of 11–35‰ (Mangum 1991).

8.2.4 *Sepioteuthis*

Sepioteuthis sepioidea occurs in the western Atlantic but other members of the genus are found in the Indo-west Pacific. All are coastal and inhabit warm, clear, tropical and subtropical waters to depths of 100 m. Most species are associated with coral reefs and in consequence are often called 'reef squids'.

The body is *Sepia*-like, stout, and slightly flattened dorsoventrally and the fins occupy almost the entire length of the lateral margin of the mantle. These features together with the large egg-size (more than 5 mm in length) and distinctive sucker dentition, separates *Sepioteuthis* from other loliginids. The largest specimen recorded is a female of 382 mm ML (Lu and Tait 1983).

Divers off the U.S. Virgin Islands, observed 4–6 *Sepioteuthis sepioidea* in close formation, and the integrity of each group was maintained as a school through characteristic postures and patterns of individuals. Displays of aggression prevented outsiders from joining existing schools (Hochberg and Couch 1971). If disturbed a deimatic response is elicited and the squid stops, flattens itself, and remains translucent except for the reddish brown rim to the edge of the fins, and two dark eyespots at the base of the fins. This squid shows disruptive coloration when swimming (Boycott 1965), and a wide variety of colour patterns (Moynihan and Rodaniche 1982).

During a 5 year period the behaviour and habits of *Sepioteuthis sepioidea* were observed in the sea, at intervals, from hatching to the adult stage (Moynihan and Rodaniche 1982). Newly hatched young are semiplanktonic during the day, and when slightly older they congregate in turtle grass flats. Medium-sized individuals and nonbreeding adults are nektonic. Most are gregarious, and they occur in large and structured groups, often associated with goatfishes, during the day, the exceptions being breeding adults and recent hatchlings. At night they scatter. Adults display numerous signals including complex body, locomotor, postural, and chromatic patterns, including ones in which the squid resembles the sea grass or the coral of its habitat and so avoids detection by predators; other displays are apparently for communication. Changes in social and spatial relations occur during reproduction when individuals exhibit courtship behaviour while still associated with large groups and then tend to form subgroups. Subsequently the subgroups become segregated, isolated, and scattered. Courting and breeding adults are attracted to coral reefs where they concentrate in water 1.5–8.0 m in depth.

During copulation the male uses the hectocotylus to 'strike' the female and deposit spermatophores in the seminal receptacle (Arnold 1965). The female may or may not be accompanied by the male as she lays the eggs in strings, to form clusters inside or under rocks, ledges, holes, or crevices, some 2 m down in water 2–3 m deep (Moynihan and Rodaniche 1982). Each string (about 38 mm long) usually contains 2–4 eggs (6 × 3 mm) in a spiral fold of jelly (Arnold 1965). The eggs take 30–42 days to develop, and they hatch in darkness. The hatchlings are 5.0 mm in mantle length and 8–9 mm in total length. At hatching the ink sac and black pupils of the eyes are conspicuous features of the otherwise semitranslucent body. Chromatophores are present in considerable numbers, relatively large in size and functional, although the chromatic patterns are simple at this stage.

Newly hatched *S. sepioidea* lives in water 0.2–1.0 m in depth, usually associated with floating vegetation above a bottom of sand, coral, and rubble mixture. It is semiplanktonic during the day. With growth the fins lengthen so that the young looks like a miniature adult and it can swim rapidly. Less time is now spent in the plankton. During the day the young congregate in shallow flats, 250–300 mm below the surface in water about 2 m in depth where there is turtle grass; they move into deeper water at night. Medium-size and prespawning adults are nektonic and go to depths of 30 m or more, notably avoiding areas with turtle grass, preferring open water at night and inshore waters during the day (Moynihan and Rodaniche 1982). Initially *Sepioteuthis lessoniana* has a basic form of attack which then leads to more specialized strategies until 4–6 weeks of age when the adult pattern is apparent (Fig. 8.11) (Kier 1996). *S. sepioidea*, irrespective of size, spends much time capturing small planktonic organisms and when feeding on schooling sardines its attacks are like those of *Loligo pealeii* (Kier 1982), and sometimes raises short arms I, as *Sepia* does (Messenger 1968). The fish, usually only a few centimetres in length, is captured with the tentacles and brought back to the mouth, bitten behind the head and held by the arms until the flesh and internal organs are devoured (Moynihan and Rodaniche 1982).

8.2.5 *Uroteuthis*

Uroteuthis has an Indo-West Pacific distribution from southern Japan to the Red Sea (Natsukari and Tashiro 1991, Vecchione *et al.* 1998). It is neritic although its vertical distribution is not known (Roper *et al.* 1984).

The first specimens were caught in the Philippines by Paul Bartsch in 1908, but only later described and named *Uroteuthis bartschi* by Rehder (1945). It has one pair of photophores on the ink sac. The maximum size recorded so far is a male of 223 mm ML.

Haneda (1963), off the island of Dumaguete in the Philippines, watched *Uroteuhis bartschi* and noted that as 'squids came floating up from the depths of a coral reef in clear seawater. . . . The light from the luminous organ above the ink sac shone out through the transparent body.' He found that if the squid was handled then light alternately appeared and disappeared. The light was emitted by two organs. These looked like pale lemon beans on either side of the ventral surface of the ink sac and each consisted of a luminous sac and opening, a lens, and a reflector. The luminous sac contained glandular tissue and bacteria, the latter in clusters or attached to the surface of the gland cells; a pure culture of the bacteria emitted a strong bluish light. A thin portion of the ink sac covers the outside of the reflector and part of the lens, and by contracting or expanding controls the passage of light from the photophore. This function of the ink sac was tested by replacing the ink with water, after which light from the luminous body shone continuously.

Each photophore of *Uroteuthis (Photololigo, Loligo) edulis (duvauceli)* is set in a cup-like reflector. The luminous sac has numerous discrete chambers containing bacteria and a narrow ciliated channel appears to connect the bacteria-containing chambers and open into the mantle cavity. On the ventral surface of the photophore there is a lens-like structure, with parallel lamellae. On the remaining sides the bacterial chambers are limited by the reflector, which separates the bacterial chamber from the ink sac (Pringgenies and Jørgensen 1994).

The life history of *Uroteuthis edulis* was reviewed by Natsukari and Tashiro (1991). It is one of the most oceanic of the loliginids found off the Japanese coast, although it does not go beyond the continental shelf. Tag and release surveys show it to migrate as some of the squids were recaptured after travelling more than 92.6 km, at average speeds varying from 0.18 to 0.55 km/h. The spawning grounds are in coastal waters, including open inlets and on the continental shelf, in depths of 30–40 m where the bottom of fine shell-sand is flat and slopes gently. The males carry spermatophores and females ovulate throughout the year. Observation in the laboratory show the male to approach the ventral or dorsal side of the female, and then swim in parallel before grasping the anteroventral margin of the female's mantle and inserting the hectocotylized arm into the cavity. The female laid egg-cases in clusters on a sandy bottom while holding her body in a perpendicular position. Each case (70–200 mm long) contains 100–400 eggs ovoid in shape (1.8 × 1.4 mm). Embryonic development takes 20 days at 20 °C. The newly hatched young has a mantle length of 2.0 mm and a total length of 4.2 mm. It swims actively in the middle depth of a tank. The juveniles feed on crustaceans, and when larger they mostly consume fish. Its life span is about 1 year. The maximum size recorded is 502 mm ML for males and 410 mm for females.

8.3 The loliginids

8.3.1 Life histories

Loligo vulgaris is a common squid along the coasts of the north-eastern Atlantic Ocean and the Mediterranean Sea. The maximum mantle length recorded is 540 mm in males and 340 mm in females of this species (Worms 1983). Its growth is rapid and sexual maturity is reached in 10–13 months. When mature it migrates vertically and horizontally in the Mediterranean (Mangold-Wirz 1963). The general pattern of migration is towards shallow coastal waters during the winter, with males arriving just before females. After copulation and spawning first the males and then the females move downwards to depths of 100 m or more. It is possible that some males survive and so may be involved in reproduction for a second or even a third season. The life span of this species is estimated as 2–3.5 years for males and 1.5–2 years for females (Worms 1983). Off the coast of South Africa the behaviour of *L. vulgaris reynaudii* was followed by tagging, filming, and applying 3D, radio-linked acoustic positioning (Sauer *et al.* 1992, 1997, Sauer and Smale 1993). These studies reveal that an apparently chaotic mass of spawning squids is well organized in both time and space. At dawn the squid move several kilometres towards the shore to small, well-defined sites near the beds of eggs on the sea floor. The squids make circling movements 50–150 m in radius, at average speeds of 180 mm/s for females and 140 mm/s for males, for several hours above an egg mass while the aggregation becomes denser as the males and females pair, mate, and lay eggs.

Naef (1928) established a series of stages in embryonic development, and these have been described for several species including *Loligo vulgaris* (Worms 1983, Blackburn *et al.* 1998) and *L. forbesi* (Segawa *et al.* 1988). Hatching is triggered in the laboratory by transition from light to dark (Paulij *et al.* 1990). The embryo is aided in its escape from the egg case by a hatching enzyme, a protease originating in the glandular epithelium of the organ of Hoyle (Denucé and Formisano 1982, Denucé 1985). In addition a transitory set of integumental cilia act to aid locomotor efforts when the young crosses the gelatinous envelope of the egg (Boletzky 1979).

In the laboratory several eggs of *Alloteuthis subulata* hatched overnight. The newly hatched squid (3.7 mm in length) was translucent except for the eyes, ink sac, and chromatophores (Fig. 8.2). The paralarva became streamlined and squid-like during movement by jet propulsion. It is capable of explosive bursts of speed and can move at up to 25× its body length per second (Gosline and DeMont 1985). The life span of this species in the English Channel is about 1 year (Rodhouse *et al.* 1988), whereas those inhabiting waters of the continental shelf off the West Sahara live only some 6 months (Arkhipkin and Nekludova 1993).

Numerous morphological, behavioural, and developmental features of *Loligo vulgaris* during its early life are summarized in Table 8.1. The chromatophore patterns of newly hatched animals have been described by Fioroni (1965), and paralarval features reviewed by Hanlon and his colleagues (1992). At hatching the oral surfaces of the upper and lower beaks have denticles (Boletzky 1971); these 'teeth' probably aid the capture and retention of prey as at this stage the arms are quite short. The development of predation in *Loligo opalescens* revealed that the initial type of attack led to more specialized strategies until, at about 4–6 weeks of age, the adult pattern of attack emerges (Chen *et al.* 1996). This indicates the sensitivity and development of the central nervous system to such encounters. By limiting the experience of the paralarva to one prey-type, its ability to capture other prey-types was subsequently much reduced.

Measurements of the volumes of various lobes of the brain of *Loligo vulgaris* at hatching and in adults show differences in their proportions (Fig. 8.3) (Wirz 1959, Frösch 1971). The higher motor centres are well developed at hatching, whereas the optic, vertical, brachial, and palliovisceral lobes are relatively small at this stage. Developmental changes in various lobes of the brain reflect changes in habits and behaviour.

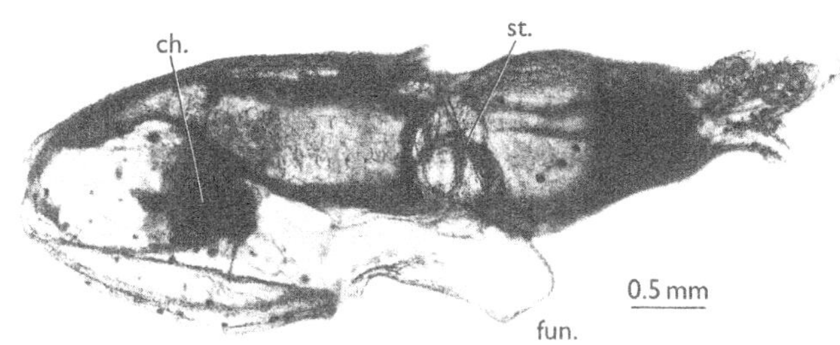

Fig. 8.2. *Alloteuthis subulata.* Newly hatched young to show its internal features when seen from the side; note the relatively huge funnel.

Table 8.1. *Loligo vulgaris*. A summary of various aspects of its early life (data from Manglod-Wirz 1963, Boyle 1983a, 1987, and this chapter)

Mature adults	Migrate-inshore shallow waters
Breeding period, Mediterranean	March–July
Spawning site	Spawning grounds at depths of 20–25 m
Bottom habitat	Sandy with rocks, covered with sand
Number of eggs per individual	3500–6000
Egg capsule length	2.7 mm
Eggs	Laid in strings (attached to substrate); each string 60–160 mm long with about 90 eggs
Egg-care	None
Embryonic developmental period	26–27 days at 22°
At hatching	
Mantle length	3.1 mm
Total length	5.0 mm
Time of hatching	At transition from light to dark
Hatchling	Translucent, with short arms
Paralarval type	Dispersive
Denticles on oral surface of beaks	Present
Buccal mass length at hatching	29% of mantle length
Habitat	In plankton for approximately 8 weeks
Response to light	Positive
Chromatophores at hatching	200 in total
Chromatic and body patterns	
At hatching	Variable
An juvenile	Variable
Locomotion of hatchling	Jet propulsion and steady swimming
Buoyancy	? No special mechanism known
Ink sac	Present and functional at or shortly after hatching
Prey in the sea	Small crustaceans
Feeding	Predatory, planktotrophic
Mode of capture in	
Posthatchling	With flared arms and tentacles and rapid jet forward
At 4 weeks	Typical adult tentacular strike
Paralysis of prey	? Not examined
Habits	
At hatching	Entirely planktonic for the first few weeks
As adults	Probably changes at 4–8 weeks of age
Receptors	
Eyes	Present
Extraocular photoreceptor organs	Present
Statocysts	Present
Olfactory organs	Present
Epidermal ciliated cells	Present
Lateral line analogue	Present
Life span	Female 1.5–2 years. Male 2–3.5 years

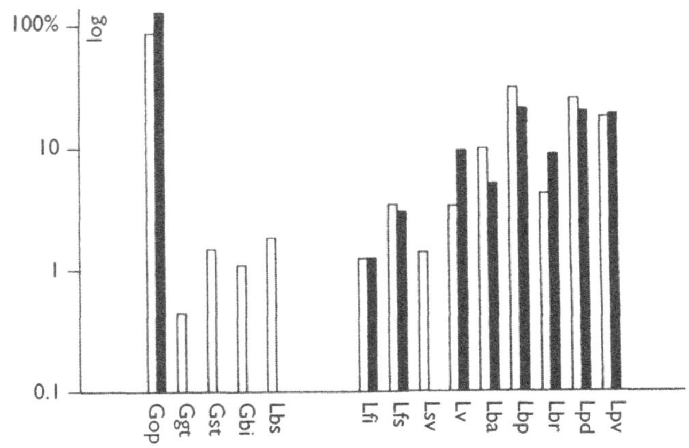

Fig. 8.3. *Loligo vulgaris* Various lobes of the brain shown as a percentage of the volume of the central nervous system, excluding the optic, plotted on a log scale. Open bars, newly hatched; filled bars, 'adult' (Wirz 1959, Frösch 1971). Gbi, inferior buccal ganglion; Ggt, gastric ganglion; Gop, optic ganglion; Gst, stellate ganglion; Lba, anterior basal lobe; Lbp, posterior basal lobe; Lbr, brachial lobe; Lbs, superior buccal lobe; Lfi, inferior frontal lobe; Lfs, superior frontal lobe; Lpd, pedal lobe; Lpv, palliovisceral lobe; Lsv, subvertical lobe; Lv, vertical lobe.

8.3.2 Mantle, fins, and locomotion

The mantle is supported internally by a chitinous plate, gladius, or pen, which is as long as the mantle. The gladius is a chitin–proteoglycan structure (42% chitin, 59% protein) in *Loligo vulgaris* (Hunt and Nixon 1981), and its ultrastructure is related to the molecular order; the chitin molecules are laid down in sheets and produce a laminated plywood-like material (Hunt and Sherief 1990).

The mantle cavity provides protection for the gills and for the orifices of the digestive, excretory, and reproductive systems and of the ink sac. Its main functions are for the movement of water into and out of the cavity for locomotion and respiration. During these activities the dimensions of the mantle can alter by as much as 21% in circumference and 27% in thickness, but with no change in length (Packard and Trueman 1974). Mantle movements during respiration and jet propulsion are the result of antagonistic actions of circular and radial muscles, in combination with the action of two layers of collagenous tunics. During locomotion water enters the cavity and the diameter of the mantle increases, mostly by expansion of the ventral region, and the water is then forcibly expelled through the funnel; the mantle diameter then decreases (D.V. Ward 1972). In resting ventilation the water flow over the gills is produced by movements of the collar flaps and associated muscles as the circular muscle fibres are inactive during respiration (Bone *et al.* 1994).

The structure and physiology of the mantle muscle has been described by D.V. Ward and Wainwright (1972) and Bone and his colleagues (1981, 1995). The mantle of *Lolliguncula brevis* (120 mm ML) has a layer of muscle, 2000 μm deep, sandwiched between two layers of collagenous fibres, an outer tunic of 100 μm and inner tunic of 25 μm (Fig. 8.4). The muscle of the mantle forms 35% of the total body weight of *Loligo vulgaris* of 350 g with a thickness of 5.2 mm (Trueman and Packard 1968). The muscle fibres are mostly circular and are divided more or less regularly into rectangular

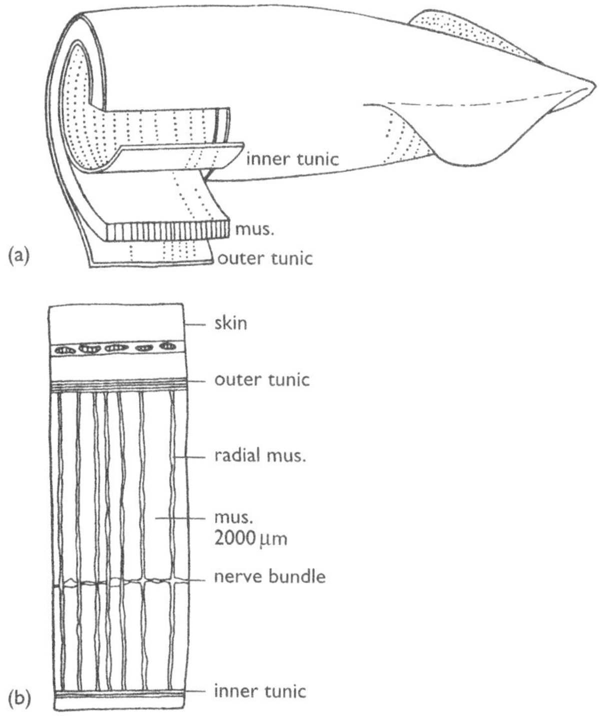

Fig. 8.4. *Lolliguncula brevis.* A diagram showing the main parts of the mantle (a), and the mantle wall in more detail (b) (D.V. Ward and Wainwright 1972).

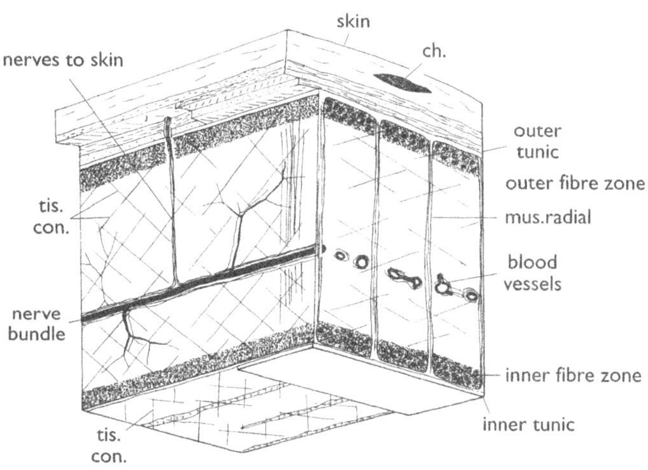

Fig. 8.5. *Loligo* to show the layers of the mantle and the disposition of connective tissue fibres in the muscle (Bone *et al.* 1981).

blocks by thin partitions of radial muscle fibres. The latter allow the mantle to become thin and the cavity to expand. These muscle fibres are opposed by the circular muscles which contract against the water within the cavity. Large amounts of collagen are disposed in oblique arrays in different planes of the mass of muscle itself (Fig. 8.5) (Bone *et al.* 1981). Collagen acts as an energy storage mechanism, the 'mantle spring' (Gosline and Shadwick 1983a, b).

The muscle fibres have a central core of mitochondria of varying diameter, surrounded by a cortical zone in which the myofilaments are in an obliquely striated array (Bone *et al.* 1995). The circular muscle comprises inner, middle, and outer zones, based on the number of mitochondria present in the central core of the fibres (Bone *et al.* 1981). The zones differ in colour, the outer is off-white to yellow, the middle white, and the inner band yellow (Mommsen *et al.* 1981). The middle layer is composed mainly of circular muscle with fibres of two types: the inner and outer bands consist of muscle fibres of up to 10 μm diameter with a large central core of mitochondria and the central band has muscle fibres of up to 5 μm diameter with only a small mitochondrial core, and the radial muscle fibres are of the mitochondria-poor type (Bone *et al.* 1981). These muscle fibre types also differ metabolically; the ratio of oxidative glycolytic enzyme activity is high in the inner and outer zones (aerobic fibres) but low in the central zone (anaerobic fibres) (Mommsen *et al.* 1981). It is probable that the giant axons supply only the mitochondria-poor fibres, and that these are also innervated by the small-fibre system since escape jets can be driven by either the small-fibre system alone or in combination with the giant fibres, but it is not yet clear how the small axons innervate the mantle muscle (Bone *et al.* 1995).

A study of the development of the circular mantle muscles of *Loligo opalescens* shows that the superficial, mitochondria-rich fibres and the central, mitochondria-poor fibres are present at the time of hatching (Preuss *et al.* 1997). The relative proportion of mitochondria-rich to mitochondria-poor fibres changes from 1 : 1 at hatching to 1 : 6 in a paralarva of 8 weeks. The changes are reflected in the muscular activity of these animals. In a hatchling of 1 week the mantle contractions have twice the relative amplitude and frequency of those in adults. This suggests that the superficial, mitochondria-poor, muscle fibres are important for locomotion at hatching but become progressively less so during the first 8 weeks of life as the central, mitochondria-poor muscle fibres become dominant for jet propulsion.

The mantle is a variable volume hydroskeleton surrounded by the inner and outer tunics of collagenous connective tissue fibres, forming alternate layers of right- and left-handed helical fibres. These fibres have a fibre angle of 27° relative to the longitudinal axis of the animal, and make the tunic rigid in the longitudinal direction. The inner and outer tunics provide for the insertion of the radial muscles of the mantle. The large amounts of collagen fibres are disposed in oblique arrays in different planes and act as an energy storage mechanism, the mantle spring, and actively allow the restoration of the original mantle volume (D.V. Ward and Wainwright 1972, Bone *et al.* 1981, Gosline and Shadwick 1983a, b, Gosline *et al.* 1983).

The speed of propulsion by jet is dependent on the velocity of the water expelled through the funnel (Table 8.2). This relies on the simultaneous maximal contraction of the mantle muscle innervated by axons in large stellar nerve bundles radiating from the stellate ganglia (Young 1939). Each stellar nerve contains one giant axon surrounded by smaller axons. The main nerve bundles pass through the mantle at the same level as the major blood vessels (Fig. 8.5). From the nerve bundles branches of the giant and smaller axons pass into and reach all parts of the mantle muscle. Branches of the giant axon

Table 8.2. Loliginidae. Speeds observed under laboratory conditions

	Body weight (g)	Body length (mm)	Speed Sustained (ms^{-1})	Burst (ms^{-1})	Direction	Source
Loligo vulgaris						
	0.003	6.0		0.16	Backwards	Packard (1969)
	3.0	65.0		1.28	Backwards	Packard (1969)
	100.0	200.0		2.08	Backwards	Packard (1969)
	160.0	280.0		2.10	Forwards	Packard (1969)
	240.0	320.0		1.76	Forwards	Packard (1969)
Loligo opalescens						
	40.0			1.00		O'Dor (1982)
	40.0		0.36			O'Dor (1982)
Loligo pealeii	100.0		0.60			O'Dor and Webber (1986)
Lolliguncula brevis						
	9.5	5.0		1.87		Finke et al. (1996)
				2.23		Finke et al. (1996)

supply the circular muscle fibres of the inner zone of the mantle. The conduction velocity rises slowly, approximately with the square root of the diameter of the nerve fibres; the largest fibres are very large, with some of 550 μm or more in diameter in *Loligo* (Pumphrey and Young 1938, Burrows *et al.* 1965). A single impulse arriving via the giant-fibre system results in an 'all or nothing' maximal response from every muscle fibre it reaches (Prosser and Young 1937, Young 1938). In this way the muscle fibres of the central zone of the mantle muscle respond as one unit to produce a maximal contraction and so empty the mantle cavity of its water in a single compressive stroke.

It has been widely accepted that the high-pressure escape jet was mediated only by the giant-fibre system, and the low-pressure respiratory movements were driven by a small-fibre system. However, more recent investigations suggest that both escape jet and slow movements involve the small-fibre system, and that the giant fibres serve to increase the amplitude of the muscle response (Bone *et al.* 1995). In *Loligo opalescens* highly coordinated activity of two motor pathways, associated with the giant and non-giant motor axons, produces all-or-none and graded contractions in the mantle muscle. The pathways appear to control different groups of circular muscle fibres. The central zone, innervated by the giant axons, generates an escape jet response, and the superficial layers of muscle fibres a slower sustained swimming, but the division of labour is probably even more complex (Gilly *et al.* 1996).

The control of jet propulsion was followed in embryos and hatchlings of *Loligo opalescens* and *Sepioteuthis lessoniana* (Gilly *et al.* 1991) (Table 8.3). The strong escape jets have been found to be driven by the small-axon system alone, with no giant axon activity, but as in adults, similar patterns of small-axon and giant-fibre control of jetting were seen. Small-axon control of jetting was present at embryo stage 26, but giant-axon control not until embryo stage 28 (embryo stages of Segawa *et al.* 1988); thus the two motor systems do not develop in perfect synchrony. The escape responses in embryos and hatchlings of *Loligo opalescens* are fast-start responses with latencies of less than the minimum value displayed by adults. Both giant and non-giant motor systems are thus functional in embryonic and hatchling squid, and both contribute to escape jetting. *Sepioteuthis lessoniana* differed, as it attained the adult-like motor patterns more quickly after hatching than did *Loligo opalescens* (Gilly *et al.* 1991).

Table 8.3. *Loligo opalescens*. Ontogeny of escape behaviour during embryonic development (embryo stages of Segawa *et al.* 1988, Gilly *et al.* 1991).

Embryo	Activities
Stage 25	Spontaneous, symmetrical mantle contractions begin
Stage 26	Eyes are a brilliant red Ink sac is visible Chromatophores on dorsal surface of head Electrical and mechanical stimulation of escape response is possible.
Stage 27	Ink is barely visible in ink sac Eyes are dark, but not black Edge of primary lid covers half of optic vesicle
Stage 28	Eye is completely covered by primary lid Mid-gut gland is visible Yellow chromatophores on arms Head retraction occurs during escape responses
Stage 29	Olfactory organ is visible as thickened disc Light flashes stimulate escape responses External yolk sac equal to length of arm Spontaneous vigorous jetting occurs
Stage 30	Hatching takes place No external yolk sac

Low-amplitude mantle radius changes occur at speeds of 0.1 m/s and are probably due to respiratory contractions; they could be driven by one of the two layers of aerobic muscle fibres. Larger contractions of the mantle occur at higher subcritical speeds, and may involve the second layer of aerobic muscle fibres. Stronger contractions at non-sustainable, supracritical, speeds need the layer of anaerobic muscle fibres as well. Escape jet thrusts differ from supracritical jets only by the added involvement of the radial muscle fibres producing hyper inflation of the mantle. At intermediate speeds the changes in pressure and jet velocity appear to depend mainly on changes in the orifice of the funnel (O'Dor 1988a,b). This author comments that the 'results indicate that squid swim continuously at variable speeds using the complex interaction of several muscle systems to produce at least four different "gaits"' and that 'a full range of mantle activity could be explained if the three layers of circular muscle and the radial muscles each received separate "all or none" innervation feedback'.

The loliginids depend largely on jet propulsion at all speeds, but these and other semibenthic squids also use their fins extensively, and all do so while hovering and manoeuvring (O'Dor and Webber 1986). The back and forth movements of these squids when schooling are due to fin movements (Bone *et al.* 1995). The fins of *Loligo forbesi* and *Sepioteuthis sepioidea* comprise a tightly packed array of muscles, essentially like those of *Sepia*, and in this regard resemble muscular hydrostats (Kier and Smith 1985, Kier 1989). Connective tissue fibres provide support for the fins and may serve as an elastic storage organ (Johnsen and Kier 1993). The maximal propulsive movements by jet are discontinuous. Analyses of ciné films and records of intramantle pressure of *Loligo vulgaris*, in a tunnel respirometer, revealed that at low speeds (0.2 m/s) the fins undulated continuously, two waves for each mantle contraction. At higher speeds a single fin wave per mantle cycle occurred, when the mantle was refilling, and may produce thrust. During rapid movements the fins are rolled tightly against the mantle (O'Dor and Webber 1986).

8.3.3 Shoaling and schooling

Many loliginids are found in groups in which the animals remain together for social reasons: such groups are termed shoals. Some species spend part, or all, of their lifetime with individuals in physical synchrony, in velocity and direction, and swim in parallel with strong cohesiveness: this is termed schooling (Hanlon and Messenger 1996). In the sea schooling behaviour has been observed in *Sepioteutis sepioidea* (Hochberg and Couch 1971) and *Loligo opalescens* (Hurley 1978). Large squids accompany schools of smaller squids in some species and are cannibalistic (Williams 1909, Hixon 1983, Summers 1983). *Sepioteuthis sepioidea* is also found in groups of mixed sizes, but there is no cannibalistic behaviour (Moynihan and Rodaniche 1982).

In the sea *Loligo opalescens* forms schools of individuals of uniform size, and their parallel organization was well developed in animals of about 120 mm ML, but less so in those of 70 mm ML (Hurley 1978). After being kept in darkness *Alloteuthis subulata* has been observed to re-form into a school on the restoration of illumination (Llewellyn 1984). Besides vision, both mechanoreception and the lateral line system may be involved in maintaining these squids in schools.

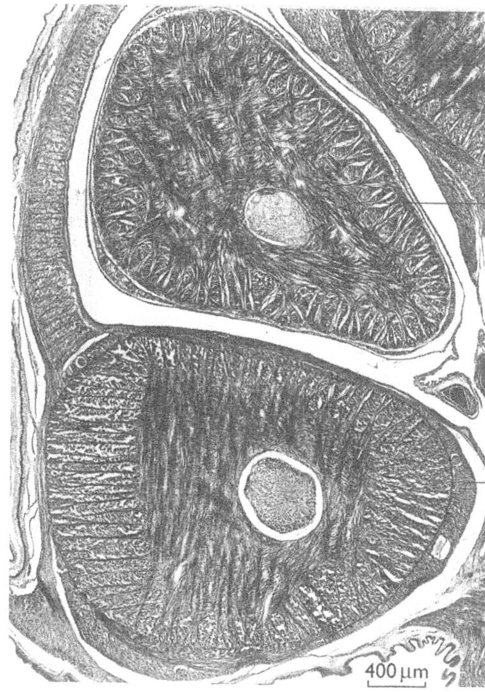

Fig. 8.6. *Alloteuthis subulata.* Transverse section to show the musculature of an arm and the stalk of a tentacle.

8.3.4 Arms and tentacles

The arms and tentacles consist largely of dense and complex muscles (Figs 8.6–8.8). They have been described from light and transmission electron microscope studies; the structure and function of the arms and tentacles have been found to differ (Kier 1982, 1985, 1991). The musculature operates on hydromechanical principles and serves as both the effector of movement and as a support system in the absence of any skeletal elements or fluid-filled cavities (Kier 1985, Kier and Smith 1985).

Each arm has an axial nerve cord with a loose connective tissue sheath. Transverse muscles surround the nerve cord, and lie in planes perpendicular to the long axis of the arm (Fig. 8.7). Orally and aborally the muscles extend out in trabeculae to insert on fibrous connective tissue sheets, and laterally on the inner member of a pair of oblique muscles on each side of the arm. Bundles of longitudinal muscle fibres run between the trabeculae formed by the transverse muscle fibres. A pair of oblique muscles, on each side of the arm, insert on the oral and aboral fibrous connective tissue sheets. The arms are manipulative but inextensible; the transverse muscle mass may resist the longitudinal compression caused by contraction of the longitudinal muscles which bend the arms (Kier 1982).

The stalk of the tentacle consists almost entirely of muscles. The axial nerve, with an artery running along one side, is surrounded by transverse muscles with trabeculae extending peripherally. Longitudinal muscles are arranged peripherally with surrounding circular, helical and superficial longitudinal muscles, enclosed by dermal epithelial tissue and epithelium (Fig. 8.8) (Kier 1982). The

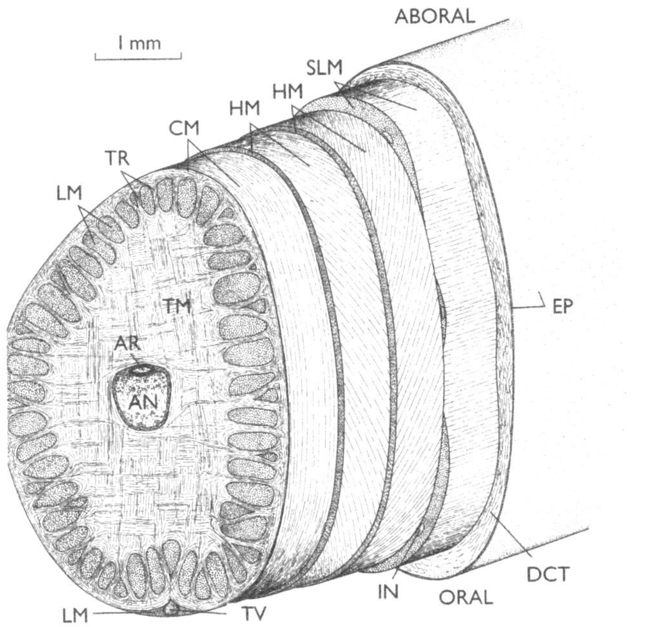

Fig. 8.7. *Loligo.* Diagram of a left arm (Kier 1982). ACT, aboral connective tissue; AN, axial nerve; AR, artery; BV, superficial brachial vein; CM, circular muscle; DCT, dermal connective tissue; EP, epithelium; HM, helical muscle; IN, intramuscular nerve cord; LM, longitudinal muscle; OCT, oral connective tissue; OM, oblique muscle; PM, protective membrane; SKLM, swimming keel longitudinal muscle; SKTM, swimming keel transverse muscle; SLM, superficial longitudinal muscle; SU, suckers; TM, transverse muscle; TR, trabeculae of transverse muscle; TV, superficial tentacular vein.

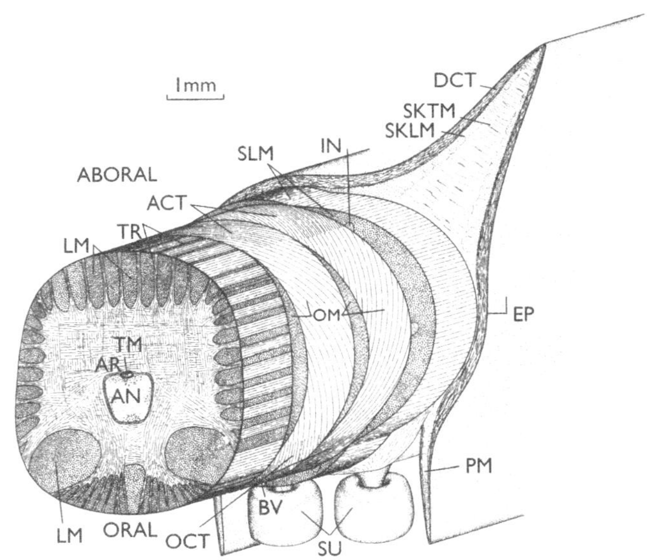

Fig. 8.8. *Loligo.* Diagram of left tentacular stalk (Kier 1982). See Fig. 8.7 for abbreviations.

longitudinal muscles, both in the deeper bundles and the superficial layer, are obliquely striated (Kier 1985, 1991). Similar fibres are found in helical patterns running either clockwise or anticlockwise, beneath the superficial longitudinal muscle layer. The helical muscles can exert torsion and may act as extensors or retractors during projection of the tentacle (Kier and Smith 1985). The transverse muscle is of specialized fibres, which show cross-striation and have short sarcomeres and thick filaments. This muscle has a high shortening speed, allowing for rapid elongation during extension of the stalk, and the club can reach the prey in 15–35 ms (Kier 1982). The specialization of the muscles has occurred mainly through changes in dimensions and arrangement of myofilament lattice and not through changes in biochemical properties (Kier and Schachat 1992).

The peduncles of the suckers of the arm and the tentacle are muscular (Fig. 8.9). The inner ring of the arm sucker has several wide, compressed teeth, pointing towards the buccal cavity, on one half and a semicrescentic plate on the other (Fig. 8.10a). The infundibulum has 4–5 rows of polygonal processes with projecting pegs (Nixon and Dilly 1977). The suckers of the tentacle are smaller and the inner ring has fairly sharp teeth around two-thirds of its diameter, and these point towards the mouth. A few polygonal processes surround the inner ring (Fig. 8.10b).

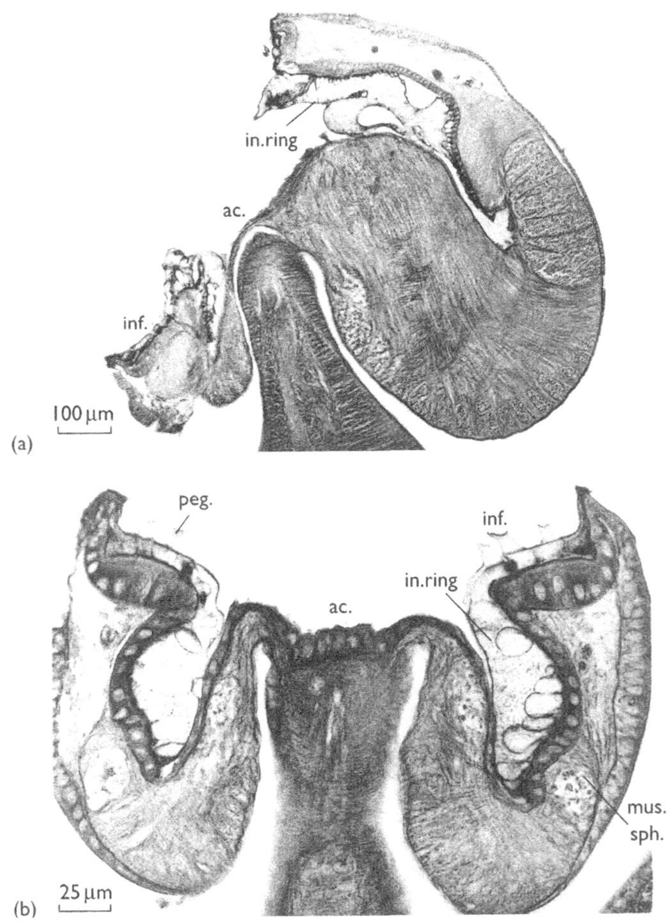

Fig. 8.9. *Alloteuthis subulata.* Sagittal sections through the sucker of (a) an arm, and (b) a tentacle (spirit blue).

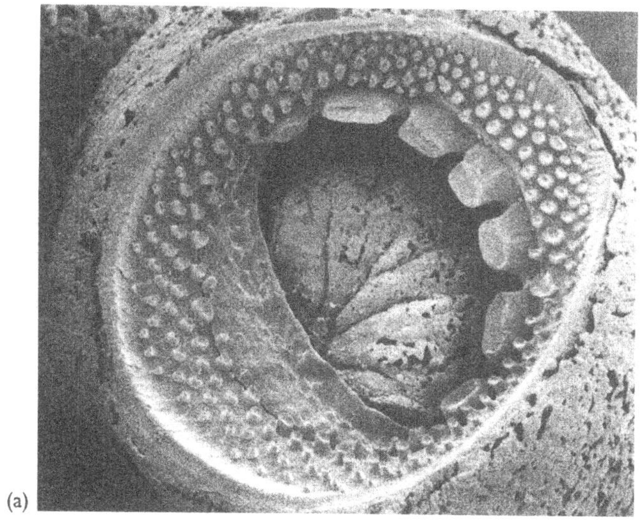

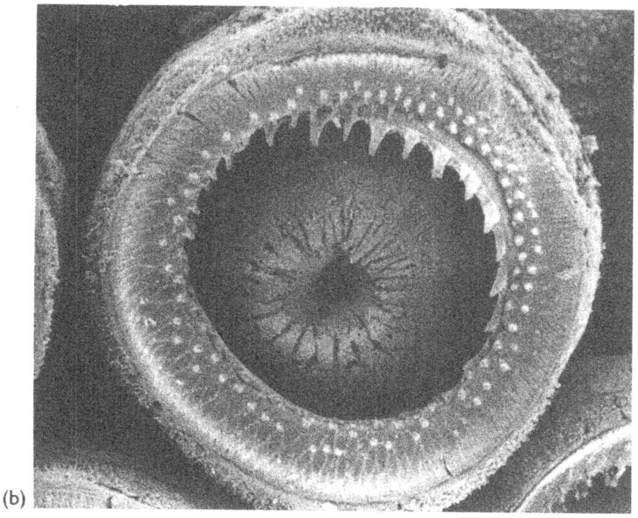

Fig. 8.10. *Alloteuthis subulata*. Scanning electron micrographs of sucker of (a) an arm (field width 567 μm), and (b) a tentacle (field width 2564 μm). The pegs on the infundibulum and the teeth on the inner rings are seen in both suckers. The pegs are more numerous and the teeth very blunt on the arm sucker (Nixon and Dilly 1977).

8.3.5 Prey capture and diet

Loliginids usually capture prey with their tentacles. The phases are sequential (Kier 1982); they were defined as attention, positioning, and strike in *Sepia officinalis* by Messenger (1968, 1977), and have been observed in several loliginids (Kier 1982, 1996, Kier and van Leeuwen 1997, van Leeuwen and Kier 1997).

The mode of attack and capture of prey was analysed in adult *Loligo*. The squid responds by rapidly turning its head to direct the eyes forward and project the arms and tentacles towards the prey. The orientation of the arms is maintained while the animal turns so that its long axis also points towards the prey. This phase lasts for 0.5–2 s and is accompanied by chromatic changes. During positioning the squid swims towards the prey with the arms enclosing the tentacles except for the slightly protruding tips of the tentacular club: this phase lasts about 1–4 s, depending on the distance to the prey (Kier 1982). Once captured the tentacles carry the prey back to the arms, which then manipulate the prey into a suitable position for its ingestion.

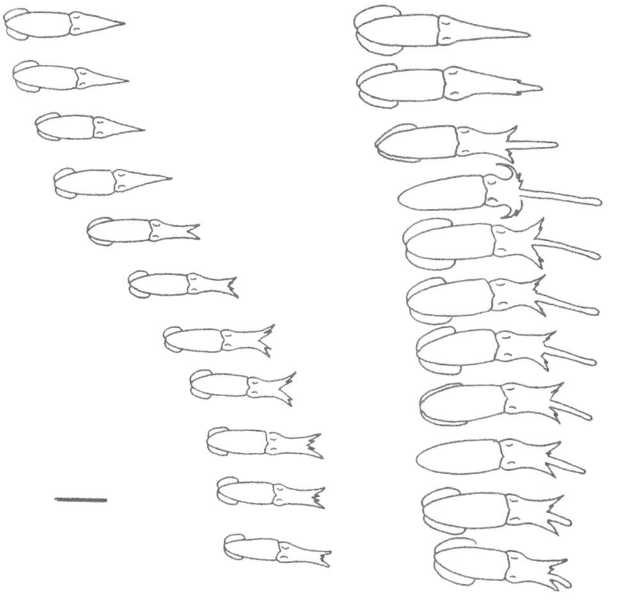

Fig. 8.11. *Sepioteuthis lessoniana*. Tracings from high-speed video sequences, taken at intervals of 20 ms, of prey-capture by young squid aged 2 weeks (left) and 4 weeks (right) (scale bar 10 mm) (Kier 1996).

The stalks of the tentacles are muscular and when projected to capture prey elongate by 70% of their resting length. Acceleration during the tentacular strike is remarkable as it can reach as high as 250 m/s^2 in as little as 20 ms (Kier and van Leeuwen 1997). This implies specialization for such rapid elongation. The ultrastructure of the transverse and associated circular muscle of the tentacles indicates the specialization significant for such rapid elongation. Whereas other muscle fibres in cephalopods are obliquely striated cells, those of the transverse muscle mass and circumferential layer in the tentacles are cross-striated cells. These cells show ultrastructural characteristics of high shortening velocity and rapid excitation–contraction coupling (Kier 1985, 1991, 1996).

The development and differentiation of the fast-contracting, cross-striated muscle cells of the tentacles and the obliquely striated muscle cells of the arms of *Sepioteuthis lessoniana* was followed at the ultrastructural level from hatching to 5 weeks of age (Kier 1996). Transverse striation of the muscle cells of the tentacle appears at about 3 weeks and by 4–5 weeks the ultrastructure found in the adult is present. During this period high-speed video recording was used to correlate these changes with the mode of prey capture. As the paralarva approached the prey (small mysid shrimp) it quickly flared the arms and tentacles to enclose the prey (Fig. 8.11). By 4 weeks of age the tentacular strike was like that of the adult.

In the sea *Sepioteuthis sepioidea* will make direct tentacular strikes on fish and it can also capture small fishes and copepods with its arms alone. It may take large numbers at the same time, and one individual was seen, under ship lights, to capture at least 50 fish in 75 min; they were held by the arms before ingestion. A fish of appreciable size was captured and held by the arms of one squid before being immobilized by a bite on the dorsal surface behind the head. Some flesh was taken from the head of the fish, and the muscles and most of the internal organs eaten; the head, skeleton, and tail were discarded in one piece. Most attacks by *S. sepioidea* in the sea appear to be successful, both in darkness and daylight (Moynihan and Rodaniche 1982).

The paralarvae of *Loligo opalescens* were recorded, by high speed camcorder, when preying on copepods (Chen *et al.* 1996). They showed progressive development of the basic attack behaviour, and this was followed by the addition of more specialized strategies. By 40 days of age the adult-like prey capture behaviour had developed. If restricted to a diet of the easily captured *Artemia* nauplii until 40 days of age then the paralarvae were unsuccessful in capturing copepods.

Loliginids feed on crustaceans and fish, and several species are cannibalistic. Where the diet has been followed, changes were found to take place with the growth of the squid (Nixon 1987). The stomach contents of *Loligo pealeii* of 10–41 mm ML contained mostly copepods; at 41–60 mm ML arrow worms, *Sagitta*, and euphausiids were eaten; at 61–100 mm crabs, shrimps, and polychaetes were consumed; and larger animals of 121–160 mm ML ate fish and squids predominately although crustaceans were still eaten. Squids of more than 160 mm ML fed only on fish and their own kind, those of 250 mm ML ate more squids than fish, and those of more than 300 mm ML were almost entirely cannibalistic. The size of the prey and of the food fragments in the stomach increased with the growth of the squid itself (Vovk 1983).

The movements of the jaw of many loliginids are visible during feeding. A series of transverse bites are made in fish and pieces of flesh (squarish in shape and about 2–3 × 2–3 mm in size) pass in rapid succession to the stomach (Bidder 1950).

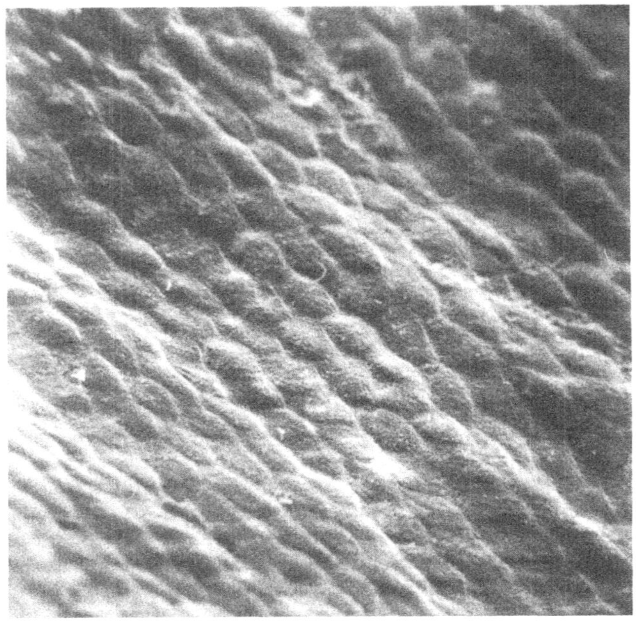

Fig. 8.12. *Alloteuthis subulata*. Scanning electron micrograph of surface of upper beak, in the crest region, showing the aligned polygonal imprints of beccublast cells (field width 80 μm) (Dilly and Nixon 1976c).

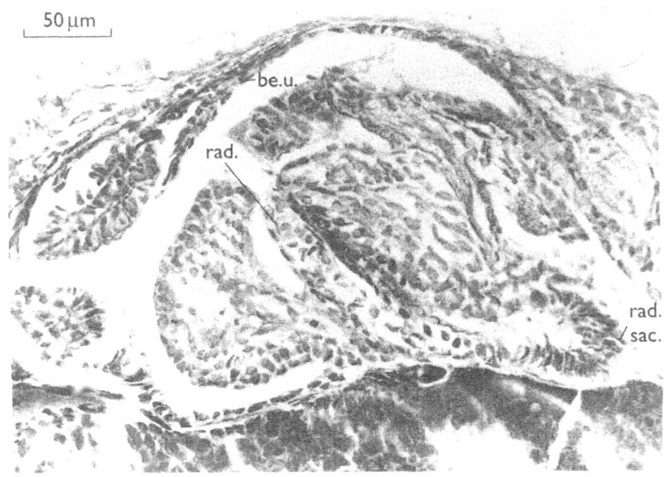

Fig. 8.13. *Alloteuthis subulata*. Sagittal section of the buccal mass of a newly hatched animal showing the radular sac and the radula emerging from it (Holmes' stain).

8.3.6 Buccal mass

The buccal mass is clearly visible as it is defined by the beaks at hatching (Fig. 8.2). Initially translucent the pale beaks darken from a small brown fleck in the lower beak (at about 20 mm ML), the coloration subsequently spreading over the lower and much of the upper beak (Clarke 1962a). At hatching the oral surfaces of the beaks bear denticles, presumably an aid in the capture of prey. The beak is chitinous and is secreted by a single layer of underlying tall columnar cells, the beccublast cells, which also serve to attach the surrounding muscles to the beak (Dilly and Nixon 1976c). The outer surface of the beak bears polygonal imprints of the beccublast cells (Fig. 8.12), and the beak so formed has a lamellar structure. The anteroposterior length of the buccal mass at hatching is about 30% of the mantle length and with growth this proportion falls to 8% or less in specimens of 120 mm ML (Morton and Nixon 1987).

Within the buccal cavity is the buccal complex comprising the radula, the radular support system, the lateral buccal palps and the salivary glands. The radula is already well formed at hatching (Fig. 8.13). It is secreted by cells in the radular sac (Gabe and Prenant 1957) and the teeth and ribbon extend from the radular sac forwards and bend acutely to terminate below and behind the bending plane (Fig. 8.14). The radula bears a series of teeth in transverse rows, each row having a central, rhachidian tooth and on either side two lateral teeth, one marginal tooth, and a marginal plate. Each newly developed tooth in the last-formed transverse row is larger in size than those in the same position and formed earlier. The number of teeth increases in each longitudinal row with the growth of the animal (Nixon 1995, 1998). The staining properties of the teeth change along the length of the ribbon, passing from the radular sac to the distal end of the ribbon. Beyond the cells secreting the ribbon, at the distal end of the radula, is a group of notably taller cells, the subradular organ (Messenger and Young 1999). These tall cells may secrete enzymes to facilitate the removal of the teeth and the ribbon of the radula. Immediately below these cells is a fold of tissue, which appears to have a mechanical role

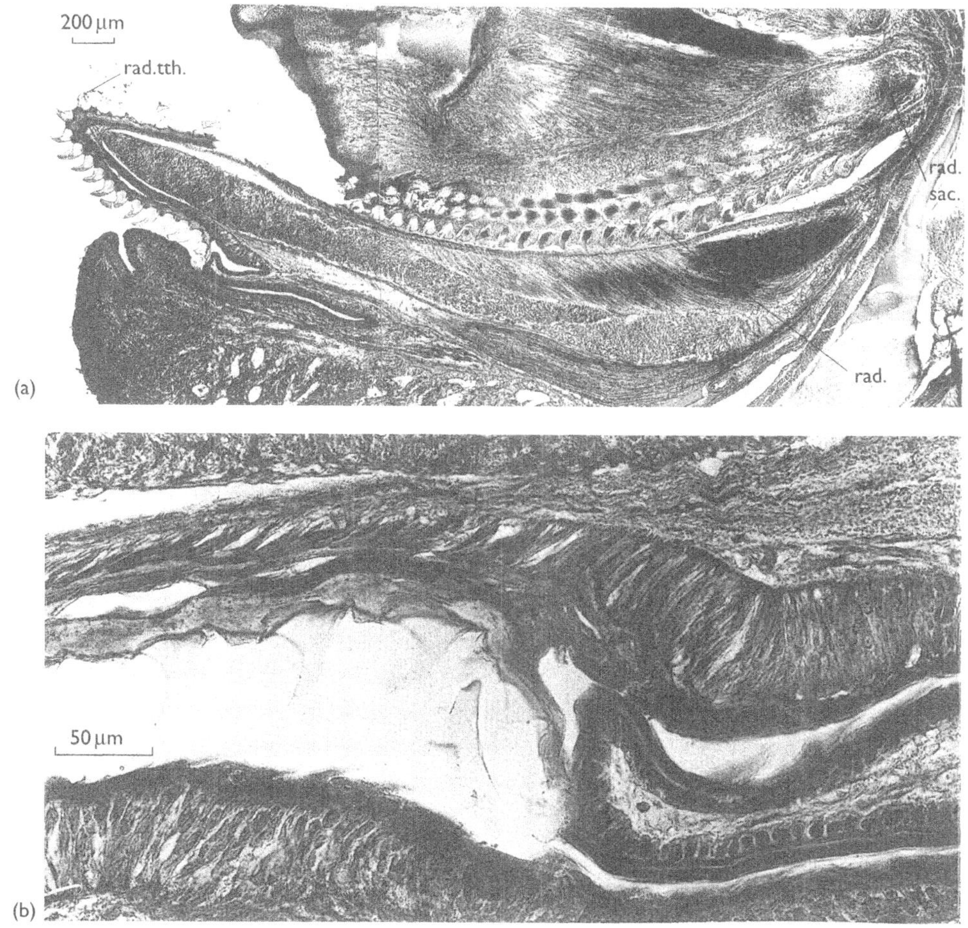

Fig. 8.14. *Alloteuthis subulata.* (a) Sagittal section showing the radular sac, radula, and subradular organ, and (b) the tissue fold which may aid in the removal of teeth from the distal end of the radula, at higher magnification.

in the removal of teeth (Fig. 8.14b). Anteriorly and on each side of the radula is a chitinous sheet, the hyaline shield, joining its upper and lower parts on either side of the bending plane.

The radular support system lies on either side of the radula. This comprises a bolster and a rod. The bolster is a roughly cylindrical mass of muscle wrapped in a sheath of connective tissue, and is attached only posteriorly and laterally and being free anteriorly. Each bolster contains an elongate structure (Fig. 8.15), or rod, composed of cells the staining properties of which differ from those of the surrounding bolster (Fig. 8.16); direct electrical stimulation causes the tip of the bolster to curve strongly upwards (Messenger and Young 1999). The bolster probably functions as a muscular hydrostat, like the arm muscles (Kier 1988). The elongation of the bolster acts to evert and raise the teeth as the ribbon approaches and passes over the bending plane, aiding in the ingestion of food material during its passage towards the oesophagus.

Lateral to the buccal complex are the lateral buccal palps. Each has a covering of chitinous material and carries numerous scattered backward-pointing teeth on the medial surface. The palps move forwards around the anterior face of the radula and then sweep backwards towards the oesophagus.

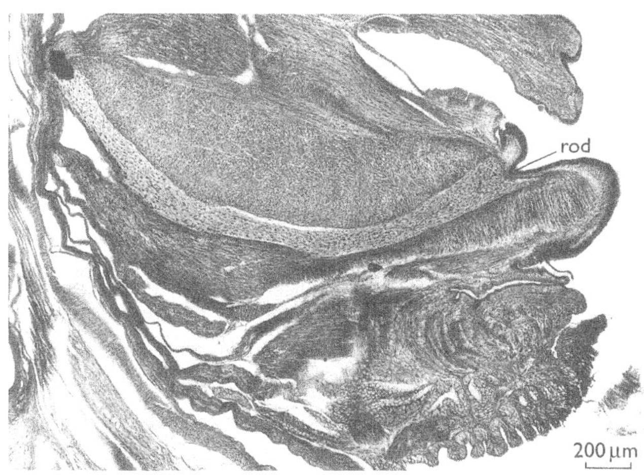

Fig. 8.15. *Alloteuthis subulata.* Sagittal section to show the bolster, and the rod which extends almost the length of the radular complex.

Between the floor of the buccal cavity and the ventral side of the radula is the salivary papilla (Fig. 8.16). This is prominent and carries the submandibular gland as well as the single duct from the paired posterior salivary glands.

8.3.7 Chromatophores, iridophores and body patterns

The rapid changes in colour displayed by cephalopods have been known to naturalists since antiquity (see Thompson 1910). A study of the physiological mechanisms of these cells was undertaken by Sangiovanni in 1819. The colour changes are controlled by complex organs, the chromatophores, which have radial muscles directly innervated from the brain.

Loliginids have a simple repertoire of patterns and for much of their time they may conceal themselves by remaining translucent, with their chromatophores retracted (Hanlon and Messenger 1996). *Alloteuthis subulata* has only yellow and red chromatophores and the body patterns it displays is limited (Cornwell *et al.* 1997). The dorsal mantle has more yellow than red chromatophores, but the latter are larger and bear more radial muscles than the former. On the ventral mantle there are more red than yellow chromatophores, and the difference in size is less marked. The significance of these findings is as yet unclear.

Our current understanding of the function and structure of chromatophores is due almost entirely to Florey and his colleagues (Florey 1966, 1969, Cloney and Florey 1968, Cloney and Brocco 1983, Florey and Kriebel 1969) who used ultrastructural and electrophysiological techniques in their studies of *Loligo opalescens*. The chromatophores of *L. pealeii* and *L. opalescens* were described by Mirow (1972a), and those of *Alloteuthis subulata* by Cornwell *et al.* (1997). The innervation of the muscle cells of the chromatophores of *Loligo vulgaris* was described by Weber (1968, 1970, 1973).

Each chromatophore organ is composed of five different cell types. These are:

- the chromatophores (pigment granules in an elastic 'sacculus')
- a set of (15–25) radial muscles
- axons
- glial cells, associated with the axons
- chromatophore sheath cells (Fig. 8.17).

The elastic sacculus containing the pigment granules is attached to the cell membrane by zonal haptosomes. The cell membrane is

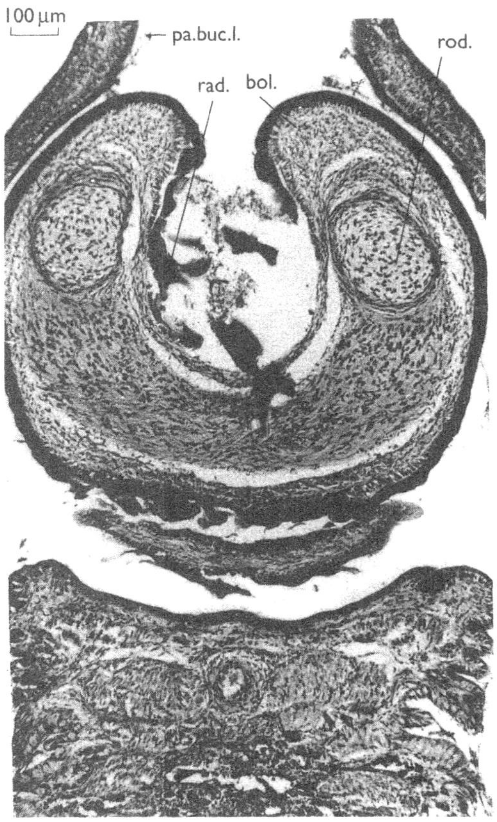

Fig. 8.16. *Loligo vulgaris*. Transverse section through the radular supports showing the bolsters and rods.

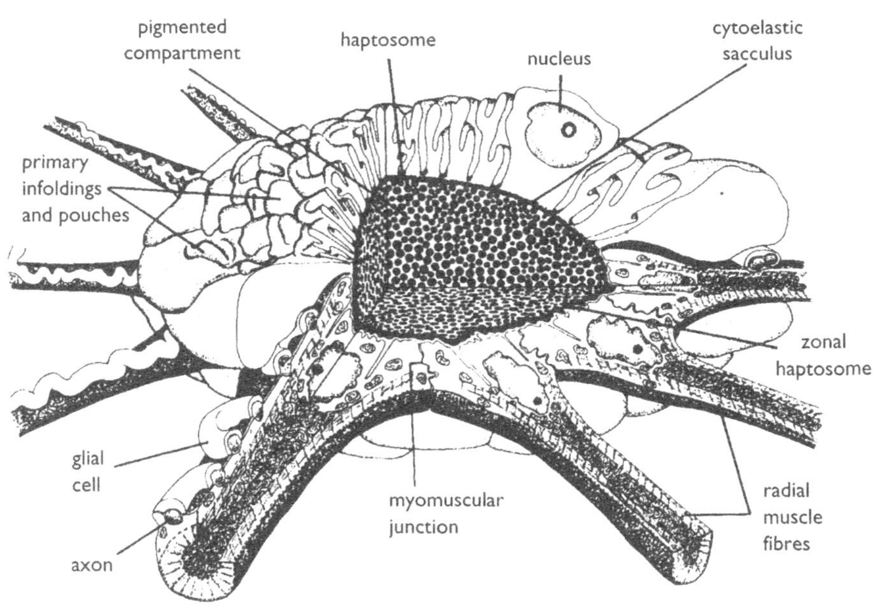

Fig. 8.17. *Loligo opalescens*. Diagram of a chromatophore, in the retracted condition, to show the arrangement of the nerve fibres, and radial muscle fibres (Cloney and Florey 1968).

attached to the radial muscle fibres by a dense basal lamina. Contraction of the muscles expands the chromatophore; when they relax, elastic forces stored in the sacculus cause it to retract. In the retracted state its cell membrane is highly folded. When the sacculus returns to its original, near-spherical, shape the pigment cell is often less than one-tenth of its expanded diameter (Florey 1969). Axons, glial cells associated with axons, and chromatophore sheath cells are present.

The chromatophores of *Loligo vulgaris*, *Alloteuthis subulata*, and *Lolliguncula brevis* receive multiple innervation (Florey 1966, 1969, Cloney and Florey 1968, Weber 1968, Mirow 1972a). Each radial muscle is innervated by 2–6 nerves running along its length. These nerves contain 50 nm electrolucent vesicles organized into numerous synapses along the muscle. Their size and appearance is consistent with their containing L-glutamate, an excitatory transmitter shown to expand the chromatophores of *Loligo vulgaris* and *Alloteuthis subulata* (Messenger *et al.* 1997). These authors show that serotonin (5-hydroxytryptamine (5-HT)) is endogenous in some of the nerve fibres innervating the chromatophore muscles, and the radial muscles bear 5-HT receptors. When topically applied 5-HT relaxes the radial muscles, causing the chromatophores to retract in loliginids, and probably has a fundamental role in their regulation although the manner in which this substance acts is not known. There is no evidence that any other substance(s) is involved in expanding or maintaining the expansion of the chromatophores of *L. vulgaris* and *Alloteuthis subulata*. These loliginids exhibit relatively few body patterns and depend for their concealment mostly on being translucent with their chromatophores retracted (Hanlon and Messenger 1996).

Lolliguncula brevis has only yellow and brown chromatophores, the yellow being more superficial and smaller than the brown ones; the latter have a fairly regular spacing, with the yellow chromatophores interspersed. This squid exhibits patterns made up of nine simple chromatic components and during calm swimming it is translucent. A hatchling *L. brevis* (2 mm ML) has 53 chromatophores on the fins and mantle, and this increases to 27 000 in an adult (100 mm ML) (Dubas *et al.* 1986). The posterior chromatophore lobe of the brain innervates the mantle and fin exclusively, and the chromatophores whose muscle fibres are controlled by motoneurons from this lobe. The motor axons to the mantle chromatophores go via the ipsilateral pallial nerve to the stellate ganglion. They pass then to the stellar nerves, which radiate over and innervate the skin of the mantle. The motoneurons of the fin chromatophores presumably go to the fin nerve, which bypasses the stellate ganglion (Young 1939).

Stimulation of individual stellar, pallial, and fin nerves allows the motor fields to be mapped from the expanded chromatophores. Considerable overlap is found between the chromatophore motor fields of adjacent stellar nerves: this was confirmed after lesions and by stimulation. When the position of the chromatophore patterns of the mantle are compared (Dubas *et al.* 1986) with the motor field, then it appears probable that subpopulations of chromatophore neurons, in several of the stellar nerves, are involved in each pattern seen on the mantle in this species.

Peripheral injection of horseradish peroxidase identified the chromatophore motoneurons in the brain of *Lolliguncula brevis*. Cells marked by a localized injection in any part of the mantle were scattered over the whole posterior chromatophore lobe. Stimulation of single motoneurons in the posterior chromatophore lobe showed the motor fields of individual chromatophore motoneurons to be compact and to include 2–60 chromatophores usually of the same colour (Dubas *et al.* 1986). A study of the brain pathways controlling the chromatophores of the same species, using cobalt iontophoresis, showed several input and output pathways of the anterior and posterior chromatophore lobes and the lateral basal lobes (Fig. 8.18). The anatomical arrangements revealed pathways which could facilitate the modification and coordination of chromatophore patterns at

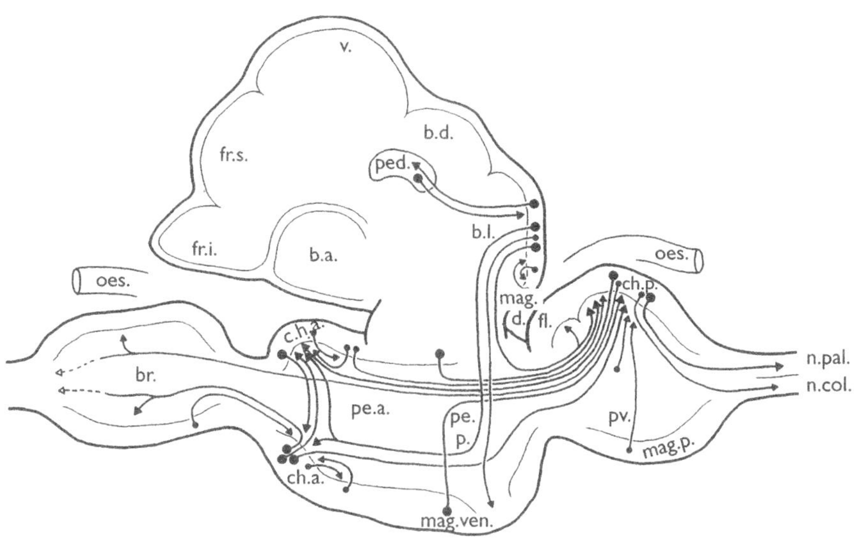

Fig. 8.18. *Lolliguncula brevis.* A diagram of the brain in sagittal section to show the pathways, on one side only, that are likely to be involved in the control of chromatophores (Novicki *et al.* 1990). The filled circles represent the origin of the pathways and arrowheads their termination. The sizes of the circles and arrowheads indicates approximately the number of perikarya or fibres filled. The fibres in the brachial lobe are difficult to follow, so the termination is shown as an open arrowhead.

every level, and several areas were found to be capable of influencing the chromatophore program (Novicki *et al.* 1990).

Below the chromatophores lie iridophores, iridocytes, or reflecting cells. These may be located in the dermis, around the eyes, on the ink sac, and sometimes in photophores (Mirow 1972b, Budelmann *et al.* 1997). Iridophores have an elongate ellipsoidal form and possess a complex structure of very thin electron-dense platelets alternating with layers of cytoplasm (Brocco and Cloney 1980). Some iridophores are physiologically active and change reversibly from a non-reflective to a reflective state (Hanlon 1982, Hanlon *et al.* 1990). Ultrastructural studies of these cells in *Lolliguncula brevis* showed that a change occurs within the platelets between these two states, and is associated with a change in colour (Cooper and Hanlon 1986, Cooper *et al.* 1990). The active cells may be controlled by acetylcholine, acting as a hormone (Hanlon *et al.* 1990).

The iridophrores are active structures which can, within a short period of time, 'switch' between a reflective and non-reflective state and thus provide a mechanism for both purposes of light reflection: camouflage and signalling. *Alloteuthis subulata* has two elongate bands of iridophores on each side of the mantle; when viewed in white light one reflects red and the other blue. As the angle of light on the iridophores becomes more oblique, then the reflected light moves further towards the shorter end of the spectrum, and becomes plane polarized to a greater or lesser extent. When viewed from above this squid can match its background illumination, and from the side the reflecting bands are highly visible. These bands reflect light which differs in intensity, polarization, and spectral composition in different directions. Their function may be in providing information on the relative position of neighbouring squid in a school. Besides these mantle bands two spots are present on the head, one above each eye. Each eyespot has three reflecting components: two are similar to those of the mantle bands, and a third appears green when in white light. The eyespot is equally bright at all angles of incidence, and it behaves like a perfect diffuser. This may act during schooling to inform neighbours of head movements; since the squid steers by turning its head, movement of the eyespot will give an early signal of changes in direction (Mäthger and Denton 2001).

The body patterns comprise the entire appearance of the animal at any given moment. Besides chromatophores and iridophores, textural, locomotor, and postural components may be added, so that a wide variety of body patterns can be exhibited by loliginids. These may be species-specific and some are sex-specific (Hanlon 1988, Hanlon and Messenger 1996). An ethogram, or catalogue of body patterns, has been compiled for *Loligo vulgaris* (Hanlon *et al.* 1994), and for *L. pealeii* (Hanlon *et al.* 1999): the former exhibited more than 20 patterns and the latter 50 patterns.

8.3.8 Learning

Most loliginid squids adapt poorly to laboratory conditions, but *Lolliguncula brevis* survives well and it has been the subject of studies of memory and visual discrimination and of habituation and dishabituation. This squid rapidly learnt to come to feed when a horizontal white rectangle was shown, but to avoid a vertical white rectangle (Allen *et al.* 1985). It was able to retain this discrimination without being shown the figures again until 9 days later; it also learnt to discriminate between a black and a white sphere, and between a white horizontal rectangle and a white sphere. Habituation was demonstrated by behavioural responses to plastic models of fish predators. The number of escape jets and the duration of neurally controlled ring patterns on the mantle of *L. brevis* were monitored by video during each trial (Long *et al.* 1989). This squid habituates to a teleost fish model and could differentiate between models, of the same size, of a teleost and a shark. Habituated responses also showed recovery, or dishabituation, when a noxious substance was presented.

8.4 Receptors and central nervous system of loliginids

8.4.1 Mechanoreceptors

The neck region is distinct and allows movement of the head relative to the body. The nuchal cartilage complex of *Lolliguncula brevis* has a set of large epidermal hair cells lying on the dorsal side of the neck and close to the midline on either side of the nuchal crest (Fig. 8.19) (Preuss and Budelmann 1995a). Each hair cell carries up to 300 kinocilia of equal length (30 μm). These cells are arranged in up to seven rows and form anterior and posterior groups. Those of the left and right anterior group are morphologically polarized in the medial direction, and those of the left and right posterior group are polarized in the anterior direction. These are primary sensory cells and are innervated by a branch of the postorbital nerve and project ipsilaterally into the ventral part of the ventral magnocellular lobe. This area of the brain also receives direct input from the statocysts and optic lobes, with further inputs from the arm and funnel nerves (Young 1976). It seems likely that it is in the ventral part of the ventral magnocellular lobe that the information from these several sources is integrated for the control of posture. The hair cells together with the associated nuchal cartilages form a neck receptor organ, and this monitors the head-to-body positions and head-to-body movements around pitch and roll axes. Several criteria indicate that these hair cells are mechanoreceptors with a proprioceptive function (Preuss and Budelmann 1995a,b).

8.4.2 'Lateral line' system

Lines of ciliated epidermal cells on the head and arms are especially noticeable in young hatchlings. They were first described as lines of cells by Naef (1928), and later shown to be sensory cells with cilia (Sundermann-Meister 1978, Arnold and Williams-Arnold 1980, Sundermann 1983, 1990).

The lines have bilateral symmetry with lines of cells all running anteroposteriorly. *Loligo vulgaris* has five lines of cells on each side of the head, two dorsal lines, two lateral, one above and one below the eye, and one ventrally placed. Each hair cell has several kinocilia, up to 12 μm in length, and an axon extending from the base. Two slightly different types of hair cell are found in this species (Sundermann 1983) and four in *L. pealeii* (Arnold and Williams-Arnold 1980).

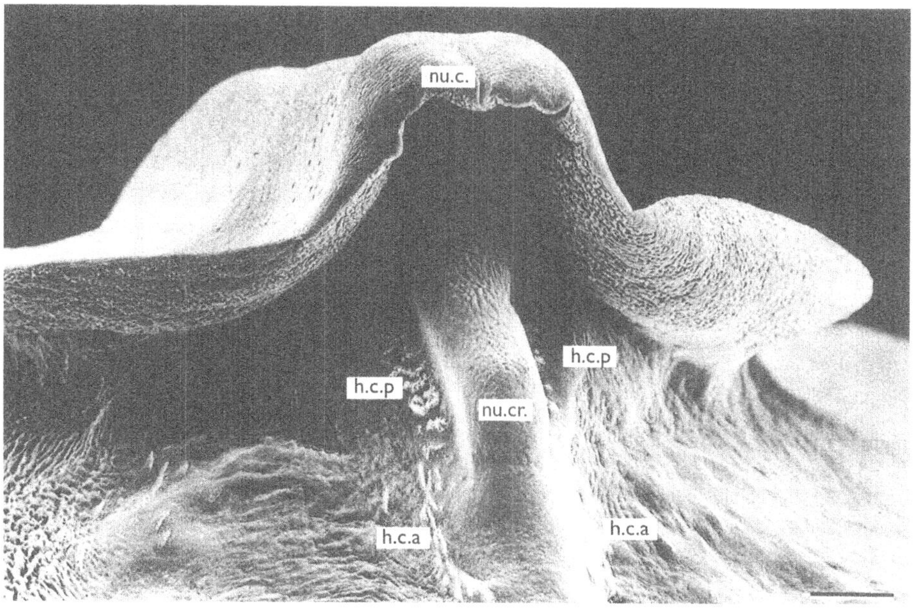

Fig. 8.19. *Lolliguncula brevis*. The dorsal neck region, the nuchal cartilage complex, with the neck receptor organ (the mantle removed) (Preuss and Budelmann 1995a). The groups of anterior (h.c.a.) and posterior hair cells (h.c.p.) of the neck receptor organ are situated on either side of the nuchal crest (nu.cr.) and covered by nuchal cartilage (nu.c.). The anterior edge of the cartilage is bent upwards as a result of fixation.

Hulet (1982) described discrete bundles of kinocilia present in hatchlings and in adult *Loligo*. Lines of ciliated cells are present in *Lolliguncula brevis* up to 40 mm ML and stimulation with sinusoidal water movements resulted in receptor potentials; this indicates a sophisticated receptor system. These cells are sensitive to local water movements (as low as 0.06 μm) and are able to detect a moving fish 1 m in body length from a distance of about 30 m (Budelmann and Bleckmann 1988, Budelmann 1994).

In loliginids the lateral line analogue system is probably involved in the detection of prey and predators, and in schooling. The system is morphologically and physiologically analogous to the mechanoreceptor lateral line system of fish and aquatic amphibians, and it has been demonstrated in *Sepia officinalis* to be used in prey capture (Budelmann *et al.* 1991).

8.4.3 Olfactory organs

There is a pair of olfactory organs. Each organ is situated on a muscular ridge of tissue, and lies posterior and ventral to each eye (Watkinson 1909). The histology and fine structure has been described for *Lolliguncula brevis* (Emery 1975) and *Loligo vulgaris* (Wildenburg and Fioroni 1989). Several types of ciliated sensory cells are present with apical dendritic processes and basal unmyelinated axons that form the olfactory nerve and make synapses in the olfactory lobe, optic lobe, optic gland, and higher motor centres of the brain (Messenger 1979a). There are two common types of sensory cell, the pyriform and the floriform receptor cells, and they are excitable. The position of these organs would allow the monitoring of water drawn into the mantle cavity during inhalation. The local application of diluted squid ink, and of L-dopa and dopamine, to the olfactory organ of a restrained *Loligo opalescens* resulted in strong escape jet responses so demonstrating that these organs can mediate detection of water-borne chemicals (Gilly and Lucero 1992). The ink of *Loligo opalescens* contains L-dopa (1.15 mmol/L) and dopamine (0.19 mmol/L), and these values are greater than those at which both compounds are found to be effective in behavioural and electrophysiological experiments (Lucero *et al.* 1994). Furthermore squid ink, L-dopa, and dopamine, inhibit spontaneous firing in primary sensory neurons isolated from squid olfactory organs (Lucero *et al.* 1992). This 'suggests that these often nocturnal schooling animals, are able to chemically sense and avoid the ink released by recently alarmed conspecifics' (Lucero and Gilly 1995) and avoid capture.

8.4.4 Extraocular photoreceptor organs

An extraocular photoreceptor organ lies behind each eye and in close proximity to the optic tract where it enters the cephalic cartilage in *Loligo forbesi*. Each comprises four or five pigmented photosensitive vesicles, usually pale orange in colour, each 200–500 μm in diameter (Cobb and Williamson 1998a). The organs form 0.19% of the total brain volume in *Loligo vulgaris* (Fig. 8.20), and 0.03% in *Lolliguncula brevis* (Maddock and Young 1987). Nerve fibres from the photosensitive vesicles enter the basal zone of the peduncle lobe. Intracellular recordings from isolated extraocular photoreceptor cells of *Loligo forbesi* showed resting potentials; flashes of white, but not red, light evoked depolarization (Cobb and Williamson 1998a). This gives some evidence for a functional system, but as yet the role of the extraocular photoreceptor organs remains unknown in loliginids, although it may serve to monitor ambient light over long periods, perhaps in relation to migratory patterns.

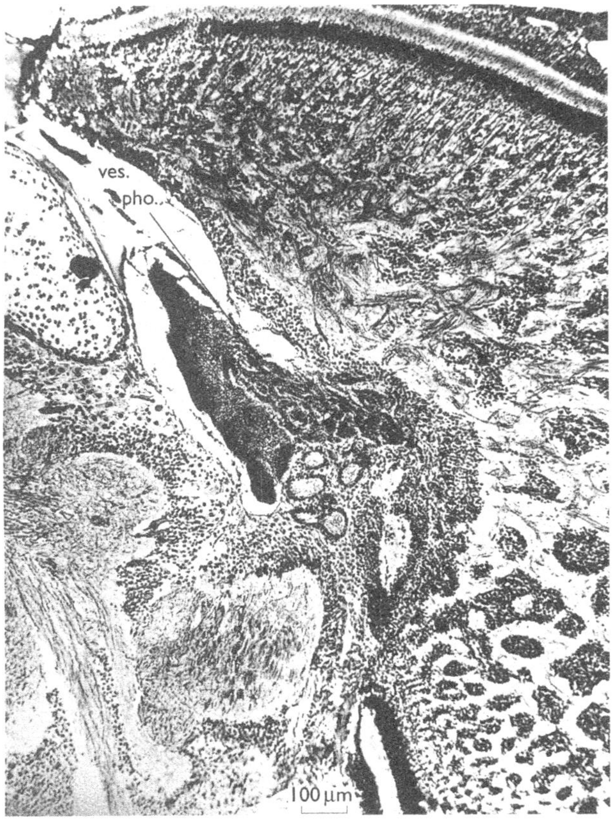

Fig. 8.20. *Loligo vulgaris*. A sagittal section showing part of the optic lobe and the photosensitive vesicles of the extraocular photoreceptor organ.

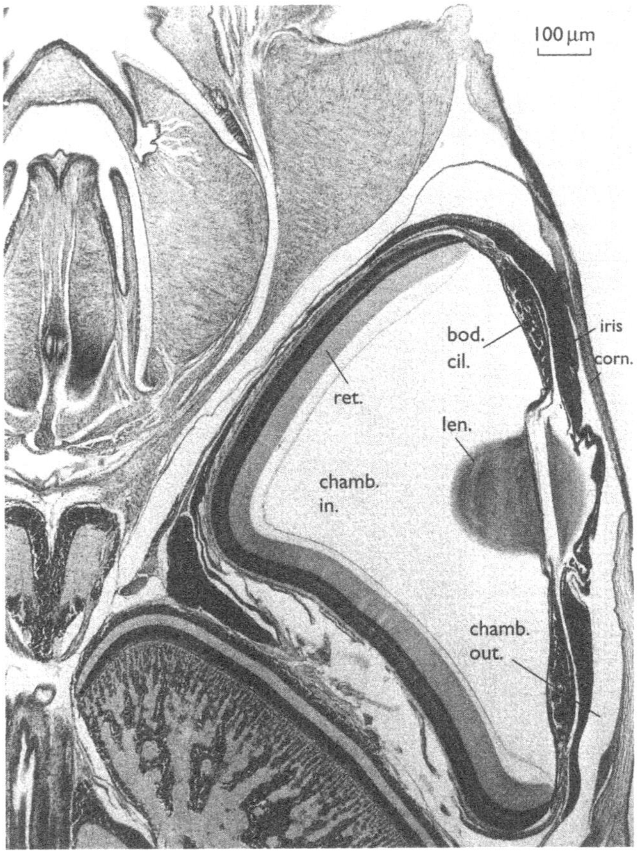

Fig. 8.21. *Loligo pealeii*. Horizontal section through the head showing the eye, part of the optic lobe, and part of the buccal mass (haematoxylin and eosin).

8.4.5 Eyes

The eyes are laterally placed and occupy much of the head (Figs 8.1, 8.21), and each lies in an orbit partly protected by cartilage. A corneal membrane covers the eye and there is a minute pore anteriorly. This is the myopsin condition and is an adaptation of the eye associated with life in coastal waters.

Movement of the eye is achieved by extraocular muscles (Williams 1909), 14 of which are present. The anterior muscles are attached to the anterior and superior faces of the eye: the former, and probably the latter, are concerned with convergent eye movements for binocular vision during fixation and the capture of prey. The anterior muscles include conjunctive muscles which cross in the midline, each presumably moving both eyes simultaneously during fixation (Fig. 8.22) (Budelmann and Young 1993). There is an additional superior conjunctive muscle in *Loligo* and *Sepioteuthis* with perhaps a similar function. The extraocular muscles are supplied by four eye muscle nerves, shown by centripetal cobalt fillings to have fibres originating from somata in the ipsilateral anterior lateral pedal lobe, which is the oculomotor centre. The main actions of most of the eye muscles is a linear movement of the eyeball; only three muscles produce strong rotation of the eyeball.

The eye superficially resembles that of a vertebrate. There is a large posterior (inner) chamber, a lens, iris, retina, choroid, sclera, and argentea (Fig. 8.21). The eye produces a perfectly focused image on the retina as the lens has a graded refractive index and a focal length approximately 2.5 times the radius (Matthiessen's ratio). The image of objects at distances from infinity to a few centimetres is of high quality. This minimizes the need for accommodation.

The structure of the retina is simpler than that found in vertebrates and contains no rods or cones. The retina (approximately 500 µm thick) comprises receptor cells, the retinal or retinula cells, supporting cells, and retinal glial, or epithelial, cells (Messenger 1981). The photoreceptive end of the retinal cell is usually referred to as the 'outer segment', although in cephalopods it is directed towards the lens. The retina can be divided into three regions: a long outer segment and a short inner segment separated by a basal membrane, and a plexiform layer below. External to the outer segment is a definite limiting membrane separating the receptors from the posterior chamber of the eye.

The retinal cells are extremely long and slender, with a cross-section of 6.0 µm in *Loligo pealeii* (Cohen 1973a). The cells are shorter centrally than peripherally and there is no equatorial strip (Young 1963a). The nuclei of the retinal cells lie below the basal membrane and their distal processes extend above it for about 320 µm. The distal segment bears two sets of parallel tubules or microvilli, at right angles to the main axis and opposite each other. The set of tubules on each side of the retinal cell is a rhabdomere. *Loligo* has some 200 000–700 000 microvilli per rhabdomere (Messenger 1981). The retinular cells are oriented with the microvilli either horizontal or vertical as the eye is

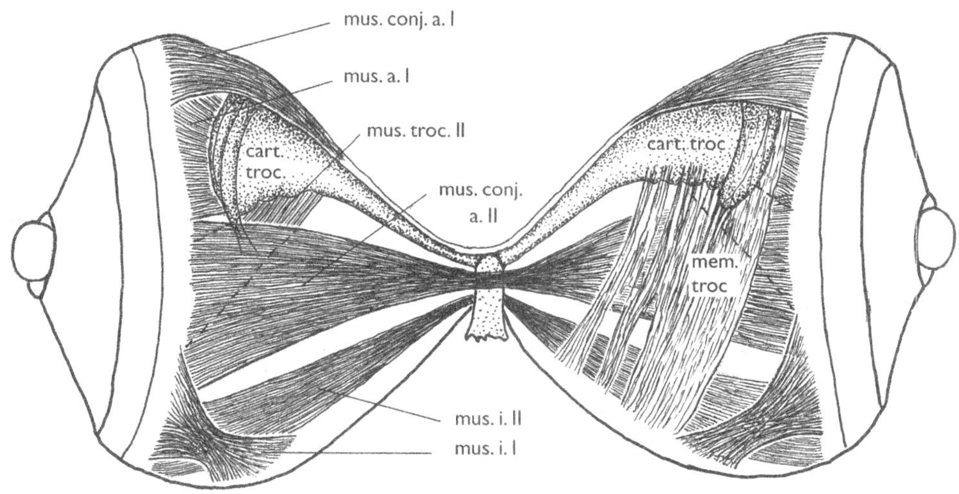

Fig. 8.22. *Sepioteuthis lessoniana.* The arrangement of the extraocular eye muscles and the trochlear cartilages on the anterior surface of the eyeballs. The trochlear membranes have been removed from the right eye (Budelmann and Young 1993).

held normally in space. This feature is thought to be associated with the sensitivity to polarized light that has been found in several cephalopods (Messenger 1991). There is a rather precise organization, with the rhabdomeres from four retinal cells forming a rhabdom. The regularity does not always hold, as one retinal cell may participate in up to eight rhabdoms (Cohen 1973a).

Within the main axis of a retinal cell are pigment granules which retreat to the base of the outer segment in darkness but migrate under conditions of light. In an intact eye, illumined from above, migration of the pigment takes place at different rates depending on the region of the retina: movement is rapid ventrally, slower dorsally, and slowest of all at the centre. Transfer to darkness results in rapid contraction during the first minute and after 10 min most of the pigment granules have reached the base (Young 1963a). The major protein is rhodopsin, which is localized in the outer segment of the retinal cell, and it has an absorption spectrum of 493 nm in *Loligo pealeii* (Messenger 1981, Saibil *et al.* 1995).

Using slices of retina of *Loligo pealeii*, electrical activity was recorded as a result of photic stimulation of the outer segment of the retinal cell, a generator potential being produced at or near the site of light absorption (Messenger 1981). In an isolated eye–optic lobe preparation of *L. opalescens* slow and fast responses were recorded from the optic nerves, the fast responses being afferent and reflecting the activity of the receptor cells (Hartline and Lange 1974, 1984).

The plexiform region of the retina is layered in the sclerad direction by imprecise zones respectively characterized by receptor collaterals, synaptic profiles, horizontal fibres, and receptor axons (Cohen 1973b). The optic nerve fibres cross in a dorsoventral chiasma in which topographical relations are strictly preserved, so that the output from the retina is re-inverted for projection on to the optic lobe (Young 1962b, 1974).

Loligo pealeii and *Sepioteuthis lessoniana* were found capable of discriminating the plane of polarization of polarized light (Jander *et al.* 1963), any particular plane being seen as a variable intensity. The microvilli have orthogonal orientation in the outer segments and adjacent photoreceptors; this, together with the physiological preference for one of two mutually perpendicular planes of polarization by single photoreceptors, led to the conclusion that 'cephalopod vision is based upon two complementary views of the world, each determined by the transformation of polarization-sensitive receptors into complementary intensity scales. A visual system based on this transformation would lead to enhanced contrast underwater and visualization of object details obscured by confounding highlights' (Saidel *et al.* 1983). This may allow apparently camouflaged fish to become visible to the squid, and so fall prey to it.

8.4.6 Optic lobes

Each optic lobe is oval in outline when viewed from the lateral side. Its long axis lies obliquely in the orbit, at an angle of about 60° with respect to the horizontal plane of the squid's body. The lobe is slightly concave on its medial side (Fig. 8.23) (Young 1974). The volume of the two optic lobes in relation to the whole brain is 336% (Table 2.6) (Maddock and Young 1987).

The optic nerve fibres enter over the lateral surface of the optic lobe and make a chiasma along a line that follows the long axis of the lobe. These fibres then spread out over the lobe and pass through an outer layer of granule cells to a plexiform zone where the majority of them end.

The organization of the optic lobe can be considered to fall into two parts. There is an outer cortex, or deep retina, where the incoming optic nerve fibres meet tangential dendrites of the second-order visual neurons, and an inner medulla composed partly of outer radial columns based on the axons of these latter neurons. The inner medulla contains third- and perhaps higher-order neurons leading to the optic tract. Both the deep retina and the medulla contain many other types of neurons. The description of cells and pathways in the optic lobe of *Loligo* has allowed some analysis of the system by which visual discrimination may be achieved (Young 1974). In the outer plexiform zone there are receptive fields of second-order visual cells

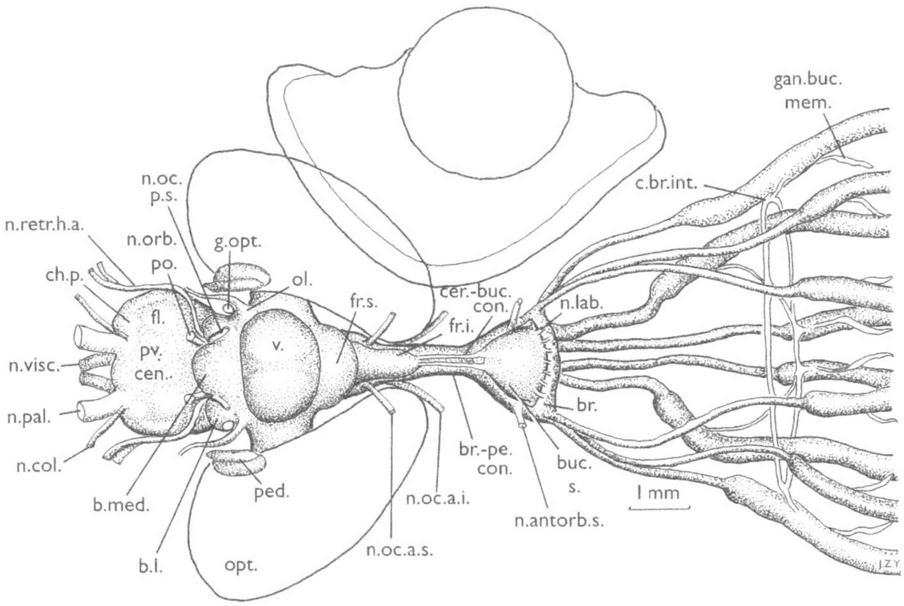

Fig. 8.23. *Loligo vulgaris* (50 mm ML). Reconstruction of the central nervous system made from tracings of transverse sections, and seen from above (Young 1976).

with specific dendritic orientations. These may serve to detect particular contours in the input. The axons of these cells then form a system of radial columns in the outer part of the medulla, linked by cells with wide-spreading dendrites. From here signals are passed by third-order fibres into the columns and axons deeper in the lobe, some directly to the optic tract, and so to the central cerebral lobes.

The deep retina of the optic lobe has thick inner and outer granule cell layers on either side of the plexiform layer. The incoming nerve fibres (processes of the retinal cells) mostly end in the plexiform zone, but a few proceed to the medulla of the optic lobe. The plexiform zone is strictly organized into the radial and tangential layers. The first radial layer contains the swollen inner ends of the optic nerve fibres, making contact with branches of the outer granular layer. The tangential layers contain the dendrites of the second-order visual neurons, whose axons pass on to form the radial columns of the medulla. These second-order cells have an unusual form in that their cell bodies lie mostly in the inner granular layer. The trunk proceeds outwards to the plexiform layer, turns to run tangentially, with spreading dendrites, and then turns in again to pass through the inner granular layer (Fig. 2.12). It is difficult to show the whole course of such a fibre. The second-order dendritic fields vary in size from less than 10 μm to nearly 1 mm in diameter (Fig. 8.24). The smaller dendrites are in the outer tangential layers, influenced by only one or a few visual cells. Their axons end in the outer part of the medulla. The axons of cells with larger fields proceed much further.

The dendritic fields of the second-order visual cells can be divided into four types:

- The smallest second-order visual cells have minute circular dendritic fields in contact with one or a few optic fibres.
- Some of the medium-sized cells also have circular fields.
- Many of the medium-sized cells have long narrow dendritic fields in the second tangential layer. These are often oriented in sets parallel to, or at right angles to, each other. Each tangential fibre carries a series of dendritic twigs. There is therefore a network of fine straight fibres running tangential to the surface of the lobe, with the greatest number in its long axis.

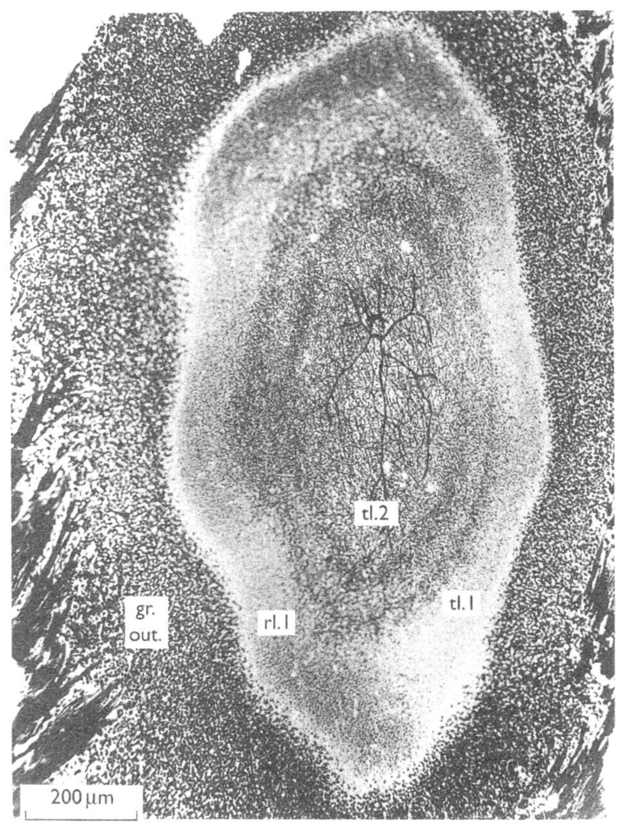

Fig. 8.24. *Loligo*. Sagittal section, tangential to the surface of the optic lobe, showing the various layers and the branches of one large second-order visual cell in the second tangential layer (retouched) (Young 1974).

- The largest of the second-order cells have enormous, oval, dendritic fields several millimetres in length. The large cell bodies are in the inner part of the inner granule cell layer and have plentiful basal dendrites in the inner plexiform layer. Their main trunks proceed to the outer plexiform layer. Each forms a flat spreading tree of branches and then arches over and turns back as an axon proceeding far into the medulla (Fig. 2.12). These remarkable cells are spaced at regular intervals and their dendritic trees form a regular series in the outer plexiform zone. The main arch between the ascending and descending trunks lies in the transverse plane. The main branches then run forwards and backwards, making huge oval dendritic trees that spread over the surface of the lobe. Each of these cells must be influenced by thousands of visual elements. These fields presumably determine the selection of response to the movements of certain contours in the visual field.

The receptive dendrites of the various types of feature detector cells lie at different levels in the plexiform zone. The smaller ones with round fields are the most superficial. Below these are the narrow, highly oriented fields and central to these are the huge dendrites of the very large cells. The impulses from these various second-order visual cells probably interact with the medulla to allow distinct responses to different visual configurations. The very numerous amacrine cells may serve to raise or lower excitability over a particular area within or around a contour. The columnar organization in the outer part of the medulla must have some significance.

The numerous efferent fibres to the retina arise from cells of the inner plexiform layer with their dendritic branches in the plexiform layers. The amacrine cells of both layers vary from those with dendrites extending only for a few microns to those with fields spreading more than 1 mm (Fig. 2.12). A single cell may have branches throughout all the layers of the plexiform zone.

Within the inner granule cell layer there is a thin inner plexiform layer, not found in octopods. This layer contains basal dendrites of the large second-order visual cells as well as fibres running for long distances. It also receives centrifugal fibres from the medulla. Immediately within this is a palisade layer of cells whose trunks proceed inwards.

The medulla of the lobe is not divided into separate regions but is organized to produce wider and wider columns of neuropil proceeding towards the medulla. Near the surface the second-order axons are grouped into columns, often of one large fibre and several smaller ones. These then become surrounded by cylinders of small cells, and the smaller axons terminate. These columns form what Cajal (1917) called (in *Sepia*) a zone of delicate cellular cartridges. The columns are richly innterconnected by large and small cells with tangential dendrites. Then, proceeding inwards, the larger axons diverge increasingly and come into contact with third-order visual neurons and with large and small multipolar cells with wide-spreading horizontal dendrites (Fig. 2.12). The effect is to give wider and wider columns with more neuropil and fibre bundles and fewer cells. All parts of the medulla receive the endings of many centrifugal fibres from the central lobes of the brain and from the opposite optic lobe.

The efferent fibres of the optic tract arise from large cells in the medulla, especially from the inner parts. The details of the arrangement of these have not been seen. Probably all of the input and output tracts proceed from all parts of the lobe. The connections can be divided into four main sets:

- Direct connections with efferent systems. These include (a) optico-magnocellular tracts, probably operating escape reactions, (b) optico-lateral basal tracts controlling the chromatophores.
- Connections with higher motor centres. These pathways lead through (a) the peduncle lobe, (b) the anterior basal lobe, (c) the median basal lobe and perhaps the dorsal basal lobes.
- Connections with centres probably concerned with learning and include (a) the superior frontal lobe and so to the vertical lobe, and (b) the subvertical lobes.
- Commissural fibres to the opposite lobe.

8.4.7 Central nervous system

The brain of *Loligo* was described by Williams (1909), and later Thore (1939), Young (1974, 1976, 1977b, 1979) and Messenger (1979a) gave detailed accounts of the central nervous system and its connections. The volumes of the various lobes of the brain have been measured in newly hatched specimens of two species of *Loligo* by Frösch (1971) and adults of the same species by Wirz (1959) (Fig. 8.3), and of three other genera by Maddock and Young (1987) (Table 2.6).

Suboesophageal mass

The suboesophageal part of the brain can be conveniently divided into the anterior, middle, and posterior suboesophageal masses, often referred to as brachial, pedal, and palliovisceral lobes (Figs 2.5, 8.25). These lobes have been described in detail by Young (1974). The anterior and middle masses are widely separated, and are joined by long tracts of fibres. The middle and posterior masses are sharply separated by the pedal arteries and a transverse septum. This septum is a sheet of fibrous tissue attached to the cephalic cartilage with orifices for the passage of nerve fibres.

The anterior suboesophageal mass contains neurons concerned with movements of the arms and suckers. The middle mass is the main centre for the control of many of the animal's movements. At the hind end of this mass is the perioesophageal lobe. This contains the first-order giant cells, the command pathway for jet propulsion, and also other large cells with axons proceeding to the arms and fins, ensuring appropriate actions of all these structures. The posterior suboesophageal mass contains final motoneurons to the fins, as well as other cells involved in movements of the mantle, the retraction of the head, and actions of the respiratory muscles of the collar. The most posterior portion of the posterior mass is concerned with the control of the viscera and of the vascular system. The skin and the muscles of the chromatophores are controlled by final motoneurons in the anterior chromatophore lobe in the posterior suboesophageal mass.

Anterior suboesophageal mass

The anterior suboesophageal mass is united with the middle suboesophageal mass by a thick bundle of fibres, the brachiopedal connective (Fig. 8.25). The anterior suboesophageal mass has two distinct

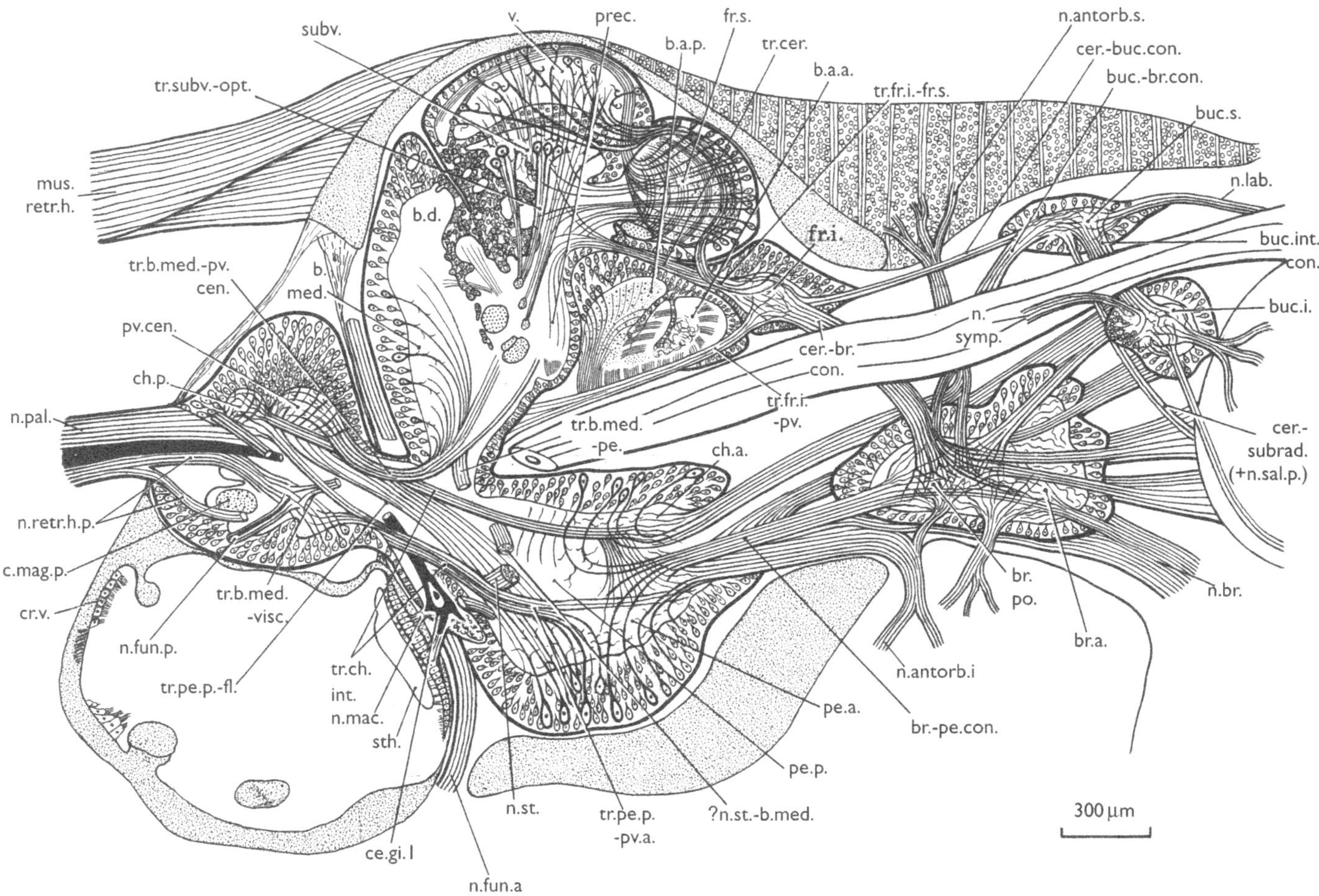

Fig. 8.25. *Loligo pealeii.* Drawing of the brain from two consecutive sagittal sections (50 μm) of the head of a newly hatched specimen. The drawing shows tracts as if seen from the medial surface of the section (Young 1976).

regions, the prebrachial and the brachial lobes, marked by walls of cells (Young 1976). However, the neuropils of the two lobes are continuous but it is not known whether the lobes are functionally distinct.

Each brachial nerve has four distinct roots:

- A dorsal root running to the supraoesophageal and buccal lobes. This contains mostly small fibres.
- A medial root, mainly of small afferent fibres, ending in the prebrachial lobe.
- A lateral root with large and small fibres. The large fibres arise from cells of the prebrachial lobe.
- A pedal root running to the brachiopedal connective and therefore making no connection with the anterior suboesophageal mass. It contains large and small fibres. Centrifugal cobalt filling of a brachial nerve showed the pattern of distribution of filled fibres and somata; this was similar in both *Sepia* and *Octopus* (Budelmann and Young 1987).

Middle suboesophageal mass

This region can be subdivided into:

- the anterior chromatophore lobes, which lie at the front
- the anterior pedal lobe just behind
- the lateral pedal lobes, which are swellings at the side and can be divided into anterior lateral pedal and posterior lateral pedal lobes
- the posterior lobe behind.

These centres control many of the animal's movements, under the higher control of the basal centres of the supraoesophageal mass.

The anterior chromatophore lobes occupy the front face of the mass. They are not clearly indicated externally, but have characteristic structure and connections. The cell wall, dorsally and ventrally, consists of cells graded in size from large (100 μm diameter) to moderate (20 μm diameter), but with no inner layer of small cells. The cells of the ventral part of the lobes are larger than those of the dorsal portion, and perhaps innervate large areas of the skin on the ventral surfaces of the head and arms where the colour patterns are simple.

The anterior pedal lobe is a large and important intermediate motor centre and receives fibres from no less than eight sources. These are:

- from the arms direct and from the anterior suboesophageal mass
- the anterior basal lobe
- the precommissural lobe
- the median basal lobe
- the posterior lateral pedal lobe

- the magnocellular lobe
- the palliovisceral lobe
- the crista statica nerve.

The fibres from the posterior lateral pedal and magnocellular lobes are very large and complex; together they constitute a set and may be concerned with initiating movements of attack with the tentacles. The whole complex of fibres has been easily followed in the small active oegopsin squid, *Pterygioteuthis* (Young 1976, 1977a).

The lateral pedal lobe occupies the upper lateral parts of the middle suboesophageal mass and consists of a large anterior lateral pedal lobe and small posterior lateral pedal lobe. The anterior lateral pedal lobe is an oculomotor centre, with large cells at the outside, and numerous small cells in the inner layers. Four nerves on each side control eye movements, and two ophthalmic nerves provide sensory and motor fibres for tissues around the eyes. There are four oculomotor nerves: anterior superior, anterior inferior, posterior superior, and posterior inferior. The two ophthalmic nerves are the superior and inferior. The posterior superior oculomotor and the superior ophthalmic nerves run together. Each oculomotor nerve has a small number of large fibres (up to 30 μm) and some also have small ones, which may be afferent fibres. The oculomotor centre receives an input directly from the crista and macula nerves and indirectly from the optic lobes via the peduncle and anterior and median basal lobes (Fig. 8.25). The posterior lateral pedal lobe is a backward continuation of the oculomotor centre, reaching to the magnocellular lobe. Its large fibres run to various parts of the pedal lobe; presumably to ensure coordinated movements of the eyes, arms, funnel, and fins, and perhaps of the tentacles when they are projected.

The posterior pedal lobe is a motor centre probably controlling the direction of swimming by regulating the position of the funnel and the action of the fins. It sends fibres direct to the muscles of the funnel and to the head retractor muscles. The lobe also sends fibres forwards to the arm centres of the anterior pedal lobe, and backwards to the centres which control jet propulsion from the palliovisceral lobe, and fin movements from the fin lobe. This important centre receives a large input from the peduncle and median basal lobes, also fibres from the crista nerve and from the magnocellular and lateral pedal lobes.

Magnocellular lobe and its connections

The region around the sides and back of the brain was called the magnocellular lobe because of its most obvious component, the first-order giant cells (Fig. 8.25) (Young 1939). The magnocellular lobe is concerned with jet propulsion but is more than a single centre of convergence on the first-order giant cells. It is not a homogeneous entity but consists of several parts functioning at different levels in the interlocking set of centres concerned with the control of movement.

There are three main parts to this lobe (Fig. 8.26):

- The dorsal magnocellular lobe contains large cells. The axons of these cells reach centres which influence the arms, funnel, fins and mantle, and are under optic and supraoesophageal influence. The dorsal magnocellular lobe is continuous with the peduncle and optic lobes and the subvertical and median basal lobes.
- The ventral magnocellular lobe contains the two first-order giant cells. These are the main command units for the propulsive jet. They receive inputs from many parts of the body and brain. The ventral magnocellular lobe is continuous in front, dorsally and medially with the posterior pedal lobe and with the lateral pedal lobe. The lobes of the two sides are joined by the ventral magnocellular commissure.
- The posterior magnocellular lobe, cylindrical in form, extends along the sides of the palliovisceral lobe with which it exchanges fibres, and ends in a posterior magnocellular commissure.

The magnocellular lobe plays a key role in initiating the actions of the jet for propulsion, but its functions are more complex than this. Besides the giant fibres themselves there are other large cells in the dorsal and ventral parts of the lobe, sending their axons to the centre of the palliovisceral lobe. Others pass forward to the pedal lobe. The dendrites of all of these large cells are mixed with those of the first-order giant cells. Perhaps they all act to produce reactions of the arms and fins as well as of the mantle in jet propulsion. In addition to these efferent pathways fibres pass from the magnocellular to the basal lobes, possibly providing an efference-copy type of feedback.

The magnocellular lobe receives fibres from all the lobes with which it is in contact, and thus signals can reach it from essentially all parts of the nervous system. Fibres (35 μm) pass in from the brachiomagnocellular connective and others come from the pallial nerve. Fibres from the optic lobe and from the macula and crista reach directly to the lobe (Fig. 8.27). Important pathways reach the ventral magnocellular lobe from the anterior and median basal lobes and the precommissural lobe (Fig. 8.28). These are partly collaterals of fibres passing to the palliovisceral lobe. This is evidently a system by which supraoesophageal influences control the movements of the animal as a whole.

Giant-fibre system

The giant axon of *Loligo* was described by Williams in 1909, and rediscovered by Young in 1936. This, together with the giant synapse, became a valuable preparation and model for the processes involved in neural conduction. The giant-fibre system is necessary for fast escape reactions and in squids includes first-, second-, and third-order giant fibres. This was described in *Loligo* by Young in 1939 and comprises a set of first-, second-, and third-order fibres. The function of this system is to ensure an escape response. A pair of giant cells in the magnocellular lobe give rise to axons that fuse so forming a bridge before making synapse with seven pairs of second-order axons. Six pairs of these axons pass directly to the retractor muscles of the head and funnel. The remaining pair of second-order axons run to the stellate ganglia, which lie on either side of the mantle, where they synapse with all of the 10–12 third-order fibres. The cell bodies of these third-order fibres lie in the stellate ganglion. Each of these giant fibres leaves the stellate ganglion in one of the stellar nerves, which also contain many hundreds of medium to small fibres, and passes to the mantle circular muscle. The third-order fibres are graded in diameter and also in their conduction velocity (Pumphrey and Young 1938). The fibre with the greatest diameter reaches to the end of the mantle and that with the smallest diameter to the part of the mantle closest to the stellate ganglion. In this manner the circular

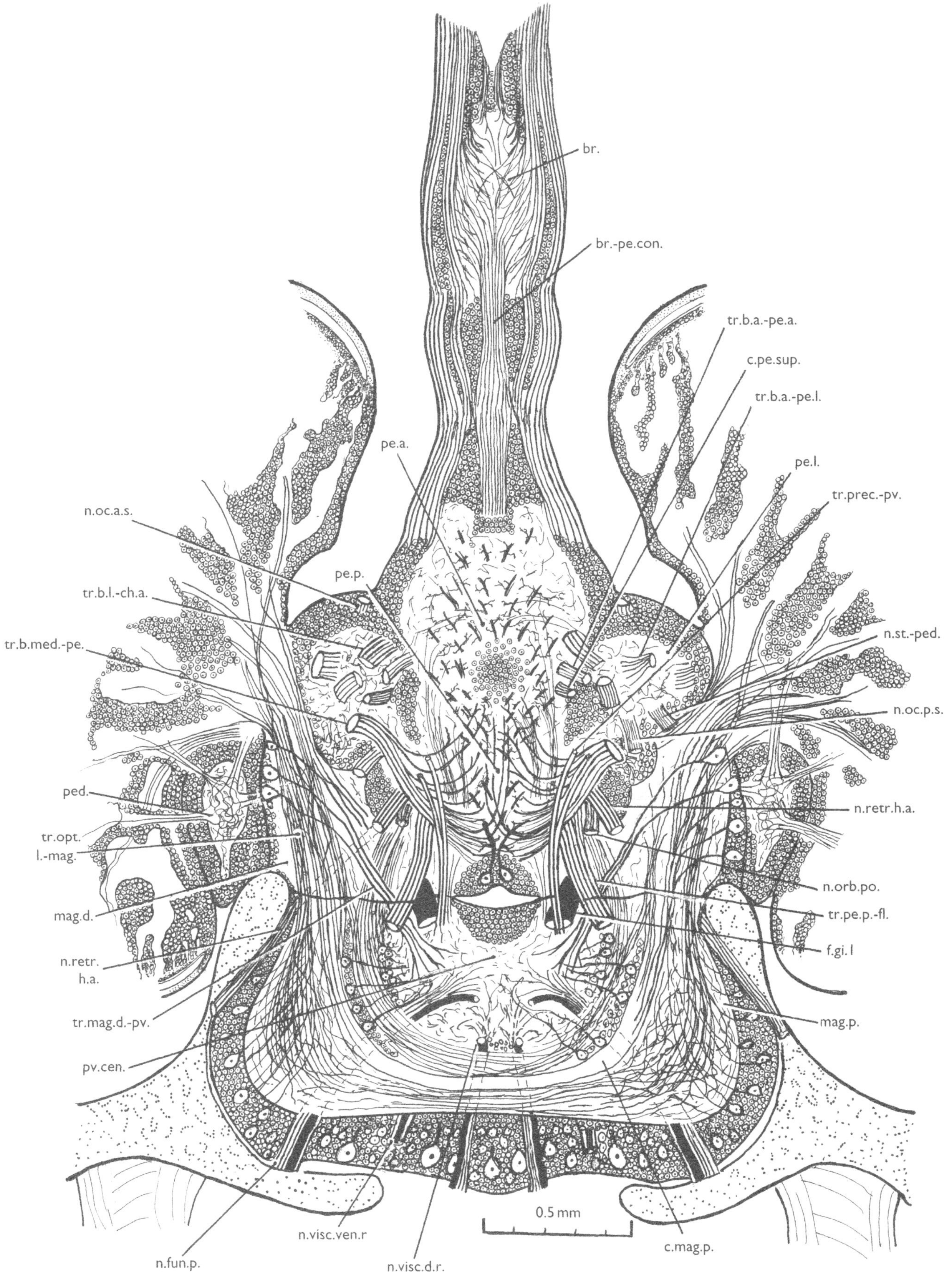

Fig. 8.26. *Loligo pealeii*. Drawing of the brain, seen from above, made from a horizontal section (50 μm) through the head of a newly hatched specimen from the same brood as in Fig. 8.25 (Young 1976).

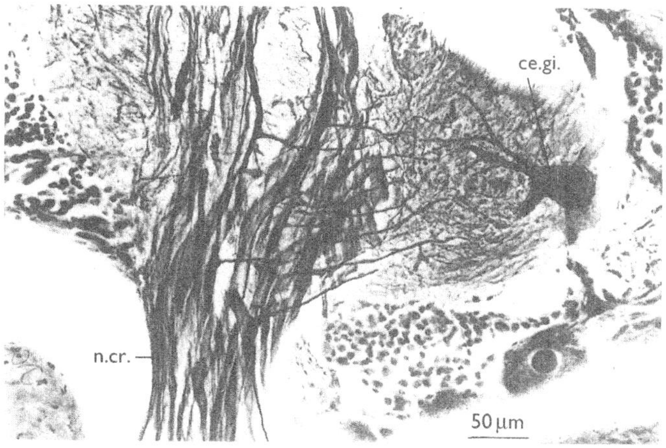

Fig. 8.27. *Loligo vulgaris.* Fibres of the *n. crista statica* showing collateral branches reaching the ventral magnocellular lobe among the dendrites of the first-order giant cell (retouched) (Holmes' stain) (Young 1976).

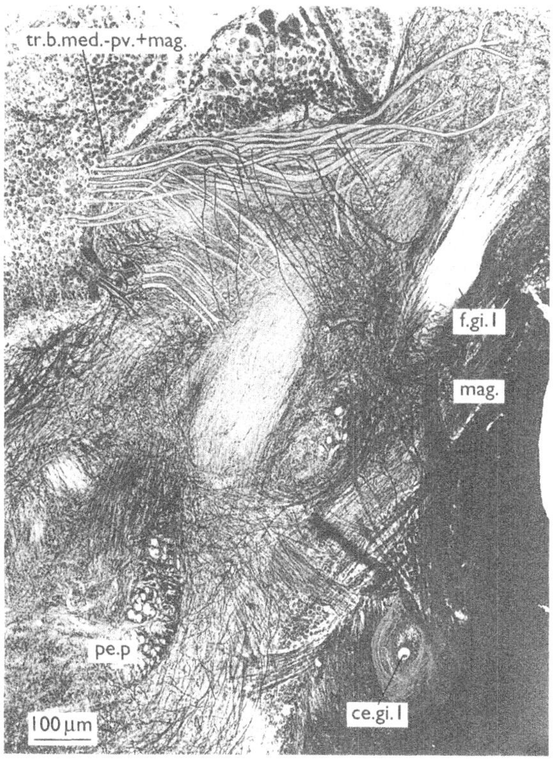

Fig. 8.28. *Loligo pealeii.* Sagittal section to show the median basal to pallio-visceral lobe tract, with its fibres sending collaterals to the magnocellular lobe (retouched) (Young 1977b).

muscles contract simultaneously to expel the largest volume of water through the funnel as a jet to ensure an escape response.

The first-order giant cells lie in the neuropil at the medial borders of the lateral part of the ventral magnocellular lobe, immediately above the maculae of the statocyst (Fig. 8.29). Each first-order giant cell has a single axon which passes backwards into the palliovisceral lobe, and there fuses with the axon of the opposite side, forming an interaxonic bridge. The diameter of the giant cell was 150 μm in

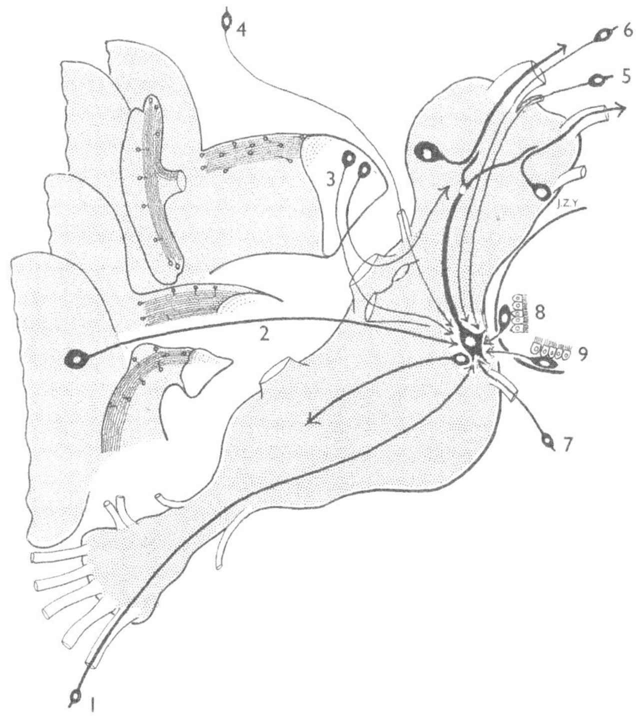

Fig. 8.29. *Loligo.* The final common path through the giant fibre system. The numbers indicate the inputs to the first-order giant cell from (1) the arms, (2) the optic lobes, (3) the median basal lobe, (4) the surface of the head, (5) the collar, (6) the mantle, (7) the funnel, (8, 9) the macula and crista. Output fibres of the ventral magnocellular lobe also go to the anterior pedal lobe and there is a third set to the dorsal magnocellular lobe (not shown). Only two of the second-order giant cells are shown (Young 1976).

a specimen of *Loligo pealeii* of 250 mm ML (Young 1939). The dendrites of the cell extend in many directions for great distances (about 750 μm in the same specimen) and are covered with terminal end feet. The ventral dendrites extend among the macula nerve fibres almost to the receptor cells (Fig. 8.30). The first-order giant axons proceed upwards and then backwards into the centre of the palliovisceral lobe, where the two axons become completely joined across the midline in an adult squid (Fig. 8.31). Evidence for the complete fusion of the axons to form a bridge is provided by light and transmission electron microscope studies of sections cut in several planes (Young 1939, Martin 1965). The stages of the process of fusion have been followed in *Loligo vulgaris* (Fig. 8.32) (Martin 1977). Such a fusion ensures simultaneous contraction of both sides of the mantle.

Behind the interaxonic bridge the first-order giant fibres diverge slightly and break up into a number of branches which make synaptic contact with the second-order axons. Each of these fibres arises from a single cell in the palliovisceral lobe. The largest of these cells are interneurons, each projecting a giant axon via the pallial nerve to the ipsilateral stellate ganglion, where it makes contact with the third-order giant axons. The other second-order giant cells are motoneurons and innervate the retractor muscles of the head and funnel. The trunks carry proximal dendrites in addition to those making contact with the first-order giant fibre. The distribution of the second-order fibres is shown in Fig. 8.31.

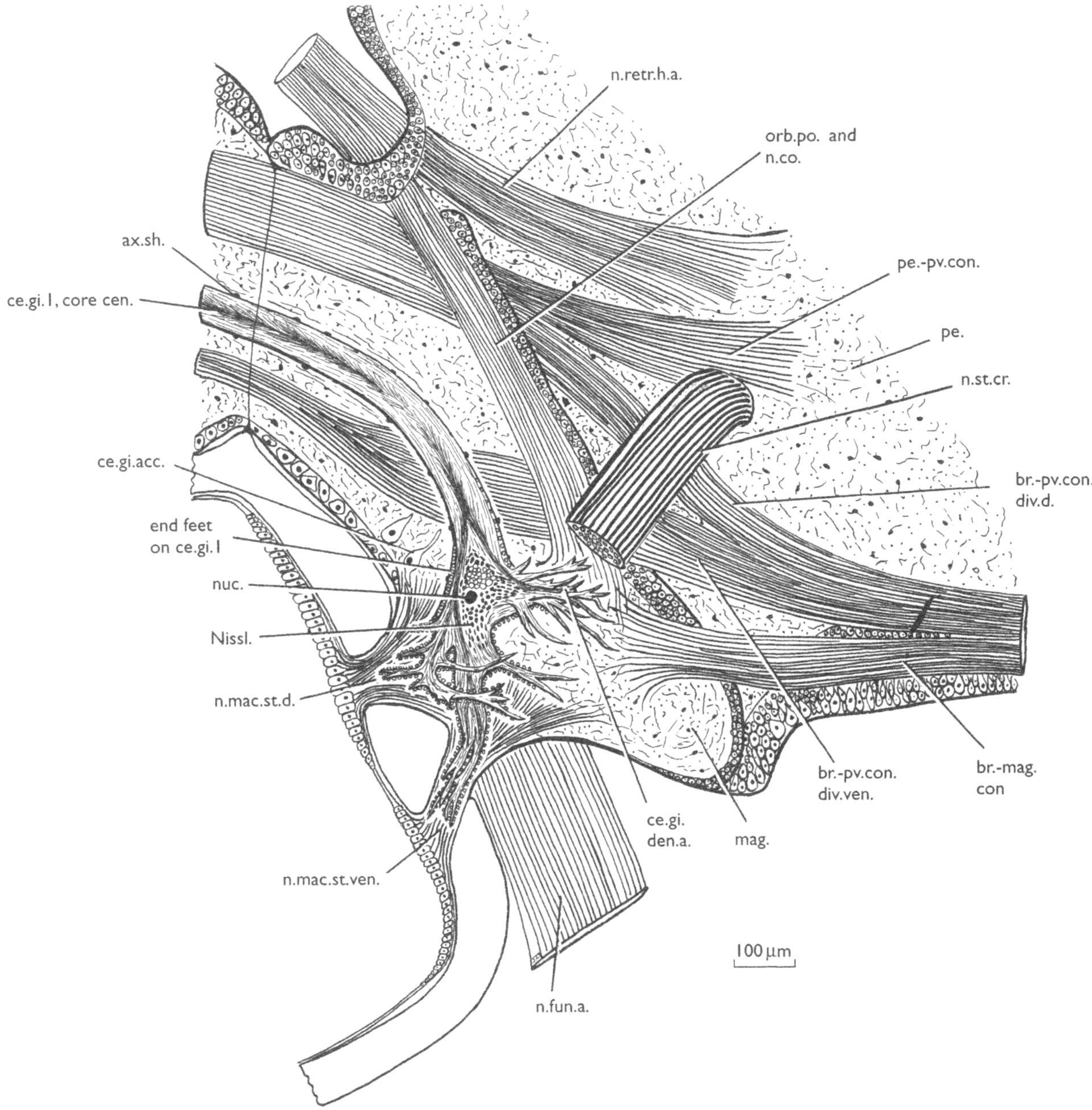

Fig. 8.30. *Loligo pealeii.* Drawing to show the structure of the first-order giant cell in an adult, made by projection from a series of sagittal sections. The giant cell is that on the right-hand side of the squid, and is seen with its lateral surface nearer to the observer (haematoxylin and eosin) (Young 1939).

The fibre that passes to the mantle, via the pallial nerve, makes synapse in the stellate ganglion with the third-order giant fibres which innervate the circular mantle muscles. These nerve fibres are unusual in that they are syncitia, formed by the union of the processes of a large number of small neurons which are aggregated in a special giant fibre lobe (Young 1939). The second-order fibre ends as some 10 branches, each making synaptic contact with one of the third-order giant nerve fibres. The contacts are made by a long series of collateral branches of the third-order fibre penetrating the sheath of the second-order fibre. These branches make close contact over about one-half of the surface of the fibre. They have narrow necks so that there is a great reduction in cross-section between the contact area and the narrow necks of the collaterals of the third-order fibre.

The giant synapse (Young 1973) lies between second- and third-order fibres in the stellate ganglion and is the largest synapse known. Martin and Miledi (1986) described the ultrastructure of the synapse which extends for over 1 mm (1125 μm in a specimen of *Loligo vulgaris* of 180 mm ML). It is a complex structure with the pre- and post-axons not simply apposed to one another uniformly. There are an estimated 15 000 synaptic contacts, and on average each synaptic contact was just over 1 μm in diameter. The structure has been examined experimentally but the nature of the transmitter(s) is still uncertain although it is probably L-glutamate (Messenger *et al.* 1995).

Isolated nerve–mantle preparations indicate that a single impulse arriving via the giant-fibre system results in a maximal response from all of the muscle fibres reached (Pumphrey and Young 1938, Young 1938)

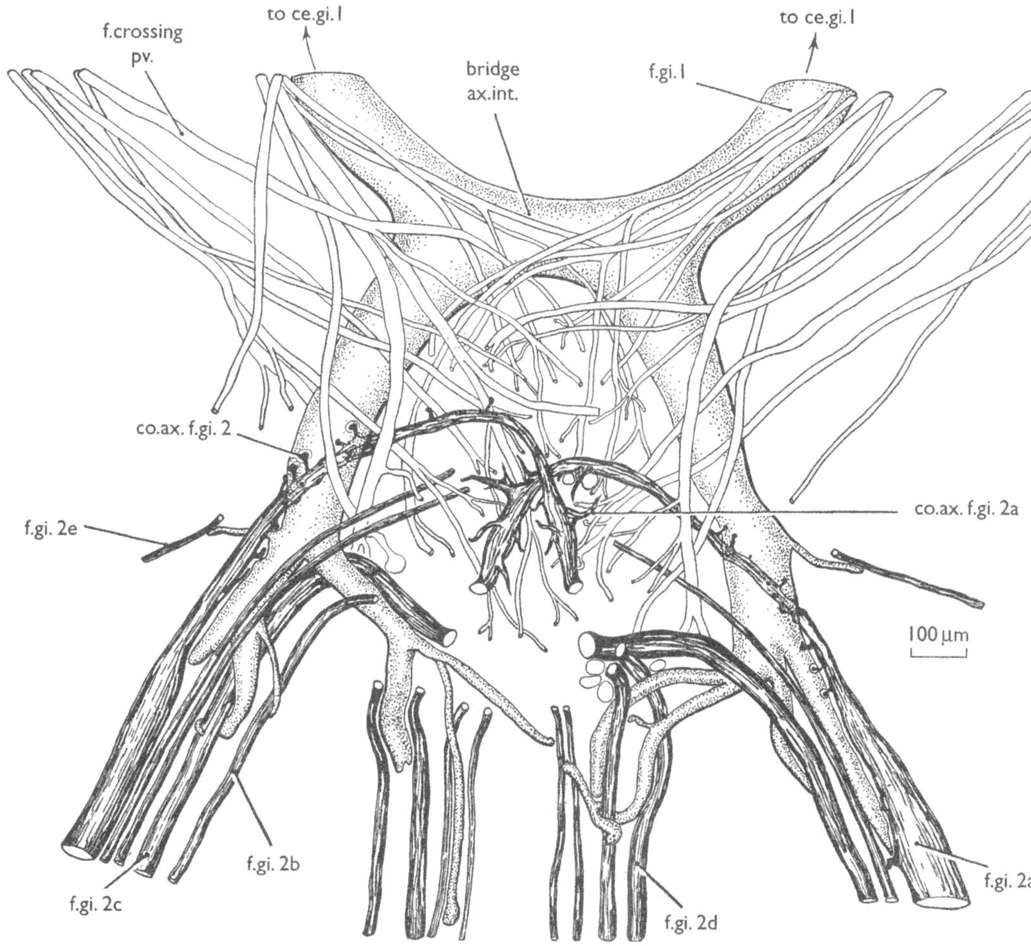

Fig. 8.31. *Loligo pealeii*. Drawing of the giant fibres in the palliovisceral lobe of a small adult (Young 1939).

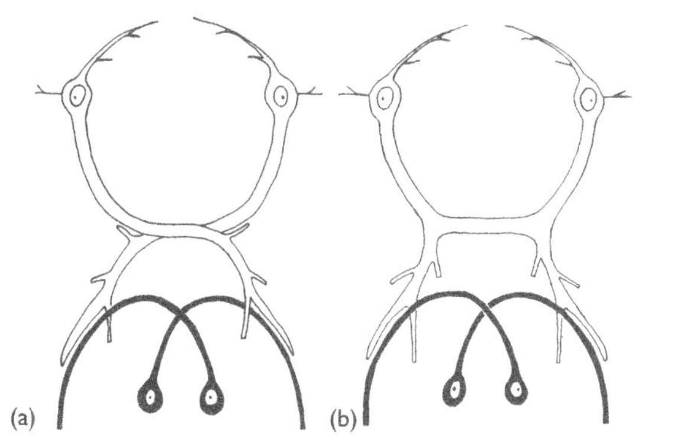

Fig. 8.32. *Loligo vulgaris*. Schematic diagrams of the course of the fibres of the giant fibre bridge: (a) A hatchling in which each axon passes to the opposite side, as they have not yet fused, and (b) in an adult (200–300 mm mantle length) after fusion (Martin 1977).

thus producing the reflex activation of an escape response. D.M. Wilson (1960) demonstrated control of muscular contraction of the mantle by a small-fibre system as well as by giant axons. The escape response of *Loligo* has been recently re-examined and the function of the giant-fibre system and found to be more complex than previously supposed (Otis and Gilly 1990). Using a flash of light as the stimulus, this triggered the activation of the giant axons, after a latency of 50 ms, and elicited a stereotyped startle-response and a powerful jet. Other stimuli produced 'delayed-escape' responses that began after 200 ms and in which the jet was controlled solely by a small motor axon pathway.

The motor function of the mantle is highly coordinated and the activity of the giant and non-giant motor axons result in the all-or-none and graded contractions respectively in the mantle muscle. Both types of axon innervate the circular muscle fibres of the mantle, and they appear to control different sets of circular muscle fibres. In one subset of circular muscle fibres there exists large sodium (Na) currents, and these fibres are proposed to employ Na-dependent action potentials to produce fast, all-or-none muscle twitches (Gilly *et al.* 1991, 1996). In paralarvae of *Loligo opalescens* Na-channel protein was detected in mitochondria-rich fibres, and these fibres assume the dominant role in jet propulsion (Preuss *et al.* 1997).

The ontogeny of the giant motor axons and the neural control of escape responses has been investigated in the embryos and hatchlings of *Loligo opalescens* and *Sepioteuthis lessoniana* (Gilly *et al.* 1991). The development of the third-order giant axons was followed together with the ontogeny of the escape response. At embryo stage 26 (Segawa *et al.* 1988) *Loligo opalescens* is capable of an escape jet driven purely by a small axon system. Later, at embryo stage 28, electrical activity is first recorded in the stellate ganglion associated with the giant axons. Maturation of these axons is accompanied by improvements in escape responses up to the time of hatching. In embryonic and hatchling *Loligo* all escape responses are fast-start responses with latencies of less than the minimum value shown by adults (50 ms). Both giant and non-giant motor systems are thus functional in embryonic and hatchling stages of this loliginid (Gilly *et al.* 1991).

The development of the circular muscles of the mantle was followed in *Loligo opalescens* (Preuss *et al.* 1997). Two fibre types, superficial mitochondria-rich and central mitochondria-poor, are present at hatching. Both types increase in size and in number during development. The central mitochondria-rich fibres increase at a higher rate and the proportion changes from 1:1 at hatching to 1:6 in an 8-week old paralarva. The proportions then remain constant into the adult stage. These differences in the muscle fibres are reflected in their activity. At 1 week of age during jet propelled swimming the paralarvae show mantle contractions with twice the relative amplitude and frequency found in adults. These findings suggest that the mitochondria-poor muscle are important in locomotion at hatching but become progressively less so during the first 8 weeks of life. At hatching the paralarvae are planktonic (see Table 8.1) but with growth they become less so and several species associate with the bottom as juveniles and adults (Boyle 1983b).

Static nerves

The crista nerves are in two parts, a smaller more dorsal branch from the anterior transverse crista and a larger branch from the other three parts of the crista (Stephens and Young 1982). The dorsal branch contains about 20 large fibres up to 40 μm in diameter. The ventral branch has some 60 large fibres. Both parts also have many smaller fibres, afferent and efferent.

The crista nerves mark the boundary between the posterior lateral pedal lobe and the ventral magnocellular lobe. The large nerve fibres divide at entry, giving a branch to the ventral magnocellular lobe in the region of the dendrites of the giant cell. Many of the small fibres then proceed to the ipsilateral lateral pedal lobe, and to the anterior and posterior pedal lobes. Other fibres cross and proceed to the same destinations on the other side.

The macula nerve fibres arise as dorsal and ventral roots innervating the two parts of the macula statica princeps, on which the statolith hangs (see Fig. 8.41). The fibres of the macula nerve spread out in the ventral magnocellular lobe among the dendrites of the first-order giant cells. From here they proceed to other parts of the magnocellular lobe and to the lateral pedal lobe.

Static nerve fibres also reach to the anterior and median basal and peduncle lobes but it is not known whether they come from the macula or the crista.

Posterior suboesophageal mass

This mass lies above the statocyst and below the oesophagus, at the hind end of the suboesophageal mass. It is clearly marked off by the passage of the cephalic arteries and a transverse sheet of connective tissue. The mass shows only slight external lobulation, but is internally subdivided into partly distinct lobes. The fin lobe and posterior chromatophore lobes are distinct anatomically and functionally, but the others are interlocking systems.

Anterior palliovisceral lobe

The front part of the posterior suboesophageal mass is occupied by the neurons that give rise to the more posterior nerves to the retractor muscles of the head and funnel. Fibres are received from the posterior pedal lobe, the anterior funnel nerve, and perhaps from the brachiopalliovisceral connective. The output from the lobe includes axons to the fin lobe, and axons from the largest cells go to make up the roots of the retractor nerves. These nerves also contain second-order fibres of the giant-fibre system.

Central palliovisceral lobe

The central palliovisceral lobe (Fig. 8.26) is one of the most difficult regions to define as it is related to so many operations. Its connections suggest that it controls the contractions of the mantle by way of the stellate ganglion. This action is related to respiration as well as locomotion and requires coordination with the actions of the retractor muscles and those of the fins, the collar, and the arms. The outputs from the central palliovisceral lobe are fibres to the mantle connective, ending in the stellate ganglion, including the second-order giant fibre, and probably several rather smaller fibres that accompany it. The accessory giant fibre probably arises separately. The smaller fibres to the stellate ganglion may include two sets, one for respiration and the other for locomotion, but as yet these have not been identified. Fibres of the collar nerves arise here, and are primarily respiratory. The neuropil shows a dense tangle of fibres, and the giant fibres, their bridge, and their synapses are conspicuous at the centre. The lobe is under many influences as it receives inputs from the magnocellular lobe, the mantle connective, from the posterior funnel nerves, the visceral nerves, fibres from the supraoesophageal lobes in the median basal to palliovisceral tract and precommissural to palliovisceral tract, and also part of the brachiopalliovisceral tract.

Visceral lobes

The lobes occupy the posterior and ventral part of the posterior suboesophageal mass, behind the posterior magnocellular commissure (Young 1976). There are thick walls of small cells with a few large cells (more than 100 μm in diameter) that give rise to the giant fibres of the visceral nerves. There is a mass of very small cells at the hind end of the magnocellular and lateral palliovisceral lobes and above the exit of the posterior funnel nerve. Bundles of fine fibres run among the small cells and form the two parts of the anterior vena cava nerves. This forms the characteristic extracortical neuropil (Thore 1939), presumably supplying cells of neurovenous tissue in the wall of the vein (Alexandrowicz 1964, 1965). These lobes contain cells whose functions are connected with internal rather than external adjustments (Young 1976).

Posterior chromatophore lobe

The posterior chromatophore lobe is clearly demarcated at the back of the posterior suboesophageal mass dorsally (Fig. 2.6). It is related to the fin lobe in front but exchanges no fibres with it. The pallial nerve enters below it and to this it sends most of its output fibres. The lobe innervates the muscles of the chromatophores and skin of the mantle and fins, by fine fibres running to them through the pallial, collar, and fin nerves without further synapse. The neuropil does not show the complications seen in those cephalopods exhibiting elaborate colour patterns. The lobe receives the lateral basal to chromatophore lobe tract, which is the main source of control of colour and is under the influence of the visual system. The lobe is joined to its fellow by a posterior chromatophore commissure, and it carries fibres from the lateral basal to chromatophore tract.

Fin lobe

This lobe occupies the dorsal part of the posterior suboesophageal mass on each side, in front of the posterior chromatophore lobe (Fig. 2.5). It lies dorsal to the magnocellular lobe and lateral to the central pallioviscera lobe. The fin lobe contains much larger cells (up to 120 μm) than the chromatophore lobe, and all are motoneurons. The neuropil is characteristic with large trunks of cells clearly apparent among the incoming fibres of the posterior pedal to fin lobe tract, which forms the main input. There is an exchange of numerous fibres where the fin lobe meets the central palliovisceral lobe medially. Some fibres reach the fin lobe from the brachiopalliovisceral tract, and a large bundle reaches the lobe from the smaller cells of the anteroventral part of the anterior palliovisceral lobe.

Supraoesophageal mass

The supraoesophageal centres (Figs 8.23, 8.25) can be divided broadly as follows:

- The basal lobes and peduncle lobes. These are higher motor centres, concerned with the regulation of movement.
- The olfactory lobes, which lie between the dorsal basal and the peduncle lobes.
- The dorsal basal and subpedunculate region, a part of the back of the supraoesophageal lobes. This region is probably concerned with reproductive, visceral, and neuroendocrinal functions. As yet there is no evidence of a central activating system in cephalopod brains, such as the reticular system of vertebrates, nor of centres for regulating internal functions, such as the hypothalamus of vertebrates. It cannot be excluded that there are areas with such functions; these may be in the dorsal basal region or in the visceral lobes, both of which are little understood.
- The lateral basal lobes are concerned with feeding by their connections with the arms and buccal lobes.
- The vertical lobe system is quite distinct and is probably (as in *Octopus*) involved in setting up visual memories.

Basal lobes

The peduncle, anterior anterior, and posterior anterior basal, median basal and interbasal lobes have some signs of systems of fine parallel fibres, recalling those of the vertebrate cerebellum (Hobbs and Young 1973, Messenger and Woodhams 1976, Young 1977b, Messenger 1979a, 1983). They have combined optic and static connections and are concerned with the regulation of motor activities, both of the whole animal and of its eye movements. The median basal lobe is more complicated than the others and may have different functions. The magnocellular lobe has several similar features of structure and connections and is sometimes considered with the others.

The peduncle, anterior anterior, and posterior anterior basal lobes have a similar structure, and may be concerned with movement in different planes. The long axes of the peduncle and optic lobes are coincident in the vertical sagittal plane and the anterior basal lobes lie in the horizontal transverse plane. The extraocular photoreceptor organs, or photosensitive vesicles, are connected with the peduncle lobe and are perhaps involved in vertical migratory movements. The anterior basal lobe is probably concerned with yawing movement since lesions to this lobe in *Sepia* (Boycott and Young 1950) and *Octopus* (Messenger 1967c) resulted in the animals making spinning movements.

Peduncle lobes

The paired peduncle lobes have been described in detail by Messenger (1979a). They lie at the sides of the brain behind the optic tracts. Each is elongate and cylindrical in shape, and partly embedded in the optic lobe. The long axis of the peduncle lobe is coincident with that of the optic lobe (Fig. 8.23).

Each peduncle lobe has two distinct regions, the basal zone and the spine. These two regions are not sharply separated and the neuropils are confluent, but their organization is quite different. The basal zone has mostly medium-sized cells (7–25 μm in diameter), the larger ones sending their trunks medially towards the central brain. The spine is characterized by its uniformly small cells (6–7 μm in diameter) and its linear organization. It is composed of two banks of fine parallel fibres running its length. These parallel fibres are partly extrinsic in origin, deriving from a variety of sources, and partly intrinsic, these last being the processes of cells of the spine itself that never leave the lobe. The connections between the two regions of this lobe are not yet understood, and in *Loligo* it is almost as if the lobe were in two separate parts (Messenger 1983).

The peduncle lobe projects to motor areas, mainly in the supraoesophageal mass (Fig. 8.33), including the lateral pedal lobe, the anterior basal, median basal, and lateral basal lobes, the interbasal lobe, the contralateral peduncle lobe, and the ipsilateral optic lobe. There is a conspicuous and disproportionately large peduncle commissure (Messenger 1979a).

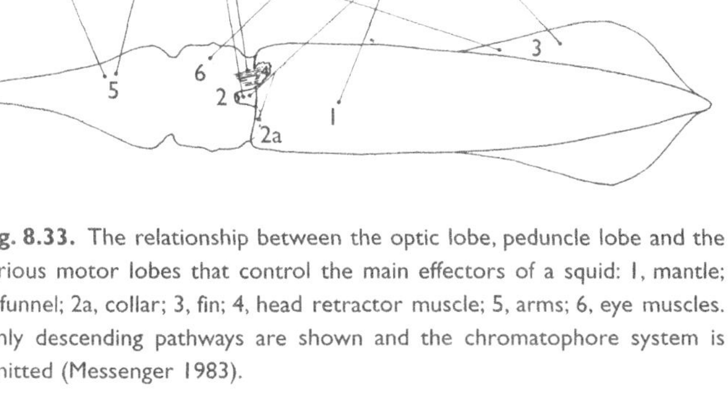

Fig. 8.33. The relationship between the optic lobe, peduncle lobe and the various motor lobes that control the main effectors of a squid: 1, mantle; 2, funnel; 2a, collar; 3, fin; 4, head retractor muscle; 5, arms; 6, eye muscles. Only descending pathways are shown and the chromatophore system is omitted (Messenger 1983).

All afferent fibres enter the peduncle lobe through the basal zone, either in tight well-defined bundles or through spaces in its medial aspect. The most conspicuous is the afferent input to the peduncle lobe that comes from the optic lobes. Fibres also enter the lobe from the statocyst as two bundles (Messenger 1983).

Anterior basal lobe
The anterior basal lobe consists of two parts set at right angles to each other, the anterior anterior and the posterior anterior basal lobes, They are separated by a narrow row of cells, which is crossed by bundles of fibres running anteroposteriorly between then. The lobes join at the sides as the lateral anterior basal lobe.

Anterior anterior basal lobe
The anterior anterior basal lobe lies in an oblique vertical transverse plane, and it slopes backwards from above in front at about the same angle to the main axis of the animal as the optic and peduncle lobes. Three parts may be recognized:

- The lateral part contains large efferent cells, with axons passing to the oculomotor and other suboesophageal motor centres.
- The anteromedian lobule sends axons to the optic lobes, and the fibres are regularly arranged in a complex commissure.
- The parallel fibre region lies behind the other two and may influence the cells of both of them. The lobe also receives fibres from the other basal lobes, both optic lobes, the statocysts and the magnocellular lobe.

The connections of the lobe with the optic tract are made by numerous bundles of fibres passing in two directions. They form a complex system with elaborate plexiform arrangements that seems to be divisible into three sets; dorsal, lateral, and ventral optic to anterior anterior basal lobe tracts. The first set runs to the midline and forms an elaborate chiasm. The second set consists of bundles all down the sides of the lateral anterior basal lobes. The third set is of very large fibres running to all the basal lobes, and occupies most of the ventral part of the optic tract. This pathway evidently serves some overriding function in either attack or retreat. Attack is likely to be the termination of a regulated process of tracking, which may end with the image of the prey, or object, acting on particular regions of the retinae and optic lobes of both sides.

Posterior anterior basal lobe
This is a more compact lobe than the anterior part. It is organized around rows of cells and fibres lying in the transverse plane, inclined so as to correspond to the plane of the optic lobe. It is not so distinctly divided as the anterior anterior basal lobe, but three components can be recognized. The main central part lies transversely, posteriorly there are the lateral lobules and dorsally there is a spine region of fine parallel fibres. The sets of parallel fibres are organized at right angle to those of the anterior anterior basal lobe.

The incoming fibres run in a series of bundles interweaving in the transverse vertical plane and they come from at least four sources: the optic lobes, the statocysts, the other basal lobes, and the magnocellular lobe. Some of the fibres end on the same side, others are crossed.

The efferent cells and fibres lie at the back and sides of the posterior anterior basal lobe. Their trunks and dendrites thus cross the bundles of incoming transverse fibres and make synapse there. The destination of these output axons are probably the same as those of the anterior anterior basal lobe. The outputs are to the lateral pedal lobe, the anterior pedal lobe, the posterior pedal lobe, the palliovisceral lobe, the other basal lobes and the optic lobes.

Posterior basal region
The region at the back of the supraoesophageal lobes is often called the posterior basal lobe. This is not a single homogeneous entity but includes six distinct centres, and although the functions of its various parts are related, many being concerned at least in part with movement of the whole animal. The parts of the posterior basal region and their probable functions are:

- median basal lobe steering and jet
- dorsal basal lobes (anterior and posterior) control centres
- interbasal lobe tentacle control centre
- dorsolateral lobes reproduction control centre
- subpedunculate lobe ocular pressure control centre
- lateral basal lobes chromatophore centre.

Median basal and dorsal basal lobes
These lobes may form a unit comparable with the peduncle and anterior basal lobes, with the dorsal basal lobe representing the spine, but there are considerable differences and this interpretation is hypothetical. The region is more extended anteroposteriorly than the anterior basal lobe. The median basal lobe occupies the whole central part of the back of the supraoesophageal lobe and joins the middle suboesophageal mass at the sides (Fig. 8.34). The dorsal basal lobes are directly above it, and consist of two areas of neuropil separated by walls of small cells. These cells each have either a single short trunk or several running in various directions. Their fibres run transversely in a meshwork through which pass the fibres of the optic connections.

The optic input to the median basal lobe comes from separate dorsal and ventral tracts. Other fibres reaching the median basal lobe come from the inferior frontal, palliovisceral, interbasal, peduncle, magnocellular, precommissural, anterior basal, and brachial lobes, and from the statocysts.

The output of the median basal and dorsal basal lobe system comes from the large cells at the back of the median basal lobe. Their axons reach to ten destinations. The output pathways include some of the largest cells and fibres of the supraoesophageal lobes. The fibres for the control of jet propulsion are especially striking. They run beside similar fibres from the precommissural lobe. Each fibre sends a large branch down to the magnocellular lobe and passes on to divide up in the palliovisceral lobe. Control of the fins is mediated through the median basal lobe by fibres passing to the posterior pedal, anterior palliovisceral and fin lobes (Fig. 8.35).

Interbasal region
This region is lateral to the other basal lobes, and lies between them and the optic lobes. It is a little understood but nevertheless important area, as afferent fibres from numerous sources converge here and are

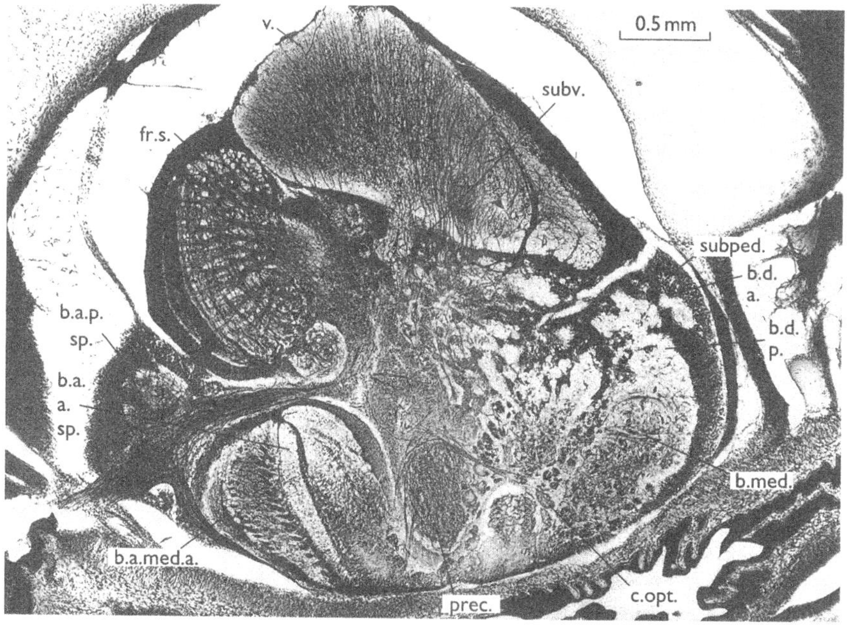

Fig. 8.34. *Loligo pealeii.* Sagittal section of the supraoesophageal mass, near the midline, to show the basal lobes of the brain (Young 1977b).

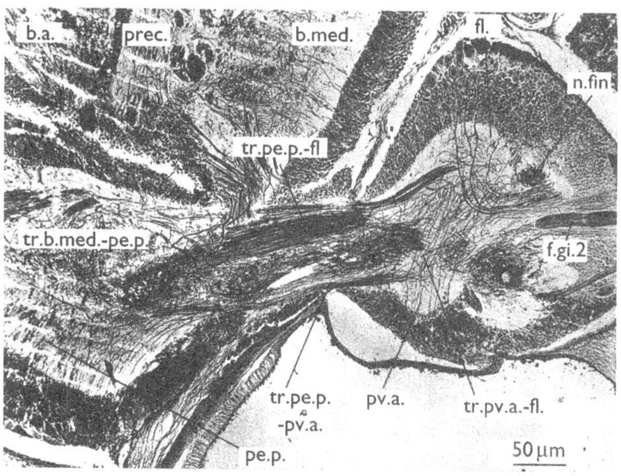

Fig. 8.35. *Loligo pealeii.* Sagittal section of a young animal showing fibres of the median basal to anterior palliovisceral and fin lobe tracts. The fin lobe lies lateral to this section but a few of its cells are shown with fibres proceeding to the fin nerve (retouched) (Young 1977b).

distributed to many destinations. Many fibres pass through this region, and it is not possible to analyse all its connections. It has lateral and medial parts separated by the lateral basal to anterior chromatophore lobe tract. The most obvious feature is the interchange in the more medial part made by an elaborate set of interweaving fibres coming from various peripheral nerves. These fibres intertwine to make a conspicuous plexus in silver preparations. The effect of this is to distribute afferent fibres to many parts of the supraoesophageal lobes, perhaps signalling trauma as a basis for avoidance reactions or learning. The more lateral part of the interbasal region receives dendrites from some of the large cells of the median basal lobe and other regions.

Dorsolateral lobes

These are a pair of small lobules at the back of the supraoesophageal mass, and are important as the source of the fibres of the nerve to the optic glands, which probably influence reproduction (as in *Octopus vulgaris*; Wells 1978). The optic gland itself lies on the lateral surface of the dorsolateral lobe, and in front is a mass of small cells of the subpedunculate and paravertical tissues. Fibres are received by the dorsolateral lobe from the optic tract, the peduncle, olfactory, and dorsal basal lobes; all may influence the development of the optic glands and so of the gonads.

Subpedunculate lobes

Across the back of the brain above the median basal lobe are three layers of lobules. These are (from the back forwards) the subpedunculate lobes 1, 2, and 3 (Young 1977b). In *Loligo* subpedunculate lobe 1 is a distinct lobe with relatively large cells (up to 12 μm in diameter), whose trunks form compact bundles of axons, which run towards the optic tract but their destination is not known. Subpedunculate lobes 2 and 3 form two rows of more irregular lobules, composed of mainly much smaller cells, which have very fine branches but perhaps carry no axon.

The main input to the whole pedunculate system is from optic nerve fibres. These come from the web of bundles in the dorsal basal lobes. The centre of each lobule of the two upper subpedunculate lobes is occupied by a dense tangle of these incoming fibres. They branch repeatedly and show many swellings.

Lateral basal lobes

These lobes lie at the back and sides of the supraoesophageal lobes in relation to the magnocellular lobes below, the olfactory and the peduncle lobes laterally, the median basal lobe medially, and the dorsolateral lobule dorsally (Fig. 8.36). They are concerned in the

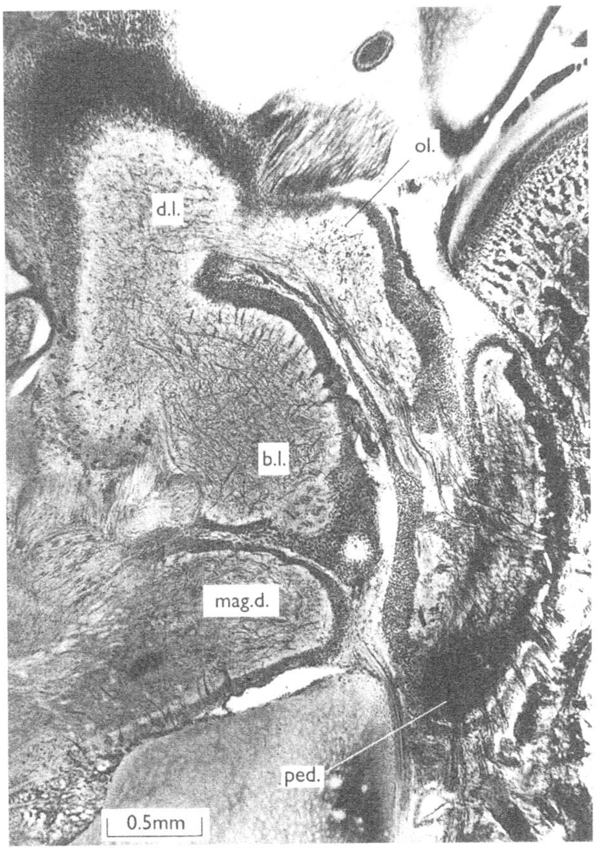

Fig. 8.36. *Loligo pealeii*. Transverse section to show the relation of the dorsolateral lobule to the olfactory and peduncle lobes (Young 1977b).

control of chromatophores (Novicki *et al.* 1990). The cells, all on the posterior wall, range in size from very small to 30 μm in diameter. The large and medium-sized cells send axons to tracts to the anterior and posterior chromatophore lobes. The smallest cells are microneurons with trunks restricted to the lobe. The input is mainly optic, from the same or the opposite optic tract. The fibres run circumferentially round the lobe across the trunks of the cells. They branch on entry to the lobe and then many times, and so each sends its influence to a large part of the lobe. A small number of fibres have been traced to the middle pedal commissure and these almost certainly come from the statocyst. Fibres also enter from the peduncle lobe.

Olfactory lobes

Each olfactory lobe is elongated, and lies medial to the peduncle lobe (Fig. 8.36). It consists of three lobules continuous with the dorsolateral lobe. Nerves from the olfactory organs enter the olfactory lobe, where they extend throughout lobe and to the dorsal basal lobe. Electrophysiological and behavioural experiments have revealed that the paired olfactory organs are chemosensitive. Living, but restrained, squid respond with vigorous jetting movements to squid ink, and to its precursor (L-dopa) as well as to quaternary ammonium ions and other potassium channel blockers (Lucero and Gilly 1995, and see 'Olfactory organs').

Vertical lobe system

The vertical lobe system forms the 'highest' part of the central nervous system (Fig. 2.5) (Young 1979) and is probably concerned with visual learning and complex behaviour, as it is in *Octopus* and *Sepia*, but as yet we have little evidence of its function in *Loligo*. Measurements of the system have been made in three genera of loliginids (Table 2.6) (Maddock and Young 1987). The vertical lobe system consists of the inferior frontal, superior frontal, vertical, and subvertical lobes (Fig. 8.37). It receives inputs from the eyes, arms, mouth, and skin receptors. The output of the system includes large neurons reaching to the control centres of the arms and mantle that initiate movements of attack or retreat; another part of the output passes back to the optic lobes (Young 1979). The vertical lobe system consists of a series of matrices in which the fibres carrying visual impulses can interact with signals of touch from the arms, of taste from the lips, and pain from various parts of the body (Young 1995a).

Inferior frontal lobe

This is the most anterior part of the supraoesophageal mass. It has been conventionally called the inferior frontal lobe (Thore 1942) and this usage is retained here. By analogy with *Octopus* it should, strictly, be called the posterior buccal lobe (Young 1971), since the lobe called the median inferior frontal in octopods is not present at all in decabrachians (Young 1979). However, there is an interchange system of fibres from the arms here, and so it may be compared with the median inferior frontal lobe of *Octopus*. The inferior frontal lobe of decabrachians serves as a main input and output channel between the arms and lips and the higher cerebral centres. It contains plexuses that ensure distribution of these fibres. There is no evidence of the contribution that its cells make to the control of behaviour.

The inferior frontal lobes of the two sides are joined in the midline where there is considerable interchange of fibres. The lobes occupy the lower part of the front of the supraoesophageal lobe (Figs 2.5, 8.25). They are narrow in front but widen out behind. The cell walls of the lobes are uneven, being thicker in front, behind, and above, but thinner at the sides. Numerous smaller cells are present with axons proceeding to the superior frontal lobe. Larger cells, mostly at the front and back, send axons to the brachial and buccal lobes, and also some to the subvertical lobe, and perhaps to the median basal lobe.

The input to the inferior frontal lobes is from several sources (Fig. 8.37) (Young 1977b). Many fibres come in the cerebrobrachial connective direct from the arms, and this has been demonstrated by cobalt filling (Budelmann and Young 1987). The fibres enter a complex plexus in which they interweave, reminiscent of the median inferior frontal lobe of an octopod. This then resolves into a series of bundles. They are joined by fibres from the cerebrobuccal connectives. Other fibres enter from behind in the cerebral tract and inferior frontal lobe tracts. These last contain fibres running between the front and back of the brain.

The output of the inferior frontal lobe is to four destinations (Young 1977b). Fibres proceed in the cerebrobrachial tract to the brachial lobes. Some pass directly to the arms. Fibres also proceed in the cerebrobuccal connectives to control the buccal mass. The inferior frontal to subvertical lobe tract carries onward signals from the combination

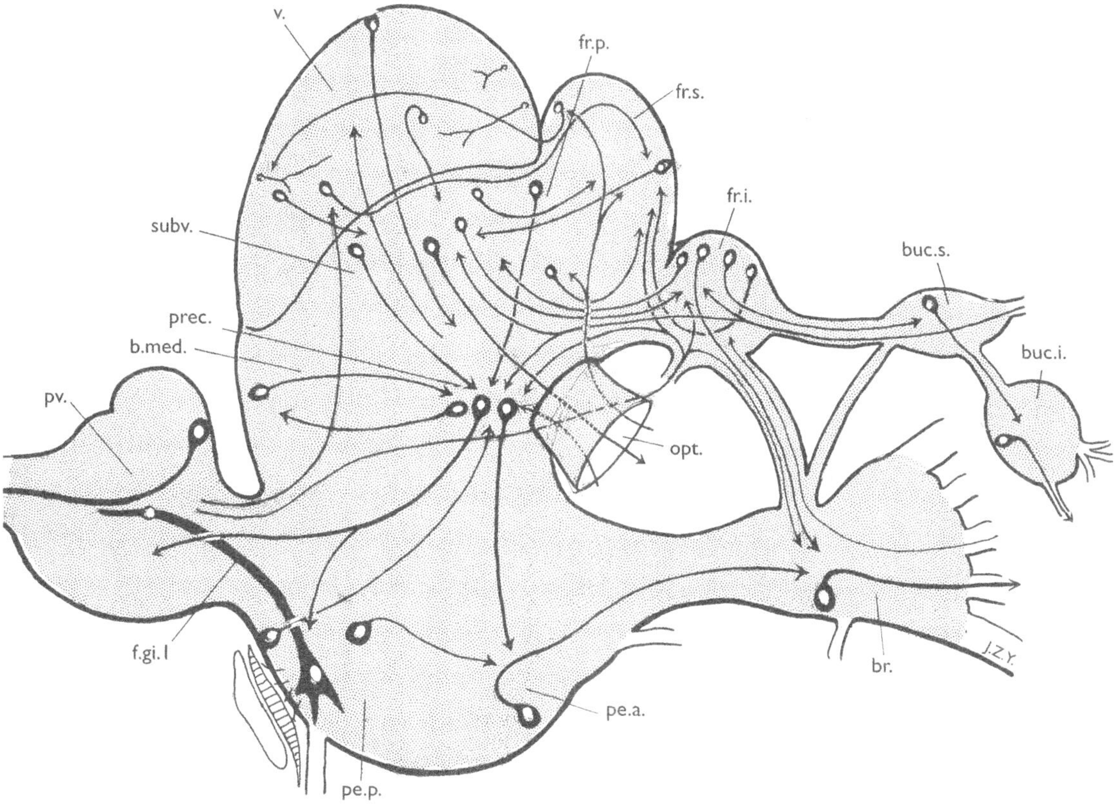

Fig. 8.37. *Loligo.* Diagram to show the inputs and outputs of the tracts of the vertical lobe system (Young 1979).

of influences that reach the inferior frontal lobe. Entrance to the vertical lobe system is thorough the interfrontal tract (inferior frontal to superior frontal lobe tract), leading to the superior frontal lobe.

Superior frontal lobe

The superior frontal lobe occupies the front of the supraoesophageal mass above the back of the inferior frontal lobe (Fig. 8.23). It is related to the optic tracts laterally, to the cerebral tract and anterior basal lobe below, and to the vertical and subvertical lobes behind. Around the back of the superior frontal lobe is a ring of rather large fibres, arranged as lobules, and these are the postfrontal lobes (see below).

The superior frontal lobe serves to receive afferent fibres from the optic lobes, to allow them to interact with those from the arms and lips to send output to the vertical lobe and to the subvertical lobe. From both of these destinations fibres pass back to the superior frontal lobe. There is thus a remarkable pair of closed circuit loops, one above the other, the lower one receiving additional inputs. The loops cross in the superior frontal lobe in a highly ordered system of layers through which the trunks of the cells pass.

The cell walls and neuropil of the superior frontal lobe show two distinct parts with different inputs and outputs. The anterior superior frontal lobe has larger cells and sends fibres to the subvertical lobe and also beyond to the precommissural and perhaps even the suboesophageal regions. The posterior superior frontal lobe contains numerous columns of small cells whose axons all run through an interweaving plexus to the vertical lobe. The superior frontal lobe contains no short axon (amacrine) cells.

The trunks of the cells of the two parts of the superior frontal lobe are similar. Each sends a single main fibre towards the centre of the lobe, giving off lateral dendritic twigs which make contact with the various layers of incoming axons (Fig. 8.38). Thus each cell samples a narrow cylinder of neuropil. The fibres that form the superior frontal to vertical tract enter the vertical lobe from the front and pass upwards and backwards immediately within its cell layer. They interweave further as they pass back to the extreme hind end of the vertical lobe.

The organization of the neuropil is similar in the two parts of the superior frontal lobe. Both receive large inputs from the optic lobes, the difference being in the relative contributions of influences from the inferior frontal, vertical, subvertical, and postfrontal lobes. Fibres from these sources of input run round tangential to the surface across the backward running trunks of the cells of the lobe, making a series of layers when seen in sagittal section (Fig. 8.38).

From the periphery of the lobe the outermost layer of the neuropil, nearest to the cell bodies, is composed of fibres of the superior frontal to vertical lobe tract. The bundles of this tract make a complicated plexus, probably allowing fibres from one part of the vertical lobe to reach back to any part of the superior frontal lobe and vice versa. Fibres from the inferior frontal lobe enter the superior frontal lobe from below by the interfrontal tract and form a second layer. The fibres of this tract divide as they enter the superior frontal lobe, one branch going to each part. The fibres from the subvertical lobe enter from below and make another layer. They are accompanied by fibres that run from the postfrontal lobe across the hilum of the lobe.

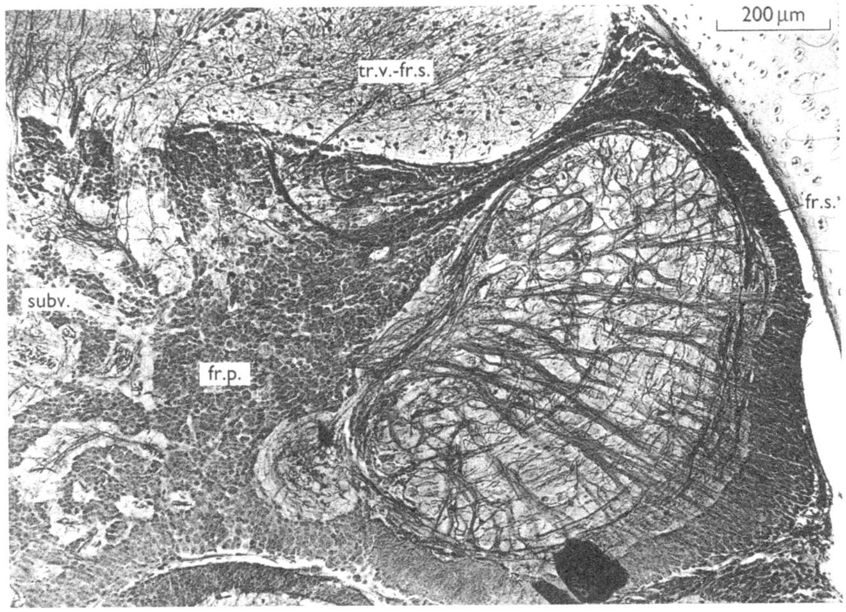

Fig. 8.38. *Alloteuthis subulata.* Sagittal section near the midline showing the vertical to superior frontal lobe tract and the layer of postfrontal cells at the back of the superior frontal lobe (Young 1979).

The fibres from the optic lobe enter laterally by three pathways and run in layers across the central neuropil of both parts of the superior frontal lobe (Fig. 8.37).

Postfrontal lobe

This is a set of cells grouped around the hilum of the superior frontal lobe. Some of the cells are large and have bushy dendrites quite different from those of the neighbouring lobes (Young 1979). The axons of these cells pass either forwards into the superior frontal lobe or back into the subvertical and precommissural lobes. Their large dendritic fields receive inputs from the optic tract and from the subvertical and the inferior and superior frontal lobes.

Vertical lobe

This is the highest part of the brain both anatomically and functionally (Young 1979). The vertical lobe is a conspicuous paired, dome-shaped structure at the top of the brain (Figs 2.5, 8.34). Each half comprises two parts with different structures yet with neuropils largely continuous, and so probably forms a functional whole. The peripheral part of the lobe has a thick cell wall with many small amacrine cells and a few large cells. The central part of the lobe occupies the middle of the dome. Its cell layer is thin, composed of a single layer of large cells, and at its centre is a mass of neuropil in which further medium and large neurons are dispersed.

The central area has large cells of a striking form. The main trunk of each cell is smooth and the first dendrites are a few short tufts attached to a swelling. At the centre of the lobe the trunk begins to curl and gives off a series of thick collateral branches, often at right angles. Some turn back to the surface, others make loops down to the base of the lobe and back again upon themselves. This gives the neuropil a very characteristic appearance (Fig. 8.39). The dendritic branches end in localized synaptic tufts of fibres tapering to fine points. These cells thus provide for collection of influence from a

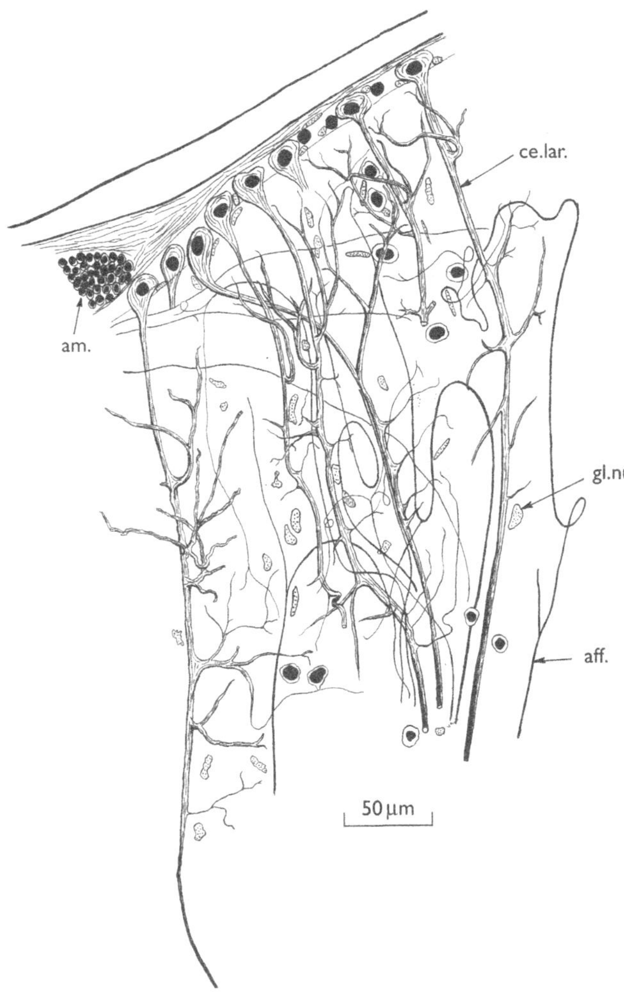

Fig. 8.39. *Loligo pealeii.* Drawing of a sagittal section of the vertical lobe showing the large cells of the central zone and some curling fibres presumed to be afferent (Young 1979).

number of widely dispersed areas. The main trunks of the cells pass through the lower surface of the lobe in bundles and either end in the subvertical lobe or turn forward to the superior frontal lobe.

The cells lying within the neuropil sample a smaller volume of neuropil. Each has a few short dendrites along the main trunk that then proceed either singly, or after division, to end in the subvertical lobe. Sometimes the part of the trunk that carries the dendrites is roughened with thousands of small indentations and protrusions, presumably occupied by incoming terminals. The cells along the floor of the vertical lobe are like those of the peripheral zone, with short dendrites.

The numerous cells of the thick layers around the more peripheral parts of the neuropil of the vertical lobe are all amacrines, each with a trunk that remains within the lobe. Shortly after entering the neuropil the trunk of each cell carries a swelling, which makes synaptic contact with a corresponding dilation on an incoming fibre from the superior frontal lobe. Passing towards the centre of the neuropil the trunks branch and their endings make bushes of 20 or more fine terminal twigs. The amacrine trunks proceed irregularly without forming bundles, making a dense network interwoven with the dendrites of the larger cells and with afferents entering from below, and also with glia fibres. Some of the larger amacrines reach from the peripheral to the central zone. The two parts of the vertical lobe probably have a similar function.

The afferent fibres to the vertical lobe from the superior frontal lobe run through a plexus near the surface with only occasional branches. Their synapses are probably made by swellings, which occur at intervals and sometimes carry small collaterals. The other input to the lobe comes from below. The fibres run up into the neuropil and branch throughout it. The final branches carry large swellings, often with twigs. These can be seen in contact with dendrites of the large cells. Each afferent probably makes synapse with many vertical lobe cells and vice versa. The source of these afferent fibres is not known. Most of them come through the subvertical lobe, they may arise there or include afferents from the mouth, arms, or elsewhere.

Subvertical lobe

This consists of a mass of neurons and islands of neuropil below the vertical and superior frontal lobes (Figs 8.34). The subvertical lobe is continuous with the precommissural lobe and with it serves to distribute the output of the vertical lobe system to the optic lobes and to other centres.

The subvertical lobe contains cells of various diameters. The large cells have spreading dendrites, evidently receiving inputs from a large number of incoming fibres. The outputs from this lobe include the subvertical to optic tract; the subvertical to precommissural tract; the subvertical to inferior frontal tract; the subvertical to superior frontal tract; the subvertical to vertical tract; and the subvertical to dorsal basal tract. Many fibres pass from the subvertical lobe to the vertical lobe but others come from more distant sources. There are also many smaller cells in the lobe but they are not like the vertical lobe amacrines, they have numerous dendrites and axons proceeding to other parts of the lobe.

The input to the subvertical lobe comes from the vertical lobe, the superior frontal lobe, the postfrontal lobe, the inferior frontal lobe, including fibres from the arms, lips, and hinder part of the body, the optic lobe, the dorsal basal lobe, and the the magnocellular and palliovisceral lobes, including possibly fibres from the pallial and visceral nerves.

Precommissural lobe

This lobe is the main output pathway from the vertical lobe system, except for the return tracts to the optic lobes. It has relatively few cells, but some are very large and probably act as command neurons producing complex functions. The limits of the region are hard to define as it lies at the centre of the supraoesophageal mass and receives fibres from many areas (Fig. 8.34).

The cell bodies of the large cells lie mainly far laterally. Dendrites from their trunks spread widely through the lobes. The axons then proceed to the anterior pedal lobe, the magnocellular and palliovisceral lobes, the interbasal lobe, the anterior and median basal lobes, and perhaps to other destinations. The large fibres to the magnocellular and pallioviscerai lobes are similar to those arising from the median basal lobe, sending collaterals to the region of the giant cell and continuing to the pallioviscerai lobe (Fig. 8.40).

The input to the precommissural lobe comes from several sources including the subvertical lobe, the inferior frontal lobe, the magnocellular lobe, the superior frontal lobe, the postfrontal lobe, the anterior and median basal lobes, the optic lobes, and the interbasal lobes.

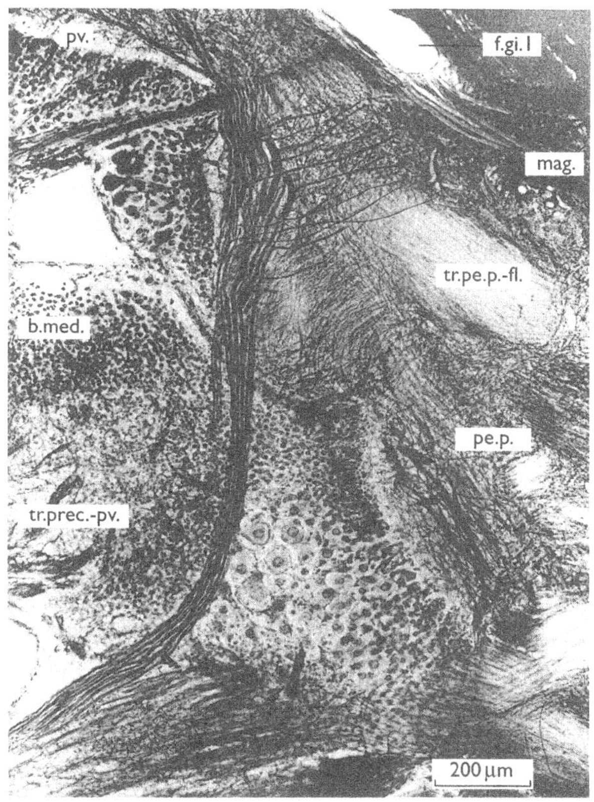

Fig. 8.40. *Loligo pealeii*. Sagittal section, retouched, to show the large fibres of the precommissural to pallioviscerai lobe tract and their collateral branches to the magnocellular lobe (Young 1979).

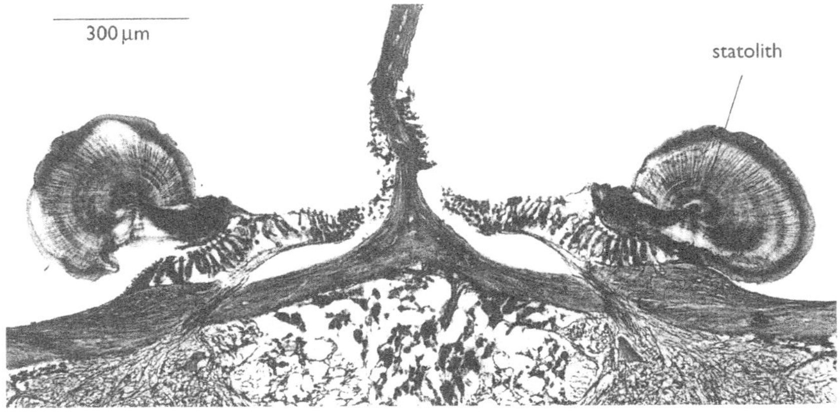

Fig. 8.41. *Alloteuthis subulata.* A section through the statoliths of a young specimen. Rings are visible in the statoliths and these can be counted to estimate the age of a specimen (Holmes' stain).

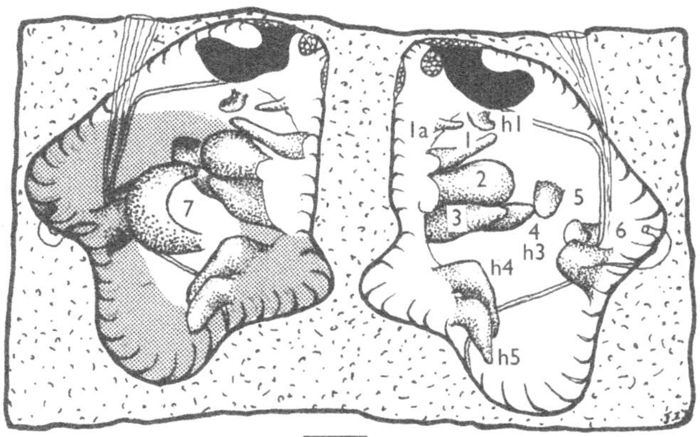

Fig. 8.42. Statocysts of *Loligo forbesii* (500 mm ML) (scale bar 1 mm) (Young 1989).

8.4.8 Statocysts

The statocysts are paired and each has an irregularly shaped cavity. Structurally and functionally two main receptor systems are present, one for the detection of gravity, the macula/statolith system, and the other for the detection of angular acceleration, the crista/cupula system. In both systems the sensory epithelia, of the maculae and cristae, are composed of many receptor cells. These transform the mechanical energy of the movements of the statoliths (Fig. 8.41) and cupulae relative to the sensory epithelia into signals that ultimately reach the brain.

The statocysts of *Loligo* were described by Stephens and Young (1982). They lie at the back of the cephalic cartilage, beneath the brain, and are separated from each other by a cartilaginous septum. The walls of the statocysts bulge outwards to form sac-like protuberances in the cartilage. Internally the walls carry cartilaginous projections of two types, straight rod-like anticristae (1–7) and hooks, or hamuli, which bend over the cristae at the point at which they change direction (Fig. 8.42). The anterior part of the statocyst is occupied by the gravity receptor organs. In each statocyst a macula statica princeps carries a statolith (Fig. 8.41), and two maculae neglectae, the inferior and superior, covered by statoconia.

The sacs of the statocyst wall form channels of irregular shape and, together with their anticristae and hamuli, serve to limit movement of the endolymph during angular acceleration in relation to the crista. This provides sensitivity appropriate to the velocity of turning by the squid (Young 1989). The crista has four sections with approximately 90° between the turns. Each section contains four central rows, flanked by three rows of hair cells and these, with their attached cupulae, respond to angular acceleration (Budelmann 1977, 1994). The crista has primary sensory cells on one side and secondary sensory hair cells on the other. Each primary sensory hair cell is about 8 μm wide and has a bundle of kinocilia protruding from the apex of the cell and an axon passing directly to the brain. Most of the hair cells of the crista are secondary sensory cells and each sends information to the central nervous system by making synaptic contact with an underlying afferent neuron. Recordings from these cells show that they are electrically coupled; this provides a mechanism for altering the sensitivity and frequency response characteristic of the organ. Inhibitory and excitatory efferent fibres both make contact with the hair cells and neurons. The output from the hair cells and the afferent neurons can be increased or decreased under direct control of the central nervous system (Williamson 1995).

At hatching the cavity of the statocyst has few interruptions; the most differentiated, and probably the only part functional at this stage, appears to correspond to the future macula statica princeps (Bülow and Fioroni 1989). The statocysts of newly hatched loliginids are readily visible, nearly spherical in shape and the anteroposterior length is more than 25% of the mantle length in the hatchling (Fig. 8.2); this contrasts with 5% in the adult. Besides the changes in size during growth, there are also changes in the shape during development of the hamuli and anticristae. Measurements of several parameters of the statocyst were subjected to discriminant analysis and placed *Loligo* amongst the non-buoyant squids, and showed it to be adapted for rapid turns (Maddock and Young 1984).

The statocyst is complex and has a level of sophistication equivalent to that of the vertebrate vestibular system (Budelmann and Tu 1997). Behavioural and physiological studies suggest that the afferent fibres of the statocyst make contact with efferent fibres and that the input is integrated with the visual input, but as yet central connections have only been shown for the oculomotor neurons in the anterior lateral pedal lobe in *Octopus* (Williamson and Budelmann 1991).

9 Chtenopterygidae

9.1 *Chtenopteryx*

Chtenopteryx (Fig. 9.1), the only genus of the family, is a warm-water form with a wide distribution (Lu and Roper 1979, Okutani 1995), occurring in Atlantic and Pacific (Roper and Young 1975) waters and in the Mediterranean (Mangold and Boletzky 1988). *Chtenopteryx* undergoes ontogenetic descent as animals of 4.4–35 mm ML were caught from 0–800 m in daylight, and 0–200 m in darkness, although some go to 600 m (Clarke 1969, Roper 1972, Clarke and Lu 1974, 1975, Lu and Clarke 1975a, b, Lu and Roper 1979). Animals of 38–62 mm ML were spread from 0–1000 m in daylight, but at night specimens of 20–50 mm ML were at 100–300 m, of 20–77 mm ML at 600–800 m, and of 42–83 mm ML at 1300–2000 m (Roper and Young 1975). A female of 54 mm ML was caught at 1100 m in the Mediterranean (Villanueva 1992b). In Hawaiian waters *Chtenopteryx* undergoes a distinct diel migration (R.E. Young 1978). Of paralarvae collected around the Hawaiian Islands, *C. sicula* was the most widely distributed (Bower *et al.* 1999).

9.1.1 *Morphology and function*

The strange fins gave rise to the name 'comb-wing' for this squid. The fins are elongated and have fin rays, with a delicate membrane between (Fig. 9.2), and a central muscular region extending the length of the fin to which dorsal and ventral membranous sheets are attached. The dorsal sheet, supported at intervals by fin rays, contains longitudinal muscle fibres running from front to back. There is a marginal muscular support structure extending along the whole of the outer edge of the fin, that is double in some places (Young 1991c). The fin rays are composed of muscles and connective tissue. The muscles are arranged in various planes. They are constrained by fibrous tissue and so act as musculo-hydrostats, serving both for support and movement, as in various tissues of some other cephalopods (Kier 1989).

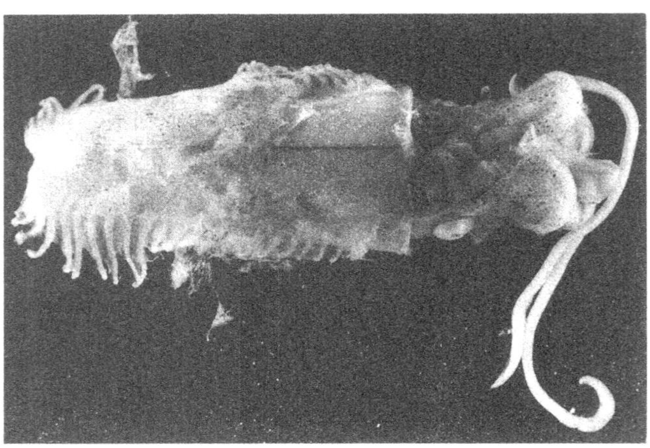

Fig. 9.1. *Chtenopteryx sicula* (32 mm ML) in dorsal view.

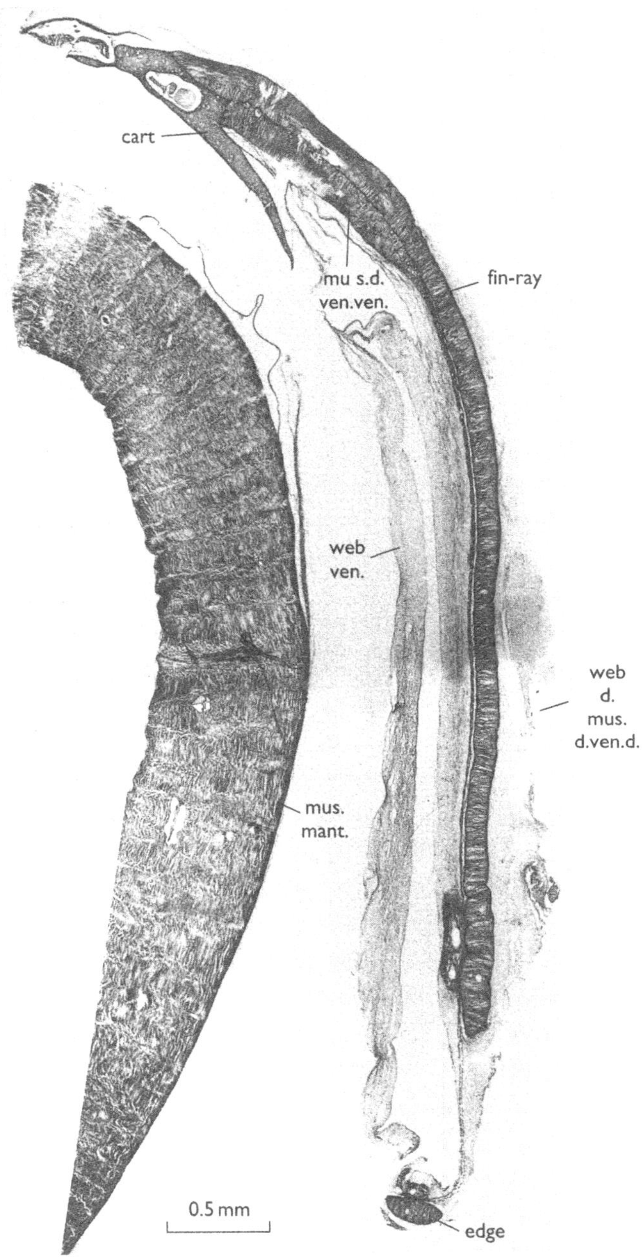

Fig. 9.2. *Chtenopteryx sicula* (36 mm ML). Transverse section of the mantle to show the mantle wall, the fin between the fin rays, its support and web (haematoxylin and eosin).

Measurements of buoyancy and the histology of the tissues indicate that *Chtenopteryx sicula* is a dense squid (Clarke *et al.* 1979). The hyaline cartilage and musculature of the head show no indication of reticulation. The mantle wall is thick and has densely organized muscles (Fig. 9.2). The funnel is prominent, muscular, and has large nerve fibres. The features of this squid indicate that it moves rapidly, and this is reflected in the statocyst whose structures resemble those which monitor fast movements and turns in various planes (Young 1989).

9.1.2 Photophores

Chtenopteryx sicula has subocular photophores with maximum emission of 425 nm (Herring 1983). A large photophore lies on the ventral surface of the ink sac, covered with a light reflecting layer, and a lens-like structure. The male develops a posterior visceral light organ at sexual maturity; this has a peak emission of 416 nm after stimulation with hydrogen peroxide (R.E. Young 1981). The wavelength of this signal is only poorly detected by potential predators, so as a sexual signal it may be safely sent (R.E. Young 1983).

9.1.3 Receptors and central nervous system

The eyes are directed laterally (Fig. 9.1). The retina is of about the same thickness throughout, except at the edges, with rhabdoms about 250 μm long. Trochlear cartilages are present and are attached to an anterior prolongation of the front of the base of the cranium. Conjunctive eye muscles are present but do not show well in our sections. The optic lobes (Fig. 9.3) form 363% of the total brain volume (Maddock and Young 1987).

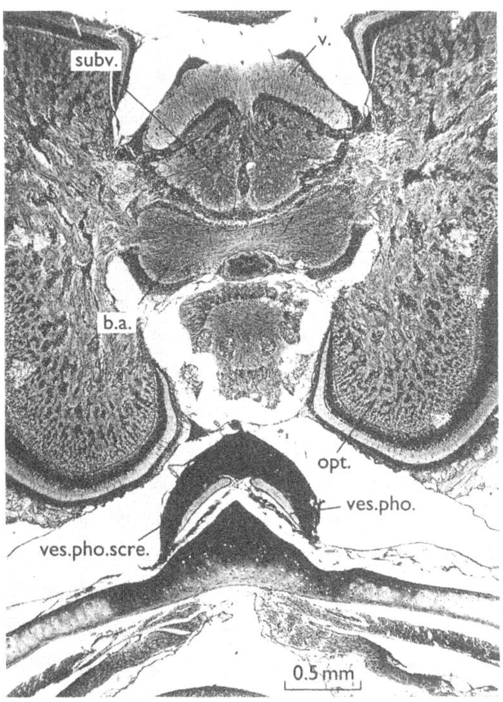

Fig. 9.3. *Chtenopteryx sicula* (36 mm ML). The central nervous system in transverse section and also shows the photosensitive vesicles which have a thick iridescent screen.

There is a thick cephalic cartilage over the central nervous system (Fig. 9.4). The brachial lobes are separated from the rest of the brain (as in *Loligo*). The ventral magnocellular lobe forms 8.6% of the total brain volume (Maddock and Young 1987), and it receives large bundles of fibres from the brachiomagnocellular connective, the anterior funnel nerve, and the posterior superior ophthalmic nerve (Fig. 9.5). This lobe has a complex neuropil filled with very large synaptic boutons and there are exceptionally large first-order giant cells (Fig. 9.6). The first-order giant axons are some 50 μm in diameter. In the palliovisceral lobe they make a chiasm in which there is a long stretch of complete fusion (Fig. 9.7). There are two large fibres in each pallial nerve, 150 μm and 100 μm in diameter. Large fibres in the stellar nerves are up to 160 μm in diameter; they arise from what appears to be a distinct giant fibre lobe of large cells (Fig. 9.8). The sections available are not adequate to be certain of the details of the giant synapse. Large fibres are present in the anterior funnel nerve, that probably arise at the back of the pedal lobe. There are large fibres

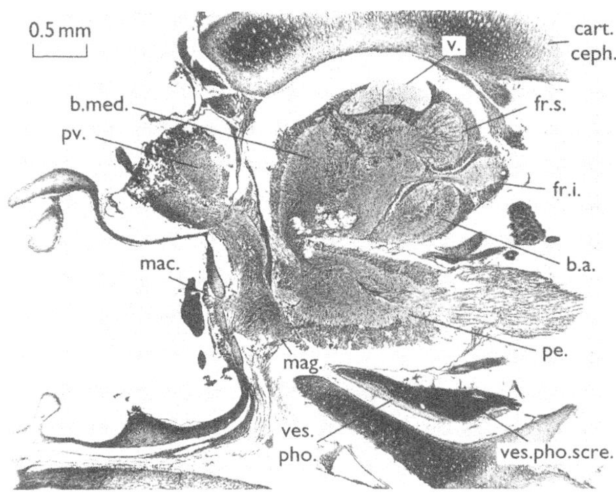

Fig. 9.4. *Chtenopteryx sicula* (35 mm ML). Sagittal section of the central nervous system, and of the extraocular photoreceptor organ and its screen. Note the size of the statocyst and the thickness of the cephalic cartilage.

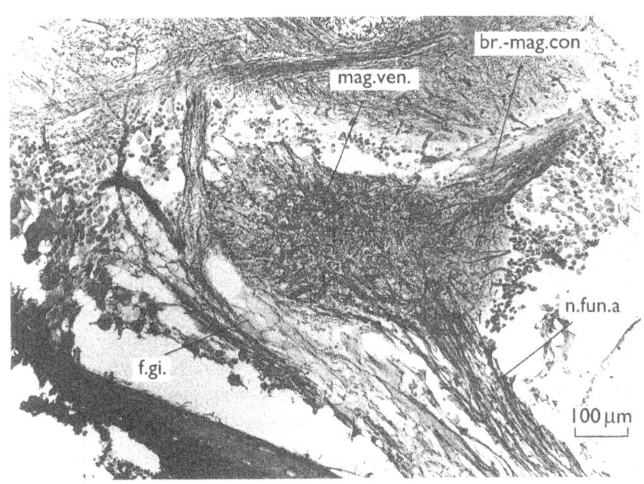

Fig. 9.5. *Chtenopteryx sicula* (35 mm ML). Sagittal section through the ventral magnocellular lobe showing giant fibres.

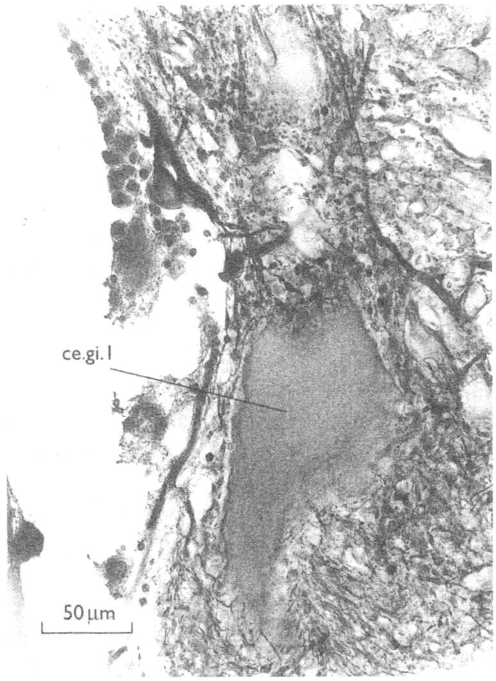

Fig. 9.6. *Chtenopteryx sicula* (35 mm ML). Sagittal section showing the magnocellular lobe with large synaptic boutons around dendrites of the giant cell.

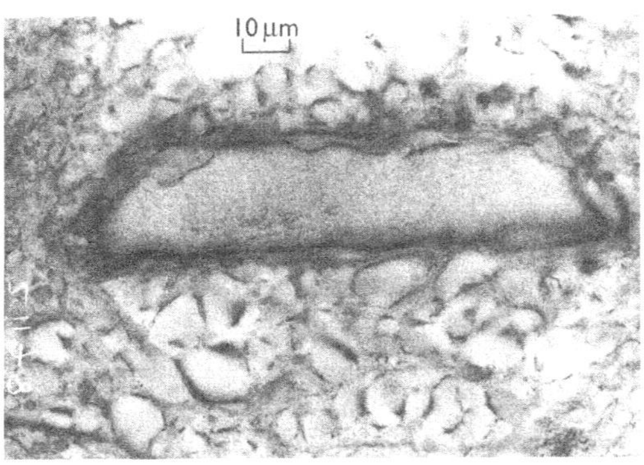

Fig. 9.7. *Chtenopteryx sicula* (36 mm ML). Transverse section to show the chiasm of the two first-order giant axons, which have fused completely.

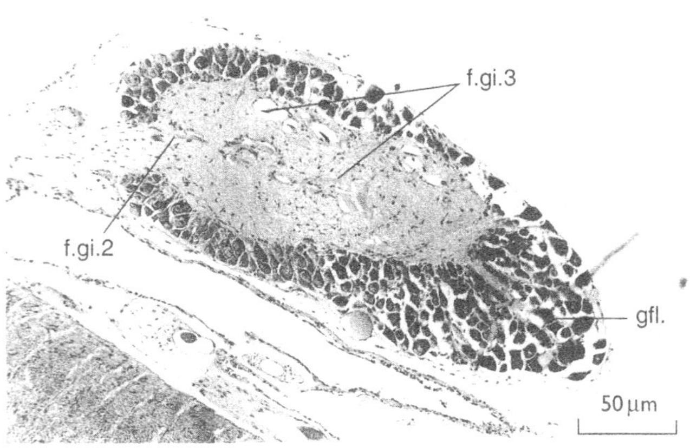

Fig. 9.8. *Chtenopteryx sicula* (36 mm ML). Stellate ganglion with third-order giant fibres, and (possible) giant fibre lobe.

Fig. 9.9. *Chtenopteryx sicula* (36 mm ML). The subpedunculate lobe at the back of the suboesophageal mass.

in the posterior pedal commissure, some of which proceed forwards (as in *Pterygioteuthis*). Two large fibres are present in each brachial cord.

The vertical lobe system forms 14.5% of the total brain volume (Figs 9.3, 9.4) and the inferior frontal system only 0.7% (Maddock and Young 1987). The subpedunculate lobes have a series of lobules filled with fine nerve fibres (Fig. 9.9) (as in *Loligo*).

The extraocular photoreceptor organs together form 4.0% of the total brain volume (Maddock and Young 1987). Each organ consists of a single, elongate, flattened, curved photosensitive vesicle, which extends from the posterior end of the supraoesophageal mass beneath the optic tract to in front of the middle suboesophageal mass (R.E. Young 1978) (Figs 9.3, 9.4). Here it meets with the contralateral vesicle, but they do not fuse. In this region there is a continuous sheath of iridophores closely associated with the organs. The convex anterodorsal wall of each vesicle is lined with a dense layer of iridophores, but light can enter through the posteroventral surface. The rhabdoms, 30–50 μm in length, are densely organized. The nerves from the photosensitive vesicles enter the peduncle lobe. The system may serve to monitor counter-illumination as it does in some other squids (R.E. Young 1977).

The statocyst is quite different from that of other oceanic squids. It is among the narrowest and the longest of all the species examined (Young 1989). The most unusual feature is a relatively very large, flattened, vertical plate which extends to the roof and lies at the first turn of the crista. This divides the front of the cavity into anterior and posterior channels. The latter is especially narrow and is walled off in front by hamulus 2 and behind by anticristae 6 and 7 and hamulus 3. The two channels are demarcated by the unusual anticristae of the medial wall; anticrista 1 is a vertical plate in the transverse plane above hamulus 1. Behind are five thin anticristae in the position of anticrista 2 and 3 of other squids (Fig. 9.10). On the floor anticrista 4 is large and anticrista 5 is small. The horizontal channel is not especially pronounced, and anticristae 6 and 7 touch but do not fuse. Hamulus 4 is markedly hooked and, like 5, is large. The vertical crista is probably tuned to rapid turns. Anticrista 4 delimits the vertical channel in front. The statocyst provides for restricted flow in all three planes and shows specialization for sensitivity to high frequencies.

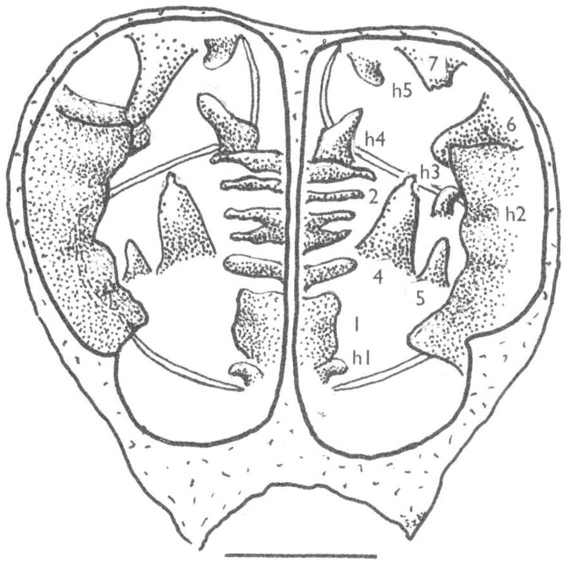

Fig. 9.10. *Chtenopteryx sicula* (32 mm ML). Anterior view of statocyst. The maculae have been removed and anticrista 7 is broken in the left statocyst (scale bar 1 mm) (Young 1989).

9.1.4 *Affinities of Chtenopteryx*

The relationship of the comb-wing squid with other families has been much debated. Roper (1969) compared Chtenopterygidae with the Bathyteuthidae and concluded that the two families share a number of features. These include the suckers on the buccal lappets, the presence of more than two rows of suckers on the arms and more than four on the clubs, and locking apparatus which is straight and simple. They differ in the attachment of buccal connectives to arms IV, but as this difference is basic it probably precludes a close relationship between the two families.

Naef (1921) noted that the Loliginidae resemble *Chtenopteryx* in many respects, and further similarities have been demonstrated by Young (1991c). The complete fusion of the axons of the first-order giant fibres is a very unusual condition and it is unlikely to be a parallel evolutionary development. The same applies to the presence of a giant cell lobe in the stellate ganglion and the single giant fibres in the stellar nerves. There are also wide differences between *Chtenopteryx* and *Loligo*, the most notable being the fins, and the suckers on the arms and the tentacular clubs.

The so-called accessory nidamental gland were considered by Naef (1921) to show similarities with that of *Loligo*. This was questioned by Roper (1969) and it has been found that this gland becomes reduced in the adult female *Chtenopteryx*, whereas it is active and carries symbiotic bacteria in *Loligo*. However, this condition does not seem to exclude affinity with loliginids, and the gland may be a 'vestige'.

Naef (1921) considered *Chtenopteryx* to be the earliest independent branch of the stem of the higher oegopsins. It is possible that it represents an early off shoot, though the conditions of the giant fibres are apomorphic.

The remarkable collection of apparently unique features in *Chtenopteryx* suggests an unusual mode of life, perhaps concerned with a capacity to live in relatively deep water although not neutrally buoyant. This may explain the divided fins, the unusual luminous organs with low emission maxima for marine organisms (Herring 1983) and the unique divisions of the statocyst (Young 1989). Long tentacles bearing numerous suckers are found in squids belonging to the 'chiroteuthid lineage', histioteuthids and brachioteuthids, all being either meso- or bathypelagic in habit (Roper 1969), and most being neutrally buoyant (Clarke *et al.* 1979). Such tentacles may allow the capture of small organisms in darkness but their prey is not known and further evidence is needed before conclusions can be reached.

10 The 'Enoploteuthid group'

10.1 Families of the 'Enoploteuthid group'

Traditionally this family has included three subfamilies, the Enoploteuthinae, Ancistrocheirinae, and Pyroteuthinae, but the last two were elevated to familial status by Clarke (1988). The three families form the 'Enoploteuthid group', and the systematics, diagnosis, and distribution of each of the species has been reviewed (R.E. Young *et al.* 1998a).

10.2 Enoploteuthidae

Enoploteuthid squids are found widely in the Atlantic, Pacific, and Indian Oceans, and two species are present in the Mediterranean Sea; their geographical distribution has been reviewed by R.E. Young and his colleagues (1998a). They are typically warm water squids found between 40°N and 40°S. The adults are mesopelagic and undergo extensive vertical migrations; the young inhabit the upper 200 m (Lu and Clarke 1975b, R.E. Young 1978).

The photophores of these squids are notable and are present on the mantle, funnel, eyeballs (Fig. 10.1), and arms. They are highly complex, and their distribution allows separation of this group into families (R.E. Young *et al.* 1992). Photophores were first described in a species of this family by Chun (1903), since when those of others have been described. The emission maxima of several species are given in Table 10.1. R.E. Young and Bennett (1988) found the integumental photophores to be of three types: simple, with a lens, or with a filter. The simple photophores are the smallest and least specialized. Each comprises:

- a basal concave reflector composed of arrays of collagen rods, which reflect light of specific wavelengths
- a cluster of highly organized photogenic crystalloids in front of the reflector
- a ring of radiating fibres of collagen, surrounding the crystalloids, guides the direction of emitted light
- a distal core partly surrounded by light guides.

Photophores with a lens have a dense ovoid cone with refractile properties (when fresh) making the term appropriate. The light guides are usually longer, and individual fibres are thicker than in the simple photophores. The basal reflector may be reduced in size. The third type has a filter, and these are the most modified and most complex of the three types of integumental photophores (Fig. 10.2) (R.E. Young and Arnold 1982). Modifications occur in the size and position of the light guide region, and in the specialization of the distal-core region.

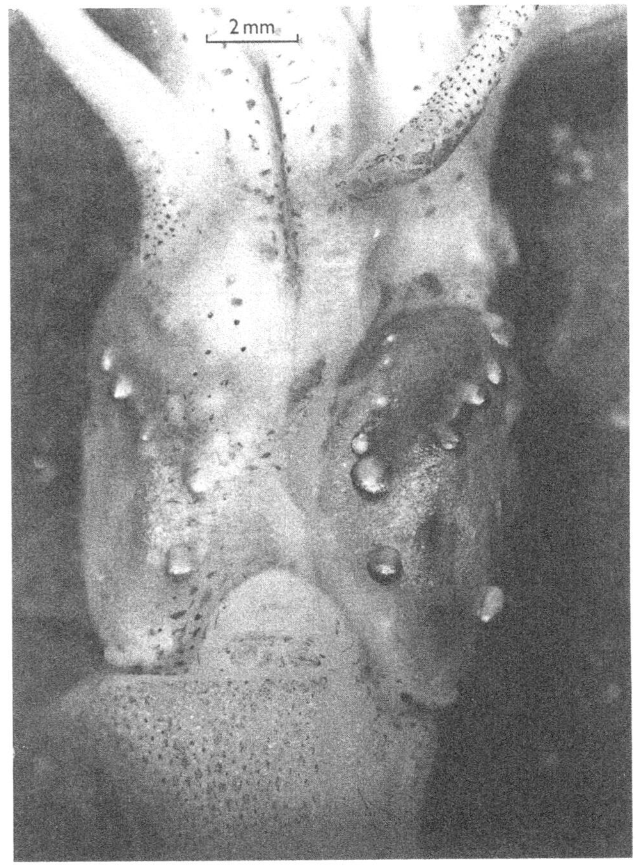

Fig. 10.1. *Abraliopsis* (23 mm ML). The ventral surface of the head to show the photophores around the eyes.

Table 10.1. Enoploteuthidae. The emission maxima of photophores

Species	Emission maxima (nm)	Photophore site	Source
Enoploteuthis higginsi	440, 480	Ventral	R.E. Young in Herring (1988)
Abralia trigonura	536, 480, 440	Ventral	R.E. Young and Arnold (1982)
Abralia veranyi	489–492 475	Ventral Subocular	Herring *et al.* (1992b)
Abraliopsis falco	421, 466	Subocular	Widder *et al.* (1983)
Abraliopsis sp.	454, 474	Tentacle	Widder et al. (1983)

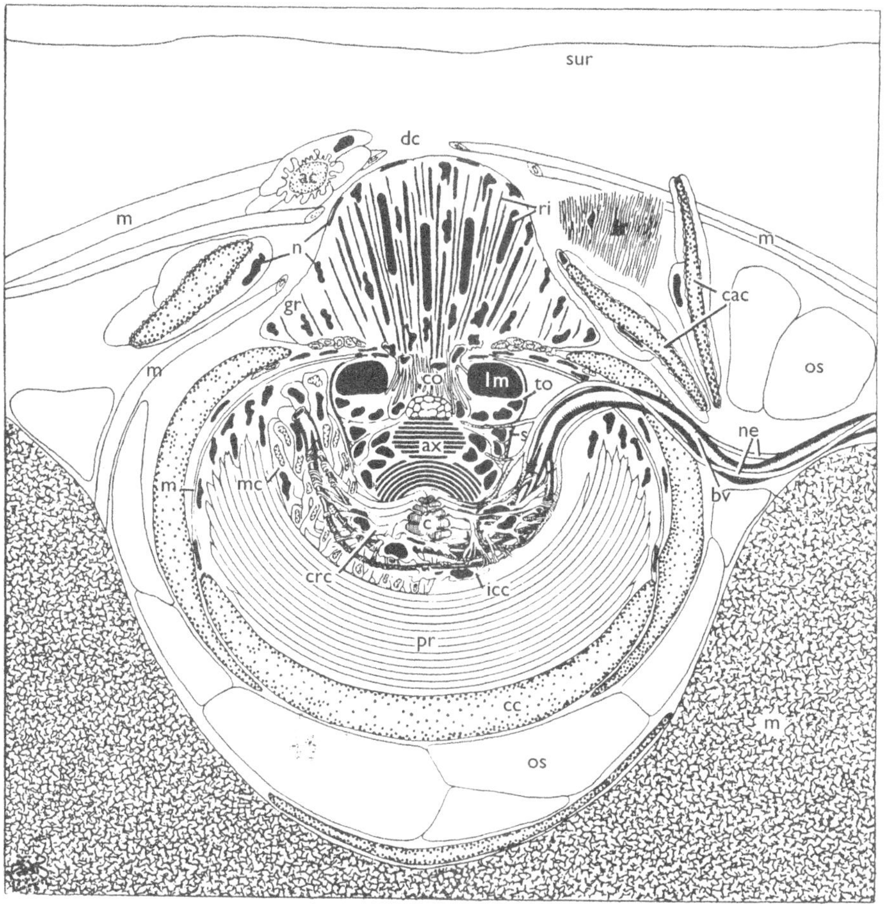

Fig. 10.2. *Abralia trigonura*. An intrinsic photophore (140–180 μm) on the ventral mantle, and one of the most complex with a filter (R.E. Young and Arnold 1982). ac, apical chromatophore; ax, axial stack; bv, blood vessel; c, photogenic cone; cac, distal cap chromatophore; cc, proximal cap chromatophore; co, core; crc, crystalloid cell; dc, distal cap; dr, distal reflector; gr, girdle; icc, immature crystalloid cell; lm, lamella; m, muscle; mc, mitochondria; n, nucleus; ne, nerve; os, orbital space; pr, proximal reflector; ri, ribbon; s, collagen sheath; sur, surface; to, torus; v, vesicles.

The function of the photophores has been demonstrated to be for counter-illumination in *Abralia* and *Abraliopsis* (R.E. Young and Roper 1977). The control of light is probably through the blood vessels, a greater flow allowing access of luciferin to the enzyme luciferase, which is produced by crystalloids. It is significant that the subpeduncular region of the brain is well developed, as it is in two other genera with photophores, namely *Lycoteuthis* and *Pyroteuthis*.

10.2.1 Enoploteuthis

Enoploteuthis is widely distributed. In the Atlantic it is found between 35°N and 40°S, and in the Pacific from 40°N to 43°S (Roper 1966, Burgess 1982, Riddell 1985). It has been captured in opening/closing nets from 0–400 m in daylight and 700–800 m in darkness (Lu and Clarke 1975a, 1975b), and off the Hawaiian Islands it migrates from depths of 500–800 m in the day to the upper 100 m at night (R.E. Young 1978). It has not been recorded from the Mediterranean Sea (Mangold and Boletzky 1988).

The eggs of *Enoploteuthis reticulata* (1.08 × 0.78 mm) are colourless, but slightly opaque with a silvery sheen and a gelatinous coating. The paralarvae are taken at 100–150 m during the day and at 0–25 m at night in Hawaiian waters. Young *E. reticulata* and *E. jonesi* are both taken in this area and can be identified by their chromatophore and photophore patterns (R.E. Young and Harman 1985b).

Each extraocular photoreceptor organ consists of several sets of photosensitive vesicles. The midventral portion is irregular and has a narrow connection with the posterior part of the organ. The latter is continuous with the dorsal part via a strand of vesicles extending over the optic lobe. Part of this strand and the posterior photosensitive vesicles contain yellow pigment (R.E. Young 1978).

The statocyst is moderately long, narrow, and deep. The lateral wall is curved and forms a horizontal canal. Anticrista 7 is tightly pressed against anticrista 6 to form a canal. The anticristae of the median row lie close together. There is no hamulus at the first turn, but the others are well developed. The flow across the vertical crista is presumably greatly restricted by the walls of the canal and the median row of anticristae (Young 1989). The statocyst is suited for rapid movement. The cartilage of the statocyst is much vacuolated and yet these squids are not buoyant (Clarke *et al.* 1979), and multiple discriminant analysis of several parameters of the statocyst also places this squid amongst the non-buoyant forms (Maddock and Young 1984).

10.2.2 Abralia

This genus is found in the Atlantic, Pacific, and Indian Oceans, in tropical to temperate regions. It is most often captured in the vicinity of steep continental slopes and has been taken in large numbers with swarms of myctophid fish (Okutani 1983d), and with the shrimp *Sergia luceas* (Kubota *et al.* 1982). In the Atlantic it has been taken in opening/closing nets at 20–60 m in the day, and at 50–100 m and 700–800 m at night (Lu and Clarke 1975a, b). Off Hawaii *Abralia trigonura* has been caught at 30–200 m at night with the majority between 50–100 m, and at 390–650 m during the day, with most at 450–560 m. In the same waters *A. astrosticta* has been caught in pelagic trawls and in benthic trawls at night, but it lives mainly near the sea floor even when migrating (Roper and Young 1975, R.E. Young 1978).

Life history

In Suruga Bay, Japan, both large (53 mm ML) and small (17 mm ML) specimens of *Abralia andamanica* have been caught during the month of December. Females, all with ripe eggs, were caught in the autumn and early winter, suggesting that spawning may take place during the same period. Nearly 8000 eggs were present in a single female, but of these only 19% were 1.2 mm or more in diameter, the remainder being 0.3–1.2 mm in diameter. Ripe eggs are translucent, yellow, and usually 0.7–1.5 mm in diameter (Kubota *et al.* 1982). In Hawaiian waters the eggs of *Abralia trigonura* are slightly greenish, and have a jelly coating, and the hatchlings possess chromatophores and photophores (R.E. Young and Harman 1985b).

Photophores

Abralia trigonura has integumental photophores (Fig. 10.2) as well as distinctive light organs on the ventral surface of the eyeball (two types), at the tips of the ventral arms, and near the posterior tip of the mantle (R.E. Young and Bennett 1988). The three types of integumental photophores, lensed, simple, filtered, occur in a ratio of 1 : 7 : 10 respectively; they are intermingled and scattered over the ventral surface of the head, arms, funnel, and mantle. Of these the filtered photophore is the most complex (R.E. Young and Arnold 1982).

One *Abralia veranyi* had more than 550 ventral mantle photophores. These comprised large photophores, arranged in 4–6 transverse rows, and smaller more numerous ones with a less regular arrangement. The bioluminescence of photophores on the ventral surface of the mantle were recorded by video and the spectral changes recorded (Table 10.1). Both large and small photophores were involved, the larger light organs giving an impression of transverse, brighter bands; no quantifiable differences could be detected in the relative intensities of the different sizes of photophores (Herring *et al.* 1992b). Further observations are needed to determine the functions of their different photophores and to unravel the complexity of their bioluminescent behaviour.

Extraocular photoreceptor organs

These organs consist of dorsal, posterior, posteroventral, and anteroventral sets of photosensitive vesicles. Each dorsal set of vesicles lies in a posterodorsal concavity of the cranium below a 'window', an area where the chromatophores are reduced in the overlying tissues. The posterior vesicles are on posterior surfaces of the optic lobes, and in front of the attachment zone of the head retractor muscles, which are translucent. The ventral photosensitive vesicles are partly encased by the cephalic cartilage and by the optic lobe, and there is a large window, devoid of chromatophores, on the ventral side of the head (R.E. Young 1978). The posterior set of vesicles contain a strong yellow pigment, and the other sets an orange pigment.

The function of the extraocular photoreceptor organs was demonstrated experimentally on board ship (R.E. Young *et al.* 1979a). *Abralia trigonura* and *Abraliopsis* were held so that the output of the photophores could be measured while light, from above or below, was directed into the photosensitive vesicles. When light was shone into the ventral photosensitive vesicles of a squid emitting light, then this decreased proportionately. The effect of downwelling light was demonstrated by shining a light from above the animal as a shield covered either the dorsal photosensitive vesicles or the eyes, or both. On removing the cover from the dorsal photosensitive vesicles the photophores showed a slow but marked increase in output; this declined when they were covered. Uncovering the eyes resulted in only a small increase in luminescence. When both the photosensitive vesicles and the eyes were uncovered there was a large and rapid rise in output by the photophores. The squid was observed with a mirror from below, and when the experimenters' eyes were adapted to the dark they could detect its silhouette to a level equivalent to 610 m off Hawaii (R.E. Young and Roper 1976). Denton and Warren (1957) suggested that the eyes of deep-sea fishes are 60–120 times more sensitive than the human eye, indicating that a fish would have a visual threshold capable of detecting the silhouettes of a squid at depths of 730–750 m.

10.2.3 Abraliopsis

Abraliopsis (Fig. 10.1) is found in the Atlantic, Pacific, and Indian Oceans and the Mediterranean Sea, from temperate to tropical regions (Nesis 1987). In Atlantic waters *A. affinis* undergoes diel migrations, animals of less than 22.4 mm ML being shallower than 60 m at night and at 10–500 m in daylight, but only squids of less than 10 mm ML were caught from 10–100 m during the day (Lu and Clarke 1975a, 1975b). *A. pfefferi*, however, appears to occur in depths of less than 200 m both day and night (Clarke and Lu 1974). In the Mediterranean it has been captured at 100, 375, and 675 m during the day and from 25–600 m at night (Roper 1972). In Hawaiian waters they undergo diel migration, the squids being in the upper 200 m at night and below 400 m during the day (R.E. Young 1978).

Life history

These squids do not exhibit seasonal changes in distribution in Hawaiian waters. They reach a maximum size of 42 mm ML, and at 26 mm ML some animals possess gonads of maximum size. Spermatophores are found in the sperm receptacle of females of 20 mm ML and more and are occasionally found in animals of 11 mm ML (Boucher 1983). The eggs (1.01 × 0.84 mm) are translucent and colourless, and have a thick, sticky, jelly coating. Paralarvae are captured in depths of 50–125 m during the day, and in the upper

60 m at night, suggesting diel migrations. At a mantle length of 2 mm the terminal photophores appear on the ventral arms, and at 4.5 mm ML the integumental ones are apparent. Chromatophores are present at hatching and these, with the photophores, provide distinctive body patterns in the early stages of the life cycle (R.E. Young and Harman 1985b).

Photophores

Ishikawa (1913), after watching *Abraliopsis* off the coast of Japan, commented that

> the swellings at the tips of the arms, as well as two or three smaller dots . . . shone so brilliantly that . . . one sees only two effulgent bodies moving in the dark water . . . next . . . in the intensity of their light are the eye organs, and then the remaining organs. The three types of organ do not always shine simultaneously; often only one or the other . . . The colour of the light (of the mantle organs) . . . is a beautiful clear blue.

Abraliopsis living in a shipboard aquarium was observed to match the overhead illumination with a downward glow, so that to the human observer the squid 'disappeared' (R.E. Young and Roper 1977, R.E. Young *et al.* 1979a). The values of maximum illumination intensity obtained for this squid indicated that the limits for concealment from predators determine the upper limits which it can inhabit during the day (R.E. Young *et al.* 1980, R.E. Young and Mencher 1980).

In the hours of darkness, in the upper 200 m of water, the intensity of downwelling light may vary over several orders of magnitude with the phases of the moon and the colour of its transmitted light changes from almost white to blue with depth. In an aquarium the water temperature was altered to correspond with those encountered by *Abraliopsis* during its diel migrations between 200–600 m; the colour of the luminescence of this squid changed with the temperature and the response could be rapid and in either direction (R.E. Young and Mencher 1980).

The detection of downwelling light by *Abraliopsis* was examined by R.E. Young and Mencher (1980). Light is able to enter the extraocular receptors via 'windows' in the integument, these are areas in which chromatophores are absent or reduced in number. There is a circular window on the dorsal surface of the head above each dorsal photosensitive vesicle, and large windows on the ventral side of the head above the funnel and below the ventral photosensitive vesicles (R.E. Young 1973b). The eyes and dorsal photosensitive vesicles were demonstrated to monitor downwelling light as when an opaque cover was placed over either pair of photoreceptors there was a marked reduction in the intensity of counter-illumination by the photophores. The ventral photosensitive vesicles probably provide the feedback mechanism for controlling counter-illumination in *Abraliopsis* (R.E. Young *et al.* 1979a).

Morphology

The cartilage of the cranium was vacuolated in six out of seven specimens examined in this study; perhaps different stages of growth are represented. The muscles of the tentacles and arms are dense and complex, as are those of the mandibular muscles which are deep and encase the thin and partly tanned beaks. Within the buccal cavity the lateral buccal palps lie on either side of the buccal complex, have a few stout cuticular teeth on their medial surfaces, and internally show reticulation. The radular supports, on either side of the radula, have a series of 'chambers' forming a type of reticulate tissue (Fig. 10.3). Transverse muscle fibres are present in the walls of these chambers. The presence of reticulate tissues is a puzzle, since ammonium ions (which have density less than that of seawater) have not been detected in *Abraliopsis* (Clarke *et al.* 1979); perhaps a different substance is involved. Measurements of the statocyst place this squid amongst non-buoyant forms (Maddock and Young 1984).

Receptors and central nervous system

The eyes lie close together; they are prominent and are emphasized by the ocular photophores (Fig. 10.1). They are joined in the midline anteriorly, and more posteriorly there are eye muscles originating from attachments in the midline. There is no indication of trochlear cartilages. The retina is thickened dorsally and ventrally, but is thinner at the centre. The outer layers of the optic lobes are correspondingly thickened ventrally and more strongly dorsally and posteriorly (Fig. 10.4).

The central nervous system is elongated, with the brachial lobes somewhat separated from the pedal lobe. The inferior frontal lobe system forms 0.9% and the vertical lobe system 11.0% of the total brain volume (Table 2.6).

No single first-order giant cell was seen in *Abraliopsis*, but several large fibres cross in the palliovisceral lobe. No outstandingly large

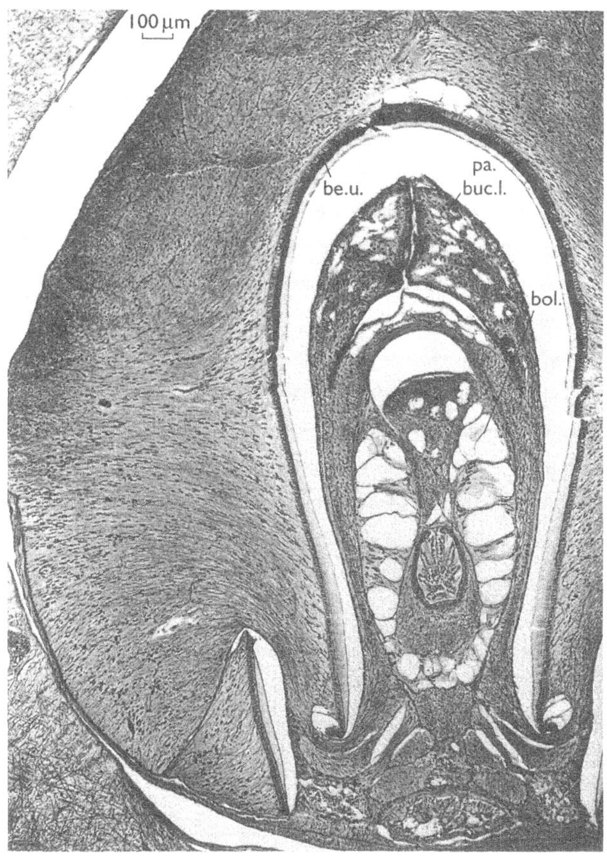

Fig. 10.3. *Abraliopsis* (34 mm ML). Transverse section of the buccal mass, showing the reticulate tissues of the lateral buccal palps and the radular supports.

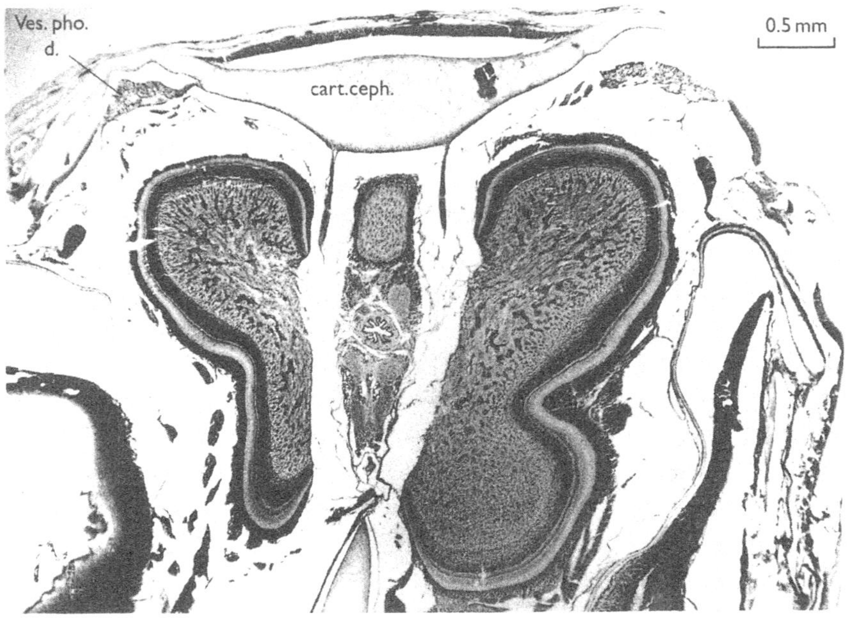

Fig. 10.4. *Abraliopsis* (34 mm ML). Transverse section through the head showing the optic lobes, retina, and the dorsal set of photosensitive vesicles.

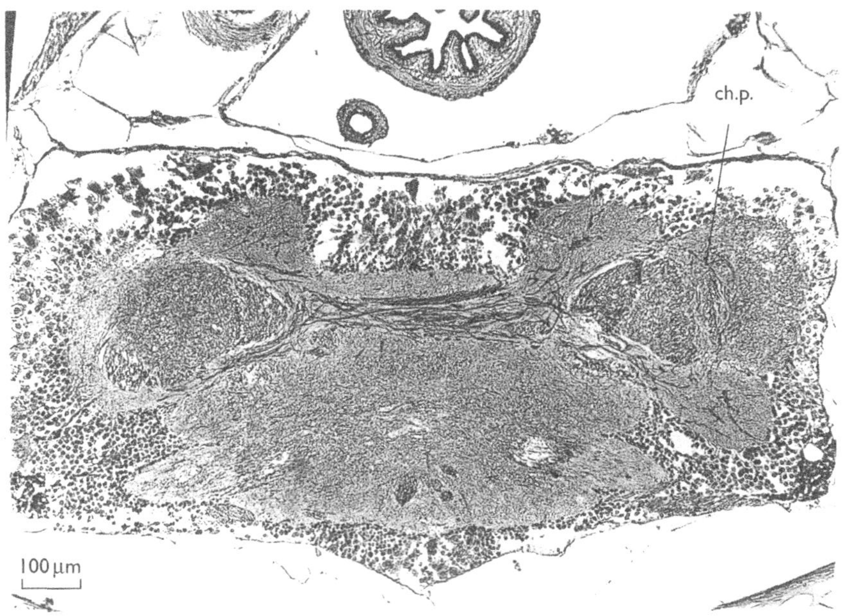

Fig. 10.5. *Abraliopsis* (34 mm ML). Transverse section showing the posterior chromatophore lobe and interchromatophore commissure.

fibres were seen in the mantle connective, stellate ganglion, or stellar nerves. The magnocellular lobe is not large or complex and no especially large fibres run forwards in the pedal lobe. The subpedunculate lobes are large but their nerves could not be traced ventrally, as they could in pyroteuthids.

The chromatophore lobes form 0.9% of the total brain volume (Maddock and Young 1987). The posterior chromatophore lobes are interconnected by a large commissure (Fig. 10.5). It is possible that these lobes are concerned with adjustment of the output of light from the photophores.

The extraocular photoreceptor organs comprise dorsal and anteroventral photosensitive vesicles and the midventral sets are enlarged and extended dorsally, but no posterior vesicles were found in two species investigated (R.E. Young 1978). A thin strand of vesicles extends between each midventral and dorsal set. None of the vesicles is pigmented. The vesicles lie within the head set in concavities of the cephalic cartilage. The anteroventral vesicles form a thin strand beginning below the peduncle lobe and extending downwards and laterally over the floor of the orbit to lie directly above the photophores of the funnel (Fig. 10.6).

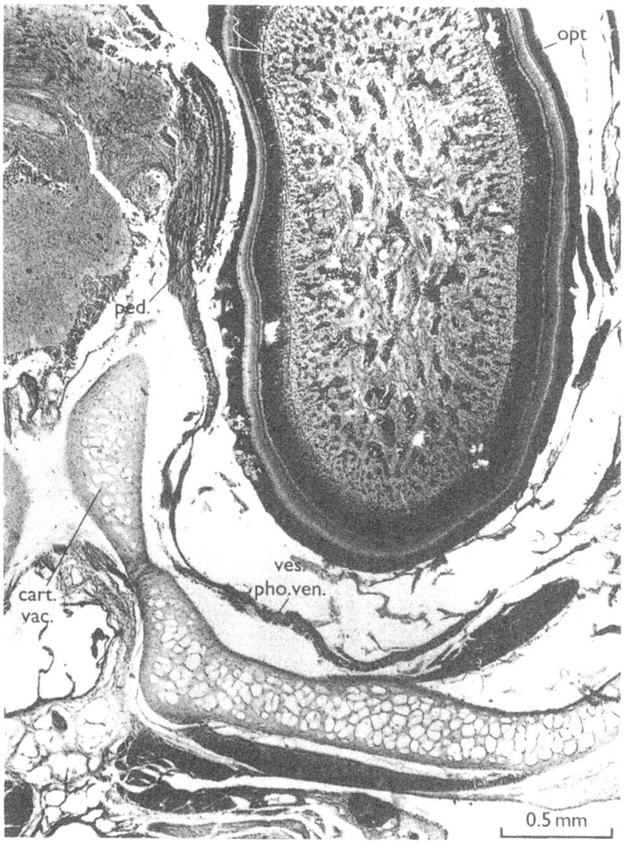

Figure 10.6. *Abraliopsis* (34 mm ML). Transverse section to show the ventral photosensitive vesicle, and its nerve entering the peduncle lobe.

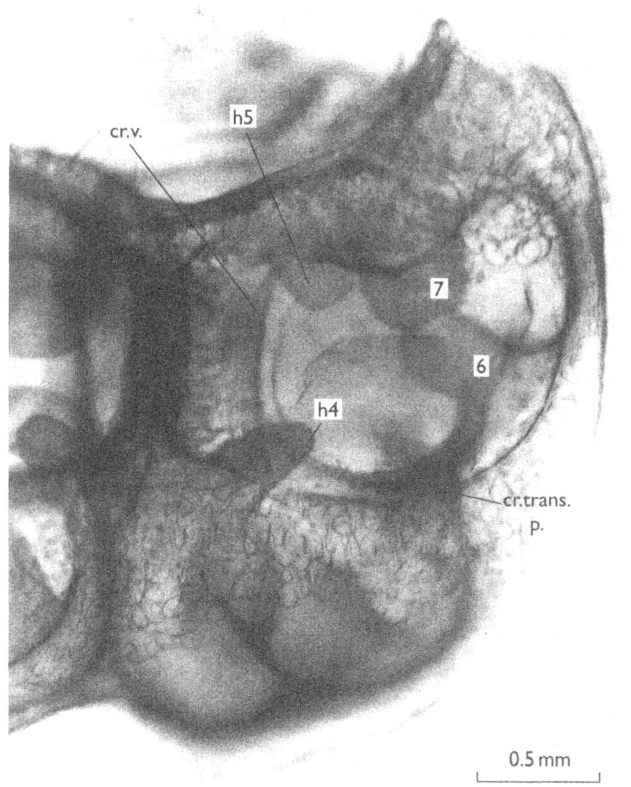

Figure 10.7. *Abraliopsis morrisi* (39 mm ML). The statocyst, in posterior view. The cartilage is reticulate (Young 1989).

The dorsal and ventral nerves of the photosensitive vesicles enter a separate lobe at the back of the peduncle lobe. The fibres of each spread out partly near the region of entry, whereas others pass up or down to spread out in the other region. Comparison between the intensity of the downwelling light and the bioluminescence produced by the squid may perhaps be made by a balance between the actions of the two regions, both of which probably send fibres separately to the main peduncle lobe and perhaps elsewhere in the brain. This arrangement is similar to that described by R.E. Young (1978). In *Abraliopsis morrisi*, the vesicles form a single compact mass embedded in the hilum of the optic lobe, and their nerve fibres enter the peduncle lobe together in one trunk from below. There is no nerve entering from above and no trace of dorsal vesicles. The photosensitive vesicles of this specimen formed 13.07% (corrected figure for 3.1% given in error in Maddock and Young 1987) of the total brain volume, but at 32 mm ML this specimen was probably immature, and some of the vesicles could be interpreted as being in the course of formation.

The statocyst is rather narrow and deep (Fig. 10.7). The cavity is greatly obstructed by the large rounded anticristae. The horizontal canal is well marked, with anticristae 6 and 7 in contact. There is no hamulus at the first turn of the crista. These characteristics are suited to monitor a life of rapid movements and turns, especially in the yawing plane (Young 1989). Multiple discriminant analysis of several parameters of the statocyst places *Abraliopsis* among the non-buoyant squids (Maddock and Young 1984).

10.2.4 Watasenia

This squid lives in the north-west Pacific from the Kurile Islands in the Sea of Okhotsk southwards to the Sea of Japan and the East China Sea (Okutani *et al.* 1987, Nesis 1987). It is primarily an oceanic squid living between depths of 255 and 1225 m. However, at maturity it moves close inshore, in spring, in shoals of many thousands to spawn. Here it subsequently suffers high mortality as it is captured and marketed, the catch being dominated by females (Hayashi 1993).

The characteristic features of *Watasenia scintillans*, the only species of the genus, include the very numerous photophores (Fig. 10.8). There are 110–180 light organs on the head in approximately four rows, and five on the ventral periphery of the eye. The mantle has some 500 arranged irregularly but leaving a bare region along the midventral line; arm III has only seven light organs, and the aboral surface of arm IV has three rows of photophores as well as three large, bulbous black light organs at the tip. The arms have hooks and each tentacle has two hooks in a single row on the club (Okutani *et al.* 1987, Nesis 1987). Males reach a mantle length of 60 mm, and females 74 mm, and their life spans are of 11–12 months and 12–13 months duration respectively (Hayashi 1993).

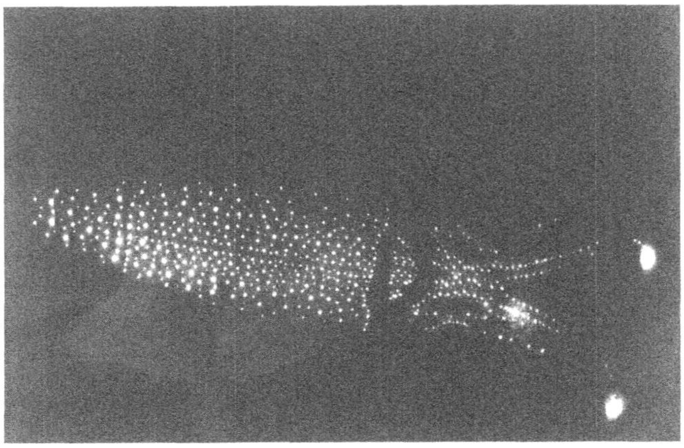

Fig. 10.8. *Watasenia scintillans*. A photograph taken by the light from the squid itself, ventral view (photograph kindly provided by F.I. Tsuji).

Life history

Watasenia scintillans takes its local name, *hotaru-ika* or the firefly squid, from its brilliant luminescent displays during the spawning season. These squids are seen near the surface, along the shores of Japan, in the spring from late April to mid-June (Haneda 1955, Zahl 1971). Young specimens of 25 mm ML have been captured in mid-October in a bay in Japan, and mature squids have been caught in February in the Pacific Ocean (Seidou *et al.* 1990). In one female some 500–1000 eggs were present in the ovary (Yuki 1985). The reproductive habits of this squid are otherwise obscure.

Photophores

Watase (1905) noted that the three oblong photophores on the ventral arm produced a light visible for a distance of some 300 mm. Sasaki (1914) described the constant expansion and contraction of the chromatophores over the large brachial organs, and noted that the light emitted was Prussian blue in colour. Zahl (1971) said this squid

> gleamed with scores of tiny blue-white lights. I tapped the glass and a flash of almost blinding intensity surged from the tips of the creature's two main arms. The flashes were immeasurably brighter than the steady lights on the squid's body. Like two ignited match heads they glowed for an instant, then faded away.

These photophores produce intense blue flashes lasting up to 20 s (Tsuji 1985). The luminescence results from an intracellular chemical reaction dependent on adenosine-5′-triphosphate (ATP) together with magnesium chloride, a soluble and an insoluble component, and molecular oxygen. The ultrastructure of these brachial photophores was described by Okada (1966).

Watase (1905) watched the pigment sacs of the chromatophores contract just as the small photophores on the body began to luminesce, thus allowing light to be emitted like 'a sunbeam shot through a tiny hole in a window curtain'; subsequently the chromatophores covered the photophores (Berry 1920). The light from these minute photophores is yellowish-white (Sasaki 1914), and is sufficient to expose photosensitive material (Fig. 10.8) (Tsuji 1985, R.E. Young and Bennett 1988). The female possesses more mantle chromatophores than the male (566–687 on a female of 70 mm ML, and 450–543 on a male of 60 mm ML), but no apparent difference in the bioluminescent display could be discerned by the observer (Sasaki 1914).

The photophores on the ventral surface of the eyeball have rod-like elements, which in turn have fibre-optic-like light-guiding properties. These elements form a dominant portion of the volume of the photophore and are aligned in such a manner that light from the bioluminescent core is directed in a cone downwards from the eye. A coplanar arrangement of lamellae in the light guides suggests that the light passing through will be narrowly restricted in both wavelength and polarization (Kawahara *et al.* 1998). These photophores are assumed to function in counter-illumination, as demonstrated in other enoploteuthid squids (R.E. Young *et al.* 1980).

Eyes

The retina of the eye is much thickened ventrally where it consists of distinct layers, one pinkish and the other yellowish in fresh material. The retina is unusual among cephalopods in that three pigments are present. A major pigment, retinal, is present throughout and has an absorbance maximum of 484 nm. The thick ventral region of the retina has two pigments. The pinkish region, about 200 μm thick, contains mainly 3-hydroretinal with an absorbance maximum of 500 nm and the yellowish region, 400 μm thick, contains 4-hydroxy-retinal with an absorbance maximum of 470 μm (Matsui *et al.* 1988, Seidou *et al.* 1990). Light and ultrastructural studies show the presence of four types of photoreceptor cells (Michinomae *et al.* 1994). The specialized portion of the ventral retina may act to maximize the information from the limited light available to these squids when in deep water (Gleadall 1994). This retina thus resembles somewhat the multibank retina of fish (Denton and Locket 1989). The visual system of *Watasenia scintillans* may be capable of colour vision and it is possible that this squid can discriminate effectively between the light it generates and that of the environment (Seidou *et al.* 1990).

10.3 Ancistrocheiridae

The members of this family have large photophores scattered on the ventral surface of the mantle, funnel, head and tentacles, and numerous minute ones scattered on the surface of the mantle, head, and arms IV (Clarke 1988, R.E. Young *et al.* 1998a).

10.3.1 Ancistrocheirus

Ancistrocheirus, the only genus of the family, is widely distributed in the temperate and tropical open waters of the Atlantic, Indian, and Pacific Oceans, and is also found in sub-Antarctic waters (Roper *et al.* 1984, Roper *et al.* 1985, Nesis 1987). Its vertical distribution is not well known but it has been recorded from 0–190 m in daylight and 0–700 m in darkness (Clarke and Lu 1975, Lu and Clarke 1975b, Roper and Young 1975, R.E. Young 1978, Lu and Roper 1979).

Life history

A gravid female (250 mm ML) was 'taken alive from the mid-layer of the ocean' at a depth of 850 m, off Japan (Okutani 1976). Its mantle was thick, soft, and flabby when touched, and covered with dark brownish-purple integument. Most of the mantle cavity was occupied by nidamental glands and eggs. The eggs were in two stages of development, the most advanced (2.5 × 1.2 mm) being golden yellow in colour. Spawning may take place on the continental slope at depths greater than 1000 m (Roper *et al.* 1985) and lasts from June to October off the coast of South Africa (Clarke 1980). This squid can reach a mantle length of 400 mm (Nesis 1987).

Ancistrocheirus lesueurii has a thick skin and a thick mantle wall with an outer, purple pigmented, gelatinous layer and an inner layer of muscle (Okutani 1976, Clarke 1980, Riddell 1985). The tissues of the mantle, fins and arms contain ammonium ions, with a lower specific gravity than seawater, so the animal is near to neutral buoyancy (Clarke *et al.* 1979, Lipinski and Turoboyski 1983).

Photophores

The photophores of this squid are dark purple or black; some have a white centre and some protrude above the surface of the skin after preservation (Clarke 1980). Bilateral symmetry is apparent in the arrangement of the photophores. One or two pairs of photophores are present on the ventral side of the head, and about five around each orbit. Very small photophores are scattered on the ventral side of the head and mantle. The larger photophores have a centrally located mass of photocytes backed by a thick reflector and encased laterally by concentrically arranged iridosomal platelets. The reflector is surrounded by a layer of chromatophores and dispersed pigment. Numerous chromatophores cover the photophore. The smaller photophores comprise a reflector, backed by pigment, and a mass of photocytes, and again a covering of chromatophores. The larger photophores are directed away from the surface of the body, the smaller ones ventrolaterally. In living animals the photophores appear silvery in white light, with a hint of blue from the larger organs. R.E. Young (1977) considered that this species does not counter-shade ventrally.

Extraocular photoreceptor organs

The extraocular photoreceptor organs of a juvenile consist of dorsal, ventral, and posterior sets of photosensitive vesicles. These form a loose association of variously shaped independent vesicles which lack any pigment (R.E. Young 1977, 1978).

10.4 Pyroteuthidae

This family of squids inhabits the epi- and mesopelagic zones of tropical, subtropical, and temperate oceanic waters. It was formerly a subfamily of the Enoploteuthidae but was raised to familial level (Clarke 1988, R.E. Young *et al.* 1998a).

10.4.1 Pyroteuthis

Pyroteuthis (Fig. 10.9) is relatively common in temperate oceanic waters between 40°N and 10°S in the Atlantic and Pacific Oceans; it also occurs in the Mediterranean Sea. In most areas it exhibits diel migration being nearer the surface at night than during the day. Off Hawaii the majority occur at 150–200 m in darkness and from 420–490 m in daylight (R.E. Young 1978), and show similar distribution patterns in other waters (Clarke 1969, Clarke and Lu 1974, Roper 1972, Roper and Young 1975, Riddell 1985).

Life history

Off the Hawaiian Islands no seasonal changes were found in the size distribution of these squids. Females of 35 mm ML or more, had spermatophores in the sperm receptacle, and those of 40 mm ML had large gonads. The ova can reach a maximum size of 1.3 mm, but only a very small percentage of those present were mature; non-yolked oocytes accounted for some 62% of the total number (Boucher 1983). These conditions suggest that spawning is intermittent (Mangold *et al.* 1993).

Photophores

The distribution of photophores is shown in Fig. 10.9. Each tentacle has six, partially embedded, photophores. In the living animal one distal and the two proximal organs appear as silvery spheres, each with a thick wall of silvery material and a greenish sphere within. This central region of photogenic tissue gives a brilliant green-blue fluorescence under ultraviolet light, and a feeble blue-green light when treated with hydrogen peroxide. The three middle photophores are larger and each has a small raised central dome and silvered internal margins (Butcher *et al.* 1982). The tentacular photophores of two species of *Pyroteuthis* have emission maxima of 463 nm and 480 nm (see Table 10.2).

The 12 ocular photophores (Fig. 10.9d) have the same basic structure and in reflected light become a beautiful iridescent green. The photogenic tissue forms a spherical cone, which is almost enclosed by the reflectors of rigid platelets of the posterior cup. In the aperture of the posterior cup is a conical plug of axial platelets, which sometimes extend to the reflectors of the anterior cap. The latter is covered by a translucent hemisphere of tissue, the distal iridosomes, in the dorsal-most ocular photophores. Each photophore lies in a cup of red-brown pigment.

The anal photophores (Fig. 10.9e) point anteroventrally. They are morphologically like the ocular ones, and have a similar iridescence. The branchial and abdominal photophores all have a similar structure (Fig. 10.9f), with a small photogenic core enclosed in a cup of reflectors. The aperture of the cup is filled with a tight bundle of silk-like fibrils running from the surface of the photogenic core to the surface of the photophore. From the surface and laterally these fibrils are diffusely reflective so the organs have a somewhat white lustre in life. The fibrils presumably act as light guides, as they have an almost hemispherical light-emitting surface.

The optical systems of these photophores comprise three types. The tentacular photophores have reflector and light guide systems of collagen. The ocular and anal ones have an elaborate organization of reflective iridosomes, those in the aperture seeming to act as interference filters. In the branchial and abdominal light organs iridosomes form the main reflective tissue but collagenous fibrils function as light guides in the apertures of these photophores. The ultrastructure of the photogenic core is similar in all of the photophores. The photocytes are the largest cells and become more numerous towards the centre of the core. They are tightly packed and the cytoplasm is almost homogeneous

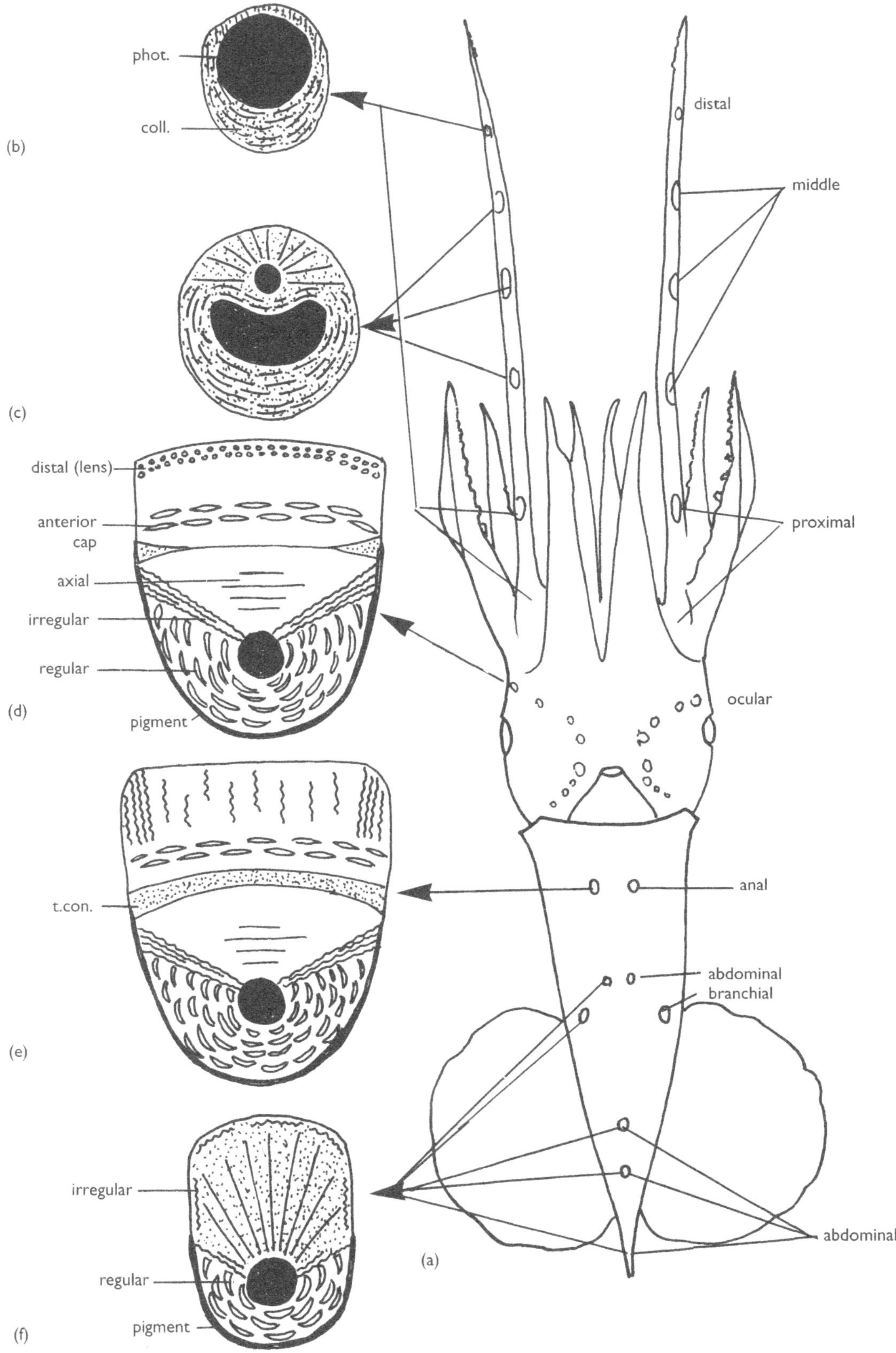

10.9. *Pyroteuthis margaritifere.* (a) An outline of the squid to show the distribution of photophores on the ventral surface. Diagrammatic representations, from an ultrasound study, of the types of photophores at different sites: (b) tentacular stalk, distal and proximal, (c) middle, (d) ocular, (e) anal, and (f) abdominal and branchial photophores (based on Butcher *et al.* 1982).

and the nuclei large and spherical. Thick sections of fresh photophores were examined with a fluorescence microscope and only the photocyte region was found to fluoresce (Butcher *et al.* 1982).

Some photophores are concerned in counter-illumination. *Pyroteuthis* can respond to changes of intensity of overhead illumination, by adjusting the intensity of its own luminescence, so that its outline is not visible when viewed from below (R.E. Young and Roper 1977).

The function of the photophores was examined in enoploteuthids and in *Pyroteuthis addolux*. The latter adopts an oblique position when illuminated from above, and on increasing the light the squid adjusts its own luminescence so that its outline is invisible when viewed from below. This squid could match the light intensities corresponding to the depth range it inhabits during the day (R.E. Young and Roper 1976). The dorsal and ventral photosensitive vesicles of the extraocular light organs are well sited to ensure a match between downwelling light and that emitted by the animal.

The buccal mass

A darkly pigmented buccal funnel surrounds the buccal mass, which is twice the length of the supraoesophageal mass of the brain (Fig. 10.10) but is narrow and the beaks thin. The mandibular muscles are dense and form a large proportion of the buccal mass. The oesophagus is moderately large where it passes through the central nervous system.

Table 10.2. Pyroteuthidae. The emission maxima of photophores in four species

Species	Emission maxima (nm)	Site of photophores	Source
Pyrotetuthis margaritifera	480	Tentacle	Herring (1983)
Pyrotetuthis addolux	463	Tentacle	E.A. Widder in Herring (1988)
Pterygioteuthis giardi	474–485, 505	Branchial	Herring (1983)
Pterygioteuthis microlampas	472, 492	Branchial	E.A. Widder in Herring (1988)

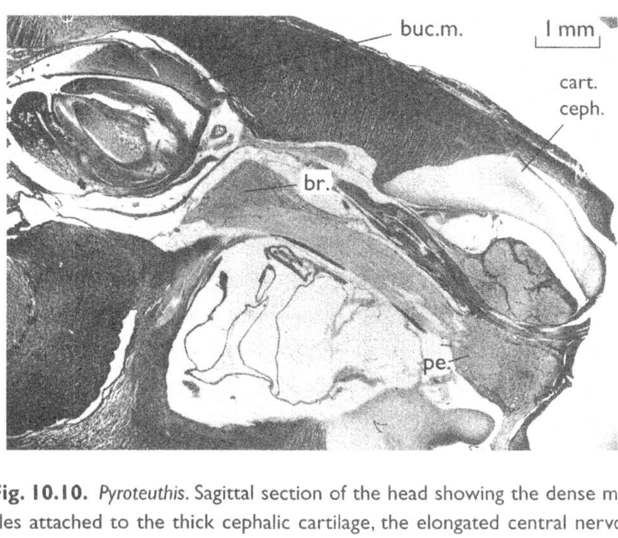

Fig. 10.10. *Pyroteuthis*. Sagittal section of the head showing the dense muscles attached to the thick cephalic cartilage, the elongated central nervous system, and the buccal mass.

Receptors and central nervous system

The eyes occupy much of the head and are directed laterally (Fig. 10.9). They lie close together, being separated only by thin membranes. The retina has rhabdoms of about equal length. The rhodopsin of *Pyroteuthis margaritifera* has an emission maximum of 480 nm (Muntz and Johnson 1978). The optic lobes form 342% of the total brain volume (Maddock and Young 1987) and the granule cell layers are thick.

The cephalic cartilage is hyaline in form and thick, especially around the central nervous system (Fig. 10.11). The musculature too is thick and there is a deep region of attachment to the anterior part of the cephalic cartilage.

The central nervous system is elongated, with the brachial lobes widely separated from the pedal lobes (Fig. 10.10). The inferior frontal lobe system forms 0.8% and the vertical lobe system 11.9% of the total brain volume (Table 2.6).

The ventral magnocellular lobe is smaller than in *Pterygioteuthis*, and its neuropil less complex. The first-order giant cells are not large and not well seen in the sections. Second-order giant fibres, such as are found in *Pterygioteuthis*, were seen at the centre of the pedal lobe, but no further details of the system could be discerned. No single outstandingly large fibre was seen in the brachial nerves, though these contain several moderately large fibres, 10 μm in diameter.

The extraocular photoreceptor organs form 17.7% of the total brain volume (Maddock and Young 1987). They comprise two sets of photosensitive vesicles: the dorsal set is embedded in the posterodorsal wall, and the ventral set in the posteroventral surface of the cephalic cartilage. The integument adjacent to both sets of photosensitive vesicles lacks pigment and so form windows for light to enter (R.E. Young 1978). The photosensitive nerves enter the peduncle lobe separately and spread out in two partly distinct lobules (Fig. 10.12).

There is a large subpedunculate lobe and nerve in *Pyroteuthis addolux* (Fig. 10.11), as there is in *Lycoteuthis*. The subpedunculate lobe is more richly vascularized than any other region of the brain.

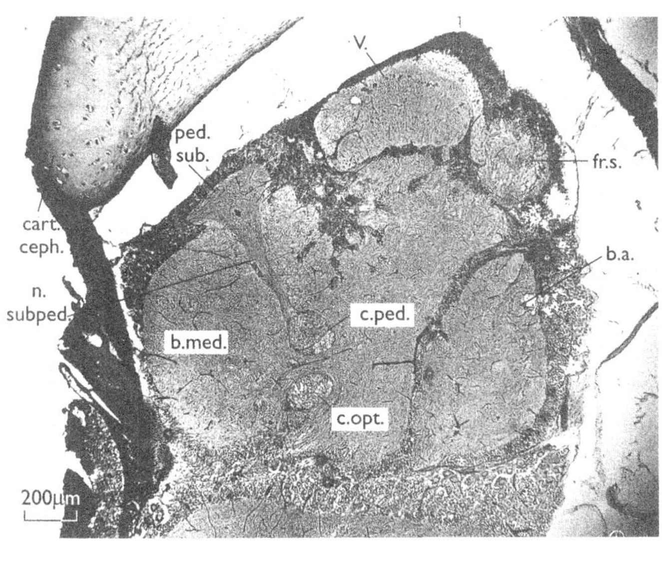

Fig. 10.11. *Pyroteuthis*. Sagittal section of the supraoesophageal mass showing the subpedunculate lobe and its nerve.

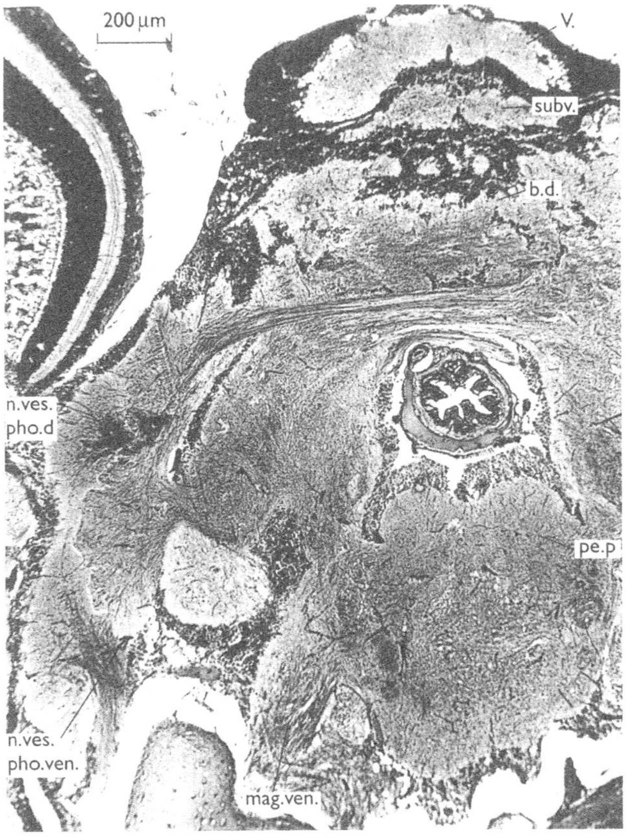

Fig. 10.12. *Pyroteuthis.* Transverse section of the central nervous system to show the nerves of the photosensitive vesicles separately entering the peduncle lobe.

The destination of the nerve fibres could not be determined beyond the level of the peduncle lobe commissure, where the nerve trunk divides.

The statocysts of *Pyroteuthis margaritifera* have the short, deep form seen in enoploteuthids (Young 1989). The anticristae are all rounded, and anticristae 4 and 5 restrict the anterior chamber of the statocyst. All five hamuli are present. Multivariant discriminant analysis of various parameters of the statocyst place this squid amongst non-buoyant forms (Maddock and Young 1984), and confirms the results of Clarke and his colleagues (1979).

10.4.2 Pterygioteuthis

These squids are found in warm waters worldwide, from just below the surface to about 600 m in depth (Roper *et al.* 1984). They undergo vertical migrations from 380–575 m during the day to between 35–200 m at night in Hawaiian waters (R.E. Young 1978), and in the Atlantic from 10–400 m in daylight and 25–200 m in darkness (Clarke and Lu 1974).

Life history

In Hawaiian waters hatchlings of *Pterygioteuthis microlampas* (<14 mm ML) were captured in the spring and larger ones (>15 mm ML) in winter, suggesting seasonal spawning (Boucher 1983). Gonads were largest in animals of 18 mm ML, and most females of 15 mm ML or more had mated as spermatophores were present in their sperm receptacles. A gravid female *P. giardi* (18 mm ML) had eggs at various stages of development in the ovary, and about 10 eggs in the oviduct (K. Mangold pers comm.). This suggests that it spawns intermittently (Mangold *et al.* 1993). *P. gemmata* is one of the smallest oceanic squids, and growth of the statolith indicates that females mature at 2 months of age and males at 1.5 months, and they live for up to 2.7 and 2.5 months respectively; its mantle length can reach 32 mm (Arkhipkin 1997). In spite of its short life span this squid is abundant in the epipelagic and mesopelagic tropical waters of the Atlantic (Arkhipkin and Schetinnikov 1989).

Photophores

Pterygioteuthis has 40–48 photophores, depending on species. There are 2 or 4 on the tentacular stalk, 14 or 15 associated with each eye, 1 on each gill, 1 on either side of the anus and 4 along the midline within the body (see Fig. 10.13). Several of these photophores have been examined with light or transmission electron microscopes (Hoyle 1902, Arnold and Young 1974), and experimental studies made to determine their function; various tissues have been tested biochemically for the presence of light-producing substances (R.E. Young *et al.* 1979b).

The photophores take several different forms and those below each eye can be divided into at least three types. Eleven are similar, each about 0.4 × 0.3 mm, lentoid in shape and approximately radially symmetrical. The central core and spherical knob is of photogenic tissue, with light-producing photocytes. The photocytes branch and ramify through the photogenic tissue and have an extensive system of microvilli (Arnold and Young 1974).

In a shipboard aquarium the nine ocular photophores together with the two anal photophores of this squid produced a blue light which glowed continuously for 15 min (R.E. Young 1977). These photophores are all ventrally sited and, with the four midline photophores, are concerned in counter-shading. In this the colour and intensity of emitted light varies to match overhead illumination equal to that encountered at a depth of 340 m during the day in Hawaiian waters (R.E. Young *et al.* 1980). This depth corresponds to the depth of captures of these squids (R.E. Young *et al.* 1982).

The control of light emission from the photophores is probably mediated by the extraocular photoreceptor organs, each of which consists of two sets of photosensitive vesicles (Fig. 10.13). The dorsal set lies behind a translucent window in the cephalic cartilage to receive downwelling light. The ventral set of photosensitive vesicles is deeply embedded in a cavity of the posteroventral surface of the cephalic cartilage. These vesicles are in a suitable position to receive light from the anal photophores when directed anteriorly. The anal photophores may conceal much of the viscera, and some of their light is probably reflected off the posteroventral surface of the head.

Stimulation of the posterior muscles of the anal photophores results in contraction and a ventral rotation of the photophore of about 45° from an orientation nearly parallel to the body axis. Light emitted directly from the photophores, together with that reflected from iridophores, is directed anteroventrally as this squid usually orients at an oblique angle to the horizontal (R.E. Young 1977, R.E. Young *et al.* 1980).

The luminescent flashing of two pairs of photophores (the branchial photophores, one on each gill, and the ocular photophores, one on each eye) of *Pterygioteuthis microlampas* and *P. giardi* was examined experimentally by R.E. Young and his colleagues (1982). Each 'gill-flasher' has a large reflector and the light output is quite broad. Muscles attached to this light organ act to allow considerable tilt and enable this squid to have some control in directing the emitted luminescence. Each 'ocular flasher' possesses a large reflector and a heavily pigmented lid which can be drawn over the face of the organ to conceal the photophore, and is withdrawn only when the light organ is stimulated. These two pairs of photophores produce an extremely bright flash, visible in a lighted room, that of the gill photophores being the more intense and each appearing as a single large white spot of light. The flash is brilliant and that from one squid is sufficient to produce a temporary blind spot in the dark-adapted eye of the human observer. The duration of the flashes was 0.3–1.3 s and 1–2 s from the ocular and gill flashers respectively. When confronted and then grasped by a large, predatory, midwater shrimp, *Acanthephyra smithi*, some of the squids emitted flashes of 3–15 s duration. These flashes of varied duration from the ocular and branchial photophores were interpreted as having a defensive function, to deter, startle, or disrupt an attack from a predator. A brilliant flash could become offensive and so allow a predator to disrupt the prey's defence and enable the predator to make a successful strike. It was suspected 'that in the dark waters of the open ocean, prey-predator strategies involving luminescent flashes are extremely complex.' (R.E. Young *et al.* 1982). The function of the tentacular photophores is not known (R.E. Young 1977).

The light emitted by the photophores is probably based on a luciferin-luciferase system since homogenized tissue of the digestive gland luminesced with considerable intensity on the addition of extract from a photophore (R.E. Young *et al.* 1979b).

Iridophores

Pterygioteuthis microlampas has an extensive array of iridophores, and much of the ventral surface of the head and adjacent funnel groove is highly silvered. There is a pair of complex chambers in the anteroventral region of the head and the basal region of ventral arms IV (Fig. 10.13). These chambers are lined with iridophores and filled with translucent extracellular material. The concave dorsal wall of the chamber is lined with a thick layer of iridophores. The flatter, ventral wall has irregularly spaced platelets, although not precisely aligned, mostly almost perpendicular to the surface of the chamber, with their long axes oriented mediolaterally (R.E. Young 1977). At the posterior edge of the chamber is a large ventral ocular photophore with its lens directed towards the chamber. A fibre-optic light probe directed into this photophore resulted in a strong ventrally directed reflection of light from the ventral iridophores of the brachial chamber. The function of this chamber is not understood.

Buccal mass

The buccal funnel is deeply pigmented (Fig. 10.14); its function is probably to hide prey organisms with photophores. There are stout papillae on the inner surface of the buccal funnel. The buccal mass has deep mandibular muscles, yet the beaks appear delicate and are only partially tanned. The bolsters form a considerable portion of the radula supports and their tissue is reticulate (Fig. 10.15).

Receptors and central nervous system

The eyes occupy most of the head (Fig. 10.13). The optic lobes appear small in contrast with the eye, and yet they form 320% of the total brain volume (Table 2.6) (Maddock and Young 1987).

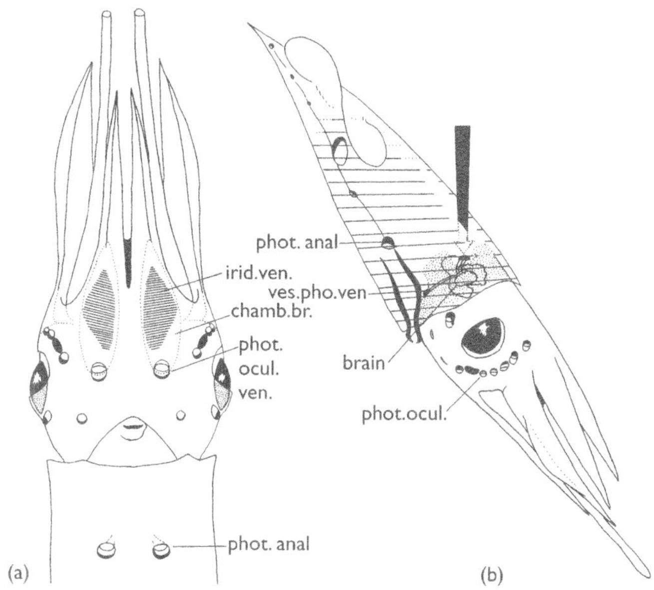

Fig. 10.13. *Pterygioteuthis microlampas*: (a) Lateral view of squid in one of a variety of attitudes observed. The arrow indicates the extraocular photoreceptor organ and represents the pathway of the downwelling light. (b) Ventral view of head to show the photophores and position of the brachial chamber with its ventral iridophores (R.E. Young 1977).

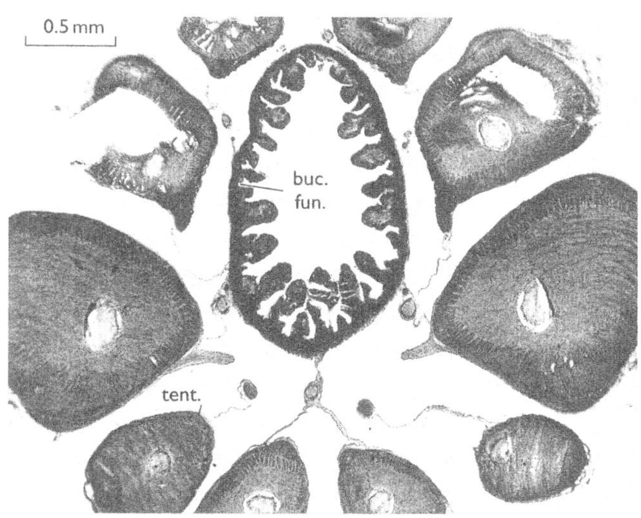

Fig. 10.14. *Pterygioteuthis*. Transverse section showing the dense musculature of the arms and tentacles, and the pigmented buccal funnel.

The central nervous system is elongated, with the brachial lobes widely separated from the rest (Figs 10.16, 10.17). The inferior frontal lobe system forms 1.3% of the total brain volume and the vertical lobe system 12.4% (Table 2.6). The neuropil of the ventral magnocellular lobe has a complex organization. There is a large brachiomagnocellular connective (Fig. 10.18). This photomicrograph also shows a large tract from the optic to precommissural lobe and two anterior giant fibres.

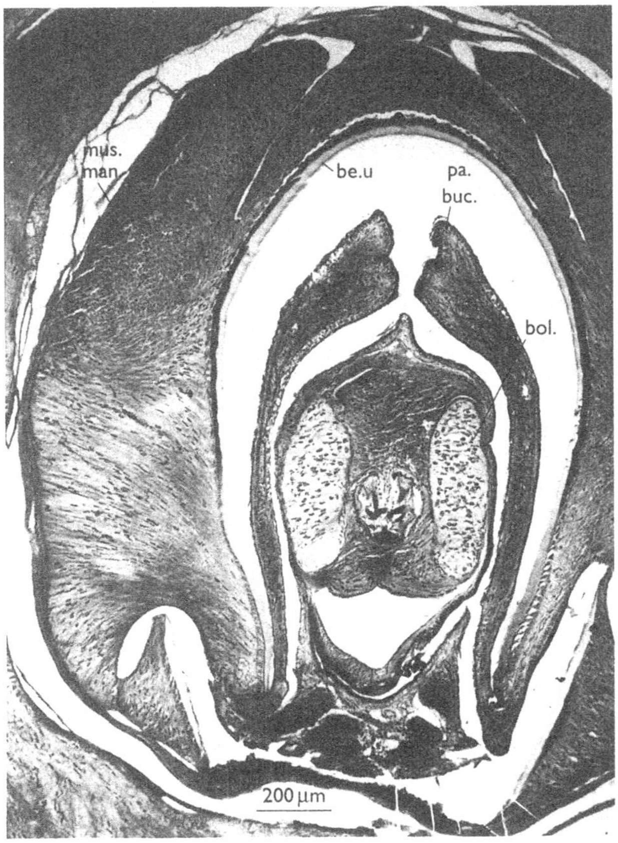

Fig. 10.15. *Pterygioteuthis.* A section of the buccal mass. The mandibular muscle is thick and mostly dense, the beaks are thin, and the tissue of the bolsters of the radular supports is pale staining.

The giant-fibre system originates from two first-order giant cells lying among the fibres of the macular static nerve. They have an elaborate dendritic system in the magnocellular lobe. Their axons proceed backwards to the palliovisceral lobe where they cross without fusion (Fig. 10.19). Beyond the crossing the branches separate and make contact with second-order giant fibres, one large one running to each mantle nerve.

The peculiarity of the giant-fibre system is the presence of a second anterior system of second-order fibres running forwards. These originate from large cells in the lateral walls of the magnocellular lobes whose axons run towards the midline. They make very close contact with the thin initial segments of first-order giant cells. The contact is so close that it is presumably synaptic (Fig. 10.20). Beyond it the two fibres of the anterior system run through the midline of the pedal lobe giving off branches here; these may be dendritic. One fibre then comes to lie dorsal to the other as they run forward above the fibres of the middle pedal commissure. Towards the front of the pedal lobe they then begin to divide. At this point the fibres are joined by further large fibres arising from cells in the ventral wall of the pedal lobe. Although not easy to follow at their origin, as they reach the front of the lobe it is possible to distinguish five especially large fibres on each side, one for each arm and tentacle (Fig. 10.21). These are third-order anterior giant fibres and the second-order fibres branch and make large synaptic contacts with them (Fig. 10.22). Besides these especially large fibres there are others not quite so large which may also receive contacts from the second-order fibres.

The five main fibres on each side continue forwards in the pedobrachial connectives and past the brachial ganglion into the brachial nerves where they lie in the lateral longitudinal columns. Their final destinations are probably the large single fibres seen in each brachial cord (Fig. 10.23). This system presumably serves to provide rapid retraction of the arms whose muscles are well developed (Fig. 10.24). It is not clear why the nerve fibres controlling them should be larger in *Pterygioteuthis* than in other squids of the enoploteuthid group.

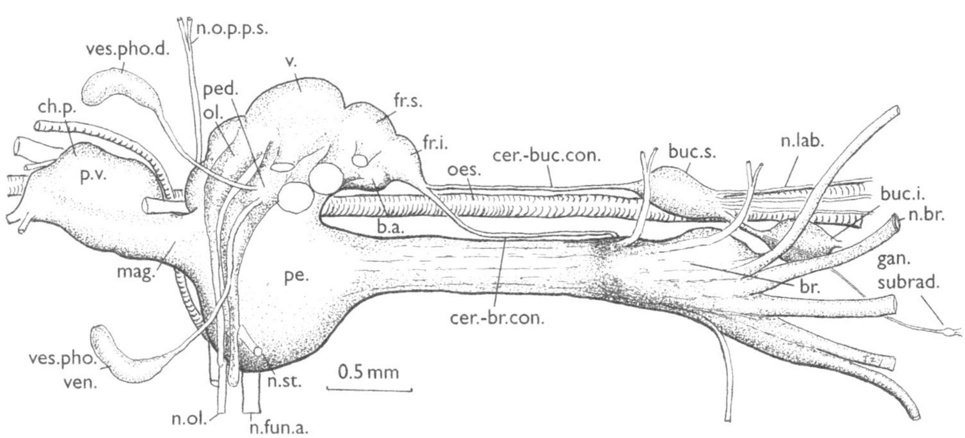

Fig. 10.16. *Pterygioteuthis.* Drawing of the central nervous system, reconstructed from serial transverse sections, in lateral view.

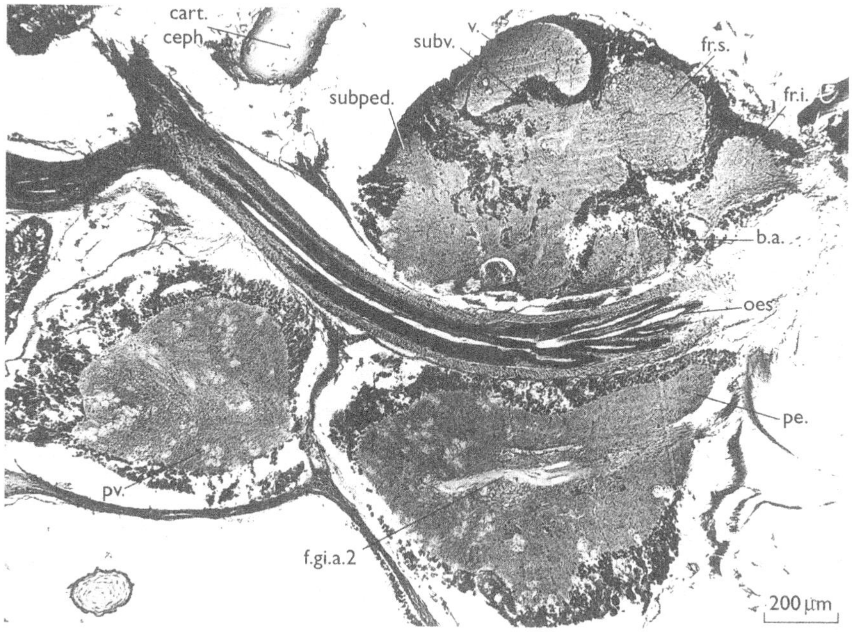

Fig. 10.17. *Pterygioteuthis*. Sagittal section through the central nervous system.

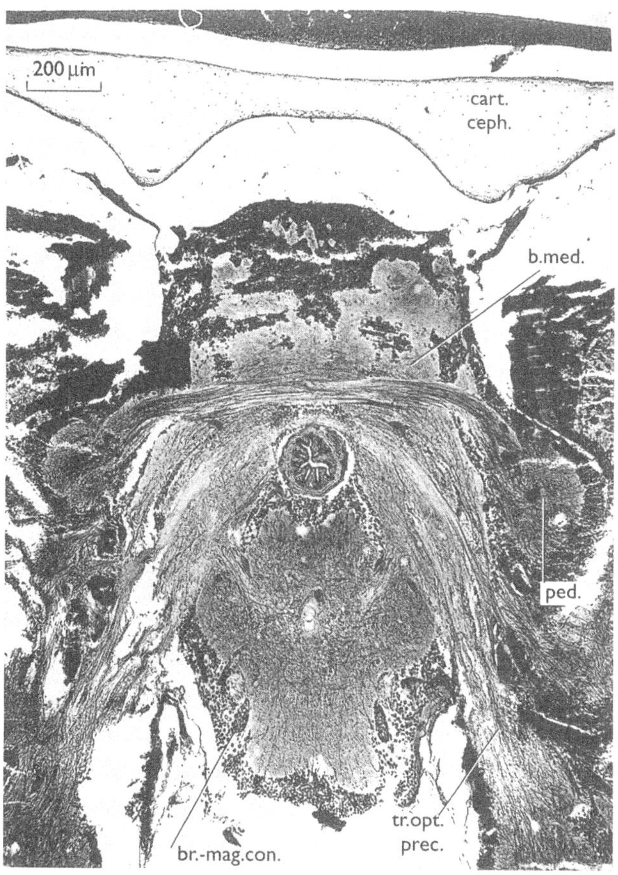

Fig. 10.18. *Pterygioteuthis*. Transverse section of the central nervous system to show the large brachiomagnocellular connective.

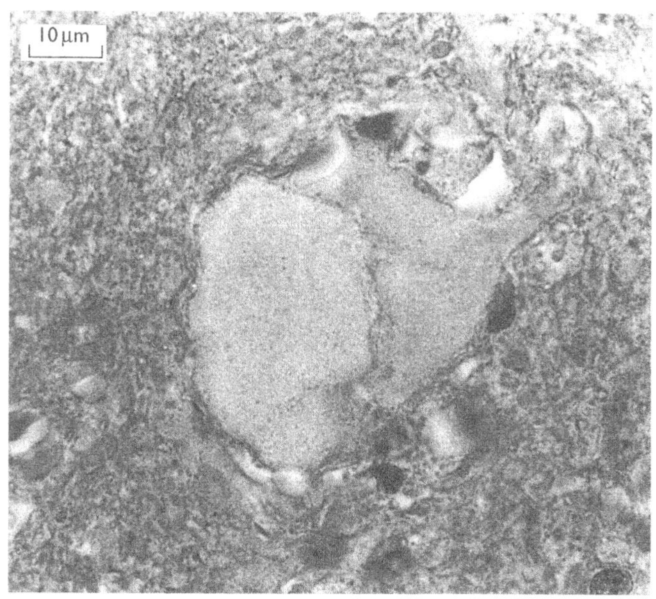

Fig. 10.19. *Pterygioteuthis*. Sagittal section of the palliovisceral lobe showing the axons from the two first-order giant cells where they cross without fusion.

No outstandingly large fibres were seen in the same positions in either *Abraliopsis* or *Pyroteuthis*.

Each extraocular photoreceptor organ comprises one dorsal and one ventral mass of photosensitive vesicles at the back of the orbits. Ventrally the vesicles are embedded in pockets of the orbits (Fig. 10.25). The dorsal set is compact, approximately disc-shaped, and tilted some 30° from the longitudinal axis of the body. This set

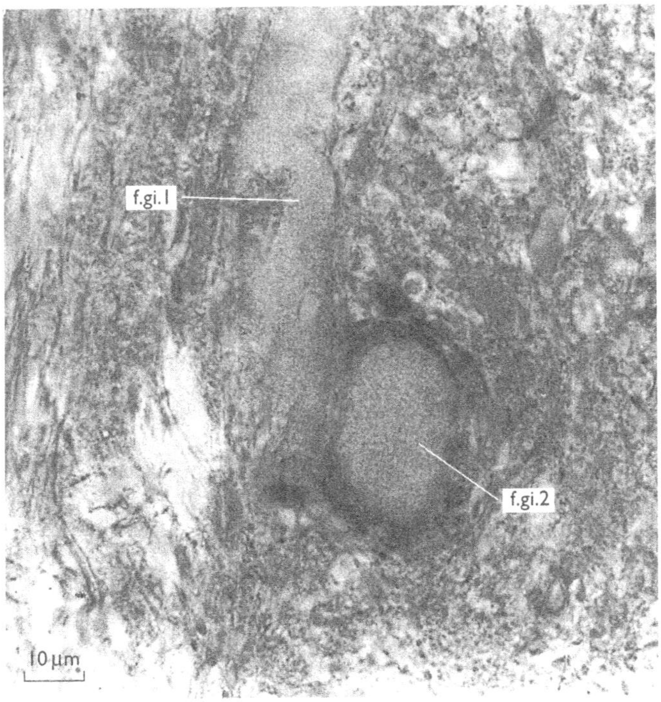

Fig. 10.20. *Pterygioteuthis.* Sagittal section showing the contact between the first-order giant axon and the anterior second-order axon.

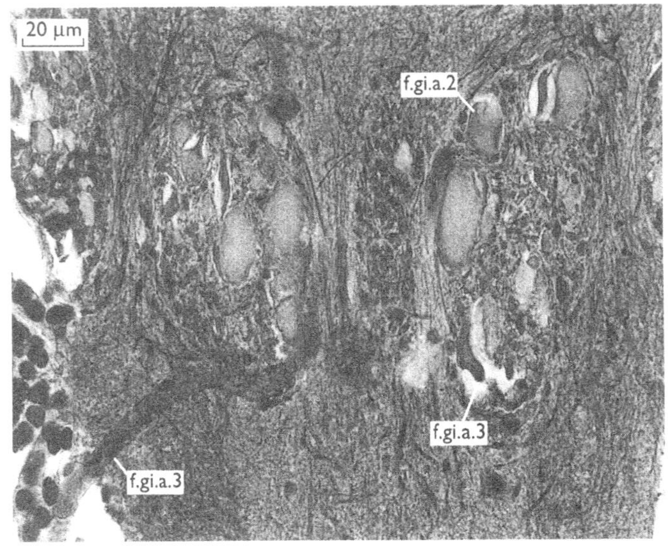

Fig. 10.21. *Pterygioteuthis.* Transverse section in which an anterior third-order fibre arises from its cell body, and runs to make contact with the second-order fibres.

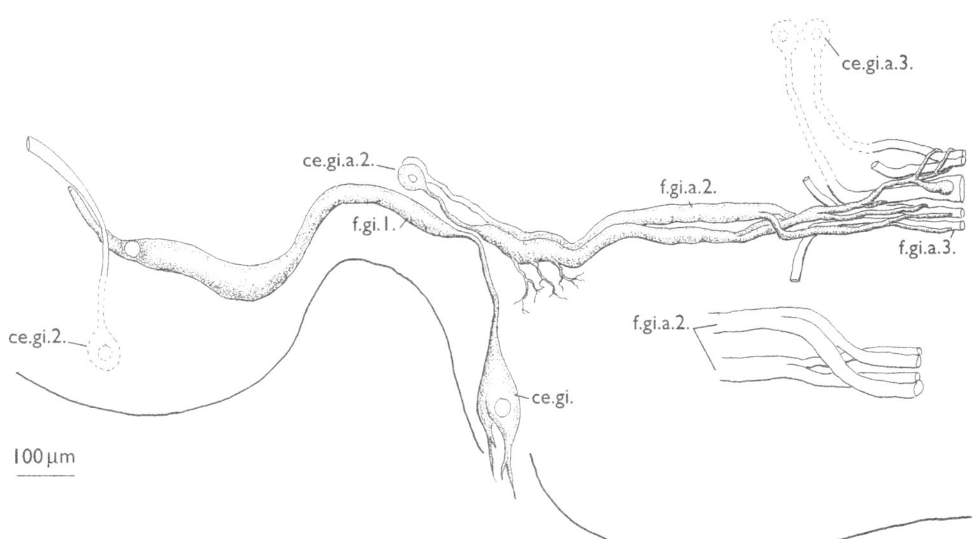

Fig. 10.22. *Pterygioteuthis.* Drawing of the anterior and posterior sets of second-order giant cells and fibres and the third-order anterior giant fibres.

consists of numerous, elongate, irregularly shaped, photosensitive vesicles, usually in a single layer. Sensory cells are distributed on all sides of each vesicle, and sensory processes are oriented mostly at right angles to the cartilage. The ventral set is thick, compact and somewhat square in outline, with a short, broad but thin extension passing along the inner surface of the cephalic cartilage. The vesicles are irregular in shape and orientation, and the sensory processes tend to lie parallel to the cephalic cartilage (R.E. Young 1977). Above the dorsal set of vesicles there are windows, which lack chromatophores and iridophores, present in the integument (R.E. Young 1978). The photosensitive nerves run into the large peduncle lobes from opposite directions. The photosensitive vesicles form 14.8% of the total brain volume (Maddock and Young 1987).

The statocyst of *Pterygioteuthis giardi* is short and deep (Fig. 10.25) (Young 1989). The anticristae are large and flat and fill much of the cavity. Anticristae 6 and 7 nearly touch, and the latter is

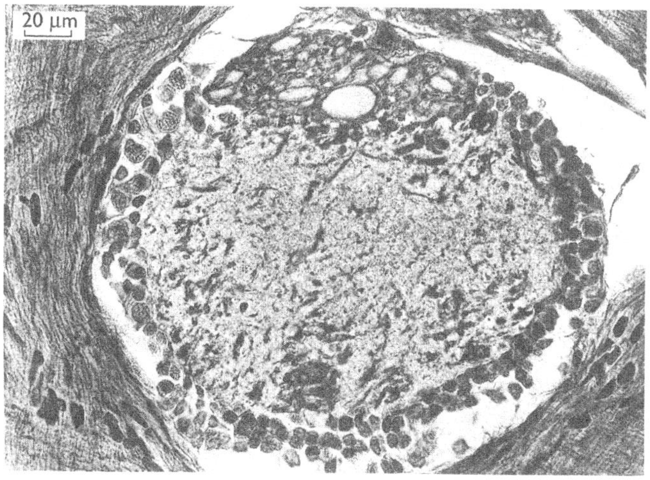

Fig. 10.23. *Pterygioteuthis*. Transverse section of brachial nerve cord with one large fibre.

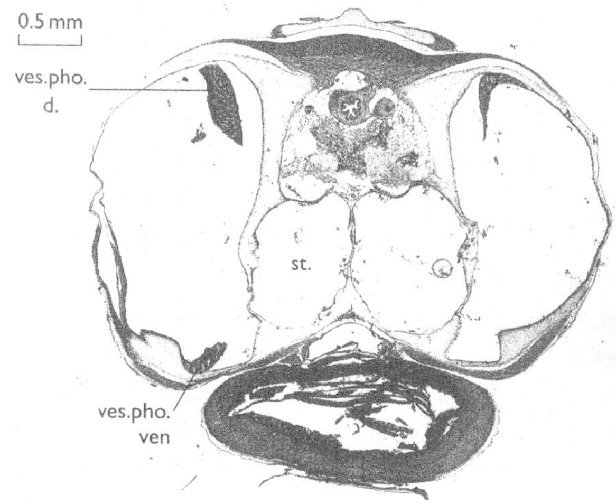

10.25. *Pterygioteuthis*. Transverse section through the head to show the extraocular photoreceptor organs with dorsal and ventral sets of photosensitive vesicles at the back of the orbit.

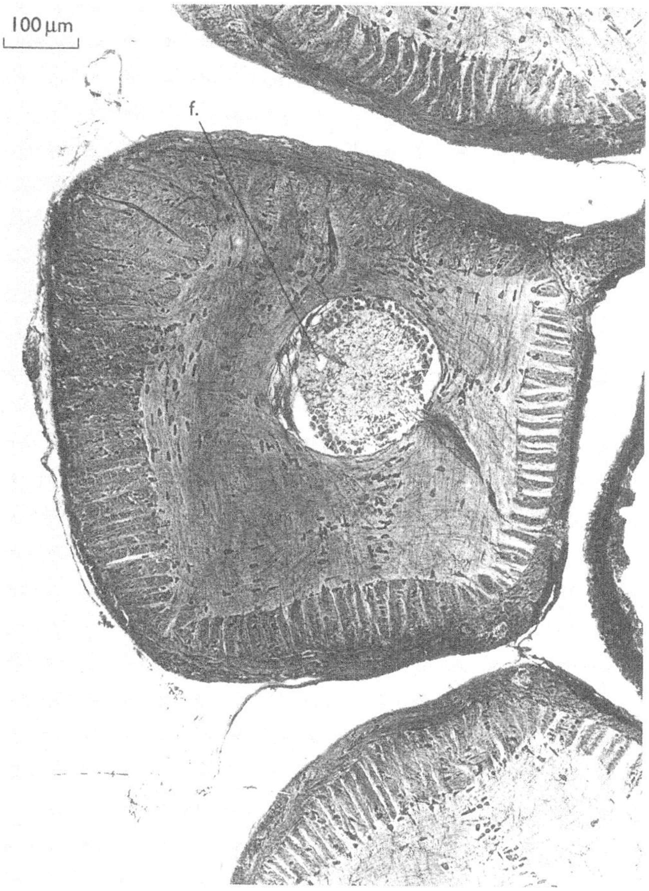

Fig. 10.24. *Pterygioteuthis*. Transverse section through the arms showing the dense musculature and brachial nerve cord.

unusual as it is curved and tapers to a point. There is a large hamulus at the first turn of the crista. This suggests that turns by the squid in the rolling and pitching planes may be especially rapid. Multiple discriminant analysis of various parameters shows that the statocyst differs from those of all other non-buoyant squids, but did not determine which measurement produced the aberration (Maddock and Young 1984).

11 Neoteuthidae

Neoteuthids are widely distributed, being found in the Atlantic, Pacific, Indian, and Southern Oceans, and the Mediterranean Sea (R.E. Young 1972b, Roper *et al.* 1985, Nesis 1987, Rodhouse 1988). They are probably mesopelagic and mesobathypelagic in habit.

11.1 *Neoteuthis*

Neoteuthis thielei has been caught in the Mediterranean Sea, in the Atlantic Ocean in opening/closing nets between 50 and 400 m, and in the Pacific Ocean (Clarke and Lu 1974, 1975, Lu and Clarke 1975b, R.E. Young 1978). The largest specimen recorded so far had a mantle length of 130 mm (Roper and Young 1975).

The eyes occupy a major portion of the head (Fig. 11.1), and there is evidence of crossing anterior conjunctive muscles. The optic lobes form 373% of the total brain volume (Table 2.6). The granule cell layers of the optic lobes are thicker in the posterior region, and the plexiform layers are well differentiated.

The brachial lobe is somewhat separated from the rest of the suboesophageal mass (Fig. 11.2). The magnocellular lobe forms 9.8% of the total brain volume and a typical giant cell is present (Fig. 11.3). There are fibres of up to 90 μm diameter in the stellar nerves. The magnocellular lobe receives a direct brachiomagnocellular tract from the brachial nerves, running below the brain (Fig. 11.4). The brachiovisceral tract also carries fibres direct from the brachial nerves, some of which pass to the magnocellular lobe. The fin lobe forms 4.2% of the total brain volume (Maddock and Young 1987). The posterior chromatophore lobe is small and was not measured. The inferior frontal system forms 2.6% and the vertical lobe system 12.1% of the total brain volume (Table 2.6).

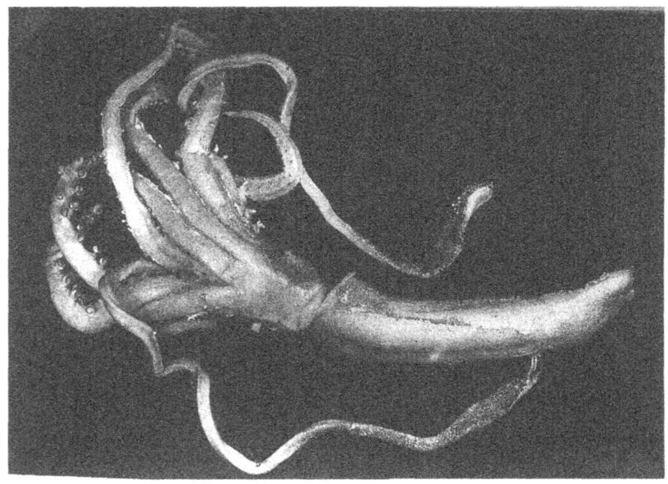

Fig. 11.1. *Neoteuthis* (104 mm ML) captured in Hawaiian waters (kindly provided by R.E. Young).

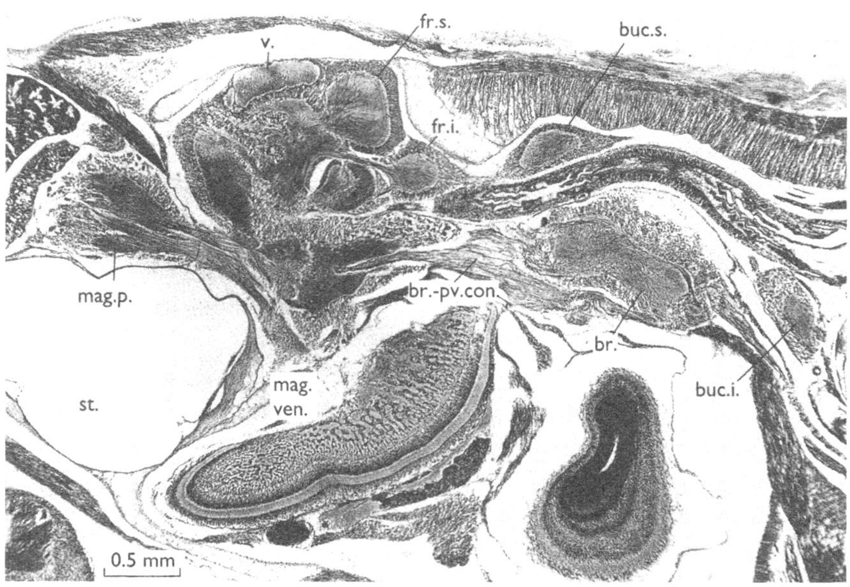

Fig. 11.2. *Neoteuthis* (23 mm ML). Sagittal section of the central nervous system.

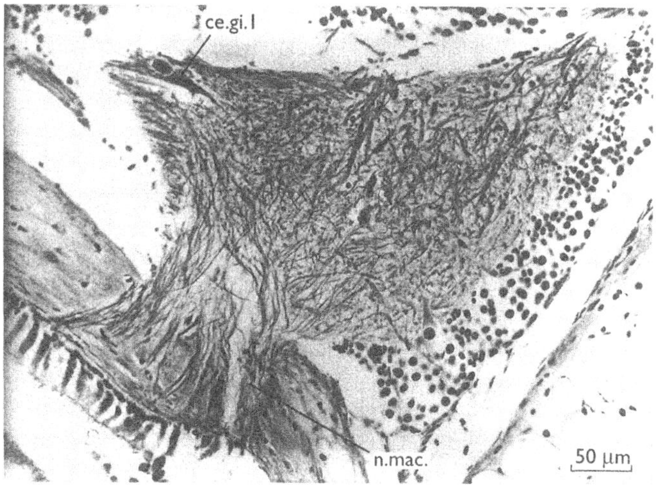

Fig. 11.3. *Neoteuthis* (23 mm ML). Sagittal section showing the magnocellular lobe and a giant cell.

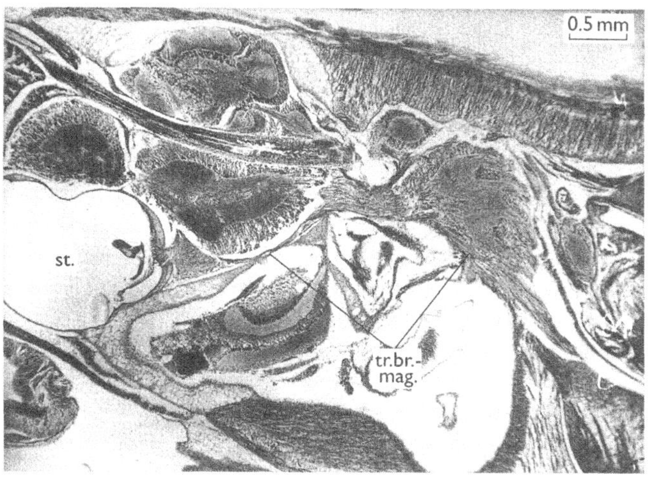

Fig. 11.4. *Neoteuthis* (23 mm ML). Sagittal section through the head showing the brachiomagnocellular tract.

The extraocular photoreceptor organ consists of a series of photosensitive vesicles (R.E. Young 1978). The ventral set arises from the medioposterior surface of the peduncle lobe and extends ventrally (Fig. 11.5). The vesicles form 0.8% of the total brain volume in a specimen of 23 mm ML (Maddock and Young 1987), but are larger in one of 104 mm ML.

The statocysts are large cavities set in a capsule of cartilage which is reticulate (Young 1989). Each cavity is little interrupted by anticristae, only 6 and 7 being large. These anticristae make contact and delineate a canal which extends for some 14% of the circumference. The hamuli are small, and one is present at the first turn of the crista.

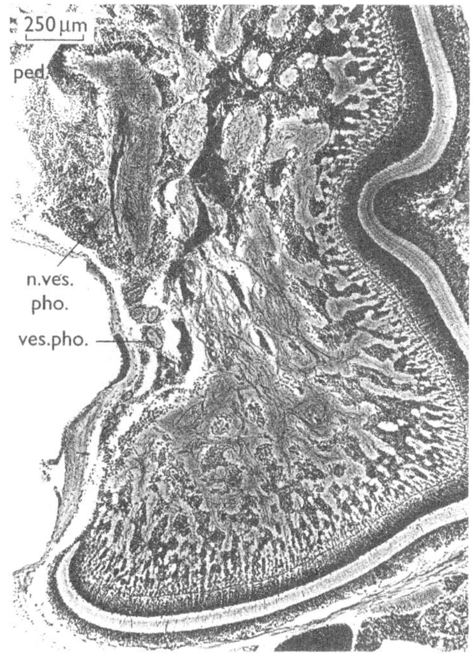

Fig. 11.5. *Neoteuthis* (23 mm ML). Part of the optic lobe, the peduncle lobe, and photosensitive vesicles.

The features of the statocyst suggest that this squid is suited to monitor only moderately rapid movements.

11.2 Alluroteuthis

This genus includes a single species, *Alluroteuthis antarctica*, which has a circumpolar distribution in the Southern Ocean where it lives mostly in depths of 750–2800 m (Filippova and Yukhov 1982, Roper *et al.* 1984, Rodhouse 1988, Kubodera 1989). Numerous specimens, including a mature female (222 mm ML), were caught in Prydz Bay at a depth of 433 m (Lu and Williams 1994b). A mantle length of 270 mm has been recorded (Nesis 1987).

Alluroteuthis preys on crustaceans (*Euphausia superba*, *Themisto gaudichaudii*), fishes and, squids (Nemoto *et al.* 1985), and the stomach of one female contained the squid *Psychroteuthis glacialis* as well as fish (Lu and Williams 1994b).

11.3 Nototeuthis

Nototeuthis dimegacotyle is found in the South Pacific Ocean and was described by Nesis and Nikitina in 1986. It has been caught from 30–60 m and 0–500 m (Nesis and Nikitina 1992). Its most characteristic feature is the presence of two enormous suckers, which occupy the whole of the central part of the tentacular club.

12 Lycoteuthidae

Lycoteuthids have a spectacular array of light organs. After observing the brilliance of the display of light emitted by living *Lycoteuthis (Thaumatolampas) diadema*, captured by the *Valdivia*, Carl Chun was so moved that he embarked on a study of Cephalopoda resulting in his magnificent monograph (1910a, 1915). These squids are known from the Gulf of Mexico, the western and southern Atlantic Ocean, the south Indian Ocean, off New Zealand, off the Kermadec Islands in the Pacific Ocean, and in Antarctic waters; their vertical distribution is poorly known, but they have been caught between 0 and 600 m, and as deep as 2195 m and 3290 m (Voss 1962a, R.E. Young 1964b, Filippova 1972, Clarke and Lu 1974, 1975, Roper and Young 1975, Toll 1983, Förch and Uozumi 1990, Villanueva and Sanchez 1993).

The family includes two subfamilies, the Lycoteuthinae with three genera, and the Lampadiateuthinae with one genus.

12.1 Lycoteuthinae

Members of this subfamily have five photophores on the ventral side of the eyeball arranged in one row, two on each tentacular stalk, and several in the mantle cavity.

12.1.1 Lycoteuthis

Lycoteuthis diadema (Fig. 12.1) has been captured in the western and eastern Atlantic Ocean, the north-eastern continental slope of New Zealand, and the southern parts of the Kermadec Ridge (Voss 1962a, Toll 1983). Its vertical distribution is not easy to ascertain since many specimens have come from the stomachs of predators, such as fishes, birds, and sperm whales (Clarke 1996a). Young specimens of *Lycoteuthis diadema* have been captured at 40–60 m and adults at 300–600 m and at 2195 m (Roper and Young 1975). As large numbers of gravid females and mature males (80–90 mm ML) were captured in two sequential net hauls, 0–100 m at 28°N 86°W, this suggests a mating aggregation (Toll 1983).

Sexual dimorphism is marked as the male has greatly elongated arms II, and a mantle drawn into a distinct tail in which there are 8 photophores; in all it possesses some 105 photophores including those on the arms. The female has subequal arms, subocular light organs, a very short tail with a single photophore in a pocket surrounded by reddish-brown chromatophores and a dorsally directed window for the emission of light, and several ventral photophores but none on the arms. The maximum recorded size is a male of 98 mm ML, and a female of 125 mm ML (Toll 1983).

Photophores

The light organs are the most striking feature of lycoteuthids and Chun (1910a, 1915) said of *Lycoteuthis diadema* that

> Its body appears to be covered with a diadem of precious stones. The middle organ of the eye shone ultramarine blue, the lateral organs with a nacreous sheen;... the posterior organs silvery white or with a nacreous sheen, except for the middle organ, which shone sky blue. A magnificent sight!

Five round subocular light organs are crowded together in both males and females. The central photophore is dark reddish-purple and the others reddish in colour (Voss 1958). The male has many photophores in the mantle, and on the head and arms. The female has 8–10 light organs in the mantle; these include a reddish one on either side of the anus, and a single large photophore in a pocket in the musculature at the apex of the fins. This last organ is surrounded

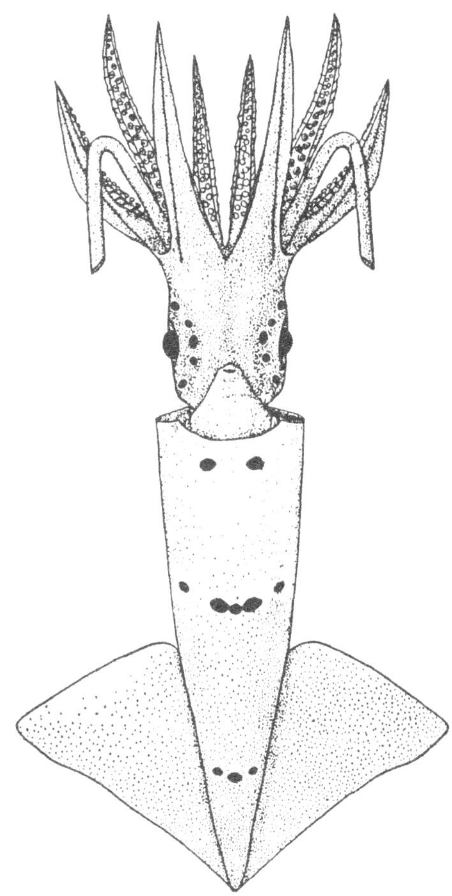

Fig. 12.1. *Lycoteuthis diadema* (65 mm ML). Female in ventral view (Voss 1962a).

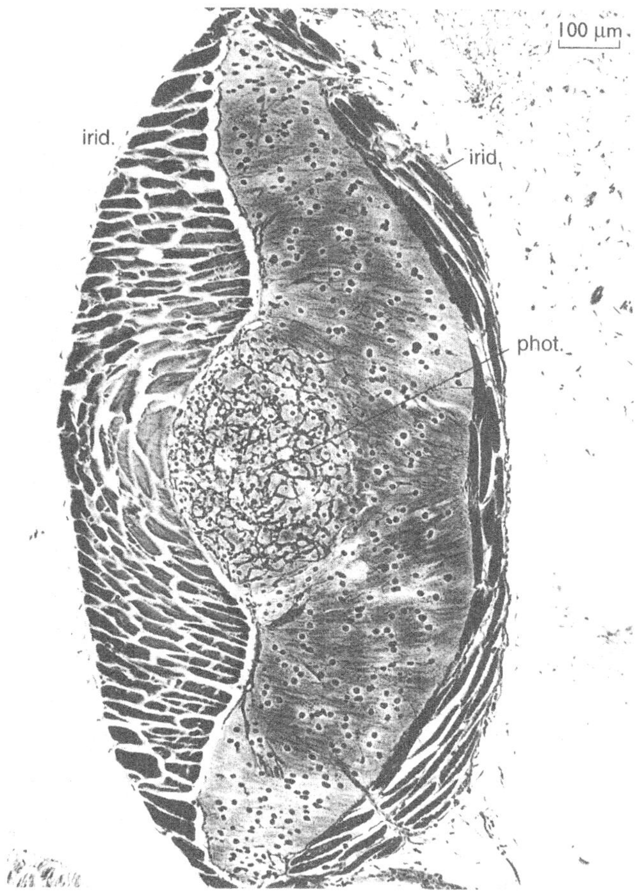

Fig. 12.2. *Lycoteuthis diadema.* Section of subocular light organ. Photomicrographs in Figs 12.2–12.10 are of same specimen (80 mm ML), stained with Bodian.

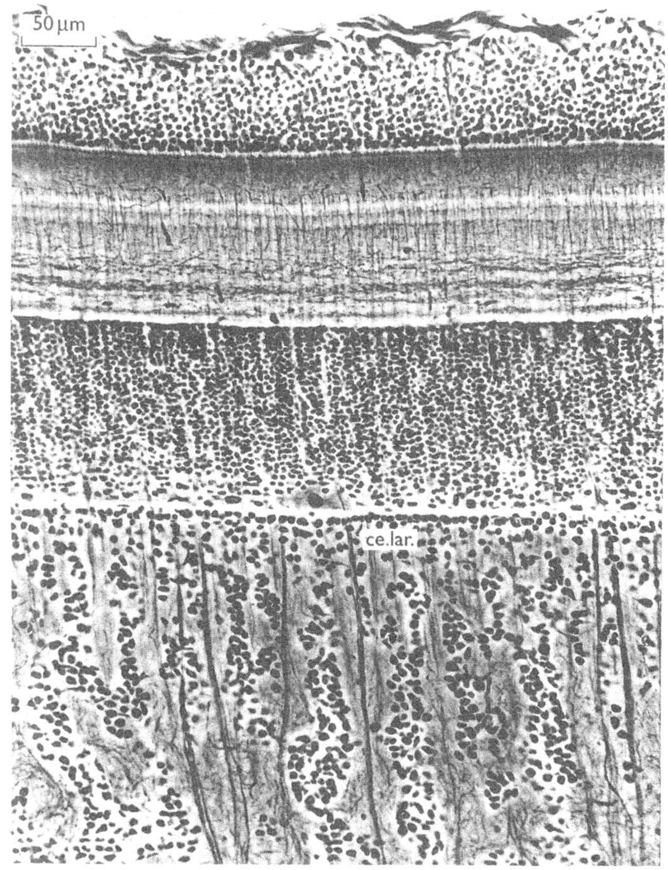

Fig. 12.3. *Lycoteuthis diadema.* The optic lobe showing the outer layers.

by a layer of densely organized dark reddish-brown chromatophores. The opening in the musculature, dorsal to the photophore, is covered by translucent connective tissue, forming a dorsally directed window for the emission of light (Toll 1983).

There is a central mass of photocytes in the light organs below the eyes (Fig. 12.2). The photocytes are in continual course of development and after being formed around the margin they proceed inwards as elongated pallisade cells. They become rounded and active in the central core where the blood vessels, entering from the surface, branch abundantly. The photocytes then proceed outwards, shrink, and presumably die at the surface. The emission of light is probably controlled by regulation of the blood supply. The blood vessels among immature photocytes are richly innervated. Beneath the photocytes are layers of flattened cells forming a tapetum. Peripheral to the photocytes are layers of 'squamous cells', as they were called by Chun (1910a), that may form a lens. There is no pigment around these organs.

Receptors and central nervous system

The retina of the eye was much folded in our sections and it was not possible to ascertain whether it varies in thickness. The longest rhabdoms reached a length of about 280 µm. The sclera of the eyeball is thickened as a marked ring of cartilage carrying the subocular luminous organs. The optic lobes form 398% of the total brain volume (Table 2.6). The granule cell layers and the plexiform layer of the optic lobe are thick, especially the inner graular layer, which has large cells at intervals around the inner border (Fig. 12.3).

The peduncle lobe is relatively large and the optic tract fibres pass to it to form parallel fibres in the spine region (Fig. 12.4). The photosensitive vesicles are in two sets and form 3.7% of the total brain volume (Table 2.6). The ventral set forms a long strand of vesicles passing from back to front on the floor of the orbit (Fig. 12.5); their nerve fibres form conspicuous bundles entering the peduncle lobe, and then spread out there. The dorsal vesicles consist of a few small sacs set amongst the optic nerves at the top of the back of the optic lobe. The nerve fibres enter the peduncle lobe from above and below, and the two sets exchange fibres. The system is thus very similar to that present in enoploteuthids.

The brachial lobes are widely separated from the pedal lobes, and large fibres run in the connectives between these lobes.

The magnocellular and pallial lobes form 17.5% of the total brain volume (Maddock and Young 1987), the ventral magnocellular lobe being prominent. The latter has a typical giant cell present (Fig. 12.6). In addition there is an anterior giant fibre on each side, originating from one of a group of large cells in the lower part of the dorsal magnocellular lobe. The largest anterior axon turns medially, crossing the axon of the first-order giant fibre, and runs towards the midline. There is probably synaptic contact at the point of crossing. The anterior fibre reaches a diameter of 100 µm and other smaller

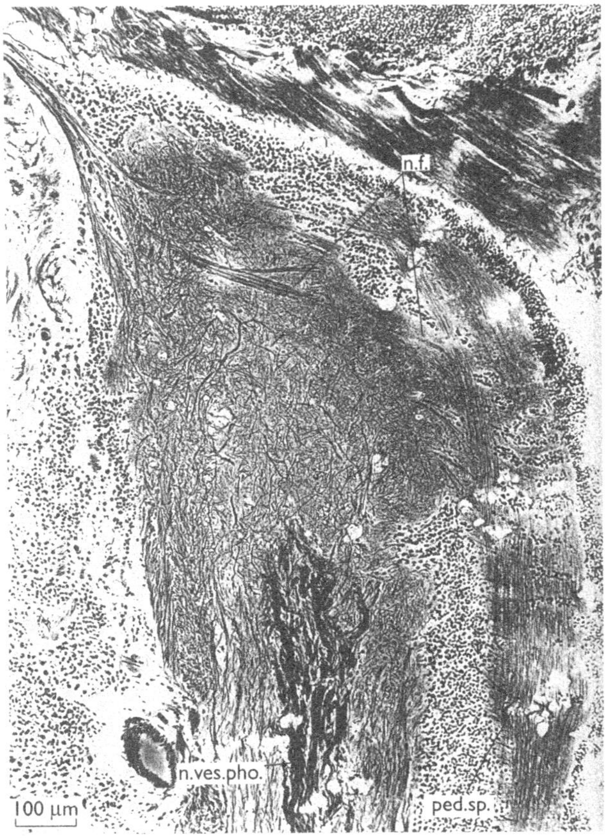

Fig. 12.4. *Lycoteuthis diadema.* Sagittal section of peduncle lobe showing fibres running between the optic lobe and the spine region, and the photosensitive nerves.

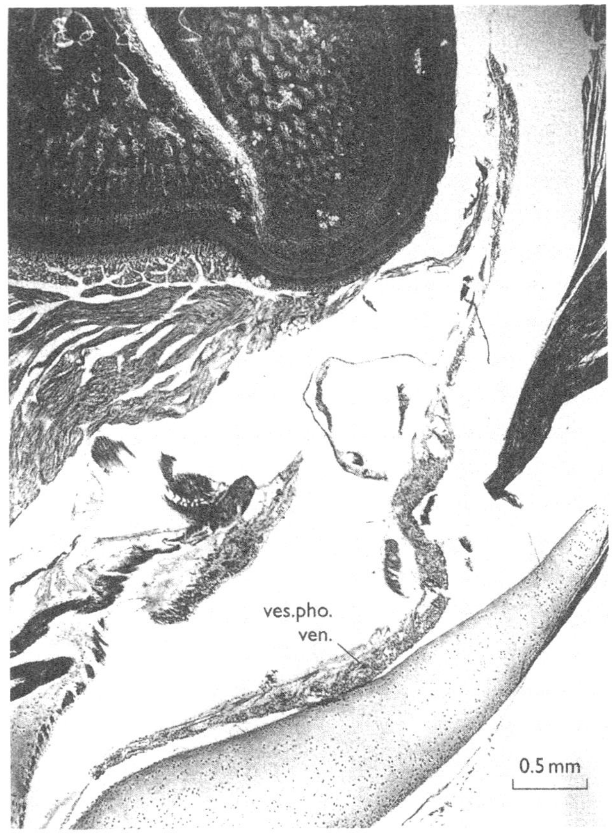

Fig. 12.5. *Lycoteuthis diadema.* Sagittal section showing the position of the ventral photosensitive vesicles.

fibres accompany it. They all cross in the posterior pedal commissure and then turn forwards to run close together, the largest fibres one above the other, in the midline above the middle pedal commissure (Fig. 12.7), and finally divide into branches in the anterior pedal lobe. This is remarkably similar to the condition found in *Pterygioteuthis*.

The inferior frontal lobe system forms 2.9% and the vertical lobe system 16.6% of the total brain volume (Table 2.6) (Fig. 12.8). The subpedunculate and other basal lobes form 12.7%; these lobes have cells of up to 50 μm diameter, and their axons form a conspicuous tract running downwards to the base of the median basal lobe. Here there is an unusual lobule, that may be called the infrabasal lobe (Fig. 12.9). It lies near the midline above the back of the pedal lobe and contains cells of up to 140 μm diameter, much the largest in the supra-oesophageal lobes. Some of the fibres from the subpedunculate lobes end here, others run on, with axons arising in the interbasal lobe, towards the pedal and palliovisceral lobes in this region. None was seen running to or from the peduncle lobe, but their presence cannot be excluded.

The subpedunculate and interbasal lobes together evidently form a system with some special function; this may be related to the displays by the numerous photophores. Subpedunculate lobes with similar features are seen in enoploteuthids. The subpedunculate region receives many optic fibres and lies close to the peduncle lobe and thus also to the influence of the photosensitive vesicles. The whole system may be concerned with regulation of the circulation, the photophores being controlled by their blood vessels, which are innervated.

The arteries of this squid are surrounded by strong circular muscles. Furthermore the cephalic artery carries a powerful sphincter, immediately behind the brain (Fig. 12.9). This is presumably capable of closing off the blood supply to the head and arms for some special purpose.

The statocyst is rather short and narrow but deep. The hamuli and anticristae are relatively large and occupy much of the cavity, so restricting the flow of endolymph. Anticristae 6 and 7 are joined together, and form a canal (Young 1989).

12.1.2 *Selenoteuthis*

Selenoteuthis scintillans was first described from a male of 32.5 mm ML, caught at a depth of 46 m in the western Atlantic Ocean (Voss 1956, 1962a). It is now known to be widely distributed in the western north Atlantic (Herring *et al.* 1985). There is some indication of diel migrations from about 300 to 800 m in the day to less than 200 m at night (Clarke and Lu 1974, Roper and Young 1975, Voss and Stephen 1992).

It was in *Selenoteuthis scintillans* that sexual dimorphism was first recognized amongst the lycoteuthids by Voss (1962a). The male, but not the female, has a round photophore on the tip of both the dorso- and ventrolateral arms. At the end of the mantle, projecting beyond

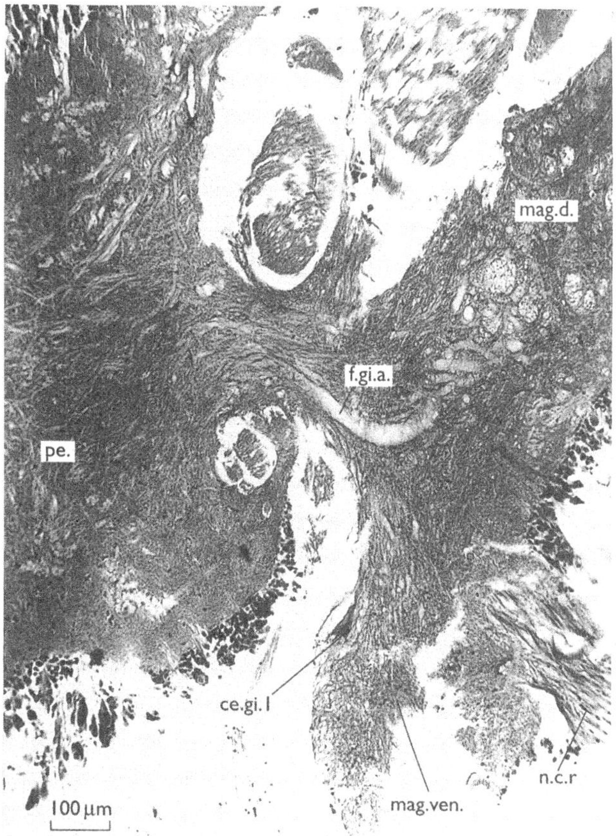

Fig. 12.6. *Lycoteuthis diadema.* Transverse section to show the ventral magnocellular lobe, the first-order giant cell, and anterior giant fibre.

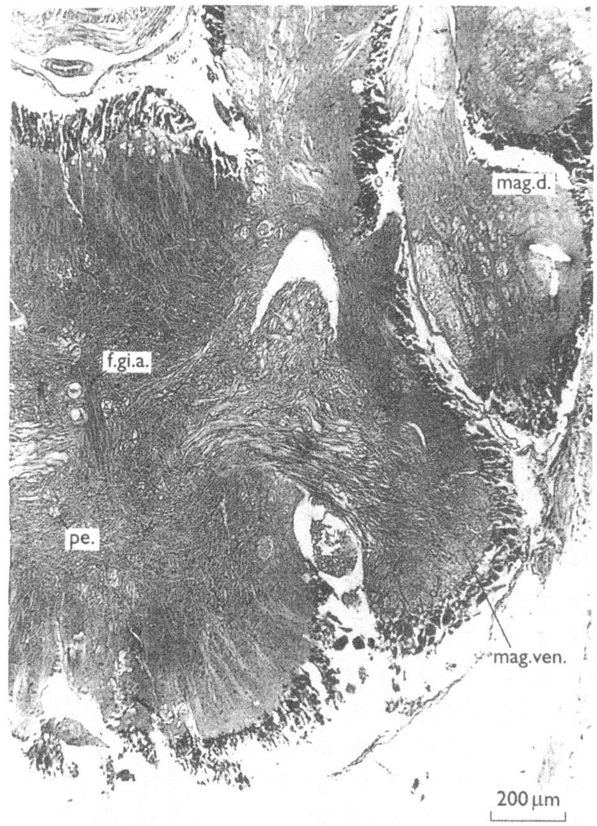

Fig. 12.7. *Lycoteuthis diadema.* Transverse section to show the two anterior giant fibres, one above the other in the midline.

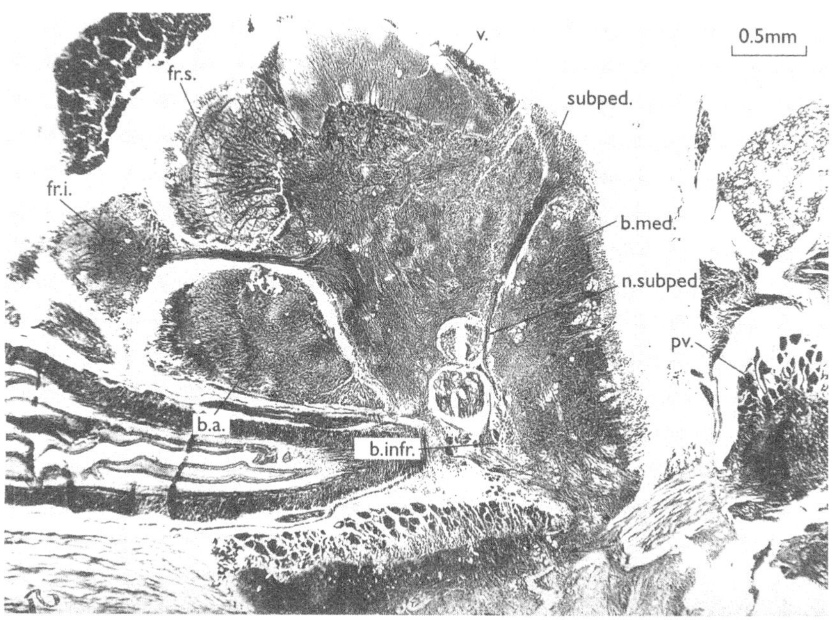

Fig. 12.8. *Lycoteuthis diadema.* Sagittal section of the supraoesophageal lobes showing the course of the subpedunculate nerve and the infrabasal lobe.

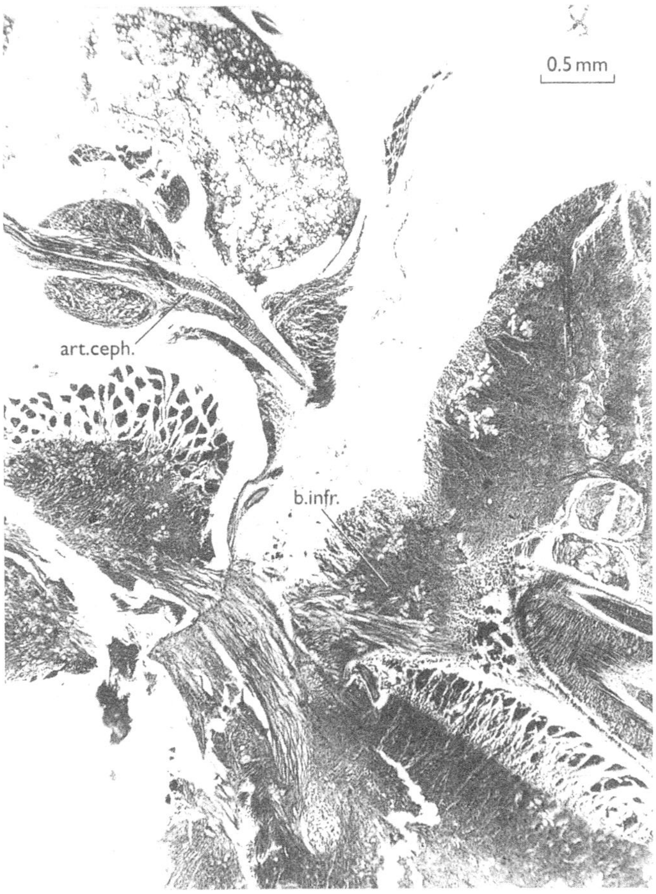

Fig. 12.9. *Lycoteuthis diadema*. Sagittal section showing the infrabasal lobe, the cephalic artery, and its sphincter.

the fins, is a large round light organ, which in the male is twice the size of that in the female. Females possess 25 photophores and additional ones are present in males (Voss 1958). An ultrastructural study revealed that the light organs are of three types. Those of the tail and some of the tentacular organs lacked optical structures (type 1), the other tentacular photophores and the abdominal ones have collagenous diffusing fibres (type 2), and the anal and ocular photophores have iridosomes but lack collagen (type 3) (Herring *et al.* 1985). The large spherical tail organ has numerous chromatophores on its surface, and these were seen to expand and contract to modify the bioluminescence of the photophore. All of the light organs fluoresced in ultraviolet light, and most glowed with increasing intensity before declining. The tail organ gave a very bright luminescence with a blue-green spectral emission, its maximum being 460 μm (Herring 1983). The function of this organ may be as a specific, possible sexual, signal, whereas the ocular and abdominal organs could be concerned with counter-illumination. Otherwise their function remains enigmatic.

12.1.3 *Nematolampas*

Nematolampas regalis was described from two animals washed on to a beach in the Kermadec Islands, in the south Pacific Ocean (Berry 1913, 1916), and no further specimens have been reported since (Voss and Stephen 1992). One of these squids (32 mm ML) had an array of at least 90 photophores, elongated thread-like ventrolateral arms with numerous embedded light organs, and a light organ near the base and another just below the club of each tentacle.

12.2 Lampadioteuthinae

Members of this subfamily have a firm, muscular, loliginiform body with broad, sagittate, terminal fins. They have five almost equally spaced, pearly photophores along each tentacle, five light organs in the mantle cavity and four subocular light organs (Berry 1916, Voss 1962a).

12.2.1 *Lampadioteuthis*

Lampadioteuthis megaleia, the only species of the genus, was described from a female (30 mm ML) washed ashore on the Kermadec Islands in the south Pacific (Berry 1916). Since that time a few further animals have been captured from the Atlantic Ocean. All have been small and one of 8.5 mm ML was caught in an opening/closing net at 210–290 m depth (Clarke and Lu 1975).

One specimen (13 mm ML) had 23 photophores, which fluoresced in ultraviolet light (Herring *et al.* 1985). One of the tentacular photophores had magenta chromatophores over its dorsal surface. The light organs were found to be of similar structure to those of *Selenoteuthis*, at both the light and ultrastructural level, although less complex. The tentacular and abdominal photophores were like type 2 of *Selenoteuthis* and the anal and ocular photophores like type 3.

13 Histioteuthidae

13.1 *Histioteuthis*

Histioteuthis is the only genus of the family (Fig. 13.1). It is pelagic and occurs in abundance in the midwaters of the oceans, from the sub-Arctic to the sub-Antarctic, and is widespread in tropical–subtropical–temperate or tropical–subtropical waters (N.A. Voss *et al.* 1998). The vertical distribution ranges from the surface to depths of perhaps 2000 m, and many species undergo diel vertical migrations (N.A. Voss 1988a). Furthermore *Histioteuthis* has been observed near the bottom from submersibles (Vecchione and Roper 1991).

13.1.1 Life history

The reproductive habits and life history of the genus are poorly known. *Histioteuthis bonnellii* is the largest species and a male of 330 mm ML has been recorded (Morales 1962). A female of the same size was gravid, the ovary and oviduct being filled with an estimated 25 000 eggs (2.3 mm diameter) (Kristensen 1980). Gravid females of *H. elongata* have been captured in February and July, suggesting a non-seasonal spawning cycle (Lu and Roper 1979). Mature males of *H. atlantica* (92–132 mm ML) have densely set photophores without lenses on the posterior part of the mantle, both dorsally and ventrally; this arrangement appears unique amongst histioteuthids (Lipinski 1993). Small specimens (5.5–20.0 mm ML) have been caught in plankton nets (Okutani and McGowan 1969).

13.1.2 Morphology

The mantle has been variously described as soft, semigelatinous, flaccid, or firm. The specimens sectioned for this study had mantle walls with bands of muscular tissue interspersed with reticulate tissue (Fig. 13.10). This tissue is also seen in the arms, head, mantle, and cephalic cartilage (Figs 13.2, 13.4, 13.9). Ammonium ions are present in *Histioteuthis*, the highest concentrations being found in the arms, followed by the mantle (Clarke *et al.* 1979). These features make it difficult to predict the orientation of the animal. The available evidence indicates that the body axis is held at about 45° from the horizontal, with the mantle uppermost (R.E. Young 1975a); some support for this comes from observations from a deep-sea submersible when one histioteuthid was seen in head-down position with its arms bent upwards (Church 1971). In the arms the muscles are largely restricted to the periphery (Fig. 13.3). The tentacles, in contrast, have quite densely organized muscles, indicating an active role in the capture of prey.

The buccal mass has stout mandibular muscles attached to the relatively thin beaks (Fig. 13.2). The tip of the papilla of the posterior salivary glands lies forward of the radula, which possesses elongated lateral teeth. Many tissues of the radular complex show reticulation (Fig. 13.4). The lateral buccal palps have teeth on their anteromedial surfaces.

Histioteuthis dofleini is reddish-brown when alive, and other species are purplish-red, a colour prevalent among animals of the

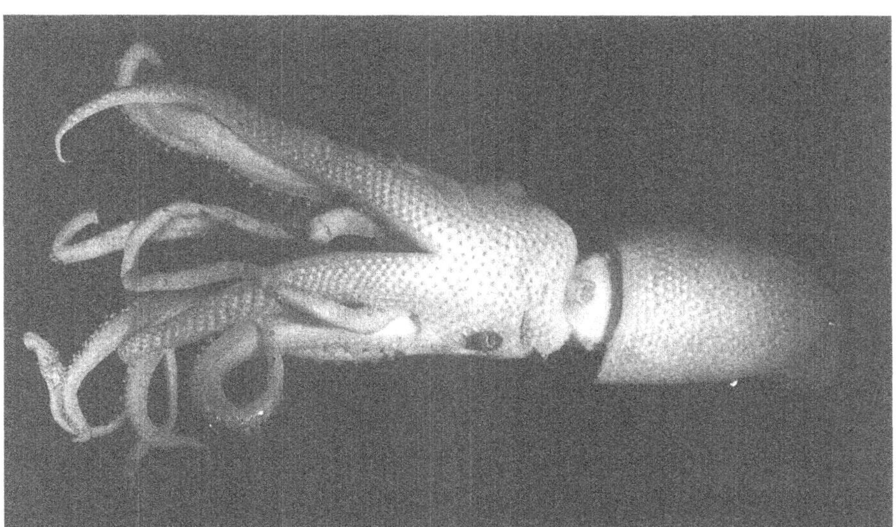

Fig. 13.1. *Histioteuthis.* Note the asymmetry of the eyes, the left being larger than the right.

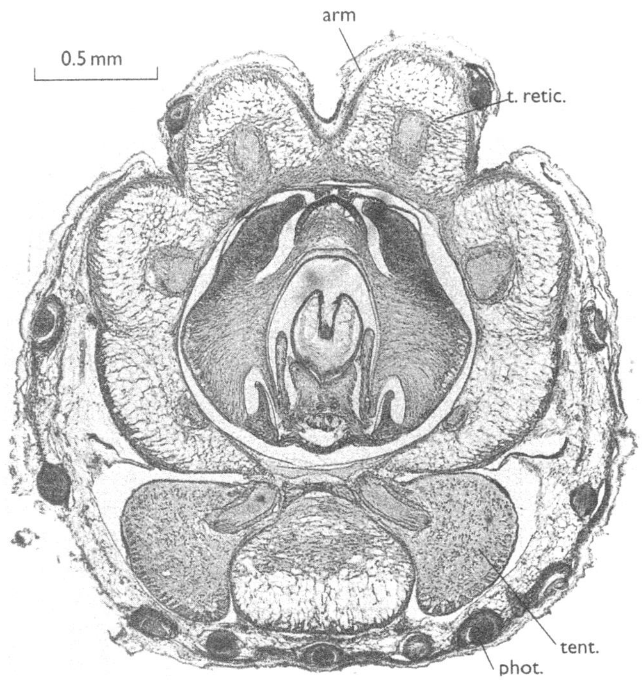

Fig. 13.2. *Histioteuthis reversa* (6 mm ML). Transverse section through the base of the arms and tentacles, the buccal mass, and photophores on the aboral surfaces of the arms.

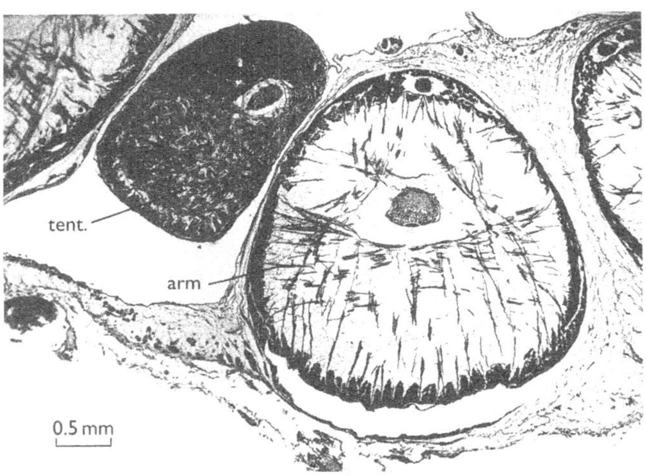

Fig. 13.3. *Histioteuthis reversa* (6 mm ML). Transverse section to show the arms and a tentacle; the latter has dense muscles and contrasts with the arms.

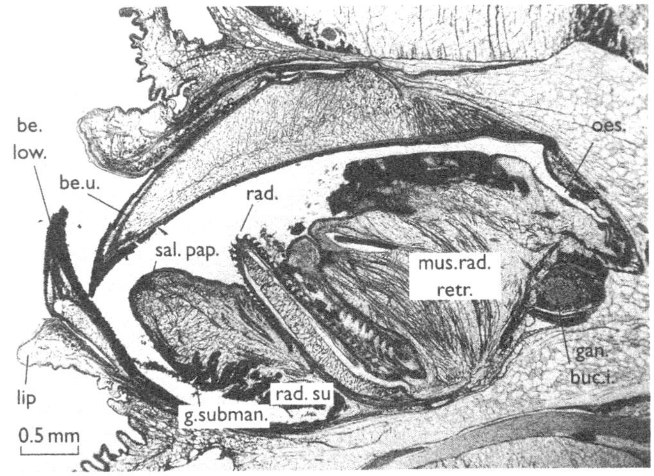

Fig. 13.4. *Histioteuthis reversa* (6 mm ML). Sagittal section of the buccal mass near the midline. The salivary papilla is relatively large.

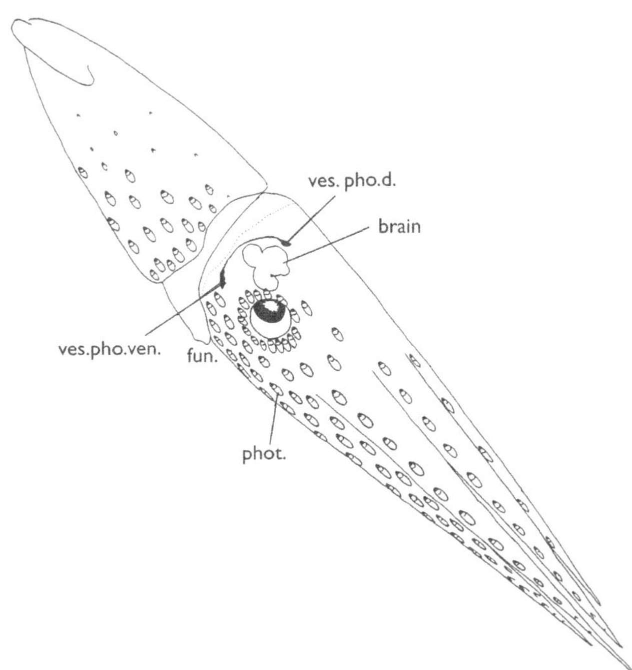

Fig. 13.5. *Histioteuthis dofleini* seen from the side, in its assumed orientation in the sea. The small right eye looks posterodorsally, and has 17 photophores arranged around its lid (R.E. Young 1977).

mesopelagic zone (N.A. Voss 1969, R.E. Young 1972b). The animal appears silvery when the chromatophores contract and the underlying iridophores are exposed. The iridescent layer is most notable on the ventral and lateral surfaces of the head, the anterior surface of the mantle and the aboral surfaces of the arms (R.E. Young 1975a).

13.1.3 Photophores

The arrangement of the photophores of *Histioteuthis dofleini* is shown in Fig. 13.5. Large light organs are concentrated on the anteroventral surface of the arms and are directed anteriorly (R.E. Young 1975a). The photophores on the ventral side of the head are large and have a close and even distribution, except ventral and posterior to the large left eye where they are lacking. Seven large light organs lie close to the anteroventral edge of the left eyelid. The smaller right eye has 17 close-set photophores along the entire circular edge of the eye lid; the light from these photophores passes anteriorly and somewhat laterally and does not enter the eye. The distribution of the photophores and their forward facing position suggests that their light would provide adequate ventral counter-shading when this squid is oriented at 45° to the body axis (R.E. Young 1977).

Each epidermal photophore of *Histioteuthis macrohista* comprises a posterior vase-shaped cup, a central 'lens' or filter, and an anterior scoop-shaped reflector (Dilly and Herring 1981). The posterior cup is almost completely enclosed in a layer of dark brown pigment that extends into the surrounding epidermis in root-like strands. The ventral external surface of the posterior cup has a window in the pigment sheath anteriorly, and this can be occluded by the expansion of a single chromatophore which lies immediately above. The posterior cup is composed of a mass of reflective material, giving it a milky-white colour in reflected light. Within the reflector cup is photogenic tissue containing photocytes, some of which have densely aggregated parallel arrays of rough endoplasmic reticulum. This abuts on to the filter, and the latter is reddish-purple in colour. The scoop-shaped reflector is translucent and its inner surface is lined with a thin layer of reflective tissue. Most of the epidermal photophores have three chromatophores over the junction of the posterior cup and the scoop-shaped reflector and two or three over its aperture.

Histioteuthis macrohista has single, greatly enlarged, elongated terminal photophores on arms I-III, and these reach 14–18% of the length of the arms. These photophores appear as tapering cylinders of white tissue whose internal surface is enclosed by a pigmented sheath, black-brown in colour; movement of the pigment can control the aperture of the photophore. Chromatophores also occur over the surface of the organs. When exposed the organs have a brilliant yellow-green fluorescence in ultraviolet light, and an emission maximum of about 490 nm. A feeble luminescence of short duration resulted from treatment of the organ with 1% hydrogen peroxide in seawater, but had no effect on its fluorescence (Dilly and Herring 1981).

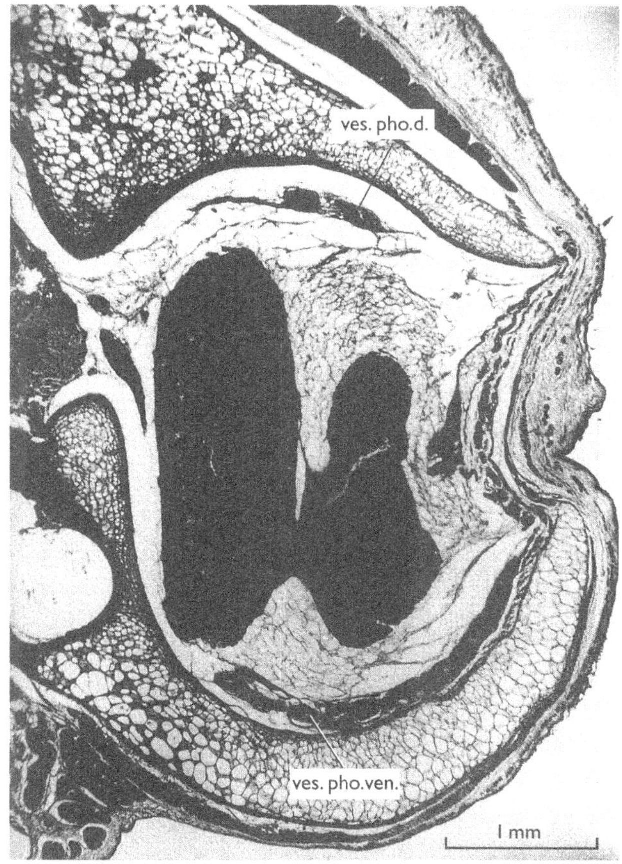

Fig. 13.6. *Histioteuthis reversa* (6 mm ML). Transverse section to show the position of the photosensitive vesicles.

13.1.4 Receptors and central nervous system

The extraocular photoreceptor organs comprise a dorsal and a ventral set of photosensitive vesicles on each side, joined by a narrow strand of vesicles (R.E. Young 1978). The dorsal and ventral sets are flat and compact, and comprise a number of vesicles of various shapes and sizes. The organ is formed of a long chain of vesicles around the back of the optic lobes and central nervous system in *H. reversa*, that swells out into larger masses dorsally and ventrally (Fig. 13.6). The vesicles form 0.2% of the total brain volume (Maddock and Young 1987). The nerves from the photosensitive vesicles enter the peduncle lobes as dorsal and ventral bundles, and spread out within the neuropil of the basal part of each lobe below the spine region (Fig. 13.7).

Some areas on the head of *H. dofleini* are devoid of chromatophores, and so form windows. One is for the ventral set of photosensitive vesicles. The nuchal crest lacks chromatophores dorsally and only scattered chromatophores and iridophores posteriorly, and so may act as a window for the dorsal vesicles. Thus the dorsal photosensitive vesicles can be exposed to downwelling light and the ventral set to light from the photophores on the anterior margin of the mantle (R.E. Young 1977).

Histioteuthis is notable for the difference in shape and size of its eyes (Fig. 13.8). Both eyes are the same size at 6 mm ML, but the asymmetry is apparent when the animals reach 14 mm ML (Wentworth and Muntz 1989). The right eye is of typical shape and has a spherical lens. The left eye is much larger and resembles a truncated cone with a curved base, and its spherical lens is about twice the diameter of that of the right eye (R.E. Young 1975a, Wentworth and Muntz 1989). The large left eye usually faces posterodorsally but is capable of some limited movement. Confirmation for this is seen in the presence of anterior eye muscles; these originate close together in the midline and the muscles undoubtedly cross. No indication of a trochlear cartilage was seen in our specimen. The right eye is directed ventrolaterally and lies slightly further forward on the head than the larger left eye. The visual axes of the two eyes lie obliquely in relation to the head. One or both lenses appear yellow in some species, yet lack coloration in other species. The spectral absorbance of the lenses from the large and small eyes show considerable variation, but the lens from the large eye always absorbs relatively more light of short wavelength than the lens from the small eye (Denton and Warren 1968, Muntz 1976, Wentworth and Muntz 1989).

The retina has the same structure as that in other coleoids, but differences are apparent in the retinal thickness, as the rhabdoms vary in length over the retinal surface (Fig. 13.8) (R.E. Young 1975a, Wentworth and Muntz 1989). Rhabdomeres have the rectilinear arrangement seen in other coleoids, but their packing varies.

When this squid is in an oblique position the large left eye looks upward so that it detects downwelling surface light under conditions of low light intensity. These conditions would be encountered by *Histioteuthis dofleini* in Hawaiian waters where it inhabits the twilight zone from about 400–500 m at night. The small downward-looking eye probably detects bioluminescence, and its compactly organized photophores may act as a searchlight (R.E. Young 1975a). The histology of the retina and optic lobes, and the characteristics of the lenses, indicate that the two eyes are suited to different tasks (Wentworth and Muntz 1989).

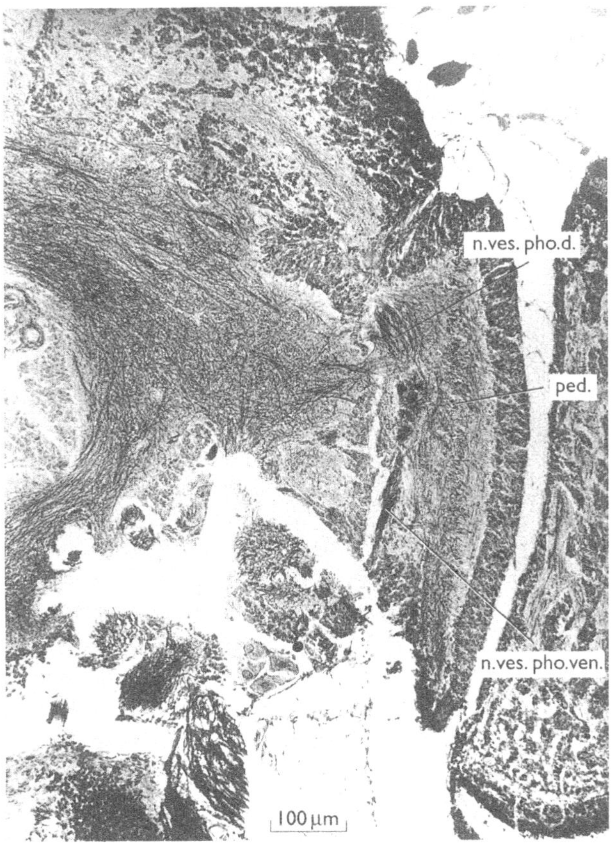

Fig. 13.7. *Histioteuthis reversa* (161 mm ML). Transverse section to show the peduncle lobe and nerves of the dorsal and ventral photosensitive vesicles.

The optic lobes, like the eyes differ in size and volume. The left optic lobe has a volume of 68.7 mm^3 and the right 29.5 mm^3 in *Histioteuthis reversa* (60 mm ML). The difference between the volumes of the optic lobes in a small specimen was even more pronounced (Maddock and Young 1987). The plexiform and granular layers are thicker in the larger left optic lobe. The volumes and histology of the two optic lobes of *H. atlanta* showed differences like those found in *H. reversa* (Wentworth and Muntz 1989).

The cephalic cartilage is notable for its thickness, especially over the supraoesophageal mass, and for its reticulate structure. The supraoesophageal mass is compact whereas the suboesophageal mass is elongated (Fig. 13.9). Of the suboesophageal lobes the pedal lobe is relatively large and forms 22.7% of the total brain volume (Table 2.6). Throughout the central nervous system there is a tendency for the lobes on the left side to be larger than those on the right. The proportional difference between several of the lobes was 1.04 to 1.20, and was statistically significant in animals of up to 60 mm ML. The exception to this was the inferior frontal lobe system which was larger on the right side, even in small specimens, the ratio being 0.60–0.68 (Maddock and Young 1987).

No distinct first-order giant cells could be found and there is no outstandingly large fibre in the mantle connective. The stellate ganglion has a single layer of cells ventrally and some smaller ones dorsally (Fig. 13.10). The stellar nerves contain two or three large fibres each but no giant fibres (Fig. 13.11).

The statocyst is relatively long and broad but shallow, and has a small volume compared with that of other buoyant squids (Young 1989) (Fig 13.12). The anticristae are elongate and rounded, and they fill much of the cavity. Anticristae 6 and 7 are in contact but are not fused. There is a horizontal canal, although only 10% of it is

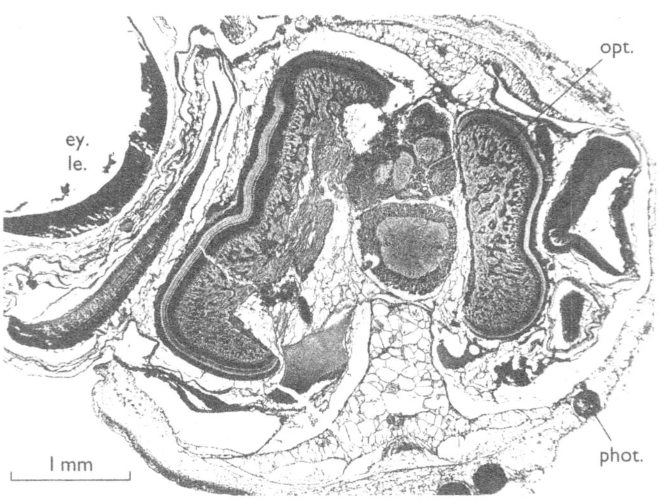

Fig. 13.8. *Histioteuthis reversa* (161 mm ML). Transverse section through the head, seen from behind, showing the optic lobes of the small right and the large left eye.

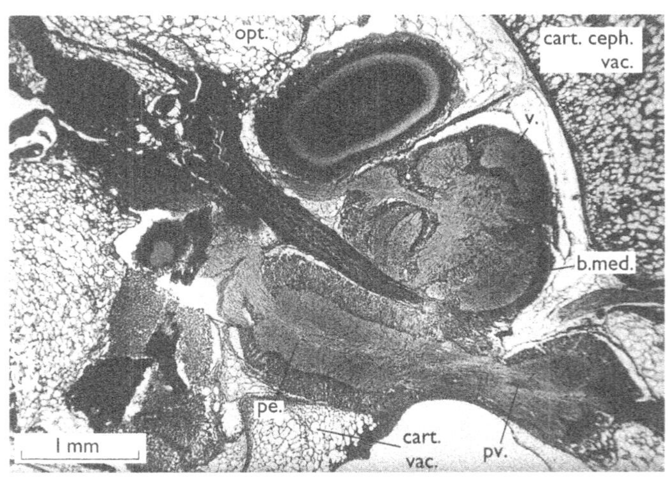

Fig. 13.9. *Histioteuthis reversa* (6 mm ML). Sagittal section through the head, near the midline, showing the central nervous system. Note the reticulation of the cartilage.

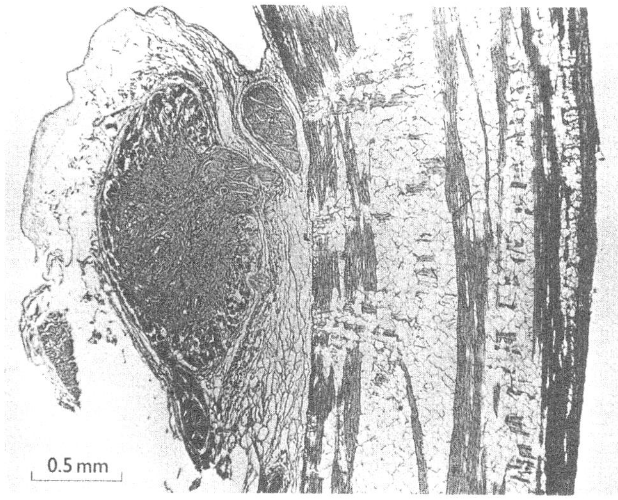

Fig. 13.10. *Histioteuthis reversa* (60 mm ML). Transverse section through the mantle to show the bands of muscle and reticulate tissue, and the stellate ganglion.

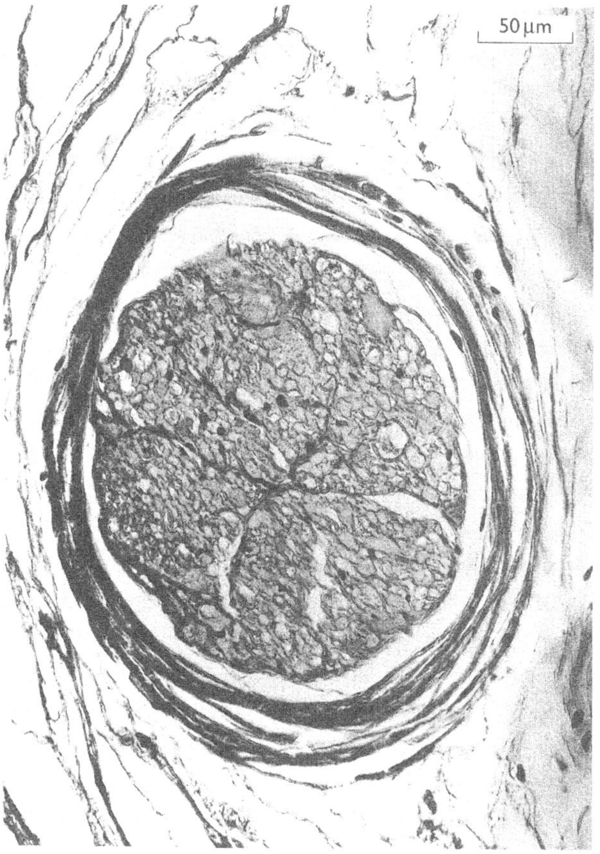

Fig. 13.11. *Histioteuthis reversa* (60 mm ML). Transverse section through a stellar nerve.

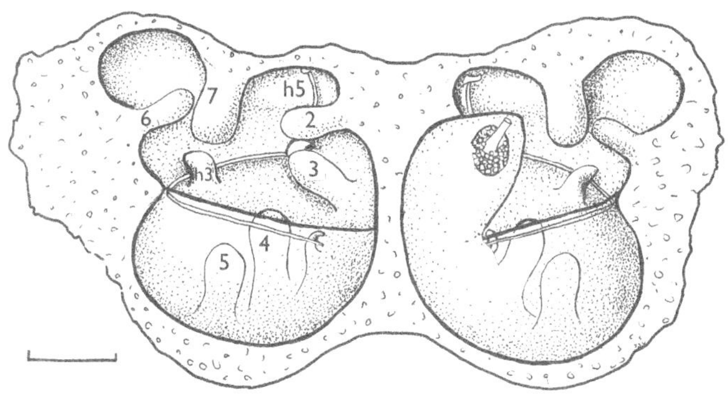

Fig. 13.12. *Histioteuthis reversa* (60 mm ML). The statocyst in anterior view (scale bar 1 mm) (Young 1989).

complete. There is no anticrista 1 and no hamulus at the first turn. Multiple discriminant analysis of the statocyst placed *Histioteuthis* amongst the buoyant squids (Maddock and Young 1984). No asymmetry was apparent in the statocysts in a small specimen of 6 mm ML. The volumes of the left and right statocysts did not differ in either *H. bonnellii* (24 mm ML) or *H. atlanta* (20 mm ML), but small differences were found in their dimensions (Wentworth and Muntz 1989); the shape of the statocysts did differ, that of the left being longer than the right.

14 Bathyteuthidae

This family of oceanic squids includes a single genus with three species. All are relatively small and dark red in colour. Their geographic distribution is broad in the bathypelagic zone of the major oceans, but they are absent from Arctic waters and the Mediterranean Sea.

14.1 *Bathyteuthis*

Bathyteuthis abyssicola is the most commonly net-caught cephalopod in Antarctic waters where the nutrient is rich and the temperature 1–5 °C (Fig. 14.1). It is captured from 100–4200 m but most frequently between 700–2000 m, seldom approaching within 1000 m of the bottom (Roper 1968, 1969, Roper and Young 1975). Small squids tend to be in shallower water than larger ones. Some diel migration may occur as specimens captured in opening/closing nets, in the North Atlantic Ocean, were found to be spread through the water column between 100 and 2000 m at night, but all were below 320 m in daylight (Clarke and Lu 1975, Lu and Clarke 1975a, 1975b).

14.1.1 Life history

Bathyteuthis abyssicola has been captured throughout the year, the specimens ranging from paralarvae to maturing, ripe or spent individuals, thus indicating little or no seasonality. The specimens captured in a single tow are usually of a widely different sizes, the only exception being those from shallow waters where paralarvae and juveniles predominate. Sexually mature females can be ripe at 40 mm ML, but more usually at 45–48 mm ML, whereas males reach maturity at 28–30 mm ML. The maximum size recorded is a female of 63 mm ML (Roper 1969).

In juveniles the photophores lie at the bases of arms I-III and are prominent, but are less apparent in specimens of 2.3 mm ML and more (Figs 14.1, 14.2). Each photophore is a small elongated sac containing

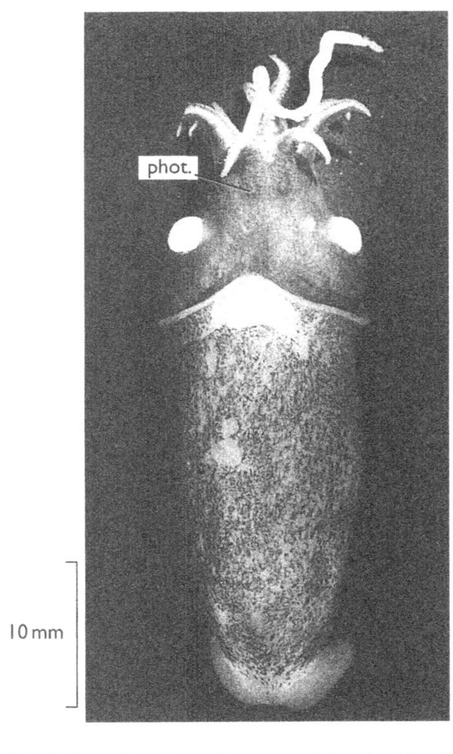

Fig. 14.1. *Bathyteuthis abyssicola* captured in the North Atlantic Ocean. Sections of this animal of 30 mm ML are shown in the following photomicrographs.

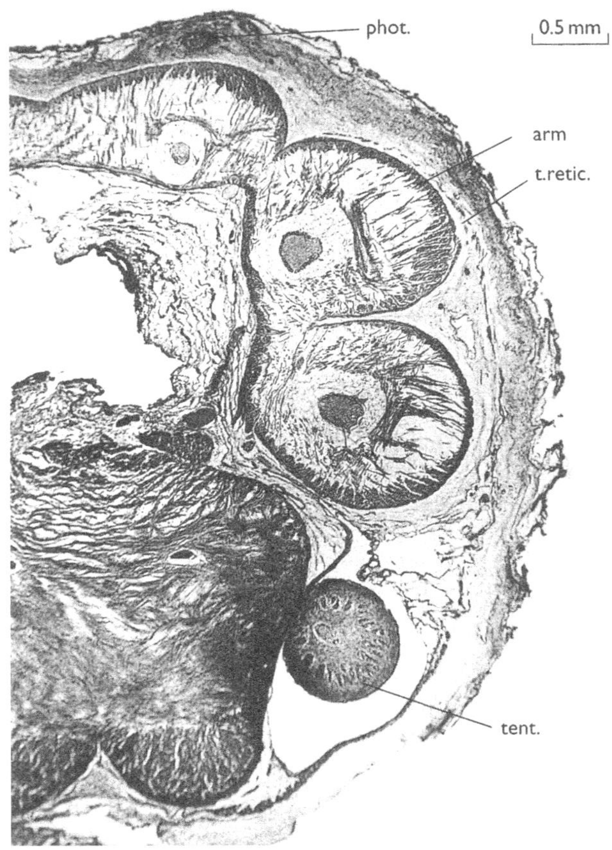

Fig. 14.2. *Bathyteuthis abyssicola.* Transverse section through the bases of the arms and one tentacle; a small photophore can be seen.

large cells, and covered with pigment but without a reflector (Chun 1910a). An area of pigment surrounds each photophore, but diminishes passing away from the light organ.

The colour of *B. abyssicola* is due to a maroon-brown pigment. There is a thin sheet of chromatophores immediately below the outer layer of reticulate tissue. Two further layers of maroon chromatophores lie beneath watery, semigelatinous tissue. Some isolated chromatophores are present in the connective tissue close to the thick wall of mantle muscle. Pigmentation is more concentrated on the head and arms than on the mantle (Roper 1969).

14.1.2 Locomotion

Bathyteuthis abyssicola has been observed to swim slowly and to hang nearly motionless in the water (Roper 1969, Roper in Denton and Gilpin-Brown 1973). The mantle possesses a deep layer of muscle, with densely organized fibres and innervated by giant fibres (see Fig. 14.5). There is also a thick layer of reticulate tissue. Two large anterior chambers in the digestive gland are filled with reddish-brown oil (Roper in Denton and Gilpin-Brown 1973). However, features of the statocyst apparently indicate that this squid is non-buoyant. Thus some features of *B. abyssicola* suggest it may be neutrally buoyant but others do not; but the various observations may have been of specimens at different stages of growth.

14.1.3 Receptors and central nervous system

The eyes, on either side of the head, are directed forwards and upwards at an angle of 45–50° from the longitudinal axis of the body (R.E. Young 1972a) (Fig. 14.1). The optical axes of the eyes are approximately parallel, indicating a wide binocular field. There is evidence for eye movement as two crossing anterior conjunctive muscles are present. The junction is by a tendon in one pair and by the continuation of the muscles across the midline in the other, as in *Loligo* (Budelmann and Young 1993). The retina is thickened ventroposteriorly with the fovea present at the centre of this region (Chun 1910a) (Fig. 14.3); in this feature *Bathyteuthis* is apparently unique, or at least amongst the cephalopods examined so far. The rhabdoms are 975 μm (46 mm ML) (R.E. Young 1972a) and these are closely packed (250 000/mm^2) (Chun 1910a).

The optic lobes form 126.9% of the total brain volume (Maddock and Young 1987) and are relatively small when compared with the eye. The plexiform zone and the granular cell layers of the optic lobe, especially the inner, are much thickened in the anterodorsal region. Here the columns of cells in the medulla of the lobe are also prominent.

The central nervous system is compact (Fig. 14.4). The brachial lobes are not widely separated from the rest of the suboesophageal

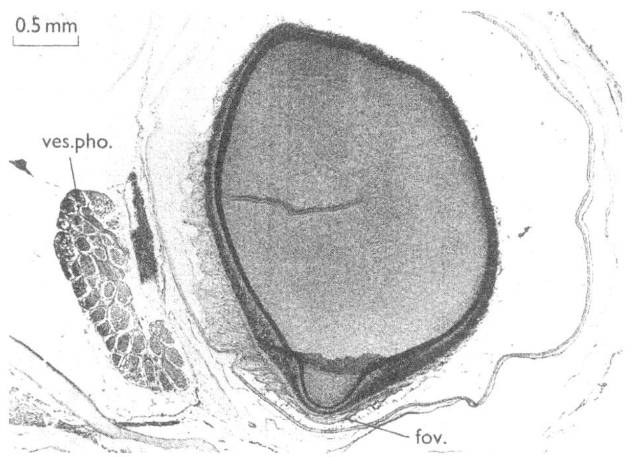

Fig. 14.3. *Bathyteuthis abyssicola*. Transverse section through the head to show the fovea of the eye and the photosensitive vesicles of the extraocular photoreceptor organ.

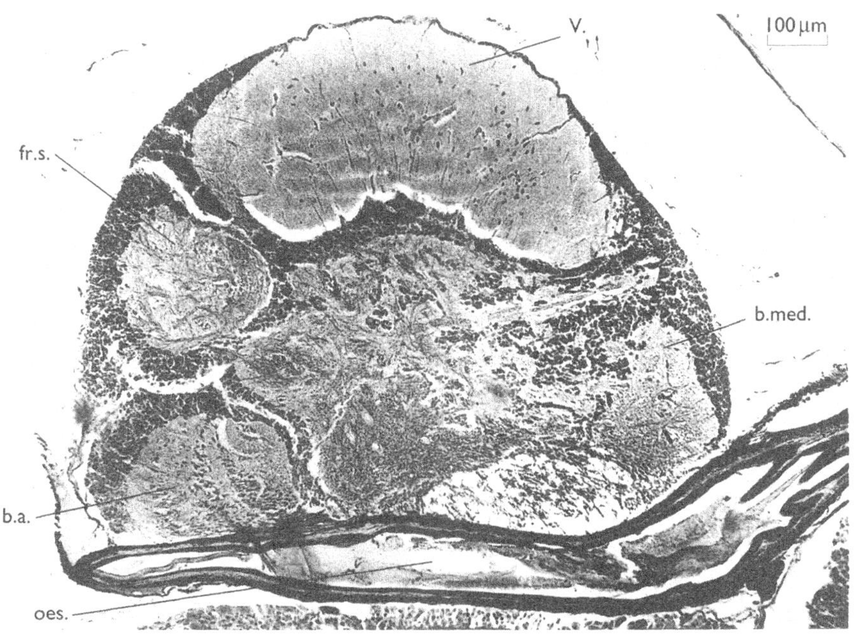

Fig. 14.4. *Bathyteuthis abyssicola*. Sagittal section of the supraoesophageal lobes of the brain.

mass, and form 8.1% of the total brain volume. This reflects the small size of the arms. The vertical lobe is a prominent part of the supraoesophageal mass, and the vertical lobe system forms 20.8% of the total brain volume. The fin lobes form 2.7% of the brain volume (Maddock and Young 1987).

The stellate ganglion (Fig. 14.5) has a large first-order giant cell, and the rest of the system is well developed, with large fibres in the stellar nerves.

The extraocular photoreceptor organ consists of a series of vesicles which were described as 'glomeruli' by Chun (1910a), who, not surprisingly, ventured no hypothesis about their nature. The vesicles lie within the confines of the orbit, located near the posteroventral surface of the eye and lateroventral to the brain. In the living animal the paired extraocular photoreceptor organs are brilliant orange in colour, and form 30.3% of the total brain volume (Table 2.6). Each organ is irregular in shape and consists of loosely packed, long, photosensitive processes extending from cells lining the vesicle. The processes vary in diameter (9–25 μm) and probably in length, but their twisted courses make measurement difficult. The lining of each vesicle includes epithelial supporting cells with irregularly shaped nuclei and scant cytoplasm, and neuron-like cells, with cytoplasm, from which the sensory processes arise. Ventral to each eye there is a lightly pigmented region of skin, or window, which lies close to the photosensitive vesicles, thus permitting any available light to enter. From each organ large nerves enter the peduncle lobe, which is quite close to the optic gland.

The statocyst has broad hamuli and anticristae. Anticrista 7 does not meet anticrista 6, and although the horizontal canal is well marked, only 10% of the circumference of the canal is complete. The anticristae of the median row divide the deep anterior part of the cavity into two parts (Young 1989). Multiple discriminant analysis of several measurements of the statocyst place this squid amongst non-buoyant forms but some distance from the centroid (Maddock and Young 1984).

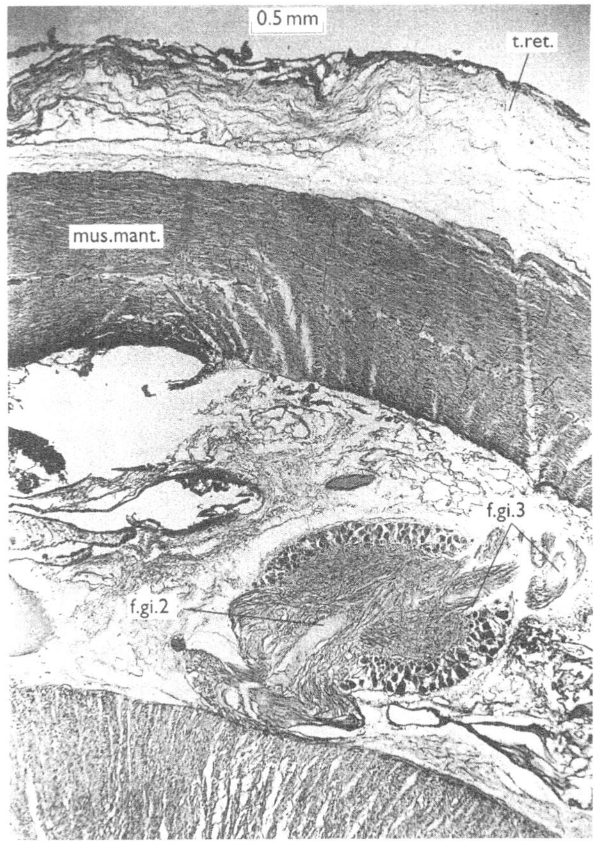

Fig. 14.5. *Bathyteuthis abyssicola.* Transverse section through the mantle to show the stellate ganglion with giant fibres, the thick layer of mantle muscle, and reticulate tissue.

15 Architeuthidae

15.1 *Architeuthis*

Architeuthis is the only genus of the family (Fig. 15.1) and it occurs widely around the world between 70°N and 65°S (Roper and Boss 1982, Nesis 1987, Roeleveld and Lipinski 1991). This squid has been seen at the surface of the sea, although usually moribund or only just alive, caught in trawls, washed ashore (Aldrich 1991, Roeleveld and Lipinski 1991), and found in the stomachs of predators such as sperm whales (Clarke 1980) and fishes (Toll and Hess 1981). Although little is known of the depths in which these animals live they have been caught recently in pelagic, semipelagic, midwater, and bottom trawls at depths of 510–550 m (Roelveld and Lipinski 1991). One tentacle was caught in a trawl sampling the water column at 500–600 m (Robison 1989). Commercial trawls off Tasmania captured two females between 500m and 1000 m (Norman and Lu 1997b). Three were caught in bottom trawls to the west of Ireland between 110 and 300 m (Lordan *et al.* 1998).

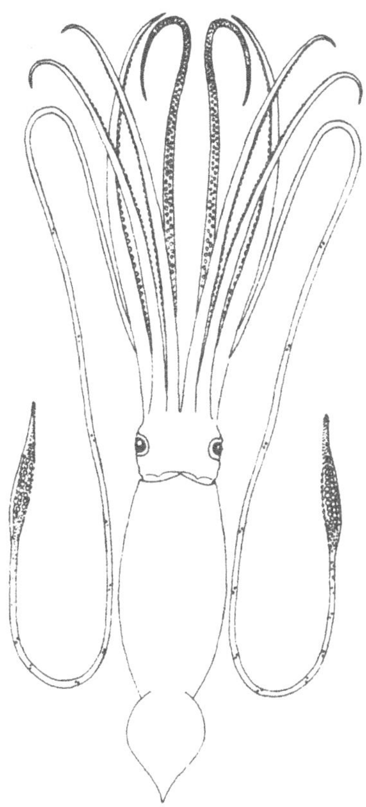

Fig. 15.1. *Architeuthis*. Dorsal view (Verrill 1880).

Whenever large animals are stranded they usually provoke comment, and this is certainly so with *Architeuthis*. Steenstrup (1857) gave records dating back to the sixteenth century, as well as of animals washed ashore in the nineteenth century. Verrill (1880) described specimens stranded along the shores of north-east America, and added notes on ones from other localities. Those from Newfoundland were described by Aldrich (1991). Roeleveld and Lipinski (1991) summarized the morphometric details of numerous animals to compare with three animals caught in southern African waters. The study by these authors has added considerably to our knowledge of the morphology of numerous internal and external structures of the giant squid.

15.1.1 Life history

The reproductive habits of this squid are little known but the males appear to mature at a smaller size than the females, the former being mature between 612 mm ML and 1100 mm ML, and the latter at 1450 mmm ML or more (Roeleveld and Lipinski 1991). One mature male of 1010 mm ML had spermatophores of 110–180 mm in length, these were covered with a rather opaque jelly (Knudsen 1957). A mature female of 1680 mm ML had round eggs, 1.2–1.8 mm in diameter, in the oviducal glands (Roeleveld and Lipinski 1991). An estimated fecundity of 10^7 eggs was given by Boyle (1986a) and is greater than that given by Roper and Boss (1982). No indication of copulation was found in the three females from southern African waters (Roeleveld and Lipinski 1991). A female with a mantle length of 2.4 m was found to have spermatophores embedded in a small shallow wound (about 10 mm diameter) about 830 mm from the mouth. Dissection revealed the remains of three spermatophores radiating about 80 mm from the wound. This mated female was submature and the ovary occupied less than 20% of the mantle cavity and weighed about 3 kg (total weight of animal was 220 kg) (Norman and Lu 1997b). Small animals of 45 mm and 57 mm ML were recovered from the stomachs of fishes (Roper and Young 1972). A small specimen, about 10 mm ML, has recently been captured 6 m below the surface and 240 m east of New Zealand (S. O'Shea pers. comm.).

15.1.2 Morphology

Ammonium has been found in some of the tissues, and concentrations of 338 mM recorded in the mantle, 228 mM in the arms, and 314 mM in the tentacles (Clarke *et al.* 1979, Roper and Boss 1982, Robison 1989). Boyle (1986a) reported that a thick oily brown fluid exuded from the digestive gland. The cartilage of the head is reticulate (Fig. 15.2). Such conditions suggest that this huge squid is near to neutral buoyancy (Clarke *et al.* 1979), and it may be a poor swimmer (Robson 1933). The fins are relatively small, their length being about 30% of that of the mantle length, and at the edges the muscles of the fin are divided into segments. The funnel is muscular but the locking apparatus appears weak.

The long arms are muscular (Fig. 15.3). The suckers are quite well spaced and relatively small (Fig. 15.4), the largest being 20–25 mm in diameter (Verrill 1880, Roeleveld and Lipinski 1991). The tentacles are long and have a short club, and the suckers are not especially notable although they are larger than those of arms.

The rostral region of the beak is darkly tanned and is in marked contrast with the surrounding pale lips. The radula has seven well-spaced teeth in each transverse row, and a somewhat ovoid marginal plate on either side (Nixon 1998). In specimens of 1820 mm and 2080 mm ML the radula measured 74 mm and 100 mm in length respectively (Dell 1970, Roper and Boss 1982). The lateral buccal palps carry numerous irregularly arranged teeth (Verrill 1880, A.C. Stephen 1962).

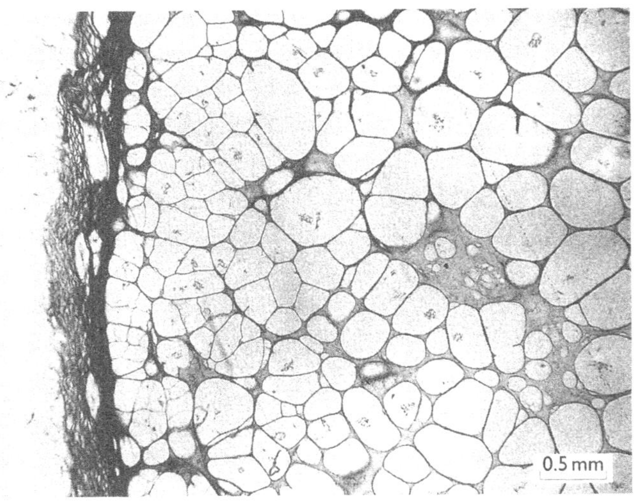

Fig. 15.2. *Architeuthis* (1620 mm ML). Histological section through the cephalic cartilage to show the extensive reticulation.

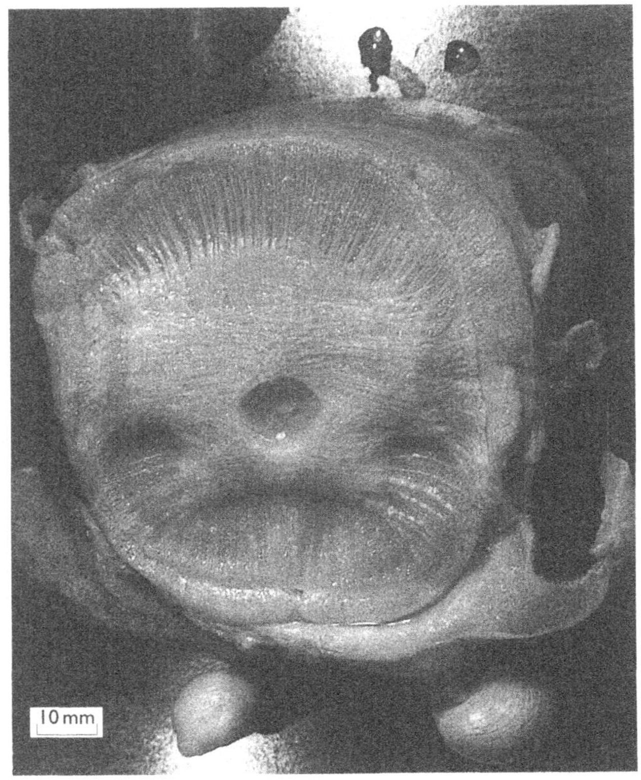

Fig. 15.3. *Architeuthis* (1570 mm ML). An arm cut transversely to show its organization, and the relatively small suckers.

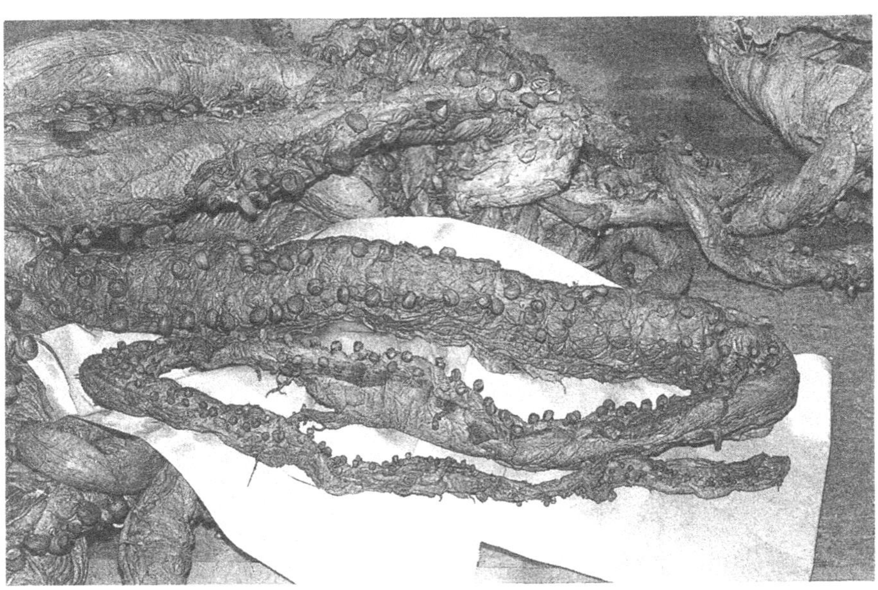

Fig. 15.4. *Architeuthis* (1620 mm ML). One arm showing the biserial arrangement and size of the suckers.

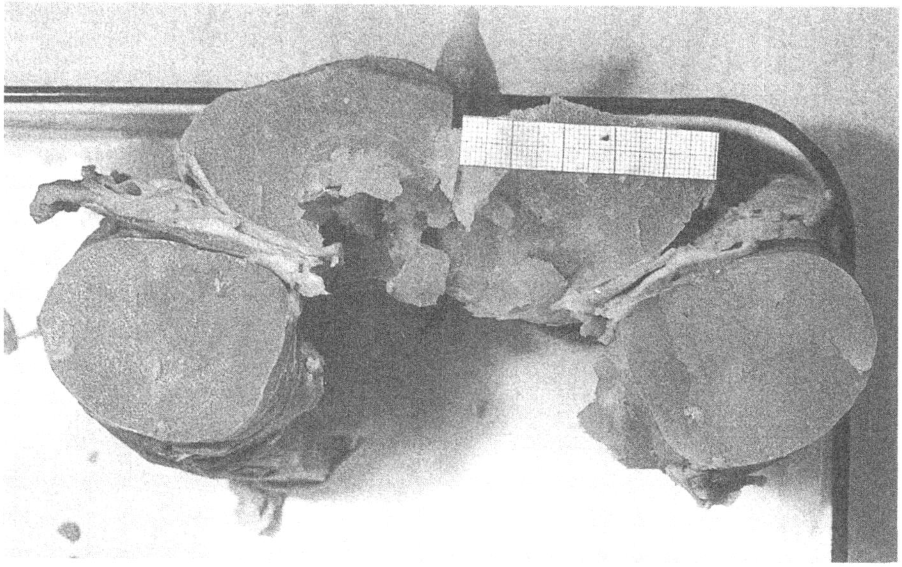

Fig. 15.5. *Architeuthis* (1620 mm ML). The cartilaginous cranium cut in the midline and the two halves turned aside (scale bar 50 mm).

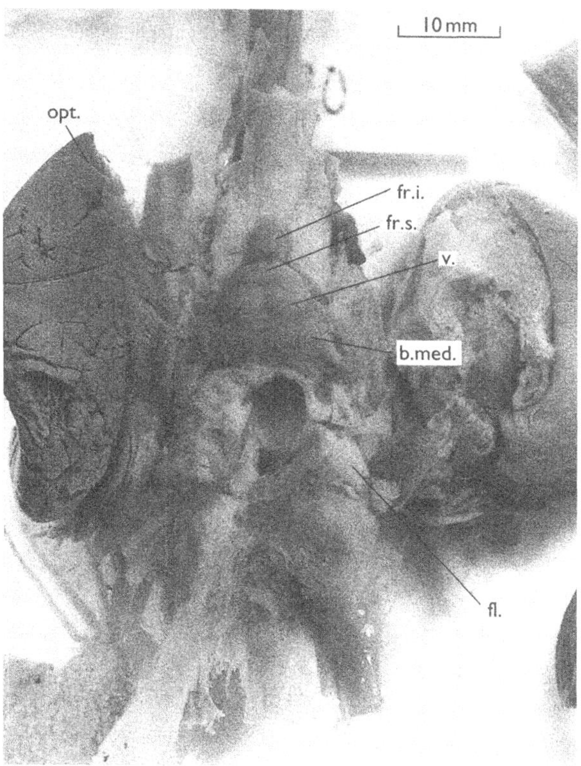

Fig. 15.6. *Architeuthis* (950 mm ML). The central nervous system and optic lobe; the oesophagus has been removed.

The oesophagus is slightly folded internally with a lining of chitin, with a thick layer of loosely organized tissue and a layer of muscle. The prey of this squid is poorly known but the remains of fish (Peréz-Gándaras and Guerra 1978, Toll and Hess 1981) and cephalopods (Roper and Young 1972) have been found in their stomachs. Three squids taken in demersal trawls had fragments from fish, crustaceans, and cephalopods (*Todarodes sagittataus, Eledone cirhosa*) (Lordan *et al.* 1998).

Table 15.1. *Architeuthis* sp. Records of the eye diameter and of the mantle length of various specimens

Mantle length (mm)	Eye diameter (mm)	Eye socket, maximum diameter (mm)	Source
1084	95		Lordan *et al.* (1998)
1170		100	Rae (1950)
1310	84		Grieg (1934)
1422		101	A.C. Stephen (1962)
1640	200		C.C. Lu (pers. comm.)

15.1.3 Eyes and optic lobes

Great difficulty was encountered when trying to find the central nervous system within the cartilage of the huge mass of the cranium, the cartilage of which was up to 50 mm thick in some places (Fig. 15.5). A smooth, ovoid opening (30–40 mm long) was present in the dorsal wall of the cranium. The brain tissue was brownish in colour after long and probably poor conditions of preservation (Fig. 15.6). Compared with the size of the animal, and the large cranium, the brain was found to be relatively very small. One specimen with a mantle length of 960 mm had a brain weighing 9.039 g, with two optic lobes of 9.25 g. Serial sections of the brains of two animals were cut, but orientation was uncertain, and so in consequence was the interpretation. The histological condition of some regions was surprisingly good; some nerve fibres stained well with Bodian's technique.

The eyes are the largest known in the animal kingdom, and yet their size is rarely recorded (Table 15.1). The optic lobes (Fig. 15.6), with the white body, lie embedded in a large mass of tissue behind the eye. The lobes have thick inner and outer granule cell layers. The outer plexiform zone has tangential layers. There is a system of columns in the medulla (Fig. 15.7).

15.1.4 Central nervous system

The various lobes of the brain have proportions similar to those found in other squids. The brachial lobes form 29.4% of the total brain volume; they are separate from the rest of the brain (Fig. 15.8) and are rather larger than in other cephalopods examined (Table 2.6) (Maddock and Young 1987). The pedal lobe (Fig. 15.9) forms 14.1% but this may be an underestimate because of damage. The magnocelluar and palliovisceral lobes form 18.4% of the total brain volume.

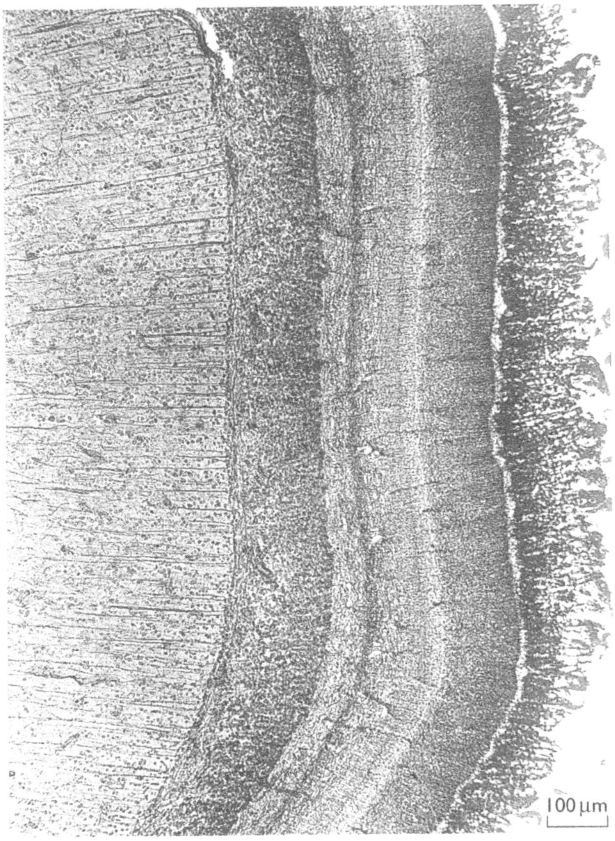

Fig. 15.7. *Architeuthis* (1610 mm ML). Transverse section of the optic lobe (Bodian).

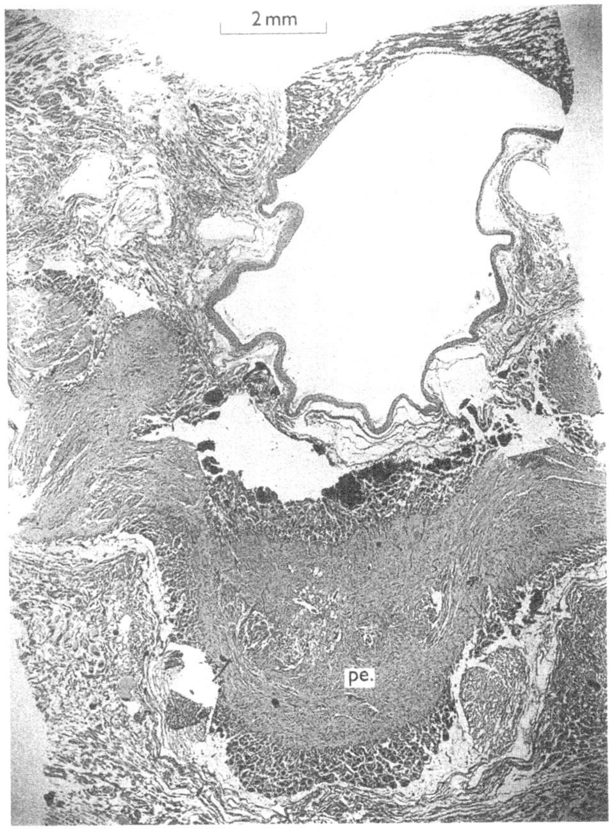

Fig. 15.9. *Architeuthis* (1570 mm ML). Oblique transverse section of pedal lobes. The space above is due to the removal of the oesophagus (haematoxylin and eosin).

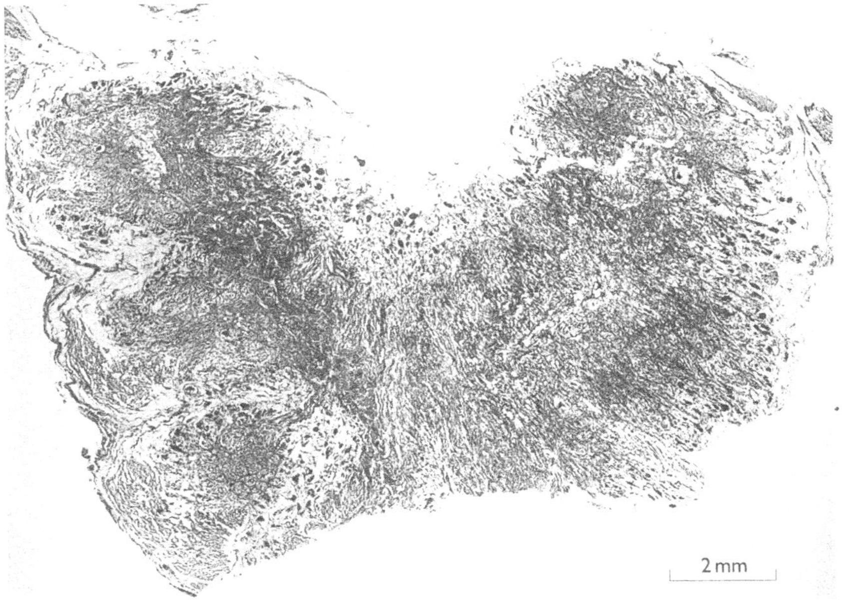

Fig. 15.8. *Architeuthis* (950 mm ML). Transverse section of the brachial lobe which has been cut slightly obliquely (Bodian).

The ventral magnocellular lobe is little developed and the ventral magnocellular commissure is a thin bundle of fibres (Fig.15.10). There are some large cells close to the midline and the largest may be a first-order giant axon (Fig. 15.11), but its cytoplasm is broken up by canals and no full section of its nucleus was seen. No especially large axon could be followed from this cell and no giant fibres were seen in the palliovisceral lobe (Fig. 15.12) or pallial nerve.

The supraoesophageal lobes are somewhat elongated (Fig. 15.13) and form 35.5% of the total brain volume (Table 2.6). The inferior frontal system forms 1.8%. The vertical lobe system is 14.1% of the total brain volume, and the vertical lobe alone is 2.93% and so is smaller than the superior frontal lobe of 3.75%. The anterior basal lobe is shown in Fig. 15.14, and the peduncle lobe in Fig. 15.15. The latter receives a large nerve from the photosensitive vesicles. These were found by taking pieces of tissue from the back of the orbit as one piece consisted of a series of undoubted vesicles with rhabdoms (Fig. 15.16). The vesicles found formed about 1% of the total brain volume, but may not represent the full extent of the tissue.

Large numbers of amoebocytes were found in the arteries and veins, and among the tissues, especially of the supraoesophageal lobes. The large motoneurons in these areas are penetrated by interweaving channels containing many nuclei. These are presumably mostly glia but include some amoebocytes. Some of these neurons are the giant cells of the magnocellular lobe (Fig. 15.11). The cells of the optic and supraoesophageal lobes have fewer of these canals and there are fewer amoebocytes in their vessels.

15.1.5 Giant fibres

Giant fibres are present in the stellar nerves (Fig. 15.17), but the distribution of the nerves to the mantle muscle is most unusual (Fig. 15.18). When the ventral mantle was cut along the midline the nerves were found to be in good condition, and only damaged where the fin nerve and mantle connective had been torn away from the head retractor muscles. The stellate ganglion is an elongated kidney-shape and is small (22 mm long and 12 mm wide) in relation to the animal. It lies 150 mm from the anterior end of the mantle and 65 mm from the midline. From the ganglion arise eight main trunks for the mantle muscles, but many of these are compound and separate out into smaller nerves. A remarkable point is that not all of the mantle is innervated by nerves arising from the stellate ganglion. The hinder stellar nerves are relatively small and short and do not run back the whole length of the mantle but break up in the muscles closer to the ganglion (Fig. 15.18). The most posterior part of the mantle

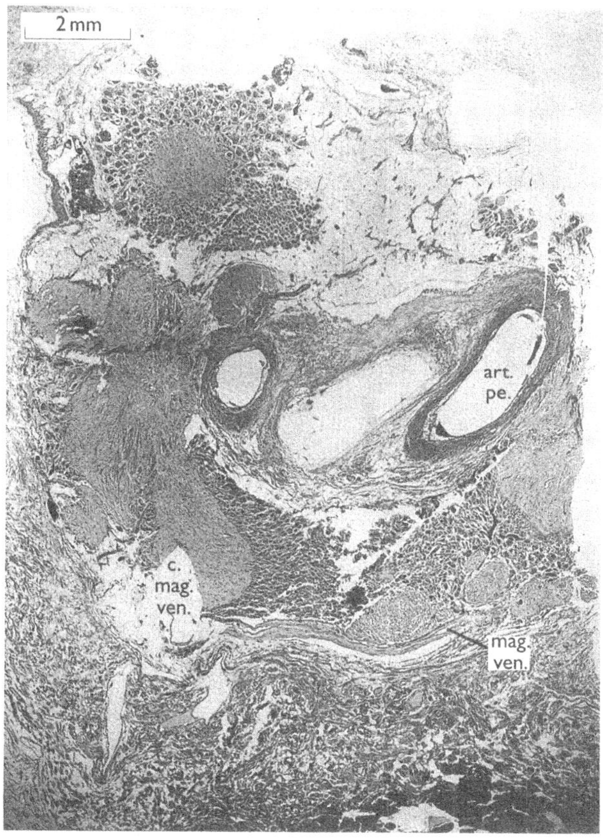

Fig. 15.10. *Architeuthis* (1570 mm ML). Oblique transverse section of the back of the pedal lobe and ventral magnocellular lobe. Note the pedal arteries with thick walls (haematoxylin and eosin).

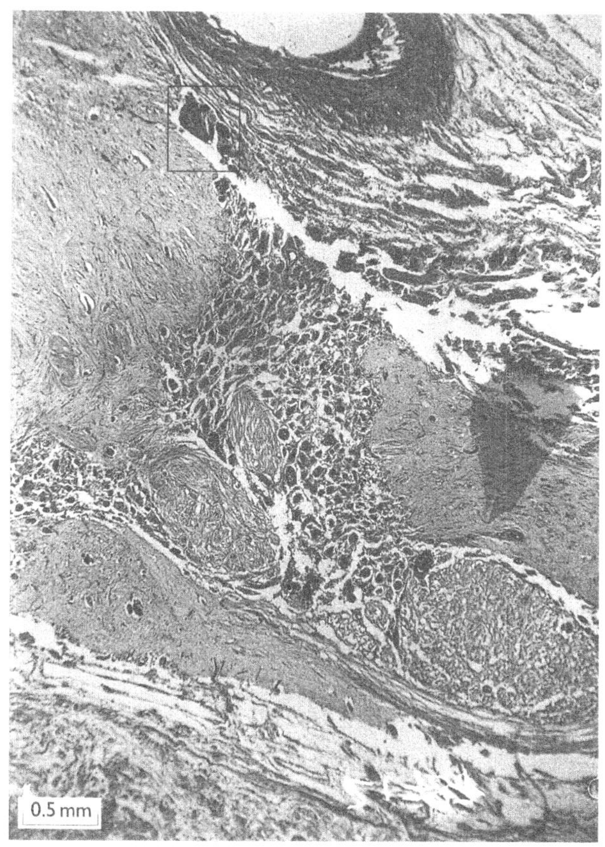

Fig. 15.11. *Architeuthis* (1570 mm ML). Transverse section of the magnocellular lobe to show region of giant cell (box) (haematoxylin and eosin).

musculature is innervated by branches of a 'median' nerve which runs with the fin nerve. Unfortunately this nerve had been torn anteriorly and its exact relation to the ganglion could therefore not be determined. It is possible that the part innervating the mantle runs from the ganglion by a path that had been destroyed, but this seems unlikely since the surface of the ganglion appeared intact, and no cut or torn branches were seen. We must therefore consider the possibility that in these squids the hinder part of the mantle is innervated directly by fibres from the central nervous system.

The giant fibres which run in these nerves are not especially large. They were examined by teasing the nerves and from cross-sections. Table 15.2 gives the largest diameters of the fibres found and the maximum distances traversed by their nerves. The median nerve was calculated to be about 620 mm long from the central nervous system

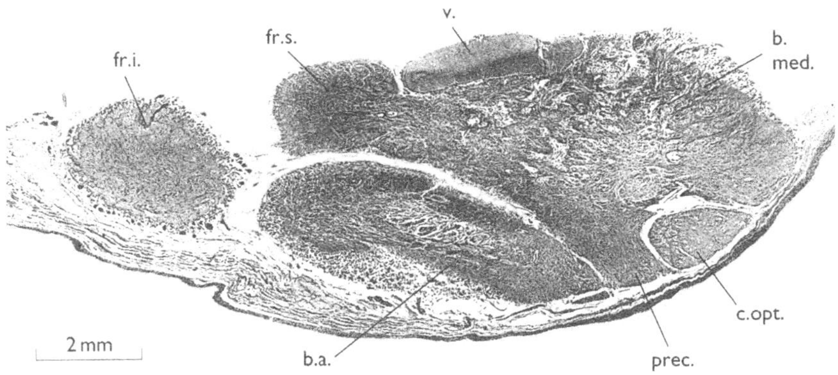

Fig. 15.13. *Architeuthis* (950 mm ML). Sagittal section of the supraoesophageal lobes (Bodian).

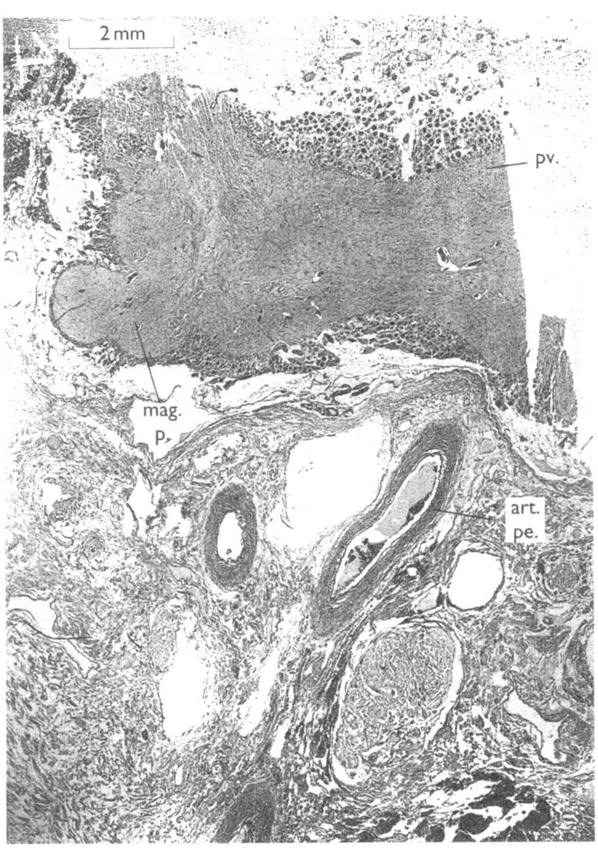

Fig. 15.12. *Architeuthis* (1570 mm ML). Oblique transverse section of the palliovisceral lobe (haematoxylin and eosin).

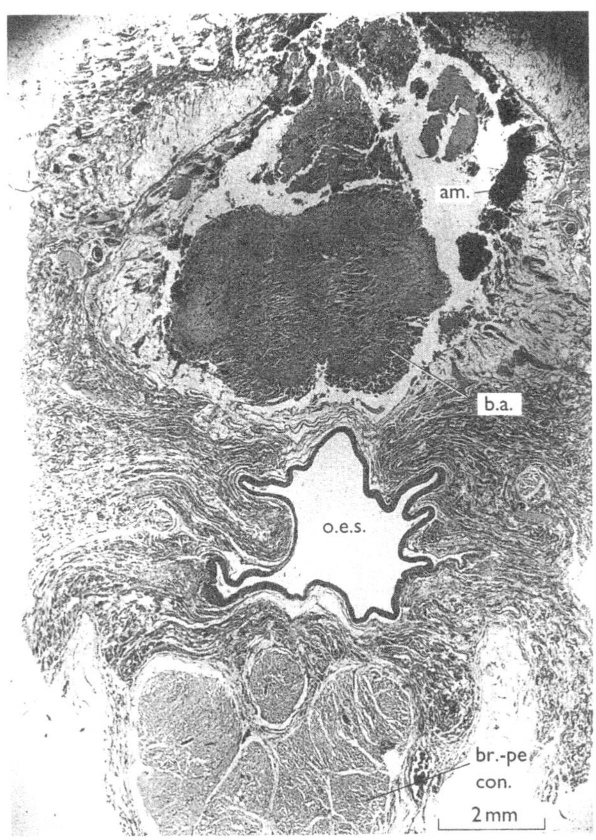

Fig. 15.14. *Architeuthis* (1570 mm ML). Transverse section of the anterior part of the anterior basal lobes. Note space left after removal of the oesophagus (Bodian).

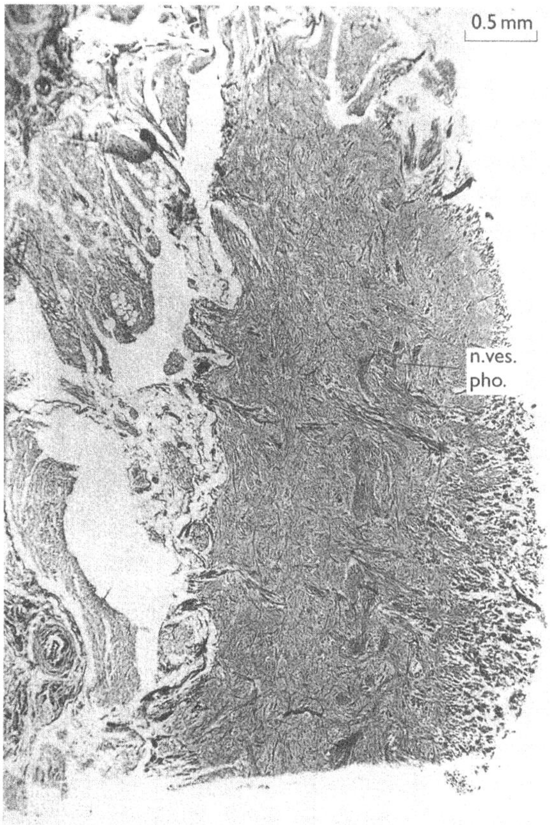

Fig. 15.15. *Architeuthis* (950 mm ML). Sagittal section of peduncle lobe, showing nerves of the photosensitive vesicles (Bodian).

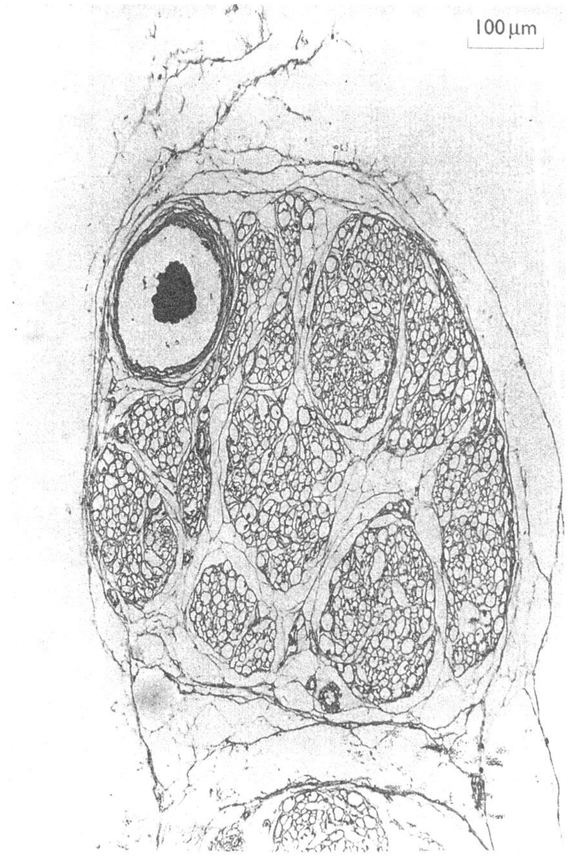

Fig. 15.17. *Architeuthis* (1620 mm ML). Transverse section of stellar nerve showing giant fibre (haematoxylin and eosin).

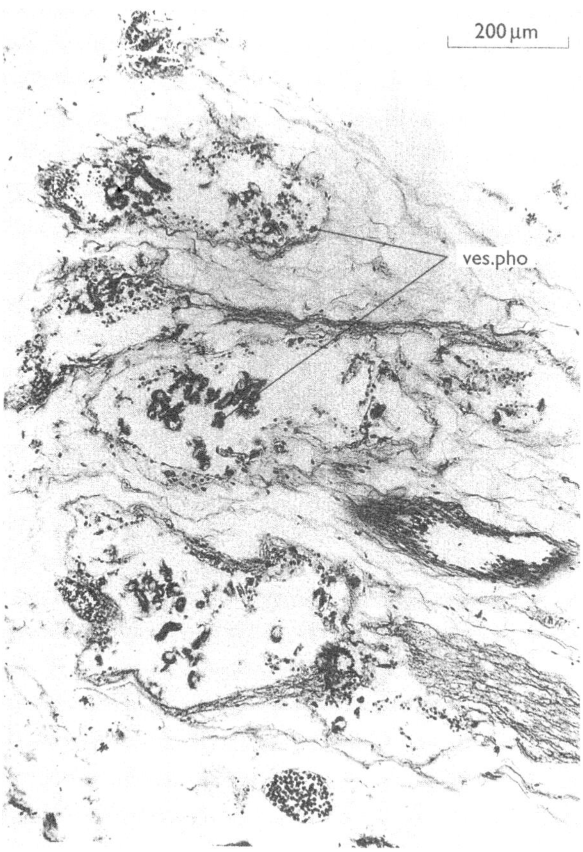

Fig. 15.16. *Architeuthis* (950 mm ML). Section through the back of the orbit to show groups of photosensitive vesicles (haematoxylin and eosin).

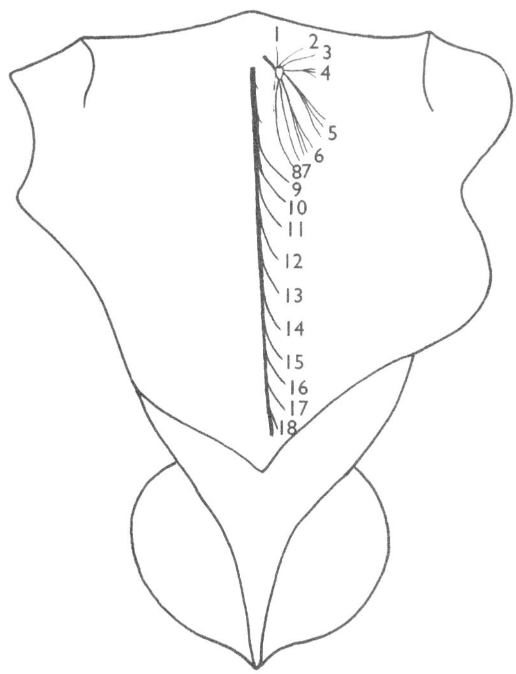

Fig. 15.18. *Architeuthis* (1620 mm ML). A dissection and drawing by J.Z. Young to show the innervation of the mantle.

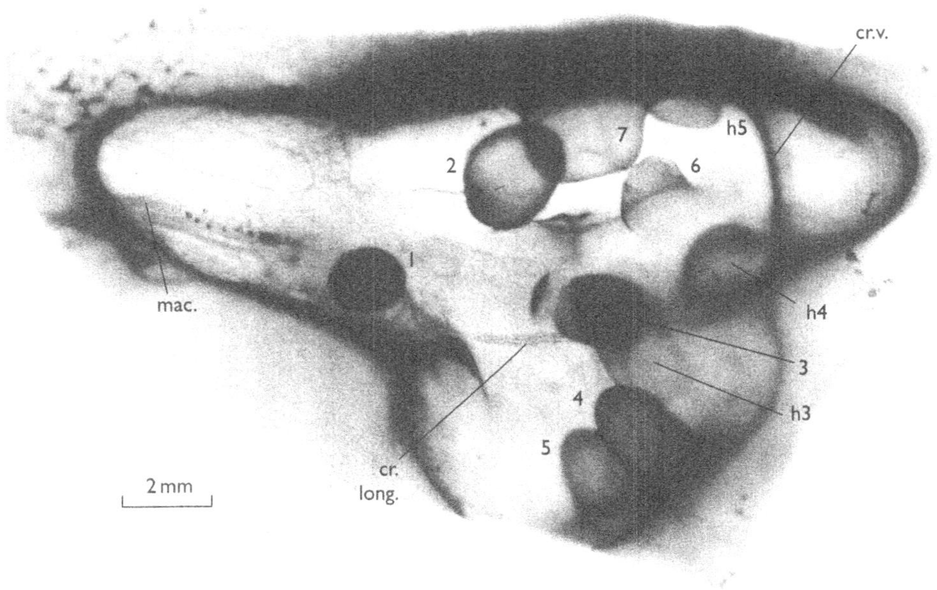

Fig. 15.19. *Architeuthis* (1570 mm ML). Lateral view of statocyst (Young 1989).

Table 15.2 *Architeuthis* sp. The lengths of the nerves shown in Fig. 15.18, measured from the ganglion to the mantle edge, and the diameters of their giant fibres

Nerves	Length (mm)	Diameter (μm)
1	140	85
2	190	100
		150
3	300	90
		75
4	400	220
5	500	140
		100
		90
6	650	260
		260
		110
		110
7	300 (?)	200

to the point of entry of the fin nerve to the muscles. Sections were cut of branch 13 leaving this nerve to the muscles and this contained fibres of 150 μm and 125 μm diameter. We therefore have the situation that some 'giant' fibres may run directly from the central nervous system to the mantle muscles.

15.1.6 Statocysts

The statocyst was described by Verrill (1880) as 'irregularly three-lobed, with several rounded papillae projecting inwards from its sides' with 'two irregular shaped otoliths'. The statocysts lie close together anteriorly but diverge posteriorly (Young 1989). Each is elongated due to the 'macula chamber' in front, it is narrow and shallow (Fig. 15.19). The posterior horizontal channel is extremely long. Anticristae 6 and 7 fit closely together but are not fused. Multiple discriminant analysis of various parameters of the statocyst place *Architeuthis* at the centre of the group of buoyant squids (Maddock and Young 1984). The strange shape of the statocyst makes comparison with other squids difficult. The elongated shape resembles the statocyst of *Mesonychoteuthis*, suggesting that it is a product of growth. The relatively unobstructed cavity of the statocyst of *Architeuthis* is probably associated with rather slow turning.

The statocyst is small relative to the mantle length of the animal (Roeleveld and Lipinski 1991). The external and internal morphology is apparently unique in many respects; the number of distinct parts is large and their relation to each other is complex (Lipinski 1997). The statoliths have growth rings, like those of other cephalopods, and those present in a female of 422 mm ML indicated an age of 153 days; if the rings are laid down daily then the average daily growth rate of the mantle was 2.76 mm per day (Jackson *et al.* 1991). The growth rings (393) of a mature female (1610 mm ML) imply that the early growth of the squid is very rapid, reaching a mantle length of 1600 mm in little over a year, but more small specimens are needed to confirm this sugestion (Gauldie *et al.* 1994). These authors found aragonite in the statolith.

16 Ommastrephidae

Ommastrephids are found in all oceans mostly between 40 °N to 40 °S, but some as far as 80 °N. They are oceanic, nerito-oceanic or neritic in habit, and are generally found from the surface to depths of 600 m but can penetrate to 3000 m. Most species undergo diel vertical migrations as well as making long-distance horizontal migrations for feeding and spawning. Several of these strong swimming squids are known to break through the surface of the water and 'fly' for some distance.

The systematics and taxonomic structure of the family and its division into subfamilies follows Wormuth (1998). The eggs are small and the hatchling is distinctive as the paralarva, or 'rhynchoteuthion', has a proboscis formed by the tentacles, which are joined longitudinally, and bear a ring of suckers at their tip (Roper and Lu 1979, Harman and Young 1985, Förch 1986). Changes in the shape and form of the rhynchoteuthion of several species have been illustrated (Wormuth *et al.* 1992), as well as an ontogenetic series of *Illex argentinus* (see Fig. 16.2).

16.1 Todarodinae

These squids are found in the Pacific, Atlantic, and Southern Oceans, from 70 °N to more than 50 °S, and in the Mediterranean Sea.

16.1.1 *Todarodes*

This genus is found on the continental slope and in oceanic waters of the North Atlantic, the western Pacific, and the Southern Oceans; it occurs from the surface to at least 1000 m (Dunning and Wormuth 1998).

Life history and habits

The life cycle of *Todarodes pacificus* was reviewed by Okutani (1983a). Ovarian eggs reach 0.8 × 0.92 mm in size, and newly hatched young have mantle lengths of 0.74–1.02 mm. The rhynchoteuthion has no ocular or visceral photophores. It has a single chromatophore between the fins on the dorsal mantle, and 14 evenly distributed photophores on the ventral mantle. The proboscis is somewhat longer than the longest arm, and the suckers on the tip are all similar in size. At 4 mm ML the proboscis begins to divide, and it separates at 7–9 mm ML (Wormuth *et al.* 1992). To reach a mantle length of 50 mm takes about 14 weeks. Sexual maturity is attained at 200–250 mm ML in females and 170–190 mm in males. A maximum size of 500 mm ML has been recorded in a female, and rather less in a male (Okutani 1983a).

Immature females will accept spermatophores from mature males, the sperm bulbs being implanted in the soft tissues around the outer lips (Okutani 1983a). A sharp arrowhead-like structure at the anterior end of the sperm reservoir ensures retention by the seminal receptacle on the buccal membrane, aided by the adhesiveness of the cement body (Takahama *et al.* 1991).

Mating has been observed on fishing grounds and in a net enclosure (Hamabe 1961). After actively swimming at the surface, around sunrise, male and female will approach and interlock their arms; the male then moves below the female so that her body is above his head. The first and second pairs of arms of the male adhere to the ventral mantle surface and the head respectively of the female. The third pair of arms entangle those of the female and the fourth pair are directed towards her mouth; this posture lasts some 3–10 s.

The spawning period of *Todarodes filippovae* is protracted in Australian waters, lasting from December to August (Dunning and Brandt 1985). In South African waters it lasts from February to August (Clarke 1980). Males of 300 mm and females of 380 mm ML are mature. Females of 440 mm and 410 mm ML, when caught, had already copulated and the larger specimen had an estimated 250 000 mature ova. A dying female was thin, and the ovary, and the nidamental and digestive glands were reduced. Spent squids occur in the Sea of Japan and their life span is considered to be of about 1 year's duration (Okutani 1983a).

Growth is rapid and diet changes take place as *Todarodes* increases in size. Animals of less 150 mm ML feed on plankton, and then until 200 mm ML they feed predominantly on crustaceans; larger ones prey mainly on cephalopods (Dunning and Brandt 1985). They feed most actively in daytime when near the bottom, at 150–200 m, rather than at night when they move into shallower water. Tagged animals had travelled horizontal distances at speeds of 0.37 m/s (32 km/day) (Okutani 1983a).

Receptors and central nervous system

The visual attack of this squid was observed in an aquarium; the sequence of actions was similar to that found in *Sepia* and *Loligo*. Most attacks were made with the eyes directed towards the bait, at a speed of approximately 500 mm/s. The bait was usually seized by widely spread arms, and after being secured was moved towards the beaks. On completion of capture of the prey the eyes returned to their initial, laterally directed, position (Flores 1983a). Discrimination tests suggest that this squid lacks colour vision (Flores 1983b), as is the case in a number of other cephalopods with the exception of *Watasenia* (Messenger 1981, 1991, Marshall and Messenger 1996).

The anatomy of *Todarodes* and of the nervous system was described by Posselt in 1891, and Sasaki in 1925. The brain is closely similar to that of *Ommastrephes*.

The giant-fibre system of *Todarodes* is well developed (Martin and Miledi 1986). Third-order giant axons measure up to 200 μm in diameter in specimens of about 160 mm ML; the second-order giant fibre in the mantle connective and ganglionic neuropil is less than 100 μm in diameter. The cell bodies of the syncitial third-order axons lie in the dorsal wall of the stellate ganglion: there is no distinct giant fibre lobe as in *Loligo*. The giant synapse also differs from that of *Loligo*. The processes of the postsynaptic axon end on a system of collaterals (2–20 μm in diameter) from the presynaptic fibre within the enlarged sheath between the two axons. In the sheath of one postsynaptic axon there were up to 10 fine presynaptic collaterals; each of these was surrounded by processes from the postsynaptic fibre that established synaptic contacts from all sides. The whole system is complex and often covered most of the circumference of the postsynaptic axons. The synaptic contacts are embedded in several layers of glial processes and extracellullar material.

The extraocular photoreceptors of *Todarodes sagittatus* consist of dorsal and ventral sets of photosensitive vesicles. The dorsal set lies near the base of the vertical lobe and its nerve fibres enter the peduncle lobe below. The ventral part consists of a grape-like structure formed by many vesicles; its long nerve bundle terminates in the peduncle lobe. Light-evoked action potentials were recorded from the nerve from the ventral set of photosensitive vesicles and their spectral sensitivity found to respond maximally at 503 nm. Extracellular recordings of light-evoked electrical responses were obtained from a dorsal photosensitive vesicle (Mauro 1977).

The statocyst of *Todarodes sagittatus* has a long narrow horizontal channel dorsally and deep vertical channels anteriorly. Anticrista 7 is extended forwards and backwards (Young 1989). Multiple discriminant analysis of several parameters of the statocyst place this squid amongst the non-buoyant forms (Maddock and Young 1984). This statocyst is suited for rapid movement and turning.

16.1.2 *Martialia*

Martialia hyadesi is an oceanic slope squid present in the South Atlantic, in the Southern Ocean, and southwards to the Antarctic Polar Front Zone, perhaps even south of this zone, and into New Zealand waters where immature animals are found (Rodhouse and Yeatman 1990, Uozumi *et al.* 1991, González *et al.* 1997, González and Rodhouse 1998).

The peak time for hatching is the austral winter. Records of small individuals from New Zealand waters (Uozumi *et al.* 1991) indicate spawning on or near the Aukland Island shelf, and from the Patagonian Shelf with spawning in the vicinity of the Falkland Islands (Rodhouse *et al.* 1992). Females of 6–12 months of age were immature, yet males of the same age were fully mature. The maximum age of captured squid was estimated, from the increment of growth rings in the statoliths, as 330 and 357 days for males and females respectively (González *et al.* 1997). A life span greater than a year has been proposed but further studies are needed. Mantle lengths of up to 370 mm have been recorded (González and Rodhouse 1998).

Occasionally huge numbers of this squid are stranded and on one occasion more than 2000 were found on the shore of Macquarie Island after persistent onshore gale-force winds. Most were still alive; they ranged from 292 to 323 mm ML, and their stomachs contained crustacean remains (O'Sullivan *et al.* 1983). The stomach contents showed that fishes and crustaceans form the major part of their diet, and cephalopods to a lesser extent (González *et al.* 1997).

16.1.3 *Nototodarus*

Members of this genus live in the tropical and subtropical waters of the Indo-West Pacific Ocean, around New Zealand and Australia, and off the Hawaiian and Midway Islands (Nesis 1987). It has been caught between 230 and 710 m during the day and from the surface to 410 m in darkness (R.E. Young 1978).

The prey of *N. gouldi* includes crustaceans, fishes, and cephalopods, and the food type differs in different months of the year. Crustaceans form the main food of smaller squids, and cephalopods of the larger ones, but fishes contribute the largest biomass of their diet. These squids feed primarily at night (O'Sullivan and Cullen 1983).

Rhynchoteuthion paralarvae have been collected over a wide area off the coast of New Zealand, on the continental shelf, from January to July. At hatching the larvae are about 0.8–1.0 mm ML. They lack chromatophores and the proboscis has eight equal-sized suckers on the extremities of the tentacles. At a mantle length of 9–10 mm the transition from rhynchoteuthion to juvenile, with the tentacles separated, was complete in most of the paralarvae (Dunning and Förch 1998). Specimens recovered up to 46 days after tagging showed that growth, of up to 45 mm per month, is rapid (Sato 1985). Females mature at about 220 mm and males at 300 mm ML; the former can reach a mantle length of 400 mm and the latter 380 mm. Large mature females are most numerous in December and January, suggesting that this is probably the time of spawning (O'Sullivan and Cullen 1983).

The extraocular photoreceptor organs of the Hawaiian species comprise dorsal, middle, and ventral sets of photosensitive vesicles on each side (R.E. Young 1978).

16.2 Illicinae

The members of this subfamily are found in the Atlantic Ocean, the Mediterranean Sea, and the Indian Ocean.

16.2.1 *Illex*

Illex (Fig. 16.1) lives primarily over the outer shelf and the continental slope in the Atlantic Ocean, between 60 °N and 50 °S, and in the Mediterranean Sea. It is neritic, or neritic and oceanic (Mangold-Wirz 1963, Roper *et al.* 1984, Nesis 1987).

Life history and habits

These squids are among the most exploited of all cephalopods caught commercially. The life history of *Illex illecebrosus* is the best known and was reviewed by O'Dor (1983). The average diameter of ripe eggs ranges from 0.77–0.82 mm in *Illex coindeti*, 0.75–0.88 mm in

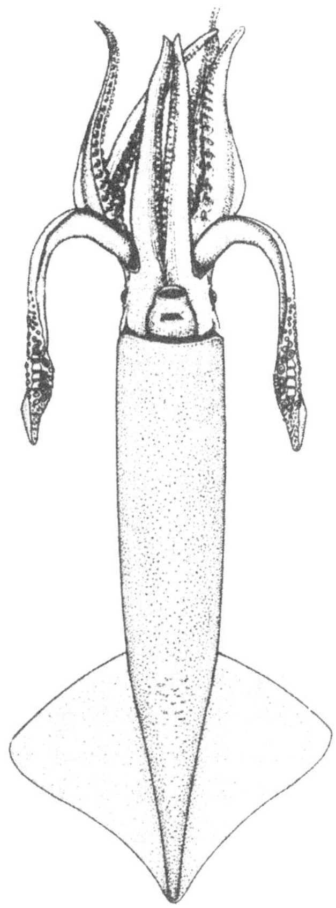

Fig. 16.1. *Illex illecebrosus* (Wormuth 1976).

I. illecebrosus to 0.96–0.97 mm in *I. argentinus*. The potential fecundity, the total oocyte count, depends on mantle length and varies from 90 000 to 800 000 in *I. coindeti* of 150–250 mm ML, from 200 000 to 630 000 in *I. illecebrosus* of 220–280 mm ML, and from 75 000 to 1 200 000 in *I. argentinus* of 360–380 mm ML. Spawning is intermittent in these squids and the production of ripe eggs decreases over time, with only 50% replacement of the initial stock of vitelline oocytes, and a decreasing volume of eggs at each release (Laptikhovsky and Nigmatullin 1993). Gonadal development begins in males of 130–150 mm and females of 100 mm ML. Mature animals, including ones of up to 340 mm ML, were present in large concentrations during a bottom trawl survey. Spawning migrations are extensive as these squids may travel more than 2000 km, at an average speed of 20 km/day. The mature female appears to pre-form a concentrated package comprising gel, eggs (1.00–1.08 mm), and broken spermatophores; once expelled, the package swells in 1–2 min (O'Dor 1983).

When the rhynchoteuthion hatches it has a mantle length of 1.1 mm and the proboscis is longer than the arms (Fig. 16.2a–c). The proboscis begins to separate when the mantle length is 3–4 mm and continues until separation into two tentacles is complete at 8–10 mm ML. There is rapid growth of the arms, suckers, and fins, while the mantle and head become narrower, before the separation of the proboscis into tentacles (Fig. 16.2) (Vidal 1994). Only a small yolk reserve is present at hatching, and O'Dor and his colleagues (1985) suggested that at least during the critical period after yolk absorption it is a suspension-feeder. Rhynchoteuthion larvae of up to 10.7 mm ML have been caught at about 250 m depth in the Gulf Stream, and juveniles occur in the slope water and adjacent upwelling regions. The young squids probably take advantage of the spring

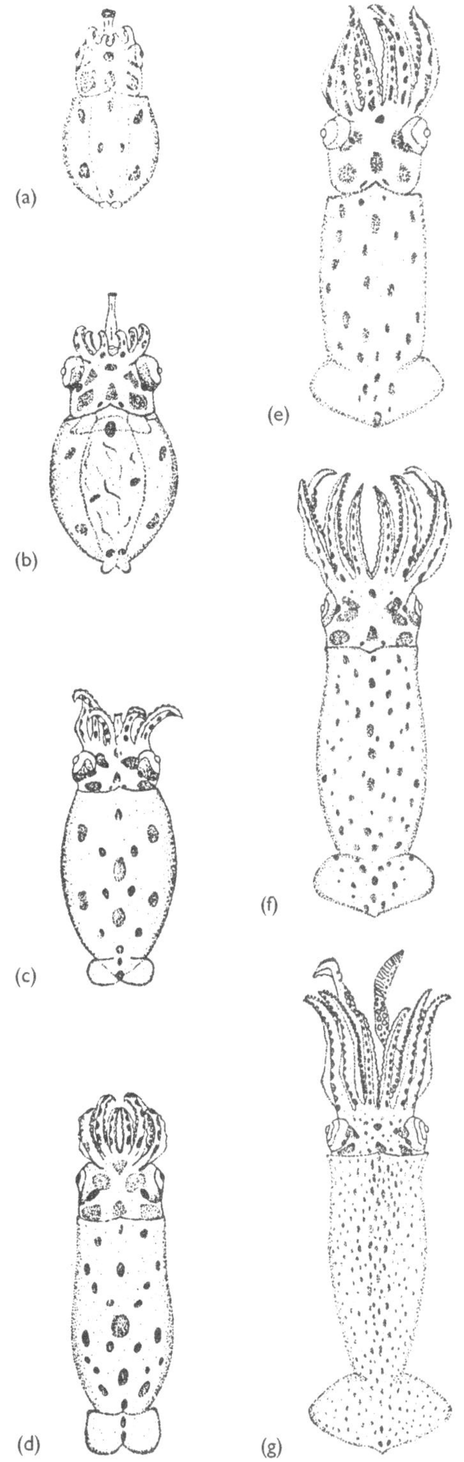

Fig. 16.2. *Illex argentinus*. A growth series of rhynchoteuthion larvae to show the changes in morphology and the chromatophore pattern, and the separation of the tentacles. The mantle lengths are 1.4 mm (a), 3.5 mm (b), 6.0 mm (c), 8.0 mm (d), 11.0 mm (e), 15.0 mm (f), and 30.0 mm (g) (Vidal 1994; with permission of Cambridge University Press).

plankton bloom. After rapid growth some move inshore in pursuit of prey; others remain on the shelf where they feed throughout the water column, and may undergo diel migrations (Vecchione and Roper 1986).

The prey of *Illex illecebrosus* includes fishes, squids, and crustaceans, the proportions varying with the seasons. Cannibalism is important among the largest squids (O'Dor 1983). One school were observed feeding on a small swarm of euphausids at the surface, and typically one squid attacked alone with a rapid head-first burst into the swarm, the arms then spread out as the prey was captured with the tentacles; the whole procedure took less than 2 s (Nicol and O'Dor 1985). The strategy for capturing fish was sometimes unsuccessful, as the latter outmanoeuvred the squid (Foyle and O'Dor 1987).

Central nervous system and receptors

The optic lobes form 287% of the total brain volume (Table 2.6). The plexiform layer has the characteristic radial and tangential layers (Fig. 16.3). The palisade layer of cells is well marked, as are the radial columns of second-order visual cells. Tangential sections show how the columns of fibres proceeding from the plexiform zone become organized into circular islands of increasing size. Each of the smaller islands has one large cell and several smaller ones.

The central nervous system was described by Richter (1913) and Thore (1939), and subjected to a biometric study by Wirz (1959). The supraoesophageal mass is compact (Fig. 16.4) except that the buccal lobe lies far forward. It forms 42.9% of the total brain volume (Table 2.6). The inferior frontal lobes occupy 1.98% and the vertical lobe system 14.2% (Maddock and Young 1987).

The dorsal and ventral photosensitive vesicles of the extraocular photoreceptor organ form 1% of the total brain volume (Fig. 16.5) (Maddock and Young 1987).

The first-, second- and third-order giant axons of the giant-fibre system can be identified in the late embryonic stages and in newly hatched larvae of *Illex*. In postlarval animals the two first-order giant axons are not fused. In the adult each axon has only an ipsilaterally descending branch which contacts the second-order fibres (Fig.16.6). At the chiasma there are two collaterals, one from each axon, that end blindly on the contralateral side. The collaterals are crossed and at the centre of the chiasma the collaterals turn around each other, from the ventral to the dorsal side, and the blind endings lie dorsally. A series of profiles through the chiasma showed that the two collaterals cross over each other and end on the contralateral side. On each side of the apposed membranes of the collaterals there are large masses of electron-transparent vesicles and the membranes show an increased electron density (Martin 1969).

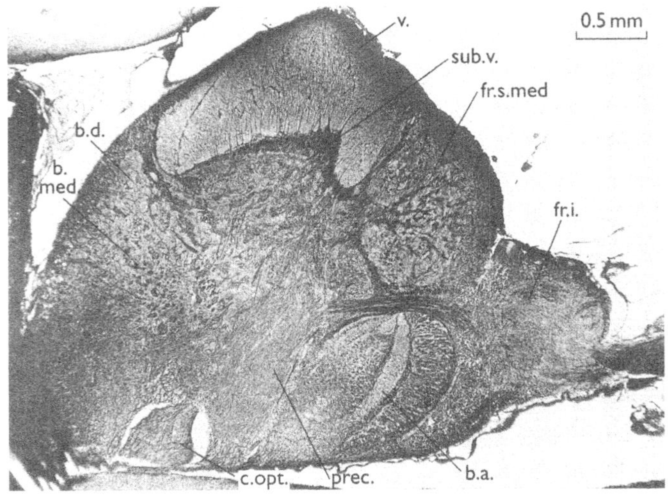

Fig. 16.4. *Illex coindeti* (175 mm ML). The supraoesophageal lobes of the brain in sagittal section.

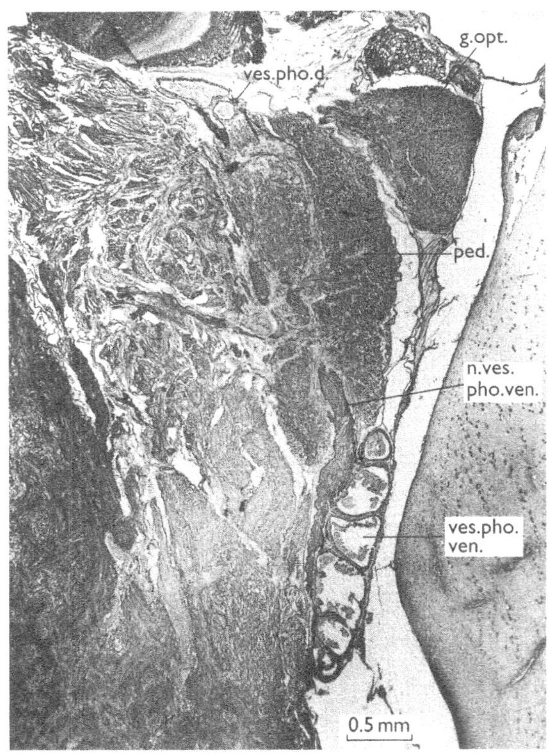

Fig. 16.5. *Illex*. Section to show the dorsal and ventral sets of photosensitive vesicles.

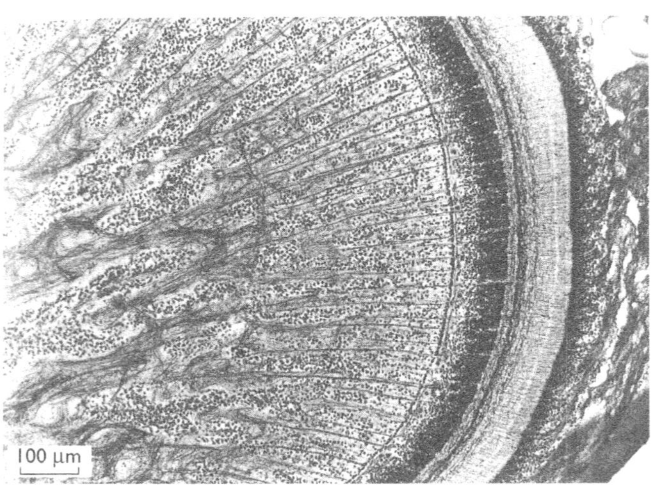

Fig. 16.3. *Illex*. Part of the optic lobe showing the well-developed plexiform layer, with its radial and tangential layers, and radial columns in the medulla.

The statocyst of *Illex coindeti* has well developed horizontal canals (Fig. 16.7). Anticrista 7 is large and extends backwards and forwards in contact with anticristae 6 and 2. The cavity of the statocyst is enlarged ventrally, allowing a large volume of endolymph to flow across the other three sections of the crista (Young 1989). Various parameters of the statocyst were subjected to multiple discriminant analysis and placed this squid amongst non-buoyant ones (Maddock and Young 1984).

16.2.2 *Todaropsis*

Todaropsis eblanae is found in the eastern Atlantic Ocean from the Shetland Islands to the Cape of Good Hope, in the Mediterranean Sea, the Indian Ocean, and off eastern and western Australia. It is usually associated with sandy or muddy bottoms, although small juveniles have been caught in near-surface waters. In the Atlantic and the Mediterranean this squid has been trawled in depths of 20–700 m but seems to be mainly confined to depths of less than 200 m in the North Sea. It has been caught in demersal trawls to 800 m in warmer waters (Wormuth 1998).

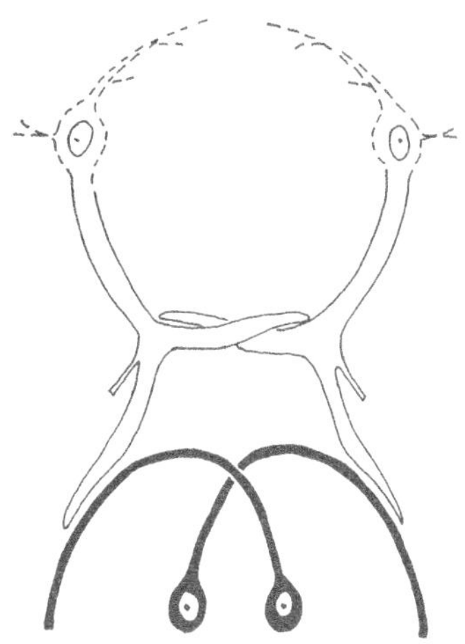

Fig. 16.6. *Illex coindeti.* Diagram to show the course of the first-order giant axons in the adult, reconstructed from serial transverse sections (Martin 1969).

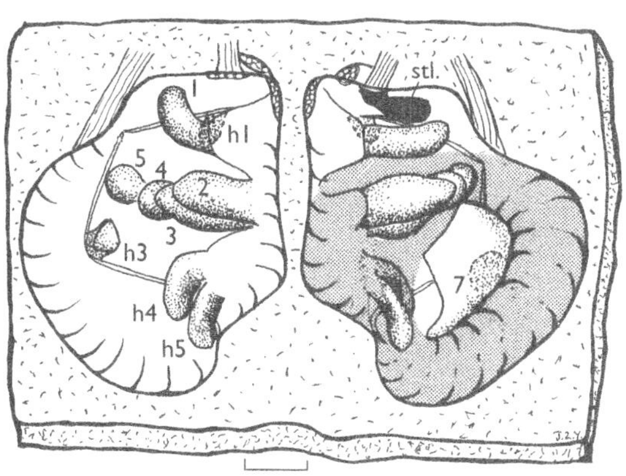

Fig. 16.7. *Illex coindeti* (270 mm ML). The statocyst seen from above, after removal of the macula on the animal's left side (scale bar 1 mm) (Young 1989).

Spawning takes place from April to June in Australian waters (Dunning and Brandt 1985). In Scottish waters mating and spawning took place from June to November and hatching from October to March (Hastie *et al.* 1994). In the Mediterranean males mature at a smaller size than females, at 120 mm and 260 mm ML respectively, and the female will produce up to 10 000 eggs, depending on her size (Mangold-Wirz 1963). The ovaries of several specimens were found to have oocytes of various sizes (up to 1.1 mm) and oviducal oocytes ranging from 1.2 mm to 1.8 mm in diameter (Rasero *et al.* 1995), suggesting it is a multiple spawner (Mangold *et al.* 1993). The largest specimens recorded are a male of 149 mm and a female of 183 mm ML (Lu 1982).

16.3 Ommastrephinae

The ommastrephines are the most widely distributed members of the family. They are found in the Atlantic, Pacific, and Indian Oceans between latitudes 60 °N and 10 °S, and in the Mediterranean Sea. Most are oceanic but the largest of all ommastrephids, *Dosidicus gigas*, has a neritic component in its distribution (Nesis 1983).

These streamlined and muscular squids are fast-moving and several species break the surface of the water. A number of observations record that they not only clear the surface but are even capable of a fairly long 'flight': one attained a height of nearly 2 m above the surface, where it remained for a distance of about 10 m (Arata 1954).

16.3.1 *Ommastrephes*

Ommastrephes bartramii has the widest distribution of all the Ommastrephidae (Dunning 1998). It lives in subtropical and temperate waters, but is absent from tropical waters. In the North Pacific this is the dominant surface-living ommastrephid, occurring in abundance off the coast of Japan. The adults occupy a broad depth range, both day and night, from the surface to at least 1500 m (Clarke 1966). It is known to be capable of 'flying', as it has been found on the decks of boats some 3–7 m above the surface of the sea (Roeleveld 1988).

Life history and habits

The male *Ommastrephes bartramii*, the only member of the genus, is mature when 320 mm ML. The females shows variation as the smallest with oviducal eggs was 420 mm ML, whereas the largest without eggs was 550 mm ML; both were from the south-west Pacific (Dunning 1998). Squids from other areas also matured at different sizes. The adults migrate to enter coastal waters, moving at speeds of up to 16 km/day, the males preceding the females. Paralarvae and small individuals have been collected in slope waters. The rhynchoteuthion larvae lack photophores. The lateral suckers on the tip of the proboscis are almost twice the diameter of the remaining suckers. The separation of the tentacles takes place at a mantle length of less than 7 mm. A distinctive chromatophore pattern is present. The life span is probably of 1 year or more in some populations (Dunning and Brandt 1985). The largest specimen recorded had a mantle length of 860 mm (Nesis 1987).

Fishes are the main prey of this squid but enoploteuthid squids (Wormuth 1976), pelagic octopods, and some crustaceans are also eaten (Lipinski and Linkowski 1988).

Central nervous system and receptors

Richter (1913) described the central nervous system of *Ommastrephes*. It is elongated with the buccal lobes lying far forward of the rest of the brain.

The stellate ganglion lies close to the midline and is considerably flattened dorsoventrally and elongated anteroposteriorly. The stellar nerves, especially the hinder ones, and the fin nerve, are extraordinarily flattened and broad. The fin nerve is also flattened, and passes laterally across the stellar nerve instead of across the ganglion. The largest fibre in the most posterior stellar nerve measured 1060 × 200 μm where it was most flattened, and 650 × 290 μm more posteriorly. The diameters of the fibres in each stellar nerve were measured, as were the distances from the ganglion to the ventral edge of the mantle. Using the data for conduction velocity provided by Pumphrey and Young (1938), it appears that the impulse would arrive at all parts of the mantle at virtually the same time, within a few milliseconds (Table 16.1). This table shows that if all fibres conducted independently of size, at either the maximum or the minimum rates, then the signal would arrive in approximately one-third of the time at the front that it would take to reach the hind end of the mantle.

The statocyst is quite broad and deep relative to its length. The anticristae are all large and smoothly rounded. Anticrista 7 is especially large and makes contact, but does not fuse, with anticrista 6 and several of those in the median row (Young 1989). Various measurements of the statocyst were subjected to multiple discriminant analysis, which placed this squid amongst non-buoyant forms (Maddock and Young 1984).

16.3.2 *Ornithoteuthis*

Ornithoteuthis is restricted to the Indo-Pacific and Atlantic Oceans. It has a nerito-oceanic distribution pattern and is found from the surface to depths of 1000 m or more, but is mostly caught between 600 m and 800 m during the day and at the surface at night. As the name implies, this squid can break through the surface of the water (Sasaki 1915).

Ornithoteuthis antillarum can change colour rapidly, turning from dark to translucent. It has a silvery sheen ventrally and laterally and on the side of the head (Voss 1957). This fast-swimming squid is

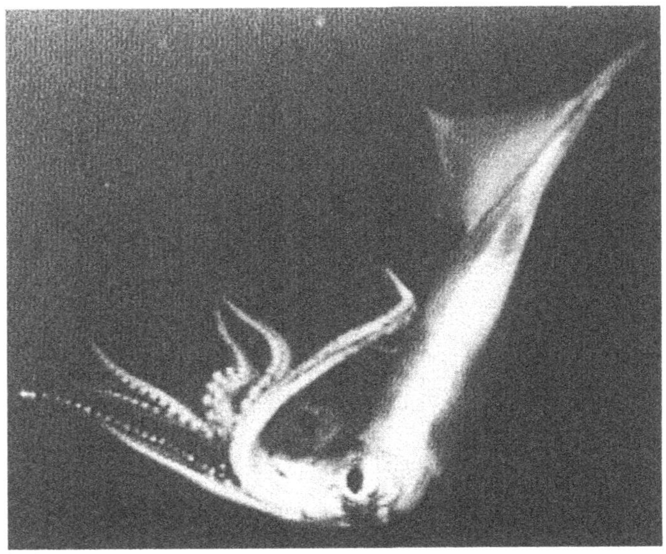

Fig. 16.8. *Ornithoteuthis antillarum* photographed, at a depth of 684 m off the Bahamas, in the 'J' posture while feeding on fish (Vecchione and Roper 1991).

Table 16.1 *Ommastrephes bartramii*. The diameter of the largest fibre in each stellar nerve and the estimated times of arrival of impulses at muscles, with the velocities calculated according to Pumphrey and Young (1938), and the arrival times with faster and slower velocities assumed independent of diameter

Stellar nerve	Length (cm)	Diameter (μm)	True velocity (ms)	Arrival time (ms)	Constant minimum velocity at 5 ms arrival time (ms)	Constant maximum velocity at 5 ms arrival time (ms)
1	14	103	6.8	21	28	7
2	18	165	9	20	36	9
3	18	165	9	20	36	9
4	24	240	11.5	21	48	11
5	24	240	11.5	21	48	11
6	24	240	11.5	21	48	11
7	26	180	9.5	27	52	12
8	27	350	14.5	19	54	13
9	28	370	15	19	56	13
10	29	385	15.5	19	58	14
11	31	470	17.5	18	62	15
12	33	555	19.5	17	66	16
13	42	630	21	20	84	20

rarely caught in trawls. It has been observed from submersibles, at depths of 100–890 m depth, on numerous occasions, and often in the so-called 'J' posture (Fig. 16.8). It has been seen 'to bend its mantle double to use the ventral arms for removal of irritants from the fins and tail' (Vecchione and Roper 1991).

Maturity is reached when the mantle length is 130 mm and 150 mm in males and females respectively, but there is considerable geographic and seasonal variation. Year-round spawning may occur in the warm waters of the East Australian Current. Small paralarvae are abundant in captures from the deeper shelf and upper continental slope waters off the coast of northern New South Wales (Dunning 1998). The proboscis is approximately equal in length to the longest arms and the tip has two suckers which are twice the diameter of the remaining six. *Ornithoteuthis volatilis* males of 237 mm ML and females of 250 mm ML are the largest yet recorded (Dunning and Brandt 1985).

16.3.3 *Dosidicus*

The Peru-Chilean jumbo squid is distributed in the eastern tropical Pacific from about 26 °S to 30 °N and 140 °W. Occasionally it is found as far north as Monterey, California, this periodic extension resulting from El Niño-Southern Oscillation events (Wormuth 1998). The horizontal distribution appears to be associated with upwelling zones and is probably related to its diet of which sardines form a significant portion (Erhardt 1991). This squid rises to the surface at night, the ascent beginning at dusk, and by dawn it has descended into midwater (Nesis 1983).

The life cycle of the jumbo squid was reviewed by Nesis (1983). Eggs have not been found at sea but mature ovarian eggs are transparent, amber-yellow in colour, and slightly ovoid in shape. Some 650 000 eggs (1.1 mm long) were found in a specimen of 500 mm ML. The paralarva (about 1 mm ML) has a few dispersed dark brown chromatophores but no photophores. It is planktonic and is found in the open sea and in nearshore waters, mostly between the surface and 100 m. Juveniles (3–5 mm ML) have been captured down to depths of 500 m. At about 1 year of age a mantle length of 350 mm is reached and 500 mm in 2 years; the largest animals are considered to be about 4 years of age. In equatorial waters females reach maturity at 200 mm and males at 150 mm ML, but in colder waters this occurs at larger sizes. Mating is in the head-to-head position and spermatophores are transferred to the buccal membrane of the female. Spawning probably takes place over or near the continental slope.

Dosidicus is a schooling squid. It is fast-moving and can clear the water with a maximum velocity of about 7 m/s (Cole and Gilbert 1970). A giant-fibre system is present, and the fibres can reach a diameter of 1500 μm. The giant axon comprises three cellular elements: a neuronal axon at the centre, a Schwann cell layer with a glial function, and fibrocytes and collagen bundles that form the endometrium (Villegas 1969, Villegas and Villegas 1984).

The statocyst is very long. The horizontal canal forms a complete tube for about 75% of its circumference in a specimen of 1000 mm ML, and it has a small volume. Hamulus 4 is large. Anticrista 7 is much elongated to form a roof to the canal, and is in contact for much of its length with anticrista 6. Anticrista 2 fits into a groove in anticrista 7 so making a roof for this part of the canal. The cavity of the statocyst of an adult allows for little flow of endolymph in the horizontal plane, although the movement of fluid is not so greatly limited in other planes (Young 1989). Multiple discriminant analysis of several parameters placed *Dosidicus* among the non-buoyant forms (Maddock and Young 1984). The statocyst is well suited to monitor rapid movement.

16.3.4 *Sthenoteuthis*

Sthenoteuthis is widely distributed. *S. pteropus* inhabits the Atlantic from 35 °N to 30 °S where the water is warm, and *S. oualaniensis* the Pacific Ocean, occurring predominantly in tropical and warm temperate surface waters (Dunning 1998). Little is known of its vertical distribution although at night the latter species is attracted to the surface by the lights from ships, and its remains have been found in the stomachs of sea birds which feed predominantly during daylight.

Sthenoteuthis oualaniensis has a darkly pigmented dorsal surface of the head and mantle, with a bluish cast, a colouring found in other near-surface dwelling coleoids. It is a swift and powerful swimmer, capable of propelling itself out of the water and on to the deck of a ship (R.E. Young 1975b). The mantle muscle has circular and radial bands of obliquely striated muscle fibres with a complex sarcotubular network spanning the area between the sarcolemma and a central sarcoplasmic core (Moon and Hulbert 1975). This muscle utilizes carbohydrate as the primary carbon energy source (Hochachka *et al.* 1975).

Life history

Off the coast of Japan adult *Sthenoteuthis oualaniensis* appear in shoals. Here their mode of copulation was observed: a male (130 mm ML) and a female (200 mm ML) swam together, the latter about a length in front of the former. Suddenly the male altered its orientation so that the two squids could grasp each other in a head-to-head position, after which they sank from the surface. All females of more than 240 mm ML had sperm bulbs implanted on the ventral side of the inner lip of the buccal membrane. Of the females examined the largest ovary contained some 250 000 eggs, and oviducal eggs of up to 0.87 × 0.72 mm (Okutani and Tung 1978). Size-frequency analyses of ovarian oocytes revealed continuous egg production throughout the adult life of this species, indicating multiple spawning (Harman *et al.* 1989). An epipelagic egg mass of *S. pteropus* was collected in the tropical eastern Atlantic, and it contained about 19 000 pale raspberry colour eggs (0.8 mm diameter) (Laptikhovsky and Murzov 1990). After spawning the female continues to feed and grow and the oocytes continue to mature. Females can reach a mantle length of 520 mm and the males 280 mm at 380 and 360 days of age respectively (Bizikov 1995).

Rynchoteuthion larvae (1.4–9.1 mm ML) have been captured from the surface to 250 m depth off Hawaii; here they are present throughout the year, with the highest abundance being in August. The changes during the growth of these larvae was followed by Harman and Young (1985). The proboscis, often very elongate, bore eight suckers of almost equal diameter on its tip. Division of the proboscis began at about 3.0–5.0 mm ML and was completed at 8.5 mm ML or more. A single round photophore was present on the ventral surface of each eye. In specimens of more than 4.0 mm ML two intestinal photophores developed between the digestive gland and the anus. Few chromatophores were present.

Photophores

The adult *Sthenoteuthis oualaniensis* has a single, large, oval photophore embedded near the anterodorsal end of the mantle. This photophore, which consists of a single layer of small organelles, is barely visible in an animal of 44 mm ML, but its size increases gradually until the mantle reaches 100 mm in length when it is apparent as an oval organ (Kishimoto and Kohno 1992), and is creamy-yellow in colour (Okutani *et al.* 1987). It produces an intense blue flash of light, which decays rapidly in intensity. The luminescence originates from numerous oval granules in the organ and the essential light-emitting components are membrane bound. The reaction is triggered by monovalent cations, such as potassium, but has an absolute requirement for molecular oxygen. The emission maximum is 455 nm (Tsuji and Leisman 1981).

A small *Sthenoteuthis pteropus*, dipnetted at the surface, emitted luminous blue flashes from the ventral surface of the arms and tentacles, as well as a pale blue, diffuse, and luminescent glow from minute subcutaneous photophores which are evenly distributed over the body. Light also emanated from the ventral surface of each eye, the intensity varying from bright flashes to a faint glow. In addition Roper (1963) noted the 'striking ability of instantaneous control of different photogenic areas' by this squid. Females, 250–380 mm in ML, have a large oval photophore on the dorsal surface of the mantle. Such an organ is also present in young females and males. The photophore emits a bright blue light of 1–2 s duration (Clarke 1965). It consists of numerous orange bodies, 1–2 mm diameter, embedded in the mantle muscle, and below is a white reflecting layer. Each is enclosed in a capsule of connective tissue comprising vacuolar tissue containing blood vessels and the photogenic body. When one individual was attacked, the whole photophore appeared to give a uniform flash with an emission spectrum of 474 nm (Girsch *et al.* 1976). This oval photophore consists of small, spherical organelles (1–2 mm diameter). Each is enclosed in a capsule of connective vacuolar tissue containing blood vessels and the photogenic body. The photogenic body is formed of a spongy matrix of capillaries with the photogenic tissue filling the interstices (Clarke 1965). The photogenic tissue forms about half of the body and in consequence lies within a few micrometres of a capillary.

Receptors

Each extraocular photoreceptor organ of *Sthenoteuthis oualaniensis* comprises dorsal, middle, and ventral sets of photosensitive vesicles (R.E. Young 1978). The latter set is formed by a series of flat, elongate vesicles. The central set is a few vesicles on the dorsal surface of the optic stalk. The dorsal set is scattered along a broad, arching nerve extending from the peduncle lobe, and along the dorsomedial margin of the optic lobe. The distribution of the vesicles suggests they may act to monitor counter-illumination.

16.3.5 *Eucleoteuthis*

Eucleoteuthis is found across the North Pacific from 20 °N to 40 °N, and in all three major oceans in the Southern Hemisphere between 20 °S and 40 °S (Wormuth 1998). The vertical distribution is poorly known but it appears in catches off Japan, and may extend to depths of 1300 m (Roper *et al.* 1984).

A pair of yellowish, ventral, longitudinal stripes of photogenic tissue are present (Okada 1968), one on each side of the midline, and broken posteriorly by a gap at the level of the maximum width of the fins. Near the anterolateral margin of the mantle there are two oval luminous patches. Two further patches lie at the base of arms IV, as well as one in the middle of these arms.

Male and female *Eucleoteuthis luminosa*, the only species of the genus, mature at between 120 mm and 130 mm ML (Wormuth 1998). Little is known of the reproductive habits but one gravid female (163 mm ML) had greatly enlarged nidamental glands, and mature eggs measuring almost 1 mm (R.E. Young 1972b).

16.3.6 *Hyaloteuthis*

This ommastrephine lives in the warm waters of the Atlantic and Pacific Oceans, between latitudes 35 °N and 35 °S (Nesis 1985, Dunning and Brandt 1985). It occurs between the surface at night and about 200 m depth during the day, and one has been caught deeper than 1700 m (R.E. Young 1978, Roper *et al.* 1984, Nesis 1987).

Hyaloteuthis pelagica is the only species of the genus and one of the smallest of the ommastrephids. Mature specimens have been captured includng one in Australian waters (46 mm ML) (Dunning and Brandt 1985), and others off Bermuda (Roper and Young 1975). Mature females (60–67 mm ML) and males (50–61 mm ML) have been found in the stomachs of *Alepisaurus ferox*, caught near New Caledonia (Rancurel 1970). One gravid female was captured in an opening/closing net between 1700 m and 2200 m during the day; its ovary was filled with mature eggs and numerous discharged spermatophores were attached to the lips surrounding the mouth. This female had dorsal, middle, and ventral sets of photosensitive vesicles in each extraocular photoreceptor organ (R.E. Young 1978).

Paralarvae are found around the Hawaiian Islands throughout the year. They have a vertical distribution between 100 m and 220 m during the day and from the surface to 150 m at night. The rhynchoteuthion has one intestinal photophore and one on the ventral surface of each eye, and the tip of the proboscis has moderately enlarged lateral suckers (Harman and Young 1985).

17 Onychoteuthidae

Most members of this family are strong, active swimmers and those of one genus are capable of 'flying'. Their geographical distribution is poorly known, but some occur widely in temperate and tropical waters in the Pacific and Atlantic Oceans, others live in sub-Antarctic waters. Their vertical range is not well known, but they are known from the surface to depths of 1000 m or more. The distributional data for each species has been summarized by Kubodera and his colleagues (1998).

17.1 *Onychoteuthis*

This genus has a circumglobal distribution between 75°N and 60°S, and is found in the Mediterranean (Roper 1972, Roper *et al.* 1984, Mangold and Boletzky 1988). *Onychoteuthis* has been seen and dip-netted at the surface at night (Rees 1949, Roper and Young 1975), and found in the morning on the decks of moving vessels. It is often called the 'flying' squid and is one of the most powerful swimmers among the oegopsin squids (Fig. 17.1), the adults rarely being caught in nets (Okutani 1974, Roper and Young 1975). From the surface it descends to depths of 1350 m or more (Clarke and Lu 1975, Roper and Young 1975, E.A. Anderson 1978, R.E. Young 1978, Lu and Roper 1979). In Hawaiian waters three members of a species complex of *Onychoteuthis banksi* migrate vertically from 50–150 m in the day to the upper 75 m at night (R.E. Young and Harman 1985a). Early stages of *O. borealijaponicus* have been caught off Japan at depths of 20–200 m (Yamamoto and Okutani 1975).

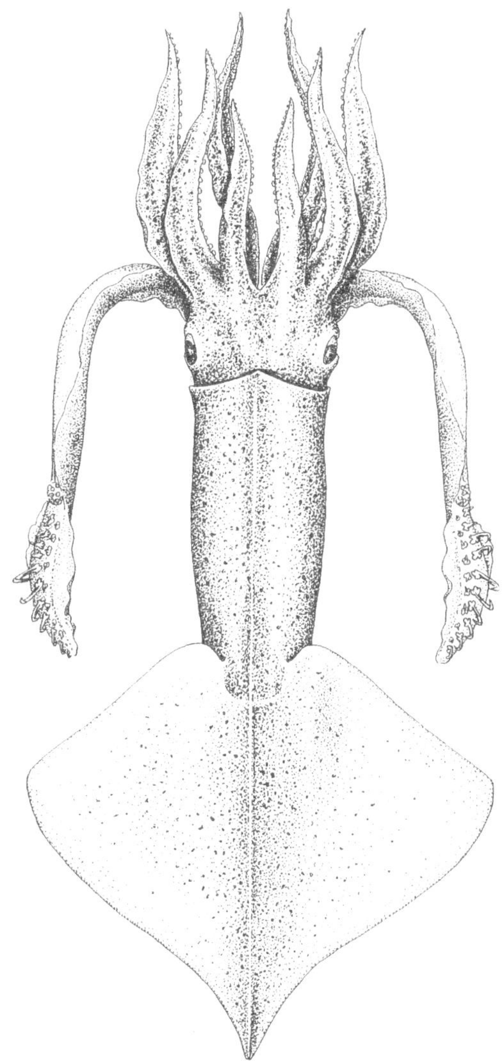

Fig. 17.1. *Onychoteuthis banksi* (Guerra 1992).

17.1.1 Life history

More than 1000 *Onychoteuthis banksi* (*Chaunoteuthis mollis*) captured off Japan, had 'sour' flesh when tasted, and all were gravid females (Okutani and Ida 1986). These squids had cuts, 100 mm long, on the left ventral side of the mantle with implanted spermatophores, and tentacles reduced to stumps 5% of the mantle length. Spermatophores, with only head and neck projecting, were embedded in the longitudinal cuts in the mantle tissue.

With specimens representing most stages of growth it is possible to follow the life cycle of *Onychoteuthis banksi*. The paralarvae (1.5–10 mm ML, up to 30 days after hatching) are abundant in the epipelagic layers of tropical and subtropical waters. They live in the upper warm layer above the thermocline and do not make any diel vertical migrations. These begin when juveniles (10–40 mm ML, 30–100 days) descend to 30–50 m depth during the day, returning to 10–30 m at night. Larger squids (40–150 mm ML, 100–260 days) penetrate to 150 m in the day, again returning to 30–50 m at night. At 200 days of age the males become functionally mature. Maturing females descend into mesoplegic waters where they possibly copulate and become mature. The maximum recorded size is 360 mm ML for females and 280 mm ML for males (Okutani and Murata 1983). The eggs range from 0.2 to 0.5 mm diameter, and the potential fecundity is some 205 000 oocytes (Arkhipkin and Nigmatullin 1997).

17.1.2 Receptors and central nervous system

The eye of *Onychoteuthis* has a bright yellow lens which absorbs strongly in the blue and ultraviolet bands (Messenger 1981). Sections of the retina show the rhabdoms apparently longer ventrally than dorsally. No trochlear cartilage or crossing muscles could be seen in the sections; perhaps this squid does not have binocular vision. The optic lobes (Fig. 17.2) form 234% of the total brain volume (Table 2.6).

The central nervous system (Fig. 17.2) is closely similar to that of *Loligo*. The vertical lobe system forms 10.5% of the total brain volume. The posterior chromatophore lobes occupy 3.53% of the total brain volume (Maddock and Young 1987). The ventral magnocellular lobe is relatively small. The first-order giant cell has not been well seen in our sections; the axons cross in the palliovisceral lobe. There is one giant fibre in each stellar nerve, the largest reaching 120 μm in diameter in *O. banksi* (ML 68 mm). The stellar nerves are arranged as in ommastrephids, the hinder ones running with the fin nerve.

The extraocular photoreceptor organs of *O. banksi* form 10.1% of the total brain volume (Maddock and Young 1987). Each organ consists of widely separated dorsal and ventral vesicles, with a thin band of vesicles close to the optic tract (Fig. 17.3). The ventral set is in the form of a narrow sheet of small sacs spread out along the dorsal surface of the ventral cartilage of the orbit. The dorsal photosensitive vesicles form a compact mass lying far forward in the dorsal wall of the orbit (Fig. 17.4). These vesicles are unusual as they almost surround an area of vacuolated tissue through which the nerve fibres pass. In *Onychoteuthis compacta* the extraocular photoreceptor organ is located on the posterior edge of the optic tract, extends ventrally and slightly laterally below the central brain along the medial side of the optic lobe, and consists of a large number of tightly packed photosensitive vesicles (R.E. Young 1978).

The dorsal and ventral photosensitive nerves are large and converge on the peduncle lobe where they enter a special part of the basal region. The fibres from above and below then overlap and interweave (Fig. 17.5). The output from this region is probably to the lateral basal lobe, controlling the chromatophores. The subpedunculate lobe lies close to the peduncle lobe, but it is not clear how the output from the peduncle lobe may control luminescence. The peduncle and lateral basal lobes lie close together, separated only by a thin layer of cells, with a narrow bridge of neuropil at one point. The posterior chromatophore lobe (Fig. 17.6) is well developed. This squid may control the emission of light necessary for counter-illumination; perhaps by a mechanism similar to that demonstrated in *Abralia* (R.E. Young and Arnold 1982).

The statocyst has large anticristae, some are pointed and others are rounded at the tips providing considerable channelling of the flow of endolymph (Young 1989). The two dorsolateral anticristae (6 and 7) come to touch in a specimen of *Onychoteuthis banksi* of 130 mm ML, so forming a well-marked horizontal canal through which endolymph must pass to reach the vertical crista. The statocyst is typical of non-buoyant oceanic squids (Maddock and Young 1984).

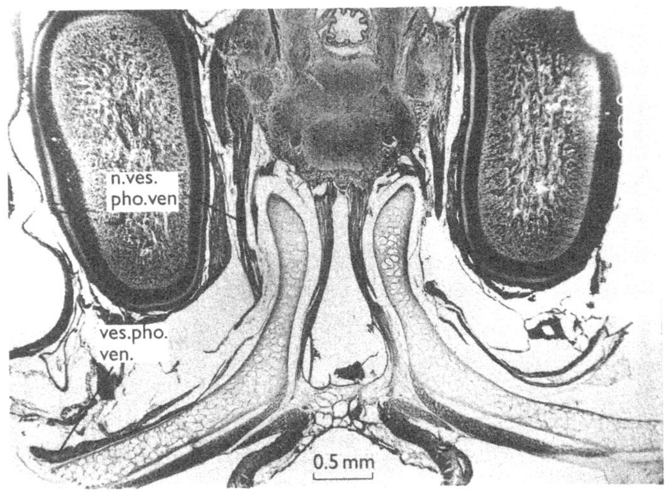

Fig. 17.3. *Onychoteuthis banksi*. Transverse section through the central nervous system showing the ventral photosensitive vesicles in the orbit. Note the reticulate cartilage.

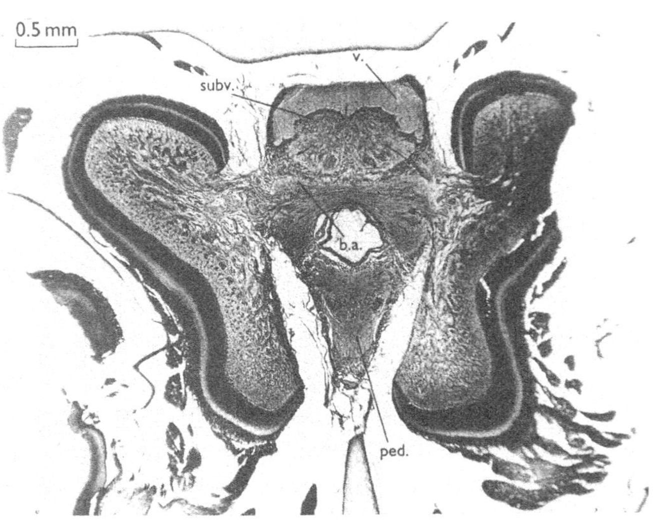

Fig. 17.2. *Onychoteuthis banksi*. Transverse section through the central nervous system and optic lobes. Figs 17.2–17.6 are all of a specimen of 68 mm ML.

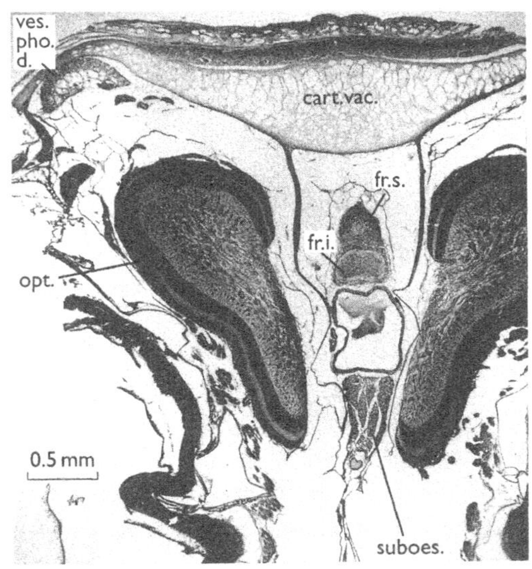

Fig. 17.4. *Onychoteuthis banksi*. Transverse section showing the dorsal photosensitive vesicles, with a 'window' above them.

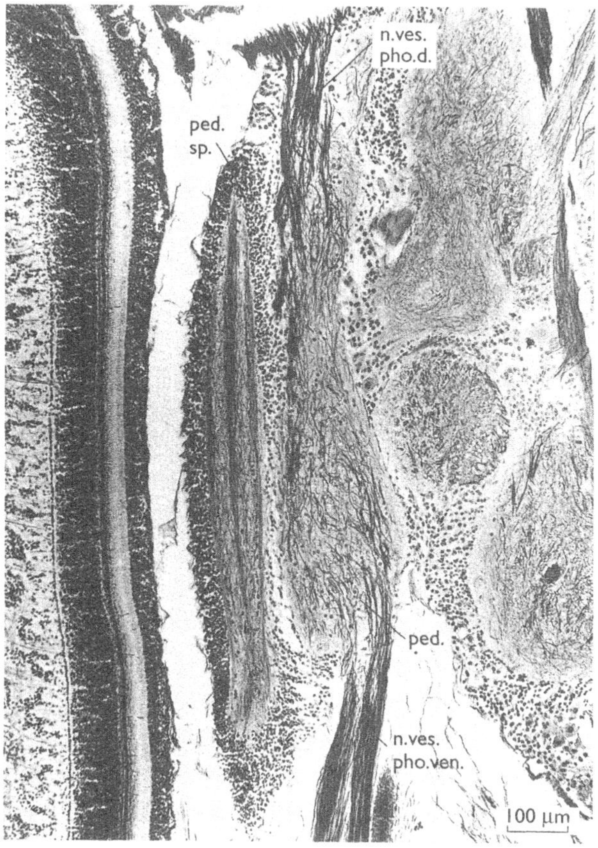

Fig. 17.5. *Onychoteuthis banksi.* Transverse section showing photosensitive nerves from dorsal and ventral photosensitive vesicles converging on the peduncle lobe.

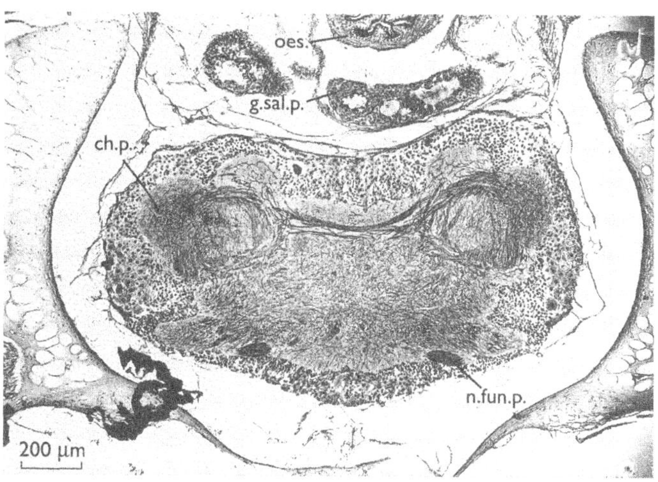

Fig. 17.6. *Onychoteuthis banksi.* Transverse section of the posterior chromatophore lobe.

17.2 *Onykia*

This onychoteuthid has been recorded from latitude 50°N to 40°S, and is probably widespread in temperate and tropical waters. It is sometimes caught in neuston nets close to the surface, day and night, and is commonly associated with floating seaweed, *Sargassum* (Roper and Young 1975, Clarke and Lu 1975). It has been recovered from the stomachs of lancetfish, *Alepisaurus ferox* (Okutani 1981).

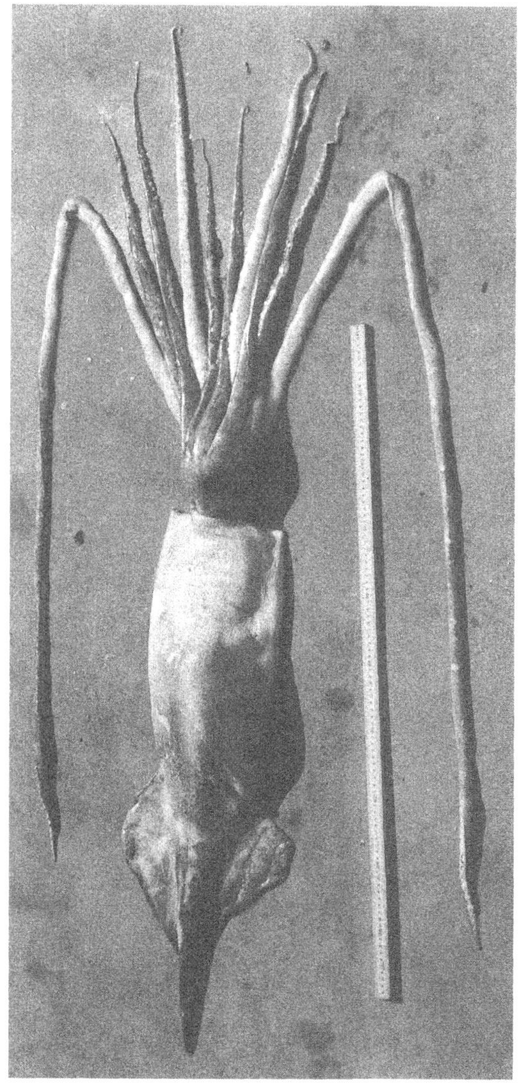

Fig. 17.7. *Moroteuthis robusta* (rule 1 m) (photograph kindly provided by F.G. Hochberg).

Onykia has a somewhat bullet-shaped mantle with a broad body, and juveniles are dark bluish in colour dorsally (Nesis 1987). Mantle lengths of up to 101 mm have been recorded (Okutani and Tsukada 1988).

A re-evaluation has linked *Moroteuthis robusta* with *Onykia carribaea* but the latter is known to be part of a species complex (Tsuchiya and Okutani 1991). Other species of *Onykia* are known but some doubt remains of their validity (Clarke 1992c, Kubodera *et al.* 1998).

The statocyst is essentially like that of a small *Onychoteuthis*, but no details were recorded (Young 1989).

17.3 *Moroteuthis*

This genus is the giant of the family, and one specimen of 2324 mm ML was recorded by Dall in 1871; more recent reports are of smaller animals (Fig. 17.7). *Moroteuthis* is found in sub-Antarctic waters and in the north Pacific Ocean (Okutani 1983b, Roper *et al.* 1984, Okutani *et al.* 1987). Little is known of its vertical distribution, but it is occasionally seen swimming at the surface, near the surf line (Hochberg and Fields 1980). It is caught on or near the bottom at depths of 100–375 m, and fragments from bottom-living sea urchins are found in the stomach (Roper and Young 1975). Captures have been made at 730–920 m depth, at 400 m, and at less than 200 m (Hochberg and Fields 1980, Roper *et al.* 1984, Bizikov and Arkhipkin 1997).

The mantle length commonly reaches 1200 mm, with a total length to the tip of the tentacles of 3350 mm. Most females mature at 400–500 mm ML and the males at a smaller size. The males possess a large penis, up to 465 mm in length. The testis increases in weight until the mantle reaches a length of 325 mm. Subsequently the testis decreases rapidly and the animal becomes emaciated; these males are, however, still reproductively active as considerable numbers of spermatophores are present within the penis. Sections of the mantle wall of juvenile and mature males have a characteristic arrangement of circular and radial muscle fibres between two collagenous tunics; in spent males the mantle wall consists of disorganized muscle fibres with 'space' between the fibres. The mantle wall of immature and maturing females has the typical arrangement of muscle fibres, but this is lost in mature and spent females and the muscle fibres are sparsely scattered. Mature ovarian ova measure 1.0×0.76 mm: the nidamental glands are very large (Hochberg and Fields 1980). This squid 'appears to fit into the category of a true terminal spawner, with spawning associated with gonad regression, tissue breakdown and death' (Jackson and Mladenov 1994).

The gladius and statolith of *Moroteuthis robusta* exhibit growth increments. The statolith is small, being 0.78–0.95 mm in length in animals of 995–1415 mm ML, but growth lines could not be ascertained as they disappear in the peripheral zone. The gladius, 931–1405 mm in length in the same specimens, has distinct incremental growth lines; assuming these to be laid down daily then the ages of these animals ranged from 413 to 520 days (Bizikov and Arkhipkin 1997).

The flesh of *Moroteuthis* tastes bitter (Hochberg and Fields 1980), and the digestive glands of two specimens (1020 mm and 1100 mm ML) contained 16.7–18.7% lipids (Hayashi *et al.* 1990). It is not certain whether this is a buoyant squid since one specimen tested was found to be dense (Clarke *et al.* 1979) though it was presumably a juvenile.

17.4 *Ancistroteuthis*

This onychoteuthid is found in west African waters, the western Mediterranean, possibly the Gulf of Mexico, and Melanesia (Nesis 1987). It is rarely caught and nothing is known of its vertical distribution. A mantle length of 300 mm may be reached, but as yet no mature specimen is known (Clarke 1980, 1992c, Roper *et al.* 1984).

17.5 *Kondakovia*

Kondakovia longimana was first described from three animals caught at 0–50 m in Antarctic waters (Filippova 1972). Further complete specimens have been recovered from the stomachs of sperm whales (Clarke 1980). One has been fished at 48 m and others found on the surface (Lu and Williams 1994a, 1994b), including one which was still alive when taken in the Ross Sea, this last being the southernmost record (Vacchi *et al.* 1994). This squid forms 50% of the cephalopods eaten by Emperor penguins in the steep slope regions of the Weddell Sea (Piatkowski and Pütz 1994).

Tissue taken from the ventral mantle wall and a tentacle of a male (355 mm ML), caught at 48 m, was found to contain an ammonium content of 329.4 mM (Lu and Williams 1994a), well within the range of concentration in neutrally buoyant squids (Clarke *et al.* 1979, Lipinski and Turoboyaski 1983, Lipinski and Jackson 1989). Sections of tissues from the same regions of this specimen showed the bundles of muscle fibres to be somewhat separated, a condition intermediate between the densely organized muscles of say *Gonatus*, and the highly reticulate tissue found in *Mastigoteuthis* (Lu and Williams 1994a).

The largest recorded specimen (estimated to be 830 mm ML) was collected from the surface of Prydz Bay, Antarctica (Lu and Williams 1994a). Evidence of growth has been derived from the lower beaks found in the stomachs of sperm whales (Clarke 1980). New methods of analysis give an estimate of at least 4–5 years longevity, and perhaps even as long as 8–10 years (Jarre *et al.* 1991).

18 Gonatidae

The members of this family inhabit cold waters including the Arctic and sub-Arctic. They extend southwards to about 35°N in the Pacific and Atlantic Oceans. Most species live in the Pacific Ocean, but two are found in the North Atlantic (Kristensen 1981a), and one in Antarctic waters to as far north as 40°N (Okutani and Clarke 1992). They have a vertical spread from the surface to 950 m but are mostly found from 400 to 800 m during daylight and above 500 m in darkness. The adults of some species show a degree of association with the bottom.

18.1 *Gonatus*

Earlier this genus was considered to include only one species, *Gonatus fabricii* (Fig. 18.1), with a worldwide distribution, but now 11 species are recognized (Kristensen 1981a, Okutani *et al.* 1988, Sweeney and Roper 1998). *G. fabricii* and *G. steenstrupia* are found in Arctic and sub-Arctic waters from the fishing banks of Newfoundland in the west to the Barents Sea in the east and as far south as latitude 45°N. The latter has a more southerly distribution and they overlap in the Irminger Basin (Kristensen 1981a). The common ancestor of these two species is considered to have migrated through the Bering Strait into the Arctic Ocean and the north Atlantic in the early Late Pliocene (some 3.5–3.2 million years before present). Somewhat later the ancestor of the notalian *G. antarcticus* migrated from the north-east Pacific into the Southern Ocean. All three species are closely related to each other (Nesis 1997).

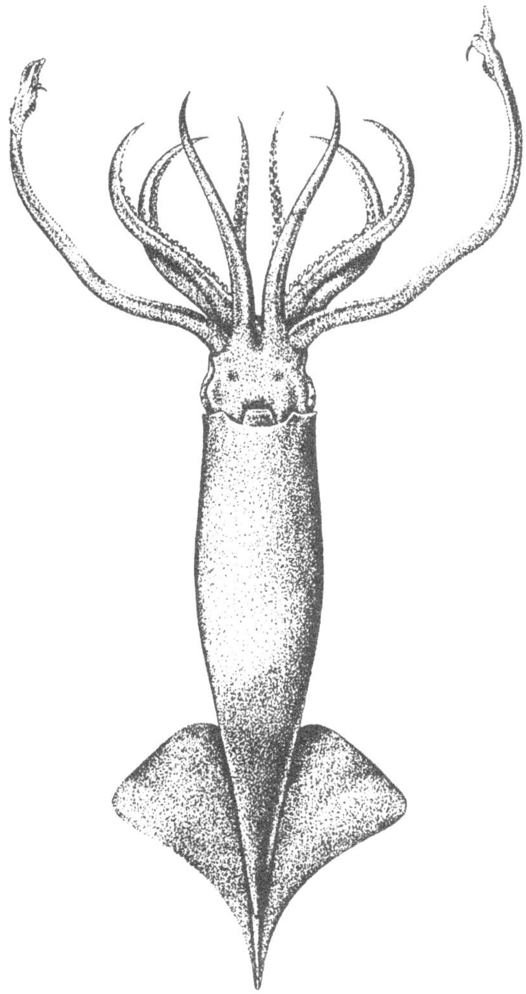

Fig. 18.1. Gonatidae. *Gonatus fabricii* (Kristensen 1981a).

18.1.1 Life history

Gonatus fabricii is the most abundant squid of Arctic and sub-Arctic waters, in both numbers and biomass. At 60°N 20°W it forms more than 70% of all the cephalopods captured (Lu and Clarke 1975a). It is a shoaling squid and lives over the continental shelf at depths of 200–300 m. Hatchlings and juveniles, 3–40 mm pen length, inhabit depths of 80–100 m; adults live above the bottom from 200 m downwards, although they have been observed near the surface at night, and a gravid female was captured at 2500–2700 m (Lu and Clarke 1975a, Kristensen 1984). South of the polar circle this squid may migrate upwards at night (Kristensen 1977b). However, along the North Atlantic Ridge and in the Sea of Irminger, it has been seen from a submersible between 420 and 1200 m and from 350–1200 m at night, suggesting little vertical migration (Moiseev 1991).

The pen length of the smallest specimens of this species yet collected is 3 mm, and this may be the size at hatching. A close relationship has been found between the pen length and body weight, and the former can be used to estimate age. The maximum recorded sizes are of a male and female with pen lengths of 293 mm and 289 mm respectively; both may have been 3 years of age (Kristensen 1982). In juveniles of 25–30 mm pen length the large central hooks of the tentacular club develop, and the adult proportions of the body are attained (Kristensen 1977a). Males reach sexual maturity at a pen length of 200 mm, and in females the ovary begins to develop at 60–90 mm pen length (Kristensen 1983, 1984). A female of 200 mm pen length was captured at the bottom at a depth of 2500–2700 m and was mature; the posterior part of the mantle was filled with the ovary, which contained an estimated 10 000 eggs. Most of the eggs were pale and elongated (1.7 mm long) but a few were mature, almost spherical (4.7 mm diameter), and orange in colour; several discharged spermatophores were present on the buccal membrane indicating copulation in the head-to-head position. The mantle and arms of this female were swollen and fleshy, the tentacles were represented by two stubs, and only scars remained of the suckers of the arms (Kristensen 1981b). Other specimens have been found in a similar condition (R.E. Young 1973a). The eggs are probably deposited on the bottom on the continental slope (Kristensen 1981b).

There are two groups of *Gonatus fabricii*, one at Disko Bugt, off Greenland, breeds during the autumn, and another group, living nearby in the Davis Strait, breeds in the spring and early summer (Kristensen 1984). These two populations have been shown to be morphologically distinct by a discriminant function analysis of some characters (Kristensen 1982).

Adult *Gonatus fabricii* can achieve near neutral buoyancy, the lift being provided by large amounts of low density oil in the digestive gland. Steenstrup (1882) commented on the volume of oil voided during dissection. Clarke (1980) recorded that the digestive glands of two female *G. antarcticus* (230 mm and 345 mm ML) were large and exuded thick yellow oil, yet that of a male (285 mm ML) was small and no oil was evident. A lipid content of 63% was recorded in the digestive gland but only 1% in muscle tissue (Kristensen 1984). One *Gonatus* was photographed from a submersible in the head-down position, so the position it adopts in life may be due to the presence of oil (Church 1971). The young, by contrast, are not neutrally buoyant (Clarke *et al.* 1979).

Juvenile *Gonatus fabricii* (7–43 mm pen length) feed on zooplankton. Copepods form nearly half of their diet, followed by euphausids, amphipods, pteropods, and chaetognaths (Nesis 1965). In subadults and adults (43–293 mm pen length) 51% of the stomachs examined were empty while amphipods were present in almost a third, and cephalopods in 13%, but the debris in the remaining squids could not be identified (Kristensen 1983). Evidently a change in diet takes place with age, the larger squids eating a higher percentage of fish and cephalopods than the subadults and juveniles (Nesis 1965, Kristensen 1984). The ability to capture fish and cephalopods may be possible only after the development of tentacular hooks.

18.1.2 Morphology

Figure 18.2 is of a small specimen of *Gonatus fabricii*, with its head withdrawn into the mantle, a characteristic of preserved specimens (Okutani and Clarke 1992). There are retractor muscles attached dorsally to the cranium that run to the mantle (Fig. 18.2), and are presumably responsible for the retractiom of the head. The muscles of the mantle, fins, and head are all dense and giant nerve fibres are present, suggesting that young gonatids can swim well and make rapid escape responses (Fig. 18.3).

The buccal mass is elongated and occupies much of the head, and its anteroposterior length is 13.5% of the mantle length in this specimen (Fig. 18.2). The mandibular and other muscles are all dense (Fig. 18.4). The salivary papilla carries the single duct from the posterior salivary glands, which lie immediately behind the suboesophageal mass of the brain in this small specimen. The oesophagus is thick-walled and relatively large in diameter.

18.1.3 Receptors and central nervous system

The retina of *G. fabricii* has rhabdoms of similar length (120 μm) in a young specimen of 20 mm ML. A trochlear cartilage is present. Two strands of muscle cross the midline, and anterior conjunctive muscle

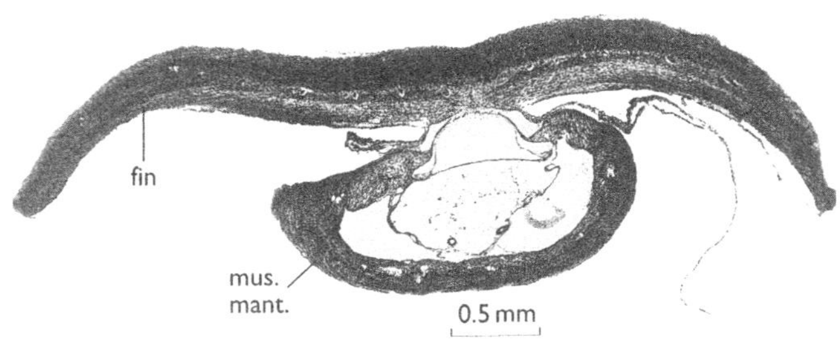

Fig. 18.3. *Gonatus fabricii* (20 mm ML). Transverse section through the end of the mantle and the fin.

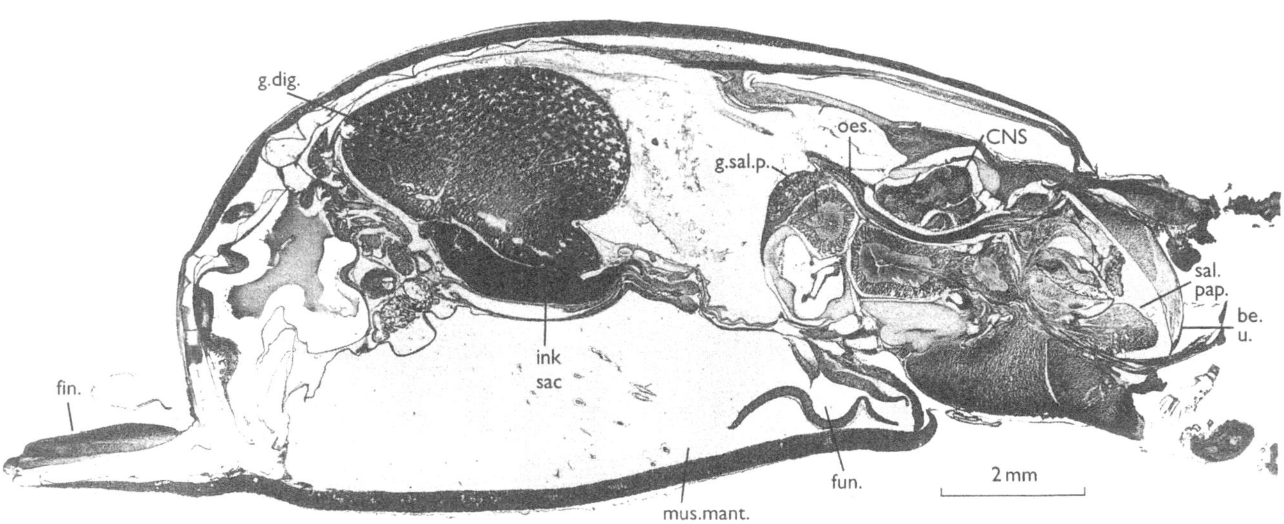

Fig. 18.2. *Gonatus fabricii* (20 mm ML). Medial sagittal section through whole animal.

I can be followed from one eye to the other (Fig. 18.5) (Budelmann and Young 1993).

The optic lobes form 392% of the total brain volume (Fig. 18.6) (Table 2.6).

The central nervous system is compact (Fig. 18.7). The supraoesophageal mass occupies 34.5% and the vertical lobe system 13.2% of the total brain volume (Table 2.6). Large efferent fibres proceed from the precommissural lobe to the pedal, magnocellular, and palliovisceral lobes, thus the brain is similar to that of *Loligo*.

The first-order giant cell is conspicuous in the relatively small ventral magnocellular lobe (Fig. 18.8). The giant axons cross in the palliovisceral lobe without fusing (Fig. 18.9). Several giant fibres are present in each stellar nerve, the diameter being larger in the longer nerves than in the shorter ones (Fig. 18.10).

Each extraocular photoreceptor organ consists of compact groups of photosensiitive vesicles on the anterodorsal surface of the optic tracts (Fig. 18.6), and form 1.3% of the total brain volume (Maddock and Young 1987).

The statocyst of *Gonatus steenstrupi* is long, broad and shallow (Fig. 18.11). The dorsolateral anticristae 6 and 7 meet with flattened surfaces to form a canal, this is complete for 16% of the circumference. Anticrista 1 is small. There is no hamulus 2, but the others are present (Young 1989). Multiple discriminant analysis of various parameters of the statocyst of a juvenile placed it among the

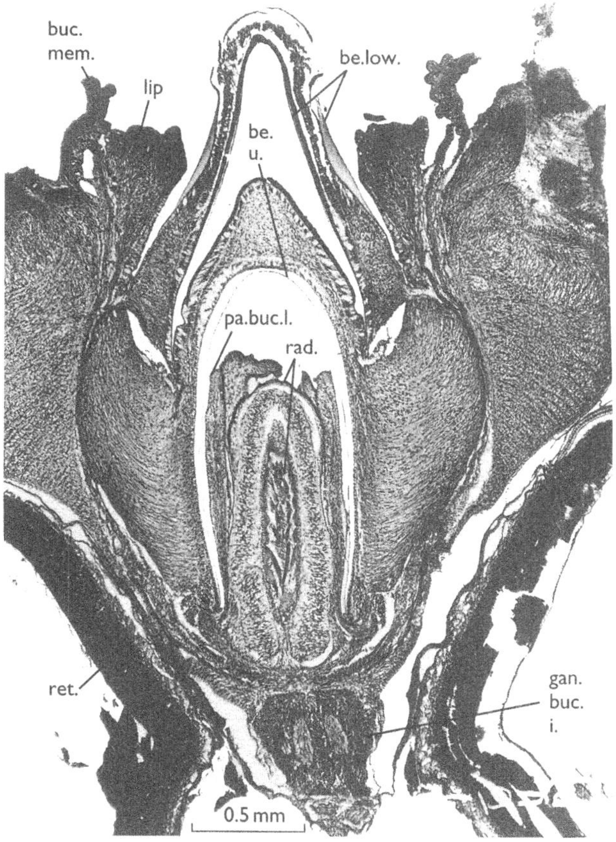

Fig. 18.4. *Gonatus fabricii* (12 mm ML). Horizontal section through the buccal mass and the inferior buccal ganglion.

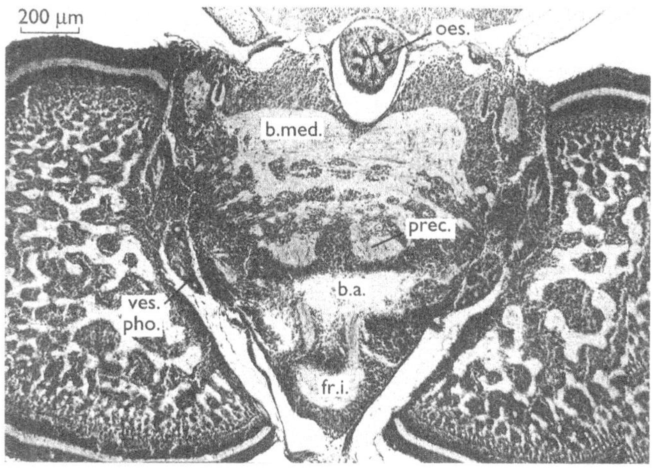

Fig. 18.6. *Gonatus fabricii* (12 mm ML). Horizontal section through the central nervous system and optic lobes. A compact group of photosensitive vesicles lies medial to each optic lobe.

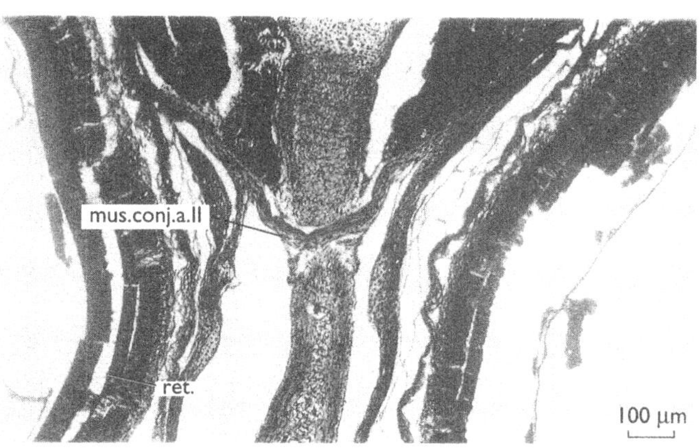

Fig. 18.5. *Gonatus fabricii* (12 mm ML). Horizontal section to show the anterior conjunctive muscle II, which can be followed from one eye to the other.

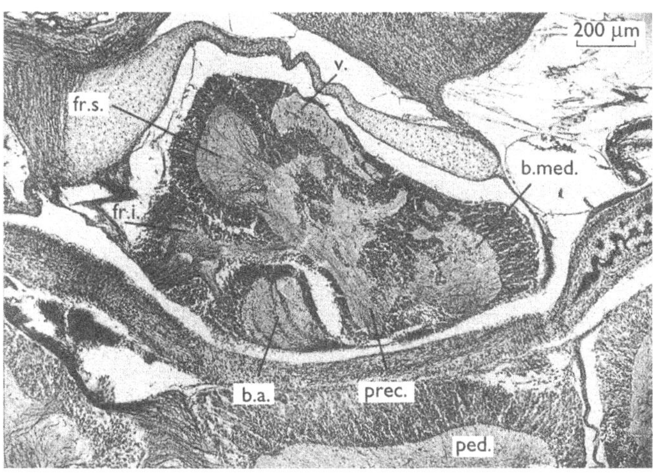

Fig. 18.7. *Gonatus fabricii* (20 mm ML). Sagittal section through the supraoesophageal nervous system.

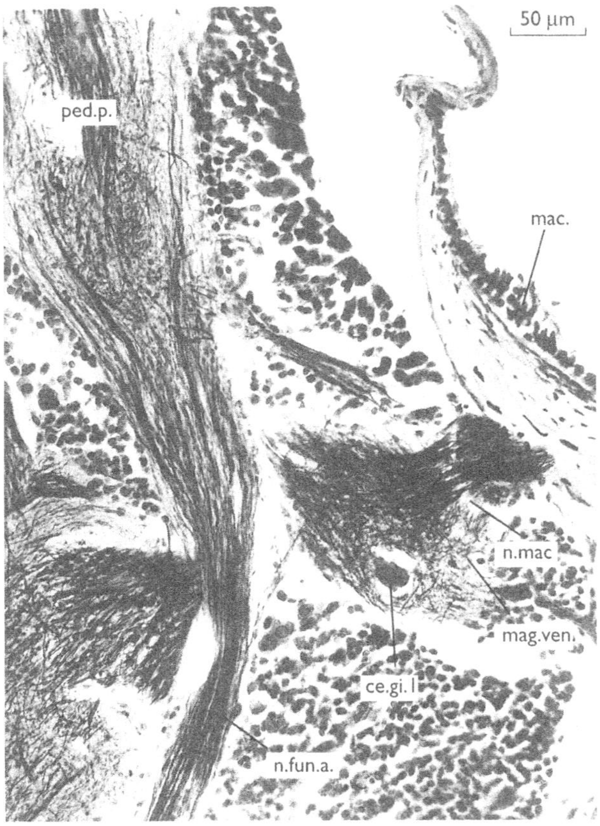

Fig. 18.8. *Gonatus fabricii* (20 mm ML). Sagittal section showing the first-order giant cell in the ventral magnocellular lobe.

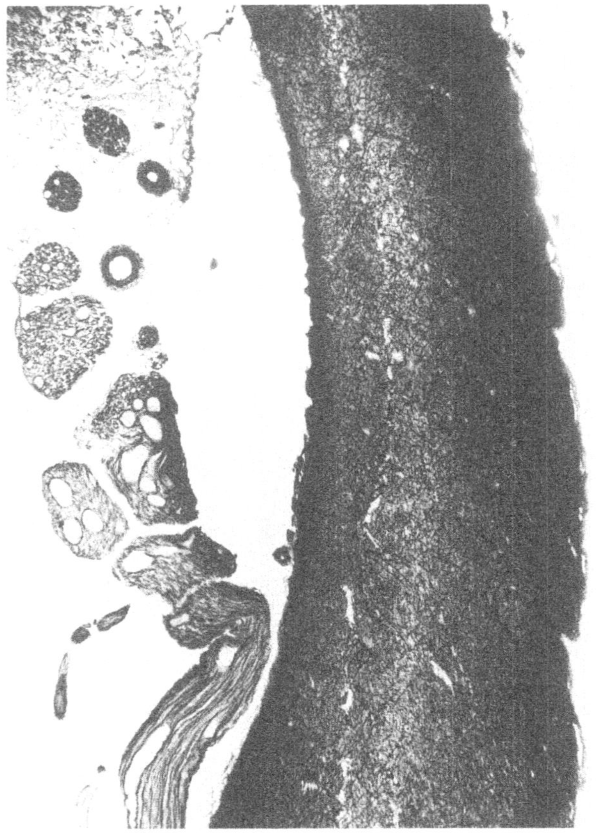

Fig. 18.10. *Gonatus* (110 mm ML). Stellar nerves and their giant nerve fibres.

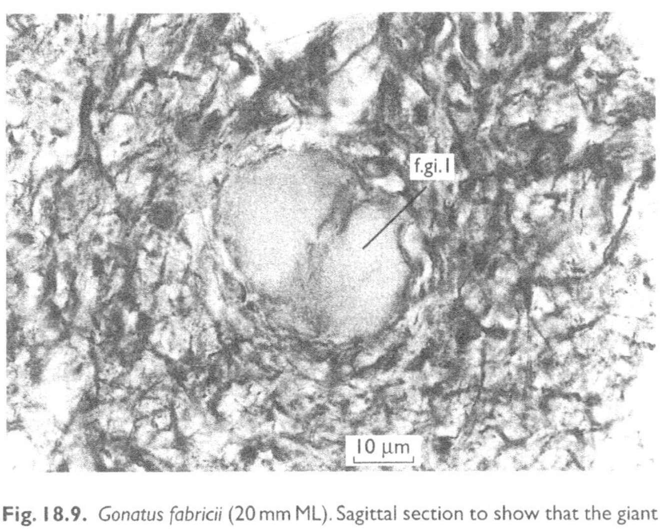

Fig. 18.9. *Gonatus fabricii* (20 mm ML). Sagittal section to show that the giant axons cross, but do not fuse, in the palliovisceral lobe.

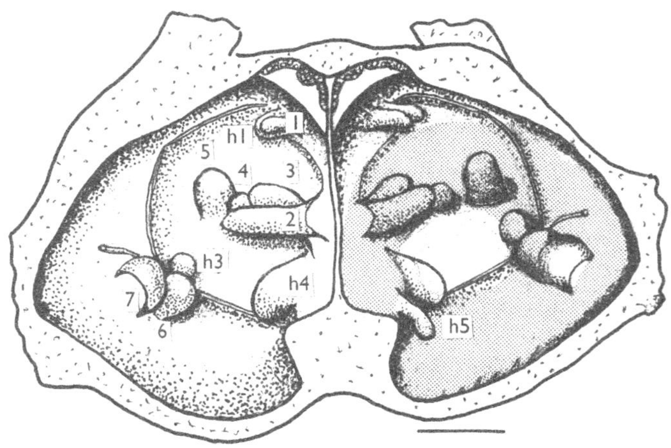

Fig. 18.11. *Gonatus steenstrupi* (85 mm ML). The statocyst seen from above (scale bar 1 mm).

non-buoyant squids (Maddock and Young 1984), this specimen had dense head, mantle, and fin muscles and a giant-fibre system, and the organization of its statocyst suggested that it was suited to measure rapid angular acceleration in all planes. Another species, *G. fabricii*, was found to be near neutral buoyancy and it is known to accumulate large volumes of oil in its digestive gland (Clarke *et al.* 1979, Kristensen 1984).

18.2 *Eogonatus*

Eogonatus tinro is found from the northern slope of the Bering Sea to the Sea of Okhotsk, the Southern Kurile Islands, and off British Columbia. It is lower epi-, meso-, and bathypelagic in habit. The adults do not ascend to the surface. Nesis (1972) described the only species of the family.

18.3 *Gonatopsis*

The members of this genus live in sub-Arctic waters, mostly in the north-west Pacific Ocean, although one species has a wider distribution across and into the north-eastern Pacific Ocean (Okutani *et al.* 1988, Nesis 1989). Their vertical distribution is little known except for *Gonatopsis borealis*, which is caught from near surface waters to a depth of 400 m at night and from 400 to 700 m in daylight; thus it undergoes diel migrations (Roper and Young 1975, Kubodera *et al.* 1983).

The genus was established for a single specimen possessing eight arms but lacking tentacles, *Gonatopsis octopedatus*, caught off the coast of Japan (Sasaki 1920). In the best-known member of the genus, *G. borealis*, the tentacles undergo abrupt degeneration at a mantle length of 10–12 mm (R.E. Young 1972b, Kubodera and Okutani 1981). A maximum size of 620 mm ML has been recorded for *G. makko*, captured in the Okhotsk Sea, making it the largest of all the gonatids (Nesis 1993b).

Spent females (240–390 mm ML) of *Gonatopsis octopedatus*, caught between 0 and 50 m, were gelatinous, their armature lost and oviducts empty; mature males of this species did not show such degeneration (Nesis 1993c, 1997).

18.4 *Berryteuthis*

The two species of this genus are found in sub-Arctic Pacific waters from Japan to Oregon and north to the Bering Sea (Roper and Young 1975, Roper *et al.* 1984, Kubodera and Jefferts 1984, Clarke 1986). *Berryteuthis magister* is a bathybenthic squid of the continental slope found between 200m and 500 m in depth in the Sea of Japan, off north-eastern Japan through the Kurile and Aleutian Islands to the Bering Sea, and off north-eastern America (Kubodera *et al.* 1983).

These squids have large rhomboidal fins at least half of the mantle length. Females, all of the species *Berryteuthis magister* (up to 267 mm ML), were caught in bottom trawls at 320 m. These animals were at an advanced stage of maturity and had bundles of sperm ropes implanted around the base of their gills. Some 5000 ovoid eggs, ranging from 1.88×1.56 mm to 5.94×4.53 mm, were present in one female, and were arranged in clusters. Copulation probably occurs in autumn or winter, and spawning in spring or summer takes place close to the sea floor (Nazumi *et al.* 1979). Recently hatched young are pelagic until reaching a mantle length of 17 mm or more; when 50–70 mm in ML they descend to live near the bottom (Okutani 1988). A maximum size of 380 mm ML has been recorded (Nesis 1987).

Observations of *Berryteuthis magister* (130–140 mm ML), from a manned submersible between 100 m and 900 m, recorded it as some 5–10 m off the bottom during the day but far above the bottom at night. They occurred mostly in groups of 5–7 squids but were also seen singly, 2–3 together, or in groups of up to 30 or more. All of the squids were oriented in the same direction and moved together, either by jet propulsion or, more usually, by fin movement (at average speeds of 0.6–0.8 m/s and 0.2–0.3 m/s respectively), thus these are schooling squids. They were mostly concentrated where the bottom relief was complex. The pelagic juveniles fed actively at night, but large squids fed more frequently in daytime (Nesis 1997).

19 Brachioteuthidae

Brachioteuthis is the only genus of the family. This small squid (Fig. 19.1) has a wide distribution as it is found in the North Atlantic, central Pacific, and Southern Oceans, and in the Mediterranean Sea (Roper *et al.* 1984, Roper and Sweeney 1992). In the Atlantic the majority have been captured from the surface to a depth of 300 m, most being juveniles (Clarke and Lu 1974, 1975, Lu and Clarke 1975a, 1975b), and others have been caught in the Gulf Stream (Dawe and Stephen 1988). Young animals have been caught in the Pacific from near the surface to about 150 m depth, most being in the upper 50 m, and these as well as larger squids show some indication of diel migration (R.E. Young 1978).

19.1 Brachioteuthis

Brachioteuthis is the most common squid to be captured near the surface in Antarctic waters and it was from here that the largest specimen yet recorded (138 mm ML) was taken (Filippova 1972).

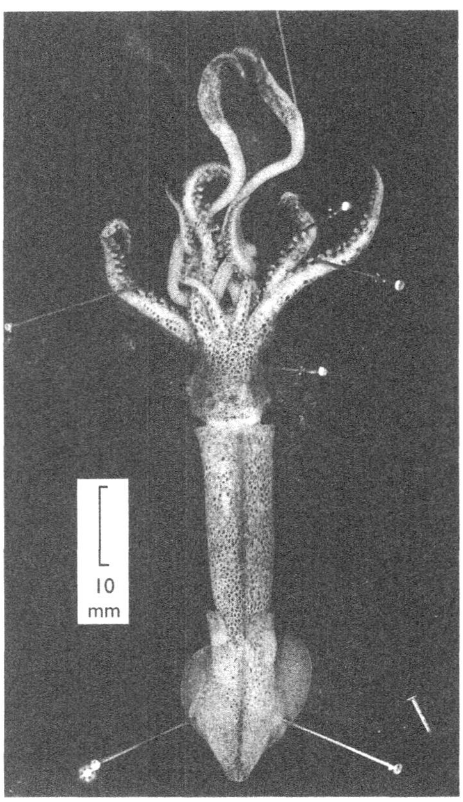

Fig. 19.1. *Brachioteuthis riisei* (52 mm ML). Specimen captured in the North Atlantic Ocean (kindly provided by M.R. Clarke).

19.1.1 Life history

A gravid female (74 mm ML) was caught in an open tow at a depth of 1125 m; it had a sperm mass embedded in the buccal membrane (R.E. Young 1978). *Brachioteuthis* is the only squid yet known to spawn individual eggs into the plankton outside the family Enoploteuithidae. The eggs are elongate and ovoid in shape (1.2 × 0.7 mm), and are not surrounded by a gelatinous envelope.

The hatchlings are readily distinguished by a long, slender neck, which is capable of extension to almost 3 mm, and of contraction, in an animal with a mantle length of 9.1 mm. The neck appears to be a muscular, fluid-filled tube, continuous with a large fluid-filled sac, or reservoir, within the mantle. When the squid is alive the neck can be quickly extended or retracted by the action of the muscles of the neck and reservoir. The tentacles are elongated and contrast markedly with the arms, which are tiny at this stage. The lower beak, but not the upper beak, has prominent teeth on the rostrum. Each eye has an ocular appendage on the anterior end of the eye, and this is a distinct, tapering extension of the silvery covering of the eye (R.E. Young *et al.* 1985). The optimum position of the animal to reduce the eye's silhouette would be with the apex of the rostrum downwards (R.E. Young 1975d).

Histological sections of a juvenile show the muscles and cartilage to be dense (Fig. 19.2). The ratio of the animal weight in seawater to its weight in air as a percentage was 2.34%, suggesting that it swims actively to prevent sinking (Clarke *et al.* 1979). The condition

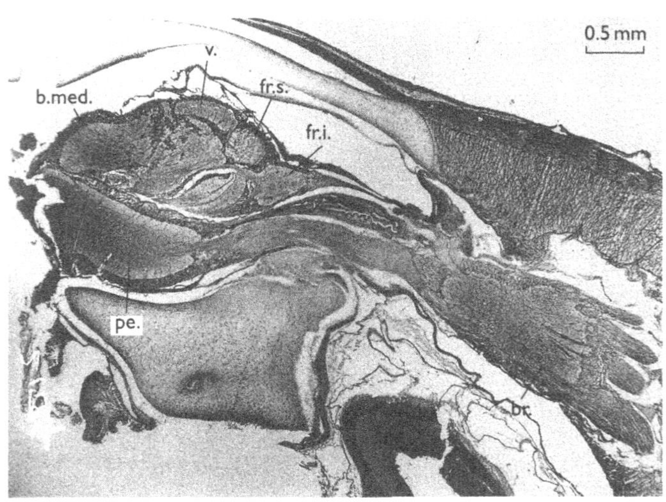

Fig. 19.2. *Brachioteuthis riisei* (52 mm ML). A sagittal section of the central nervous system.

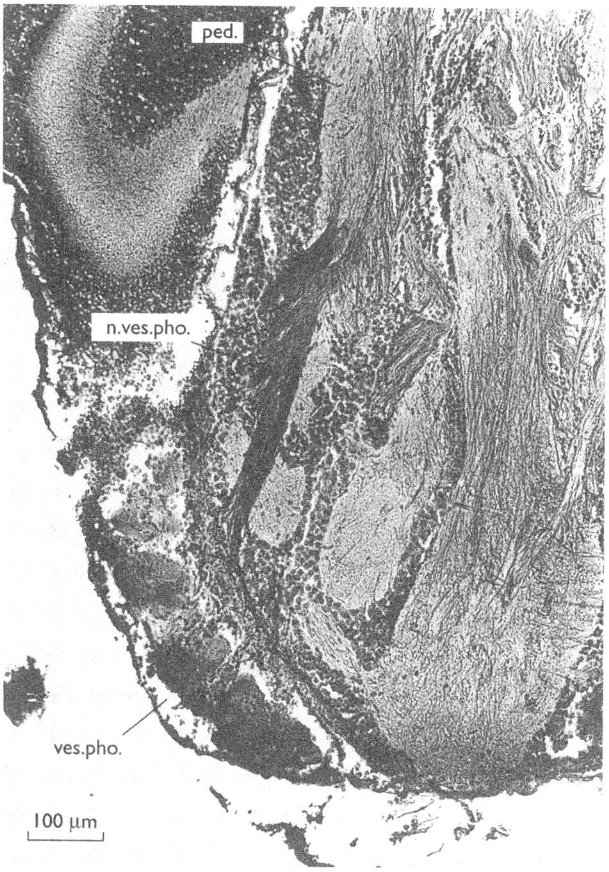

Fig. 19.3. *Brachioteuthis riisei* (52 mm ML). A sagittal section showing the photosensitive vesicles and peduncle lobe.

of the statocyst also indicates that this squid is not buoyant (Maddock and Young 1984).

19.1.2 Receptors and central nervous system

The retina of the eye has a uniform population of rhabdoms, 240 μm in length.

The cephalic cartilage is thick and hyaline in form. The central nervous system is elongated, with the brachial lobes widely separated from the rest (Fig. 19.2). The inferior frontal lobe is large relative to the superior frontal and vertical lobes. The magnocellular and palliovisceral lobes are missing from the sections, so we have been unable to make any measurements. No giant fibres were seen in the sections of the mantle connective and stellate ganglion.

The extraocular photoreceptor organ consists of a large mass of photosensitive vesicles (R.E. Young 1978). The vesicles extend down behind and below the optic tract as a series of partly separated sacs (Fig. 19.3). Their nerves run to the peduncle lobes, where the fibres interweave.

The statocyst of the juvenile is small but relatively wide in relation to the mantle length, and is nearly equal in all directions. Anticristae 6 and 7 touch but do not fuse, and define a distinct horizontal canal. The anticristae of the median row are unusual as anticristae 2 is smaller than 3, and anticrista 5 is broad. There is no anticrista 1 and no hamulus at the first turn. The other hamuli are small and equal in size (Young 1989). Multidiscriminant analysis of the statocyst places this squid amongst the non-buoyant forms (Maddock and Young 1984).

20 Lepidoteuthidae and Pholidoteuthidae

20.1 Lepidoteuthidae

This family was created by Pfeffer (1912) for *Lepidoteuthis grimaldii*. Later *Pholidoteuthis* and *Tetronychoteuthis* were added (Roper *et al.* 1969), but there are numerous and considerable differences between *Lepidoteuthis* and these last two other genera (Adam 1950, Clarke and Maul 1962). Clarke (1980, 1992a) proposed that the family should again be restricted to include only *Lepidoteuthis grimaldii*, and this position is retained in the present study.

The features characteristic of the family, and of the genus and species include the absence of tentacles in the adult (Clarke 1992a). There are rhomboidal dermal cushions (scales) on the mantle, except posteroventrally between the fins. The fins have an ovoid form with the long axis along that of the body, and they lack scales. The largest specimen recorded had a mantle length of 100 mm (Nesis 1987). The paralarval stage of *Lepidoteuthis grimaldii* (9.7 mm ML) lacks the characteristic dermal structures but slightly larger specimens do possess them (R.E. Young 1991). Tentacles are present at this stage and they possess peculiar clubs, which are similar to those of the paralarva of Octopoteuthidae; this is one feature that suggests a relationship between the two families (R.E. Young *et al.* 1998b).

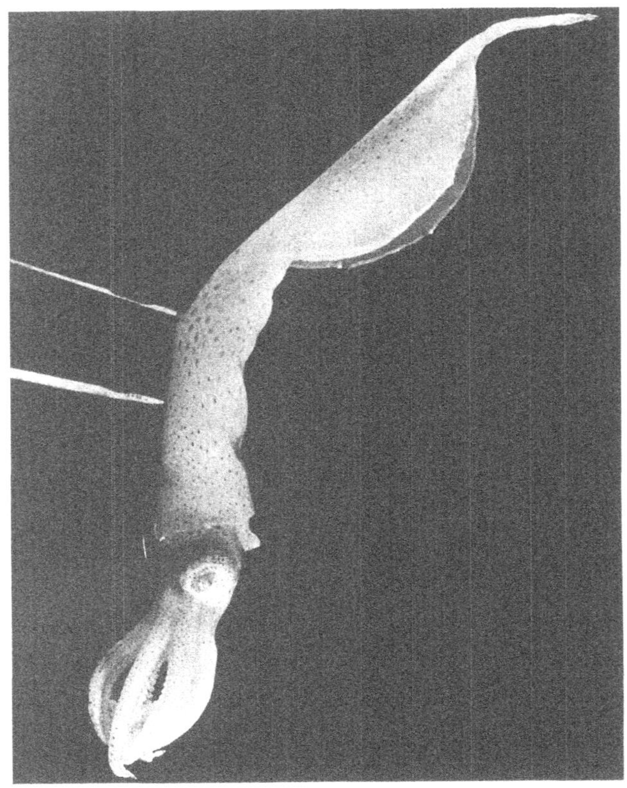

Fig. 20.1. *Lepidoteuthis grimaldii* photographed while still alive after capture. The tips of the forceps are approximately 7 mm apart (Clarke 1964).

20.1.1 Lepidoteuthis

Lepidoteuthis (Fig. 20.1) has been found in the stomachs of sperm whales caught in the Atlantic Ocean between 40°N and 40°S, the southern Indian Ocean (Clarke 1980), and Japan (Okutani et al. 1976), and of fishes caught in the south-west Pacific Ocean (Rancurel 1970). Young animals have been captured in opening/closing nets in the North Atlantic at depths of 23–50 m and 100–270 m (Clarke 1964, Clarke and Lu 1974, Lu and Clarke 1975b) and in Australian waters (Zeidler 1987).

The most striking feature of *Lepidoteuthis grimaldii* is the presence of many relatively large 'scales'. Roper and Lu (1990) investigated these structures and suggested that they are better termed dermal cushions. They are diamond- to hexagonal-shaped, 1.5 × 2 mm in a specimen of 180 mm ML, and 10 × 12 mm in a squid of 755 mm ML. The cushions are diagonally arranged and attached posteriorly so that their configuration permits a smooth passage for the body when moving tail first through the water. The cushions are somewhat gelatinous and if pressed will rebound to their original form. The internal structure is reticulate with small chambers of irregular size and shape, and thin membranous walls. A fragile epidermal layer envelops each cushion and continues as the outer epithelium of the mantle. Roper and Lu (1990) suggest that the primary function of the dermal cushions is as a buoyancy mechanism in conjunction with the reticulate tissue of the mantle. It is possible that the cushions may act as vortex generators (Bone 1972).

A juvenile (60 mm ML) was observed in a shipboard aquarium (Fig. 20.2) (Clarke 1964). The mantle was translucent, apart from orange chromatophores on the mantle and head that gave a faint colour. The digestive gland, visible through the mantle, was bright red. The squid continually attempted to move downwards by rapid and powerful beats of the large and highly mobile fins. The mantle wall moved little while the funnel moved gently in and out of the cavity. These latter actions were presumably for respiration as they did not propel the animal. The funnel is large but has 'paper thin walls and a wide external opening' (Clarke and Maul 1962).

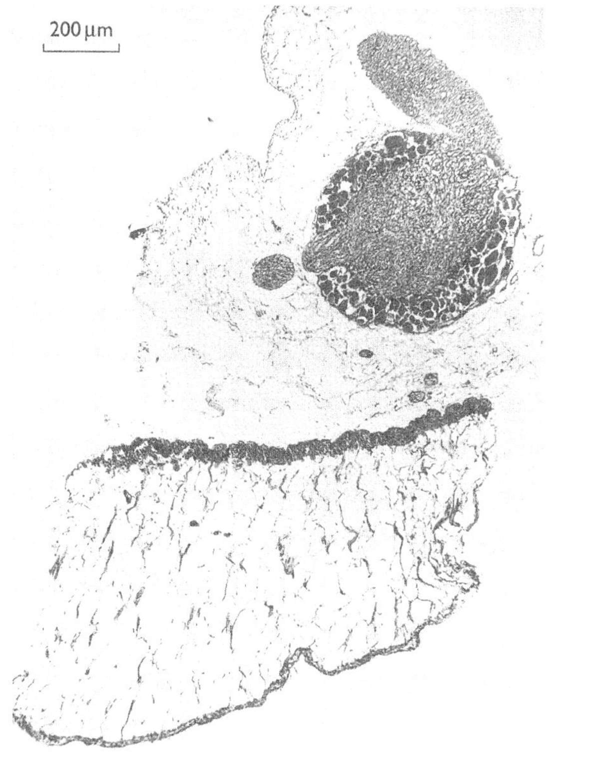

Fig. 20.2. *Lepidoteuthis grimaldii.* An oblique transverse section through mantle and stellate ganglion. Note the reticulate tissue in the mantle wall (haematoxylin and eosin).

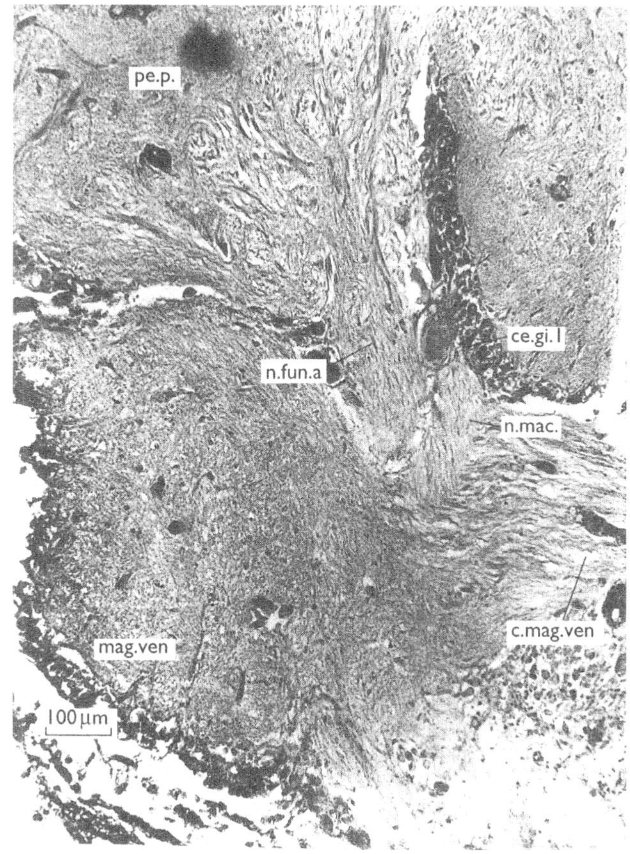

Fig. 20.3. *Pholidoteuthis adami* (460 mm ML). An oblique transverse section showing magnocellular lobe and first-order giant cell (haematoxylin and eosin).

Histological sections of the mantle wall and stellate ganglion reveal no especially large nerve fibres in the mantle connective, stellate ganglion, or stellar nerves (Fig. 20.3). The mantle has thin strands of longitudinal muscles, circular muscles, a thick layer of reticulate tissue with scattered muscle fibres, and a layer of circular muscles. Clarke *et al.* (1979) considered that the mantle tissues contained ammonium ions to provide buoyancy, and this was confirmed in *Lepidoteuthis grimaldii* by chemical analyses (Lipinski and Turoboyski 1983).

20.2 Pholidoteuthidae

The family includes one genus and its characteristic features include small dermal cushions on the mantle, except on the ventral surface between the fins. The tentacles are moderately long with slightly expanded clubs bearing tetraserial suckers. The suckers are compressed and have a characteristic slit-like opening with prominent peg-like teeth on the infundibulum (Clarke 1992b).

20.2.1 Pholidoteuthis

This squid was first caught in the Pacific Ocean (Adam 1950), but has since been captured in the Gulf of Mexico, and found in the stomachs of sperm whales caught in the Atlantic Ocean (Clarke 1980), and of fishes in the south-west Pacific Ocean (Rancurel 1970). It has been captured at depths of 20–230 m, and can occur in vast schools in the upper part of the Gulf of Mexico where it appears at the surface several hours after sunset (Voss 1956).

The small, soft dermal cushions on the mantle of Pholidoteuthis adami are irregularly rounded with a pentagonal-like outline (0.5 mm diameter); the cushions have continuous thin-walled, irregularly shaped chambers which appear vacuolated. The mantle is thick and spongy and, as in the dermal cushions, reticulate tissue is present (Roper and Lu 1990). The presence of reticulate tissue may be an indication of neutral buoyancy in this squid.

Central nervous system and receptors

In the central nervous system the median inferior frontal and vertical lobes form a relatively large proportion of the supraoesophageal mass. The dorsal basal lobe consists of numerous islands of neuropil and small cells surrounding the origin of the nerve to the optic gland. The small cells are continuous with a large mass of paravertical tissue, which is penetrated by neuropil, and so establishes the neural origin of this tissue. The suboesophageal lobes are elongated with the anterior mass lying forward and separated from the rest. The ventral magnocellular lobe is relatively small. The first-order giant cells are small (Fig. 20.3). Giant fibres cross in the palliovisceral lobe and a large one enters the mantle connective. The fin lobe of the brain is relatively large (Fig. 20.4).

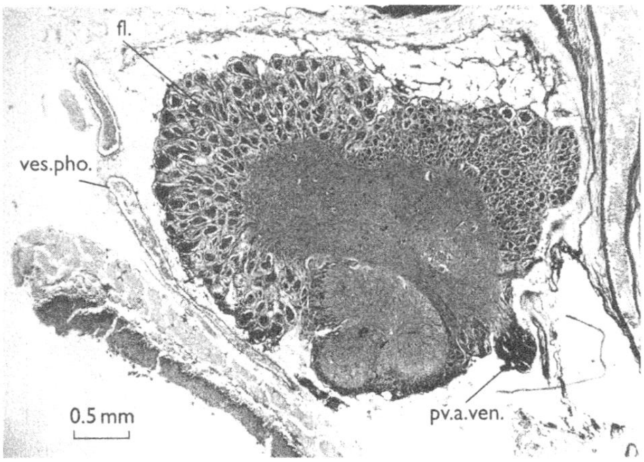

Fig. 20.4. *Pholidoteuthis adami* (460 mm ML) An oblique transverse section of the pallioviscera lobe and photosensitive vesicles (haematoxylin and eosin).

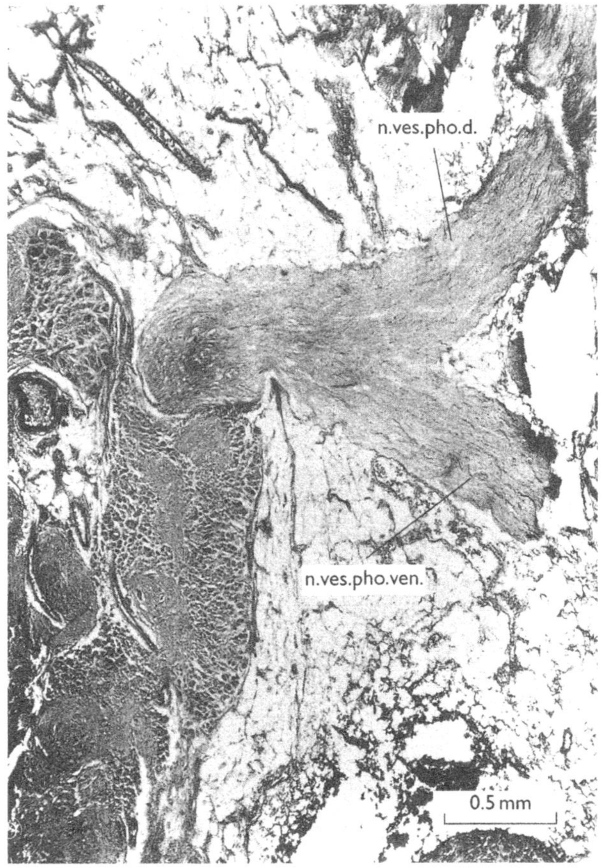

Fig. 20.5. *Pholidoteuthis adami* (460 mm ML). Oblique transverse section to show the nerves of the photosensitive vesicles (haematoxylin and eosin).

The extraocular photoreceptor organ consists of photosensitive vesicles running continuously from far dorsally above the optic lobe, around the fin lobe, to below and lateral to the palliovisceral lobe (Fig. 20.4). The vesicles are large open sacs, wider above, and narrow at the sides. The photosensitive vesicle nerves do not enter separately from above and below, but together dorsally (Fig. 20.5), and they then proceed to divide and their fibres interweave as they pass ventrally through the peduncle lobe. The photosensitive vesicles may respond to downwelling light, as this squid has no photophores.

The statocyst is short, broad, and deep. The horizontal canal is well marked. There is a deep channel anteriorly. The hamuli and anticristae have blunt, rounded tips. Anticristae 6 and 7 are in contact but are not fused; anticrista 7 has an unusual triangular form (Young 1989). Multidiscriminant analysis of the statocyst places this squid amongst neutrally buoyant species (Maddock and Young 1984), and this has support from the presence of reticulate tissue in the mantle and dermal cushions.

21 Octopoteuthidae

The squids of this family occur in the Atlantic, Pacific, Indian, and Southern Oceans and *Octopoteuthis* also in the Mediterranean Sea (Clarke 1980, 1983, Roper *et al.* 1984, Roper *et al.* 1985, Stephen 1985). The young are found in surface waters and the adults inhabit the deep sea.

Both genera lack tentacles for the greater part of their lives and when *Octopoteuthis* was first described the characters of both octopods and squids seemed apparent in the animal, hence the name. The two genera have very large fins and are neutrally buoyant.

21.1 *Octopoteuthis*

This squid (Fig. 21.1) is found worldwide from 56 °N to 50 °S, in subtropical and temperate waters. Young specimens of up to 20 mm ML occur in near-surface waters, primarily from 0 to 300 m depth (Stephen 1985, Stephen and Jefferts 1992) and adults are meso- to bathypelagic. They have been captured at various depths down to 800 m (Roper and Young 1975, Lu and Clarke 1975b). One species, *O. nielseni*, undergoes diel vertical migrations in Hawaiian waters (R.E. Young 1978), where many become prey to the sperm whale (Clarke 1980).

21.1.1 *Morphology*

Reticulate tissue is present in many parts of this squid including the cartilage (Fig. 21.2), and the thick mantle wall, which has only narrow layers of muscle (Fig. 21.3). The mantle serves mainly for respiration and produces little propulsive effect, and furthermore the wall of the funnel is thin. The fins are very large and muscular (R.E. Young and Roper 1977), and provide the main locomotor force. *Octopoteuthis* is neutrally buoyant and concentrations of 280–500 mmol/L ammonium

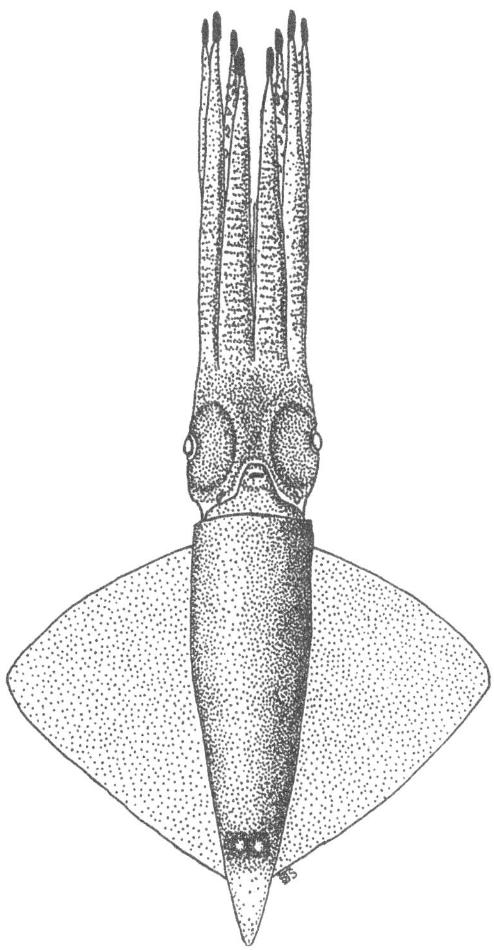

Fig. 21.1. *Octopoteuthis.* A ventral view (original drawing by S.J. Stephen).

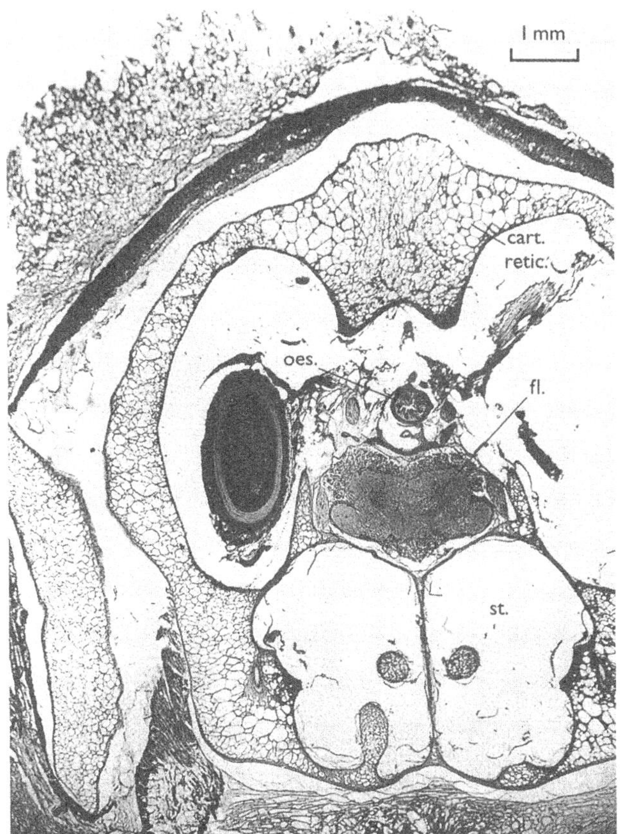

Fig. 21.2. *Octopoteuthis danae* (61 mm ML). A transverse section through the head showing the prominent fin lobe of the brain and the extensive reticulation of the cartilage. Figs 21.2–21.5 are all of this specimen.

have been recorded in the mantle tissue (Clarke *et al.* 1979, Lipinski and Turoboyski 1983).

The buccal mass and the radular support system has much reticulate tissue, with small bundles of muscle fibres (Fig. 21.4). The salivary papilla has isolated glandular patches, some of which open ventrally. The posterior salivary glands are well developed and the single duct opens on the anterior face of the prominent salivary papilla. The anterior salivary glands lie in the lateral buccal palps. The long, narrow oesophagus leads to a large pointed stomach.

Two types of photophores are present (Herring *et al.* 1992a). The first has a complex capillary network, with elastic walls and a thick reflector and is found on the head, body, and arms; the second type lacks a capillary core but has a dense matrix containing paracrystalline assemblies and occurs on the tips of the arms.

21.1.2 Receptors and central nervous system

The eyes are relatively large and prominent. The vacuolation of the cartilage of the head makes it difficult to discern the extent of the trochlear cartilages. Muscle fibres cross in the midline but it is not clear whether they belong to one conjunctive muscle or two.

The optic lobes form 288% of the total brain volume (Maddock and Young 1987). The granule layers of the optic lobes are thin and uniform all over (Fig. 21.5).

The central nervous system is elongated, with the anterior sub-oesophageal mass far in front of the rest of the brain. The fourth brachial nerves are very small, reflecting the condition of the tentacles. The remaining brachial nerves each include a few large fibres.

The magnocellular lobes form 8.6% of the total brain volume, but no single first-order giant cells or fibres were seen. There are large cells in the ventral magnocellular lobe but all are within the main rows, none is isolated in the neuropil. In the palliovisceral lobe several large fibres cross in the commissure. There are no outstandingly large fibres in the mantle connective. The stellate ganglion and nerves contain large cells and fibres but none is of outstanding size. The fin lobes are relatively large (Fig. 21.2) and occupy 8.8% of the total volume of the brain, and form a larger portion of the brain than those of any other oegopsin squid measured except *Discoteuthis* (Table 2.6). The posterior chromatophore lobes form 1.5% and the vertical lobe system 15.4% of the total brain volume.

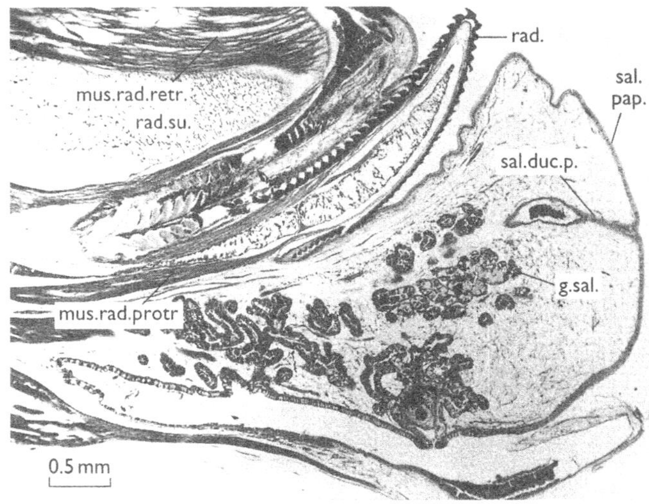

Fig. 21.4. *Octopoteuthis danae.* A sagittal section through the buccal mass to show the radula and the large, salivary papilla with scattered glandular patches in otherwise amorphous tissue.

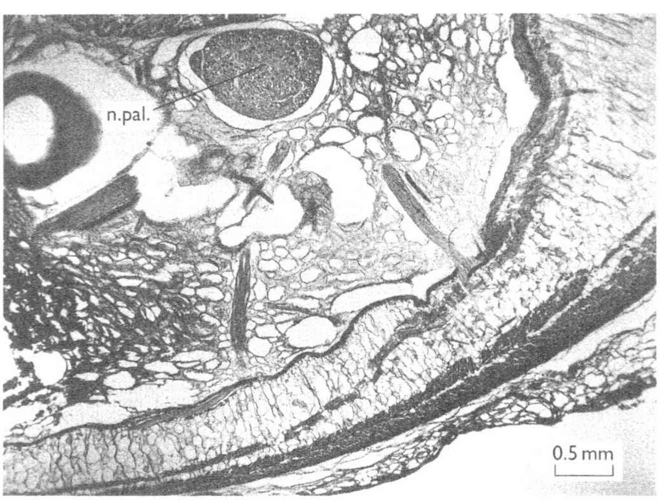

Fig. 21.3. *Octopoteuthis danae.* A transverse section through the mantle to show the reticulate tissues crossed by strands of muscle fibres.

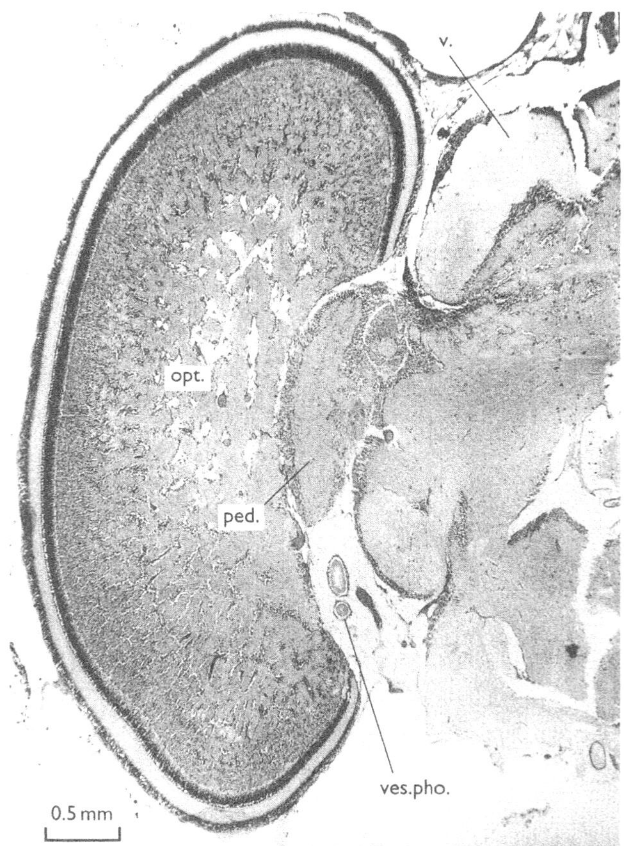

Fig. 21.5. *Octopoteuthis danae.* A transverse section through the central nervous system showing the optic lobe and the vertical lobe system.

The extraocular photoreceptor organs consist of central and ventral sets of photosensitive vesicles on either side, joined by a long narrow vesicle; together they form 3.7% of the total brain volume. In *Octopoteuthis nielseni* the central set comprises many small lobules around the optic tract; the ventral vesicles are fewer and larger and lie packed together in a special chamber in the orbital cartilage (R.E. Young 1978). The nerve fibres from the vesicles enter the basal peduncle lobe. *O. nielseni* is capable of counter-shading and the photosensitive vesicles probably serve to control this as in other midwater cephalopods (R.E. Young and Roper 1977).

In the statocyst the anticristae, although large and rounded, do not fill the whole cavity. Anticristae 6 and 7 are close to each other but do not touch; the horizontal canal is therefore demarcated for less than one-third of its circumference (Young 1989). These conditions are similar to those found in other cephalopods in which the fins are the main organ of locomotion, such as *Sepia* and *Discoteuthis*, but there is no hamulus at the first turn. The large anticristae suggest a capa-city for rapid turns but nevertheless the dimensions of the statocyst fall into the same category as that of other buoyant squids (Maddock and Young 1984).

21.2 *Taningia*

This genus is cosmopolitan, primarily tropical to subtropical (Roper *et al.* 1985, Nesis 1987), and is known from some regions of all oceans (Roper and Vecchione 1993). In the Atlantic Ocean it has been caught in depths of 25–300 m (Clarke and Lu 1974, Lu and Clarke 1975b).

Taningia danae is considered to be monotypic (Vecchione and Roper 1992). The animal is a rich maroon colour with pigment almost evenly distributed, although slightly darker dorsally (Roper and Vecchione 1993). The body is thick, soft, and gelatinous. One specimen of 640 mm ML had a mantle 60 mm thick, and its fin was 50 mm thick. Much shrinkage takes place; one animal weighed 25.4 kg when fresh but only 8.6 kg after a few months in preservative (Clarke 1967). The tissues have a concentration of 755 mmol/L ammonium, providing lift for neutral buoyancy (Clarke *et al.* 1979, Lipinski and Turoboyski 1983).

There is a photophore on each arm II, and it is superficially similar to that of *Octopoteuthis* although the surface can be occluded by densely pigmented lateral lips (Herring *et al.* 1992a). In a female of 1400 mm ML the arm tip photophore reaches a length of 60 mm, and has a thick layer of photogenic tissue overlying the reflector layer (Clarke 1967). The photogenic core is composed largely of densely staining acellular material through which strands of cytoplasm run (Herring *et al.* 1992a). The axial nerve extends to the tip of the arm and supplies branches to the photogenic layer and related muscles (Clarke 1967). When a hand was moved slowly across a shipboard aquarium containing a living squid the action resulted in bright, simultaneous flashes of green-blue light from the two arm tip photophores, lasting a fraction of a second; occasionally the photophores glowed with fluctuating intensity for 1–7 s (Roper and Vecchione 1993). The luminescent emission has a maximum at 475–480 nm (Herring *et al.* 1992a). Photophores, located on each side of the intestine, glowed continuously for periods in excess of 15 min and illuminated a broad circular area in the mantle cavity, in an animal of 60 mm ML (Roper and Vecchione 1993). The function of these photophores is for counter-illumination (R.E. Young and Roper 1976, 1977).

Spermatophores have been found embedded near the base of the aboral side of the arms of females, and also on each side of the mantle close to the anterior edge of the fin (Clarke 1967, Okutani 1974). The largest recorded specimen is a mature female 1600 mm ML, trawled at 260 m (Roper and Vecchione 1993). Clarke (1967) examined a female of 1400 mm ML and found more than 250 000 eggs, all of a similar size, in the ovary. Spawning may take place near the bottom as young animals are caught in nets between 100 and 300 m (Clarke and Lu 1974).

22 Thysanoteuthidae

The large size and rhomboidal fins of this oceanic, epipelagic squid make it easy to observe in the sea. It has a wide distribution in the Pacific and Indian Oceans and is common in the Sea of Japan, and is found in the Atlantic Ocean, and the Mediterranean Sea (Barnard 1934, Rees and Maul 1956, Voss and Erdman 1959, Bruggen 1962, Okutani 1966, Clarke 1969, Filippova 1970, Clarke and Lu 1974, Yamamoto and Okutani 1975, Morales 1981, Okiyama 1993). Its vertical distribution is poorly known as it is infrequently caught, the exception being captures in opening/closing nets at 25–85 m in daylight (Clarke 1969, Roper and Young 1975).

22.1 *Thysanoteuthis*

The family includes only *Thysanoteuthis* (Sweeney and Roper 1998). *T. rhombus* can reach a mantle length of 800–900 mm (Nigmatullin *et al.* 1991). Its mantle is muscular (see Fig. 22.3) and the fins, which extend the entire length of the mantle, are the major agents in locomotion. When caught in set nets these squids move by 'fluttering the fins' (Nishimura 1966). G.L. Voss (pers. comm.) said after watching from the deck of a boat that 'with their fins spread widely they soar like no other squid'. They show low activity and swim with undulating movements of the fins, but are capable of making powerful jet propulsive movements of short duration, as well as sharp, jerky movements when in danger (Nigmatullin *et al.* 1991). Schools consisting of about 20 of these squids have been observed close inshore at the surface of the water in Madeira (Rees and Maul 1956).

Macroplanktonic and micronektonic organisms, never more than 10% of the absolute length of the squid, were found in the stomachs of several specimens (Nigmatullin *et al.* 1991). An animal of 790 mm ML was washed ashore and its stomach contained the remains of a mantis shrimp, *Squilla armata*, a fish, and a squid, *Ornithoteuthis volitans* (Roeleveld and Pheiffer 1987).

22.1.1 Life history

In the coastal waters of Japan they have been observed 'swimming side by side, keeping a close relation with each other, as if... controlled by a complicated psychophysiological mechanism.' (Nishimura 1966). Each such pair consists of a male and a female of about the same size. Maturation begins at 400–500 mm ML in males and at 500–600 mm ML in females, when approximately 6–9 months of age. The reproductive system of both sexes has been described. The male has a hectocotylus which differs markedly from that of other squids, and has two oval enlargements which are probably glandular. The Needham's sac is long and narrow and contains 10–15 large spermatophores; there is also a long, thin penis (terminal organ). During copulation the male holds the spermatophores by the unarmed suckers of the hectocotylus and the blade shaped appendage of the dorsal protective membrane. After mating the female is often found with marks around the buccal conus and 3–6 attached spermatophores (Nigmatullin *et al.* 1991). The female has small oviducts, set in a tight zigzag, in comparison with the very large ovary and oviducal glands. This squid shows a pattern of repeated partial spawning (Arkhipkin *et al.* 1983a, b).

Adults have been found spawning at the surface during the warmest period of the year in the most productive areas of the oceans (Sabirov *et al.* 1987). Egg masses float and two have been found off the shores of Japan at a depth of about 2 m (Fig. 22.1) (Misaki and Okutani 1976, Suzuki *et al.* 1979, Sabirov *et al.* 1987). One sausage-shaped mass was 1000 mm long and 150 mm in diameter, it was translucent and gelatinous with the eggs arranged in a single layer in a continuous spiral, with two rows of eggs about 10 mm apart (Misaki and Okutani 1976). Shortly before hatching the egg mass loses its buoyancy and sinks (Sabirov *et al.* 1987).

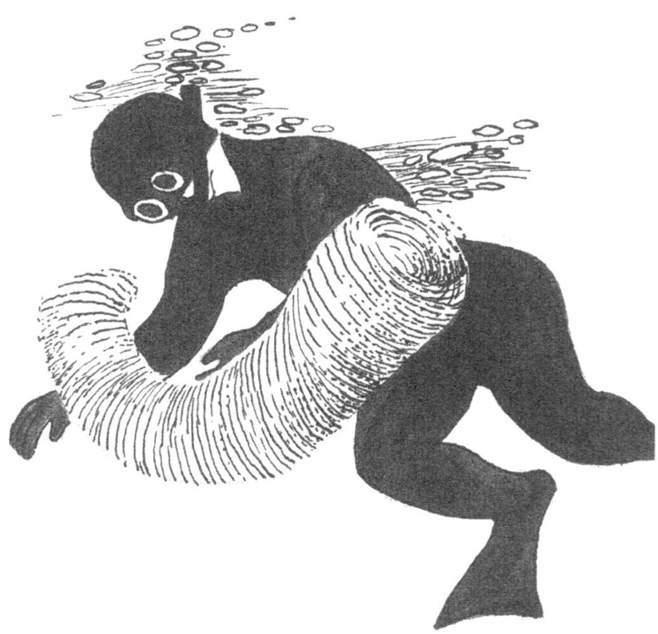

Fig. 22.1. *Thysanoteuthis rhombus*. A floating egg mass (130 cm long and 20 cm across) photographed while held by a diver in Japanese waters (Suzuki *et al.* 1979) (drawing by Raymonde Parsons).

Young, hatched from eggs maintained in an aquarium, were 2 mm ML at hatching. Each had a dome-shaped mantle and tiny fins, with densely organized mantle chromatophores which changed the little squid from pink to brown (Misaki and Okutani 1976). Recently hatched young, caught in the south-western waters of Japan, have been described and the early stages of growth illustrated (Yamamoto and Okutani 1975, Sabirov *et al.* 1987, Stephen 1992). This squid lives for 2 years (Nazumi 1975, Nigmatullin *et al.* 1991).

22.1.2 Central nervous system and receptors

Figure 22.2 shows the central nervous system. The vertical lobe system forms a prominent portion of the supraoesophageal mass, and is well developed.

The extraocular photoreceptor organ consists of some six small photosensitive vesicles on each side, lying at the base of a mass of paravertical tissue (Fig. 22.2). The peduncle lobes do not have any conspicuous bundles of darkly staining fibres. A portion of the retina of the eye showed rhabdoms of 130 mm.

The mantle connective has a giant fibre, seen in Fig. 22.3 indicating the presence of a giant-fibre system. Other giant fibres were found to be weakly stained in the tissues available.

The statocyst has five prominent hamuli. The anticristae of the central row have 'square' tops leaving wide channels for flow across the anterior transverse crista and longitudinal crista. The horizontal endolymph channel is unusual as anticrista 7 is elongated, front and back, and curved but does not touch anticrista 6, thus leaving a rather wide canal (Young 1989). Multiple discriminant analysis of the measurements of the statocyst place this squid amongst the non-buoyant forms (Maddock and Young 1984). The statocyst differs from those of the fast-moving ommastrephids but resembles those of *Sepia* and *Chtenopteryx* in having a hamulus at the first turn of the crista (h_2). These last two decabrachians, like *Thysanoteuthis*, have fins which extend along the length of the mantle.

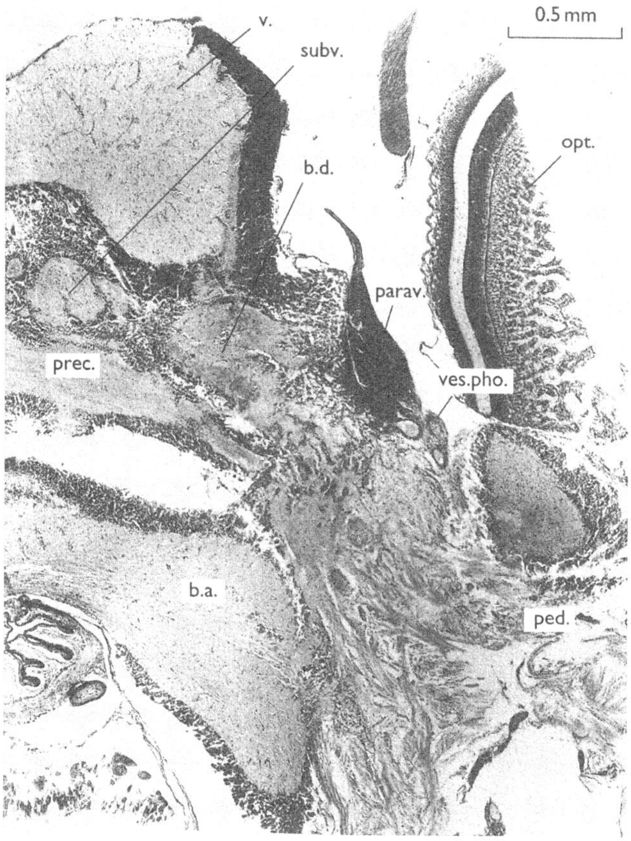

Fig. 22.2. *Thysanoteuthis rhombus.* A transverse section of the supraoesophageal mass and part of the optic lobe.

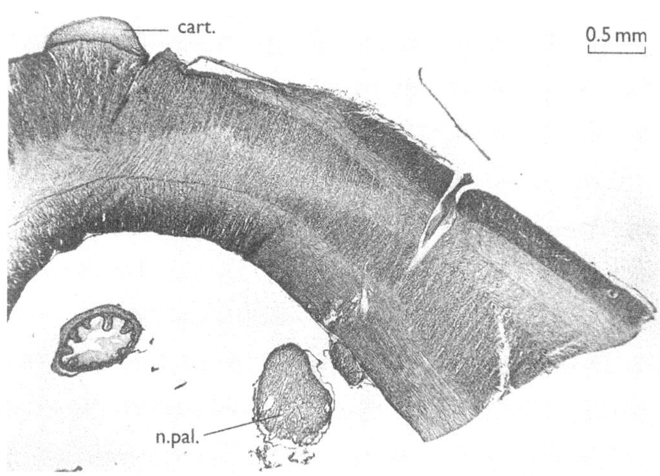

Fig. 22.3. *Thysanoteuthis rhombus.* Section through the mantle to show the mantle connective with giant fibre.

23 Cycloteuthidae

The family was established by R.E. Young and Roper (1969a), and the main characteristic is the pair of large round fins. Its members are widely distributed.

23.1 *Cycloteuthis*

Cycloteuthis was the first member of this family to be described (Joubin 1919). It is found in the north Atlantic, Indian, Pacific, and Southern Oceans, and has been caught in opening/closing nets from 55 to 800 m (R.E. Young and Roper 1969a, Roper and Young 1975, Clarke and Lu 1974, 1975, R.E. Young 1978).

Photophores are present around the eyes, and one on the ink sac (Jefferts and Roper 1992). The size distribution of lower beaks present in the stomachs of sperm whales showed a peak between February and August, which led Clarke (1980) to suggest that spawning takes place mostly from September to November and is followed by death.

23.2 *Discoteuthis*

The geographic distribution of *Discoteuthis* is in the Atlantic, and south Pacific Oceans, from depths of 0 to 700 m (R.E. Young and Roper 1969a, Roper and Young 1975).

The fins are muscular and can attain the full length of the mantle. The mantle wall of *Discoteuthis discus* is thick with a central, gelatinous layer between inner and outer sheets of circular muscle (R.E. Young and Roper 1969a). Photophores are present in the integument (Jefferts and Roper 1992). The cranial cartilage is reticulate (Fig. 23.1), and so too are the tissues of the arms, suggesting that these squids may be neutrally buoyant (Clarke *et al.* 1979), and some confirmation comes from the form of the statocysts (Maddock and Young 1984).

The brain is elongated with the brachial lobes far forward. The magnocellular lobe forms 6.8% of the total brain volume (Table 2.6), and its ventral portion is enlarged (Fig. 23.1). There is a distinct giant-fibre system with first-order giant cells whose axons cross in the palliovisceral lobe, but we have no preparations of the stellar nerve. The fin lobe forms 13.1% of the total brain volume and dominates the posterior suboesophageal mass (Fig. 23.2). The posterior chromatophore lobes are small. The anterior basal lobe receives fibres mainly from the ventral part of the optic lobe in the oblique section in figure. The peduncle lobe is complex and it receives dorsal and ventral nerves from the photosensitive vesicles (Fig. 23.1).

The extraocular photoreceptor organ is a broad, flat irregular ventral lobe located between the posteroventral surface of the optic lobe and

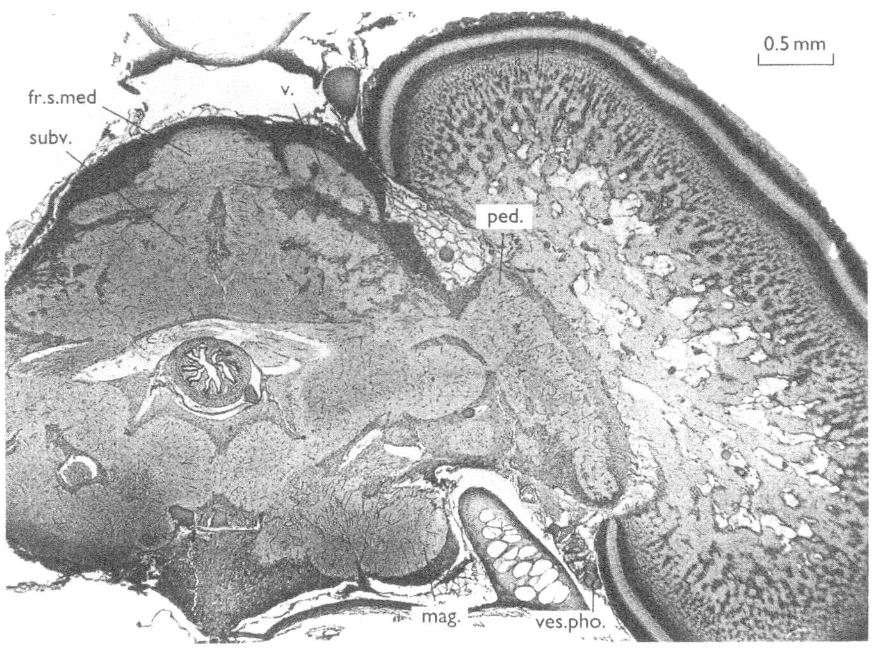

Fig. 23.1. *Discoteuthis discus* (51 mm ML). Oblique transverse section of the central nervous system.

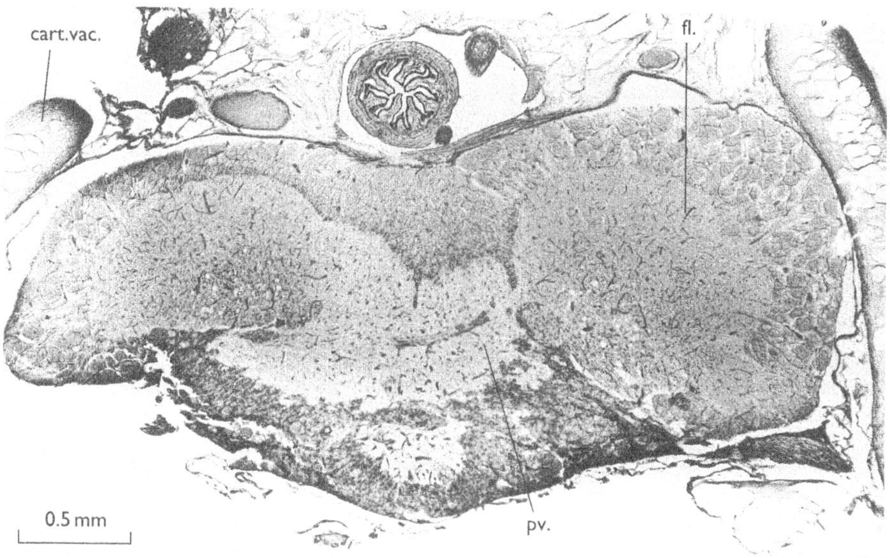

Fig. 23.2. *Discoteuthis discus* (51 mm ML). Oblique transverse section of the suboesophageal mass which is dominated by the fin lobe.

the cephalic cartilage (R.E. Young 1978). There may be a series of isolated vesicles or irregular strands of vesicles, and these form about 8% of the total brain volume (Maddock and Young 1987).

The statocyst of *Discoteuthis discus* (51 mm ML) is short and narrow, and very deep (Young 1989). The anticristae and hamuli are relatively large. Anticristae 6 and 7 do not quite meet but anticrista 7 is extended longitudinally, making a channel estimated to be complete for one-third of the circumference. Kölliker's canal is large. The features of the statocyst suggest the need to monitor rotatory movements in several planes. The dimensions of the statocyst indicate that this squid is neutrally buoyant (Maddock and Young 1984).

24 The 'Chiroteuthid lineage'

Five families are considered to be related, the Chiroteuthidae, Mastigoteuthidae, Joubiniteuthidae, Promachoteuthidae, and Batoteuthidae. They possess tentacles which bear extremely small suckers, and have a distinctive paralarva, or doratopsis. R.E. Young (1991) termed the group the 'Chiroteuthid lineage', to which he considers the recently described family Magnapinnidae may also be related (Vecchione and Young 1998).

24.1 Chiroteuthidae

Chiroteuthids are widespread in all oceans from sub-Arctic to sub-Antarctic waters, and are meso to bathypelagic though their young are found from near the surface to depths of perhaps 700 m.

24.1.1 *Chiroteuthis*

Chiroteuthis (Fig. 24.1) was described by Chun (1910a) as 'one of the most splendid deep-sea Cephalopoda'. Members of this genus are caught to depths of 2000 m, the paralarvae mostly between the surface and 700 m, and juveniles from 50 to 1000 m; some species make diel vertical migrations (Clarke and Lu 1974, 1975, Roper and Young 1975, R.E. Young 1978).

Life history

The doratopsis, or paralarva, has a long tail, which extends beyond the fins (Fig. 24.2). The tail is fragile and is usually broken or lost during capture, or it may be autotomized (Vecchione and Roper 1991). *Chiroteuthis* (43–60 mm ML) was observed before capture by remote operated vehicle in the Pacific and Atlantic Oceans in depths of 280–470 m. The tail was complete, or nearly so, and it had several knob-like pouches spaced along its lateral edges, filled with clear fluid. The posture adopted by a doratopsis is typically oblique to nearly vertical, with the tail pointed downwards, and the rest of the animal including the arms and tentacles extending upwards, or with arms and tentacles spread (Fig. 24.3). The doratopsis shows several, simple chromatophore patterns. Observations show that the fins undulate slowly, and to escape the squid turns downwards to jet rapidly away, leaving behind it a long, thin, pseudomorph of ink, approximately its own size (Vecchione *et al.* 1992).

The juvenile differs markedly from the doratopsis. The head is wider than the mantle, and is largely occupied by the laterally placed eyes. The arms are long, especially arms IV. The tentacles are exceptionally long. The maximum recorded size is 270 mm ML for *Chiroteuthis calyx* captured in the Sea of Okhotsk (Nesis 1993b).

Fig. 24.1. *Chiroteuthis veranyi* (51 mm ML, length of arm IV 165 mm, length of tentacle 328 mm) captured in the north Atlantic Ocean (photographed after preservation by P.N. Dilly).

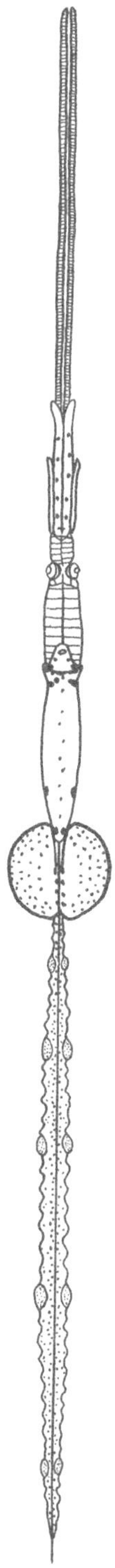

Fig. 24.2. *Chiroteuthis* (44 mm ML). Drawing of doratopsis in ventral view. It was captured by remote operated vehicle in the Pacific (Vecchione *et al.* 1992).

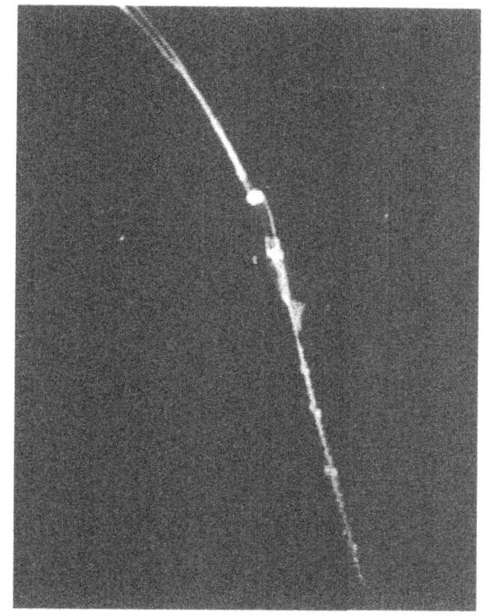

Fig. 24.3. *Chiroteuthis*. Doratopsis photographed from a remote operated vehicle (Vecchione *et al.* 1992).

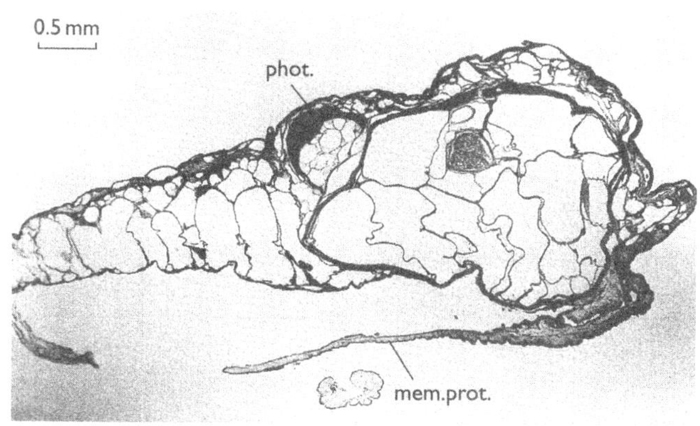

Fig. 24.4. *Chiroteuthis* (51 mm ML). Transverse section through arm IV, the protective membrane, and a photophore, and showing the extensive vacuolation of the tissues.

Morphology

Chiroteuthis is gelatinous and analysis of tissues from the neck, head and arms, and mantle shows the presence of ammonium, the highest concentration being in the long neck of the doratopsis. In juveniles the large ventral arms IV have the most buoyant tissues (Clarke *et al.* 1979). Reticulate tissue is present in the cephalic cartilage (see Fig. 24.15), wings of the conus of the gladius, the buccal complex (see Fig. 24.12), the arms where it extends into the protective membranes, and the photophores (Fig. 24.4). The muscle tissue is restricted often to narrow, peripheral bands.

The tentacles are extremely long and the muscles densely organized (Fig. 24.5). Each tentacular sucker is set on a peduncle with a long, cylindrical basal part, and a collar (Fig. 24.6). The sucker has a thin layer of muscles encircling a stout inner chitinous ring (Fig. 24.7). The inner ring has a recurved median tooth on the distal margin and 10–12 smaller teeth.

The tentacular stalk has a series of pigmented pads along its length, but their function is not known. They were described by Chun (1910a) as glandular knobs. Each small, oval pad is loosely attached to the stalk of the tentacle, and surrounded by pigment and a collar (Fig. 24.8). The collar is mainly of circular muscles, richly innervated. At the centre of the pad is a series of tubular structures, each consisting of a single tall cell with a basal nucleus (Fig. 24.9). Many nerve fibres lie at the base and some proceed among the cells. The lumina between the tubes contain amorphous material at the tip,

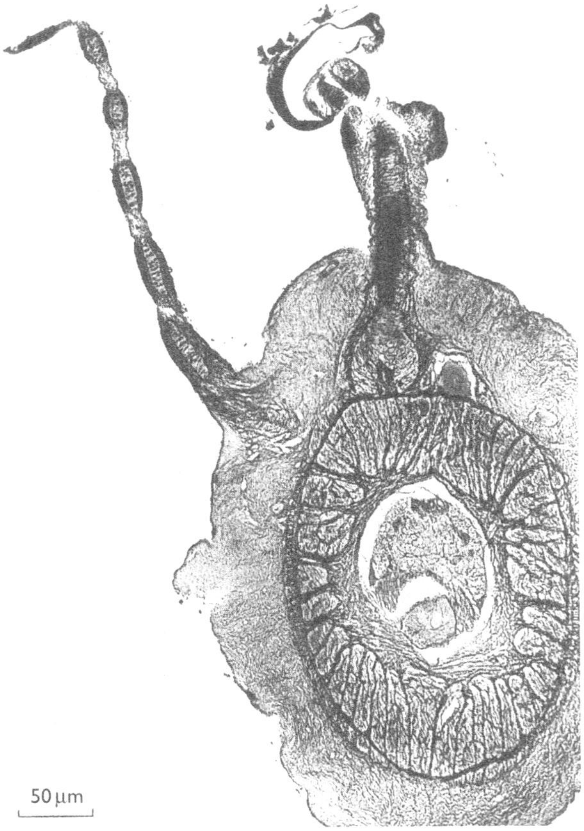

Fig. 24.5. *Chiroteuthis* (51 mm ML). Transverse section through the tentacular club, sucker and keel. The tentacle is muscular and in marked contrast with arm IV in Fig. 24.4.

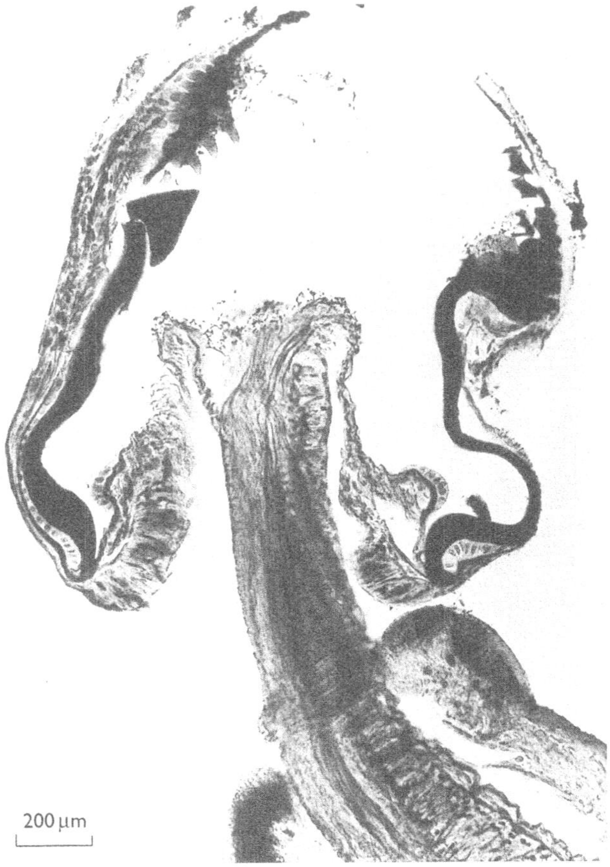

Fig. 24.7. *Chiroteuthis*. Longitudinal section through a tentacular sucker. Figs 24.7–24.12 and 22.14 are all of the same specimen of 51 mm ML.

Fig. 24.6. *Chiroteuthis* (68 mm ML). Tentacle showing the suckers and protective membrane. Each sucker is borne on a very long stalk and has a frill, or collar.

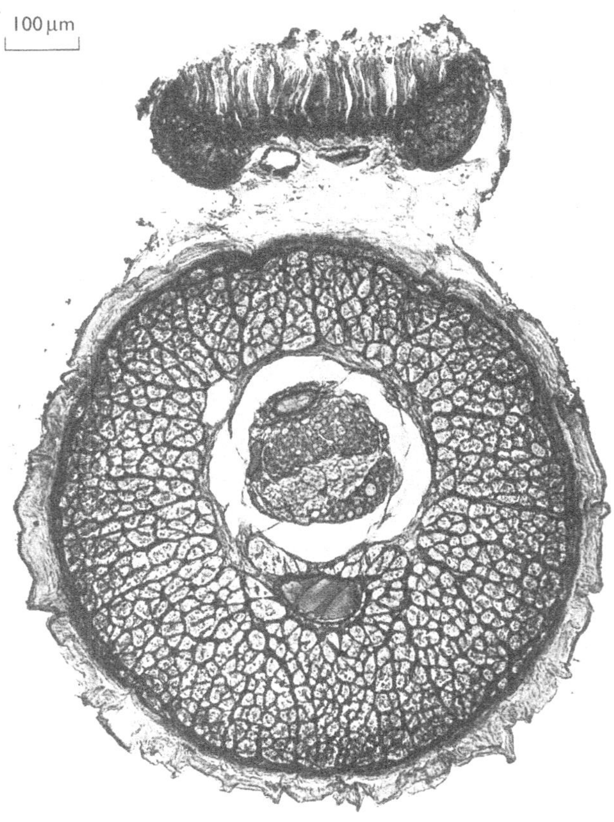

Fig. 24.8. *Chiroteuthis*. Transverse section through tentacle stalk and pad.

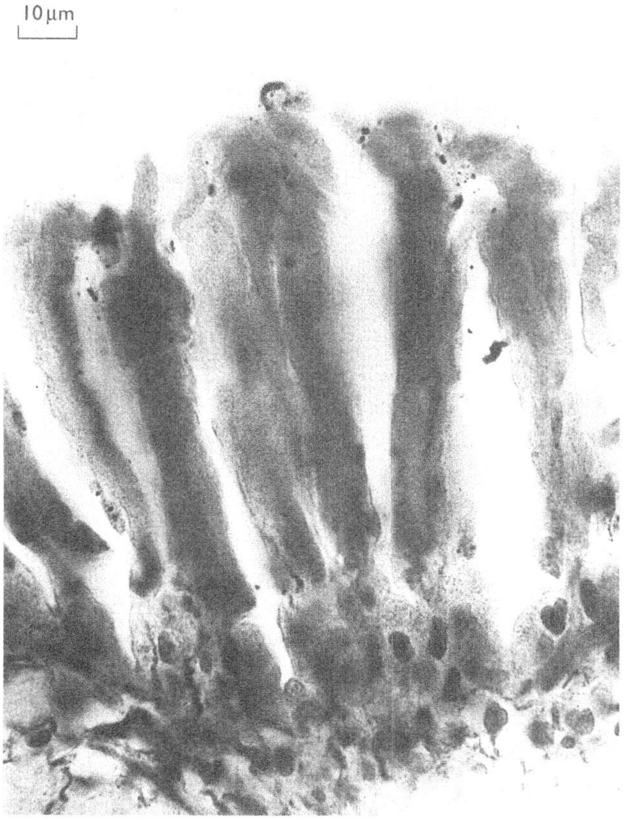

Fig. 24.9. *Chiroteuthis*. Pad of tentacle to show the tubular processes and rich innervation of the circular muscles.

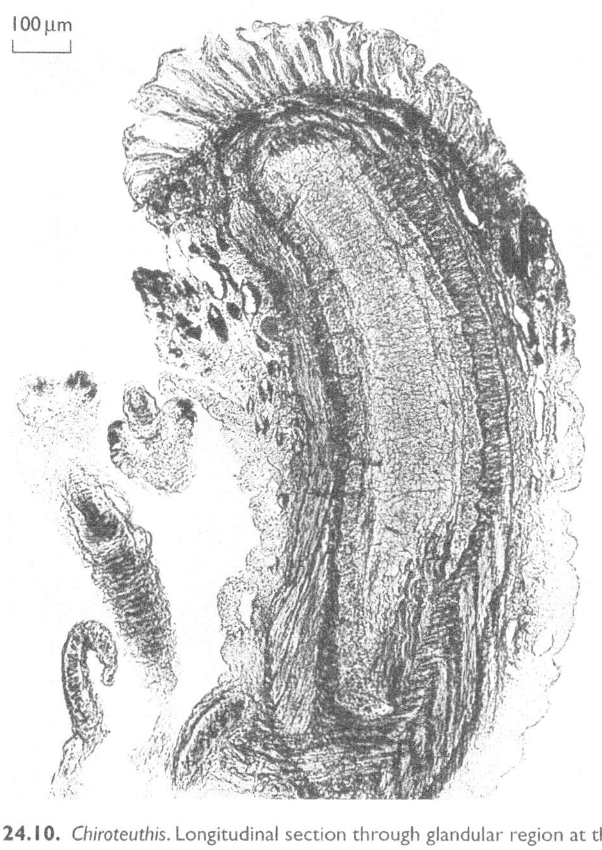

Fig. 24.10. *Chiroteuthis*. Longitudinal section through glandular region at the tip of the tentacle.

staining darkly with silver. Large chromatophores lie below each pad. At the distal tip of the tentacle there is a further pad of tissue of similar structure (Fig. 24.10). The tentacular nerve cord remains large to the tip of the tentacle, and contains a mass of neuropil into which pass the processes of very large nerve cells (Fig. 24.11). Around the outside is a dense network of nerve fibres. The complex network of nerve fibres and neuropil suggests that there is a sensory system, which may serve to attract, detect, and seize prey, and perhaps even immobilize it.

The buccal funnel extends forward as membranes between the arms, and its inner surface is densely pigmented and violet in colour (Chun 1910a). It is well supplied with nerves arising from the superior buccal lobe. The buccal mass forms a relatively large proportion of the head in the small specimen examined in this study. The upper beak is stout and the mandibular muscles are dense. Within the buccal cavity many of the tissues are reticulate, including the lateral buccal palps and the bolsters which support the radula (Fig. 24.12).

The diet of this squid is poorly known. Of 385 *Chiroteuthis imperator* caught in shrimp trawls (20–160 m depth) the majority had empty stomachs but some contained recognizable debris from fishes, crustaceans, and cephalopods, in that order (Kubota *et al.* 1981). These authors noted that this squid, which may be neutrally buoyant and inactive, is nevertheless capable of preying on a diverse spectra of organisms including fast-swimming fishes.

Photophores

Photophores are present along the tentacular stalk, arms IV, on the eyes, and on the ink sac. The tentacles of *Chiroteuthis imperator* may reach a length of 2 m and have 'luminous organs, like many small lamps on one thread' (Haneda 1955). Arms IV have as many as 55 brilliant, spherical photophores along the protective membrane (Fig. 24.13). Each light organ consists of a dorsal cup of reflective tissue. At the centre of the cup there is a mass of irregular cells, well supplied with capillaries, which run into it from below, through the vacuolated tissue filling the centre of the cavity (Fig. 24.14). These

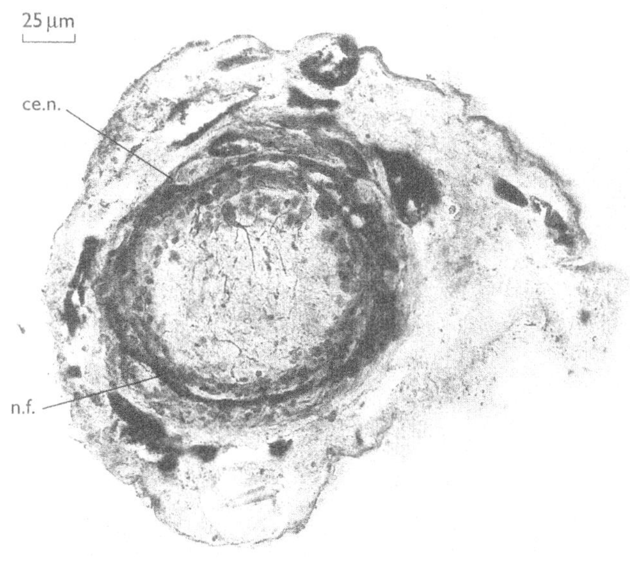

Fig. 24.11. *Chiroteuthis*. Distal tip of tentacular club in transverse section showing large nerve cord and nerve cell.

photophores are covered by a layer of dark brown chromatophores. Light from the photophores presumably passes through the translucent tissues of the arms (Fig. 24.4). *Chiroteuthis imperator* has 'numerous lens-shaped organs... arranged in three rows on the ventral surface of the eye and which glitter intensely and resemble precious stones on the dark background of the eyeball' (Chun 1910a). A pair of oval photophores are embedded on the ventral surface of the ink sac, each 7 mm long in animal of about 186 mm ML (Kubota *et al.* 1981).

Receptors and central nervous system

The eyes form a considerable portion of the elongated head and those of *Chiroteuthis veranyi* were described as 'almost as large as the visceral sac' (Chun 1910a). The retina has rhabdoms of up to 500 μm in length, and those at the centre may be shorter than those peripherally. Transverse sections show the presence of conjunctive eye muscles indicating binocular vision, as in *Mastigoteuthis* and *Joubiniteuthis*. There is no trochlear cartilage (Glockauer 1915). The optic lobes form 366% of the total brain volume (Figs 24.15, 24.16) (Table 2.6).

The central nervous system, like the head, is elongated, and the anterior part of the suboesophageal mass is widely separated from the rest. There is a direct brachiomagnocellular nerve running outside the brain: interestingly, it is shown by Chun (1910a) but not labelled in his figure. This nerve is smaller than in *Mastigoteuthis* and correspondingly the magnocellular lobe is smaller and does not show a rich network of fibres (Fig. 24.15). The magnocellular lobes together with the palliovisceral lobe form 20.3% of the total brain volume in *Chiroteuthis*, whereas they form 40.1% in *Mastigoteuthis* (Maddock and Young 1987). The large magnocellular lobes in both genera may be connected with the function of the elongated tentacles in the detection, recognition and capture of prey. No first-order giant cells were seen, nor any especially large fibres in the palliovisceral lobe, pallial nerve, or stellar nerves. Movement is probably with the fins, rather than by jet propulsion, as was observed in *Mastigoteuthis magna* (Roper and Vecchione 1997).

The inferior frontal lobe system forms 1.8% and the vertical lobe system 13.8% of the total brain volume (Table 2.6). The chromatophore lobe forms 0.2% of the total brain volume (Maddock and Young 1987). The chromatophores themselves may, by expansion, be concerned in the control of light emission from the photophores.

The extraocular photoreceptor organs are elongated, and each lies below the peduncle lobe within the inturned fold of the optic lobe

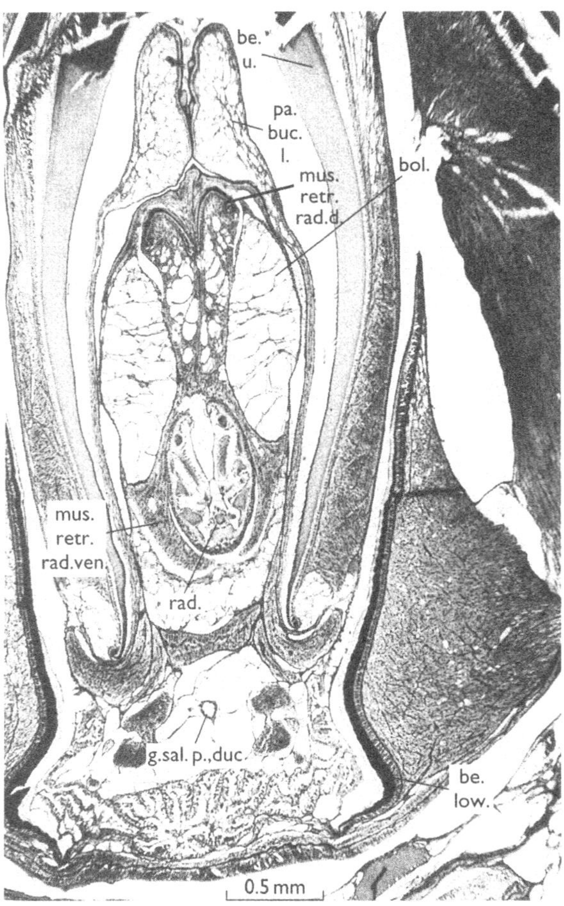

Fig. 24.12. *Chiroteuthis*. Transverse section through the buccal mass showing the vacuolation of the lateral buccal palps and bolsters.

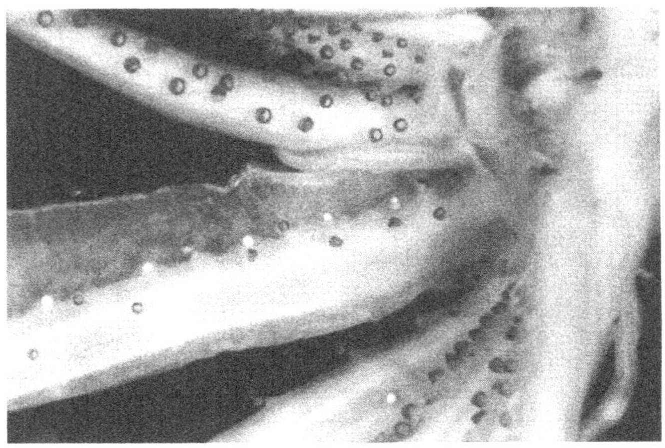

Fig. 24.13. *Chiroteuthis* (68 mm ML). The luminous organs and protective membrane of arm IV.

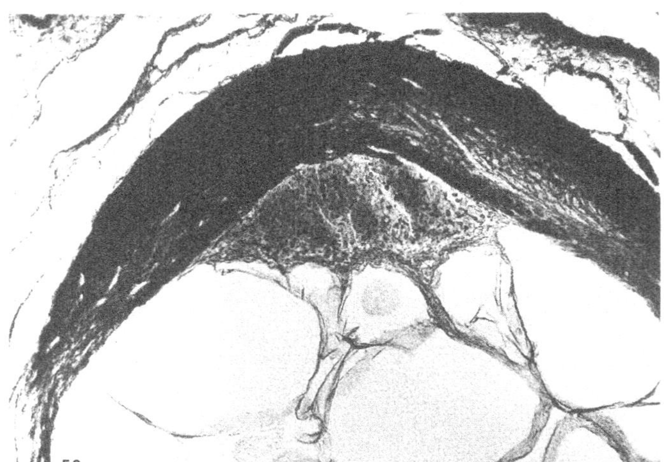

Fig. 24.14. *Chiroteuthis* (51 mm ML). Photophore of arm IV.

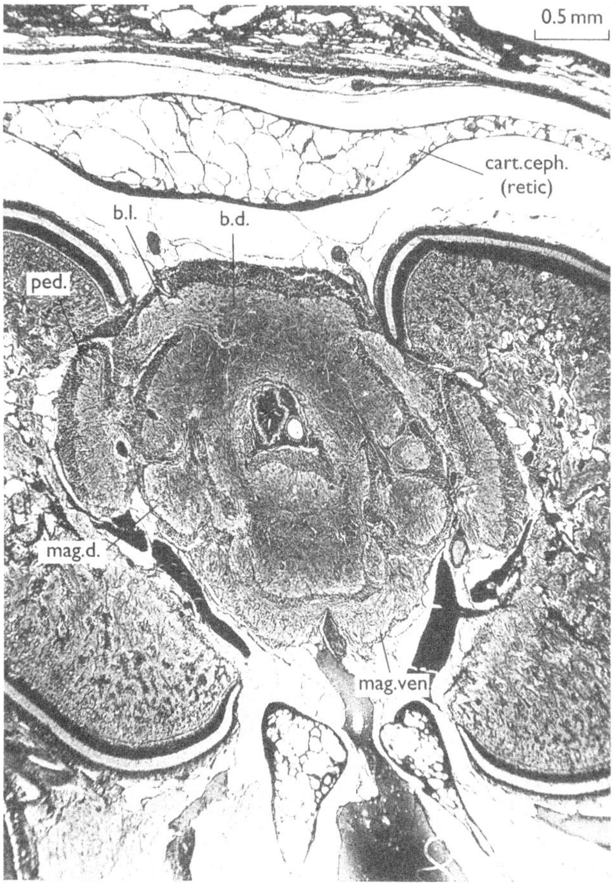

Fig. 24.15. *Chiroteuthis*. Transverse section through the central nervous system.

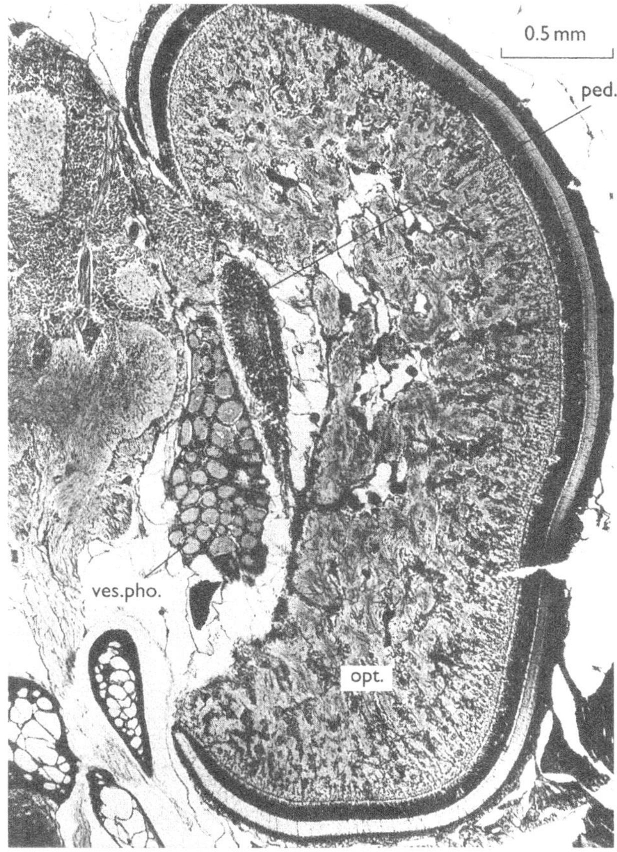

Fig. 24.16. *Chiroteuthis*. Transverse section to show the optic lobe and the photosensitive vesicles.

(R.E. Young 1978). Each organ contains many photosensitive vesicles and together they form 0.8% of the total brain volume (Table 2.6). The vesicles are in essentially the same position as those of *Mastigoteuthis* except that they are more compact and extend less far ventrally. They may receive light from the photophores below the eyes. The photosensitive nerve enters the peduncle lobe from below and divides into fine branches near the base of the spine (Fig. 24.16).

The statocyst is broad, and shallow at the back but deep at the front, with dorsal and ventral chambers. All five hamuli are present but are small. The conditions suggest that the crista provides sensitivity for slow turning in all directions. The presence of hamulus 2 is unusual, and it is not found in *Mastigoteuthis*. Anticrista 1 is absent but 2–5, although small, occupy their usual positions, and anticrista 6 is minute. Two protuberances are present in the roof, 7 and 7a (Young 1989). There is no enclosed horizontal canal. The statocyst clearly belongs to a squid which is buoyant (Maddock and Young 1984).

24.1.2 *Planctoteuthis*

Planctoteuthis (*Valbyteuthis*) (R.E. Young 1991) has been caught in the Atlantic Ocean between 25°S and 46°N, and in the eastern Pacific Ocean from 40°S to 35°N (Roper and Young 1967). Specimens of 3–57 mm ML inhabit depths of 0–1300 m in the North Atlantic (Clarke and Lu 1974, 1975, Lu and Clarke 1975b). It has been captured between 750 and 1200 m in Hawaiian waters (Roper and Young 1975) and from 1550 to 2000 m off the Californian coast (R.E. Young 1972b).

A gravid female was caught off southern California, 32°N and 120°W, at 2000 m depth. Its mantle (76 mm long) was almost entirely filled with a mass of mature eggs, 1.5 mm in diameter, and large nidamental glands (R.E. Young 1972b). During development *Planctoteuthis* shows few changes, although the neck does become reduced.

Histological examination shows many tissues to be reticulate, so this squid is probably neutrally buoyant (Clarke *et al.* 1979).

24.1.3 *Asperoteuthis*

Asperoteuthis is found in the tropical North Pacific (Lu 1977, Okutani 1995).

A strange chiroteuthid was described by Lu in 1977 (*Chiroteuthis acanthoderma*) but it is now considered to be the type species of *Asperoteuthis* (R.E. Young 1991). Its surface is entirely covered with numerous small, conical, cartilaginous tubercles. These are especially closely arranged on the head, mantle, and aboral surfaces of the arms, but are sparse on the oral surfaces of the arms. Changes during growth are probably similar to those in *Chiroteuthis*, except that most photophores are absent and the relative arm length shows larger changes (R.E. Young 1991). Each tentacle has a pad at its tip and was described by Lu (1977); it is similar to the pad found in *Chiroteuthis*.

24.1.4 *Grimalditeuthis*

Grimalditeuthis bonplandi is the only species of the genus (Fig. 24.17) It is rare but apparently widely distributed as it has been reported from the Atlantic Ocean, at 11°N 20°W and 18°N 25°W (Clarke and Lu 1975, Lu and Clarke 1975b), the Straits of Florida (Cairns 1976), the Gulf of Mexico (Voss 1956), the Pacific Ocean, in the Santa Catalina Basin and off Hawaii (R.E. Young 1972b, 1978), and in Japanese waters (Okutani 1975). In the Atlantic Ocean animals of 39–77 mm ML have been captured in opening/closing nets at 200–250 m and 900–1020 m during the day, and at 200–1500 m at night (Clarke and Lu 1975, Lu and Clarke 1975b). Off Hawaii animals of 2–24 mm ML have been caught between the surface and 300 m; ones of 20–98 mm ML were caught between 180 and 300 m depth in darkness, and from 750 to 1300 m in daylight; the two largest animals captured were also at the greatest depth (R.E. Young 1978, 1991).

Life history

The doratopsis of *Grimalditeuthis* is almost identical with that of *Chiroteuthis*, small specimens of the former being separable from other chiroteuthids by the number of chromatophores and eye shape. *Grimalditeuthis* retains its tail at metamorphosis.

Our specimen has a neck behind the head and a brachial pillar in front (Fig. 24.17), and both elongated regions are crossed by strands of tissue but are otherwise translucent (Fig. 24.18). There is no cephalic cartilage around the central nervous system.

Receptors and central nervous system

The eyes are laterally placed on the long, cylindrical head. The eye is relatively small with a diameter of 5.4 mm in an animal of 67 mm ML (Fig. 24.17). The retina has rhabdoms of about 75 μm in length at the centre but as long as 120 μm around the periphery. There is much pigment concentrated in the base of the rhabdoms at the centre, but this is dispersed in the peripheral rhabdoms.

The optic lobes form 712% of the total brain volume (Table 2.6). The granular and the palisade layers are well marked, but the columns of neuropil and fibres are short and there are large venous spaces.

The optic tract is long (Fig. 24.18) and the peduncle lobe lies in an unusual position, immediately above the extraocular photoreceptor organ, close to the brain and far from the optic lobe (Fig. 24.19). Such separation of the optic lobe from the peduncle lobe is rare, although it occurs in two octopodids, *Alloposus* and *Eledonella*, and in some cranchiid squids. The optic glands are well developed and highly vascular in our specimen, perhaps indicating approaching maturity (Fig. 24.20). Nothing is yet known of reproduction in *Grimalditeuthis*.

Each extraocular photoreceptor organ consists of a small cluster of photosensitive vesicles attached to the posteroventral margin of the optic tract, and together they form 0.4% of the total brain volume (Table 2.6).

The suboesophageal mass of the brain is elongated and the brachial lobe lies far forward. From this lobe the brachial nerves emerge as two strands and accompany the oesophagus forward along the pillar (Fig. 24.18), and each strand divides into four brachial nerves just behind the buccal mass.

The magnocellular lobe is prominent (Fig. 24.19) and receives large tracts from in front and behind. No giant fibres were seen either in the central nervous system or in the stellate ganglion. Each stellate ganglion is reduced to a single layer of cells ventrally and a very few cells dorsally (Fig. 24.21).

In the supraoesophageal mass the superior frontal and vertical lobes are well developed. The vertical lobe system forms 15.6% of the total brain volume, and the inferior frontal system 1.8% (Table 2.6).

The statocyst has a cavity with little obstruction and distinct antero-posterior and dorsoventral chambers The crista is set at the centre with space around it, except anteriorly. All five hamuli are present, but are small; there are no other anticristae (Young 1989). Multiple

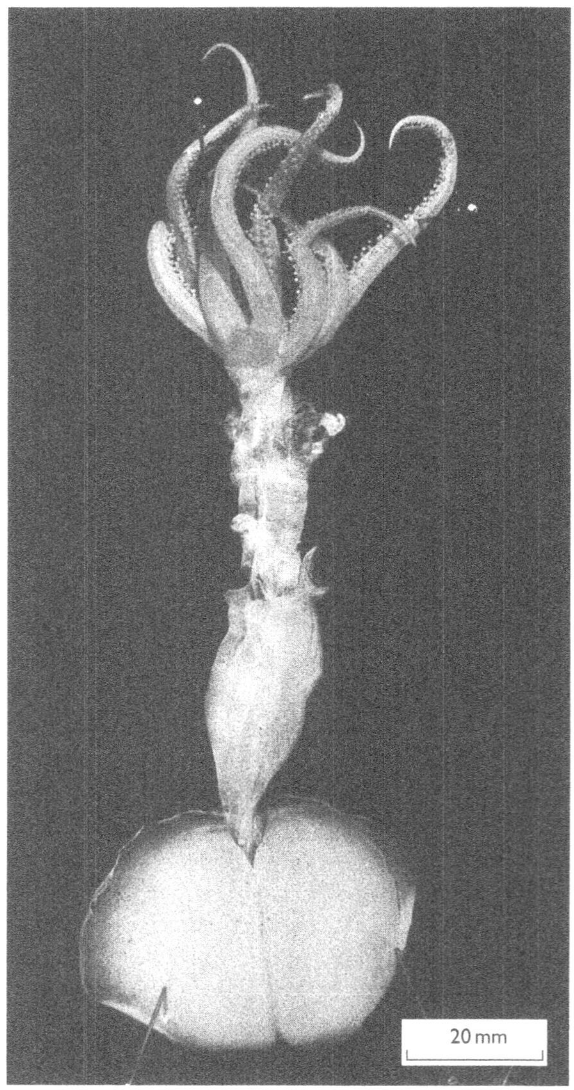

Fig. 24.17. *Grimalditeuthis bonplandi* (67 mm ML, measured without the accessory fin and tail). Ventral view. Figs 24.18–24.21 are all of this animal.

discriminant analysis of various parameters of the statocyst place *Grimalditeuthis* among the buoyant squids (Maddock and Young 1984). The statocyst is suited to monitor slow turning in all planes.

24.2 Mastigoteuthidae

Mastigoteuthids have been caught in all oceans except the Arctic Ocean. They are meso- and bathypelagic in habit, although some species are associated with the bottom (Nesis 1987, Roper and Vecchione 1997).

24.2.1 *Mastigoteuthis*

Mastigoteuthis schmidti (Fig. 24.22) is widely distributed between 60°N and 50°S (Clarke 1966, Clarke and Lu 1974, 1975, Lu and Clarke 1975a, 1975b, R.E. Young 1978).

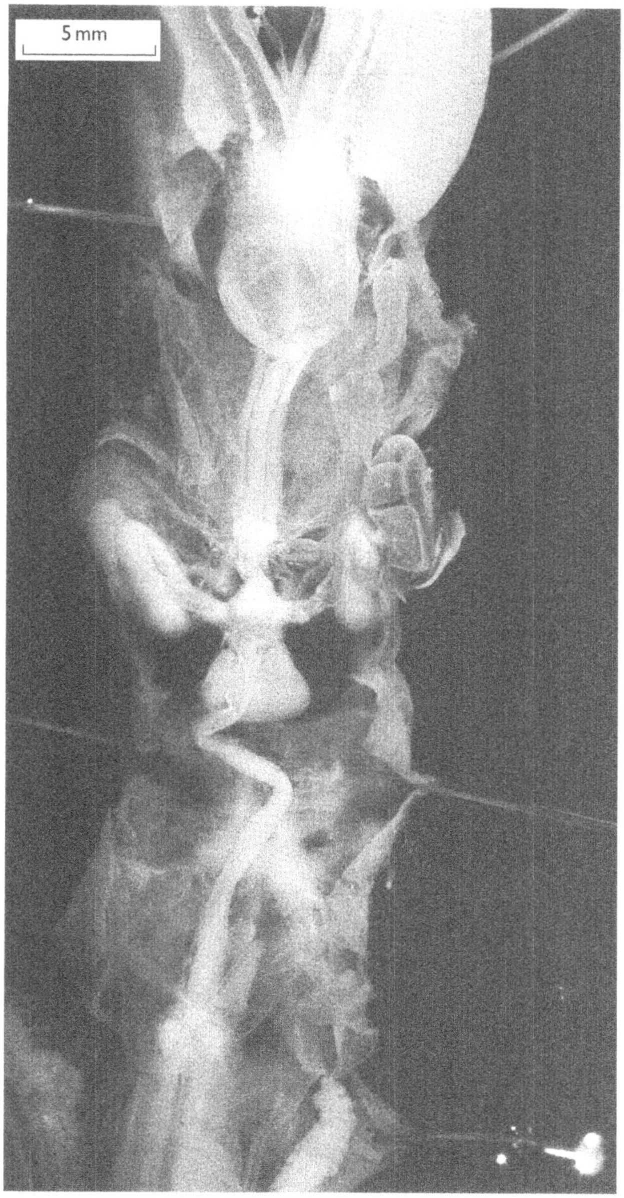

Fig. 24.18. *Grimalditeuthis bonplandi*. Dissection of the neck and pillar to show the organization of the central nervous system, optic tract, and brachial nerves.

Mastigoteuthis has a cylindrical mantle which tapers rapidly to a tail. Small photophores are scattered over the animal, mainly ventrally, and some species possess a large light organ below each eye (R.E. Young 1972b). Verrill (1882) said the skin shows spots of light like tiny eyes and when alive the squid is a magnificent dark red (Chun 1910a). *Mastigoteuthis schmidti* is soft-bodied and many of

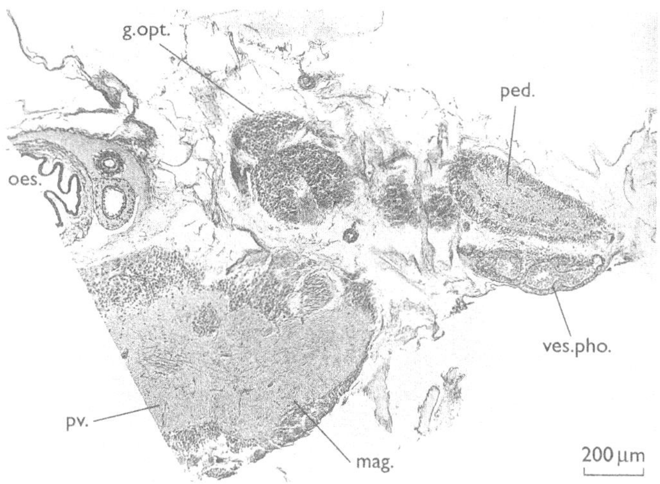

Fig. 24.19. *Grimalditeuthis bonplandi*. Transverse section through the suboesophageal mass and the peduncle lobe, photosensitive vesicles, and optic gland.

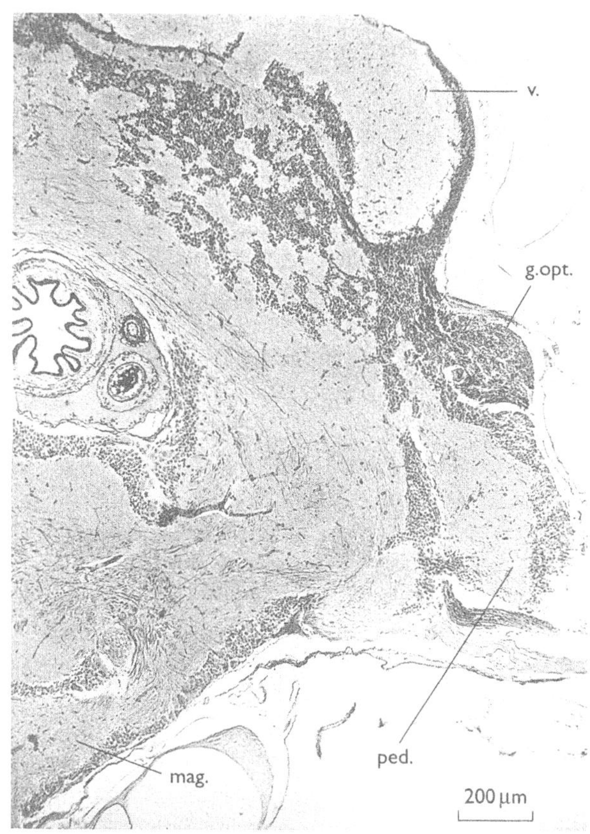

Fig. 24.20. *Grimalditeuthis bonplandi*. Transverse section through the central nervous system.

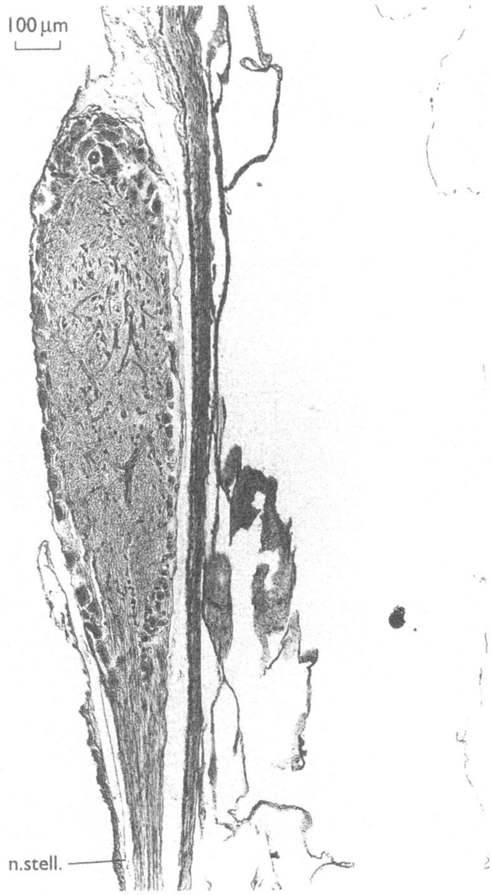

Fig. 24.21. *Grimalditeuthis bonplandi.* Longitudinal section of the stellate ganglion.

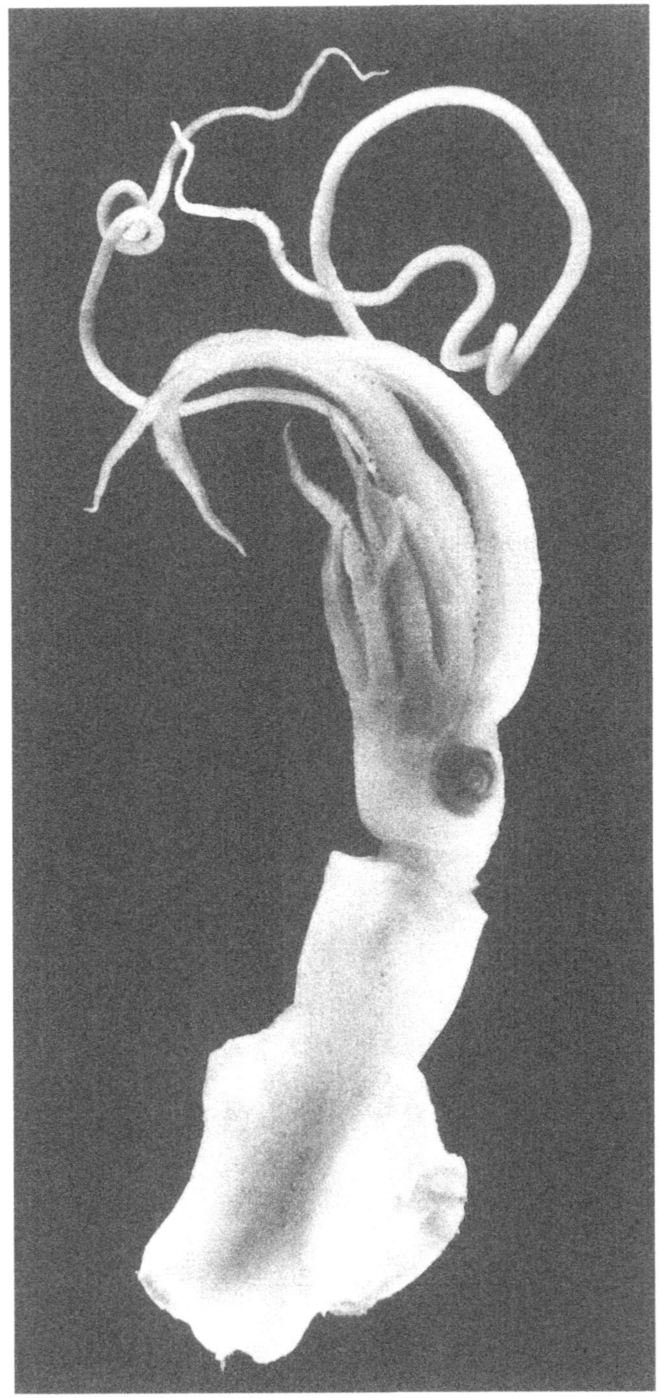

Fig. 24.22. *Mastigoteuthis schmidti* (44 mm ML).

its tissues are reticulate, including the mantle (Fig. 24.23), fins (Fig. 24.24), arms (see Fig. 24.27), and cephalic cartilage (see Fig. 24.31). The large ventral arms are more buoyant than the rest of the animal (Clarke *et al.* 1979). These conditions were considered to result in a head-up posture but this is not so as observations in the sea show that the squid maintains a head-downward posture by constant manoeuvring with the fins (Roper and Vecchione 1997).

The inner and outer layers of the mantle muscles are separated by reticulate tissue (Fig. 24.23). The muscles are weak and are used mainly for respiration (M.R. Clarke pers. comm.). The surface of the mantle, head, and arms of *Mastigoteuthis cordiformis* and *M. hjorti* bear tubercles of elastic fibrocartilage. The tubercles and underlying skin of *M. cordiformis* are reticulate, whereas only the underlying skin of *M. hjorti* is reticulate as the tubercles have quite a dense organization (Roper and Lu 1990). The fins are nearly circular, and are large and strong. They extend in some species for more than 60% of the ML. They form the main propulsive organ and their movements are powerful during locomotion (Clarke pers. comm.). Double sinusoidal waves were seen to pass forwards or backwards along the fins (Roper and Vecchione 1997). The fins can roll together and enwrap the mantle, one fin overlapping the other along the midline; any water trapped between the mantle and the fins is expelled, almost simultaneously, anteriorly and posteriorly. This action gives a low-velocity, jet-like, propulsion. A short tail extends behind the fins and includes the end of the conus of the gladius. The tail has longitudinal muscle fibres (Fig. 24.24).

Arms and tentacles

The arms are large, especially the ventral pair, which are twice as long as the others, and can be longer than the mantle (Dilly *et al.* 1977). The suckers are stalked, and are arranged in two longitudinal rows. Each sucker has an inner ring with a number of sharp teeth pointing towards the mouth (Fig. 24.25). The surrounding infundibulum has

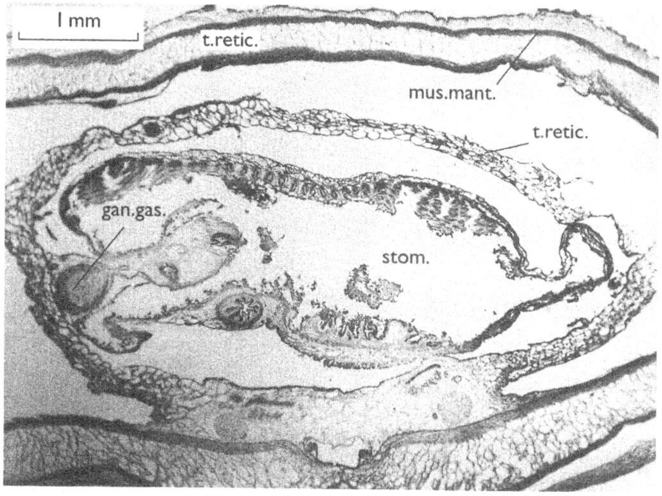

Fig. 24.23. *Mastigoteuthis schmidti* (41 mm ML). Transverse section through the mantle and stomach.

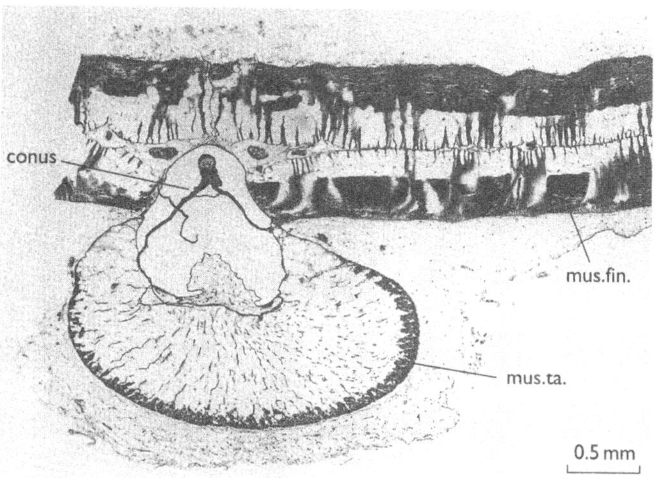

Fig. 24.24. *Mastigoteuthis schmidti* (41 mm ML). Transverse section through the fin and tail to show the closed conus (haematoxylin and eosin).

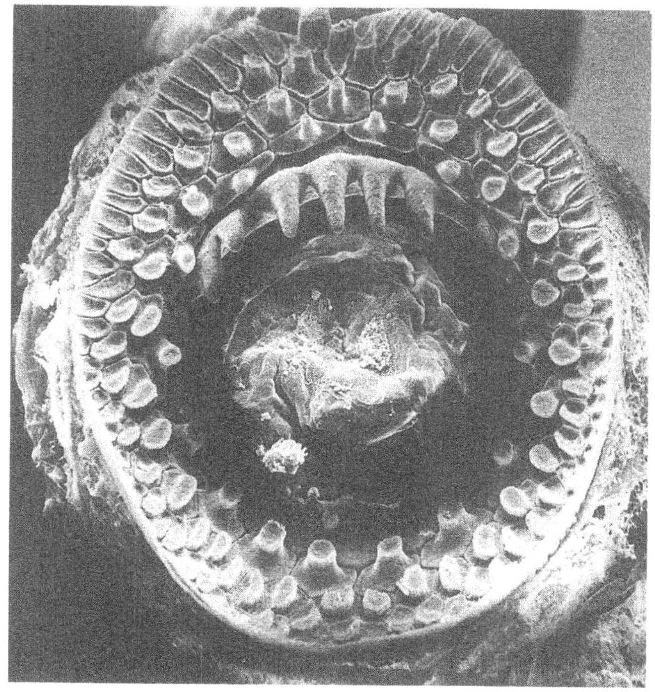

Fig. 24.25. *Mastigoteuthis schmidti*. Sucker from an arm to show the polygonal processes with projecting pegs and sharp teeth on the inner ring (field width 2000 μm).

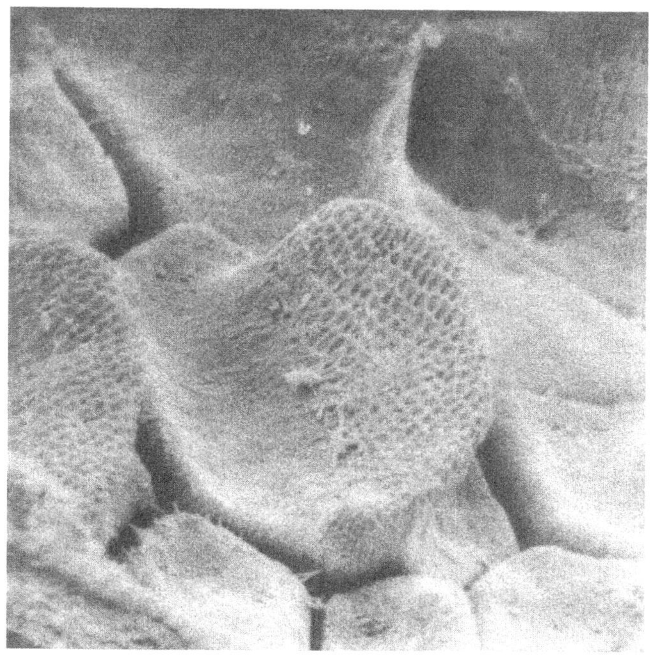

Fig. 24.26. *Mastigoteuthis schmidti*. The same sucker as in previous figure, but at higher magnification (field width 8 μm).

two or three rings of polygonal processes with pegs, each with many minute pores on its surface that may serve for adhesion (Fig. 24.26) (Nixon and Dilly 1977). Much tissue in the arms is reticulate, and the muscles and axial nerve form only a small portion (Fig. 24.27). The large ventral arms are highly reticulate and contain ammonium, which gives lift (Clarke *et al.* 1979). Further buoyancy is provided by the digestive gland which contains oil and lies far forward in the mantle (Rancurel 1971).

The tentacles are extremely long and slender, and they are covered with many minute suckers (Fig. 24.28). In the sea the tentacles can elongate rapidly to twice their length (Roper and Vecchione 1997). The axial nerve cord is prominent and is surrounded by dense longitudinal muscles (Fig. 24.29). The tentacles are often found isolated and clinging to the meshes of fishing nets (Rancurel 1971, M.R. Clarke pers. comm.). In the sea the tentacular suckers felt sticky to touch and had to be firmly pulled from the observer's skin (Roper and Vecchione 1997). The distal two-thirds of the tentacle is covered by innumerable minute suckers, each about 0.1 mm in diameter. Each sucker is set on a stalk and has rings of pegs, very similar to those of the arm suckers (Fig. 24.30) (Nixon and Dilly 1977). The muscles in the tentacular suckers can affect strong adhesion to

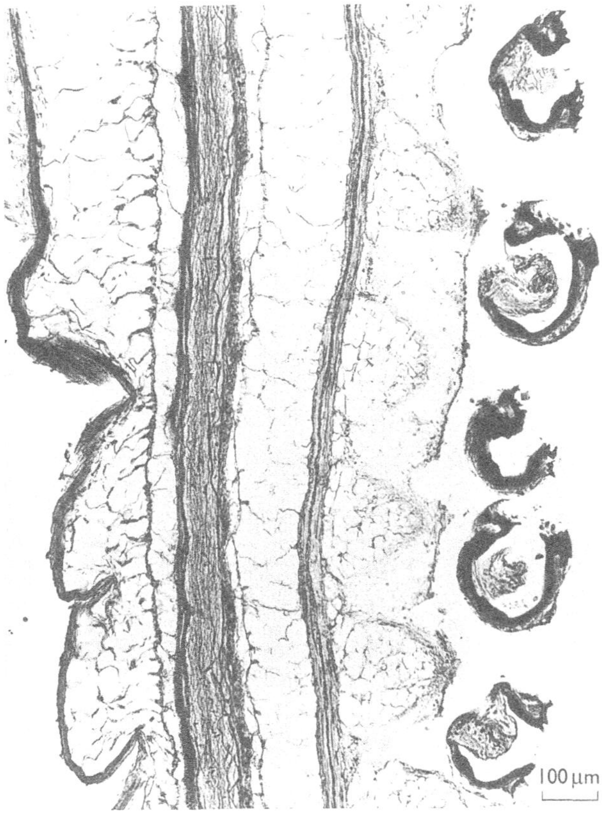

Fig. 24.27. *Mastigoteuthis schmidti.* Longitudinal section through an arm, revealing the extent of the reticulate tissues.

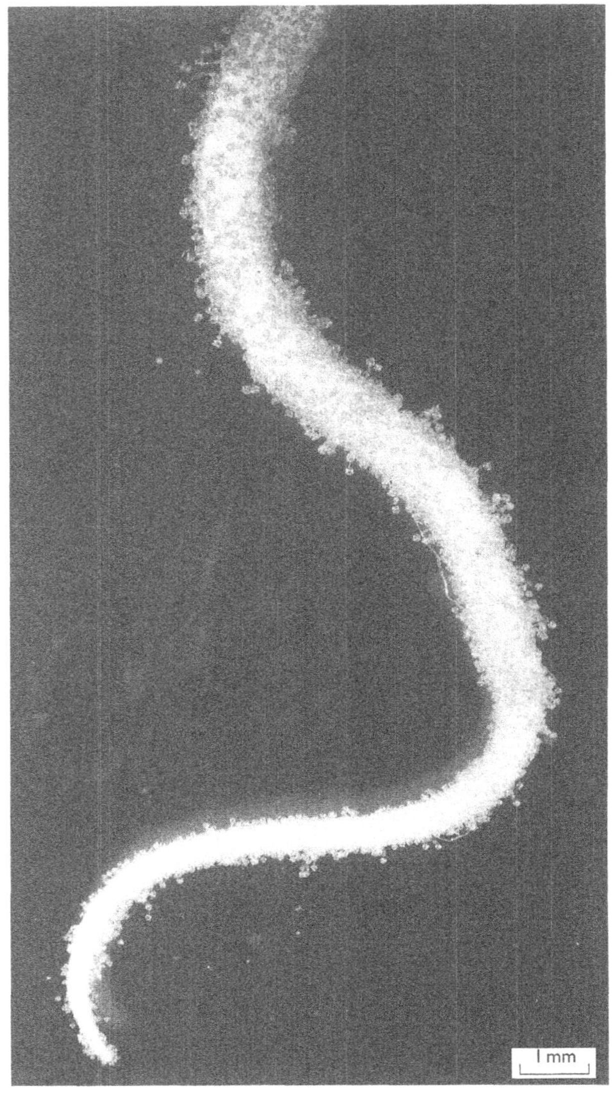

Fig. 24.28. *Mastigoteuthis schmidti.* Part of a tentacle and its minute suckers.

human skin (Roper and Vecchione 1997). The tips of the tentacles skim along the sea floor suggesting the presence of receptors in the suckers. *Mastigoteuthis* sp., photographed from a submersible, shows the extremely long tentacles emerging from their tentacular sheaths; these are formed by the enrolled lateral membranes of ventral arms IV, with which the tentacles appear to be in continuity. The tentacles, held apart by the ventral arms, are trailing close to, or over, the substrate (R.E. Young *et al.* 1998b). This would allow the very small and very numerous suckers to detect and capture small bottom-associated zooplankton.

Buccal mass

The buccal mass is longer than the central nervous system and is approximately one-tenth of the mantle length of the specimen in Fig. 24.31. Many tissues are reticulate, but the mandibular muscles are mostly dense. Stout teeth are present on the lateral buccal palps, which also carry the anterior salivary glands. The duct of the posterior salivary glands opens on the salivary papilla, which has glandular patches on its surface. The buccal membrane has a folded, glandular inner surface and may serve to entangle prey. Verrill (1882) found fragments of small crustaceans in the stomach of one specimen, and Rancurel (1971) found antennae from crustaceans larger than the squid itself. A deep layer of reticulate tissue covers the viscera, and separates the coelomic and mantle cavities (Fig. 24.23).

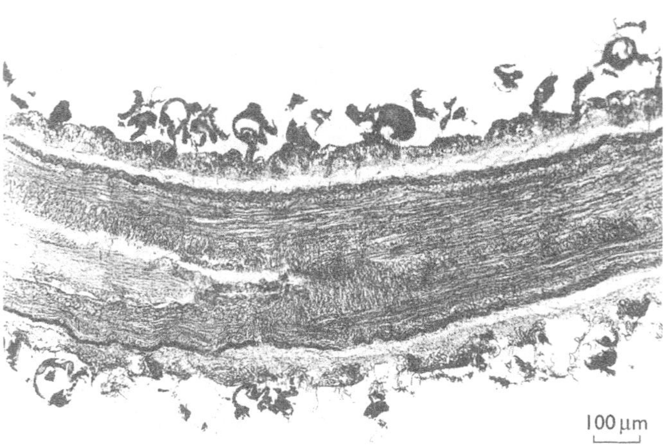

Fig. 24.29. *Mastigoteuthis schmidti* (44 mm ML). Longitudinal section of a tentacle showing its dense tissues that are in marked contrast with those of the arm shown in Fig. 24.27.

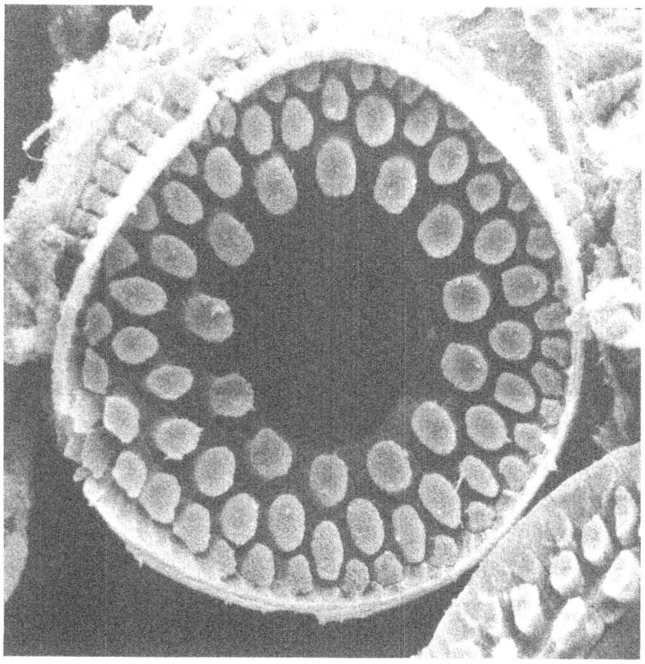

Fig. 24.30. *Mastigoteuthis schmidti* (41 mm ML). Sucker from tentacle (field width 55 μm).

The living animal

Hypotheses developed from morphological studies (Dilly *et al.* 1977) were tested by observations of *Mastigoteuthis magna* in the sea. Video recordings, made off Cape Hatteras at 700–1000 m, usually revealed this squid to be close to the bottom and never in the photic zone. From a very small size this species displayed a characteristic 'head-down' posture, in which the animal hovered by constant manoeuvring with the fins. The mantle was held vertical, the head oblique, arms I–III held at approximately 45°, while the long tentacles hung vertically, this was termed the 'tuning-fork' posture. An animal would maintain this position for up to 15 min and even remained as the submersible moved away. The vertical orientation was held by the varied actions of the fins, these were often complex depending on the correction needed. Recordings showed that the tips of the tentacles and their shadows converged on the substrate. In one sequence when the bottom fell sharply away from beneath the animal the tentacles elongated very rapidly to double their previous length (Roper and Vecchione 1997). Another *Mastigoteuthis* was photographed in the same 'tuning-fork' attitude at 885 m depth, off Hawaii (R.E. Young *et al.* 1998b).

Receptors and central nervous system

The eyes are directed laterally in the adult and have a diameter of up to 15% of the mantle length. In the retina the rhabdoms have an almost uniform thickness of 200 μm (Fig. 24.32). The retina is folded in one of

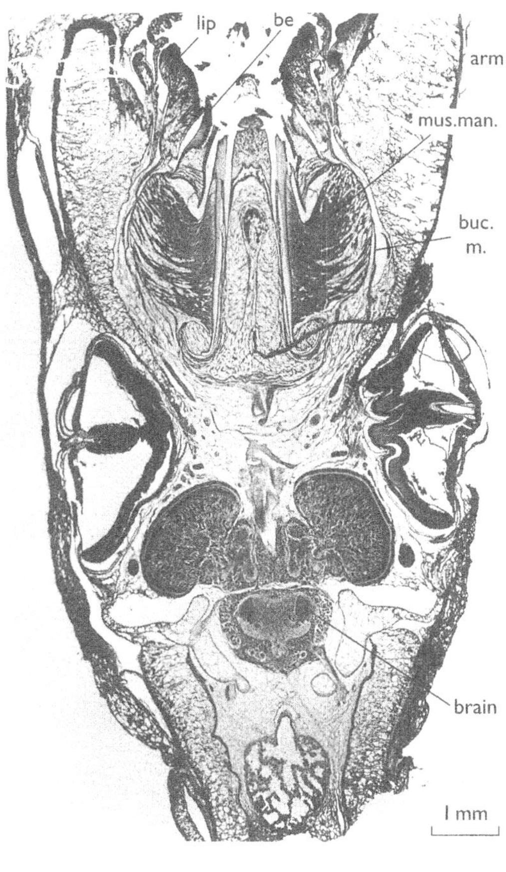

Fig. 24.31. *Mastigoteuthis schmidti* (41 mm ML). Horizontal section through the head to show eyes, optic lobes, the central nervous system, buccal mass, and bases of two arms.

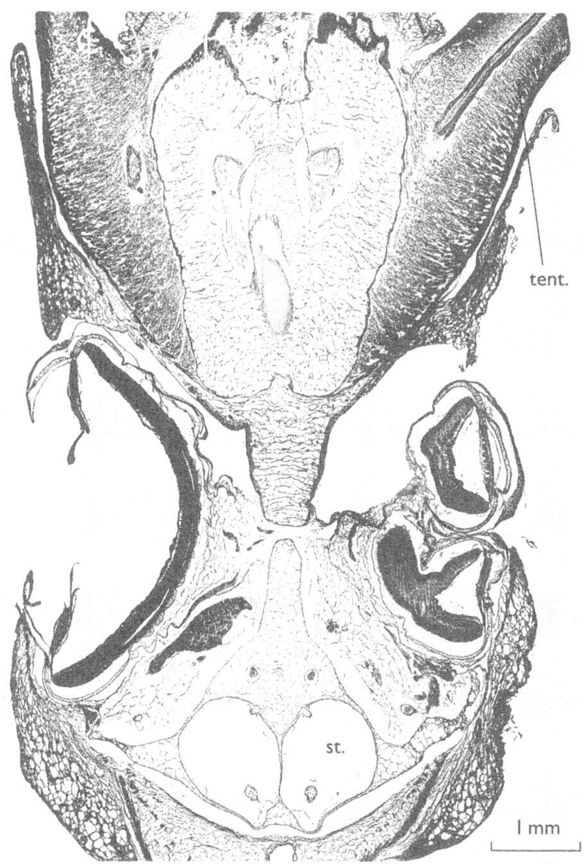

Fig. 24.32. *Mastigoteuthis schmidti* (41 mm ML). Horizontal section through head, eyes, statocysts, and bases of the tentacles.

our specimens, but this is probably an artefact. The rhabdoms range from 135 to 200 μm in length, being longest posterodorsally, but there is no marked equatorial thickening. A trochlear cartilage is present, and two conjunctive eye muscles cross in the midline (Fig. 24.33). The presence of such muscles suggest that this squid is capable of binocular vision (Budelmann and Young 1993).

The optic lobe is set on an axis with an oblique angle (Fig. 24.34). The inner and outer granule cell layers are thick (Fig. 24.35). Together the two optic lobes form 146% of the total brain volume (Maddock and Young 1987).

The most remarkable feature of the brain is the enormous development of the ventral magnocellular lobe, which is as large as the vertical lobe (Maddock and Young 1987). This lobe receives a massive innervation direct from the arms and tentacles called the brachiomagnocellular nerve (Fig. 24.34). This term distinguishes the nerve from the brachiomagnocellular connective, which runs within the brain, as it does in other decabrachians. The ventral magnocellular lobe is larger than the vertical lobe system and is an elaborate centre for influencing swimming activities (Dilly *et al.* 1977). The ventral magnocellular lobe has a rather thin outer layer of cells but a densely interweaving network of fibres in the neuropil (Fig. 24.36). This tangle of fibres resembles that in the median inferior frontal lobe of *Octopus*, even to the presence of many pieces of degenerating fibres, which may arise from a continual process of shedding sucker discs. The magnocellular lobe receives fibres from the macular nerve behind and the crista nerve enters laterally. The output fibres of the ventral magnocellular lobe pass to the large centres in the pedal and palliovisceral lobes that control the fins. There may also be efferent fibres to the arms and

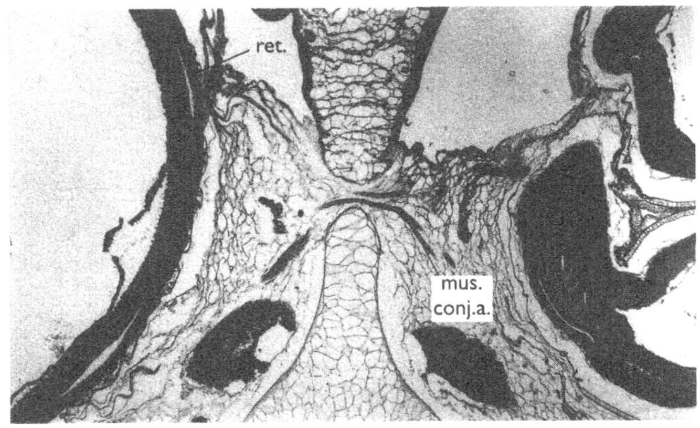

Fig. 24.33. *Mastigoteuthis schmidti*. Horizontal section showing the anterior conjunctive muscles of the eye.

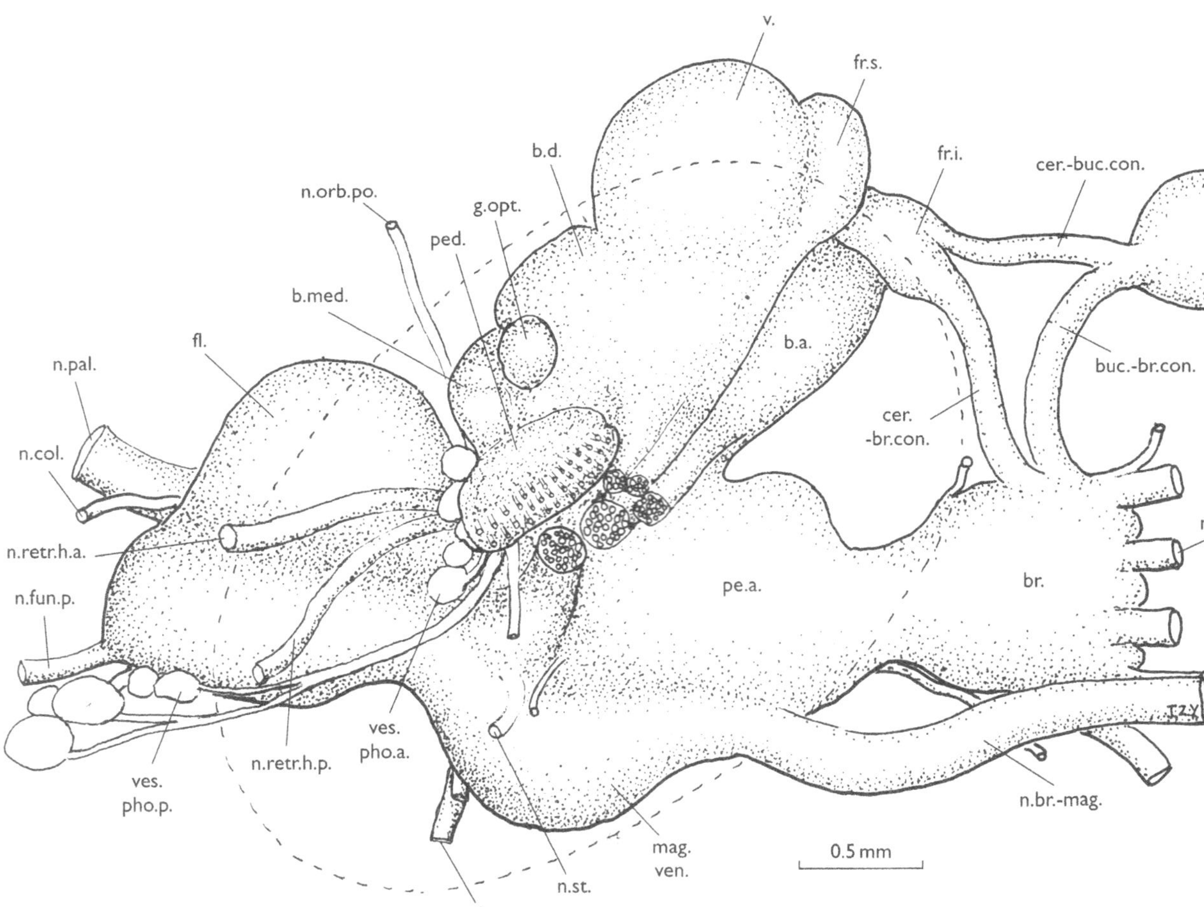

Fig. 24.34. *Mastigoteuthis schmidti* (60 mm ML). Drawing of the lateral view, reconstructed from histological sections, to show the central nervous system and photosensitive vesicles. Note the large ventral magnocellular lobe. The broken line indicates the position of the optic lobe (Dilly *et al.* 1977).

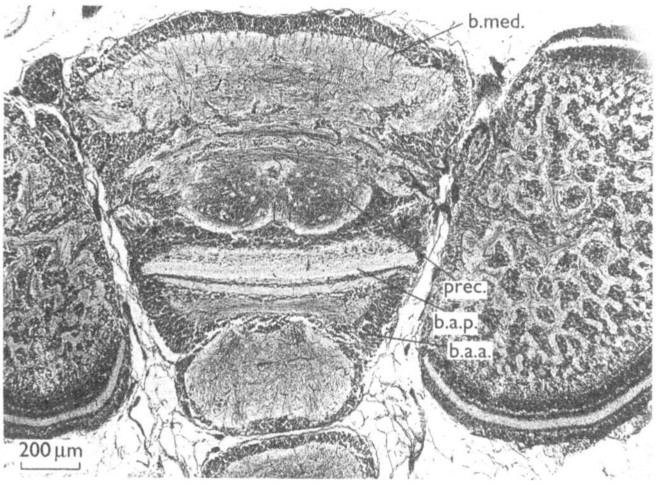

Fig. 24.35. *Mastigoteuthis schmidti* (41 mm ML). Horizontal section to show the basal lobes.

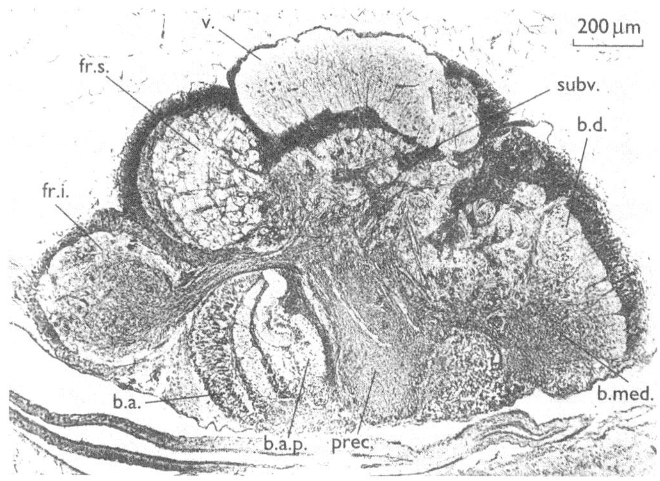

Fig. 24.37. *Mastigoteuthis schmidti* (70 mm ML). Sagittal section through the supraoesophageal mass of the central nervous system.

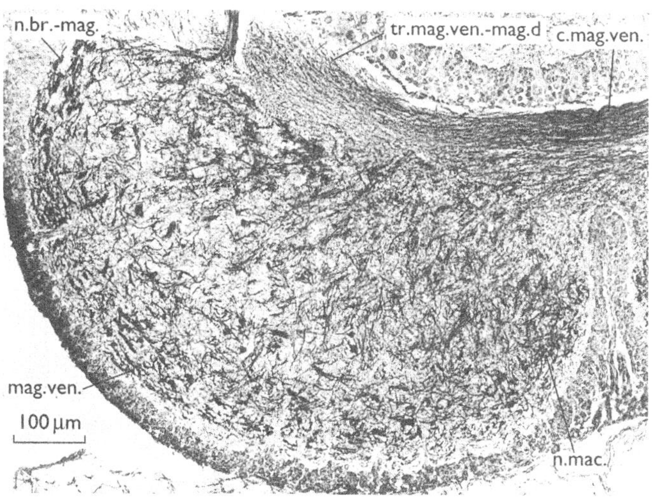

Fig. 24.36. *Mastigoteuthis schmidti* (60 mm ML). Transverse section of the magnocellular lobe.

tentacles. No giant cells have been seen in this lobe but there is a reduced giant-fibre system, though its elements are small. There are fibres up to 8 μm diameter at the centre of the palliovisceral lobe. The pallial nerve contains fibres of various sizes, of which one or two fibres are distinctly larger than the rest (of up to 30 μm diameter). There are no especially large fibres in the stellate ganglia or nerves.

The supraoesophageal lobes are characterized by an inferior frontal lobe system (2.6%) that is larger than the superior frontal lobe (1.3%) (Maddock and Young 1987) (Fig. 24.37). This is the reverse of the condition in *Loligo* and other more visual squids, and is an indication of the importance of the tactile input. The superior frontal and vertical lobes resemble those of *Loligo*, but the vertical lobe is relatively small. The vertical lobe system forms 6.4% of the total brain volume. The peduncle lobes lie in an oblique vertical plane and show a development of parallel fibres that is perhaps greater than in *Loligo*. The fibres in the anterior part of the anterior basal lobe run vertically at the sides and across the centre, and those of the posterior anterior basal lobe run transversely (Fig. 24.35). The median basal lobe is the other part of the supraoesophageal mass concerned with locomotion, probably operating steering about the yawing axis and propulsion with the fins and the funnel.

The extraocular photoreceptor organs comprise anterior and posterior sets of photosensitive vesicles. They lie behind and medial to the peduncle lobe in several species of *Mastigoteuthis* (Dilly *et al.* 1977, R.E. Young 1978). The posterior set of vesicles is a string of some five sacs (Fig. 24.34). The rhabdoms are rather short and irregular. The most posterior vesicles lie at the back of the orbit, below the eye. The nerve from the photosensitive vesicles enters the peduncle lobe from below as a single trunk.

The statocyst has a cavity divided by blunt anticristae and there is a well-defined horizontal channel with anticristae 6 and 7 almost in contact, but not fused, making a closed section (Dilly *et al.* 1977). This suggests the need to monitor quite rapid turns in the yawing plane. The statocyst reaches 6% of the mantle length in specimens of up to 41 mm ML. The dimensions of the statocyst place *Mastigoteuthis* amongst the buoyant squids (Maddock and Young 1984). Kölliker's canal is large and long, and has a broader region lined with cilia. All the cilia of each cell unite to form a blade (Dilly *et al.* 1977). The region around the opening of the canal is covered by epithelium, some cells of which carry short, blunt cilia

24.3 Joubiniteuthidae

The squids of this monotypic family are found in the North and South Atlantic and the western North Pacific. Their vertical distribution is from the surface to 1500 m (Roper and Clarke 1992).

24.3.1 *Joubiniteuthis*

Joubiniteuthis occurs between 10°N and 40°N, and also at 30°S in the Atlantic Ocean and off Japan and Hawaii in the Pacific Ocean

(R.E. Young and Roper 1969b, Okutani and Kubota 1972, R.E. Young 1978). Most have been caught in open nets or found in the stomachs of predators, but seven were captured in opening/closing nets, four by day at 0–200 m and 600–800 m, and one each at 0–120 m, 800–895 m, and 1250–1500 m during the night in the north Atlantic (Clarke and Lu 1974, 1975). There is too little material to be certain of the extent of either the horizontal or the vertical distribution of this squid.

Morphology

Joubiniteuthis portieri (Fig. 24.38) has numerous small chromatophores on the mantle. These are more densely organized on the dorsal surface than on the ventral surface, and there are large, elliptical ones on the lateral surfaces. The tail-like extension of the mantle has a thin integument with many small chromatophores, although the gladius is visible as a central ridge (Okutani and Kubota 1972). The fins are thick and gelatinous medially. The mantle has outer and inner layers of muscle, and these are separated by a thick layer of reticulate tissue (Fig. 24.39). The cartilage of the head region is also vacuolated (see Fig. 24.41). The body shape and texture, together with histological evidence, led Clarke *et al.* (1979) to suggest that this squid is near to neutral buoyancy; and this is supported by the morphometry of the statocysts (Maddock and Young 1984).

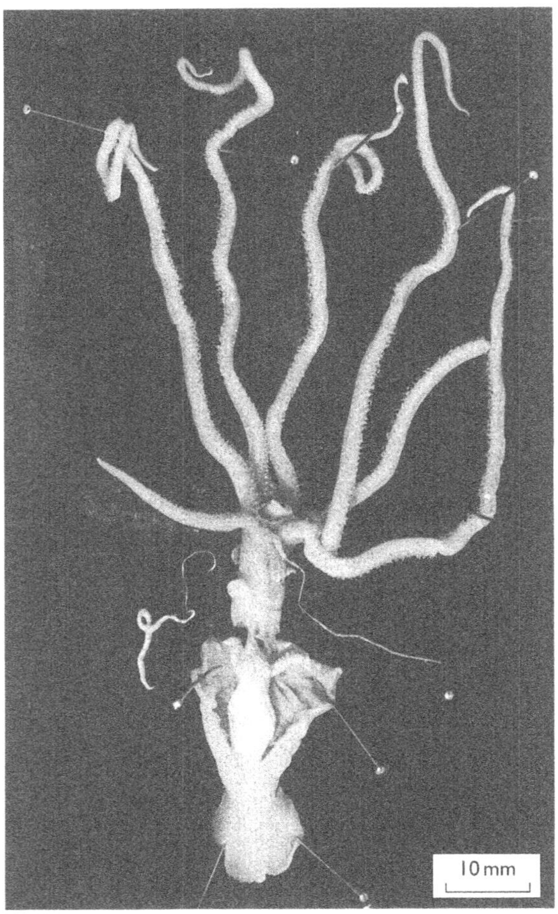

Fig. 24.38. *Joubiniteuthis portieri* (36 mm ML, measured without the tail). Ventral view. Figs 24.39–24.42 are all of the same animal.

The buccal mass occupies much of the head, and the rostral tips of the beaks protrude through the lips. The rostrum of the upper beak is sharp, that of the lower is blunt (Clarke 1986). The radula has simple teeth, the rhachidian tooth and lateral teeth 1 have small cusps, the marginal teeth are longer than the others (R.E. Young and Roper 1969b).

Receptors and central nervous system

There is no special differentiation of the retina of the eye, although the rhabdoms range from 150 to 200 μm in length. Serial sections show the presence of conjunctive eye muscles, as in *Mastigotetuhis*. In sagittal section bundles of crossing fibres and the inner end of one trochlear cartilage, attached to the base of the cranial cartilage, are visible. Both series of sections show several strands of muscle or tendon, loosely attached to the trochlear cartilage but with no clear differentiation into the typical sets found in loliginids (Budelmann and Young 1993). The presence of the crossing fibres suggests that *Joubiniteuthis* has binocular vision.

The optic lobes form 133.8% of the total brain volume (Maddock and Young 1987). The granule cell layers of the optic lobe are well differentiated (Fig. 24.40).

The central nervous system is shown in Fig. 24.41. In the suboesophageal mass the brachial (17%) and magnocellular lobes (12%) form large proportions of the total brain volume. The magnocellular lobe receives a large tract direct from the brachial nerves, in addition to a contribution from the brachiomagnocellular connective (Fig. 24.42). This arrangement is similar to that found in

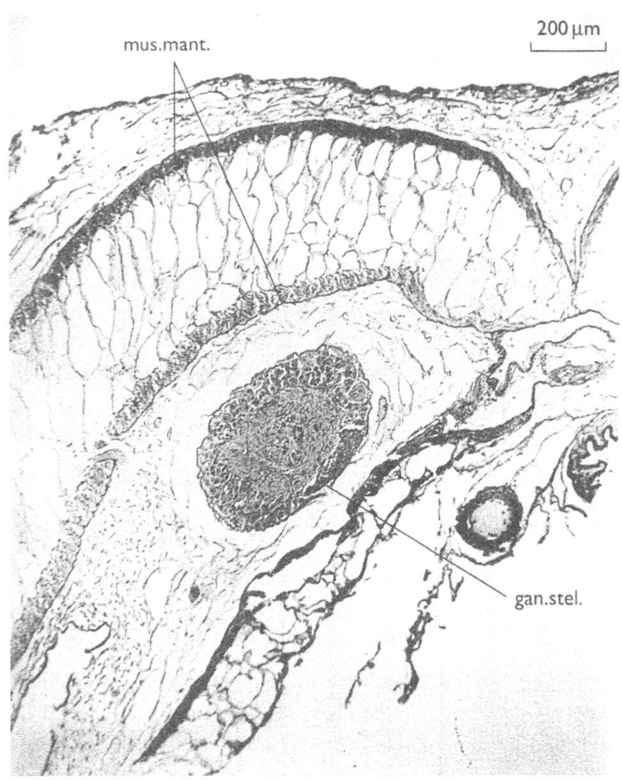

Fig. 24.39. *Joubiniteuthis portieri*. Transverse section to show the stellate ganglion, and the muscle layers of the mantle separated by reticulate tissue.

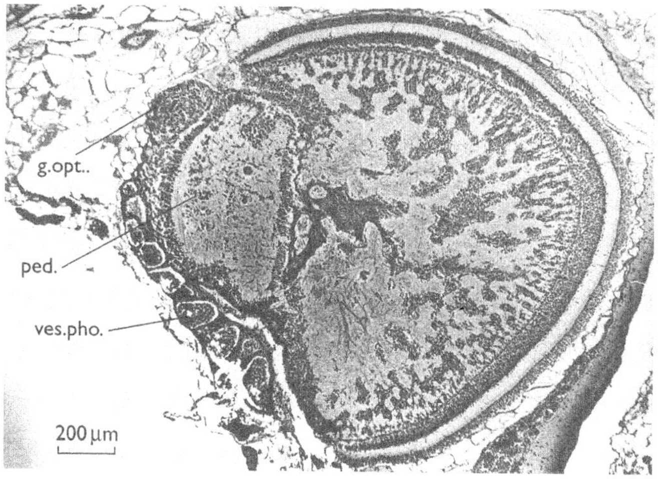

Fig. 24.40. *Joubiniteuthis portieri*. Section through the optic lobe, optic gland, peduncle lobe, and photosensitive vesicles.

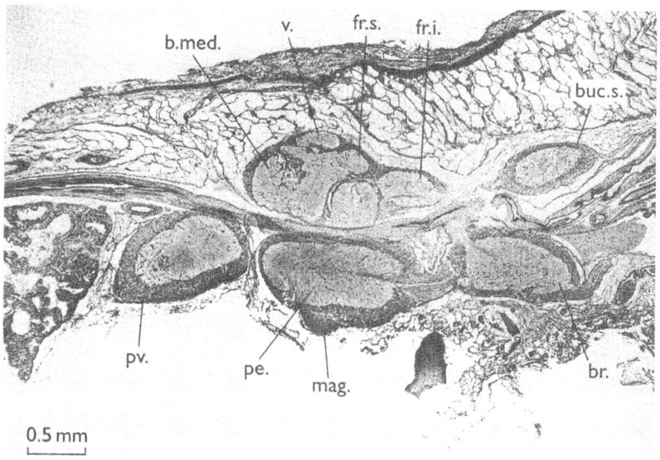

Fig. 24.41. *Joubiniteuthis portieri*. Sagittal section through the supra- and sub-oesophageal masses of the central nervous system. The inferior frontal lobe is prominent when compared with the superior frontal and vertical lobes.

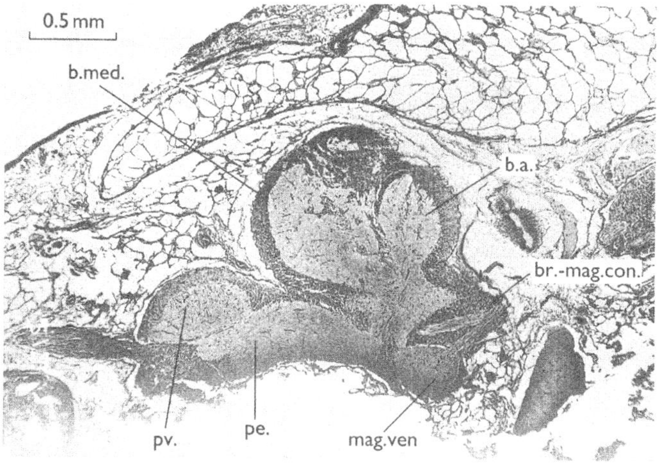

Fig. 24.42. *Joubiniteuthis portieri*. Sagittal section through the brain to show the prominent ventral magnocellular lobe.

Mastigoteuthis, which also possesses long, whip-like tentacles bearing many small suckers.

In the supraoesophageal mass the inferior frontal lobe system occupies 4% of the total brain volume, and so is a prominent part in our specimen (Fig. 24.41). This reflects the importance of tactile information from the slender, delicate arms and tentacles and their many small suckers. The vertical lobe system forms 7.7% of the total brain volume; a small proportion compared with that in most of the other squids measured (Maddock and Young 1987). However, these measures of the various lobes of the brain may also be a reflection of the stage of development of our specimen.

The parallel-fibre lobe system occupies 13.1% of the volume of the brain, much of this being taken up by the peduncle lobe (Fig. 24.40). This lobe receives the large nerve from the compact extraocular photoreceptor organ. This organ consists of a number of globular photosensitive vesicles, which are attached to the posteroventral margin of the optic stalk (R.E. Young 1978), and it forms 3.3% of the total brain volume (Table 2.6). The peduncle lobe is large and, in spite of the small optic lobes, suggests that light is monitored via the photosensitive vesicles, perhaps to allow the animal to correct its depth.

No first-order giant cells or axons were seen, nor any other signs of a giant-fibre system. There are no outstandingly large fibres in the mantle connective or stellate ganglion (Fig. 24.39). The largest fibres in the stellar nerves are 10 µm in diameter, suggesting that movement is mainly by the actions of the fins. The fin lobes form 3.6% of the total brain volume, a somewhat similar proportion to that in other chiroteuthids (Table 2.6).

The statocyst has only small hamuli present at all three turns of the crista and at either end. The only other interruption of the statocyst cavity is a slight thickening of the roof (Young 1989). Kölliker's canal is relatively large. Multiple discriminant analysis of the measurements of the statocyst place *Joubiniteuthis* amongst other buoyant ones (Maddock and Young 1984), and the condition of the statocyst and the delicate, needle-like tail indicate that this squid does not move rapidly.

24.4 Promachoteuthidae

The few specimens captured so far indicates a wide distribution in the North and South Atlantic and Pacific Oceans. Their vertical distribution show deep-living habits.

24.4.1 *Promachoteuthis*

This is one of the strangest and rarest of all the cephalopods and was first caught in the north-west Pacific Basin by *H.M.S. Challenger*, in an open net fished to 3600 m where the bottom was 6000 m. This specimen was described by Hoyle (1885b, 1886) who named it *Promachoteuthis megaptera*, the generic name being a rendering of 'Challenger'. Two more specimens were captured off southern Chile in the south-east Pacific Ocean, one from between 0 and 1830 m and the other at 1391 m depth (Roper and Young 1968). A further

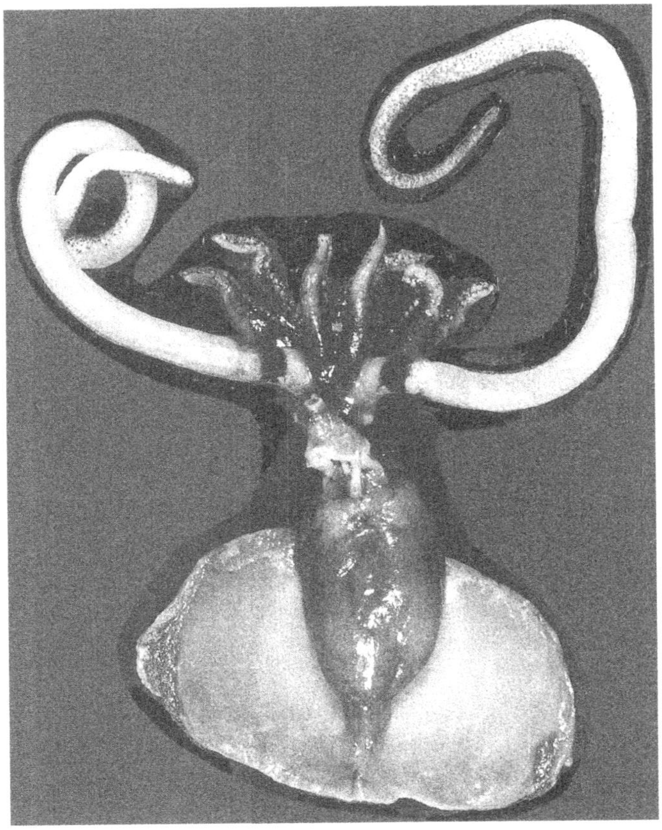

Fig. 24.43. *Promachoteuthis megaptera* (54.2 mm ML). A female captured in 1980 at 31°N 147°E (photograph kindly provided by T. Okutani).

specimen (Fig. 24.43) was caught in the north-west Pacific Basin in a discrete depth tow at 2750 m, where the ocean bottom was 6000 m (Okutani 1983c). The largest specimen so far recorded is a female of 112 mm ML captured at 2650 m in the North Atlantic Ocean (Toll 1982). A mature male has been captured (N.A. Voss 1988a, 1992). These few records suggest that *Promachoteuthis* is a bathypelagic squid living well above the sea floor.

This family shares with the other four families of the 'chiroteuthid lineage' the apparent loss of the club of the tentacle (R.E. Young 1991). The tentacles are notable for their thickness and the presence of minute suckers. The body texture is semisolid, gelatinous, and the mantle stout, with thick walls. The fins are large and broad. The eyes are relatively small. The arm suckers are small and arranged in two rows.

24.5 Batoteuthidae

The representative of this family is found in Antarctic waters but its vertical distribution is poorly known.

24.5.1 *Batoteuthis*

Batoteuthis skolops is the only species of the genus and family and it has been caught in Antarctic waters of the Atlantic, Pacific, Indian, and Southern Oceans, at depths of 366 m to 2525 m in open nets (R.E. Young and Roper 1968).

Batoteuthis has a small head, which is narrower than the long, slender mantle, the posterior end of which is drawn into an extremely long, needle-like tail. The fins are small and short. The funnel is relatively huge. The tentacle lacks a club and bears six longitudinal rows of stalked suckers (R.E. Young and Roper 1968).

24.6 Magnapinnidae

R.E. Young (1991) described an unusual squid, which had been captured in Hawaiian waters (33°49′N, 121°51′W) between 0 and 200 m, and called it 'bigfin'. Further specimens were identified from collections and allowed the description of a new species, which was considered to represent a new genus and family (Vecchione and Young 1998).

24.6.1 *Magnapinna*

Magnapinna pacifica has a small mantle with a thin muscle wall. In contrast the fins are notably very large with a nearly circular outline, and are separate throughout their length. The distal tips of the arms and the tentacles are unique among known extant squids in that they are vermiform and lack suckers. The arms all show an abrupt transition from the sucker-bearing part to slender vermiform filaments which are devoid of suckers and membranes. The tentacles are short, thick and robust, and their suckers small; they also lack the teuthid tentacular club as do other members of the 'chiroteuthid lineage'. Chromatophores are scattered over the head, funnel, dorsal mantle, dorsal and ventral surfaces of the fins, and the aboral surfaces of the arms and tentacles.

Several features of *Magnapinna* indicate a relationship with families of the 'chiroteuthid lineage' (R.E. Young 1991) although distinct from their paralarvae. All have large terminal fins, although those of the former are relatively much larger. The brachial crown of *Magnapinna* differs in the presence of vermiform filaments on the tips of the arms and tentacles. Somewhat similar bare arm tips are present in some paralarval octopodids (Hochberg *et al.* 1992) including *Octopus vulgaris* (see Fig. 31.7); their function has not been investigated but they may be chemosensory.

25 Cranchiidae

Cranchiids are found worldwide except in Arctic waters. They range in size from small to very large and have radiated into every zoogeographical region, and a wide variety of habitats with epi-, meso-, and bathypelagic representatives. The success of cranchiids is apparent from the number of genera and species belonging to the family, they are also the most numerous in the total catch of oceanic squid in the Antarctic, Atlantic, and Pacific Oceans, and in the biomass of its members estimated from those consumed by predators, most notably sperm whales (Clarke 1980). In the North Atlantic, at 60°N, cranchiids form 18% of the diet of this mammal, and in Antarctic waters *Mesonychoteuthis* forms as much as 77% by weight. Sperm whales are also responsible for the capture of the largest specimens of this family yet known, as nets take only smaller animals (Clarke 1983).

The family includes 13 genera, more than any other oegopsin family (Table 1.1); some genera are speciose. It is divided into two subfamilies, Cranchiinae and Taoniinae. N.A. Voss (1980, 1985, 1988b) revised the genera and reviewed the systematics, biology, and biogeography of the family.

Photophores are present in all genera of the family. They include ocular, digestive gland, and brachial end-organ photophores. Examination of the morphologies of the ocular light organs of the cranchiins and taoniins showed that they differ considerably although serving the same function, that of ventral camouflage or counter-illumination (R.E. Young 1977). In cranchiins the light organs act to obliterate the silhouette of the eyeballs by illuminating numerous small ventrally directed photophores, whereas taoniins use their elaborate reflectors to spread the light over the entire projected ventral surface of the eyeball. The brachial end-organs in adult females are, however, very similar and have a glandular structure (Herring *et al.* 2002). The comparative anatomy of the photophores thus supports the separation of the two groups.

25.1 Cranchiinae

Cranchiins are all found in tropical and subtropical waters and have a circumglobal distribution (N.A. Voss 1980). The young live in near surface waters and may undergo ontogenetic descent.

25.1.1 *Cranchia*

Cranchia scabra has a circumglobal distribution in tropical and subtropical waters between 44°N and 37°S (R.E. Young 1972b, N.A. Voss 1980). The majority are concentrated in the upper 200 m but some are found to depths of 1000 m (Clarke and Lu 1974, Lu and Clarke 1975b).

Tubercles

The mantle is covered with tubercles of various sizes, shape and ornamentation (Sasaki 1929). At 30 mm ML they range from small, relatively simple nodules (20 μm across), to large and complex ones (300 μm across) (Fig. 25.1); at 53 mm ML the tubercles reach a width of 660 μm. The tubercles are of hyaline cartilage (Chun 1910a, Person 1969). Growth occurs around the periphery of small tubercles, and mainly at the apices of the conids in the large tubercles (Fig. 25.2) (Dilly and Nixon 1976b). The epidermis is adherent to the tips of the tubercles (Fig. 25.3), like that of *Galiteuthis glacialis* (McSweeny 1978). The thin epidermal sheath, enveloping the mantle, is attached to and supported like a canopy by the tubercles, leaving spaces which were described as 'empty' by Roper and Lu (1990). These so-called empty spaces would presumably be fluid-filled in life, but as yet no investigations have been made. The tubercles may give protection in life as this squid has been observed to retract its head into its mantle, and so with the chromatophores expanded it looks like a large orange (R.E. Young 1972b). A predator would first encounter the tubercles and so deter it, or the tubercles may be involved in buoyancy in conjunction with the subepidermal chambers.

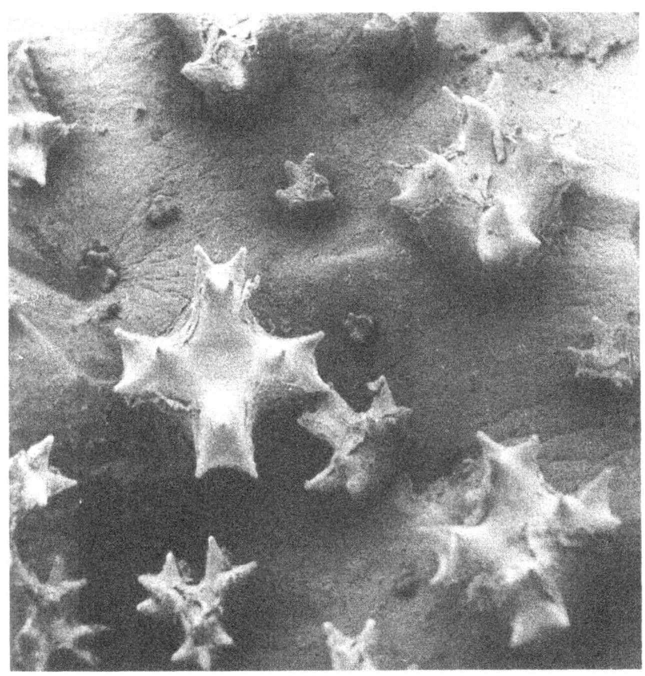

Fig. 25.1. *Cranchia scabra.* Scanning electron micrograph of skin showing tubercles, ranging from small and newly emerged to large and fully formed ones (field width 900 μm) (Dilly and Nixon 1976b).

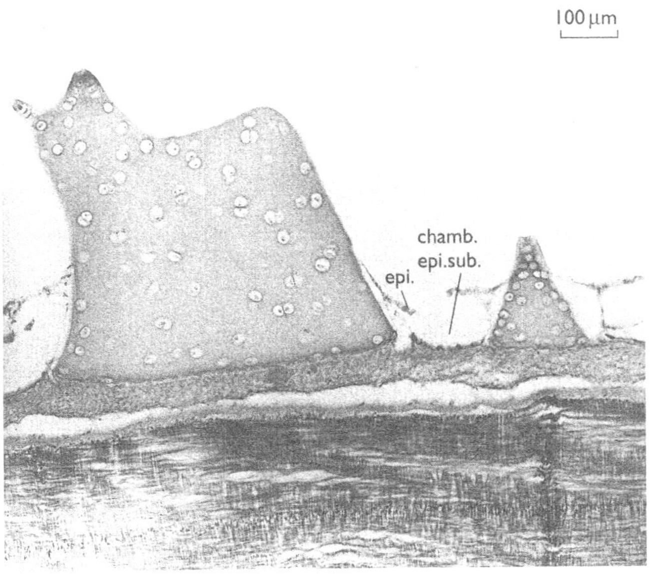

Fig. 25.2. *Cranchia scabra.* A section through two tubercles to show the relative densities of the dividing cells in the cartilage (Holmes).

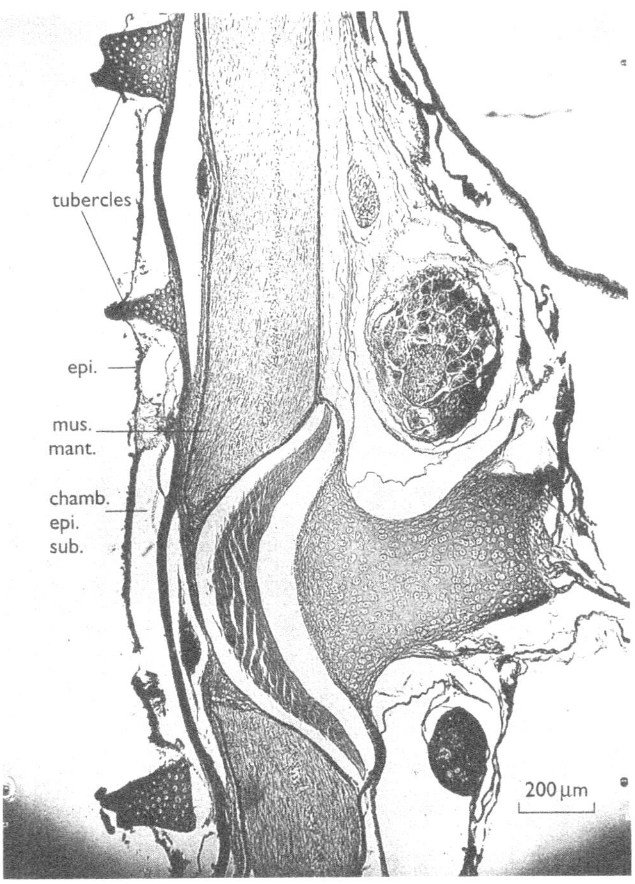

Fig. 25.3. *Cranchia scabra.* Section through the mantle wall to show the tubercles, epidermal sheath and chambers, and the stellate ganglion (Holmes).

Locomotion

Observation of this neutrally buoyant squid, in a shipboard aquarium, revealed it to change position and orientation relative to light and gravity by continual and rapid movements of its fins. The animal survived for about 2 h during which time the mantle contracted once or twice with a resulting backward movement of 3–4 body lengths. The mantle remained deeply indented for a few moments, only slowly regaining its globular shape (Clarke 1962b). The coelomic fluid has a high concentration of ammonium (412 mmol/L) (Denton *et al.* 1969).

Receptors and central nervous system

The eyes are not as pronounced as in other cranchiins, but the presence of a large trochlear cartilage, a strong tendon and conjunctive muscles suggest that *Cranchia* is capable of binocular vision (Budelmann and Young 1993). The optic lobes (Fig. 25.4) are notable for being the largest among the cephalopods measured, forming 609% of the total brain volume (Table 2.6).

In the central nervous system the supraoesophageal lobes occupy 52.4% of the total brain volume, the vertical lobe system 16%, and the inferior frontal lobe system 2% (Maddock and Young 1987). The magnocellular lobe is of moderate size and together with the palliovisceral lobe forms 16% of the total brain volume (Table 2.6). There is a complete giant-fibre system, although the largest fibre is only 20 μm in diameter.

The extraocular photoreceptor organs are multivesicular and form an elongate array (N.A. Voss 1988b). Each organ lies lateral to the peduncle lobe and comprises about 12 small vesicles, forming 7.2% of the total brain volume (Table 2.6), and with rhabdoms which can reach 560 μm (Fig. 25.4).

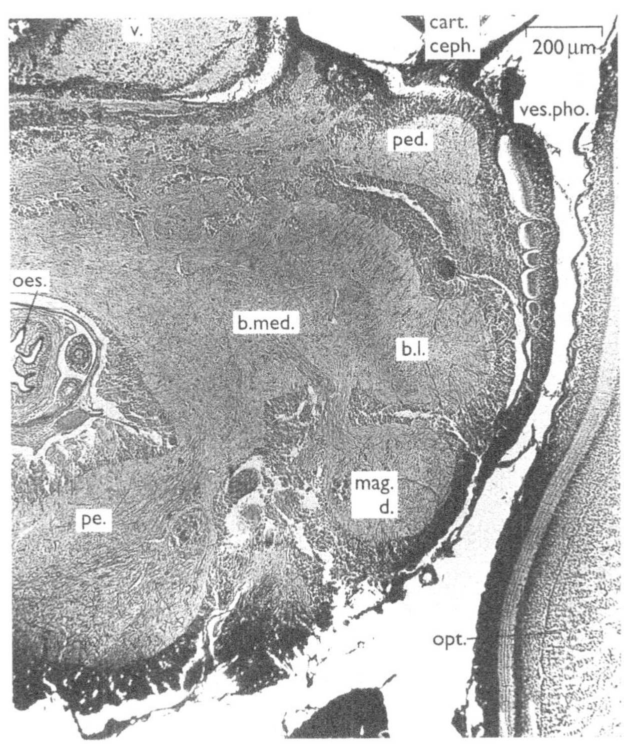

Fig. 25.4. *Cranchia scabra* (55 mm ML). Transverse section to show the central nervous system and the photosensitive vesicles.

The statocyst, in anterior view, has an indentation below the large dorsal sac. Three hamuli are present, at the second and third turns of the crista and at the dorsal end. In a specimen of 125 mm ML 33 straight anticristae were present, but there were fewer in smaller animals. The anticristae are in three parallel rows forming a spiral in the medial, ventral, and dorsal walls, and a large one lies close to Kölliker's canal. The dorsal set is complete and presumably acts to disturb the flow of endolymph (Young 1984, 1989). Multiple discriminant analysis of various measurements of the statocyst places *Cranchia* amongst the buoyant squids (Maddock and Young 1984), confirming observations of living animals (Clarke 1962b) and the analysis of coelomic fluids (Denton *et al.* 1969).

25.1.2 *Liocranchia*

Liocranchia reinhardti has a circumglobal distribution, but *L. valdivae* is restricted to the Indian and Pacific Oceans, in tropical and subtropical waters. Both species undergo ontogenetic descent from the surface to depths of 1200 m (Lu and Clarke 1975b, R.E. Young 1978). There is no evidence of diel migration.

Life history

A recently spawned female was captured in Hawaiian waters, at 775 m depth, with the remains of what appeared to be sperm reservoirs attached to the inner wall of the mantle, close to the funnel. The ovary was depleted, the nidamental glands gelatinous and swollen, and the muscles of the mantle, fins, head, and arms were flaccid (R.E. Young 1978). The size of the larva at hatching is not known but specimens of 2.5 mm ML have been captured in the Atlantic Ocean (Lu and Clarke 1975b). Considerable change takes place in its shape and form with growth. Young animals of up to 7.8 mm ML are globose with two almost circular fins. As the mantle elongates the fins gradually merge into a single fin (Clarke 1969). Ontogenetic descent begins at quite a small size in the Atlantic Ocean (Lu and Clarke 1975b), but is not found in those living in Hawaiian waters (R.E. Young 1978). *Liocranchia reinhardti* grows rapidly until reaching a mantle length of 200 mm in the tropical waters of the Atlantic Ocean (Arkhipkin 1996).

Concealment

Liocranchia is translucent except for the sessile eyes, the visceral nucleus, and the chromatophores when expanded. The visceral nucleus, comprises the digestive gland and ink sac, is elongated and conical in shape. A reflective sheath covers the surface of the visceral nucleus, with an outer thick layer of iridophores laterally and ventrally, and thinning dorsally. The iridophores are flat platelets (115 nm thick). The inner layer is of small reflective 'rods' (149 nm thick) of electron-dense platelets. These are elongate, half cylindrical in cross-section, and arranged parallel to the long axis of the visceral nucleus (Seapy and Young 1986). The rods are structurally almost identical to the light guides described in the ocular photophores of *Bathothauma* (Dilly and Herring 1974). In *Liocranchia* the rods effectively reflect light at lateral angles and may supplement lateral reflection. When the longitudinal axes of the visceral nucleus and of the eyes are oriented vertically, pointed ends directed downwards, then the ventral silhouette of the animal is minimal. In life these organs adjust rapidly, accurately, and consistently to such an orientation, and thus conceal the squid from predators (Seapy and Young 1986).

Receptors and central nervous system

The extraocular photoreceptor organs are paired and in *Liocranchia valdivae* each consists of three elongated vesicles (R.E. Young 1978). In *L. reinhardti* each organ is formed by a linear array of 20–25 vesicles lying between the central brain and the optic lobe on either side (Fig. 25.5); the arrangement of the vesicles is such that light could be received over an arc of almost 180° (Messenger 1967a). Each vesicle consists of a pigmented cup, with rhabdoms of 200 μm and 260 μm in animals of 120 mm and 126 mm ML. The rhabdoms are organized parallel to the long axis of the cup, and the cell bodies occupy the peripheral end of the cup. Light thus passes through the cells to reach the light sensitive region, as in the vertebrate retina. The nerve fibres from each sac of photosensitive vesicles pass separately to the peduncle lobe where the bundles all join and interweave.

Giant cells and the giant-fibre system are well developed (Fig. 25.6). Each pallial nerve has a large second-order giant fibre (50 μm diameter). In the stellar nerve the largest fibre is only 20 μm in diameter.

The statocysts are similar in shape to those of *Cranchia*, and likewise possess 5 hamuli and about 16 anticristae on each side. A row of four anticristae on the medial wall is continued by 5 ventrally, and so divide

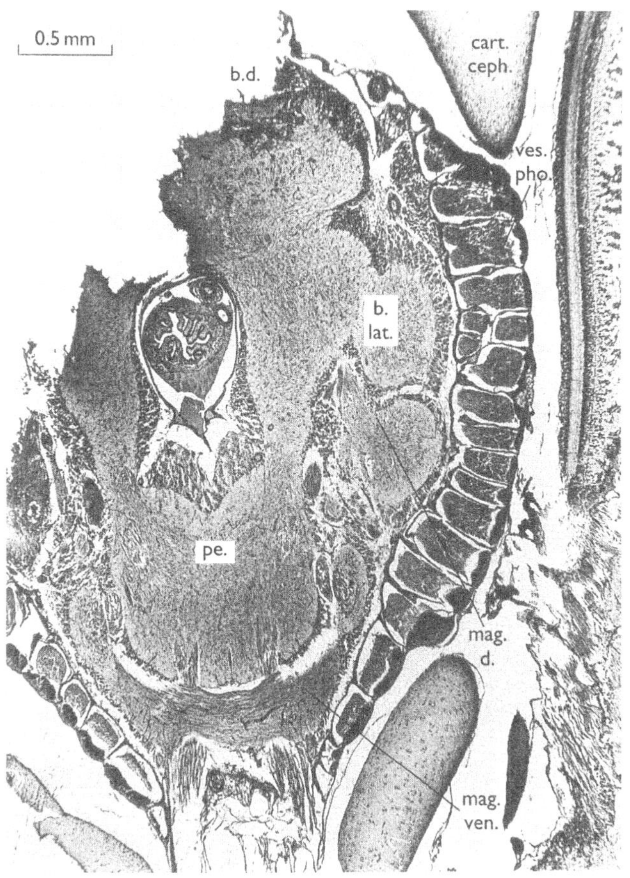

Fig. 25.5. *Liocranchia reinhardti* (126 mm ML). Transverse section showing the central nervous system and elongate array of photosensitive vesicles.

the cavity transversely. In the roof there is a row of 4; this set approximately delimits the horizontal canal. Three smaller protuberances in the dorsal wall are not in line with the others, suggesting a function for disturbing the flow of endolymph.

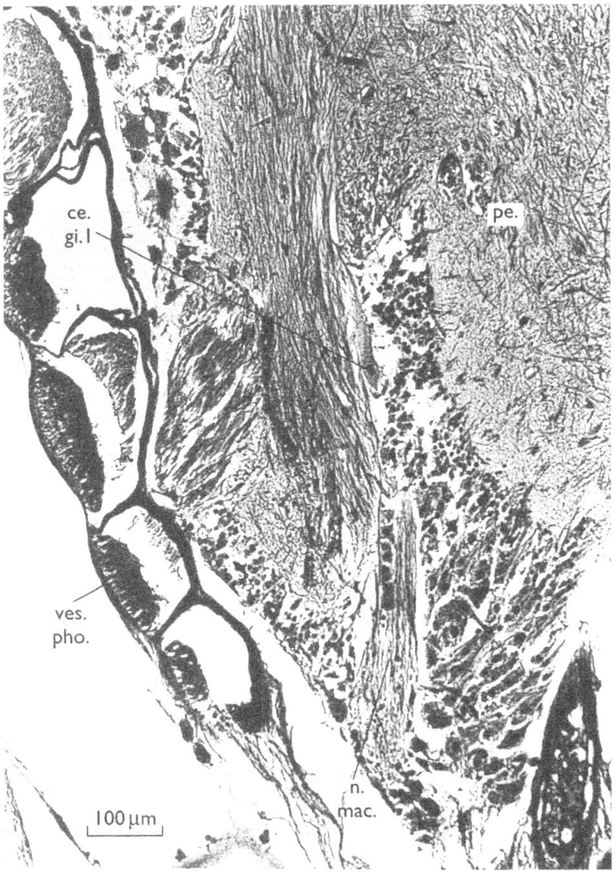

Fig. 25.6. *Liocranchia reinhardti* (126 mm ML). Transverse section to show first-order giant cell.

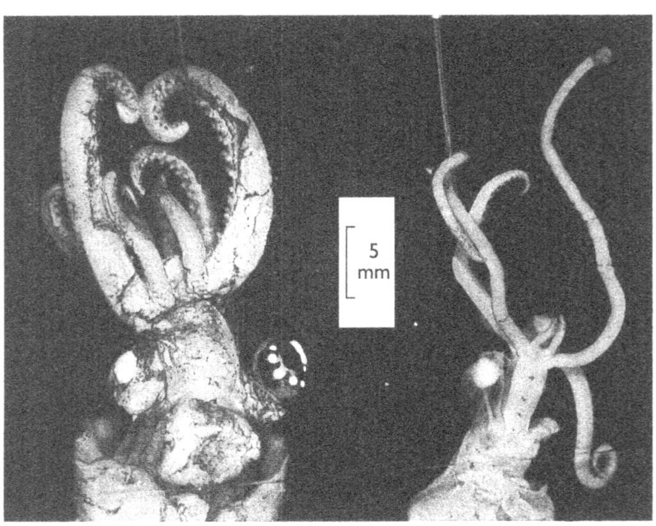

Fig. 25.7. *Leachia*. The eye stalk is long in the small specimen (46 mm ML) and short in the larger animal (80 mm ML).

25.1.3 *Leachia*

This cranchiin has a circumglobal distribution in tropical and subtropical waters (N.A. Voss 1980), and is found from the surface to depths of 2000 m or more (R.E. Young 1975c).

Life history

Leachia pacifica undergoes an abrupt descent from near surface waters to depths of some 2000 m off Hawaii, when the mantle reaches 80% of its maximum length. The sudden vertical migration is correlated with rapid developmental changes, and most notably the eyes become sessile. In addition the brachial end organs develop, and the nidamental and oviducal glands enlarge. Females caught at depths of more than 1000 m had discharged spermatophores embedded in the dorsal surface of the head and mantle, indicating that mating had taken place. Some of these females may have spawned, since they were flaccid and gelatinous, and yet they differed little in mantle length from the largest of the paralarvae, suggesting an abbreviated adult life (R.E. Young 1975c).

Brachial end organs develop only in females at the onset of maturity. The end organs are presumed to be photophores as they are large and covered by pigmented muscular flaps, which would allow luminescence to be concealed or exposed (R.E. Young 1975c). Capture records of near mature females show them to be solitary, with an estimated minimum separation of 61 m. The density of the females is within a reasonable order of magnitude for sexual attraction by bioluminescence to be feasible, especially in depths of 1000 m or more where background bioluminescence is low and predators are scarce (R.E. Young 1983). Males do not develop brachial end organs but they do descend from near surface waters to depths greater than 1200 m as they mature sexually.

Leachia retains a long arm pillar until the mantle length reaches at least 56 mm, at which size the tentacles and the arms are relatively long (Figs 25.7, 25.8). The buccal mass is visible as it lies anterior to the bases of the arms and tentacles. The oesophagus is about 5 mm in length before it reaches the central nervous system; it is deeply folded in the arm pillar.

Receptors and central nervous system

The eyes of *Leachia* are stalked, and remain so until the squid descends into deep water, until when at depths of 1000 m or more they become

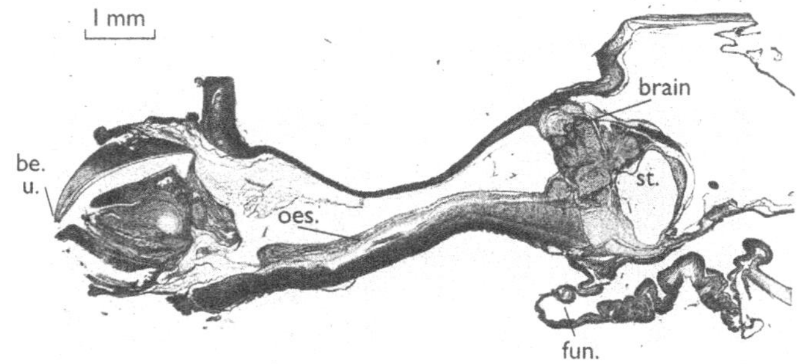

Fig. 25.8. *Leachia dislocata* (56 mm ML). Sagittal section through the head and buccal mass which are separated by the long arm pillar.

sessile (R.E. Young 1975c). In adults the eyes are relatively small, being 5.5% of the mantle length. At 80 mm ML the eyes carry 5–25 subocular photophores (Fig. 25.7) (N.A. Voss 1980). These photophores have emission maxima of 475–500 nm (Herring 1983).

The optic lobes are separated from the central nervous system by long optic tracts. Together the lobes form 284.9% of the total brain volume (Table 2.6) (Maddock and Young 1987).

The cephalic cartilage is short anteroposteriorly, and thick dorsally. It is hyaline in form, and the chondrocytes are especially noticeable. The central nervous system is compact and there is no separation of the anterior suboesophageal mass even in the smallest specimen sectioned (Figs 25.9, 25.10). The magnocellular and palliovisceral lobes form 20.4%, and the vertical lobe system 12.4% of the total brain volume (Table 2.6).

The extraocular photoreceptor organs lie on the posteroventral surface of the peduncle lobe. Each organ is oval and consists of several closely packed, cup-shaped vesicles (Fig. 25.11). A dark brown pigment covers much of the anterior surface of each organ, and this becomes more extensive in adults. There is a pigment in the walls between some of the vesicles so that light can only be received from behind. The vesicles are minute in the paralarva and are small even in the adult (R.E. Young 1978). The cell layers are rather thick and irregular; the rhabdoms increase from 35 μm in a specimen of 46 mm ML to 200 μm at 126 mm ML. The photosensitive organs formed 7.5% of the total brain volume in one specimen (Table 2.6).

The statocyst forms a prominent portion of the head of a larva (Fig. 25.8). In a subadult *Leachia danae* (200 mm ML) 5 hamuli and 13 straight anticristae, some very small, are present and are arranged in ventral, medial, and dorsal sets. The ventral and medial anticristae are unusually long, thin, and pointed, and some bear accessory protuberances. The statocyst of an adult male *L. dislocata* (132 mm ML) has a much less obstructed cavity, with only three hamuli and two small anticristae on the medial wall (Young 1984, 1989). Multiple discriminant analysis of several parameters of the statocyst places *Leachia* amongst the buoyant cephalopods (Maddock and Young 1984).

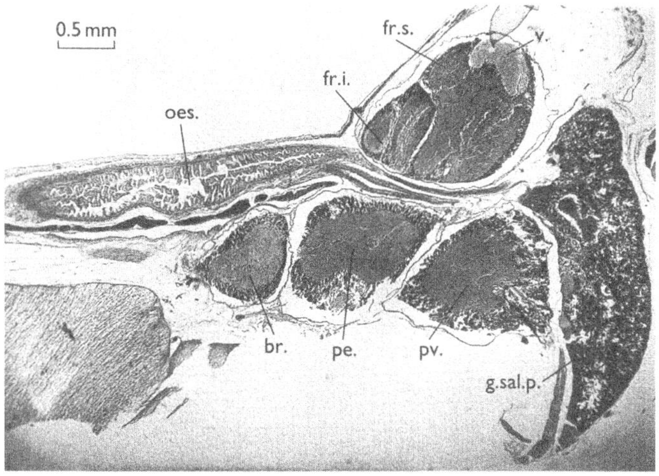

Fig. 25.9. *Leachia dislocata* (132 mm ML). Sagittal section through the central nervous system, oesophagus and posterior salivary gland (Bodian).

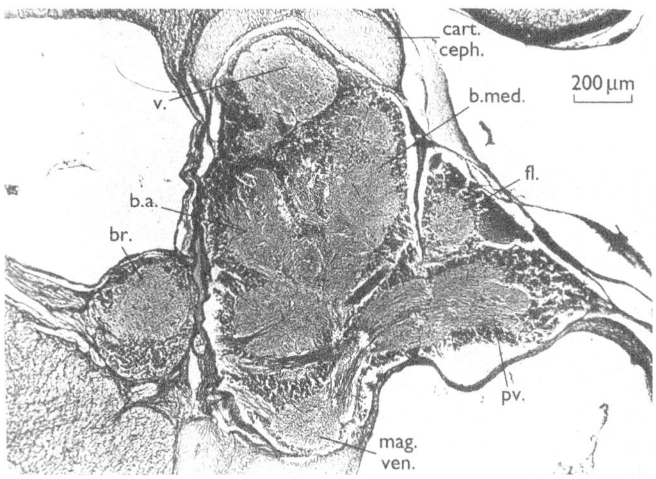

Fig. 25.10. *Leachia dislocata* (56 mm ML). Sagittal section through the central nervous system.

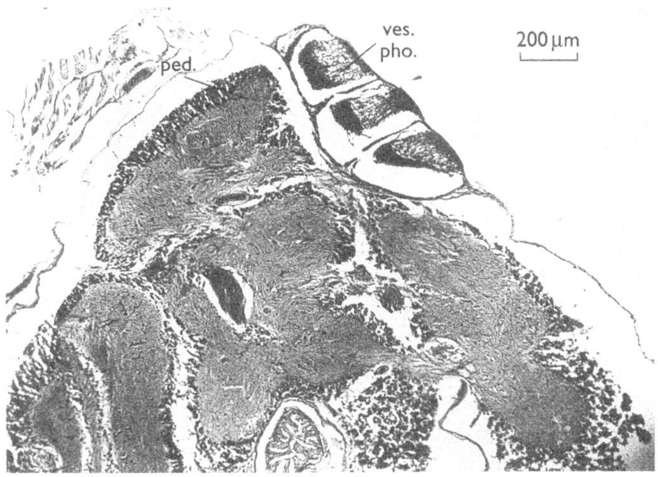

Fig. 25.11. *Leachia dislocata* (132 mm ML). Section through photosensitive vesicle.

25.2 Taoniinae

The Taoniinae have wider geographical distributions than the cranchiins. All are circumglobal, mostly in tropical and subtropical waters, and some extend into temperate and subpolar Antarctic waters. A restricted circumglobal distribution is seen in *Egea* in tropical waters, and *Mesonychoteuthis* in Antarctic waters. Taoniins are found from near surface waters to depths of 1400 m, and perhaps deeper. Many genera undergo ontogenetic descent.

25.2.1 *Helicocranchia*

This genus (Fig. 25.12) has a circumglobal distribution in tropical and subtropical waters (N.A. Voss 1980). In the Atlantic Ocean it is found to 40°N, and from the surface to 500 m in depth; there is also evidence of ontogenetic descent (Clarke and Lu 1974, 1975, Lu and Clarke 1975a, b), and this is also seen in Hawaiian waters where it is caught between 100 and 1200 m (R.E. Young 1978).

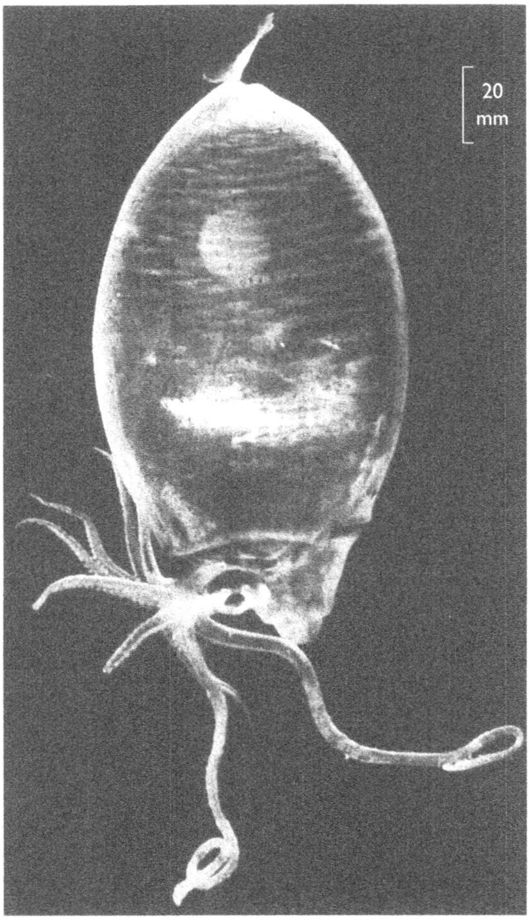

Fig. 25.12. *Helicocranchia pfefferi* photographed while still alive, with its head downwards (Denton *et al.* 1969).

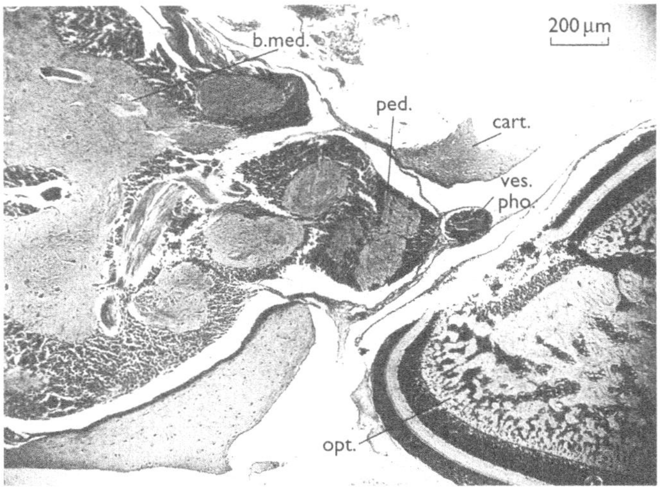

Fig. 25.13. *Helicocranchia pfefferi* (50 mm ML). Oblique transverse section of the central nervous system showing the small photosensitive vesicle (haematoxylin and eosin).

Receptors and central nervous system

The eyes have muscular lids. The retina is uniform, except at the periphery where it thins. The subocular photophores appear quite prominent in an animal of 44 mm ML and the light produced has emission maxima of 500–505 nm (Herring 1983).

The optic lobe is in close contact with the eye. The lobes form 288.4% of the total brain volume (Maddock and Young 1987). The columns of cells in the optic lobe are short. The peduncle lobe is separated from the optic lobe, even in a specimen of 50 mm ML. Few fibres of the optic tract enter the peduncle lobe. The spine region of the lobe is small.

In the central nervous system (Fig. 25.13) the vertical lobe system forms 12.6% of the total brain volume. The vertical lobe alone is relatively large, being 6.5%, whereas the superior frontal lobe is 2.2%. The magnocellular and palliovisceral lobes together occupy 34.8% of the total brain volume (Table 2.6). The ventral magnocellular lobe is connected with the pallial nerve by a large tract. A fully developed giant-fibre system is present in the brain but we have no suitable sections of the stellar nerves to examine. The cephalic cartilage is hyaline.

The extraocular photoreceptor organs are sited on the posterior surface of each peduncle complex (R.E. Young 1978). Each organ consists of a small oval vesicle with short rhabdoms 40 µm in length in a squid of 44 mm ML. Together the two organs form only 0.5% of the total brain volume (Maddock and Young 1987).

The statocyst of a subadult *Helicocranchia papillata* has no hamuli and just two anticristae of unusual pyramidal shape present in the roof (Young 1984, 1989). Multidiscriminant analysis of various measurements of the statocyst indicate that this squid is buoyant but at some distance from the centroid (Maddock and Young 1984). The evidence for buoyancy is supported by the presence of ammonium ions in the coelomic fluid (Denton *et al.* 1969).

25.2.2 *Bathothauma*

Bathothauma lyromma has a circumglobal distribution in tropical and subtropical waters (N. A. Voss 1980). It is found from near the surface to depths of 1400 m, and perhaps deeper (Aldred 1974, Clarke and Lu 1974, Lu and Clarke 1975, R.E. Young 1978). There is evidence of an ontogenetic descent but not of diel vertical migration.

Growth

Remarkable changes in shape and form take place during the growth of this squid from 4 mm ML to a subadult. Initially the arms are short and the tentacles long and with many suckers; they surround the buccal mass and are all at the end of a long arm pillar (Fig. 25.14). This pillar is gradually lost so that the buccal mass, arms, and tentacles come to lie immediately in front of the eyes. The eye stalks are long in the smallest specimen examined, but lengthen still further until the mantle reaches a length of 80 mm before regressing until the mantle reaches 110 mm in length. The eyes change from an elongated ovoid to oval, and then become almost round in a large animal, and they come to lie close to the central nervous system (Fig. 25.15). The maximum size so far recorded is of a female of 205 mm ML (Aldred 1974).

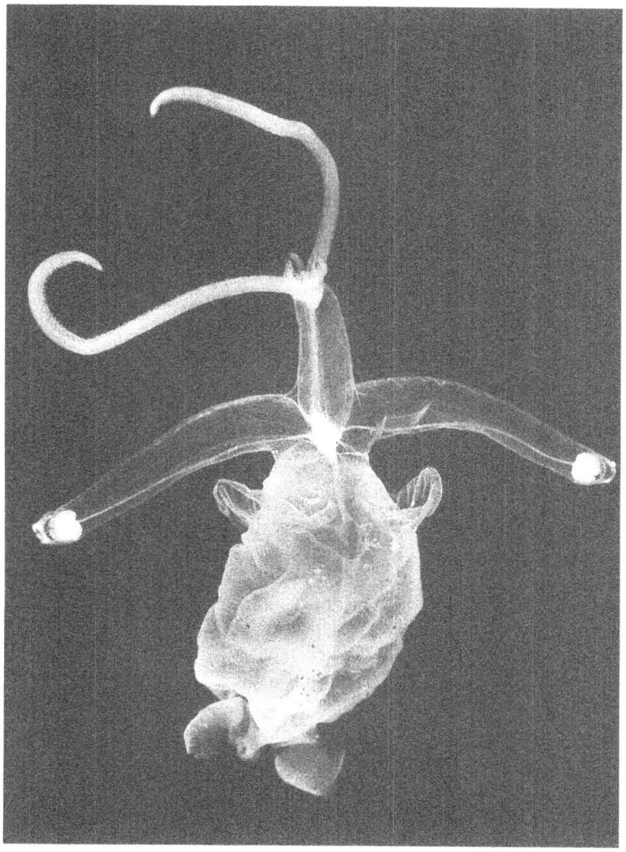

Fig. 25.14. *Bathothauma lyromma*. A young specimen of 12 mm ML with the eyes at the end of stalks, with the buccal mass and the arms and tentacles at the end of a long arm pillar.

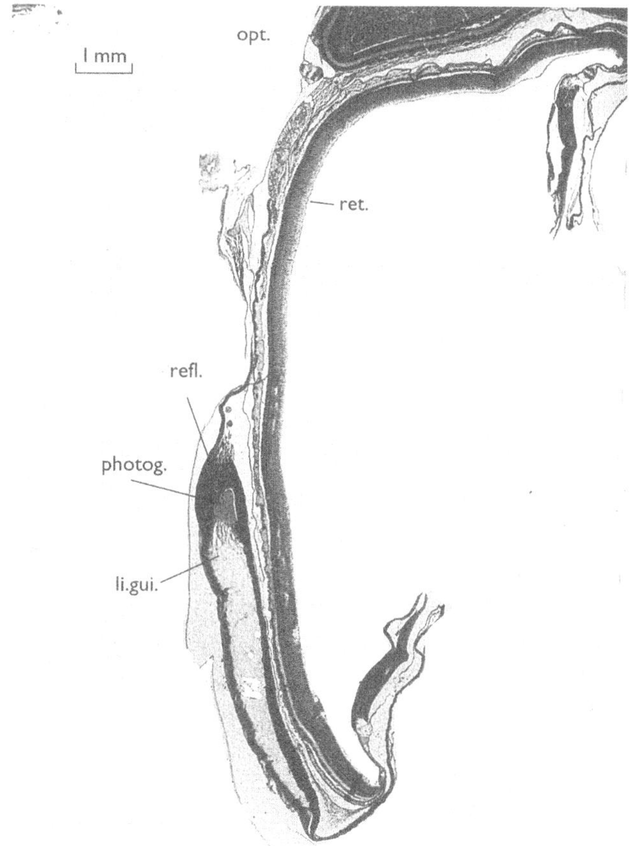

Fig. 25.16. *Bathothauma lyromma* (150 mm ML). Section through eye and photophore.

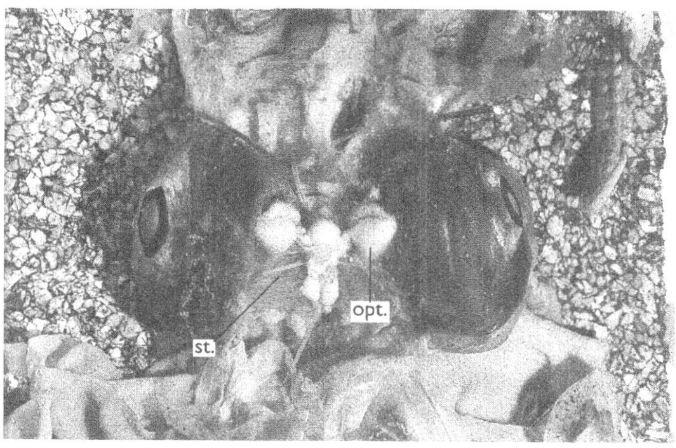

Fig. 25.15. *Bathothauma lyromma* (195 mm ML). Dorsal view of head to show eyes, central nervous system and statocysts exposed after dissection.

Photophores

A subocular photophore is visible in an animal of 35 mm ML, and is pear-shaped at 63 mm ML. The medial edge of the photophore is an iridescent gold when the animal is first captured, and it is covered with reflecting platelets. The area of light emission has a dull white surface, and internal to this is a wide transparent region backed by the silvered inner margin of the photophore. At the top of the photophore the core of photogenic tissue is embedded in a highly silvered reflector. On exposure to ultraviolet light the core gave a feeble yellow-green fluorescence. After treatment with hydrogen peroxide a green-blue light was emitted over the whole of the dull white surface (Dilly and Herring 1974). Emission maxima of 475–505 nm have been recorded from several specimens (Herring 1983, 1988).

The photophore comprises a photogenic core, a reflector and a light guide region (Fig. 25.16). The photogenic core lies deep inside the photophore, and it has cells (up to 150 μm in length) filled with rod-like refractive bodies, highly organized photogenic crystalloids. The reflector layer is a mass of iridophores surrounding the photogenic core, except in the region of the light guides. The light guide region consists of elongated, modified, iridophores which extend from the ventral surface of the photogenic tissue to the surface layer of the photophore. These cells are bounded by iridosomal platelets. The blood vessels supplying the photogenic tissue are surrounded by many small diameter nerve fibres; these may provide some mechanism to control light emission (Dilly and Herring 1974).

Receptors and central nervous system

In a young *Bathothauma* (Fig. 25.17) the retina of the eye appears to be in three parts with rhabdoms of different lengths. Where the retina abuts on to the optic lobe the rhabdoms are 120 μm in length, and

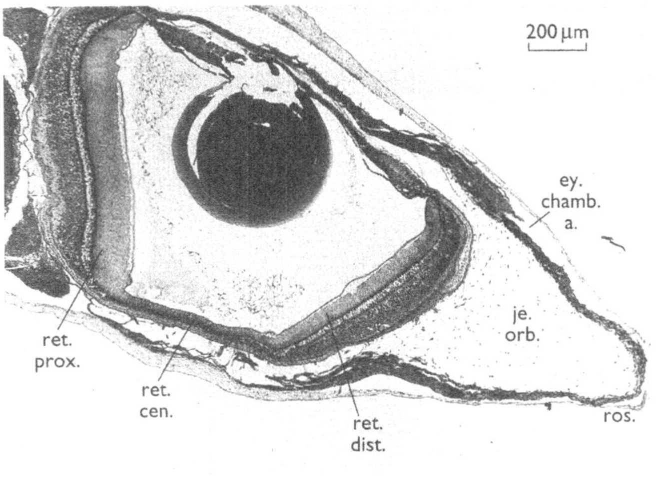

Fig. 25.17. *Bathothauma lyromma* (12 mm ML). Transverse section of eye and ocular appendage, to show the retina and the thinner central part (Young 1970a).

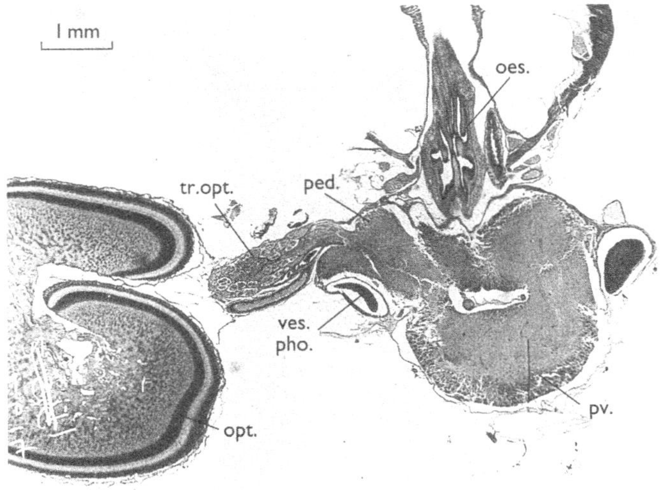

Fig. 25.18. *Bathothauma lyromma* (145 mm ML). Horizontal section of central nervous system and optic lobe (haematoxylin and eosin).

the distal ones 70 µm (Young 1970a). In a larger animal the rhabdoms appear to be of almost uniform length (Fig. 25.16).

The stalked eye carries a distal prolongation, the ocular appendage or rostrum. This appendage is filled with loose connective tissue, jelly like material, with a sparse population of large, branched cells (Fig. 25.17). The jelly is surrounded by dense layers of fibres, shining iridescent when seen from the outside. The appendage has an aperture at its tip. The function of the appendage remains obscure (Young 1970a).

The optic lobes in a larva lie close to the eyes and far from the central nervous system; the long optic tract is visible through the transparent stalk (Fig. 25.14). With growth the eyes become sessile, the optic tract becomes relatively short, and the optic lobes lie closer to the brain (Fig. 25.15). The lobe is indented where the optic tract enters and becomes deeply infolded (Fig. 25.18) so that it almost forms two distinct parts (Young 1970a). The optic lobes occupy 367.6% of the volume of the central nervous system (Maddock and Young 1987).

The peduncle lobe lies close to the central brain (Fig. 25.19). In the larva and the adult there is only a small contribution of fibres from the optic tract to the peduncle lobe. The reduction of this path is correlated with the relatively small size of the peduncle lobe and its spine, as well as the short columns of cells in the optic lobe.

Each extraocular photoreceptor organ is a flat, oval vesicle located on the posteroventral surface of the peduncle complex (Fig. 25.18). They form 6.1% of the total brain volume (Maddock and Young 1987). There is no screening pigment associated with these organs (R.E. Young 1978). Rhabdoms of 150 µm were present in the photosensitive vesicles in an animal of 195 mm ML.

In a subadult the central nervous system is compact with all of the suboesophageal lobes close together. The vertical lobe system forms 16.4% of the total brain volume. The ventral magnocellular lobe is relatively small but the giant-fibre system is well developed with first-order fibres crossing in the palliovisceral lobe (Fig. 25.20). There are one or two giant fibres in each stellar nerve.

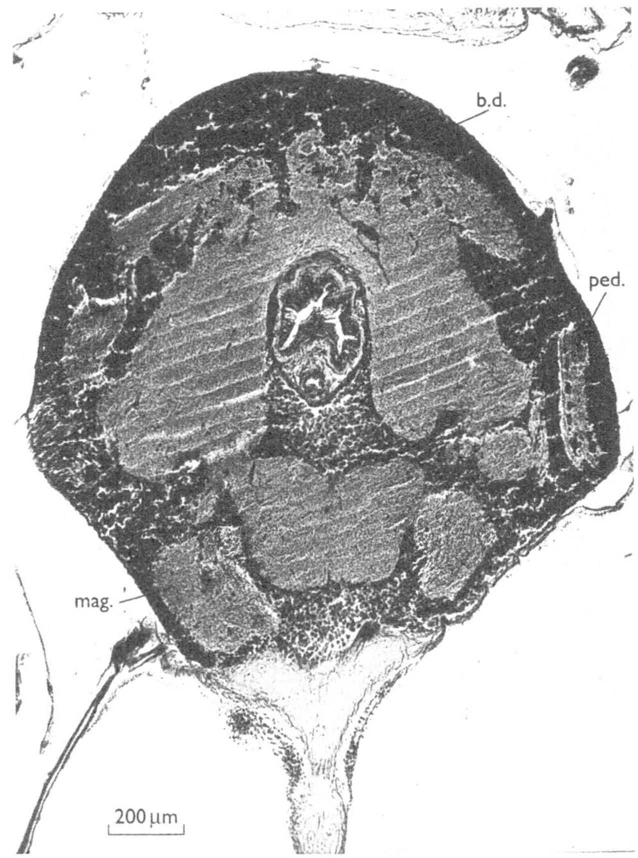

Fig. 25.19. *Bathothauma lyromma* (12 mm ML). Transverse section of the central nervous system, with the closely associated peduncle lobe.

The statocysts of subadult *Bathothauma* (145–195 µm ML) are simple (Young 1984). The cavity is broad and large, and lacks internal division as it is without obstructions. Measurements of the statocyst place this squid amongst the buoyant cephalopods (Maddock and Young 1984). From the condition of the statocyst

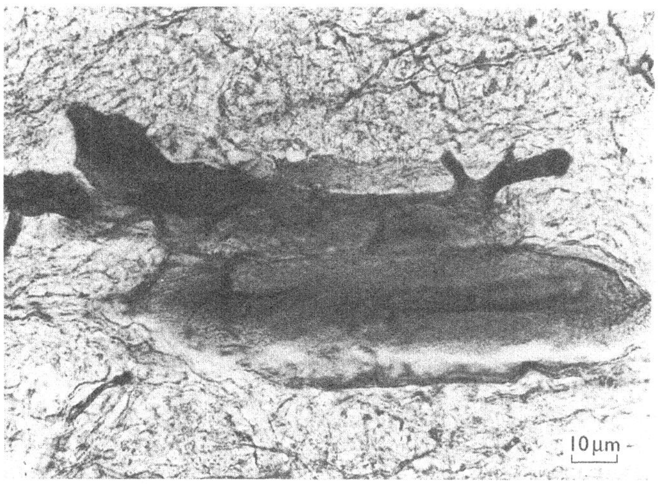

Fig. 25.20. *Bathothauma lyromma* (25 mm ML). Transverse section showing first-order giant fibres crossing in the palliovisceral lobe and giving off branches.

N.A. Voss (1988b) suggested that this squid has low mobility. The fins of *Bathothauma* are small and their muscles weak, and the fin lobe of the brain forms only 1.8% of the total brain volume (Maddock and Young 1987). These features are all indicative of slow movement, and so it is strange that it has a giant-fibre system (Young 1989).

25.2.3 *Sandalops*

This taoniin has a circumglobal distribution in tropical and subtropical waters and is found from near surface waters to depths of more than 2000 m (N.A. Voss 1980). It does not undergo diel vertical migration but does descend into deeper waters with growth (R.E. Young 1975d). Specimens of up to 110 mm ML have been recorded (Nesis 1987).

Photophores

A large photophore covers most of the ventral surface of each tubular, upwardly directed eye of the juvenile, with a smaller light organ whose ventral edge lies in an indentation of the larger one. Most of the photophore is occupied by rod-like iridophores, composed of long concentric ribbons of iridisomal platelets. A hemispherical photogenic body lies in the centre of a dorsal bulge at the medio-posterior end of the larger photophore. The flat dorsal surface of this body abuts against a thick, dorsal reflector. Light guides originate between the light-producing body and the dorsal reflector. A dense mat of light guides form a concave reflector over the ventral surface of the photogenic body and functions as a broad-band reflector. The dorsal reflector is partly encircled by a strip of iridophores, the fibrous reflector (R.E. Young 1977). Except for this zone a heavy pigment covers the outer surface of the dorsal and fibrous reflectors. In the adult the ocular photophores are greatly enlarged, the larger one covering much of the posterior ventral surface of the eye (R.E. Young 1975d).

When alive the animal is transparent apart from the eyes and visceral organs. The eyes are shielded ventrally by the photophores which provide ventral countershading. Luminescence has not been observed in *Sandalops* at sea, but a 'white' light directed into the ocular photophore of a freshly caught specimen resulted in a blue glow in the dorsal half of the fibrous reflector (R.E. Young 1977). The opaque digestive gland and ink sac complex maintains a vertical orientation and in this way reduces the problem of counter-shading. The sides of the eyes have a strong purple-gold metallic sheen; this would be effective when the eyes are oriented vertically (R.E. Young 1975d).

Receptors and central nervous system

The eyes undergo considerable change in shape and form as the animal grows; this has been correlated with depth of capture in *Sandalops melancholicus* (R.E. Young 1975d, 1977). Young animals (up to 28 mm ML) were caught above 400 m possessed compressed eyes, each with an ocular appendage. Juveniles (30–52 mm ML) captured at 425 m had tubular eyes. The eyes of a larger animal (66 mm ML) taken at 675 m were in a state of transition from tubular to almost hemispherical form. Such eyes were also seen in animals of 100 and 105 mm ML taken at depths of 800 m to 1075 m.

Behind the strange eyes is a trochlear cartilage and two pairs of anterior conjunctive muscles. One pair, conjunctive muscle I, cross via a tendon below the brachial lobe and pass laterally in a groove of the trochlear cartilage until the latter divides into two branches; here the tendon separates into numerous muscle fibres, which attach to the eyeball. The crossing of conjunctive muscle II is just below the tendon and the muscle fibres spread ventrally. The system is thus very like that of *Loligo* (Budelmann and Young 1993), and forms a mechanism for directing the eyes forward.

The optic lobes form 218.6% of the total brain volume (Maddock and Young 1987).

In the central brain the magnocellular and palliovisceral lobes form 38.9% of the total brain volume (Table 2.6). There are large tracts from the brachial and pallial nerves to the magnocellular lobe. The vertical lobe complex forms 8.3% of the total brain volume, and so is small when compared with that of other cranchiids measured. The fin lobe is 3.2% of the volume and has only large cells. The first-order giant fibres are fused in the midline of the palliovisceral lobe.

The extraocular photoreceptor organs consist of a pair of single, bilobed, vesicles located on the ventral surface of the peduncle complex. The vesicles increase in size between the juvenile and adult stages (R.E. Young 1977, 1978). They form 54.1% of the total brain volume (Maddock and Young 1987). The rhabdoms are 200 µm in length.

The statocyst is a little more complex than that of *Bathothauma* and *Helicocranchia*, but nevertheless still simple. It has one hamulus and eleven straight anticristae; the latter are all small, except two larger ones in the position of 2 and 6 (Young 1984, 1989). Multiple discriminant analysis of various measurements of the statocyst places *Sandalops* among the buoyant squids (Maddock and Young 1984).

25.2.4 *Liguriella*

This taoniin is found in the tropical waters of the Indian and West Pacific Oceans, and the subtropical and northern sub-Antarctic waters of the Atlantic, Indian and Pacific Oceans. Its vertical distribution is from subsurface waters to depths of more than 1000 m (N.A. Voss 1980). Specimens of up to 240 mm ML have been recorded (Nesis 1987).

The extraocular photoreceptor organs are monovesicular, bilobed, and moderately large in size (N.A. Voss 1988b). In our specimen of *Liguriella podophtalma*, of 138 mm ML, the vesicle is an oval sac (900 × 700 μm) with rhabdoms 350 μm in length.

Liguriella has a compact central nervous system. The ventral magnocellular lobe is relatively large and is directly connected with the pallial nerve. It also receives a large brachiomagnocellular tract. Giant fibres cross in the palliovisceral lobe. The histological sections of our specimen are incomplete and did not allow measurements to be made.

The statocyst cavity shows little subdivision. There are only two hamuli and two anticristae, one is in the midline and the other close to Kölliker's canal (Young 1984, 1989). Multidiscriminant analysis of several parameters indicate that this squid is neutrally buoyant (Maddock and Young 1984).

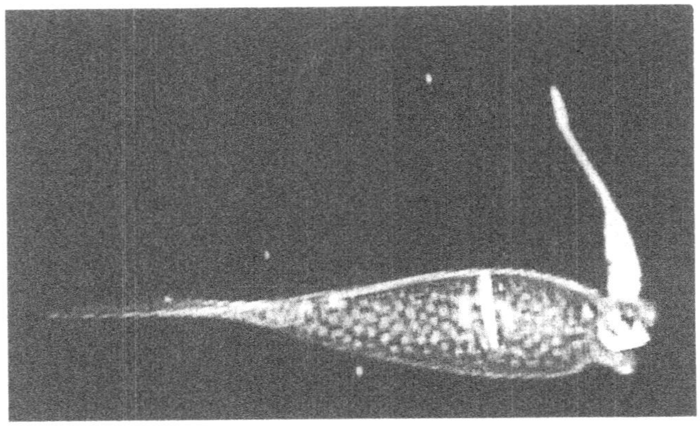

Fig. 25.21. *Taonius pavo* photographed in the sea (684 m depth) in the 'cockatoo' posture (Vecchione and Roper 1991).

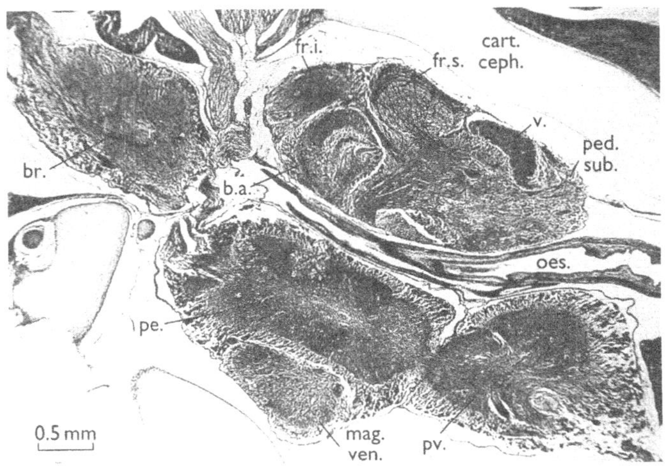

Fig. 25.22. *Taonius* (283 mm ML). Sagittal section of the central nervous system (Bodian).

25.2.5 *Taonius*

This genus is circumglobal from sub-Antarctic to sub-Arctic waters (N.A. Voss 1980). Its members are found from just below the surface to depths of more than 2000 m, but are mostly concentrated below 400 m (R.E. Young 1978). The juveniles inhabit depths of 500 m to 700 m, and the adults descend to greater depths. *Taonius pavo*, recorded from a submersible between 830 and 862 m, was seen in characteristic 'cockatoo' posture (Fig. 25.21).

Receptors and central nervous system

The eyes have long stalks in the young, but in the juvenile they are close to, and aligned along, the head so that they are anteriorly oriented and tubular in form. In the latter the retina is divided into a thick main portion (180 μm) and a very thin (25–50 μm) accessory retina. In adults the eye is almost hemispherical and oriented anteriorly; the retina is uniform (290 μm) except at the periphery where it thins rapidly (R.E. Young 1975d).

The changes in the eyes with the growth take place during the descent from near surface to much deeper waters. Two pairs of anterior conjunctive muscles are present and a large trochlear cartilage. The tendon of conjunctive muscle I crosses in a channel below the brachial lobe. Conjunctive muscle II is a large bundle of muscle fibres; one part of it runs between two cartilages and the other part spreads out ventrally. This condition is somewhat similar to that found in *Sepia* that ensures convergence of the eyes (Budelmann and Young 1993).

The optic tract is relatively long. Fibres diverge from the tract to the peduncle lobe and there intertwine with large bundles from the photosensitive vesicle. The peduncle spine is small and receives fibres from the plexus of the two inputs. The optic lobes form 448.7% of the total volume of the central nervous system (Table 2.6).

The cephalic cartilage is hyaline and thick, especially dorsal to the central nervous system. The brain is elongated in the two large specimens sectioned, one of 283 mm ML and another of 540 mm ML. The sub- and supraoesophageal masses are shown in Fig. 25.22. The brachial lobe lies some way front of the other lobes of the sub-oesophageal mass, and it forms 18.4% of the total brain volume (Table 2.6). The ventral magnocellular lobe is connected directly with the pallial nerve (Fig. 25.23). This lobe also receives a bundle of small fibres from the macula. The anterior basal lobe is a notable part of the supraoesophageal mass, and the parallel fibre lobe system as a whole forms 14.2% of the total brain volume (Maddock and Young 1987).

The vertical lobe occupies 4.41% and the superior frontal lobe 3.66% of the total brain volume (Table 2.6). The latter sends many fibres to the subvertical lobe and to the vertical lobe (Fig. 25.23). In two of our specimens there are large patches of degeneration in the vertical lobe, with an invasion of blood sinuses. These degenerative patches cannot be artefacts since they are present in two females captured at different locations and separated by 11 years. There is no indication of degeneration in any other part of the brain, and none in a large mature male examined.

The subpedunculate lobe is relatively large and sends a tract of fibres towards the optic tract; this is perhaps concerned with the control of the output of the photophores.

The giant-fibre system is well developed with one large fibre (70 μm in diameter) in each stellar nerve (Fig. 25.24).

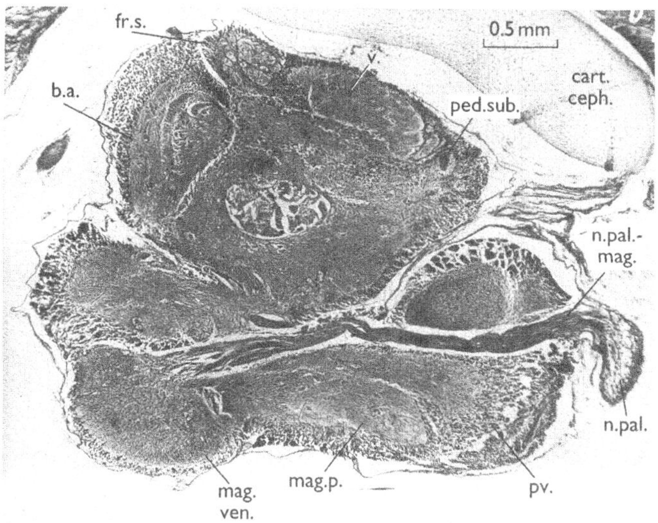

Fig. 25.23. *Taonius* (283 mm ML). Sagittal section of the central nervous system to show (a) the palliomagnocellular nerve, and (b) the vertical lobe and area of degeneration (Bodian).

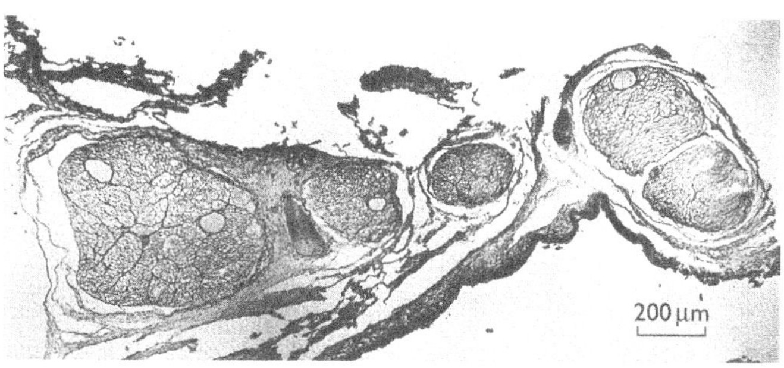

Fig. 25.24. *Taonius pavo* (540 mm ML). Section through stellar nerves.

The paired extraocular photoreceptor organs lie on either side on the posteroventral side of each peduncle complex. No screening pigment was found in the animals R.E. Young (1978) examined. The photosensitive vesicles form 47.8% of the volume of the central nervous system (Maddock and Young 1987). They have rhabdoms of 400 μm in a specimen of 283 mm ML and of 500 μm in one of 540 mm ML.

The statocyst has only 3 hamuli and 17 straight anticristae. The anticristae form a continuous curved row with minute ones ventrally, then proceeding up the medial wall and then dorsolaterally. Their size varies irregularly. There is a protuberance close to Kölliker's canal (Young 1984, 1989). Multiple discriminant analysis of various parameters of the statocyst places *Taonius* amongst the buoyant squids (Maddock and Young 1984).

25.2.6 *Galiteuthis*

Galiteuthis occurs circumglobally in tropical, subtropical, temperate, and subpolar regions, and in Antarctic waters (N.A. Voss *et al.* 1992b). It exhibits some diel movements and ontogenetic descent, and has been captured in nets to depths of 2000–2400 m (N.A. Voss 1980, McSweeny 1978, Roper *et al.* 1985, Rodhouse and Clarke 1986). *Galiteuthis* is one of the most abundant of the pelagic cephalopods found south of the Antarctic Polar front (Piatkowski and Hagen 1994).

Living *Galiteuthis*, observed in a shipboard aquarium, was normally quite active moving around for much of the time, but would hang motionless for long periods when the only movements were largely restricted to the tentacles and arms, with occasional fin movements. One squid moved about 300 mm backwards during an escape response. No respiratory movements were seen although a constant stream of water issued from the funnel (McSweeny 1978). One squid (identified as *G. phyllura* by R.E. Young 1972b) was photographed in the San Diego Trough, at 1100 m, in the head down position with arms and tentacles held closely around the funnel (Church 1971).

Galiteuthis bears numerous cartilaginous tubercles scattered over the surface of the mantle of subadults, mostly anteriorly and dorsally. Eventually the tubercles become thick, pillar-like, and rounded distally, and can measure up to 0.4 mm in diameter and 0.6 mm in height. The epidermis of the mantle is supported by these tubercles, fusing at their rounded shoulders, leaving a subepidermal space which is presumably filled with fluid (McSweeny 1978).

Life history

One gravid female (496 mm ML) had about 24 sperm reservoirs attached to the mantle surface close to the anterior margin. The reservoirs (each 0.30 mm diameter and up to 27 mm in length) were either firmly attached along their length, or deeply embedded in the mantle wall. Several had penetrated the mantle and were projecting into the inhalant chamber. The passage of one reservoir was indicated by a scar on the mantle wall. Movement through the wall may be aided by enzymes or the breakdown of the mantle muscle. The musculature of the mantle wall was soft, fragile, and greatly reduced and consisted of isolated strands in a gelatinous matrix. A large number of eggs of uniform size (4.0 × 2.8 mm) were present. These eggs were significantly larger than those in a female of 397 mm ML, indicating their rapid development before spawning (McSweeny 1978). Two further females (425 and 475 mm ML) were captured at the surface of an ice hole over a depth of 1900 m. Both were gelatinous, devoid of tentacles, had empty stomachs, and empty spermatangia distributed in the dorso-anterior part of the mantle. One large mass of eggs appears to mature in a short time, while the others degenerate; ripe eggs measured 3.3 × 2.5 mm (Nesis *et al.* 1998).

Several hundred *Galiteuthis glacialis* were captured to the south of the Antarctic Polar Front, in opening/closing nets at 300–400 m in daylight, and 200–300 m in darkness. This is the upper zone of warm deep water beneath the Antarctic surface layer. The captures were dominated by specimens of less than 20 mm ML, the remainder were up to 60 mm ML (Rodhouse and Clarke 1986). In the early stages of its life cycle it is caught from the surface to 300 m, and one large specimen was taken at 800 m (Piatkowski and Hagen 1994). Specimens of only 3 mm ML as well as a mated female (493 mm ML) were caught near Heard Island (Lu and Williams 1994b). The largest specimen recorded as yet is of *G. phyllura* of 2750 mm ML (Nesis 1993b).

Receptors and central nervous system

The eyes of young *Galiteuthis glacialis* are borne on short, stout stalks. At 15 mm ML each eye is oval and occupies more than half of the stalk, but at 50 mm ML the eyes are sessile, and by 100 mm ML are round; eventually the eyes occupy most of the head (McSweeny 1978). A trochlear cartilage and two conjunctive muscles are present, suggesting that this squid has binocular vision. The retina has rhabdoms of uniform length of 300 μm in a specimen of 297 mm ML. A crescent-shaped ocular light organ lies on the ventral surface of each eye, and it has connective tissue fibres which act as light guides. The light produced by these organs has an emission maximum of 506 nm in *G. phyllura*, 510 nm in *G. glacialis* and 495 nm in an un-named species (Herring 1983).

The optic lobes are well differentiated (Fig. 25.25), and relatively very large, forming 467% of the total brain volume (Table 2.6).

The central nervous system is compact with the brachial lobes not widely separated from the pedal lobe (Fig. 25.26). The inferior frontal lobe is simple whereas the median superior frontal lobe is complex. The vertical lobe system forms 11.9% of the total brain volume (Table 2.6).

The magnocellular lobe is simple but has a large anterior commissure. Giant cells were not seen. The magnocellular lobe is probably without a direct connection with the pallial nerve. There is a small giant fibre chiasma in the palliovisceral lobe. We have no sections of stellar nerves. The fin lobe has only large cells.

Each of the paired extraocular photoreceptor organs consists of a single, crescentic-oval, vesicle lying on the posteroquadrant of the optic stalk. The vesicles of *Galiteuthis phyllura* have a thick rind and rhabdoms 550–600 μm in length (R.E. Young 1972b, 1978). In *G. glacialis* these vesicles (Fig. 25.25) form 12.2% of the total brain volume (Maddock and Young 1987). The rhabdoms measured up to 300 μm in our specimen of 297 mm ML. The photosensitive nerves run in several small bundles to the peduncle lobe, passing on both sides of the peduncle spine.

The statocyst is much less subdivided and the butterfly shape less marked than that of *Taonius* and of *Mesonychoteuthis* (Young 1984). Five relatively small hamuli and a few small anticristae are present in subadult specimens. The sacs above and below the crista are large, the statolith is small. Multiple discriminant analysis of various measurements of the statocyst place *Galiteuthis* amongst the buoyant squids (Maddock and Young 1984). The statocyst appears to be suitable for monitoring slow turns, although this squid is capable of escape movements (McSweeny 1978).

25.2.7 *Mesonychoteuthis*

Mesonychoteuthis hamiltoni (Fig. 25.27) reaches a large size. One of 1170 mm ML was taken in a haul sampling a depth horizon from 2300–2000 m, and is the only adult female recorded so far (Rodhouse and Clarke 1985). It is restricted to Antarctic waters, where the subtropical convergence may act as a boundary. Its vertical distribution is little known, although a large specimen was caught in an opening/closing net at 2000 m (Clarke 1986), and others from the surface to 500 m depth.

The smallest specimens (4.8–27.9 mm ML) caught in nets were apparently newly hatched since fragments of the chorion remained attached to their mantles. Many changes in shape and form take place during the early stages of the life cycle (McSweeny 1970, Rodhouse and Clarke 1985). A further specimen of 1050 mm ML was captured in a haul sampling a depth horizon of 2000–2200 m (M.R. Clarke, pers. comm.). These squids are known to reach 2500 mm ML (N.A. Voss 1980). From the sizes of beaks recovered from the stomachs of sperm whales it has been estimated that *Mesonychoteuthis hamiltoni* may exceed 4000 mm ML (Clarke 1986).

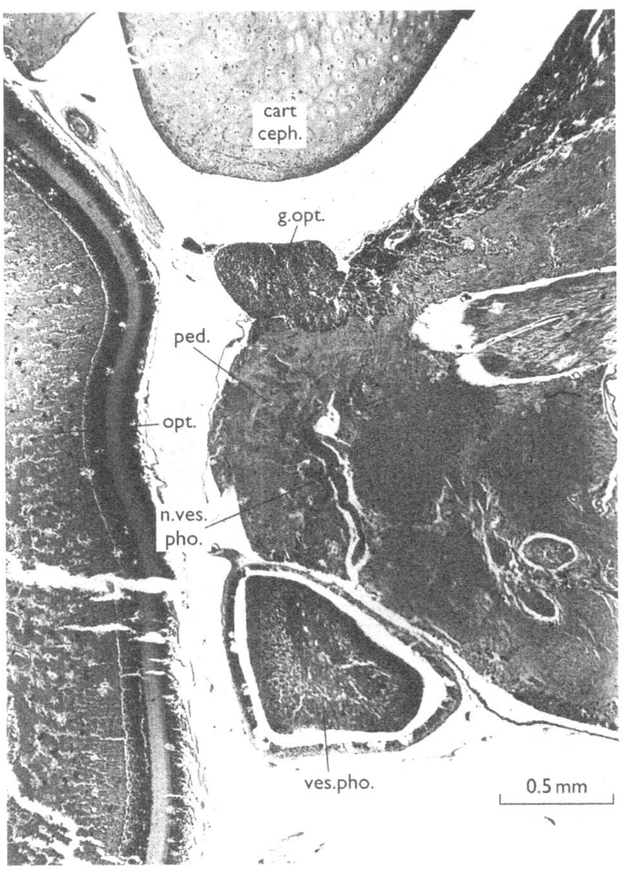

Fig. 25.25. *Galiteuthis glacialis* (297 mm ML). Transverse section showing part of optic lobe, photosensitive vesicle, and optic gland (haematoxylin).

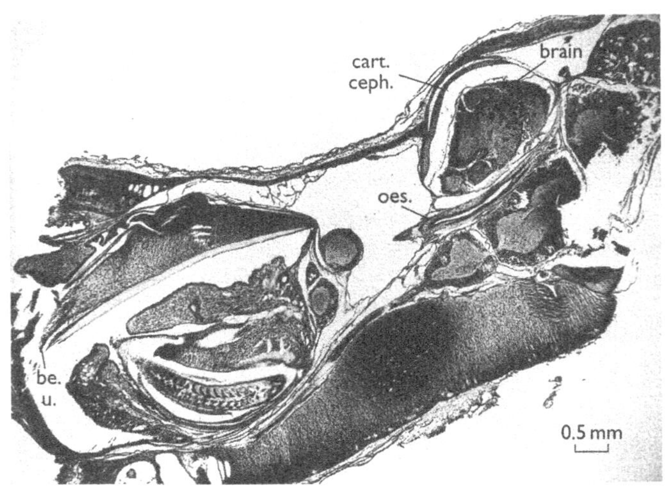

Fig. 25.26. *Galiteuthis*. Sagittal section of the central nervous system and buccal mass, close to the midline.

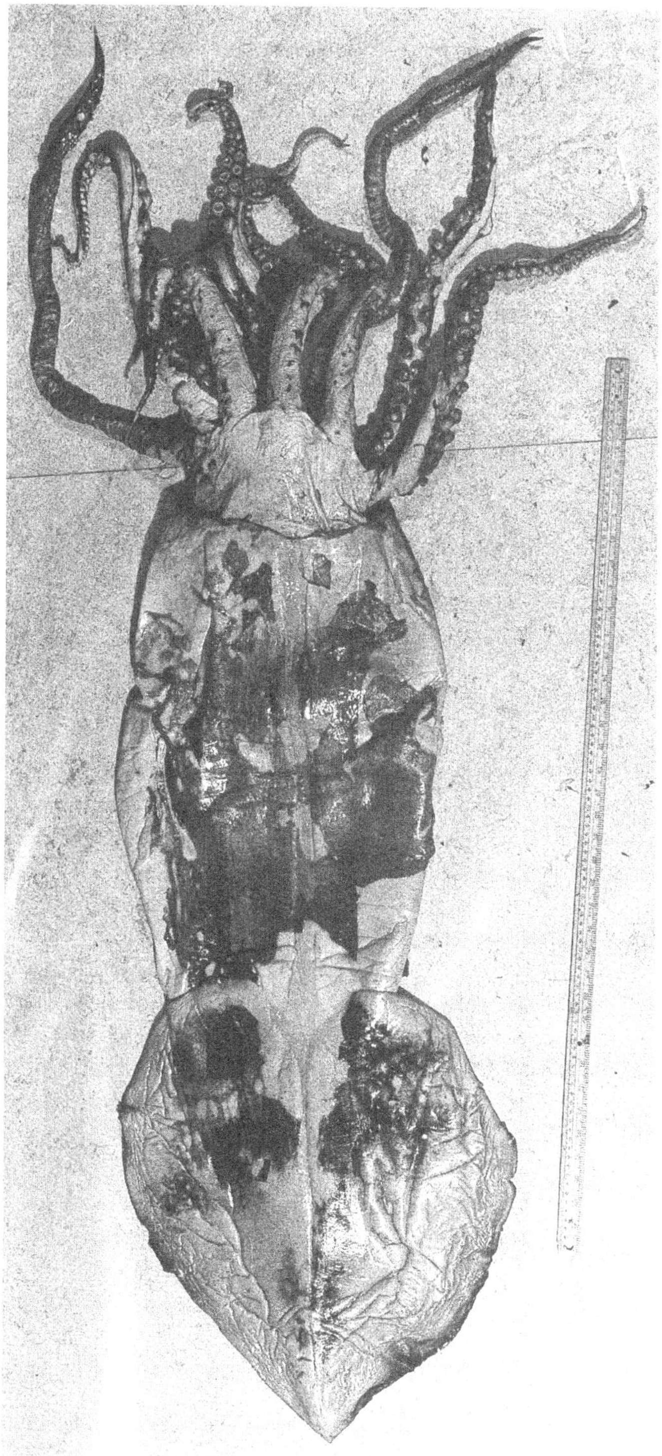

Fig. 25.27. *Mesonychoteuthis hamiltoni* (1170 mm ML). Dorsal view of a large specimen captured in a net (photograph kindly provided by M.R. Clarke).

Stellate ganglia and statocysts

Serial sections of the stellate ganglion show the presence of unusually thick layers of large and small cells, both dorsally and ventrally (Fig. 25.28). There are many glial cells, both in the cell layers and neuropil. The pallial nerve contains a fibre 130 μm in diameter, which divides on entering the ganglion (Fig. 25.29). Each stellar

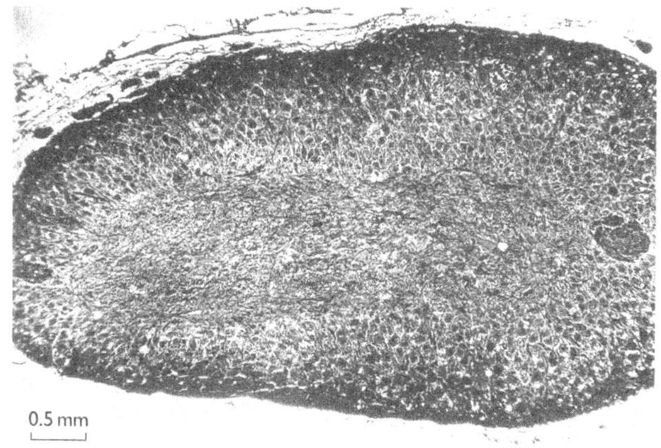

Fig. 25.28. *Mesonychoteuthis hamiltoni* (1170 mm ML). Section of the stellate ganglion.

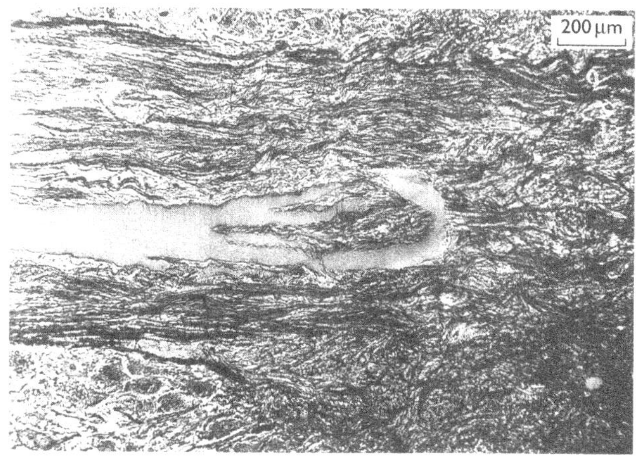

Fig. 25.29. *Mesonychoteuthis hamiltoni* (1170 mm ML). Pallial nerve entering the stellate ganglion.

nerve contains one or more giant fibres reaching 300 μm, but most are smaller.

The statocyst has a butterfly like form when viewed from in front in the specimen of 1170 mm ML. Five large hamuli are present. There are 21 straight anticristae in a single row with the lowest, ventrally, close to the hind end of the longitudinal crista. The row then turns up the medial wall and curves laterally and backwards across the roof, ending lateral to the dorsal end of the vertical crista. The cavity is thus divided into anterolateral and posteromedial parts by the row of anticristae (Young 1984, 1989). Multiple discriminant analysis places this squid amongst the buoyant forms, but at a short distance from the majority (Maddock and Young 1984).

25.2.8 *Egea*

This genus is usually found in the tropical waters of the Atlantic, western Pacific, and Indian Oceans and occasionally in subtropical waters (N.A. Voss 1980). It lives in subsurface waters down to depths of 800 m and perhaps to 2000 m (Clarke 1969, N.A. Voss 1974, Clarke and Lu 1975). There is some confirmation that it goes to considerable depths, as it is consumed by sperm whales (Clarke 1980).

Receptors and central nervous system

The eyes are a prominent part of the head. A large trochlear cartilage is present and two anterior conjunctive muscles, which cross below the brachial lobe. Conjunctive muscle I runs in a groove of the cartilage and conjunctive muscle II is large and spreads out ventrally (Budelmann and Young 1993), and so, like some other cranchiids, *Egea* is probably capable of binocular vision. The optic lobes form 236.7% of the total brain volume (Table 2.6).

Above the central nervous system the cephalic cartilage is notably thick (Fig. 25.30). The brachial lobe forms 9.7% and the pedal lobe 15.9% of the total brain volume; the former is somewhat separated from the latter. The magnocellular lobe occupies a considerable part of the suboesophageal mass and has a large connection with the pallial nerve. There are giant fibres crossing in the palliovisceral lobe. The vertical lobe system occupies 16.1% of the total brain volume (Table 2.6).

The extraocular photoreceptor organs comprise a pair of large, elongate, crescentic-oval sacs with a scalloped posterior margin. They lie at the sides of the back of the brain and fit closely to the contours of the cephalic cartilage (R.E. Young 1978). In the specimen of 198 mm ML they form 44.2% of the total brain volume (Maddock and Young 1987). Each monovesicular sac has a continuous cavity filled with a mass of curled rhabdoms, 500 μm in length in a squid of 198 mm ML. Pigment develops on the dorsolateral surface of the sac only when the animal reaches 35 mm ML (R.E. Young 1978). The bundles of nerve fibres from the photosensitive vesicles make indentations in the walls as they converge on the peduncle lobes, which they enter laterally across the base of the vertically oriented spine. The bundles proceed medially to anastomose in the more dorsal part of the basal region of the lobe, close to the lateral dorsal basal lobules from which the nerve to the optic gland arises.

The statocyst is remarkable as the cavity is interrupted by 38 anticristae, the largest number found among the cephalopods examined

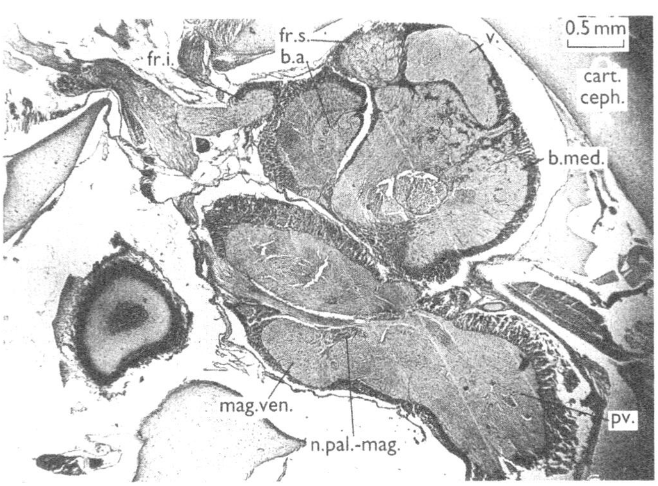

Fig. 25.30. *Egea inermis* (198 mm ML). Sagittal section through the central nervous system (haematoxylin and eosin).

(Young 1984, 1989). The anticristae are straight, and many are long, thin, and pointed, whereas others are small protuberances. They lie on the floor, medial wall, and roof, but are not in a single row. The row on the floor is almost at right angles to that in the medial wall and roof, and in each row there are large central fingers flanked by smaller ones on either side. This arrangement presumably results in disturbance in the flow of endolymph in all three planes, and the five relatively large hamuli restrict the effective length of each section of the crista. The hamuli and the statolith are also large. These conditions are appropriate to a high degree of mobility. Multiple discriminant analysis of various parameters of the statocyst place this squid amongst the buoyant ones (Maddock and Young 1984).

25.2.9 *Megalocranchia*

These moderately large to large cranchiids have a circumglobal tropical and subtropical distribution (N.A. Voss 1980). Paralarvae are found from the surface to about 200 m day and night, juveniles 40–50 mm ML occur in the upper 200 m at night and at 600–800 m in the day, and subadults in depths of more than 2000 m during the day and at less than 700 m at night. Mature and perhaps spent females float on the surface at night, suggesting that spawning is in shallow water (N.A. Voss *et al.* 1992b).

The mantle of the larva has a mucous coat, but in subadults its wall is thin and leathery after preservation (N.A. Voss 1980). In a living squid (40 mm ML) there is a thick layer of gelatinous tissue between the mantle and the surface epithelium. Attempts to hold this squid after its capture proved frustrating as, apart from the head and tail, the surface was extremely slippery (Seapy and Young 1986).

The young (40 mm ML) are translucent when alive, apart from the eyes and visceral nucleus, which comprises the digestive gland and ink sac, and the chromatophores when expanded. The eye is laterally compressed and the ventral surface has an ocular appendage, giving it an inverted tear-drop shape. The digestive gland is compact, narrow, and conical with an external reflective layer. For concealment this cranchiin relies on its translucency, and on the reflective surfaces of the eyes and visceral nucleus to hide the opaque structures. When swimming the stalked eyes and visceral nucleus are adjusted rapidly and accurately to a vertical position in response to changes in the orientation of the body, ensuring that when seen from below any shadow is minimal (Seapy and Young 1986).

Receptors and central nervous system

The eyes in the young are stalked, but become sessile as the squid grows and form prominent bulges on the head.

The optic lobes are the smallest among the cranchiid squids measured, being 166.1% of the total brain volume (Table 2.6). The inferior frontal lobe system forms 4.8% and the vertical lobe system 10.5% of the total brain volume. The size of the former suggests that *Megalocranchia* makes especial use of afferent fibres from the arms and tentacles.

The brachial lobes are somewhat separated from the middle suboesophageal mass. They form 15.7% of the total brain volume and are second in size amongst the cranchiids measured (Table 2.6). The magnocellular lobe probably has no direct connection with the pallial nerve, though it is difficult to be sure of this from our transverse sections. There are giant fibres crossing in the palliovisceral lobe.

A pair of extraocular photoreceptor organs is present. Each organ develops from a narrow, crescentic band along the posterior surface of the peduncle lobe in a young specimen and later forms a broad elongate vesicle extending from the surface of the peduncle complex and then ventrally until making contact below the brain. These organs are heavily pigmented (R.E. Young 1978). In our specimen of 335 mm ML the organs are more than 7 mm long and 1.2 mm wide; they form 63.8% of the total brain volume and are the largest among the cephalopods measured (Maddock and Young 1987). The organs form narrow sacs on each side of the brain. The nerves enter the peduncle lobe (Fig. 25.31) as in *Egea*. Unfortunately the organs are much broken up in the histological sections but have rhabdoms 250 μm long, in an animal of 335 mm ML, that are rather regularly arranged in the transverse plane. The cell bodies lie along the outer walls of the sacs and there is pigment over their inner ends; light presumably passes through the cells to reach the photosensitive layer.

The statocyst has 5 hamuli and 19 straight anticristae, some very small, but many others long and thin. Fourteen of these form a row beginning ventrally, continuing up the medial wall and then backwards dorsally. Two other anticristae lie separately in the dorsal wall, and there is a large protuberance close to the opening of Kölliker's canal (Young 1984, 1989). The subdivision of the statocyst suggests that these cranchiins are capable of rapid movements. This is supported by observations of living animals by Seapy and Young (1986) in a shipboard aquarium, and other features related to mobility (N.A. Voss 1988b). Multiple discriminant analysis of various parameters of the statocyst place *Megalocranchia* among the buoyant squids (Maddock and Young 1984).

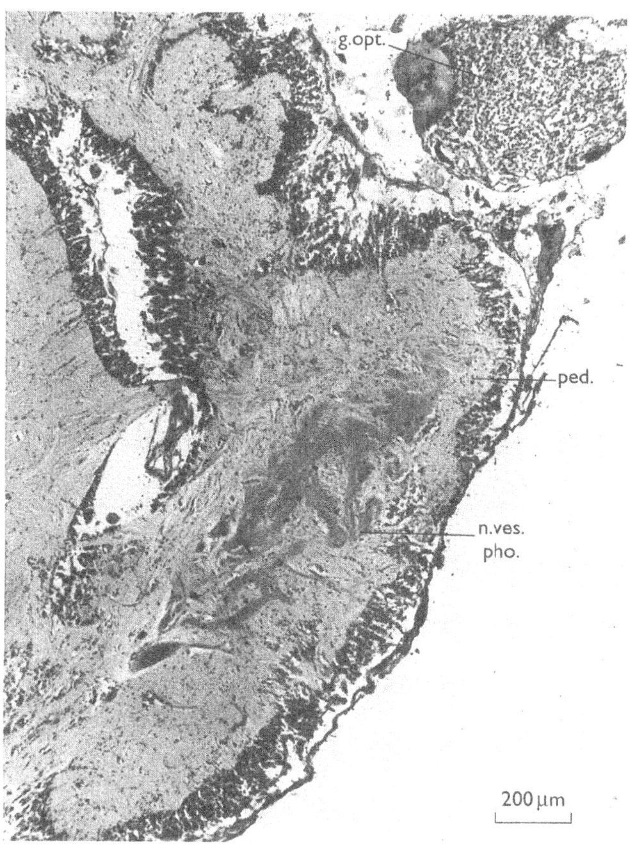

Fig. 25.31. *Megalocranchia* (335 mm ML). Transverse section to show nerves of photosensitive vesicles in peduncle lobe.

25.2.10 *Teuthowenia*

Three closely related species with discrete, ecologically distinct distribution patterns comprise the genus. *Teuthowenia megalops* is largely confined to sub-Arctic and temperate waters of the North Atlantic Ocean, between 68°N and 30°N (Nixon 1983), but is very occasionally found in the Mediterranean (Sánchez 1985). *T. maculata* inhabits eastern tropical waters of the Atlantic from 22°N to 20°S and *T. pellucida* is circumglobal between 30°S and 60°S (N.A. Voss 1985). *Teuthowenia* is found from surface waters to depths of 1000 m, and even to 2700 m.

Life history

Teuthowenia megalops (Fig. 25.32) is one of the few oceanic squids for which something is known of the life cycle (Table 25.1). The smallest specimens captured so far are 3 mm in ML. Growth has been followed and the changes that take place are almost imperceptible until the near-mature subadult stage is reached. There is marked sexual dimorphism and males mature when 196 mm ML but females not until more than 380 mm in ML. The female develops brachial end organs on all of the arms, as well as numerous glandular processes between the largest suckers on all arms. Though as yet no mature female has been captured, one of 315 mm ML contained some 70–80 000 eggs, all of similar size (0.6 × 0.5 mm) (Dilly and Nixon 1976a, Nixon 1983). The male too displays marked changes at maturity, especially of its arms (N.A. Voss 1985).

Maturity-related morphological changes in *Teuthowenia pellucida* appear to take place in a short period of time. N.A. Voss (1985) speculated on the probable functions of some of these changes and their role during courtship and mating. Mature females may attract a male by the bioluminescence of the brachial end organs, effective at a distance of 50–100 m (R.E. Young 1975c), or perhaps over a shorter distance by one or more pheromones, produced by gland-like organs on the dorsum of the mantle. On reaching a female the male may grasp her and insert the tips of arms I and II into her mantle chamber. These arms are attenuated and have specialized suckers which adhere to the surfaces of the inner linings and organs of the female; subsequently this specially modified region detaches from the arms to become encysted. The male, with the long penis extending out of the opening of its mantle, deposits the sperm reservoirs of the ejaculated spermatophores on to the anterodorsal mantle surface of the female, where they become embedded in the mantle. They then penetrate the inner wall and release sperm into the mantle cavity. Fertilization probably takes place when the eggs leave the oviducal gland. The eggs are of near-uniform size, and may be deposited in gelatinous masses. The animals mature, mate, and spawn in near surface waters, and the hatchlings undergo ontogenetic descent to midwater and beyond (N.A. Voss 1985).

Behaviour

In the sea (750 m depth) *Teuthowenia megalops* was seen to adopt a 'cockatoo posture', as also seen in *Taonius pavo* (Figs 25.21, 25.32).

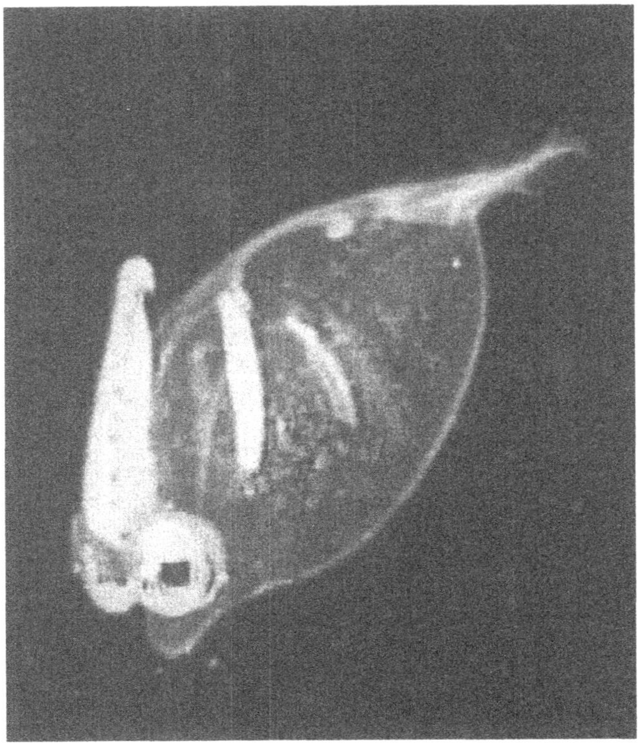

Fig. 25.32. *Teuthowenia megalops* in the sea, recorded on videotape at a depth of 789 m, in the characteristic 'cockatoo' posture (seen also *Taonius pavo*, Fig. 25.21). It is translucent except for the large opaque eyes and the visceral nucleus. The latter is in a vertical position to provide optimum concealment (Vecchione and Roper 1991).

The squid rotates, head-up or head-down, with the digestive gland as the fulcrum; the gland is maintained in a vertical position regardless of the orientation of the mantle and head (Vecchione and Roper 1991). In a shipboard aquarium the squid, if chased, would retract the fins, the head, and, finally the tentacles into the mantle cavity to form a ball (Dilly 1972). A similar balling phenomenon has been seen in living *Cranchia* (R.E. Young 1972b).

Photophores

Growth of the eyes is accompanied by the development of subocular light organs. The photophore below each eye is initially a simple crescent shape, but subsequently divides to become quite complex, and can be asymmetrical (Dilly and Nixon 1976a). The photogenic tissues of these photophores closely resembles that present in *Bathothauma* (Dilly and Herring 1974). From the photogenic core there is a system of well-oriented fibrils, acting as light guides, leading to the surface. The subocular photophores have an emission maxima of 490–505 nm (Herring 1983).

Receptors and central nervous system

The eyes are a striking feature of the living animal (Dilly 1973). They are initially on short stalks, but later are incorporated into the head. With growth the eyes increase in diameter to form 8–20% of the mantle length in animals of 3–180 mm ML (Dilly and Nixon 1976a, N.A. Voss 1985). The retina is uniform in specimens of up to 40 mm ML, with rhabdoms 300 μm in length. Two pairs of anterior conjunctive muscles are present and cross beneath the brachial lobe of the brain. There is no trochlear cartilage and the two

Table 25.1 Cranchiidae. A summary of various aspects to show the changes during the life cycle of *Teuthowenia megalops* to the subadult stage. There is no sharp demarcation as the external changes are gradual and almost imperceptible (data from Dilly and Nixon 1976a, Nixon 1983, N.A. Voss 1985)

Stage	Eggs	Paralarva	Postparalarva	Juvenile	Subadult
Form	Flask-shaped	Translucent, globular	Translucent, elongated	Translucent except, eyes and visceral nucleus	Probably as juvenile
Size (mm)	1.7 long	3 (ML at hatching)	7–70 (ML)	70–180 (ML)	380 (gladius length)
Mantle shape		Round, sac-like	Elongated, narrowing posteriorly	Tapers posteriorly to slender cone	Tapers to slender cone
Fins		Small, rounded, terminal	Rounded, extend 30% ML	Extend to middle of mantle	Extend > 55% ML
Tentacles		Elongated, 120% ML	Changes from 80% to 40% ML Club develops	Less prominent: 40% ML	About 40% ML
Arms		Only visible when magnified	Develop, III > II > IV > I	III > II > IV > I	III > II > I > IV
Brachial end organs					Develop on all arms of female
Chromatophores		Few, large, and discrete	Small number, large, discrete	Innumerable small ones	Innumerable small ones
Eyes		Stalked, 10% ML	Stalked to sessile, 12% ML	Sessile, 16% ML	Sessile
Subocular light organs		Not visible	Develop at 22 mm ML	Present	Present
Gonads		Not visible	Just apparent at 22 mm ML	Developing 50% ML	Not recorded
Depth (m)	Unknown	Mostly at 0.25, a few to 1000	0–200, a few to 900	0–1000	>1000

muscles proceed together to the eyeball, enclosed in a sheath of fibrous tissue.

The optic lobes, infolded slightly medially, are well differentiated. They form 407% of the total brain volume in an animal of 265 mm ML (Maddock and Young 1987). The optic tract is elongated in an animal of 40 mm ML as the eye is carried on a short stalk (Fig. 25.33). The peduncle lobe is far from the optic lobe, and close to the photosensitive vesicle of the extraocular photophore.

The central nervous system lies just behind the large buccal mass in small specimens (Fig. 25.34) (Dilly and Nixon 1976a). In the supraoesophageal mass the vertical lobe system is well developed (Fig. 25.35), and forms 10.9% of the total brain volume in a specimen of 265 mm ML (Maddock and Young 1987). The anterior and median basal lobes are also well developed.

The ventral magnocellular lobe is characteristically prominent. It receives a large tract direct from the tentacles and arms running independently of the brachiopalliovisceral connective. There are fibres 6 μm in diameter in the ventral magnocellular lobe that branch repeatedly and spread tangentially (Fig. 25.36), crossing the trunks of the fibres arising from cells of the lobe. These join into bundles running to the pallial nerve, and are probably efferent fibres. Many fibres from the ventral magnocellular lobe proceed to the posterior magnocellular lobe (Fig. 25.35), and to the palliovisceral lobe. Here they come into relationship with the neuropil of the swimming centre, whose fibres proceed to the fin lobe (Fig. 25.37). It seems possible that the ventral magnocellular lobe provides an integration centre

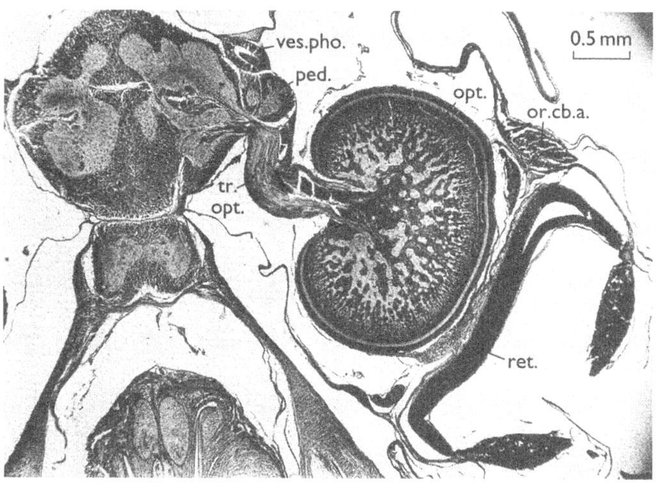

Fig. 25.33. *Teuthowenia megalops* (41 mm ML). Horizontal section through the central nervous system near the midline.

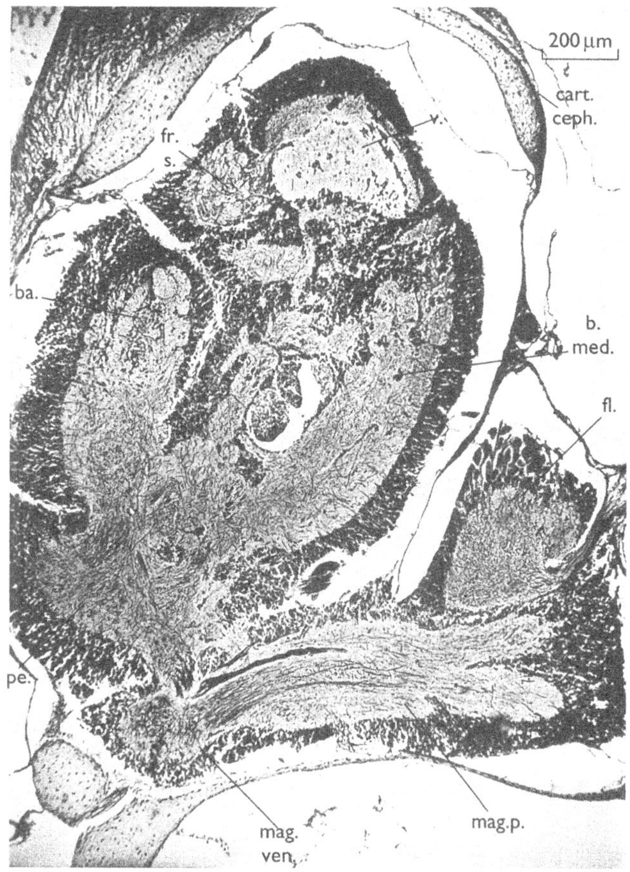

Fig. 25.35. *Teuthowenia megalops* (40 mm ML). Sagittal section of central nervous system showing fibres from ventral to posterior magnocellular lobe.

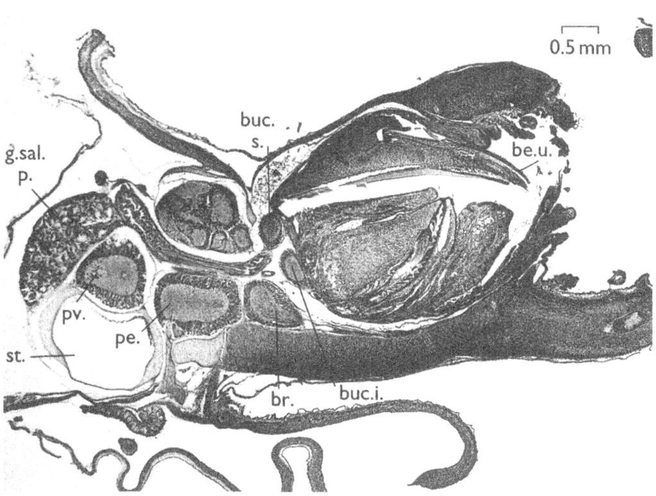

Fig. 25.34. *Teuthowenia megalops* (40 mm ML). Sagittal section through the head region showing the close apposition of the brain and buccal mass.

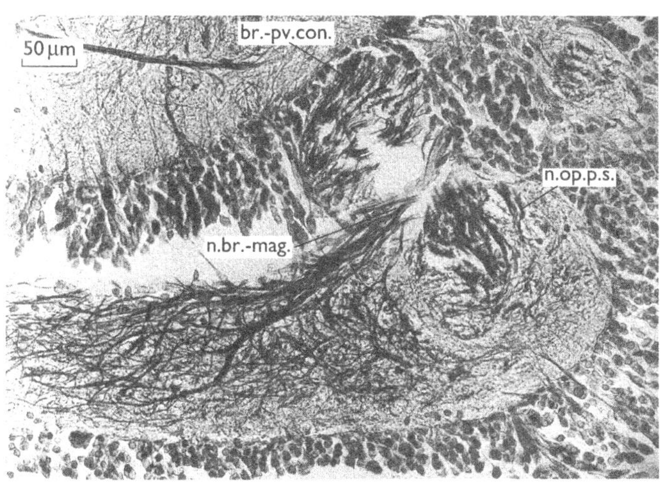

Fig. 25.36. *Teuthowenia megalops* (45 mm ML). Transverse section through the magnocellular lobe showing large fibres.

for the direct control of the mantle and fins, following stimulation from the arms and tentacles. In addition this lobe receives a bundle of small fibres from the surface of the head by way of the posterior superior ophthalmic nerve (Fig. 25.36). The lobe is thus also concerned with escape reactions of the giant-fibre system. This system contains typical first-order giant cells (Fig. 25.38). Their axons proceed as usual to the palliovisceral ganglion where they cross without fusing in the animal of 40 mm ML (Fig. 25.39).

Giant fibres arise in the stellate ganglion by the fusion of several axons (Fig. 25.40). Each stellar nerve contains one or more giant fibres, the largest being only 40 μm in diameter in a specimen of 134 mm ML.

The photosensitive vesicles of the extraocular photoreceptor organs are simple and small in the paralarva (13 mm ML) but in a subadult (265 mm ML) the sacs are 4 × 3 mm and form 42.3% of the total brain volume (Table 2.6). The sacs lie behind the brain and extend up above the peduncle lobes, to which their fibres run (Fig. 25.33). They end close to the lobe that gives origin to the nerve to the optic gland. The rhabdoms of the vesicles range from 44 μm in length to 500 μm in squids of 13 mm to 265 mm ML respectively.

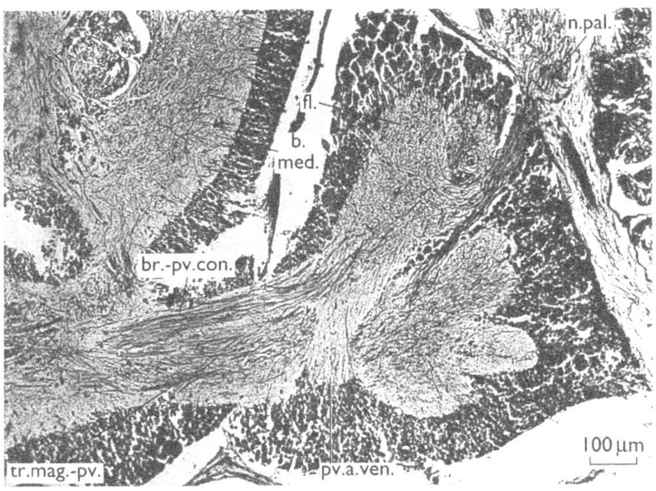

Fig. 25.37. *Teuthowenia megalops* (40 mm ML). Sagittal section through the fin lobe.

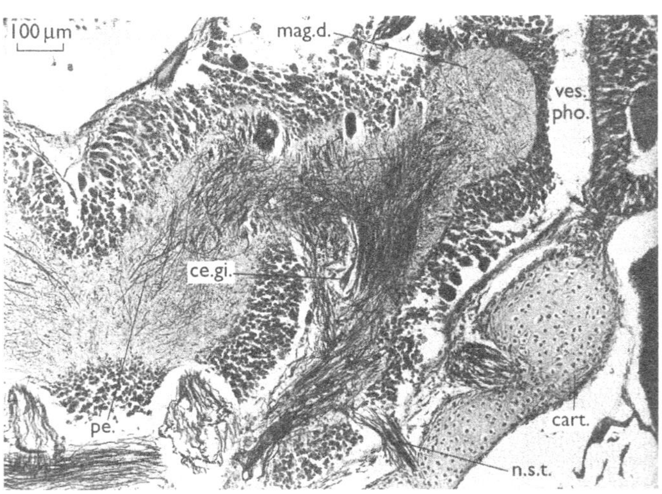

Fig. 25.38. *Teuthowenia megalops* (45 mm ML). Transverse section of magnocellular lobe showing a giant cell.

The statocyst is surprisingly little divided, even in the largest individual examined (265 mm ML). There are only three hamuli and the single straight anticrista is a large protuberance close to the opening of Kölliker's canal (Young 1984, 1989). Multiple discriminant analysis of various parameters of the statocyst place this taoniin amongst the buoyant squids (Maddock and Young 1984); this is supported by buoyancy measurements of the animal (Clarke *et al.* 1979).

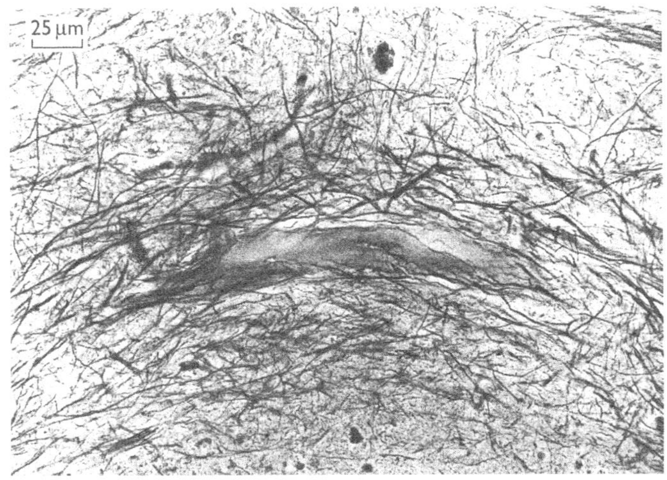

Fig. 25.39. *Teuthowenia megalops* (41 mm ML). Horizontal section of palliovisceral lobe to show axons of first-order giant cells crossing without fusing.

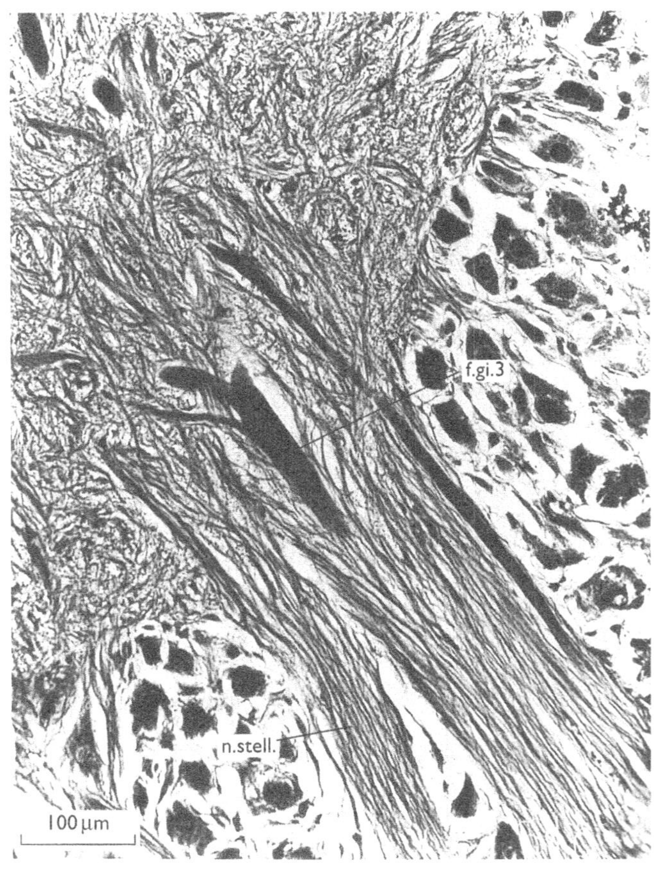

Fig. 25.40. *Teuthowenia megalops* (41 mm ML). Horizontal section showing fusion of several axons in the stellate ganglion to form third-order giant fibres (Holmes).

26 Vampyroteuthidae

26.1 *Vampyroteuthis*

Vampyroteuthis infernalis (Fig. 26.1) is the only living representative of the family and of the order Vampyromorphida. The vampire squid is bathypelagic and has a worldwide horizontal distribution between latitudes 40°N and 40°S. Its vertical distribution is 100–3000 m, but mostly between 600 m and 1500 m, and it seldom rises above 600 m (Pickford 1946). In the Atlantic Ocean it has been caught in opening/closing nets mainly between 700 m and 1250 m, day and night, and less often down to 1500 m (Clarke and Lu 1974, 1975; Lu and Clarke 1975a). In the Pacific Ocean the vampire squid is caught between 100 m and 1200 m, and well above the bottom, and is therefore presumably pelagic throughout its life. The larvae occupy deeper water and young of less than 20 mm ML have been caught below 900 m (Roper and Young 1975). There is evidence that the vampire squid is restricted to waters with a low salinity (34.4–35.4{‰}), and where the temperature is 2–10°C. *V. infernalis* is thus largely limited to the upper deep waters of tropical and subtropical seas (Pickford 1946, 1952). Recently it has been found associated with seamounts, apparently a result of passive advection by currents (Nesis 1993a), and three adults have been caught in benthic trawls at 970–1290 m around Walters Shoals, a seamount in the south-western Indian Ocean (Nesis 1994).

The first specimen to be captured was named by Chun in 1903 and has been called the little black octopus or the vampire squid. As these ambiguous names suggest, features of both octobrachians and decabrachians are apparent, besides which *Vampyroteuthis* possesses unique features. These include the development of larval fins, which are resorbed after the adult pair begin to develop, and long slim retractile filaments, lying between arms II and III. The filaments are the modified second dorsal pair of arms (Robson 1932, Pickford 1936, 1938, 1940, R.E. Young 1967) whose nerves exchange fibres with the other arms through the interbrachial commissure before joining the brain (Young 1977a).

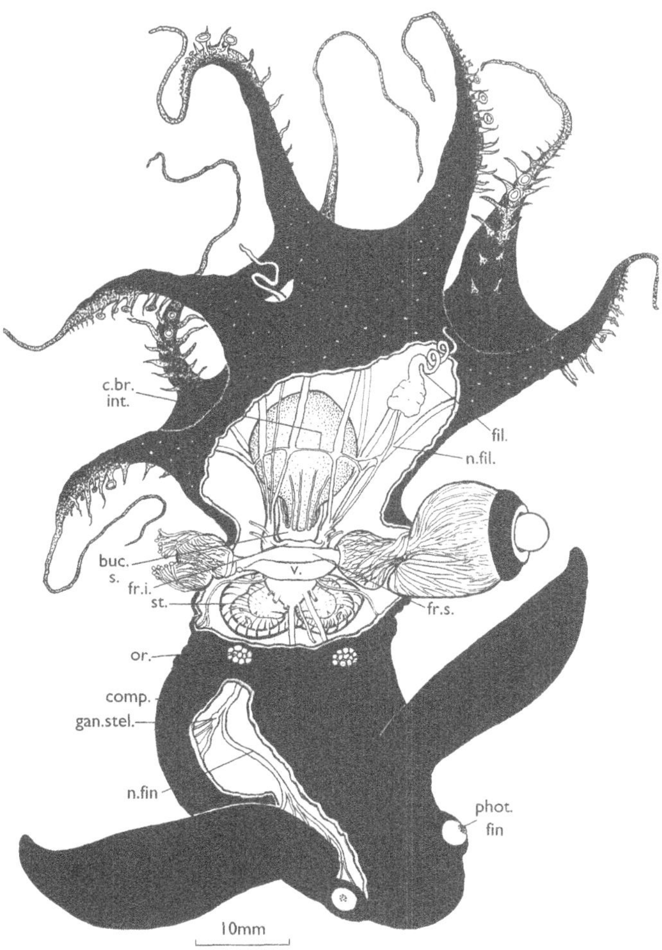

Fig. 26.1. *Vampyroteuthis infernalis*. Drawing of dissection of whole animal. The photophores have been added from other illustrations (Young 1977a).

26.2 Life history

One female (63 mm ML) was found with ovarian eggs at various stages of development; the largest female examined (100 mm ML) had recently spawned and one coelomic egg (4 mm in diameter) had been retained (Pickford 1957). Spawning was apparently not seasonal and this, with the condition of the ovarian eggs, suggests that *Vampyroteuthis* spawns intermittently (Mangold *et al.* 1993). The female matures at 78 mm ML and the male at 44 mm ML (Pickford 1949a, 1949b, 1957). The eggs are spherical and are pelagic.

At hatching the young has a single pair of fins; a second pair develops, and then the first pair of fins is resorbed leaving only the second pair (Fig. 26.2) (Pickford 1949a). One hatchling (8 mm ML) was caught in a trawl, between 1050 and 1300 m. It possessed a large internal yolk sac, and was embryo-like as the mantle was not fused to the head as in older animals, its filaments were broad, short and arm-like, and did not reside in pockets (R. E. Young *et al.* 1998b). A paralarva (9 mm ML) was collected by a submersible, at 830 m depth. Its morphology, except for the oblique orientation of the larval fins, was much more like that of larger specimens than of the 8 mm ML hatchling (R.E. Young and Vecchione 1999). The young are pelagic until a mantle length of 20 mm is reached; at this size they are found in depths of more than 800 m (Roper and Young 1975).

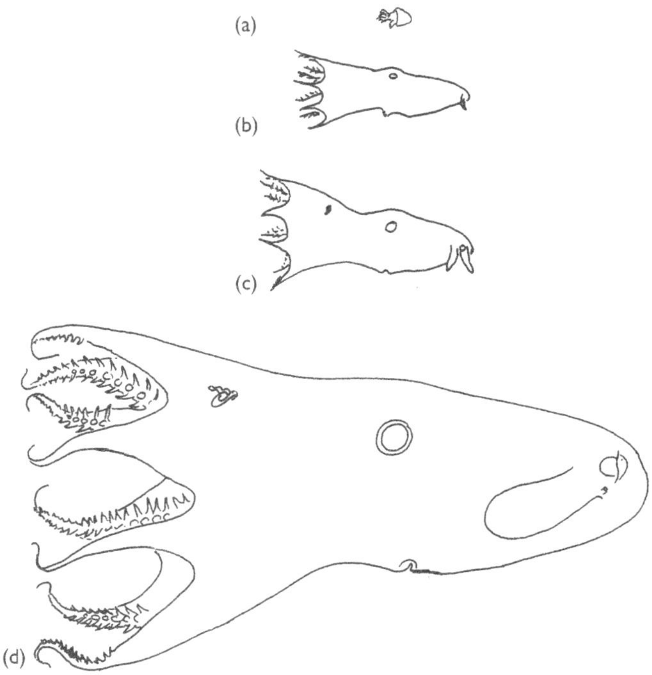

Fig. 26.2. *Vampyroteuthis infernalis.* Outline drawings of animals of (a) 5.5 mm ML, (b) 17.0 mm ML, (c) 18.0 mm ML, and (d) an adult 55 mm ML, to show changes with growth (Pickford 1949a).

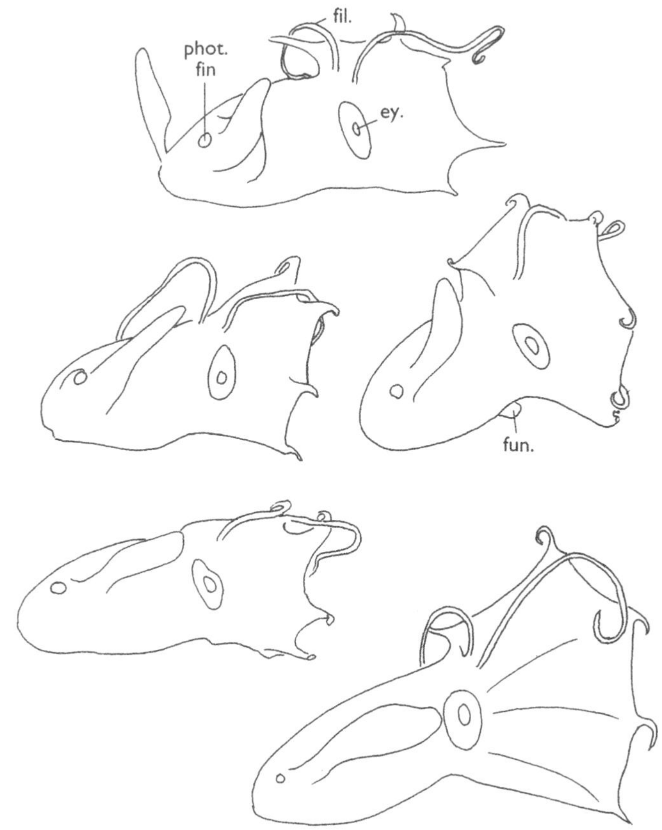

Fig. 26.3. *Vampyroteuthis infernalis.* Outlines of the animal as it moved in an aquarium (traced from a film kindly lent by R.L. Zimmer).

Females tend to be larger than males at maturity. The oviducts are paired and have wide, fluted openings. No nidamental glands have been found. Postlarval females have a pocket-like invagination in front of each eye, and sections of the sac showed a simple non-glandular epithelium. The right sac in one mature female was found to contain a coiled spermatophore (Pickford 1946). In one of our specimens the receptacle has a complex structure and several projections from the floor of the invagination, with a loose internal organization and a covering of muscle fibres. A further projection rises at the centre of the invagination with similar internal features but lacking muscle fibres. Smaller papillae, around the central projection, have a glandular epithelium.

The male has a penis on the left side, but lacks a hectocotylus. Spermatophores, about 20 mm long, are stored in spermatophore sacs which project freely into the mantle cavity (Pickford 1946). An ultrastructural study of the spermatozoa shows the retention of several primitive features. Late spermatids and mature spermatozoa show a strong resemblance to developing spermatids of *Octopus* (Healy 1989, 1990b).

26.3 Locomotion

A short film of *Vampyroteuthis infernalis* (taken in a shipboard aquarium by R.L. Zimmer and lent to us by R.E. Young) showed the animal swimming slowly backwards, apparently by the action of the fins. The arms were sometimes held together, or spread outwards, with their terminal parts projecting backwards (Fig. 26.3). The long filaments coiled and twisted. They were protruded for very considerable distances and then withdrawn, and these actions occurred several times during the few minutes of the film. The filaments were pale and contrasted with the dark mantle. R.E. Young (in Roper and Brundage 1972) described three modes of locomotion:

- by the fins, with the arms forward in the direction of movement
- rapid swimming or escape response with a medusoid action of the arms and web
- 'slow swimming with water ejection through funnel and fins spread as stabilizers'.

A video-recording, taken at 670 m depth from a submersible (P.J. Herring, pers. comm.) showed *Vampyroteuthis* swimming with screw-like actions of the fins aided apparently, on some occasions at least, by jet propulsion. The slender distal tips of the arms showed pale as they trailed at the end of the arm-cone. When the arms opened the suckers were white, so contrasting starkly with the darkness of the arms and web. The pair of filaments extended an astonishing distance, some five or more times the total length of the animal. The eye, in spite of its relatively large size in relation to the mantle, appeared capable of some movement. This is supported by the presence of conjunctive muscles which indicate binocular vision.

26.4 Arms and filaments

The eight arms are of similar length, although arms I and II are slightly longer. The arms are joined by an extensive web. The ends of

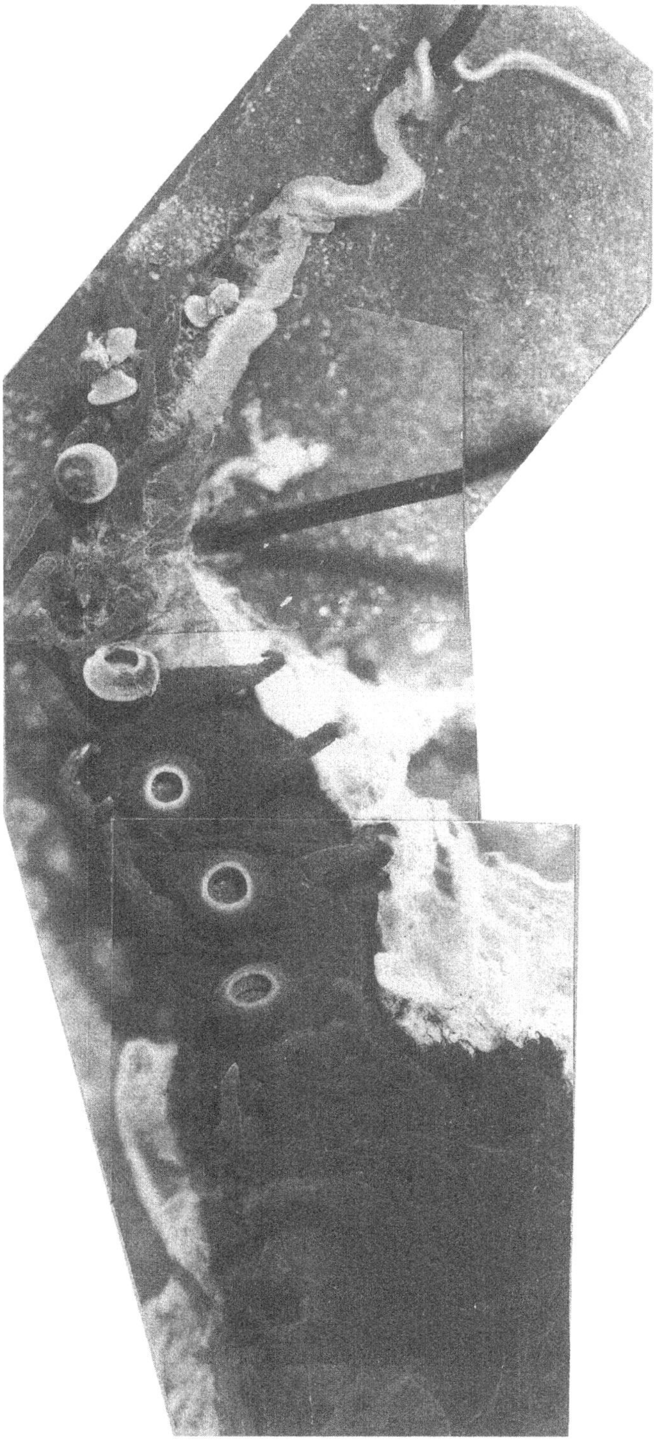

Fig. 26.4. *Vampyroteuthis infernalis.* An arm to show the uniserial suckers and paired cirri.

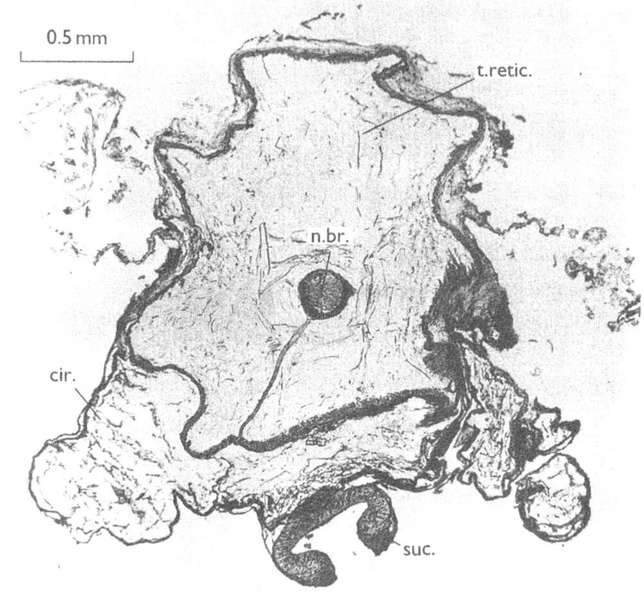

Fig. 26.5. *Vampyroteuthis infernalis.* Transverse section through an arm, a sucker, and a pair of cirri (much contracted). Most tissue is reticulate except the sucker and narrow band of muscle in the arm.

the arms are extremely slim. Each arm has a single, longitudinal row of up to 21 suckers (Fig. 26.4). There may be 10 pairs of primary cirri before the first sucker, after which paired cirri alternate with each sucker. The arms have few muscles, but much loosely organized tissue, and the sucker has dense radial muscles but lacks a sphincter (Fig. 26.5). Chemical analysis of an arm showed it to consist of 94% water and 3.22 mmol/L ammonium (Robison 1989).

The suckers closest to the mouth are sessile (Fig. 26.4), those further along the arm have a peduncle, and more distally there are papillae which later develop into suckers (Pickford 1949a). A scanning electron microscope study showed that the fully formed sucker is like an urn with a straight neck surrounding the opening; the acetabulum is wide but shallow in preserved material (Nixon and Dilly 1977). The rim of the sucker has complex radiating folds and hillocks, on the apices of which are pores, each with a bundle of cilia (0.1 μm diameter and 5 μm long). Nothing is known of the function of the cilia but they are well sited to sample the surrounding water. The presence of cilia on the rim of a sucker has not been found among other cephalopods so far examined.

In life the cirri have been observed to stretch outwards and away from the arms. Internally the cirri are highly reticulate (Fig. 26.5), but there is no division into compartments as in *Cirrothauma*. At the tip of the cirrus there are pores between the cells through which cilia protrude (Nixon and Dilly 1977). Abundant secretory cells as well as 'olfactory-like' cells have been found in this region of the cirrus (R.E. Young pers. comm.).

The paired filaments emerge from pockets and can be projected a great distance (Fig. 26.3). The nerve of each filament is joined to the circumoral commissure ring in the same way as are the nerves from the other eight arms; this arrangement of the nerves shows that the filaments are to be regarded as the second pair of arms (Pickford 1940, Young 1977a). The pocket of one of our specimens contained a crustacean, perhaps part of its diet.

The filament nerve is smaller than those of the other arms. In the basal portion of the filament the nerve is surrounded by cells, like a typical brachial nerve. More distally it consists only of a large bundle of fibres running to the tip. The base of the filaments has a thin epithelium and an irregular central stroma. The terminal portion has a thick epithelium, dense cellular stroma and some muscle fibres near the tip. It has a large vascular sinus.

26.5 Buccal mass

The buccal mass forms a considerable portion of the head and is relatively large when compared with the brain and even the mantle (Figs 26.1, 26.6) (Young 1977a). The beak is black and the rostrum of the upper beak is quite sharp, but that of the lower is short and protrudes forwards only slightly (Pickford 1949a, Clarke 1986). The mandibular muscles have a loose texture and relatively few muscle fibres. The lateral buccal palps lie on either side of the radular complex. The radula has seven teeth in each transverse row, all similar in form with a single cusp (Nixon 1998). On either side of the radula are the bolsters (Messenger and Young 1999); these sacs are largely devoid of tissue but have a loose mesh of fine fibrils crossed by a few sparse strands of transverse muscle fibres.

The salivary papilla carries the duct from the small, single posterior salivary gland, which lies behind the brain (R.E. Young 1964a). The duct runs forward to open at the tip of the papilla (Fig. 26.6). Tubular glands surround the duct and open by several pores near the free tip. Behind is the huge submandibular gland, which opens into the buccal cavity.

A pair of subradular ganglia lie on the dorsal side of the base of the salivary papilla. They are more widely separated than in *Octopus* and do not seem to be in any special relationship to the small posterior salivary gland duct. The preservation is not good enough to show where their nerve fibres end, except that they lie among muscles which surround the openings of the subradular glands.

The anterior salivary glands lie on the posterior surface of the buccal mass. A duct from each gland runs forward to open at the front of the lateral buccal palps. The cuticle is thickened a little on the anteromedial surface of the palps. Posteriorly the lateral buccal palps unite to form the entrance to the large oesophagus. The cuticle is continuous and forms a rather thick, but translucent lining to the oesophagus. The epithelial cells of the oesophagus, however, contain dark pigment granules, serving to make the oesophagus almost black (Fig. 26.6) (R.E. Young 1964a). A crop is present and here too the epithelium is pigmented as it is in the stomach, but the cuticle remains translucent. The stomach has very thick walls of loosely organized tissue containing muscle fibres.

The feeding habits of *Vampyroteuthis* are known only from isolated reports of crop or stomach contents. Larval crustaceans and copepods, and 'most commonly, what appeared to be the spermatophores from copepods' have been found (R.E. Young 1964a). Copepods were present in the crop of one of our specimens.

26.6 Mantle, gladius, funnel, fins, and skin

The mantle is fused to the broad head and its aperture is very wide. The mantle muscles are weakly developed and are separated by much gelatinous tissue. The main muscle fibres are circular, and attached to the gladius dorsally, and are continuous ventrally. The mantle is supported by a large structure embedded dorsally in the mantle, the gladius.

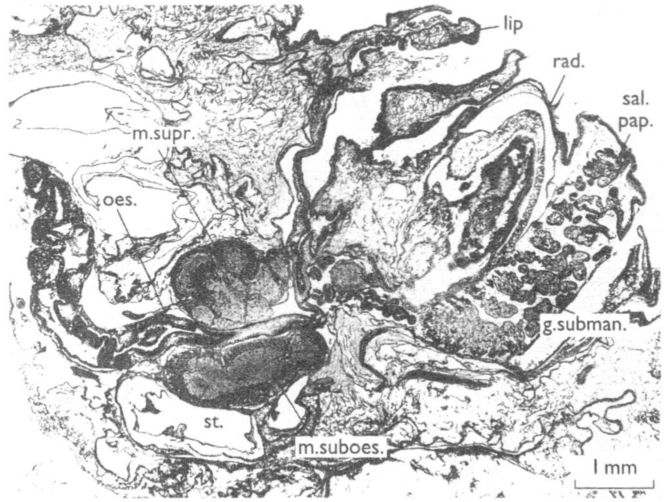

Fig. 26.6. *Vampyroteuthis infernalis*. A medial sagittal section through the head and buccal mass.

The gladius lies in a shell sac, which extends from the neck backwards to the apex of the body where it bends ventrally. The gladius is thin and translucent, with a broad pro-ostracum, the free forwardly projecting region of the middle plate. It is rounded anteriorly and is narrowest where the margin bends outwards into the lateral plates. These plates widen as they diverge and at the point of greatest width the direction of curvature is reversed sharply. This region marks the junction of the conus vane with the lateral plate (Pickford 1949a).

The centre of the conus has a small cavity, which communicates with the posterior end of the coelom by a narrow duct. This duct and cavity presumably represent the siphuncular canal of *Spirula* (Pickford 1936) and *Nautilus*, although no chambered phragmocone remains. The cavity is surrounded by a series of groups of cells, which appear to be secretory. Each group is centred round a space but no ducts were seen leading to the main cavity. The latter contains some debris and amoebocytes. Nothing is known of the function of this system.

Vampyroteuthis is the only living cephalopod to possess a gladius with a broad median field, and it most nearly resembles that of fossil loligo-sepiids except that it is not calcified (Donovan 1977). In the larva the gladius is longer than the mantle, but it becomes progressively relatively shorter as the animal grows. At first the anterior margin of the mantle projects freely over the back of the neck. In the adult the union of the head and neck appears complete but the anterior margin of the shell sac projects forward below the skin, which conceals it, and is connected to the nuchal cartilage by short, slender nuchal muscle fibres (Pickford 1949a).

The funnel is short and largely sunk in gelatinous tissue. It is moved by a large retractor and other extrinsic muscles, but there is no median pallial adductor muscle. Within the funnel are circular and sphincter muscles.

In juveniles and adults the animal possesses one pair of fins which lie posterolaterally on the mantle. Initially there is a posterior pair of fins, and then a second pair develops anteriorly (Fig. 26.2). Finally at metamorphosis the hinder pair are resorbed, leaving only a small pit and the anterior pair of adult fins which are nearly as long as the mantle (Fig. 26.1).

The adult fins are paddle-shaped, elongated laterally, and narrow at the base, and have a horizontal position (Fig. 26.1). The underside of each fin is light pink in colour. The muscles of the fin are probably attached not to the gladius but to a pair of oval fin cartilages lying above it. The levator fin muscles pass from one fin to the other across these cartilages (R.E. Young 1964a). Depressor fin muscles also run between the fins and there are other accessory muscles both above and below. A film and video-recording show the fins to move up and down with a curling, screw-like action (Fig. 26.3).

The outer surface of the skin has a very thin pavement epithelium. Beneath is a layer of collagenous connective tissue, well supplied with blood vessels, perhaps serving for cutaneous respiration since the gills are small (Pickford 1940). There are many chromatophores but they lack attached muscles. The chromatophores contain black and brown pigments, the pigment granules being spread out in a single layer or in thin protoplasmic strands. The animal, when alive, is jet black except for the red of the eyes (from the retina), which gives it a 'terrifying appearance'. There is red pigment on the suckers, cirri, and filaments. The whole of the inside of the mantle is a pale iridescent pink, as are the linings of the coelomic cavities.

26.7 Photophores

Three types of structure have been considered to be luminescent. These include a pair of dorsal fin-base organs, numerous, small, simple organs over the surface of the head, arms, and body, and a pair of composite organs on the dorsomedial surface of the head at the level of the eyes (Pickford 1949a). Recently the function and structure of all of these organs has been investigated by Herring and his colleagues (1994).

Each of the fin-base organs (Fig. 26.1) has a dark, crescentic 'eyelid' across the posterior surface that can close the aperture (Joubin 1920). The dark pigment cells act to modulate the emission of light, as they can expand and contract, more or less in synchrony with a period of 1 Hz (Herring *et al.* 1994). These authors showed the organ to be seated in a deep, black pigmented, pit. The upper half of the organ consists of a transparent region. Hydrogen peroxide treatment results in luminescence with a peak emission at 460 nm, and a broad bandwidth. In a living animal the two fin-base organs emitted a brief flash after gentle prodding.

The photogenic tissue of the fin-base organ has a somewhat uniform appearance. The cells resemble a syncitium (Brunnmüller 1929). At the ultrastructural level most of the cytoplasm consists of a complex assemblage of membranous vesicles and sinuous filamentous profiles, sometimes forming microtubular assemblies. Associated with the region of very fine filaments is the wide distribution of very small, densely stained proteoglycan particles.

Below the photogenic tissue is a thick reflector with a highly silvered reflecting surface. Adjacent to the photogenic tissue is a thin layer of closely aligned radial fibres overlying the main reflector. Thick and thin fibres are present, but deeper in the reflector the fibres are more uniform, arranged in layers with differing orientation, with some regularity in packaging. Both layers are formed of collagen fibres which have a regular banding. This photogenic tissue in *Vampyroteuthis* is not like that found in the photophores of any other cephalopod so far examined (Herring *et al.* 1994).

The numerous small organs scattered over the surface of the head, arms, and body (Fig. 26.1) were found to fluoresce, but they did not luminesce when treated with hydrogen peroxide (Herring *et al.* 1994). The organs appear as very small reflective spots, often set in an area devoid of pigment. Each consists of a tight mass of cells with distinct boundaries and fairly uniform granular cytoplasm. A proximal fibrous layer provides the reflective properties of the organs and this is composed of collagen, like that of the fin photophores. The cells of the core and the collagen fibres of the reflector of the fin-base organs and the scattered small organs show structural similarities, suggesting that they do have a bioluminescent function.

Herring *et al.* (1994) found that the composite organs did not respond to hydrogen peroxide, and did not fluoresce. Sections showed the core to consist of a number of tubules bound by a mass of membranous material, a few nuclei in the middle of the nodules, and only clear cellular organization at the peripheral wall of the capsule. There is no fibrillar structure and no reflector layer in these organs. These authors suggest that there is no justification in continuing to regard these structures as photophores. The organs each have a clear window in the overlying skin to allow the entry of light in an otherwise pigmented epidermis; structurally they are similar to the photosensitive vesicles of the extraocular photoreceptor organs of other cephalopods. Pickford (1949a) found that the number of nodules remains similar in the composite organs throughout life, but individual nodules increase in size with the growth of the animal. Furthermore the organs are longer in females than in males, but females also reach a larger maximum size.

26.8 Eyes

The laterally placed eyes are relatively large (Figs 26.1, 26.2). There is a clear space around the lens, and this presumably gives it an 'aphakic' character (Munk and Frederiksen 1974, Young 1977a); such a gap can enlarge the area of the retina that will receive an image at full aperture. The retina has rhabdoms up to 420 μm long (Fig. 26.7) at the centre, thinning gradually in all directions towards the periphery. The optic nerves collect into a few large bundles. Two specimens we have examined show somewhat different conditions. In neither of them is there evidence of clear crossing of the fibres in the dorsoventral chiasma, as is seen in other coleoids with the exception of cirroctopods. In one specimen (48 mm ML) the optic nerves are rather short with some irregular interweaving, whereas in another specimen (65 mm ML) they are longer and run uncrossed through the centre of a pink 'white body'. In other specimens the 'white body' forms a thin layer around the optic lobe.

There is a conjunctive eye muscle crossing the midline allowing movement of both eyes. These crossing muscles differ considerably from those found in decabrachians but there is no doubt of their presence. They occupy precisely the usual position in which they are found in decabrachian coleoids, below the brachial lobe (Budelmann and Young 1993). The muscles consist of bundles of elongated fibres.

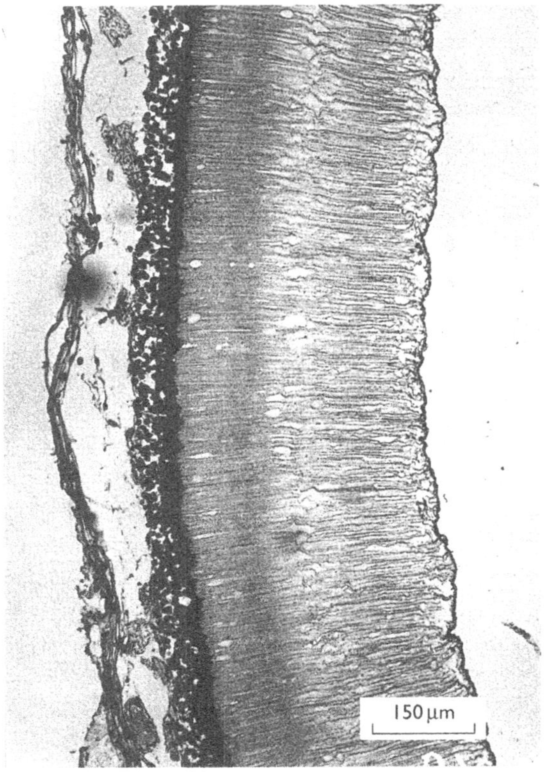

Fig. 26.7. *Vampyroteuthis infernalis*. Part of the retina.

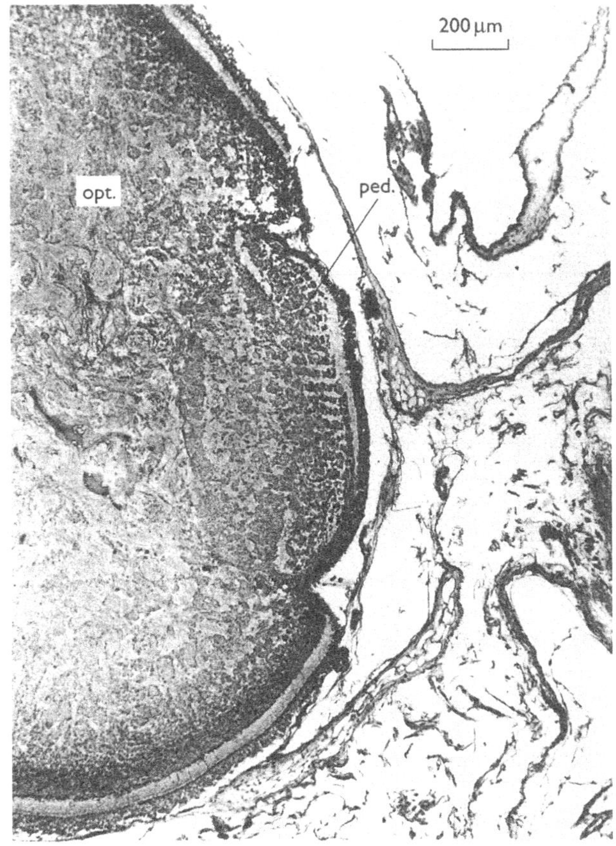

Fig. 26.8. *Vampyroteuthis infernalis* (27 mm ML). Sagittal section of the optic and peduncle lobes.

There are well-marked trochlear cartilages, which are reticulate. The cartilages have a fibrous sheath and some of the crossing muscle fibres appear to join them. The main bulk of the muscle fibres proceed alongside the cartilage. The cartilage and muscles reach the eyeball together and appear to insert on it. There is apparently only one crossing muscle, corresponding to anterior conjunctive II rather than to the tendon of conjunctive muscle I. The presence of this muscle shows that the eyes must sometimes be used in unison.

26.9 Optic lobes and optic tract lobes

The optic lobes form 378.4% of the total volume of the brain (Table 2.6). The lobes have outer and inner granule cell layers but the former is thin and the latter poorly marked off (Fig. 26.8). There is no sign of an inner plexiform layer or palisade cells, such as are found in decabrachians but not in octobrachians. The main plexiform layer is thin and almost absent from some sections. There is no differentiation of the cells into columns in the medulla. The groups of cell islands fill the whole inner part of the lobe and continue into the optic tract. These cells are numerous and mostly of a uniform and moderate size, presumably with considerable influence on the behaviour of the animal.

The optic tract lobes are set at an angle of about 60° to the horizontal anatomical plane. The peduncle lobe is elongated dorsoventrally and its spine region is faintly indicated externally. There is a well-marked olfactory lobe. The optic gland lies ventrally behind the olfactory lobe. The peduncle lobe is large and the spine has two rows of parallel fibres running anteroposteriorly (Fig. 26.8). The rows of parallel fibres are rather widely separated by large groups of small cells. This gives a regular segmented appearance in the sagittal plane. In the basal peduncle lobe the lateral part of the neuropil is interrupted by many neurons. More medially, the nerve from the photosensitive vesicele enters and forms a dense plexus. The connections of these vesicles are essentially as they are in decabrachians. There are few commissural fibres joining the two peduncle lobes; indeed, the whole optic commissure is small.

26.10 Central nervous system

The eight arm nerves are united by an interbrachial commissure of typical form (Fig. 26.1). The nerve from each filament lies between the nerves of arms I and II, and gives off a small branch that joins the main commissural ring. Each filament nerve can be traced back, like the other brachial nerves, towards the magnocellular lobe (Fig. 26.9).

The brain is exceptionally compact, more so than in many octopods, and much more so than in most decabrachians. No sharp divisions can be seen externally between the three main suboesophageal masses. The eight arm nerves enter the brain and join to form the prebrachial part of the anterior suboesophageal mass (Fig. 26.10). There is a short brachial lobe, not sharply marked off from the middle suboesophageal mass. Many bundles of the brachial nerves can be followed back to the magnocellular lobe (Fig. 26.9).

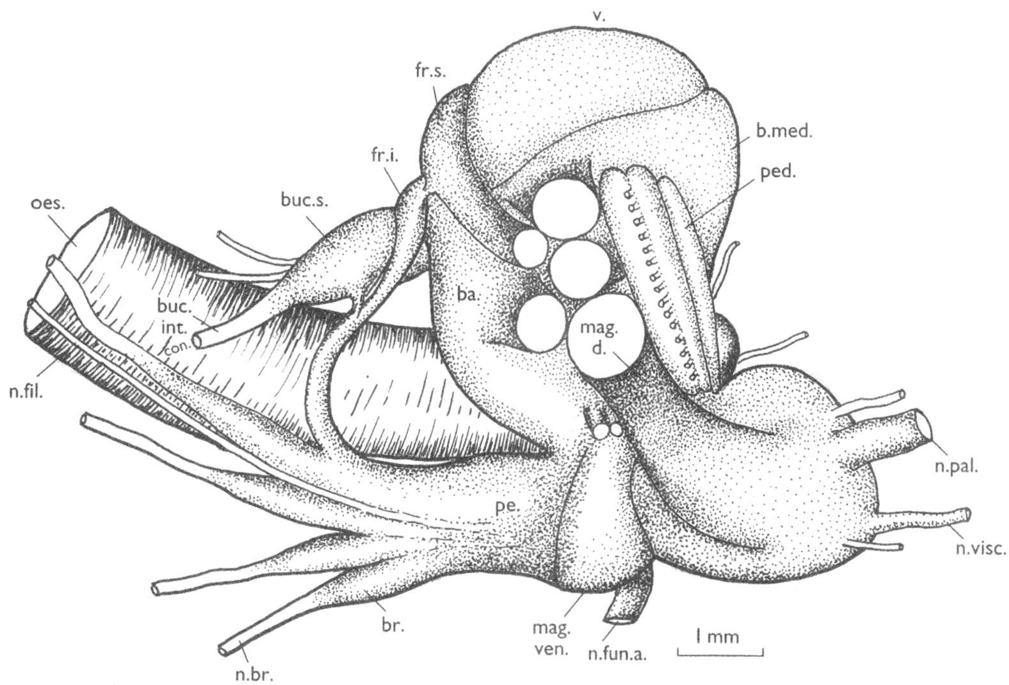

Fig. 26.9. *Vampyroteuthis infernalis*. Drawing of lateral view of dissected brain (Young 1977a).

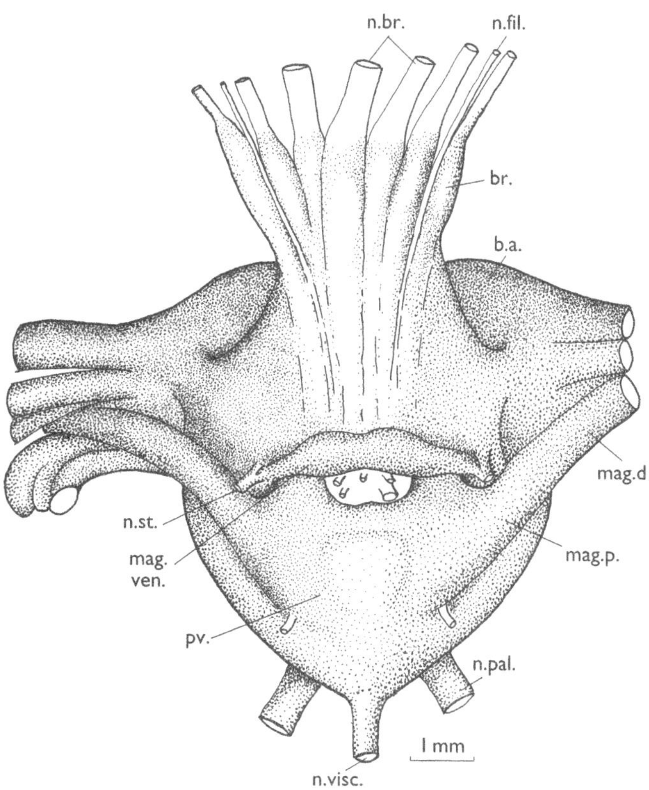

Fig. 26.10. *Vampyroteuthis infernalis*. Drawing of the central nervous system in ventral view.

The pedal lobe is broadly expanded laterally in the region of the oculomotor centre. The ventral magnocellular lobe runs round the ventral end of the pedal lobe as a well marked band of tissue (Figs 26.9, 26.10). In the midventral line it forms a pair of swellings, from which bands proceed to join the dorsal magnocellular lobe behind the entry of the static nerves. The dorsal magnocellular lobe is a conspicuous band, and it proceeds backwards to merge into the posterior suboesophageal mass. The latter is of rather simple form, with none of the lobes well marked externally.

The union between the suboesophageal and supraoesophageal regions is made by smaller anterior and larger posterior connectives, separated by a wide lateral cerebral foramen. The superior buccal lobe is united to the supraoesophageal mass as in octobrachians, not separated from it as in decabrachians. A large and long interbuccal connective joins it to the inferior buccal lobe (Fig. 26.9). The cerebrobrachial connective runs partly to the superior buccal lobe but much of it passes on dorsally to expand into a small pair of inferior frontal lobes. These in turn are joined by interfrontal tracts to the superior frontal lobe, which is a broad band of tissue running across the front of the supraoesophageal lobes and joining the optic tracts at the sides. The superior frontal lobe is marked off from the vertical lobe. No external divisions of the vertical lobe are apparent; there is some indication of folding at the back. The vertical lobe is not sharply marked off from the subvertical or basal lobes.

26.10.1 Supraoesophageal mass

The superior buccal lobes are prominent and are attached to the rest of the brain, but almost completely separated from the posterior buccal lobe (Fig. 26.9). The inferior frontal lobe system forms 5.6%

of the total brain volume (Table 2.6). The fibres of the cerebro-brachial connective enter a posterior buccal lobe, whose upper part constitutes a median inferior frontal lobe. Here a few bundles of fibres cross, but there is no distinct lateral inferior frontal lobe and no mass of interweaving fibres. At the back of the posterior buccal lobe there are groups of small cells, clearly representing the subfrontal lobe. The subfrontal lobe is unusual in receiving connections from the posterior buccal lobe rather than from the median inferior frontal lobe, as in octobrachians. Fibres run direct from the sub-frontal to the subvertical lobe. The posterior buccal lobe receives a large palliovisceral to inferior frontal lobe tract. It also sends several interfrontal tracts to the superior frontal lobe. This imperfectly divided posterior buccal lobe shows a simplified version of the octo-brachian condition. It could be either a fore-runner or a secondary simplification. The posterior buccal lobe opens behind to the sub-vertical lobe, which is much subdivided by islands of cells. There is no distinct precommissural lobe.

The superior frontal lobe forms a prominent part of the brain, with many interweaving and crossing fibres (Figs 26.9, 26.11). The vertical lobe system forms 27.5% of the total brain volume (Table 2.6), but it is not divided into five lobules (Fig. 26.12). Its structure may be described as intermediate between decabrachians and octobrachians, perhaps being closer to the latter. The outer cell layers are thick over the whole surface (as in octrobrachians) but the neuropil is extensive, with scattered cells (as in decabrachians). There are islands of cells and neuropil in the dorsal basal region, and below is a distinct median basal lobe. The anterior basal lobe is well developed, with regions of parallel fibres and small cells across the midline.

26.10.2 *Suboesophageal mass*

The bases of the arm nerves join to form prebrachial lobes, which merge behind as brachial lobes. The two filament nerves pass on either side between the roots of nerves of arms I and II (Fig. 26.13). The pedal lobe is almost cut off from the brachial lobe by a wall of cells. Large brachiomagnocellular and brachiopalliovisceral connectives run along its ventral side. The contributions from the various brachial nerves merge together at the front of the pedal lobe but the filament nerve remains distinct (Fig. 26.13–26.15), and is still recognizable to its entry into the magnocellular lobe.

The fibres from the arms enter the ventral magnocellular lobes at the sides and then interweave in a plexus (Fig. 26.9). The fibres of this plexus form a regular series of bundles near the periphery and also curl round to make a further network at the centre of the ventral magnocellular lobe (Fig. 26.16). Fibres from the plexus spread and branch in the neuropil and are presumed to be afferents. They make contact with the processes of the cells of the lobe. These collect into bundles which proceed backwards to form larger bundles of fibres which pass to the dorsal and posterior magnocellular lobes (Fig. 26.17). Here they meet a tract of brachiomagnocellular and brachiopalliovisceral fibres, many running through direct from the arm nerves. The two sets of fibres spread out through the magnocellular and palliovisceral lobes and seem to proceed partly to the pallial nerves, while a bundle passes up far laterally to the interbasal region. The posterior magnocellular

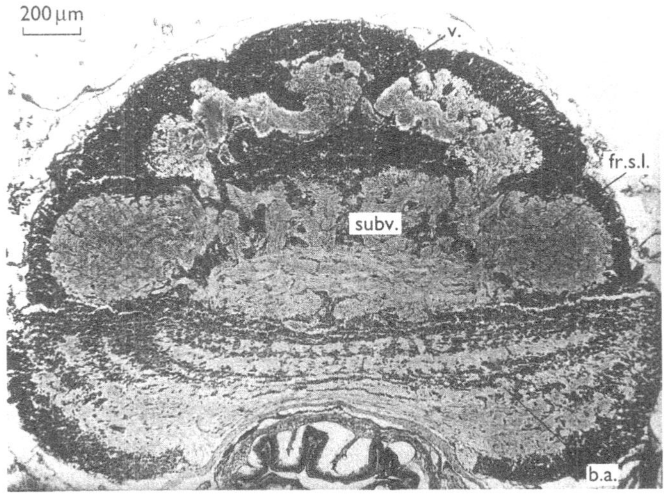

Fig. 26.11. *Vampyroteuthis infernalis.* Transverse section to show the lateral superior frontal lobe, the vertical, subvertical, and anterior basal lobes.

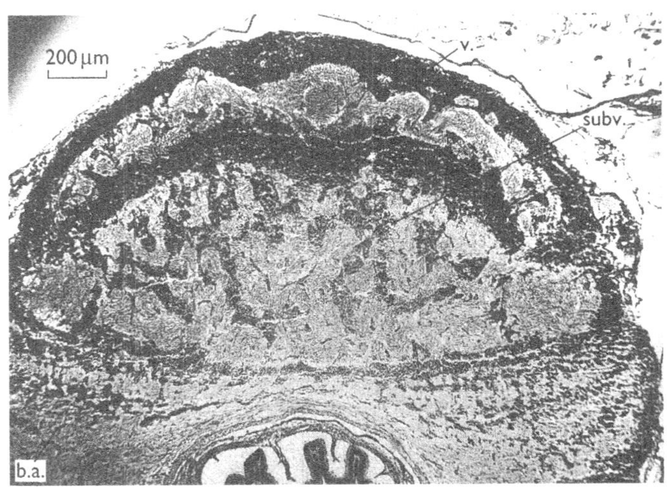

Fig. 26.12. *Vampyroteuthis infernalis.* Transverse section of the vertical, subvertical, and anterior basal lobes.

lobe contains some notably large cells but the destination of their axons is not clear. There is no unique first-order giant cell, and no large axons cross in the palliovisceral lobes. There are no especially large fibres in the stellate ganglion or nerves (Fig. 26.18). The magnocellular lobe thus has an elaborate organization recalling that of *Mastigoteuthis*, which possesses long tentacles with innumerable small suckers.

There is a large fin lobe and this receives a tract from the dorsal part of the posterior pedal lobe (Fig. 26.19) and another from the interbasal region, which is evidently an important area for motor coordination, as it is in *Loligo* (Young 1976).

The fibres of the statocyst nerves spread widely to the pedal and other lobes and some can be seen to reach to the anterior basal lobe (Fig. 26.19).

26.11 Stellate ganglia

No sign of giant fibres has been seen in the stellate ganglion or in the stellar nerves (Fig. 26.18). *Vampyroteuthis infernalis* exhibits an

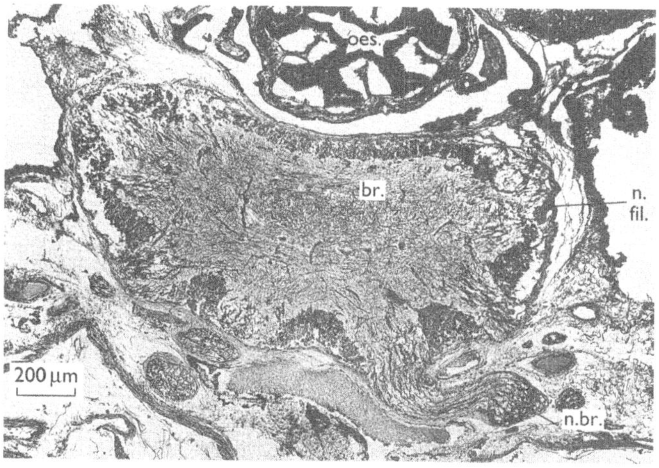

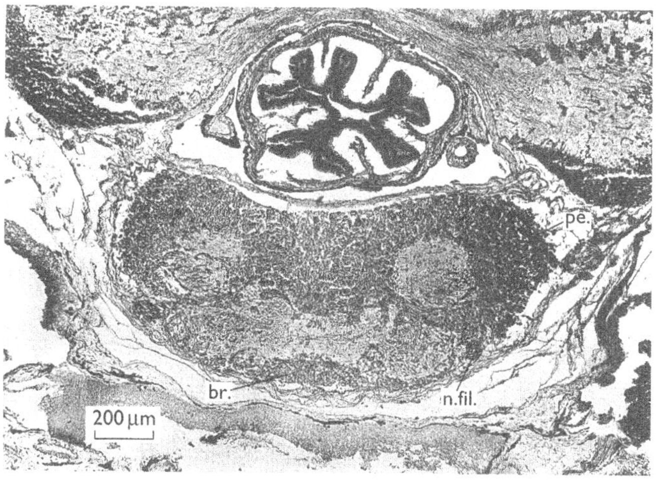

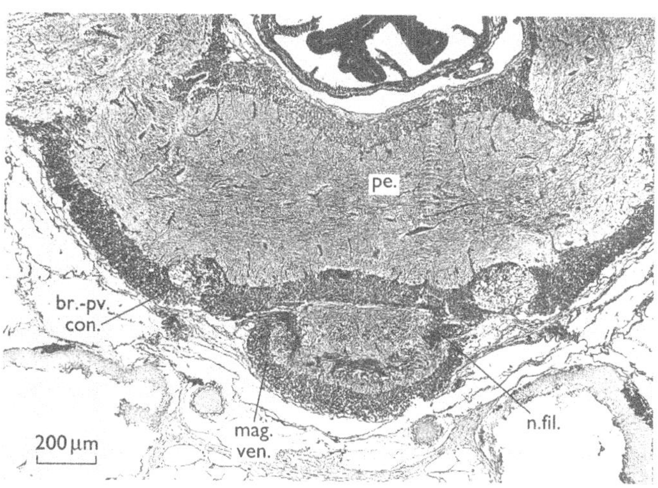

Figs 26.13–26.15 *Vampyroteuthis infernalis*. A series of transverse sections of the brachial and pedal lobes to show the course of the filament nerves to the magnocellular lobe.

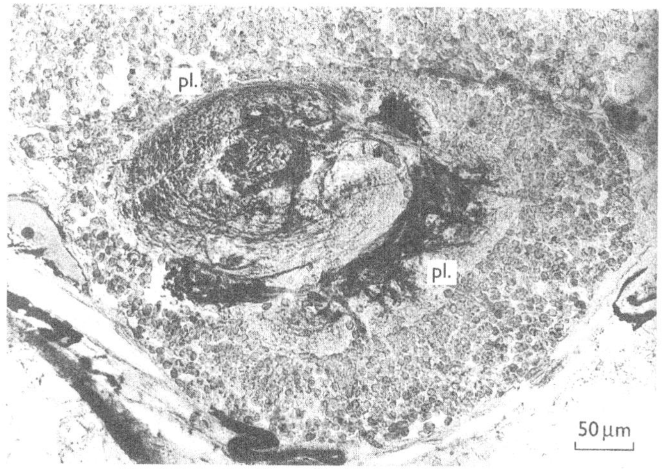

Fig. 26.16. *Vampyroteuthis infernalis* (27 mm ML). Sagittal section of magnocellular lobe to show the plexus of fibres from the arms.

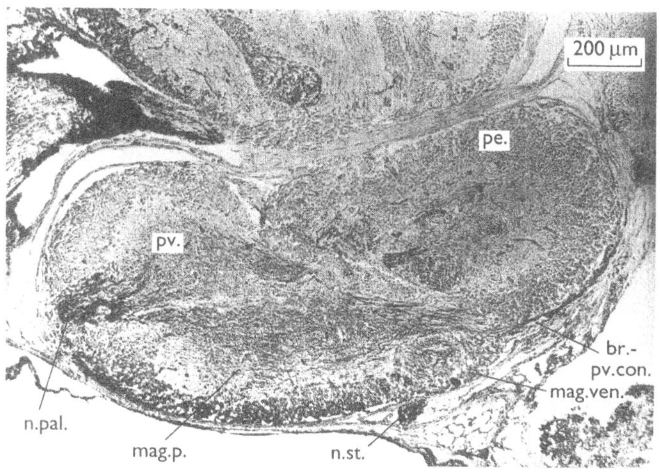

Fig. 26.17. *Vampyroteuthis infernalis* (27 mm ML). Oblique sagittal section of the magnocellular and palliovisceral lobes.

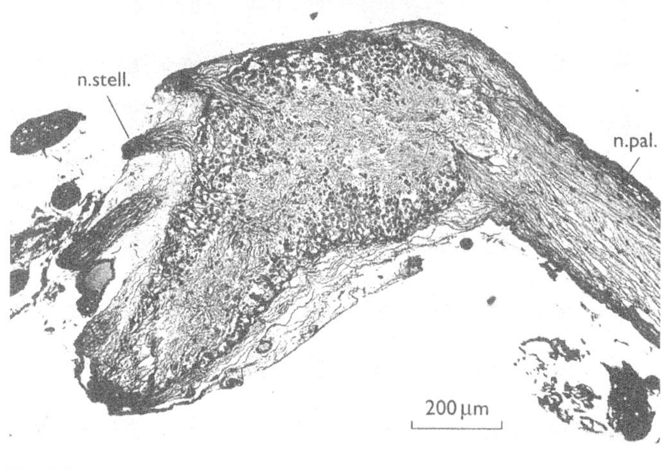

Fig. 26.18. *Vampyroteuthis infernalis*. Stellate ganglion and pallial nerve.

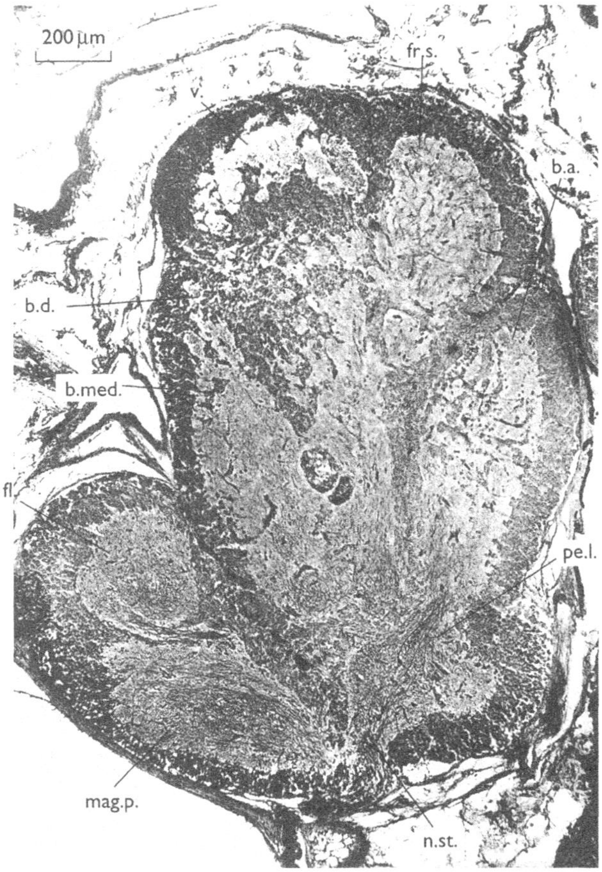

Fig. 26.19. *Vampyroteuthis infernalis* (27 mm ML). Sagittal section showing the vertical and basal lobes, and the static to anterior basal lobe tract.

escape response (R.E. Young in Roper and Brundage 1972, P.J. Herring pers. comm.) but the mantle muscles appear weak, and the question remains whether it has lost the giant fibres, or has never had them (Young 1977a).

26.12 Extraocular photoreceptor organs

Each extraocular photoreceptor organ in one specimen (65 mm ML) comprised about 30 round, white, photosensitive vesicles (Figs 26.20, 26.21). The vesicles are bright orange when fresh and form rows extending away laterally on either side of the vena cava, and thus lie in the ventral body wall, below and behind the statocyst and above the funnel (R.E. Young 1972a, 1978). The body wall above these vesicles is deeply pigmented with black. Posteriorly the tissue separating them from the mantle cavity is translucent. Nerve fibres from the vesicles join the posterior funnel nerve, a branch of which can be followed to the peduncle lobe. R.E. Young (1972a) suggests that they could detect light within the mantle cavity, perhaps emitted by luminescent prey. The fact that the nerve reaches to the peduncle lobe implies that the vesicles are concerned in monitoring movements of the whole animal. The vesicles occupy a position between that found in the squids, where they are on the optic stalk, and the octobrachians, where they are found on the stellate ganglion. Nerves from the

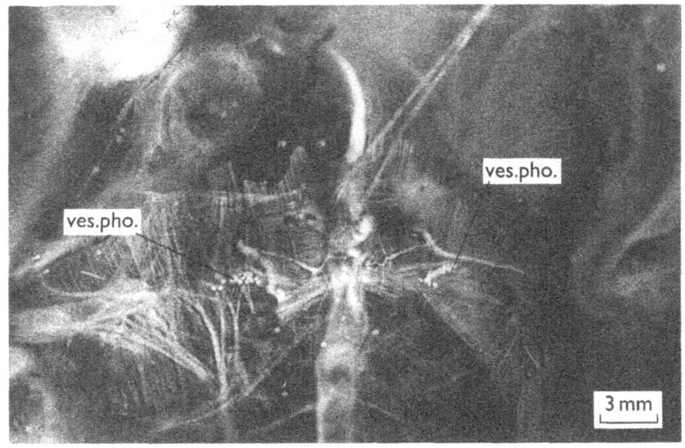

Fig. 26.20. *Vampyroteuthis infernalis* (65 mm ML). Dissection to show the photosensitive vesicles.

Fig. 26.21. *Vampyroteuthis infernalis* (27 mm ML). Sagittal section of the central nervous system to show the photosensitive vesicles.

vesicles join the posterior funnel nerve, a nerve from which joins the peduncle lobe.

26.13 Statocysts

The statocysts show a curious combination of features present in those of Octobrachia and Decabrachia. The statocyst is large (Fig. 26.1), and is a feature of coleoids which move and turn slowly (Young 1989). There are inner and outer sacs, with perilymph between, as in octobrachians but never in decabrachians. The macula is divided into macula

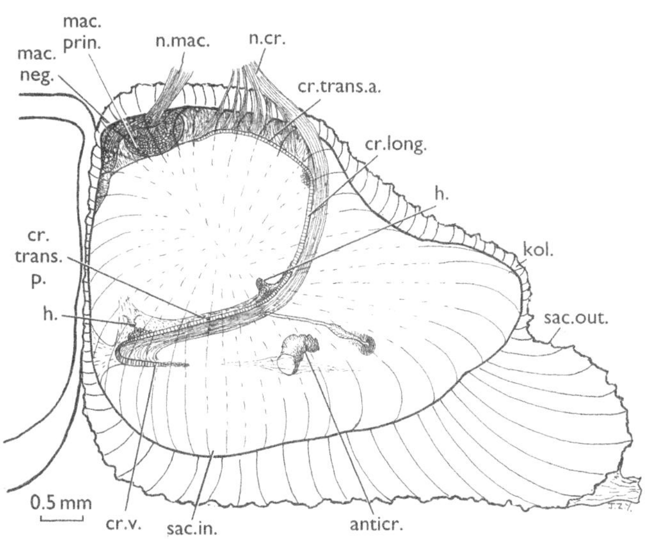

Fig. 26.22. *Vampyroteuthis infernalis.* Drawing of a statocyst from a young animal with only two small hamuli and one anticrista present, in dorsal view (Stephens and Young 1976).

princeps and a single curved macula neglecta (Fig. 26.22). This condition is intermediate between that found in the two major groups.

The crista has four sections, oriented as in decabrachians but not sharply separated. The sections are not subdivided as they are in octobrachians. There are hamuli at two of the turns of the crista, an entirely decabrachian feature. Eight anticristae restrict the flow of endolymph; they are arranged approximately as in non-buoyant decabrachians such as *Loligo.*

Kölliker's canal is large, and its duct opens at the tip of a large anticrista. The lining of the wall of the inner sac contains many hair cells whose base is prolonged in various directions. Their nerve fibres form a network and bundles of these fibres join the crista. Parts of the network cover the anticristae. The function of these hair cells is not known (Stephens and Young 1976).

This set of characters is puzzling both functionally and phylogenetically. The large anticristae are presumably related to pelagic life, involving turning in three planes. They are formed from the wall of an inner sac.

26.14 Affinities with other coleoids

Vampyroteuthis infernalis possesses features which are present in Decabrachia and Octobrachia, but they are so mixed that it is difficult to decide on its ancestry although a number of people have discussed this problem (Pickford 1940, 1957, R.E. Young 1967, Young 1977a, Fioroni 1981, Bandel and Leich 1986, Healy 1989, 1990b, R.E. Young *et al.* 1998b).

Vampyroteuthis has 10 appendages, evidently a fundamental feature of coleoids (see Doyle *et al.* 1994, Donovan in press c). One pair of these appendages is modified into filaments, which can be projected and subsequently retracted into a pouch. The remaining four pairs are arms which bear suckers and paired cirri. The latter have a similar arrangement to those present in Cirroctopodida.

The gladius is like that of a Jurassic phragmoteuthid (Bandel and Leich 1986) as well as that of a Recent cirroctopodid, *Cirrothauma murrayi* (Aldred *et al.* 1983). The latter is probably the nearest living relative of *Vampyroteuthis infernalis* and has evidently evolved independently from some stage close to the separation of the two groups, perhaps at the end of the Triassic.

The features of the brain of *Vampyroteuthis infernalis* are similar to those of cirroctopods with compactness of lobes but rather diffuse inferior frontal and vertical lobe systems, without the definite lateral inferior frontal and subfrontal lobes found in octobrachians. There is little similarity of the brain of *V. infernalis* with that of decabrachians where the lobes are more widely sepearated. The inferior frontal system is like that of a rather simple octobrachian. We must assume that the *Vampyroteuthis* stock evolved this system after leaving the main stem, perhaps then going on to give rise to Octobrachia. The ventral magnocellular commissure was carried on from the ancestral condition and later lost in Octobrachia.

The statocysts show an extraordinary assemblage of characters, and insofar as there are similarities then these are with the cirroctopods. Hamuli and anticristae are in positions approximately similar to those in decabrachians, and the crista gives three turns and has four sections. The two sacs of the statocyst are an octobrachian feature but the complex macula and multiple anticristae (Young 1989) must be characters carried on from an earlier ancestry, and indeed these are still partly present in cirroctopods.

Characters of *Vampyroteuthis infernalis* that are definitely octobrachian include the presence of cirri along the arms, the web between the arms, and suckers without stalks or chitinous inner rings. The spermatozoa are simple but do show some resemblance to part of the spermatid of *Octopus* (Healy 1989). There is no rod in the bolster of the radular complex, as in decabrachians (Messenger and Young 1999) The number of lamellae in the gills is reduced (Pickford 1940). The conjunctive eye muscles are like those found in decabrachians (Budelmann and Young 1993).

Several features appear to be unique to *Vampyroteuthis.* The suckers possess cilia but apparently lack cuticle (Nixon and Dilly 1977). The structure of the photogenic tissues 'Provides little guidance to the rather diffuse relationships of the Vampyromorpha' (Herring *et al.* 1994) as it is unlike that of the bolitaenid octopod *Japetella,* and there are no apparent links with the photogenic tissues in that of any sepiid, sepiolid, or teuthid decabrachians investigated so far. Collagen fibres are found as reflectors in some photophores of some of the enoploteuthid and lycoteuthid squids (R.E. Young and Arnold 1982, Herring *et al.* 1985), and in *Spirula spirula* (Herring *et al.* 1981). *Vampyroteuthis* is so far apparently unique in having such a thick reflective structure, formed of collagen fibres. Indeed, Herring *et al.* (1994) consider that collagen may have been the first reflective material in ancestral cephalopods, which would indicate an early separation of Vampyropmorphida from other coleoid cephalopods.

Vampyroteuthis infernalis is in many ways like a phragmoteuthid ancestor of Decabrachia but has begun to diverge in the octobrachian direction. It retains an early form of gladius, with a siphuncle, as well as other features, but has evolved unique characters. As a survivor from the past it shows how little the organization of coleoid cephalopods has changed during their evolution.

27 Cirroctopodida

27.1 Cirroctopodids

Three families of finned octopods form the order. Their members occur in all oceans, the Cirroteuthidae between latitudes 75°N and 60°S in the Atlantic, Pacific, and Arctic Oceans, the Stauroteuthidae in the north-west Atlantic Ocean, and Opisthoteuthidae throughout the Atlantic and Pacific Oceans and Antarctic waters associated with the continental slope and seamounts (Nesis 1993a). Most have been recorded at great depths, cirroteuthids from 2000–4300 m, stauroteuthids from 460–2500 m, and opisthoteuthids from 100 m to 7279 m, the latter is the world depth record for a cephalopod (Voss 1988a,b). The abundance of cirroctopods ranges from 1.1 to as many as 98/km^2 (Roper and Brundage 1972, Pearcy and Beal 1973).

The first octopod with fins to be found was described by Eschricht in 1836 and named *Cirroteuthis mülleri*. It was captured at Jacobshavn, Greenland, and a coloured illustration of this specimen was published by Reinhardt and Prosch in 1846. Cirroctopodida possess a pair of large blade- or paddle-shaped fins and cirri along the arms (Figs 27.1, 27.2), and these two external features readily separate the members of this order from the Octopodida. The morphological features of the family and their bathypelagic distribution has been discussed by Nesis (1987), Voss (1967, 1988a), and Guerra and his colleagues (1998).

Underwater observations and midwater captures show that cirroctopodids swim freely near to, but mostly above, the ocean floor (Roper and Brundage 1972, Pearcy and Beal 1973, Vecchione and Young 1997, Villanueva *et al.* 1997, Guerra *et al.* 1998) and may be said to be mainly suprabenthic, as they associate with but do not make contact with the sea bed (Fig. 27.3). The exception to this is the Opisthoteuthidae whose members may spend some time on the substrate and are perhaps better termed benthopelagic (Fig. 27.4) (Vecchione and Roper 1991). One cirroctopodid was seen, from a submersible at 2800 m, to be hovering just above the sea floor, with the web in an inverted umbrella-like position (Boletzky *et al.* 1992). When touched this animal gave a spectacular reaction in which it became transformed from an inverted umbrella-shape to a balloon-shape, all within 15 s; the curvature of the arms changed so that the web expanded between them and the cirroctopodid appeared more like a pumpkin. The distal tips of the arms disappeared beneath the web and the balloon continued to expand for a further 10 s. A coordinated arm flexion, moving back from the arm tips as a peristaltic wave, resulted in the tips re-appearing, the grooves between the web sectors disappearing and the body again exposed above the now shrinking web. The arms, which had remained immobile for 43 s, began to lift and the animal swam away, with its arms trailing behind. These authors suggested that the balloon response may be a defence mechanism against a potential predator.

The texture of finned octopods is gelatinous and observations of these animals hovering, apparently suspended, above the bottom, suggests that they are neutrally buoyant. This was confirmed in an adult cirroteuthid which was found to be neutrally buoyant shortly after its capture by a submersible at 3351 m (Villanueva *et al.* 1997). Thus movement by cirroctopods can be by the fins, by umbrella-like medusoid drifting, and by ballooning.

The anatomy of these animals has been described in *Opisthoteuthis depressa* (Ijema and Ikeda 1895, Meyer 1906), *Grimpoteuthis umbellata* and *Stauroteuthis syrtensis* (Ebersbach 1915), and *Cirrothauma murrayi* (Chun 1913, Aldred *et al.* 1983). The behaviour and habits of these animals during reproduction are as yet not known. However, it is apparent from the presence of a few large eggs and numerous small ones in the ovaries of four species, from two different families, that they spawn over a considerable period of time (Mangold *et al.* 1993).

27.2 Cirroteuthidae

This family has a distribution between latitudes 75°N and 60°S in the Atlantic, Pacific, and Arctic Oceans (Voss 1988b). Its members usually live in deep water, but above the bottom; a very small number have been found at the surface, perhaps as a result of upwelling (Voss 1988a).

27.2.1 *Cirroteuthis*

Cirroteuthis mülleri is known from the north-west Atlantic, Baffin Bay to Iceland, off Jan Meyen Island and off Oregon, and it has been caught from the surface to depths of 2300 m (Voss 1988a, b).

The reproductive organs were exposed and the central nervous system dissected in the first specimen captured of *Cirroteuthis mülleri*; it was a gravid female and, as the illustration by Reinhardt and Prosch (1846) shows, the ovary contained eggs at various stages of development, including a few outstandingly large ones. Recently a mature male, captured by a submersible at 3351 m over the mid-Atlantic Ridge, had a mantle length of 220 mm and a total length of 1300 mm (Villanueva *et al.* 1997).

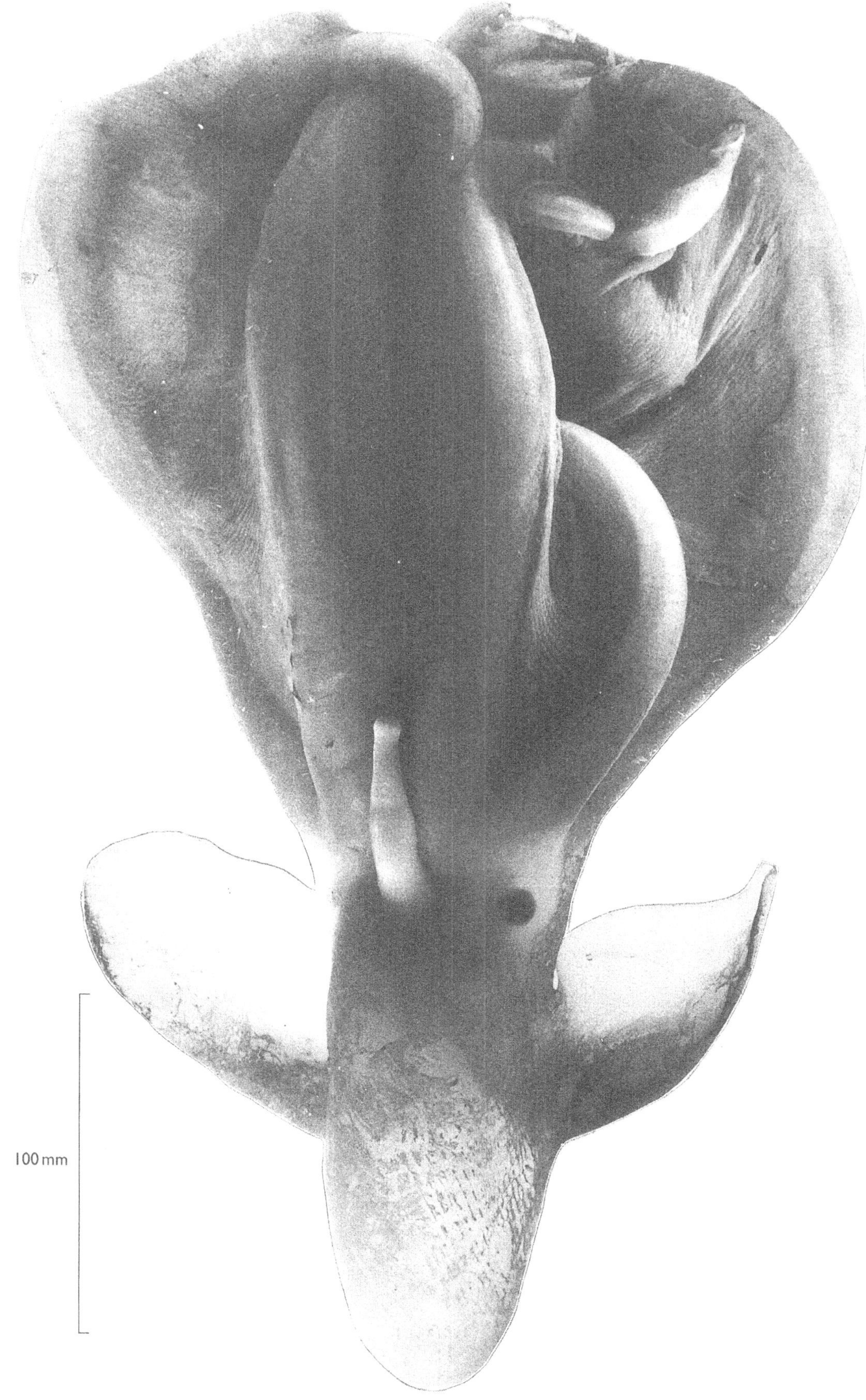

Fig. 27.1. *Cirrothauma murrayi* (130 mm ML) photographed immediately after capture (Aldred *et al.* 1983).

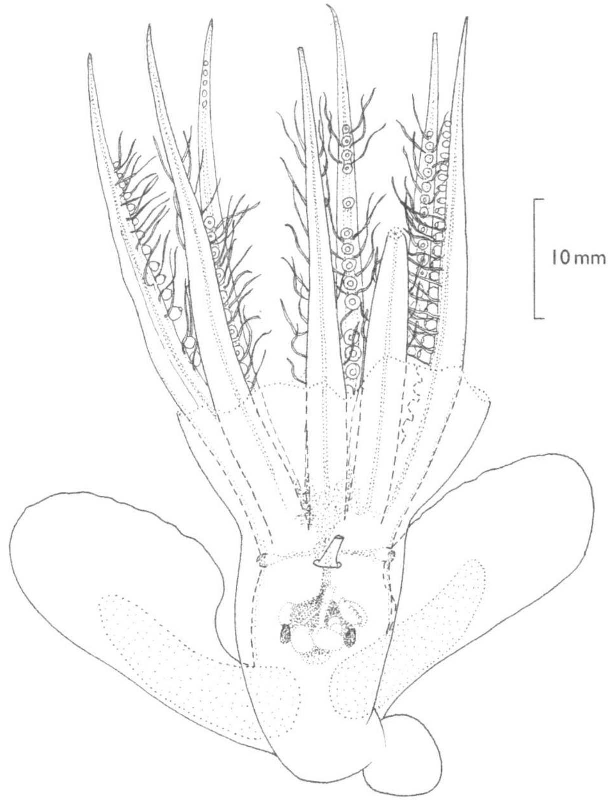

Fig. 27.2. *Cirrothauma murrayi* (21 mm ML). Drawing to show the large, mainly translucent fins and the elongated cirri on the arms (scale bar 10 mm) (Aldred *et al.* 1983).

Fig. 27.4. *Opisthoteuthis agassizii*. A photograph made from a video-recording of an animal at 580 m depth, off St Vincent (Vecchione and Roper 1991).

Fig. 27.3. Cirroctopodida. The shadows of the animals in these photographs show that they are close to, but above, the sea floor. The animals were photographed at 3970 m (a), 3620 m (b), and 5030 m (c) (Roper and Brundage 1972).

Five sightings of *Cirroteuthis* have been recorded from submersibles, at depths of between 2702 and 4527 m; two were drifting umbrella-like 1–3 m off the bottom, and three were fin-swimming 1–7.5 m off the bottom with strokes of 4–30/min (Villanueva *et al.* 1997).

27.2.2 *Cirrothauma*

Cirrothauma murrayi (Fig. 27.1) has been caught in the Atlantic and Pacific Oceans, in the Arctic, and in Antarctic waters. Some specimens have been taken in opening/closing nets, at 4505–4515 m depth, not far off the bottom; several are known from midwater and some considerable distance above the substrate (Aldred *et al.* 1983).

Cirrothauma murrayi is a bulky, gelatinous creature, and when freshly caught the adult is red-brown virtually all over its body and arms. Pigmentation is of two types: an outer layer with numerous single pigment granules, and beneath it a layer containing pigment granules in well separated spherical clusters. No chromatophores were seen. A small specimen (Fig. 27.2) of only 21 mm ML was translucent and nearly colourless except for part of its alimentary canal, which was deeply pigmented.

Arms, suckers, and cirri

The arms form a major portion of the animal, especially in the adult (Fig. 27.1). Each arm has a single row of suckers alternating with paired cirri. The arm is gelatinous and its tissues are traversed by internal muscle bundles. A transverse septum divides the arm into inner (oral) and outer (aboral) compartments. The division of the arm into two partially independent, flexible, tapering tubes may give greater control over movement. This suggests that the movement is in the oral–aboral plane; important in extending and curving the arms into the umbrella position, and holding them together to streamline the animal during fin-swimming.

The strange suckers caught the attention of Chun (1913) who described them as 'minute suckers poised on long spindle-shaped and clumsy stalks of gelatinous substance... and evidently devoid of the sucking pit'. Each arm is 12–17 mm in length in specimens of 105–150 mm ML, with approximately 40–60 suckers, each 0.8–4.0 mm in diameter. Most of the suckers lack an aperture and instead the surface is like a cap (Fig. 27.5). The cap is cuticular and has pegs, those around the periphery being a little larger than the others (Fig. 27.6). The pegs have pores some of which display an exudate on the surface, perhaps mucus. The first six or seven aboral suckers have a small orifice (1.9–3.1 mm in diameter) and one is shown in Fig. 27.7. The infundibular cuticle has a radial arrangement and pegs are present on the upper part. The pegs have regularly arranged pores approximately 0.2 μm in diameter.

The suckers are almost certainly functional, as the pores may allow adhesion of the infundibular surface to a foreign surface, with each tiny pore acting like a capillary. The hypothesis proposed by Nixon and Dilly (1977) predicted that only a relatively small infundibular surface area, and hence a small number of pores, would be needed by deep-sea octopods to achieve functional attachment. If the effectiveness of the system is dependent on pressure then this suggestion appears to be supported by the conditions found in *Cirrothauma* which inhabits considerable depths.

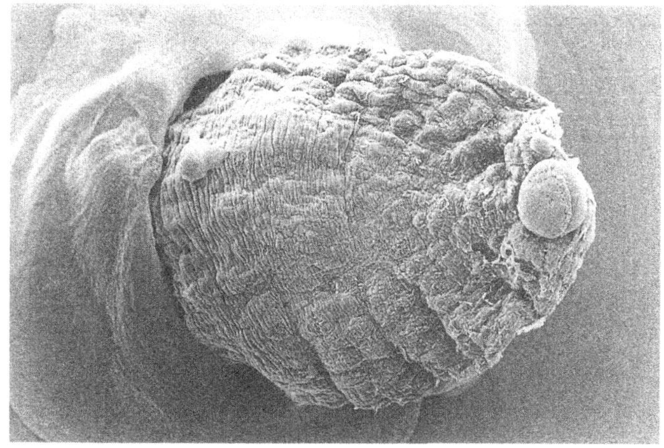

Fig. 27.5. *Cirrothauma murrayi.* Scanning electron micrograph of peduncle with sucker showing the cap-like structure (field width 4166 μm) (kindly provided by K.P. Ryan).

There is a small organ in the stalk of each sucker that Chun (1913) suggested could be luminous. We were at first inclined to agree with this interpretation. However, better material showed this organ to be a complex ganglion with large nerve cells (Aldred *et al.* 1984). The significance of these large cells and the strange neuropil is not known. Particularly puzzling is the presence of a collagenous cap, which Chun (1913) likened to a reflector.

The paired cirri range from 0.5 to 37 mm in length in preserved specimens of 105 and 155 mm ML. In the sea the cirri extend some distance away from the arms (Fig. 27.3a), and their length is especially notable in the small animal (Fig. 27.2). Each cirrus is formed of an inner core of reticular tissue divided by septa (Fig. 27.8). There is a rich blood supply with a few vessels running the length of the cirrus. This arrangement is similar to that of other soft-bodied invertebrates with a 'hydrostatic skeleton' (Elder 1980, Kier 1988). The cirri, in their outheld position, are well suited to receive chemical or tactile stimuli from potential prey or predators.

Interbrachial web

A deep and delicate interbrachial web joins the arms and extends along most of their length. The primary web is joined directly to the proximal half of the arms and at their tips. The distal part of the arm, except the tip, is connected to this primary web by a thin segment-shaped secondary web. This arrangement is seen in a photograph taken at great depth (Fig. 27.3a) (Roper and Brundage 1972).

Mantle, funnel, fins, and fin support

The mantle is relatively short, and the posterior part behind the fins is taken up by a soft mass of gelatinous tissue. The mantle wall has two thin layers of muscles separated by a thick layer of gelatinous tissue. Each layer of muscle is divided into outer longitudinal and inner circular muscle fibres. The mantle adductor muscle consists of a narrow membranous strip, and is apparently weak. The stellar nerves in the mantle have no giant nerve fibres.

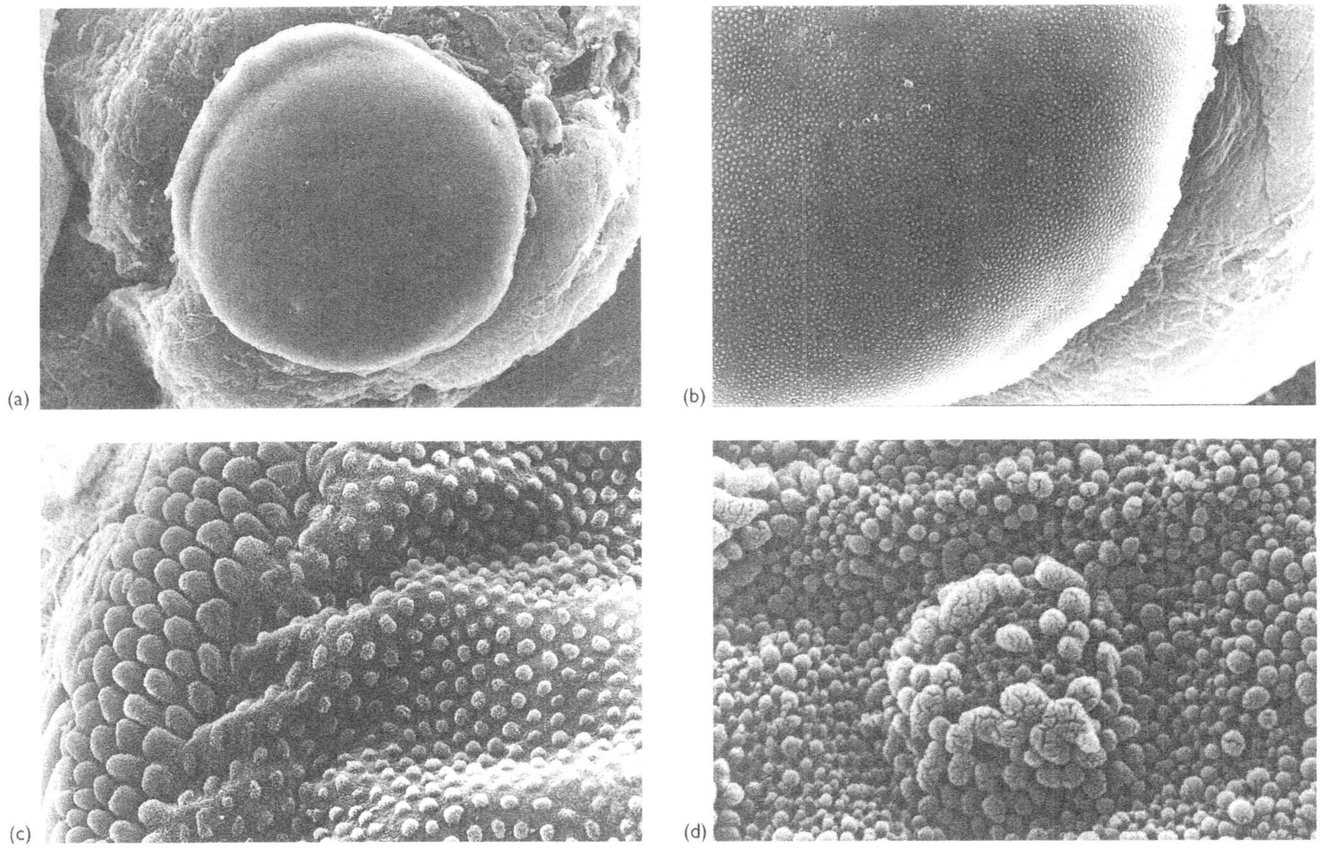

Fig. 27.6. *Cirrothauma murrayi.* Scanning electron micrographs of cap-like structure of sucker: (a) The cap (field width 830 μm), and (b) at higher magnification (field width 310 μm) showing the presence of small pegs over most of its surface with slightly larger ones at the periphery. (c) The small pegs of the cap with larger ones to the left (field width 70 μm), and at higher magnification (d) to show the pores on a small peg (field width 5 μm) (kindly provided by K.P. Ryan).

The funnel is extremely long and slender (Fig. 27.1), and reaches 84% of the mantle length in a female of 220 mm ML. The mantle aperture surrounds the funnel closely leaving only a small space, presumably for inhalation. These features of the mantle and funnel indicate that propulsion by jet is not strong.

The fins are relatively huge in comparison with the mantle in the young specimen (22 mm ML) and even in a larger specimen (130 mm ML) each fin almost equals the length of the mantle (Figs 27.1, 27.2). Attachment of each fin is to the anterior half of the mantle. The posterior edge of the fin is thick, and it is here that the major muscles lie. The leading edge of the fin is thin and almost completely translucent. This clear area would be directly over the eyes when the fins are folded around the body during locomotion.

The fin support is an exceptionally large and complex structure. It is translucent and in shape it resembles a butterfly with open wings (Fig. 27.9). The wings are concave and are the site of attachment of the fin musculature.

Heart

The heart has an unusual arrangement as it has two muscular chambers, joined by a short narrow duct. The first, or ventral, chamber is small and has a rounded somewhat triangular form. Arising from the ventral chamber is a single gonadal artery and two smaller vessels to the branchial hearts. The dorsal chamber is crescent-shaped and gives rise to two pallial arteries on the left side, and on the right it narrows to form the anterior aorta. The anterior chamber has developed from the swollen part of the aorta, together with the basal part of the pallial arteries (Aldred *et al.* 1983). The significance of this extra chamber is not understood but it may improve the blood flow, perhaps needed by the large fins and long arms during locomotion.

Buccal mass

The buccal mass is relatively large and so the beak encloses a spacious cavity. The buccal complex consists of a prominent salivary papilla and long tongue (Fig. 27.10). The salivary papilla carries a single duct from the median salivary gland. The terminal part of the salivary duct is toothed as is the papillary shield around its orifice. These features are characteristic of the terminal tip of the duct from the posterior salivary gland and the surrounding papillary shield in *Octopus vulgaris* and *Eledone cirrhosa* (Nixon 1979a, b). Thus the median salivary gland of the buccal complex of *Cirrothauma murrayi* (Aldred *et al.* 1983) and other cirroctopodids (Ebersbach 1915) represents the posterior salivary gland of Octopoda. The anterior position of this salivary gland of cirroctopodids is the phylogenetically original position.

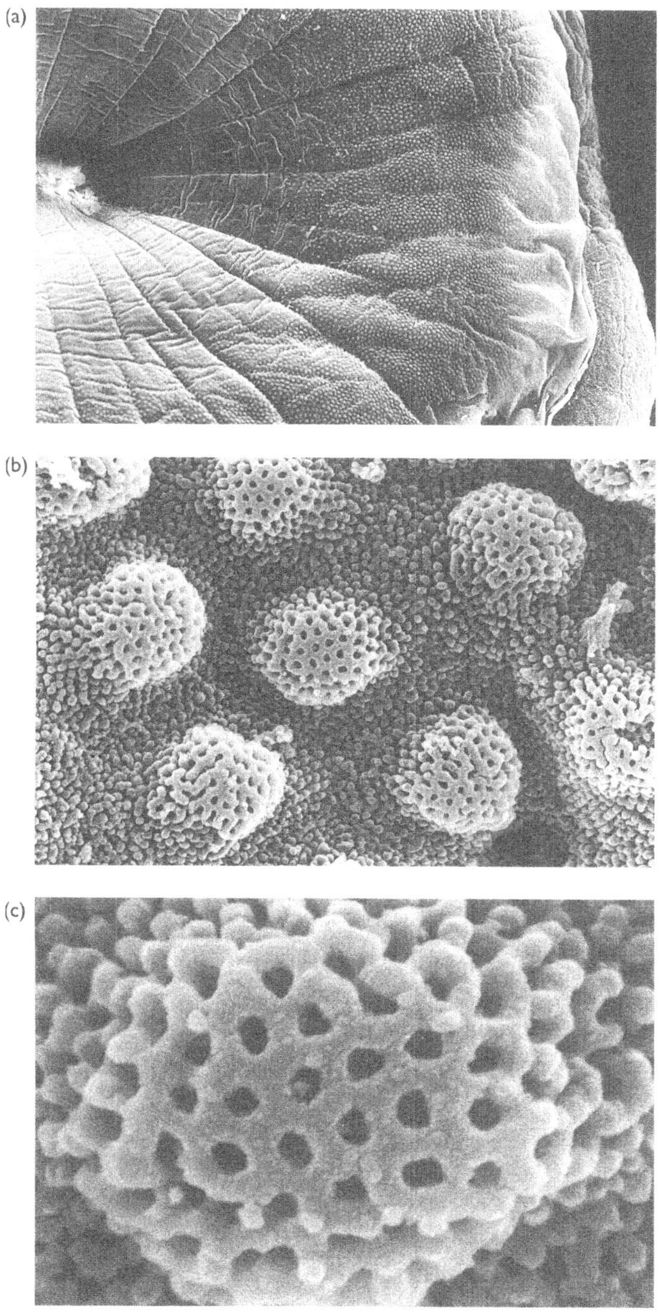

Fig. 27.7. *Cirrothauma murrayi.* Scanning electron micrographs of one of the suckers with a small aperture: (a) The cuticle with the radial arrangement of the upper part of the infundibulum (field width 530 μm), (b) the pegs at higher magnification (field width 12 μm) and (c) the pores on a peg (field width 3.2 μm) (kindly provided by K.P. Ryan).

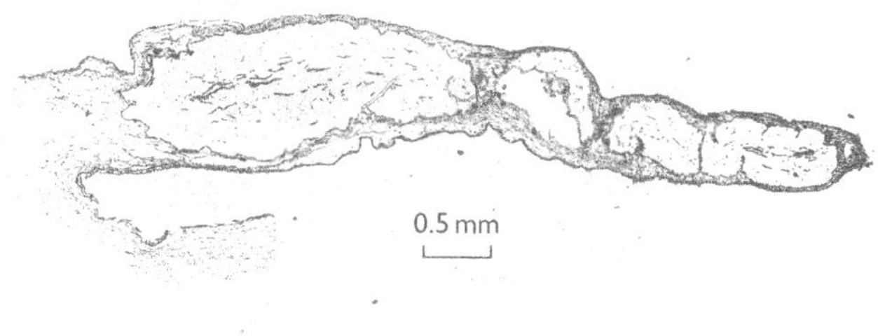

Fig. 27.8. *Cirrothauma murrayi* (155 mm ML). Longitudinal section of a cirrus, showing the septa (Masson) (Aldred *et al.* 1983).

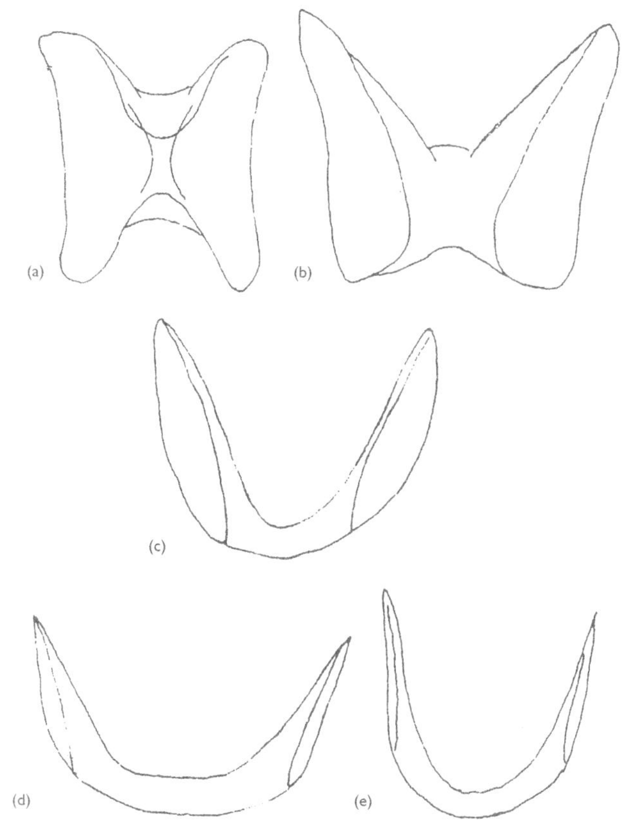

Fig. 27.9. The fin support (shell or gladius) of Cirroctopoda. Cirroteuthida: (a) *Cirroteuthis mülleri* and (b) *Cirrothauma murrayi*; Stauroteuthidae: (c) *Stauroteuthis syrtensis*, and Opisthoteuthidae: (d) *Opisthoteuthis agassizii* and (e) *Grimpoteuthis* sp. (Hoyle 1886, Ebersbach 1915, Voss 1956, Aldred *et al.* 1983).

The nerve of the posterior salivary gland in *Octopus vulgaris* runs forwards from the central nervous system and then backwards; this confirms the migration of the gland posteriorly, to a position behind the brain (Young 1971). There is a submandibular gland at the base of the salivary papilla that opens into the buccal cavity.

The tongue lies behind the salivary papilla and has a cuticular covering which is thrown into a series of transverse folds. Within the tongue is a pair of sac-like supports, one on each side of the presumed vestigial radula (Fig. 27.11). Each sac has fine branching fibrils in an amorphous matrix.

The cuticle lining the buccal cavity, and covering the buccal complex, continues along the oesophagus covering long finger-like projections. Similar projections are found in the stomach, where the cuticle becomes thick and is ridged on the two lateral walls. These lateral walls have thick muscles although the dorsal and ventral walls are thin. The stomach presumably would be capable of breaking

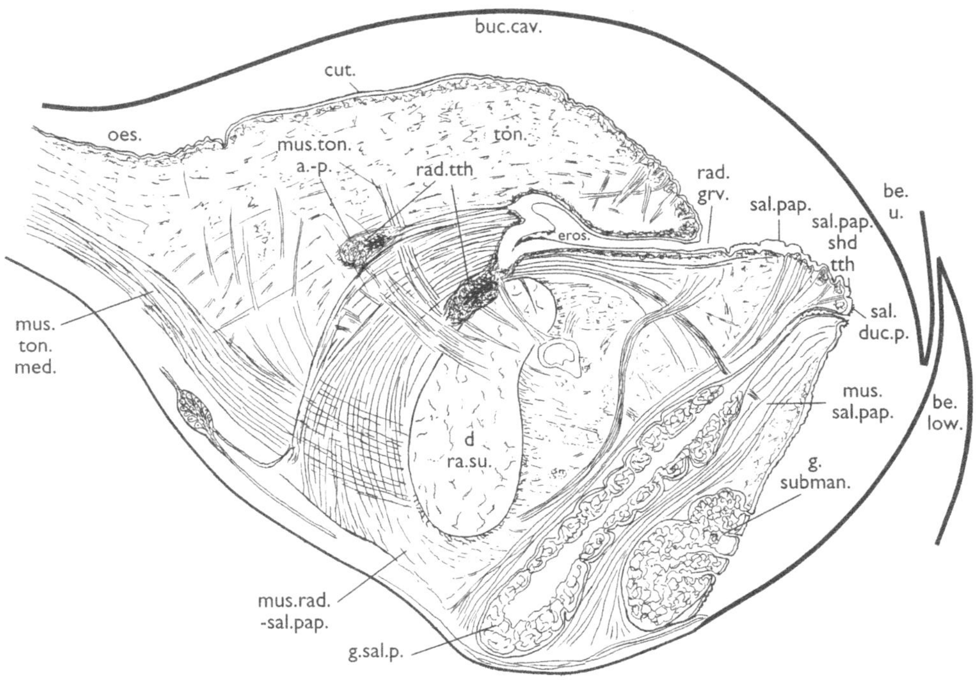

Fig. 27.10. *Cirrothauma murrayi* (155 mm ML). Drawing of the buccal mass reconstructed from serial sections near the midline (Aldred *et al.* 1983).

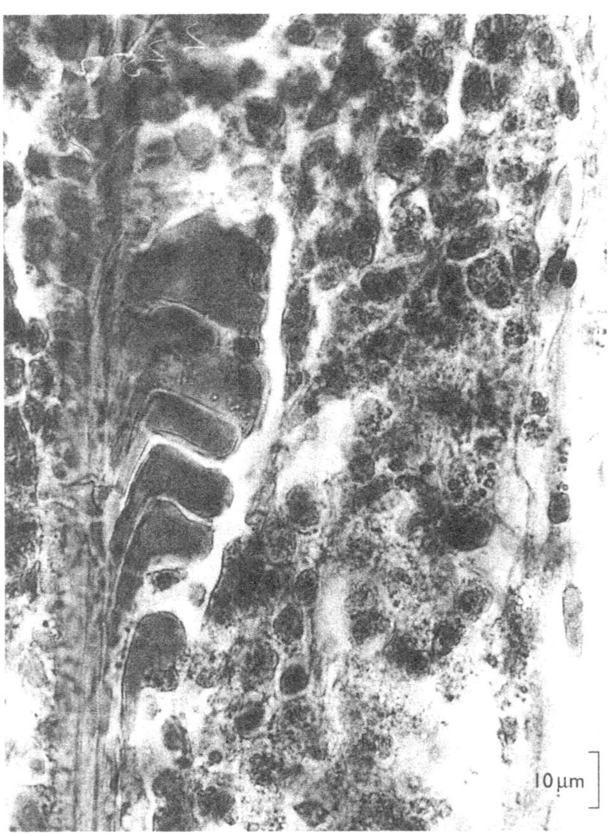

Fig. 27.11. *Cirrothauma murrayi* (155 mm ML). Posterior part of the 'radula' with tooth-like projections lying among amoebocytes (Aldred *et al.* 1983).

down relatively large particles, such as could pass along the oesophagus (4 mm diameter in a specimen of 155 mm ML). No debris was found in the stomachs examined. The dorsal side of the stomach and the oesophagus are deeply pigmented, suggesting a need to obscure any light emitted from ingested luminous prey. The stomach leads into the caecum, which is less muscular. The intestine and distal end of the rectum are devoid of pigment.

Reproduction

The male has a simple duct leading from the testis to a vas deferens with thick, folded walls. This leads to a seminal vesicle which is filled with a mass of small packets of sperm (Fig. 27.12). Each packet contains aggregations of spermatozoa with the heads arranged at the periphery and the tails at the centre. From the seminal vesicle the duct then leads through a voluminous accessory gland to a simple penis and a single opening. The structural simplicity and the small size of the sperm packet explain the uncomplicated reproductive tract of the female.

The female has a thin-walled ovary situated posteriorly in the mantle, below the fin support. Anteriorly the ovary narrows and gives rise to a single (left) thin-walled oviduct. The oviduct opens into the oviducal gland, which is divisible by its coloration into two parts. Internally both parts of the oviducal gland are highly folded. The proximal part of the gland functions as a spermatheca and contains several packets of sperms (Aldred *et al.* 1983). The gland secretes a substance which forms the outer coating of the egg case as it passes through the central cavity (Boletzky 1978). Fertilization is known to be internal as sperm packets were present in the oviducal

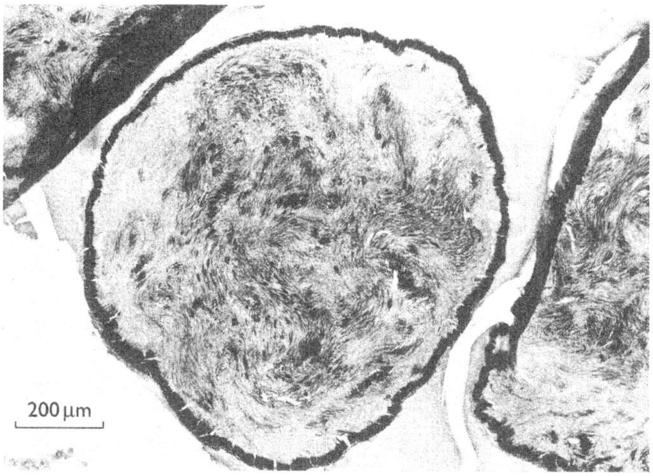

Fig. 27.12. *Cirrothauma murrayi* (155 mm ML). Section showing sperm packet in seminal vesicle (Aldred *et al.* 1983).

gland. The method of copulation is not known, since the male has no hectocotylus.

The ovaries in two of our specimens contained ovoid eggs. Approximately 200 were present in our specimen of 155 mm ML. The eggs varied widely in size, the largest being 9 mm long (Fig. 27.13). An egg lay on each side of the oviducal gland, the proximal one being 11.3 × 8.2 mm and the distal one 13.8 × 8.4 mm; the latter had a tough, brown outer coat and the rudiments of a stalk at one end (Aldred *et al.* 1983). The eggs mature only in small numbers and are presumably laid singly or only a few at a time, so it is an intermittent spawner (Mangold *et al.* 1993). Voss (1988a) reported one egg of *Cirrothauma murrayi* to be 15.5 mm in length with a capsule 22.5 mm long. The site of spawning is not known, although the presence of a stalk suggests that the eggs may be attached to an object or to the substratum.

Descriptions of the embryos of cirroctopods have, without exception, included a comment on the large size of the fins at this stage (Verrill 1885, Robson 1932, Boletzky 1982). This is so in *Cirrothauma murrayi* as the fins remain disproportionately large until a size of at least 21 mm ML is reached. The largest specimen recorded had a mantle length of 220 mm (Aldred *et al.* 1983).

27.2.3 *Froekenia*

The first specimen was captured at 6°N 81°W. In a note written immediately afterwards, Hoyle (1904) said it was like *Cirroteuthis*, but without a web. Further captures of specimens in good condition show the genus distinguished by the absence of a web (Guerra *et al.* 1998). *Froekenia* is a lower bathyal species and has been found living near the slope of guyots and seamounts (Nesis 1993a). Mantle lengths of up to 110 mm have been recorded (Nesis 1986).

27.3 Stauroteuthidae

There is only one genus and species in the family, and it is found in the north-west Atlantic Ocean, from Nova Scotia along the coast of

Fig. 27.13. *Cirrothauma murrayi*. The ovary showing eggs at various stages of development (Aldred *et al.* 1983).

North America to the north of the Bahamas, at depths of 460–2500 m (Voss 1988b).

27.3.1 *Stauroteuthis*

Stauroteuthis syrtensis was recorded on video from a submersible at 782 m off Cape Hatteras, Noth Carolina. When first approached its body was in a horizontal position with the web inflated to a bell-shape, and the arms separated from the primary web by the extended membranes of the secondary web (Vecchione and Young 1997). This posture was maintained by a gentle sculling motion of the fins, rotating at the fin base by up to 90°. The secondary web contracted slowly, pulling the arms aborally into the primary web. The arms and web were then brought together, beginning proximally, and progressing distally in a wave-like contraction, and so expelling the water from within the web, with the fins wrapped ventrally round the mantle. This allowed the animal to glide some distance after which the bell-like posture was resumed. This sequence of actions was repeated as the submersible again approached. When the web was inflated the cirri were elongated and extended laterally against the web. Ballooning like that observed by Boletzky *et al.* (1992) was seen briefly.

The animal observed by Vecchione and Young (1997) was captured and preserved, after which it appeared rather different from when it was alive since the gelatinous subcutaneous layer appeared to have increased in thickness. The mantle wall is about eight times thicker posteriorly than anteriorly, with a thick gelatinous core and thin radial fibres. The mantle closely surrounds the funnel, which can be withdrawn into it. The oesophagus and stomach of this specimen contained the remains of small crustaceans, mostly copepods. A honeycomb of secretory glandular tissue surrounds the lips and it is suggested that the secretions may entrap planktonic prey. Anatomical examination revealed that the fins have distinct regions. The proximal part has a flexible, sponge-like, central core that flattens against the gladius. The central core has thick bundles of muscle dorsally and ventrally, oriented parallel to the long axis of the fin, and these insert on the core and the gladius sac. Distally the fin is formed mainly of two thin layers of muscle, oriented transverse to the long axis.

A transverse section of an arm shows the presence of much reticulate tissue with a narrow band of muscle around the periphery (Fig. 27.14). The sucker is deep and its muscle dense, but the peduncle is reticulate as well as the cirri. The rim of the sucker has complex folds. The infundibulum is very small if, as it is in other octopods, the sphincter muscle is the demarcation between the infundibulum and the acetabulum (Nixon and Dilly 1977).

The central nervous system of a 'cirroteuthid' was described by Young (1977a), but this specimen was subsequently identified by G.L. Voss as *Stauroteuthis syrtensis* from photographs. The statocyst was earlier investigated by Ebersbach (1915).

27.4 Opisthoteuthidae

Opisthoteuthids occur throughout the Atlantic and Pacific Oceans and in Antarctic waters, in depths from about 100 m to below 7000 m (Voss 1988a,b).

The family includes *Opisthoteuthis* and *Grimpoteuthis* with some 20 species, thus making it the most speciose family of the order (Voss 1988a, b, Voss and Pearcy 1990).

27.4.1 *Opisthoteuthis*

Members of this genus are found between 50°N and 45°S in all oceans, and mostly in depths of 100–1000 m (Voss 1988a,b), the deepest record being 2250 m (Verrill 1885). They are found associated with the continental slope and with seamounts (Nesis 1993a), and may live just above or in contact with the bottom (Vecchione and Roper 1991).

When 39 specimens of *Opisthoteuthis californiana* were captured, mostly between depths of 686 and 823 m, off the west coast of Canada and America, from 59°N 145°W to 45°N 124°W (Pereyra 1965), the rate of capture was 0.09–1.00/h, suggesting that this cirroctopod is not unduly rare as Berry (1949, 1952, 1955a) had surmised. Pereyra (1965) observed *Opisthoteuthis californiana* in an aquarium. The animal appeared dark reddish-brown in colour, bell-shaped in form, and its ear-like fins quite noticeable. When at rest only the basal parts of the arms were in contact with the bottom. Medusoid movements were initiated by extreme opening of the umbrella, the oral surface taking on a convex shape with the arms curved slightly upwards; the umbrella was then closed by a powerful downward stroke which reduced the opening into the concavity. These actions forced the animal upwards and their repetition kept it above the bottom. The frequency of the umbrella stroke was 1 per 2–4 s, and the fin stroke in the ratio of 3 to each umbrella stroke. The main function of the fins was of balance and orientation, but they also moved in a coordinated manner and augmented the medusoid movement. The animal could move in all directions but this depended on its attitude at the time the umbrella closed. Stimulation produced a rapid upwards escape movement. Vecchione and Roper (1991) observed *Opisthoteuthis agassizii*, from a submersible (586 m), apparently resting on the substrate (Fig. 27.4).

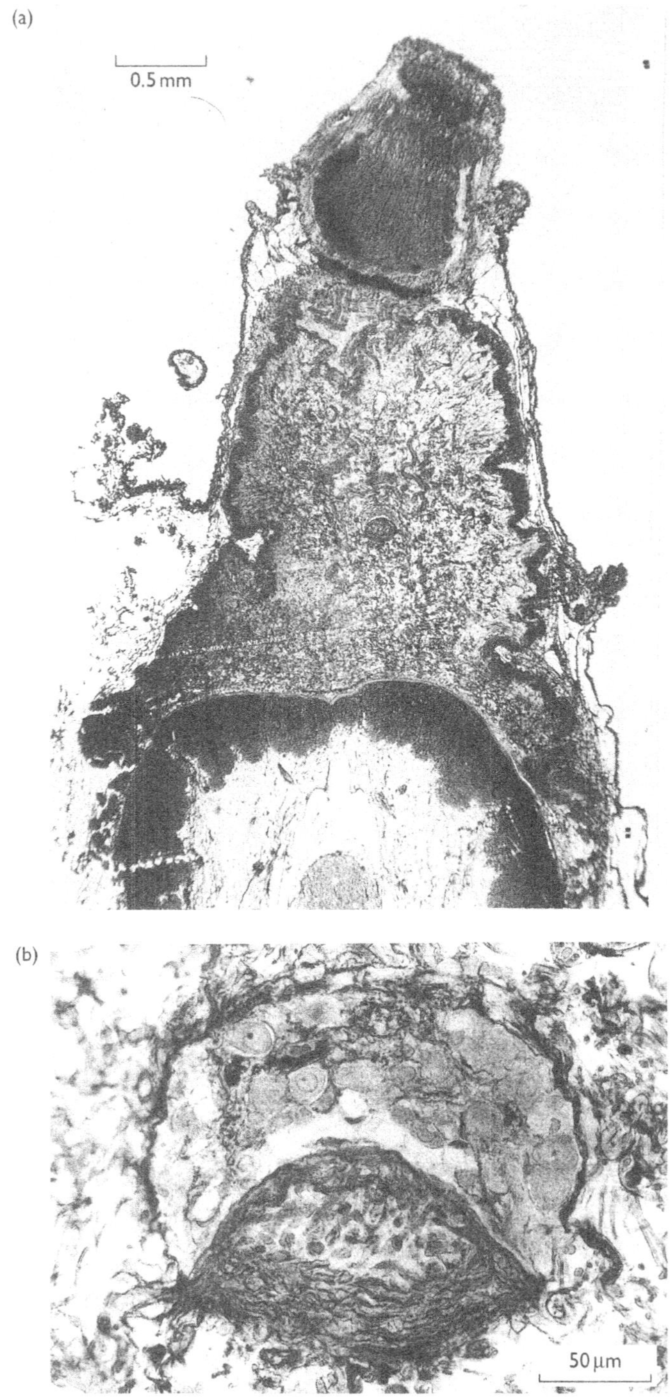

Fig. 27.14. *Stauroteuthis syrtensis*: (a) A section through part of the arm to show the axial nerve cord, the pale-staining reticulate tissue and narrow band of muscle, and the sucker peduncle with its deep and muscular sucker; (b) the complex sucker ganglion in the peduncle at higher magnification.

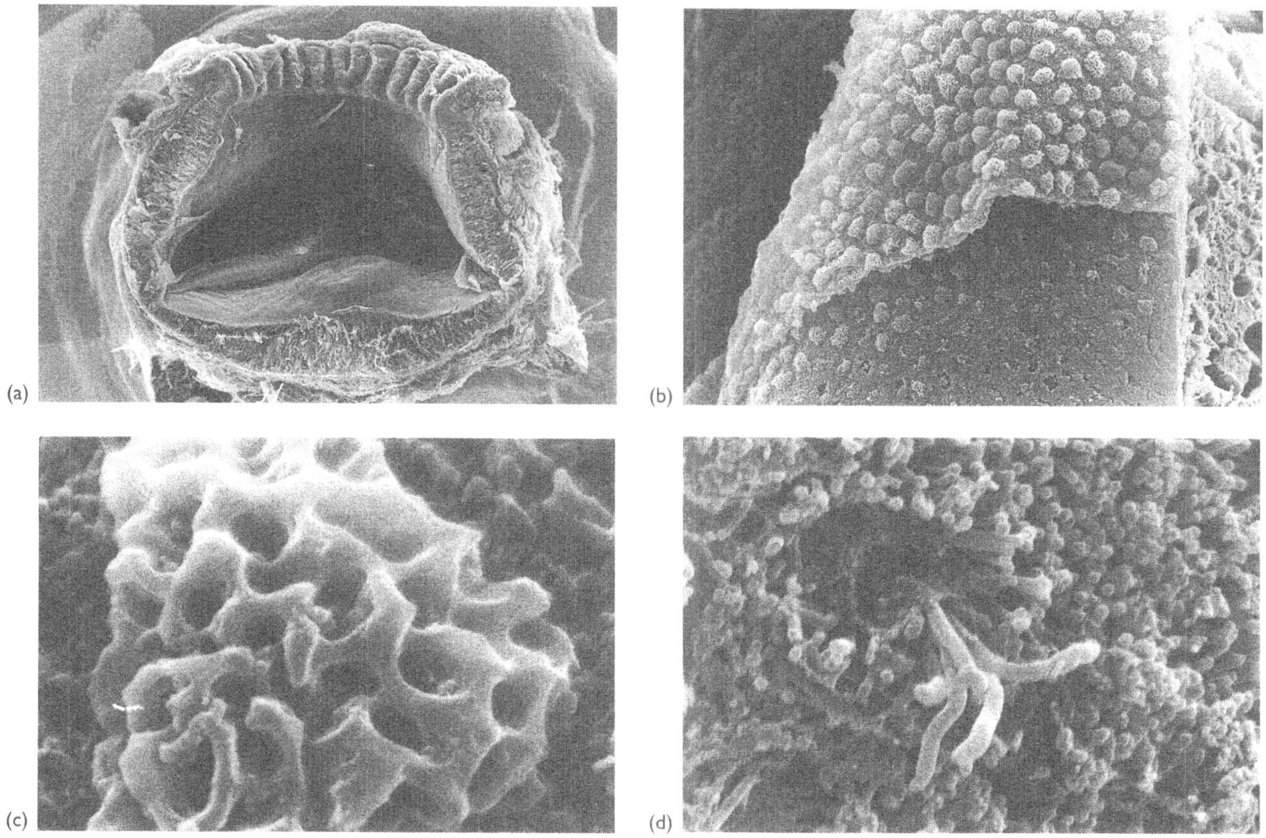

Fig. 27.15. *Opisthoteuthis agassizii.* Scanning electron micrographs of the sucker (a) in hemisection (field width 3333 μm), (b) part of the infundibular cuticle with pegs (field width 62 μm), (c) a peg at higher magnification (field width 2.5 μm) and (d) cilia present in the rim (field width 5 μm) (kindly provided by K.P. Ryan).

Opisthoteuthis has well-spaced suckers along the arms. In hemisection the infundibular region is seen to be narrow and the acetabulum deep and wide at its base. The muscles of the acetabulum are deep (Fig. 27.15). The infundibular surface has radial folds and is covered with pegs. The cuticle is shed, as it is for example in *Octopus vulgaris* (Nixon and Dilly 1977), and a new cuticle is seen already formed below. The pegs are rounded and have pores. Putative sensory structures are present on the rim.

Opisthoteuthis agassizii reaches maturity at 95 g in the male, when a few enlarged suckers are present at the base of the arms and margin of the web (Voss 1988a). Females are mature at 190 g, but growth continues and fecundity is highest in the largest animals; a weight of more than 1600 g has been recorded. Ovarian eggs vary widely in size ranging from 0.2 mm to 6.9 mm in length and from 209 to 833 in number. The period of egg maturation and of spawning is long and is a continuous process in *O. agassizii, O. vossi* (Villanueva 1992a), and *O. californiana* (Berry 1952, Pereyra 1965), as it is in *Cirrothauma murrayi* (Mangold *et al.* 1993).

The stomachs of eight *Opisthoteuthis californiana* contained predominantly crustacean remains, mostly small whole mysids about 7 mm long, large copepods 10 mm in length, and a small crangonid shrimp about 22 mm long (Pereyra 1965). Debris in the stomachs of *O. agassizii* and *O. vossi* of less than 500 g comprised mostly of benthic gammarid amphipods, whereas polychaetes dominated in individuals of more than 500 g. Other prey included whole copepods, ostracods, malacostracan crustaceans, the setae and mandibles from polychaetes, and whole or pieces of gastropods and bivalves, and foraminiferans (Villanueva and Guerra 1991).

The statocyst has inner and outer sacs. The crista is divided into four parts by rather sharp turns. Near to the last turn there is an anticrista (Young 1989).

27.4.2 *Grimpoteuthis*

Grimpoteuthis includes 10 or more species (Kubodera and Okutani 1986, Sanchez and Guerra 1989, Voss and Pearcy 1990). The distribution of this genus is wide, being found in the Atlantic, Pacific, and Southern Oceans, between latitudes 50°N and 75°S (Voss 1988b, Voss and Pearcy 1990). Captures are from near the surface to great depths, including one in the Cayman Trough at 7279 m, so far a world depth record for any cephalopod (Voss 1988a), and others from abyssal plains (Nesis 1987). A total length of some 4 m can be reached; such a giant was photographed from a submersible in the Galapagos Trench in the Pacific Ocean (Voss 1988a).

A small *Grimpoteuthis*, recorded on video from a manned submersible, was sitting on a hard substrate at a depth of about 1500 m, off Hawaii. The arm tips were tucked beneath the web, the mantle bent ventrally so that the animal lay horizontal with the bottom, with the fins held close to the body. Another specimen left the sea floor as the submersible approached and swam, 1–2 m above the bottom, by strong fin movements (one complete cycle about every 3 s) before again settling, this time on soft substrate, its posture like that of the other animal (Vecchione and Young 1997). Four animals, at 2702–4527 m, were seen initially to be resting on the bottom, and they remained until approached by the claw of the submersible. On being touched they crawled away or took off from the bottom and swam with 12–30 fin strokes/min (Villanueva *et al.* 1997).

In the statocyst the crista runs in a curve with three rather sharp bends. There are three cartilaginous knobs, two close together near the dorsal end of the crista and one near the last turn (Young 1989).

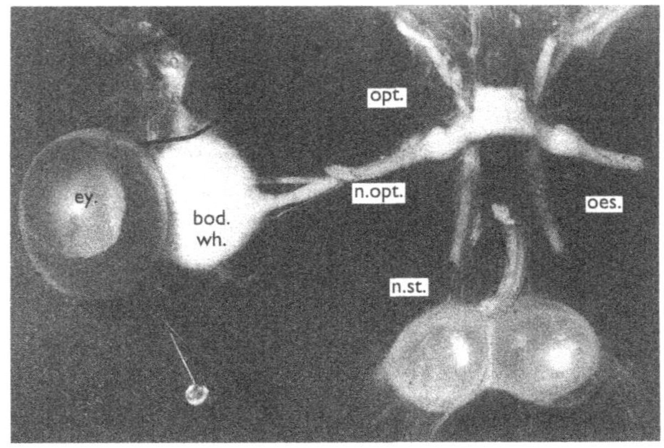

Fig. 27.16. *Cirrothauma murrayi* (155 mm ML). Dissection showing the eye, central nervous system, and statocysts.

27.5 Central nervous system and receptors of cirroctopodids

The central nervous system of cirroctopodids retains many primitive features. It also shows specialization in the reduction of the visual system and the development of the tactile system, and so provides valuable tests for hypotheses about the functions of various lobes of the brain. Indeed, the simple conditions found provide clues to the understanding of the brains of all cephalopods.

The tissues of cirroctopodids are so translucent, even after preservation, that the nervous system is readily visible with little dissection (Fig. 27.16). Furthermore there is no cephalic cartilage. The brain appears as two narrow rings around the oesophagus, and when seen from the side (Fig. 27.17) it recalls the condition in *Nautilus* (Young 1965a). In *Opisthoteuthis* the central nervous system is flattened asymmetrically. The supraoesophageal lobes are pushed forwards and the posterior part of the suboesophageal lobe backwards. The buccal mass lies skewed to the right, below the optic lobe, which is compressed to a thin sheet.

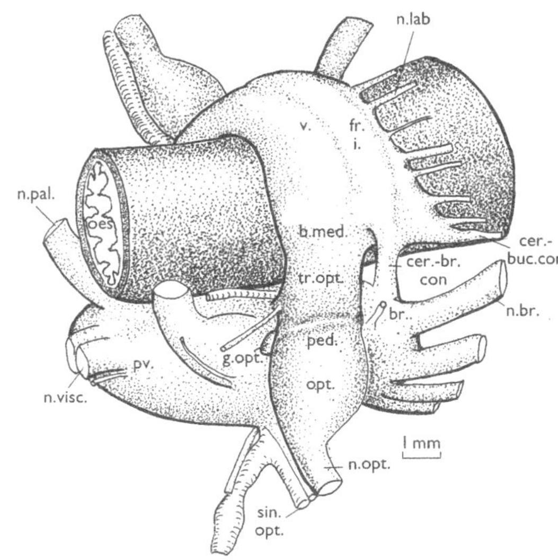

Fig. 27.17. *Cirrothauma murrayi* (155 mm ML). Drawing of the central nervous system (Aldred *et al.* 1983).

27.5.1 Suboesophageal lobes

The three parts of the suboesophageal mass lie close together (Fig. 27.18). The whole mass forms 75.6% of the total brain volume in *Cirroteuthis*, 74.3% in *Cirrothauma*, 76.8% in *Grimpoteuthis*, but only 62% in *Opisthoteuthis* (Table 2.6). The anterior and posterior masses are especially large, whereas the middle mass is concentrated and its anterior part is little developed. In *Cirrothauma* the lateral pedal lobe is large and receives the main input from the statocyst and the tracts from the supraoesophageal lobes. In cephalopods with well-developed eyes this region includes the oculomotor centres, but in *Cirrothauma* the eye muscles are reduced. The development of the lateral pedal lobe is therefore presumably related to its other function, namely the combined control of the funnel and of the arms. The lateral pedal lobe has proved difficult to interpret; it lies between many other lobes and should perhaps be treated as an 'interbasal' lobe (Young 1977a). This condition in *Cirrothauma* emphasizes the importance of this region. It sends fibres forwards to the brachial lobes, down to the posterior pedal lobe and back to the palliovisceral lobe.

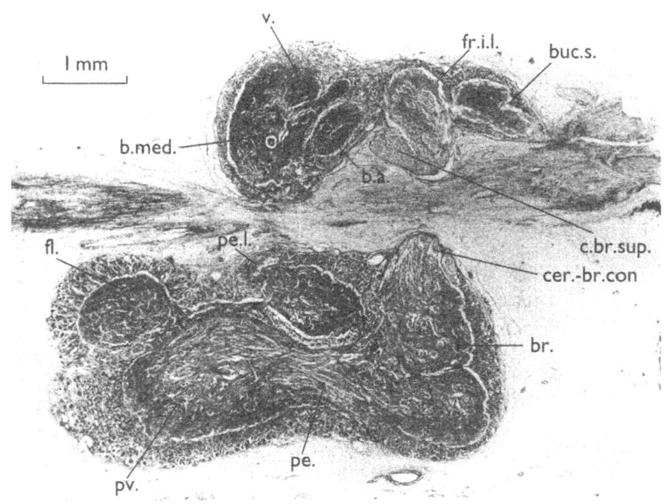

Fig. 27.18. *Cirrothauma murrayi* (155 mm ML). Sagittal section of the central nervous system (Aldred *et al.* 1983).

The magnocellular lobe is relatively reduced in all the cirroctopods, but serves as a main pathway from the optic lobe (Fig. 27.19). Bundles of fibres of neurovenous tissue issue from its ventral surface in *Cirrothauma* but not in *Stauroteuthis.* This neurovenous tissue is abundant and ends in the walls of the large ophthalmic vein. There is no ventral magnocellular lobe in these animals. The posterior magnocellular commissure is small. Although no conspicuous giant cells have been identified, there are some large fibres in the pallioviscerai lobe and in the stellar nerves.

The posterior suboesophageal mass carries a large fin lobe containing a uniform population of cells (Fig. 27.20), as in decabrachians. There is a large central palliovisceral lobe giving rise to the pallial nerves. The visceral nerves arise by two roots, above and below the magnocellular commissure. There is no trace of chromatophore lobes. The posterior part of the palliovisceral lobe contains many small cells and neurovenous tissue.

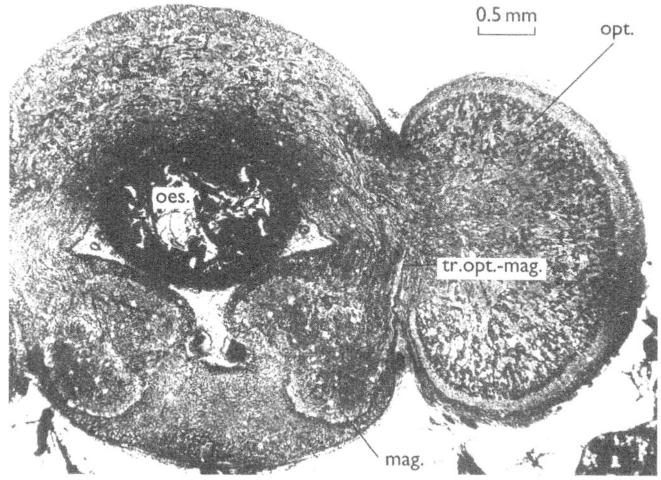

Fig. 27.19. *Stauroteuthis syrtensis.* Transverse section of the central nervous system.

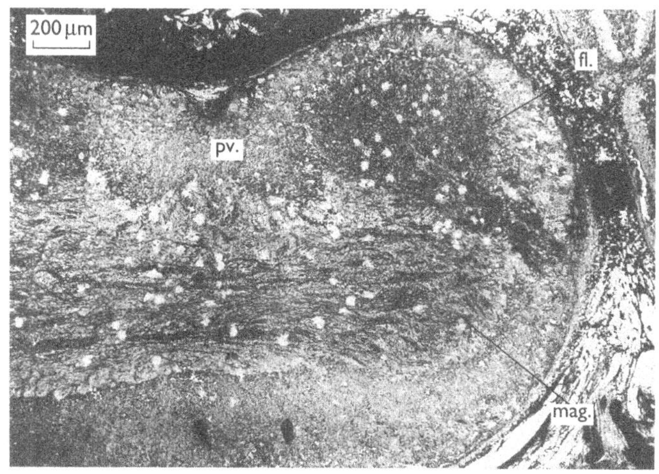

Fig. 27.20. *Stauroteuthis syrtensis.* Transverse section of the posterior suboesophageal mass.

27.5.2 *Eyes*

The eyes lie on either side of the head, and are usually set far from the brain. Long optic nerves lead to relatively small optic lobes, which are joined by short optic tracts to the central nervous system.

The eyes of *Opisthoteuthis*, *Stauroteuthis*, and *Grimpoteuthis* are relatively large, but they are small in *Cirroteuthis* and minute in *Cirrothauma.* The optic nerves are different from those of other octobrachians in that the fibres join to form a few thick bundles (Fig. 27.16). These run to the optic lobes directly, without forming the dorsoventral chiasma that is present both in octopods and decabrachians. In *Grimpoteuthis* (160 mm ML) the eye was 27 mm in diameter; the retina was greatly thickened at the centre where the rhabdoms reached 600 µm, decreasing gradually to 100 µm at the periphery. There is little pigment, and this is nearly all concentrated at the base of the cells. In *Cirroteuthis mülleri* the eye was 7 mm in diameter and the retina of a more uniform thickness of 400 µm, decreasing only little at the periphery (Fig. 27.21).

The eye of *Cirrothauma* is so reduced that the animal was described by Chun (1911) as 'ein blinder Cephalopod'. The eye is better developed than this suggests, being 13 mm and 17 mm in diameter in specimens of 130 mm and 155 mm ML respectively. The eye is simple and is a nearly spherical cup covered with a thin transparent window with no trace of lens, ciliary body or iris (Fig. 27.16). The back of the cup is formed by a pigmented retina, continuous in front with the transparent cornea. The rhabdoms of the retina are better developed than is implied in the classical account by Chun (1913), although at the periphery they may appear as isolated rods, but the back of the eye is occupied by rhabdoms of conventional form. However, in one specimen of *Cirrothauma* a considerable part of the retina shows an extraordinary condition of folding. Stages can be seen in which the retina appears to be in the process of crumpling to make vesicles, lined by rhabdoms (Aldred *et al.* 1983). In some of these sacs the rhabdoms are isolated from each other, reminiscent of the 'flames' described by Chun (1913). In the sacs so formed the rhabdoms are similar to those in an epistellar body or other extraocular photosensitive vesicle. The centres of the sacs often contain masses of amoebocytes, apparently removing the tips of the rhabdoms. But these regions of the retina do not give the impression of total degeneration. Indeed, by adding to the available area of rhabdoms they may actually increase the sensitivity of the retina, assuming that it serves only to detect the presence of light or flashes from bioluminescence.

27.5.3 *Optic lobes*

The optic lobes form 38.6% of the total brain volume in *Cirroteuthis*, 13.5% in *Cirrothauma*, 48.3% in *Grimpoteuthis*, and 15.9% in *Opisthoteuthis* (Table 2.6). The optic lobe has a simple structure (Fig. 27.22). There is an outer plexiform layer and central medulla, but no well-marked inner and outer granule cell layers in *Cirrothauma*, although they are moderately developed in the other cirroctopods. The cell islands and fibre tracts of the medulla of the optic lobes of *Cirrothauma* are simple and show no signs of arrangement into radial columns.

The optic lobes of *Cirrothauma* are the focus for many connections with the rest of the brain. There are five connections:

- large brachio-optic tracts, perhaps mainly afferent
- optic-magnocellular tracts, which may be mainly efferent, perhaps serving for escape reactions (by analogy with *Octopus* and decabrachians)

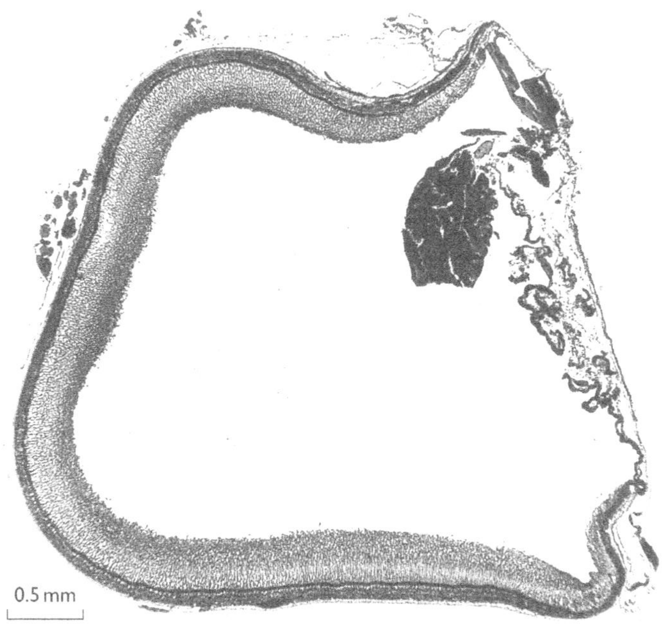

Fig. 27.21. *Cirroteuthis mülleri*. Section of the eye.

- connections with the peduncle lobes, and anterior and median basal lobes; these probably constitute one single set regulating posture and movement
- connections with the vertical lobe system
- a thin optic commissure joining the two optic lobes.

27.5.4 Origin and organization of supraoesophageal lobes

The simplicity of the optic lobes and their close proximity to the central brain in cirroctopods makes it possible to detect how these lobes and all the higher centres have arisen. The basic plan of cephalopod ganglia has been analysed in the stellate ganglion of *Octopus* (Young 1972). In the dorsal part of the ganglion, which receives afferent fibres, motoneurons are mixed with amacrine cells, whose axons do not leave the ganglion but make contacts in a special neuropil close to the surface. We can now see that the optic lobes and all the higher supraoesophageal centres are special developments of such a surface layer of neuropil. In *Stauroteuthis* the outer layers of the neuropil of the magnocellular lobe are in continuity with the ventral plexiform layers of the optic lobes (Fig. 27.23). Moreover, on the dorsal side of the optic tract the 'spine' of the peduncle lobe is similarly continuous with the plexiform zone, as it is in *Grimpoteuthis* (Fig. 27.24). A little further forward the system of islands of cells in the dorsal basal lobe can be seen as a continuation

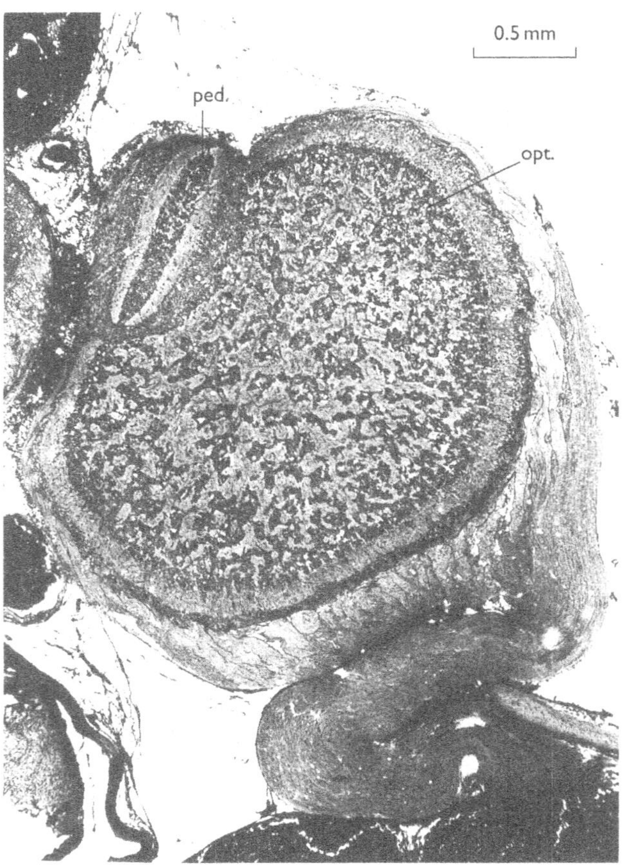

Fig. 27.22. *Stauroteuthis syrtensis*. Transverse section of the optic lobe and peduncle lobe.

Fig. 27.23. *Stauroteuthis syrtensis*. Transverse section to show the transition between the optic and magnocellular lobes.

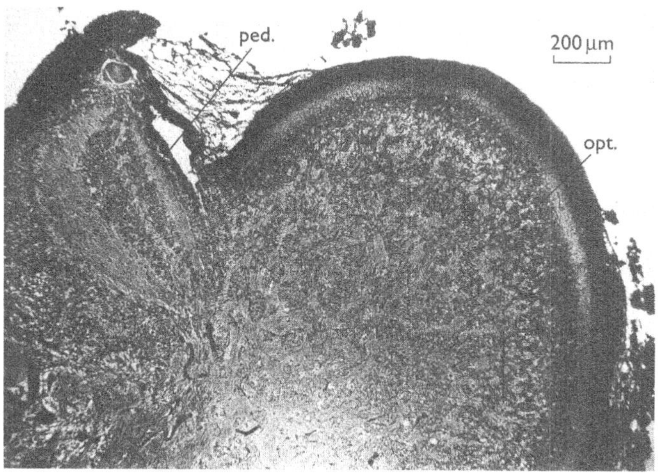

Fig. 27.24. *Grimpoteuthis glacialis.* Transverse section of optic lobe and peduncle lobe.

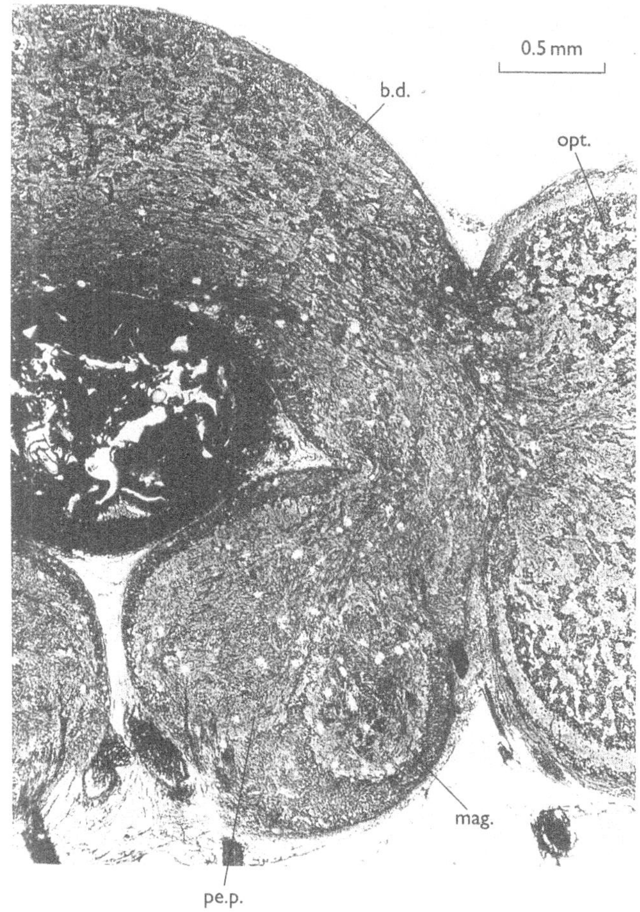

Fig. 27.25. *Stauroteuthis syrtensis.* The continuity between the cell islands of the optic and dorsal basal lobes is seen in this transverse section.

of the islands of cells and neuropil of the optic lobe (Fig. 27.25). The anterior basal lobes are similarly continuous with the optic lobes.

It has been recognized that the ventral parts of the anterior and median basal and peduncle lobes contain the efferent cells and fibres (Young 1977a, 1979, Messenger 1979a). These presumably represent the more ventral, motor, parts of the stellate ganglion that are continuous with the dorsal small-celled region. Similarly, the precommissural lobe is the motor part of the vertical lobe system.

The nature of the major supraoesophageal lobes thus becomes clear for the first time. It has previously been recognized in *Octopus vulgaris* that the tactile system of the inferior frontal lobes is derived from a similar specialization of the superior buccal lobe system (Young 1965b). The great complexity of the visual parts of the supraoesophageal lobes has prevented full recognition of their nature until seen in the simplified form present in cirroctopods. The similarity of the organization of the optic and supraoesophageal lobes has also been noticed in *Nautilus* (Young 1965a).

27.5.5 Peduncle and basal lobes: visuo-static system

These lobes all have the same organization, but are set in different planes, as in other cephalopods. They are well developed in cirroctopods, in spite of the reduced vision. In *Cirrothauma* static fibres can be traced to some of the basal lobes, but unfortunately the details of these pathways are not clear. Each of these lobes contains a set of fine parallel fibres, the 'spine', and a basal zone of large efferent cells and fibres. The major part of the output of all these lobes goes to the lateral pedal lobe from which there are pathways to the fins, mantle and funnel, and arms.

The peduncle lobes are conspicuous bands of tissue running down the back of the optic tract and forward over its dorsal side (Figs 27.22, 27.24), and therefore lie largely in the vertical transverse plane. The anterior anterior basal lobe has parallel fibres organized in the vertical sagittal plane, those of the posterior anterior basal lobe in the horizontal plane.

The organization of both the median basal and the dorsal basal lobe is similar to that of the neighbouring peduncle lobe, with larger fibres ventrally and finer ones dorsally. But it is not clear that these are arranged in parallel bundles. Numerous fibres reach from the median basal to the lateral pedal lobe. A lobe between the peduncle and median basal lobes lies in the position of an olfactory lobe, but no nerve was seen connected with it.

There are large well-vascularized optic glands lying below the optic tract in *Cirrothauma*. No nerve fibres were seen in them, but they lie very close to the tissues of both the optic and the 'olfactory' lobes.

27.5.6 Buccal and inferior frontal lobes: tactile system

The superior buccal lobe serves for the control of the buccal mass. Behind this lobe is the inferior frontal lobe, which together with the superior buccal lobe forms a prominent part of the supraoesophageal mass (Figs 27.26, 27.27). The inferior frontal lobe system forms 6.3% of the total brain volume in *Cirroteuthis*, 10% *Cirrothauma*, 9.2% in *Opisthoteuthis*, and 9.3% in *Grimpoteuthis* (Table 2.6). In Cirroctopoda the superior buccal lobes are united with the other supraoesophageal lobes, a feature characteristic of octopods, and are typical motor ganglia, with large cells in the outer layer and small cells next to the neuropil. These lobes receive labial nerves from the lips and are joined by connectives to the inferior buccal

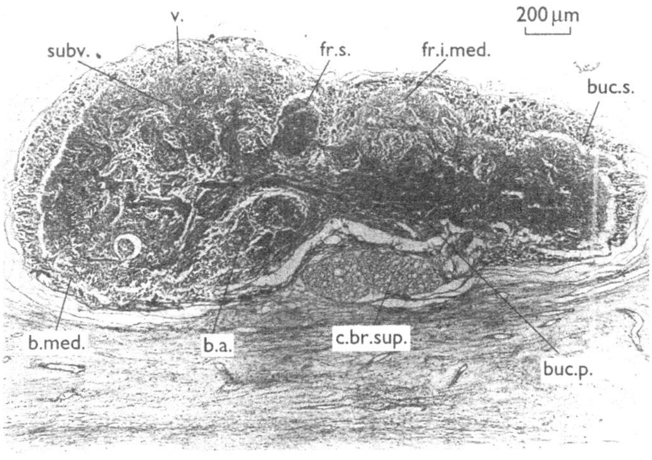

Fig. 27.26. *Cirrothauma murrayi* (130 mm ML). Sagittal section of supraoesophageal lobes.

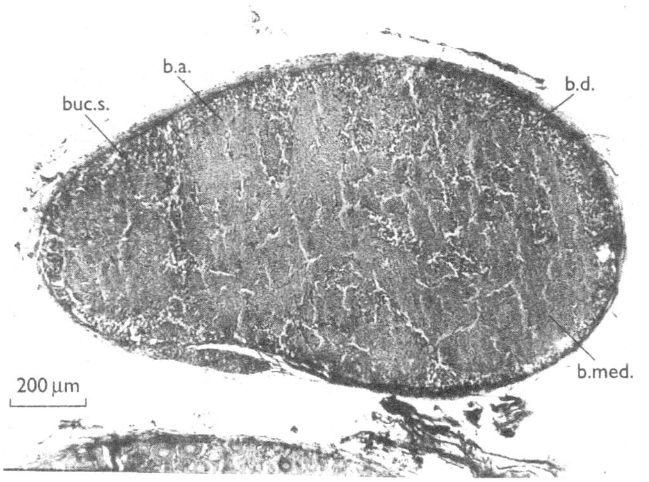

Fig. 27.27. *Cirroteuthis mülleri.* Sagittal section of supraoesophageal lobes cut rather far laterally.

ganglion. Numerous fibres enter from the arms but the major part of the cerebrobrachial connective runs to the inferior frontal lobe system.

Fine afferent fibres pass up the side to a median inferior frontal lobe, as in *Octopus*. In *Stauroteuthis* the superior buccal lobe continues backwards as a posterior buccal lobe from which thick fibres pass downwards in the median part of the cerebrobrachial connective to the centre of the brachial lobe (Fig. 27.28). From the median inferior frontal lobe fibres pass into a large zone of islands of small cells and fine fibres; this is obviously the subfrontal lobe. The arrangement is similar in other genera examined (Fig. 27.29).

The posterior buccal lobe receives a large tract from the subvertical lobe, presumably carrying the influence of the vertical lobe system (Fig. 27.26). It also receives a connection from the median basal lobe region, as in other cephalopods, allowing integration of tactile activities with those of the mantle and viscera.

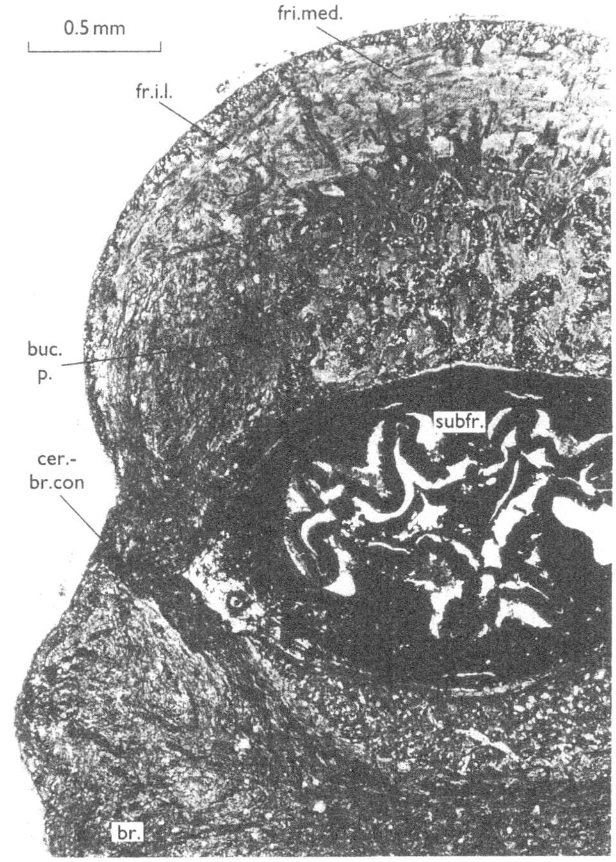

Fig. 27.28. *Stauroteuthis syrtensis.* Transverse section of the inferior frontal lobe system.

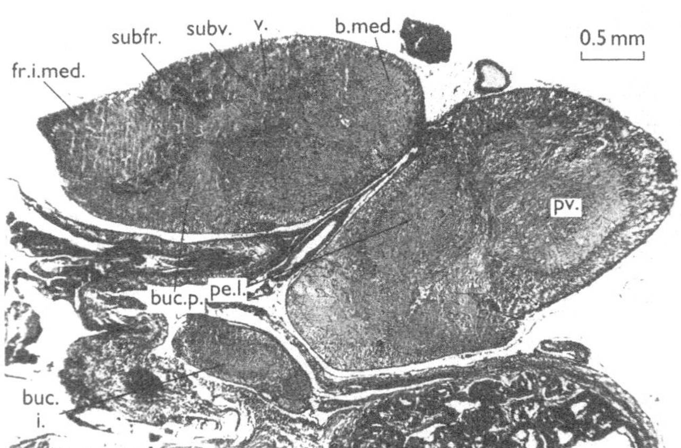

Fig. 27.29. *Opisthoteuthis agassizii.* Sagittal section, somewhat oblique, showing the central nervous system.

27.5.7 Vertical lobe system

The superior frontal lobe of *Cirrothauma* can be recognized as a very small lobule behind the median inferior frontal lobe (Fig. 27.26), from which it receives fibres. A few fibres come from the optic lobe in *Stauroteuthis*, but none has been seen in *Cirrothauma*. Behind the superior frontal lobule there lies a region of small cells and islands of neuropil, presumably constituting a vertical lobe in *Cirrothauma* and *Opisthoteuthis* (Fig. 27.29). There is no sign of division into gyri as in octopods nor is there a large neuropil as in decabrachians. The cell islands continue down in the midline almost to the oesophagus. There is no clear division into subvertical or precommissural lobes. However, bundles of fibres proceed down to the posterior buccal lobe and also to the lateral pedal lobe. Large bundles of fibres also run to the optic lobes, but their direction of conduction is not clear.

The vertical lobe system is less developed in cirroctopods than in any other cephalopod examined. This system forms 4% of the total brain volume in *Cirroteuthis*, 2.9% in *Cirrothauma*, 3.4% in *Grimpoteuthis*, and 3.8% in *Opisthoteuthis* (Table 2.6). Its functions are perhaps mainly concerned with tactile operations including memory, which it is known to influence in octopods.

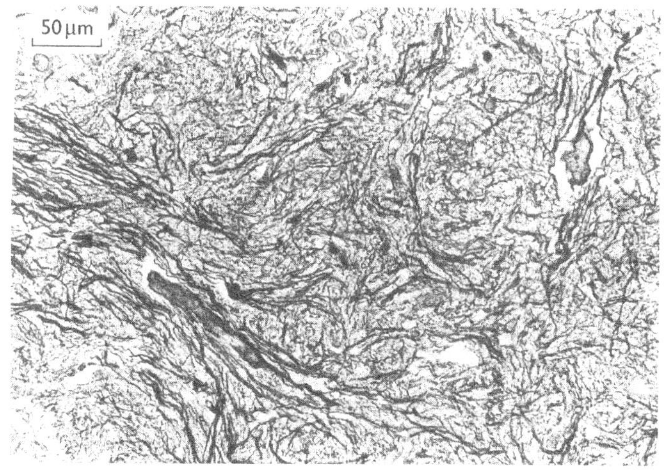

Fig. 27.30. *Stauroteuthis syrtensis*. Section of stellate ganglion showing large fibres.

27.5.8 Stellate ganglia and epistellar bodies

The pallial nerve includes a large fin nerve, which continues past the stellate ganglion, exactly as it does in *Vampyroteuthis* and in decabrachians.

The stellate ganglia are remarkable for the presence of some specially large nerve fibres in *Stauroteuthis* (Fig. 27.30), but not in *Cirrothauma*. The ganglia have distinct dorsal and ventral parts, with small cells present only in the dorsal part, as in *Octopus*.

The epistellar body is a large irregular sac with several lobules, and a diameter of 1.5 mm. It occupies the whole of the back of the ganglion in *Cirrothauma* (Fig. 27.31). The irregular, closely packed arrangement of the rhabdoms resembles that found in the sacs of photosensitive vesicles of decabrachians. In *Cirrothauma* there are pigmented amoebocytes covering the tips of the rhabdoms, and a mass of debris at the centre. The epistellar nerve can be seen in *Stauroteuthis* entering the neuropil (Fig. 27.32). It runs towards the pallial nerve but cannot be followed into it with certainty.

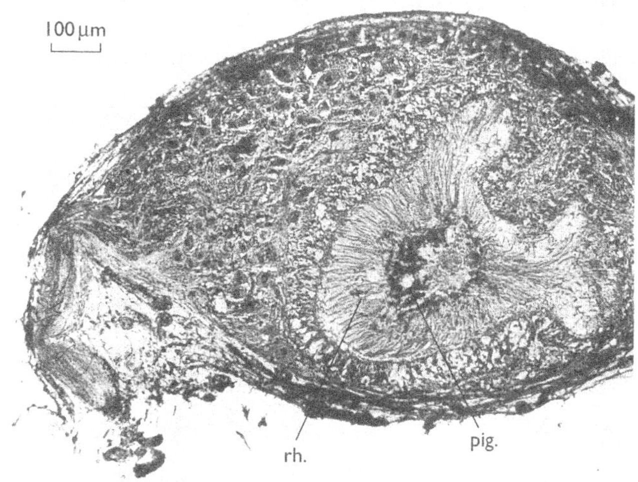

Fig. 27.31. *Cirrothauma murrayi* (155 mm ML). Epistellar body.

The significance of these photosensitive vesicles inside the mantle remains obscure (Nishioka *et al.* 1966, Mauro 1977). The epistellar bodies may serve to detect the presence of luminous matter in the water entering the mantle, and then to ensure its removal. It is of note that the oesophagus, in all of the cirroctopodids examined, is heavily pigmented, presumably to ensure that light from luminescent prey consumed is screened.

27.5.9 Giant nerve fibres

Cirroctopodids are the only eight-armed coleoids in which traces of a system of giant fibres has been found. The fibres are not very large and their passage has not been easy to determine. They seem to be a distinct set following the plan that is so well developed in decabrachians; they may represent reduced ancestral stages of it. There are no obvious single giant cells but there are large cells in the lower parts of the magnocellular lobe, especially in *Stauroteuthis* where, in the neuropil close to these large cells, there are several fibres proceeding forwards and apparently reaching the brachial lobe. Large fibres are seen in this

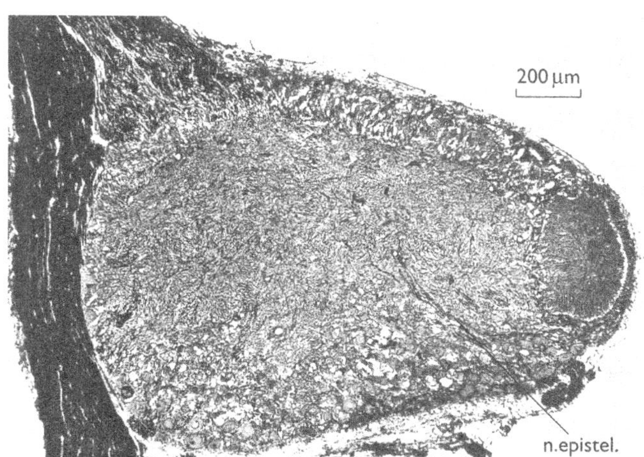

Fig. 27.32. *Stauroteuthis syrtensis*. Epistellar body.

region in *Cirrothauma*. This recalls the anterior giant-fibre system of some decabrachians, such as *Pterygioteuthis*. The posterior giant fibres cannot be followed far in our sections of either cirroctopodid although there are some moderately large fibres in the mantle connective, but they are undoubtedly present in the stellar nerves of *Stauroteuthis* (Fig. 27.30). There are one or two in each nerve, reaching 50 μm diameter in the ganglion, but are not readily visible further peripherally. They could not be found in the ganglion or nerves of *Cirrothauma*.

27.5.10 Statocysts

The statocysts of cirroctopodids are relatively large, especially in *Cirrothauma*, and they show some characteristics both of octopods and decabrachians. They are transparent sacs lying free at some distance from the brain. There are inner and outer sacs, with perilymph between (Fig. 27.33), as in octopods. In some cirroctopodids there are several processes comparable with the anticristae of decabrachians (Ebersbach 1915). *Stauroteuthis syrtensis* probably has four anticristae (Young 1989), and *Grimpoteuthis glacialis* three. Comparison with the statocysts of decabrachians is not easy, but two of the anticristae of *Grimpoteuthis* lie at the ends of the vertical part of the crista and may possibly represent hamuli 4 and 5; they are quite unlike the single anticrista of *Octopus*. The crista of cirroctopods is not divided into sections as it is in *Octopus*. The macula of *Opisthoteuthis* (Fig. 27.34) is oval and that of *Grimpoteuthis* and *Cirrothauma* is an elongated oval. There are no maculae neglectae.

In *Cirrothauma* the statocysts are a pair of large, semitransparent sacs lying some distance behind and below the brain (Fig. 27.16). The sacs are nearly spherical, but are flattened against each other medially. The outer sac is attached to the surrounding tissues by several bundles of muscle fibres. The inner sac almost fills the outer (Fig. 27.33), unlike the condition in *Octopus* where there is abundant perilymph (Young 1960b).

The macula is a greatly elongated oval area round the anterior wall of the inner sac, and is unlike that found in other cephalopods. A very small statolith occupies only a fraction of the surface of the hair cells. This statolith is an aggregate of numerous fine grains. It may have become reduced in size by fixation, but statoliths with the same form were present in all four statocysts of our two specimens. A few isolated particles were seen scattered over the surface of the maculae.

Round the edge of the macula is a web of elongated cells with fine processes spreading around the pavement epithelial cells that line the sac. These are in some ways similar to cells seen in *Vampyroteuthis* (Stephens and Young 1976), but they seem not to carry hairs. They send fibres towards the macula and may serve to anchor it to the wall. There are no signs of the accessory maculae neglectae so prominent in decabrachians.

The crista of *Cirrothauma murrayi* is a continuous ridge running round a large part of the circumference (Young 1989). It lies in the sagittal plane that is parallel to the macula, except for the first portion which is more nearly in the transverse plane. The first part shows some possible sign of subdivision into sections, as in *Octopus*, but the rest of the sensory epithelium seems to be continuous. With this orientation the greater part of the crista is arranged to detect angular acceleration in the yawing plane, that is to say rotation about a dorsoventral axis. If the animal often lives suspended head downwards the crista would serve to monitor its rotations above the bottom. The short first section of the crista, lying in the transverse plane, monitors dipping movements, but there is no horizontal component to prevent rolling movements.

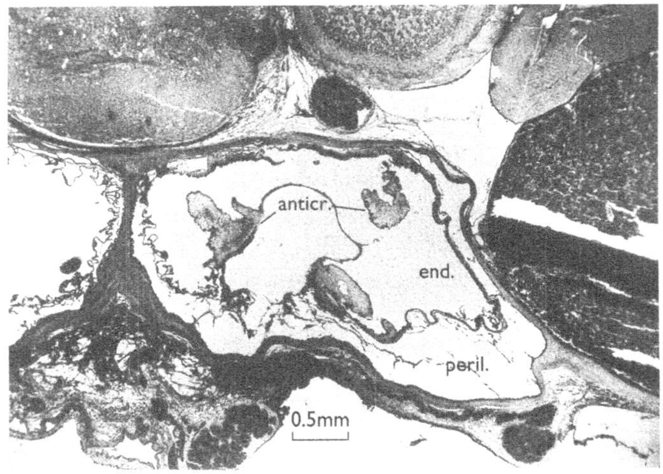

Fig. 27.33. *Stauroteuthis syrtensis*. Transverse section of statocyst showing several anticristae. The wall of the inner sac has become folded.

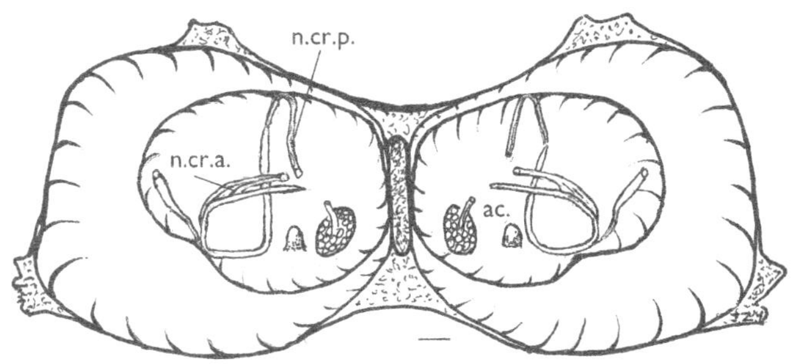

Fig. 27.34. *Opisthoteuthis*. Statocyst in dorsal view (scale bar 1 mm).

The capacity of the statocyst cavity suggests that movements of the animal must be slow, requiring a considerable inertial mass of fluid to provide high sensitivity. The crista runs round the circumference of the sac with sizeable spaces medial and lateral to it. On the more medial side of the crista, is the single large anticrista, a large knob of cartilage, close to the outer end of Kölliker's canal. This is comparable to the position of the more expanded anticrista of *Octopus*. Kölliker's canal is a complex duct, lined by long cilia, and with a special epithelium around the opening.

The statocysts of *Cirrothauma*, *Grimpoteuthis* and *Opisthotetuthis* show similarities to the characteristics of those of buoyant squids.

28 Amphitretidae

28.1 *Amphitretus*

Amphitretus pelagicus is the only member of the family, and is known from just a few specimens. It is a pelagic, oceanic octopod found in the Pacific and Indian Oceans between 10°N and 40°S. Its vertical distribution is little known except that it is captured far above the bottom. Two specimens were caught in opening/closing nets in Hawaiian waters, one (35 mm ML) at 240–260 m (Roper and Young 1975), and another (40 mm ML) in less than 350 m (R.E. Young 1978). The largest specimens recorded include a female of 87 mm and a male of 40 mm ML (Thore 1949).

28.2 Behaviour

Ijema and Ikeda (1902) observed one *Amphitretus pelagicus* while it was still alive (Fig. 28.1). These authors commented that

> It was soft, transparent and... almost jelly fish like... the animal was an elongate bell,... A thick layer of a delicate, colourless and perfectly transparent tissue covered up alike the body, the head and the umbrella... over 20 mm thick at the posterior end and 8–10 mm at the sides.

The outer layer of tissue made the animal smooth and slippery to touch. Small yellow-brown chromatophores were present on the surface of the gelatinous coat, with larger and more numerous ones close to the mantle muscles, on the brachial muscles and around the eyes. The animal exhibited changes from 'intensely red' to 'nearly colourless'. The eyes were of 'brilliant coloring' being on

> the bulbous inner end... a shining dark purplish or bluish ground-color, dotted with madder-brown chromatophores. The conical outer end was marked with rings of reddish, purplish and of a light color.

The tubular-shaped eyes only just reached the surface of the gelatinous layer. After fixation the specimen became 'a shabby object, incomparable to its formed beauty' (Ijema and Ikeda 1902). The gelatinous layer was reduced to a soft skin with a wrinkled surface, and the animal had shrunk from 190 mm to 148 mm in total length.

R.E. Young and his colleagues (1998b) observed one living animal in a shipboard aquarium. It was translucent apart from its tubular eyes, gills, stomach, and caecum. The stomach and caecum adhere to the long tapering digestive gland, and together form the slender, elongated digestive tract complex. The whole complex rotates to a vertical position when the octopod is oriented horizontally. In this way any shadow cast by the digestive tract complex is minimized, the reflective coverings further reducing its opaqueness, thus providing concealment from predators below.

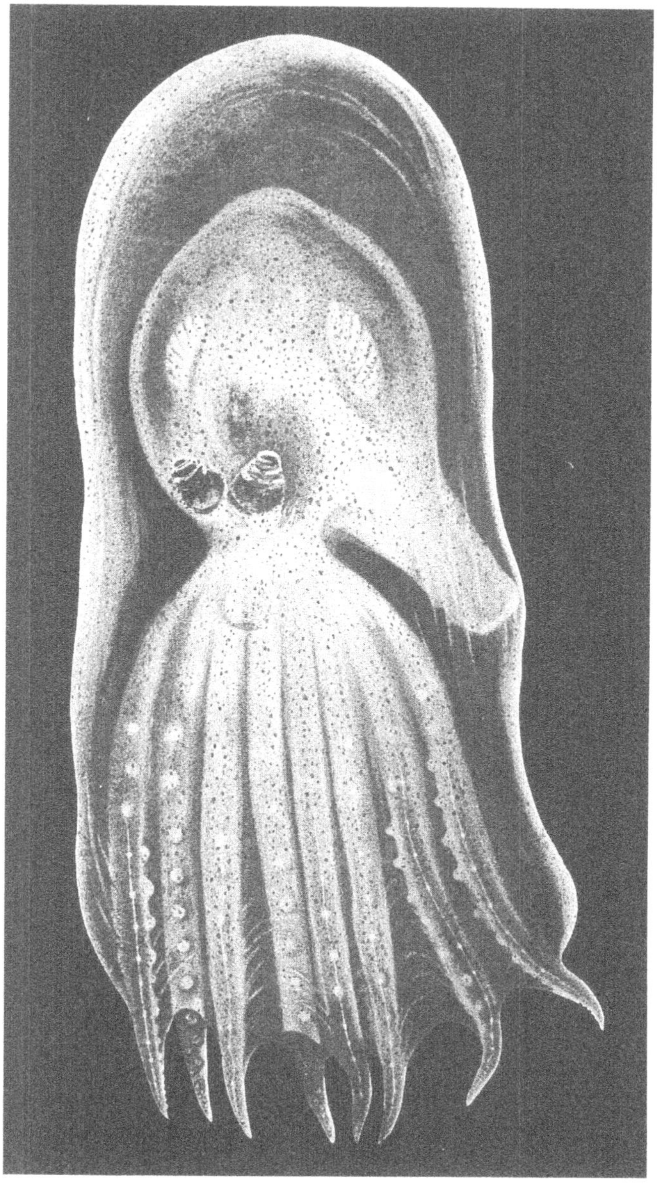

Fig. 28.1. *Amphitretus pelagicus.* Photograph of the water-colour illustration made while the specimen described by Ijima and Ikeda (1902) was still just alive. The total length of the animal, including the gelatinous coat, was 190 mm.

28.3 Receptors and central nervous system

Each eye is tubular in shape and is supported by the surrounding thick gelatinous coat. After fixation the eye collapses so the elongated wall becomes deeply folded (Fig. 28.2). The collapsed wall of the eye

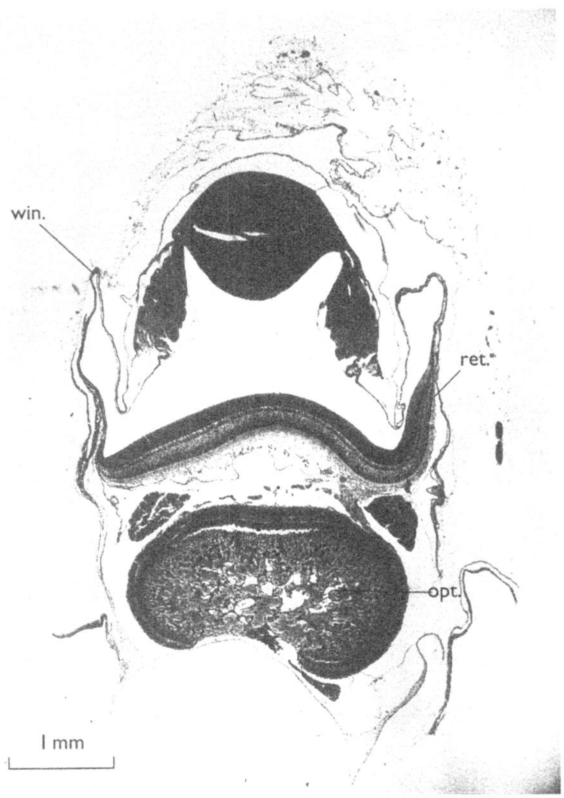

Fig. 28.2. *Amphitretus* (37 mm ML). Transverse section through the head, to show the eye and optic lobe. The posterolateral extension of the retina can be seen on the right.

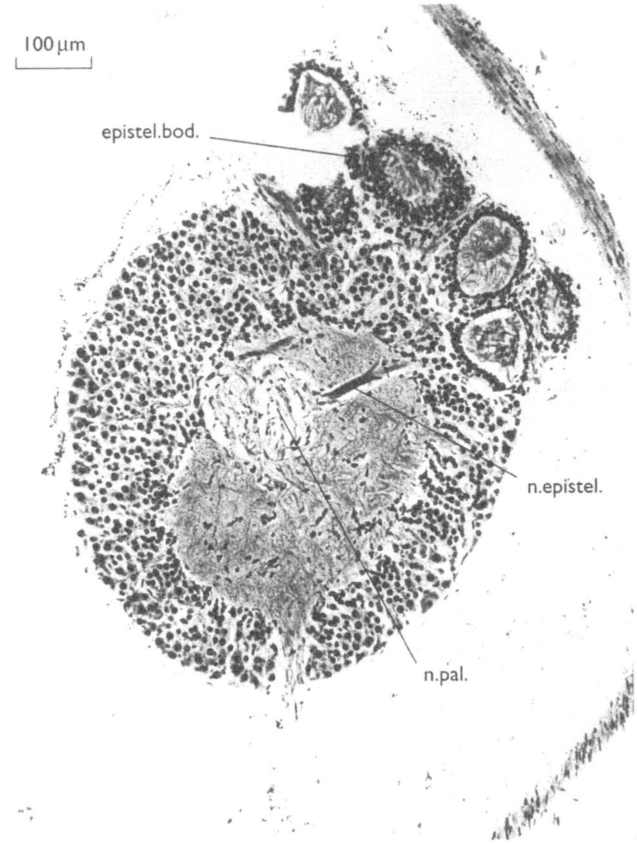

Fig. 28.4. *Amphitretus*. Section through the stellate ganglion and photosensitive vesicles of the epistellar body lying on its anterior wall.

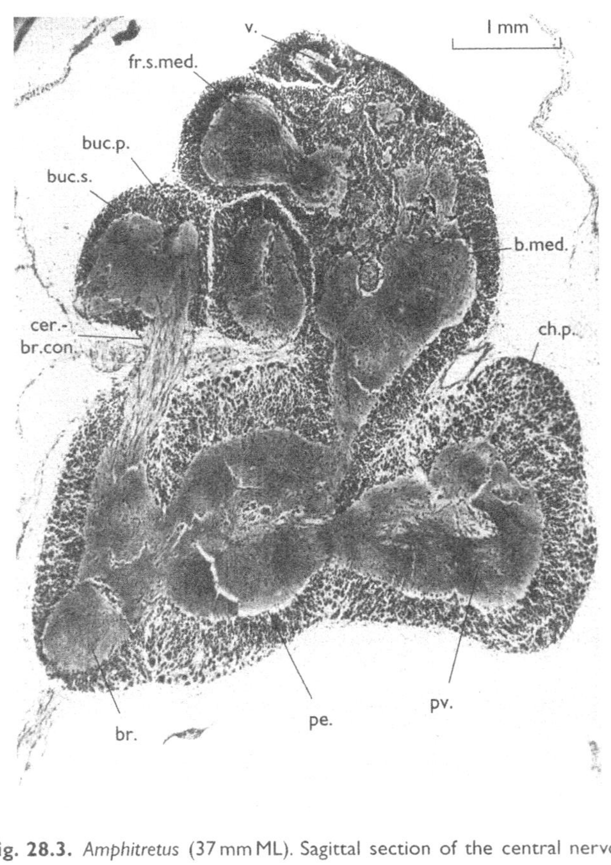

Fig. 28.3. *Amphitretus* (37 mm ML). Sagittal section of the central nervous system.

is thin, and is probably largely translucent. Light could enter through this tissue, forming an aphakic window, and reach the retina (Young 1991a). The lens occupies much of the diameter of the eye and bulges prominently, especially after fixation. It is deep and should be capable of providing a good retinal image. The light entering the aphakic window would, in contrast, only allow a record of light or of luminescent flashes to be received by the retina. The retina is a simple cup, slightly extended posterolaterally. The rhabdoms are 180 μm long, narrow and closely packed. The layers of supporting cells, with the cell bodies of the rhabdoms and the nerve fibre layer, reach a depth of 600 μm at the centre.

The optic lobes form 146% of the total brain volume and are smaller than in the bolitaenids measured (Table 2.6, Fig. 28.2). Internally the lobes show a marked arrangement into columns. The peduncle spine runs across the optic tract and turns down behind it.

The central nervous system is compact (Fig. 28.3). Most notable are the chromatophore lobes which form 3.8% of the total brain volume, and the brachial lobe which forms 14% (Table 2.6). The latter is the largest found amongst the meso- to bathypelagic octopods measured (Maddock and Young 1987). There is a prominent cerebrobrachial connective, fibres from which proceed to the lateral and median inferior frontal lobes. The median inferior frontal lobe is small and is barely differentiated from the posterior buccal lobe, and there is a small amount of subfrontal tissue in the midline. The vertical lobe system forms almost 11% of the total brain volume; of this the median superior frontal lobe occupies nearly 3% and the vertical lobe almost 4.5%.

One extraocular photoreceptor organ, or epistellar body, is present on each stellate ganglion, and takes the form of 'a large complex of a dozen or more generally circular vesicles which cover most of the anterior wall of the ganglion' (R.E. Young 1978). The dominance of this part of the ganglion is seen in Fig. 28.4. The epistellar nerves clearly join the pallial nerve (mantle connective) running towards the brain. This is the first anatomical evidence of the course taken by these nerves.

The sac of the statocyst of our specimen was broken and no measurements could be made. The macula and crista were recovered intact and are similar to those of *Vitreledonella*. The crista makes a continuous curve and is divided into nine sections; the first has two irregular rows of hair cells and the second a single row. The first two sections are served by an anterior crista nerve in the form of distinct bundles. This is a somewhat different arrangement from that of *Vitreledonella*. Kölliker's canal has a wide outer section. There is no evidence of an anticrista (Young 1989). The condition of the statocyst suggests that this octopod moves by using its arms, and by jet propulsion using its relatively huge funnel. The gelatinous coat presumably provides some measure of buoyancy.

29 Bolitaenidae

The bolitaenids are pelagic octopods with a wide distribution in tropical to temperate waters. *Eledonella* and *Japetella* have been caught worldwide between latitudes 50°N and 43°S. The young are found in the upper few hundred metres and then undergo ontogenetic descent to become mesobathypelagic in habit as adults (Thore 1949, Nesis 1987).

29.1 *Bolitaena*

Bolitaena microtyla is caught only rarely. It is a tropical–subtropical, mesobathypelagic octopod (Nesis 1987).

The body is light purple in colour and the oral surface of the web dark purple. The ovary of one female (20 mm ML) was filled with eggs 2.5 mm in diameter (Thore 1949). The female carries the eggs on her arms until they hatch (Nesis 1987). Thiele (1915) showed a series of chambers across each arm with lamellae or parallel walls, and the illustration shows a close resemblance to the chambers seen in the histological sections of *Japetella* (Figs 29.3, 29.4). This suggests that there is probably a similar system in *Bolitaena* for achieving near neutral buoyancy.

29.2 *Eledonella*

Eledonella pygmaea is the only member of the genus (Fig. 29.1). It is found in tropical and subtropical waters between latitudes 38°N and 42°S (Thore 1949). Young (5–15 mm ML) have been caught at depths of about 200 m or below 600 m, larger animals (25 mm ML or more) from 975–1425 m, and brooding females between 800 m and 870 m.

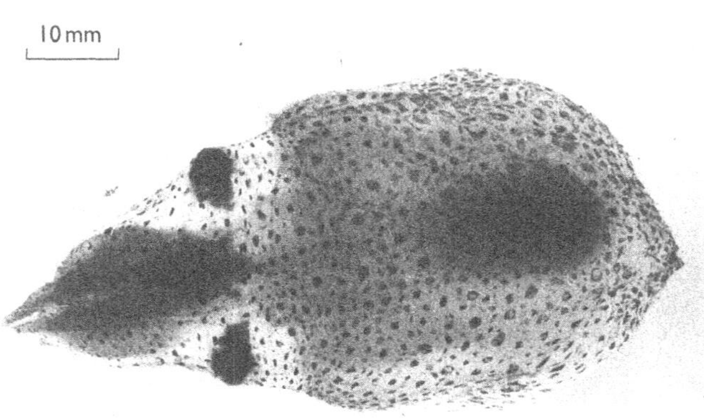

Fig. 29.1. *Eledonella pygmaea* (45 mm ML) seen from above.

Several males nearing maturity were captured at 1200–1425 m (R.E. Young 1978).

29.2.1 Reproduction and photophore

Sexual dimorphism is apparent in maturing females with the development of a circumoral light organ (Robison and Young 1981). The posterior salivary glands in the male grow at five times the rate in the female (Voight 1995). Mature females become heavily pigmented with numerous chromatophores on the mantle and the aboral surface of the arms and web; those on the oral surfaces of the arms and web are more densely pigmented. The tissues of the arms and web become thicker at this time. Gravid females of up to 48 mm ML have been caught (R.E. Young 1978). A mature female (30–35 mm ML) (recorded as *Bolitaena microtyla* by R.E. Young 1972c, but since referred to *Eledonella* by Hochberg *et al.* 1992) was captured in an opening/closing net between 830 and 975 m, off Oahu, Hawaii. Its arms were arched only to converge near their tips and all were joined by the web, except the two ventral ones (R.E. Young 1972c). The arms and web formed a capacious pouch, presumably to hold offspring as several were attached to the head and body of this female, while others were present in the same trawl. The buccal mass of this female was almost sealed by the oral surface of the buccal crown, with only a small, short, canal to the lips remaining. In the digestive tract the posterior salivary glands were small and flaccid, the crop had collapsed, the stomach was contracted, and the digestive gland was small. The condition of the digestive tract indicates that feeding had ceased. The suppression of feeding would prevent cannibalism which could otherwise occur with young in such close proximity to the mouth. The hatchlings associated with this mature female were 2 mm ML, their arms were short, the eyes elliptical and directed forward. The ovary of this adult was depleted except for a single egg (2 × 1.5 mm) (R.E. Young 1972c), and in another mature female (40 mm ML) an oviducal egg of 3 × 1.8 mm was present (Rancurel 1970). The eggs are cemented together in clusters and brooded within the brachial web (Hochberg *et al.* 1992).

Maturing females, of more than 50 mm ML, have a bright yellow, lobulated, circumoral organ, which produces a very bright, greenish luminescence after the application of hydrogen peroxide (Robison and Young 1981). The distinctive shape of the light organ was recognizable at a distance of 3 m. The organ lies beneath a transparent integument on the oral surface and forms a thick ring around the mouth, with short extensions on to the bases of the arms. The aboral surface of the organ was covered by the dark pigmentation of the oral region. Histological sections showed that besides numerous blood vessels and muscles a single cell type predominated. Besides this luminescent organ, luminescent activity was observed in homogenates of the digestive gland (R.E. Young *et al.* 1979b). Development of the circumoral organ was followed in several specimens. One female (22 mm ML) had small ova (0.14 × 0.12 mm) but only a trace of the light organ; another (37 mm ML) had larger ova (0.5 × 0.17 mm). In a larger female (45 mm ML) the circumoral organ was orange in colour and the ovary packed with ova (1 mm long). Males examined showed no trace of such a light organ, even in one close to maturity with a well-developed hectocotylus, enlarged suckers, and a large penis (Robison and Young 1981).

29.2.2 Diet

The diet of *Eledonella* is not known, but the crop of one of our specimens was filled with a mass of tissue cut into pieces, including arms and other parts of a smaller bolitaenid, and a few intact copepods. This material may have been consumed while the net was being hauled in, but does demonstrate that this octopod is capable of biting through soft flesh and of ingesting whole, small crustaceans.

29.2.3 Receptors and central nervous system

The eyes are set widely apart (Fig. 29.2a). Part of the wall of the eye lacks pigment and could serve as an aphakic aperture (Young 1991a).

The optic lobes are deeply folded. Together the two lobes form 226% of the total brain volume (Maddock and Young 1987). The optic tracts carry the peduncle lobe and optic gland, which can lie near the optic lobe or, with growth, at some distance from it. The spine of the peduncle lobe lies along the axis of the optic tract.

The supraoesophageal lobes form 49.1% of the total brain volume (Table 2.6). The vertical lobe consists of five small gyri, and forms 2.75% of the total volume of the whole brain, and the superior frontal lobe forms 3.63%. The inferior frontal system is 5.38% of the volume.

The extraocular photoreceptor organs each consist of a single, spherical photosensitive vesicle attached to the posterior margin of the stellate ganglion (R.E. Young 1978).

29.3 *Japetella*

Japetella diaphana is found all round the world between latitudes 50°N and 35°S, but thrives best where there are well-established currents (Thore 1949). In the Atlantic Ocean it is found from 11°N to 40°S (Clarke and Lu 1974, 1975, Lu and Clarke 1975a, b). The young are found in epipelagic and mesopelagic zones and they undergo ontogenetic descent, as larger animals inhabit depths of 600 to some 4000 m (R.E. Young 1978, Hochberg *et al.* 1992). There is no evidence of diel migration.

29.3.1 Buoyancy

This octopod is gelatinous and a freshly captured specimen weighed 88 g in air and 0.12 g in seawater. The ratio (weight in seawater/weight in air × 100) gave a value of 0.136, a figure similar to that found in squids which are close to neutral buoyancy, and so indicating that the bolitaenid is at least partially buoyant (Clarke *et al.* 1979). Pieces of the arms and mantle were found to be positively buoyant and the liquid expressed from a fresh piece of mantle was isosmotic with seawater with a sulfate concentration about half that of seawater (Denton and Gilpin-Brown 1973). Reticulate tissue is present in the arms, head, and mantle, and in some places this develops into large chambers. In the arms there is separation of the chambers from the large axial nerve cord, which is surrounded by a wall enclosing loosely organized tissues (Fig. 29.4). The muscle tissue of the arms is restricted to peripheral bands. The suckers in contrast have deep, dense muscles, and shed the infundibular cuticle (Fig. 29.4) as *Octopus* does (Nixon and Dilly 1977).

Multiple discriminant analysis of measurements of the statocyst places *Japetella* close to the buoyant squids (Maddock and Young 1984). This supports other evidence that this octopod is close to neutral buoyancy and a living animal when observed 'hung passively horizontal in the water, with the arms and head retracted into the gelatinous mantle' (Herring *et al.* 1987).

29.3.2 Reproduction

Reproductive habits are poorly known. Females with ovarian eggs of 1.3 × 0.4 mm, and 3 × 1 mm have been captured. One gravid female (60 mm ML), caught at 1050 m, had a sperm mass embedded

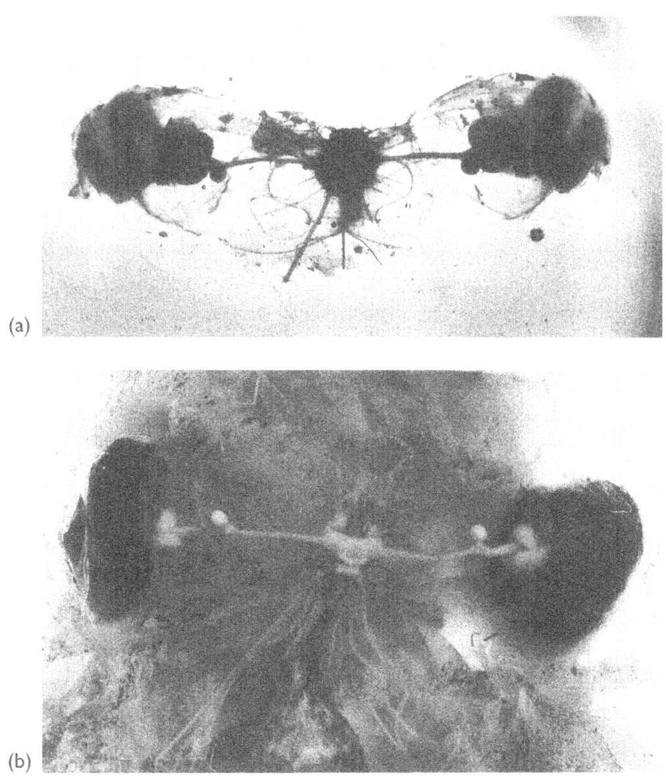

Fig. 29.2. *Eledonella pygmaea* (45 mm ML) in which the eyes and the central nervous system have been exposed (a), and *Japetella diaphana* (78 mm ML) (b).

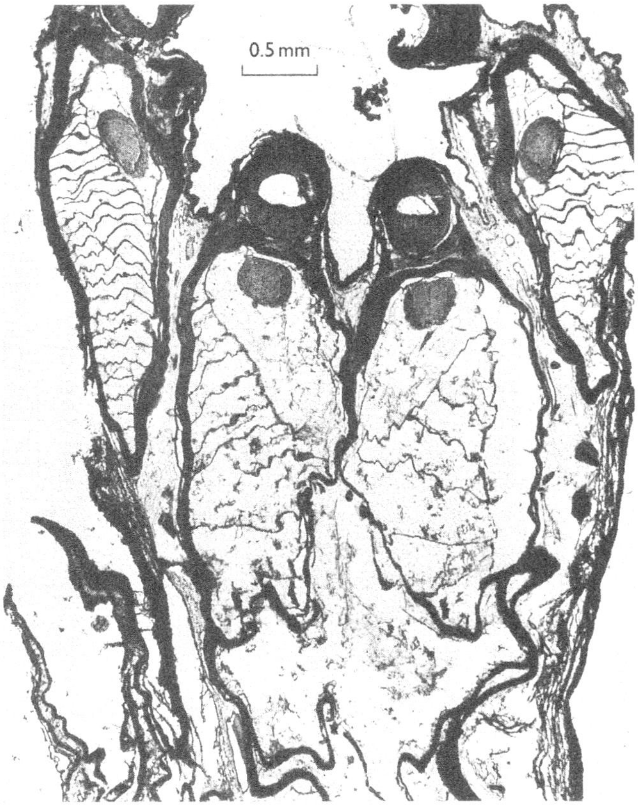

Fig. 29.3. *Japetella diaphana*. Longitudinal section through the arms to show the large chambers.

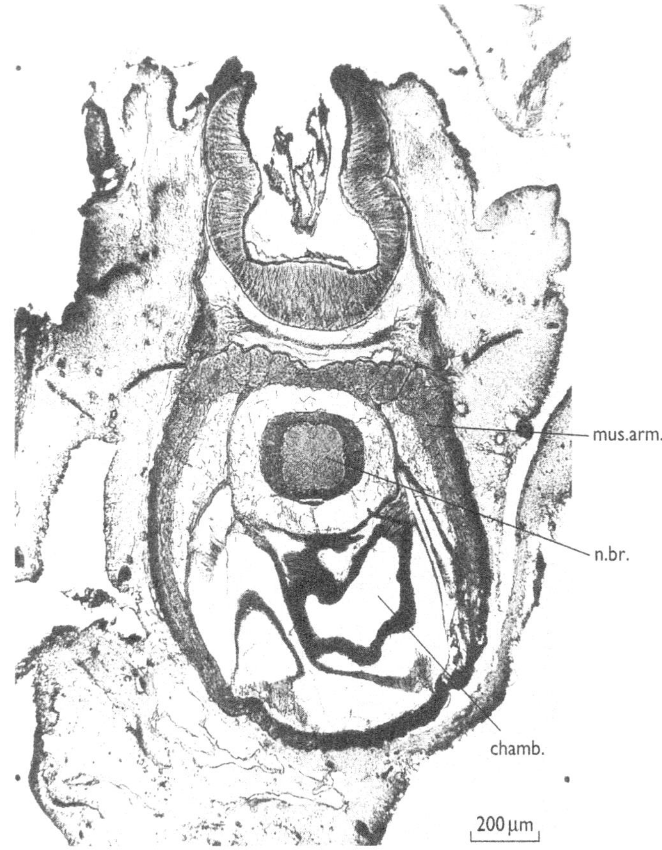

Fig. 29.4. *Japetella diaphana* (22 mm ML). Transverse section through an arm to show the arrangement of the chambers and axial nerve cord. The sucker is shedding the cuticle as a disc.

between arms II and III and possessed ovarian eggs about 1.1 mm in length (R.E. Young 1972c, 1978, Robison and Young 1981). The female broods the eggs, which are cemented together in clusters, in the brachial web, while remaining in depths of 700–800 m (R.E. Young 1978). The eggs may all hatch at the same time, or in batches, as 44 specimens 6–12 mm in length were captured in a single haul far from any coast, and the smallest specimen (2.5 mm ML) had a sticky gelatinous coat (Thore 1949). Other young have been identified and illustrated (Hochberg *et al.* 1992).

Females approaching sexual maturity become deeply pigmented, lose their silvery tissues, and develop a circumoral light organ. This organ is thick (5 mm wide) and lobular, and in living animals it is pale lemon in colour (Robison and Young 1981, Herring *et al.* 1987). After capture the circumoral organ of one female was treated with dilute hydrogen peroxide and the ring glowed steadily with a greenish-yellow light (emission maximum of 505 nm) (Herring 1983). In a shipboard aquarium females were observed in several postures; in the most active the head was held beyond the mantle with the upper arms flexed forwards and upwards and the lower arms forwards and downwards, so exposing the circumoral ring. The circumoral organ develops from a muscular ring in females of 40–56 mm ML. Cellular proliferation follows with a gradual degradation of the muscle tissue, and fine blood vessels are common although there are few nerve fibres. The light organ entirely lacks accessory optical structures (Herring *et al.* 1987). The directionality of the light emitted from the ring is determined by the position of the surrounding arms and web. The function of this organ remains enigmatic but it seems most likely to be a sexually significant signal, as both gravid females and near mature males have been captured in depths of 1050–1065 m (R.E. Young 1978).

29.3.3 Diet and buccal mass

Analysis of the digestive tracts of specimens of up to 20 mm ML showed the presence of calanoid, euphausiid, and decapod crustaceans, molluscs, chaetognaths, and fishes in their diet (Passarella and Hopkins 1991).

The beaks are broad, their walls are thin and delicate, and the rostral edge tanned (Clarke 1986). There is no rostrum, instead the rostral edge of the lower beak is toothed. The radula is wide and its teeth multicuspidate. The buccal mass forms a large proportion of the head (Fig. 29.5), and in this small specimen its length is 23% of the mantle length. The musculature of the radular complex is dense and contrasts with the surrounding tissues of the head and arms.

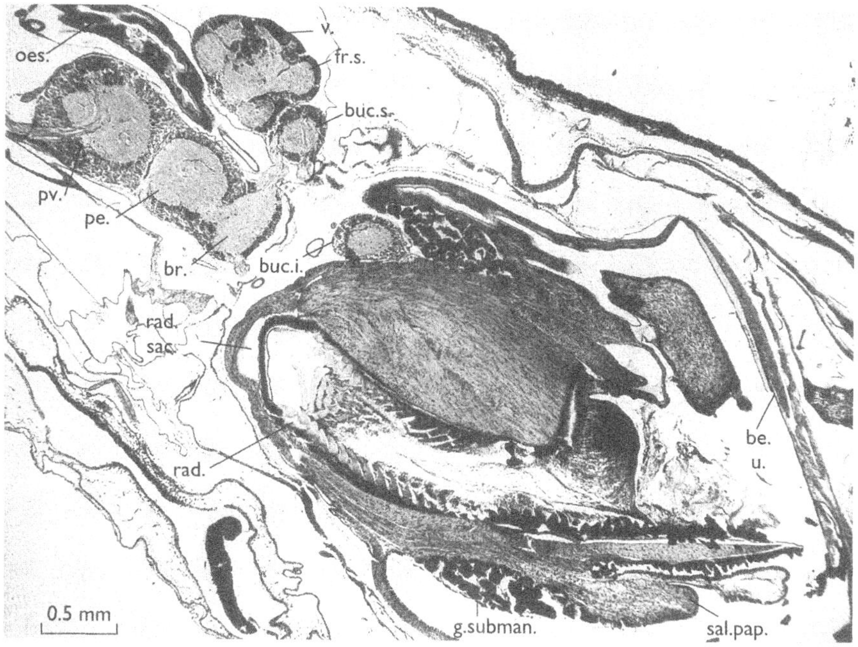

Fig. 29.5. *Japetella diaphana* (22 mm ML). Sagittal section showing that the buccal mass forms a large proportion of the head, whereas the central nervous system is relatively small.

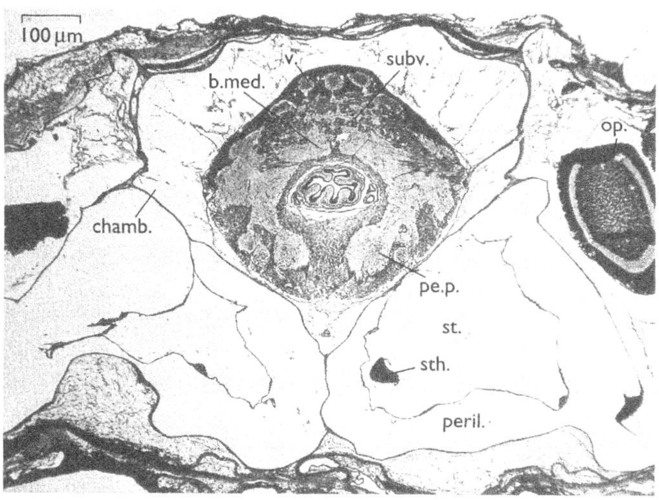

Fig. 29.6. *Japetella diaphana* (22 mm ML). Transverse section through the head showing the large chambers radiating above the central nervous system, and the statocysts.

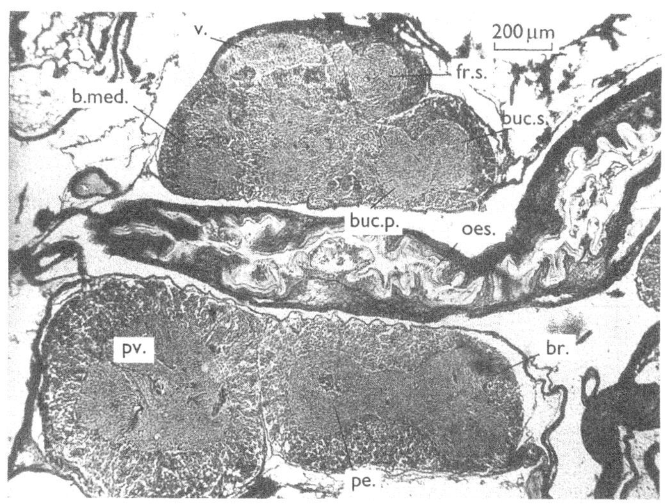

Fig. 29.7. *Japetella diaphana* (22 mm ML). Sagittal section through the central nervous system.

29.3.4 Receptors and central nervous system

The eyes are widely separated and the optic tracts much longer than in *Octopus vulgaris*. The optic lobes are partly divided, the dorsal lobule being smaller. The peduncle lobe and optic gland lie apart from the optic lobe (Fig. 29.2b). A large tract of fibres from the ventral part of the optic lobe bypasses the peduncle lobe and runs directly to the central nervous system.

The central nervous system has a series of deep chambers around it dorsally and laterally (Fig. 29.6). The brain itself is compact and lies just behind the large buccal mass. The superior and posterior buccal lobes form a prominent part of the brain (Fig. 29.7). The median inferior frontal lobe is small and a few small cells represent the subfrontal lobe (Young 1977a). The vertical lobe system is quite well developed and forms 11.3% of the total brain volume (Table 2.6) (Maddock and Young 1987).

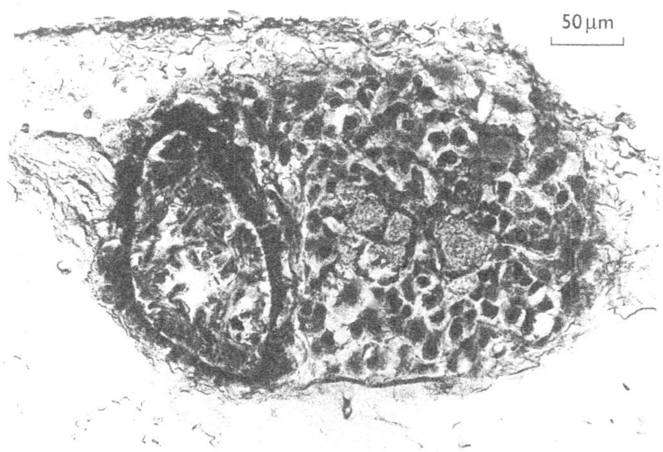

Fig. 29.8. *Japetella diaphana.* Transverse section of the stellate ganglion and epistellar body.

The mantle muscles are thin and the stellate ganglion is small. The epistellar body is a spherical vesicle attached to the posterior margin of each stellate ganglion (Fig. 29.8).

The statocyst is large and is enclosed in a larger perilymphatic sac (Fig. 29.6). The crista runs in a continuous curve across the front, along the side and then up the back of the sac; it is divided into nine sections, as in *Octopus*. There is no anticrista (Young 1989).

30 Vitreledonellidae

30.1 *Vitreledonella*

Vitreledonella richardi is the only species of the genus and family. It is gelatinous and glass-like in appearance when alive (Fig. 30.1). It is widely distributed in all oceans between 50°N and 40°S, but is absent from the Mediterranean. It has been captured in opening/closing nets at depths of 50–1000 m (Clarke and Lu 1975, Lu and Clarke 1975a, b). Small specimens of less than 20 mm ML are found between 50 and 300 m, and larger animals in depths of 100–1000 m (Clarke and Lu 1974, Land 1992).

30.2 Life history

Joubin (1937) found 540 larvae, each about 5 mm long, present in the mantle of a female whose total length was 330 mm when measured just after capture. This octopod may be viviparous, or the young may be held within the brachial web. Alternatively, hatching perhaps takes place in depths of 1000 m or more, and the young then migrate upwards together as 889 recently hatched paralarvae were captured in a single haul at 310–400 m (Lu and Clarke 1975a). Two frames from a video-tape taken in sequence from a submersible, at a depth of 909 m off Hawaii, showed one *Vitreledonella* engulfed within the web of another perhaps engaged in courtship or mating, or becoming prey (R.E. Young *et al.* 1998b). The largest female captured so far had a total length of 450 mm and a mantle length of 110 mm (Thore 1949).

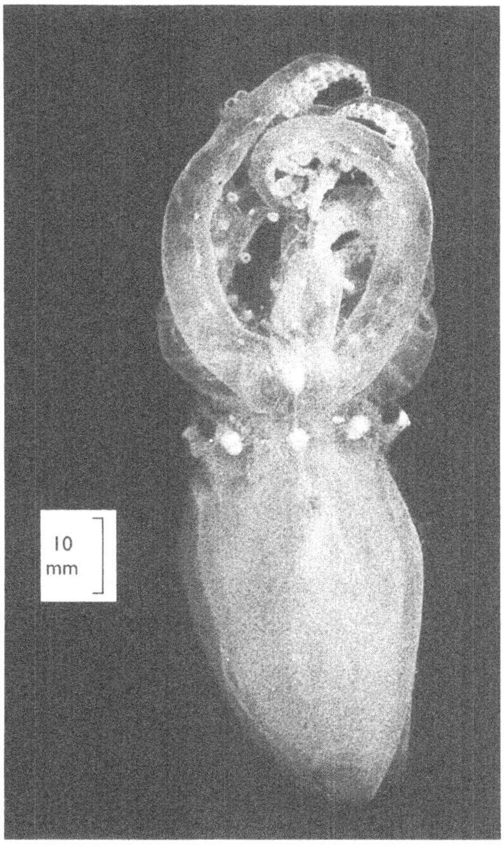

Fig. 30.1. *Vitreledonella richardi* (55 mm ML) caught in the North Atlantic at 195–295 m depth. The central nervous system, optic tracts, optic lobes, and arm ganglia are all readily visible through the translucent tissues.

30.3 Morphology

The buccal mass is wide, and lies some distance in front of the central nervous system. The beak is flattened anteriorly and is translucent except at the oral edge, which is amber in colour. The radular sac protrudes prominently behind the buccal mass (Fig. 30.2). Each transverse row of the radula has a rhachidian tooth with a long central cusp, a small lateral tooth 1 and large lateral tooth 2, and the marginal tooth has a long cusp (Nixon 1998).

The longest arms are about three times the length of the mantle (Thore 1949). The tissues are translucent, except the axial nerve cord and ganglia which are readily visible, as are the nerves going to the suckers and their ganglia (Figs 30.1, 30.3). The sucker has a squarish orifice after preservation, and it is deep and the musculature dense. The infundibulum takes the form of a shallow encircling band, 100–150 μm deep, with little apparent division into sectors. The pegs on the infundibulum are conical and somewhat sparsely distributed (Fig. 30.4). Each peg has numerous minute pores, similar in size to those found on the suckers of *Octopus vulgaris*. The infundibular cuticle is sloughed off at intervals, like that of *Octopus* (Nixon and Dilly 1977).

The eyes and the digestive gland are opaque and contrast with the rest of the animal (Fig. 30.1). The dorsal surface of the eye is darkly pigmented and the anterior surface has a reflective coating. The digestive gland is long and spindle-shaped and it too has a reflective covering. When the animal is oriented with its body axis 45° from the horizontal the longitudinal axes of the eyes and of the digestive gland are parallel, and vertical. This has the effect of reducing the size of the opaque parts of the animal when it is viewed from below in silhouette; this camouflage strategy is seen in other midwater animals (Land 1992). Confirmation that the digestive complex is reflective and becomes vertical when the animal moves at an angle of about 45° in the sea, and is seen in a photograph of a specimen in the Atlantic Ocean (R.E. Young *et al.* 1998b).

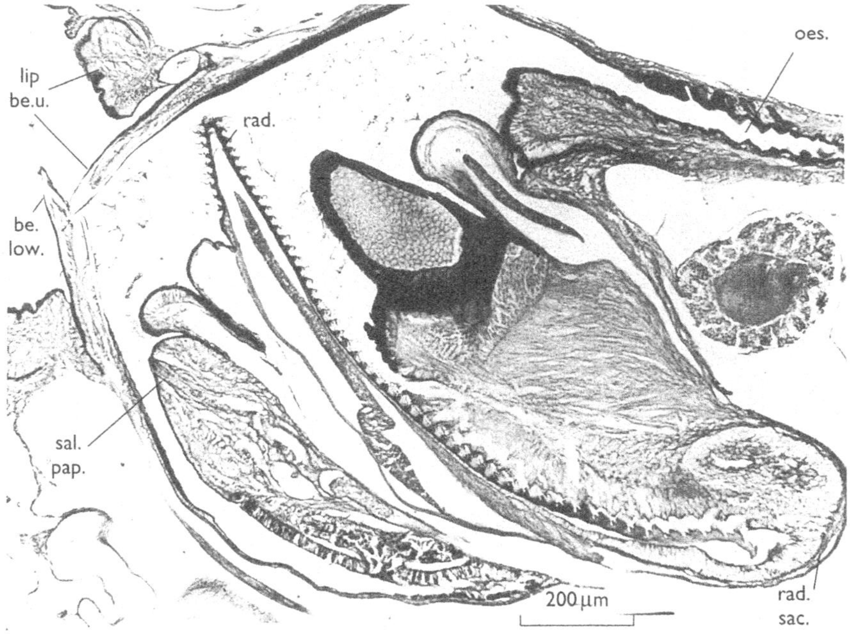

Fig. 30.2. *Vitreledonella richardi* (17 mm ML). Sagittal section of the buccal mass (Spirit blue).

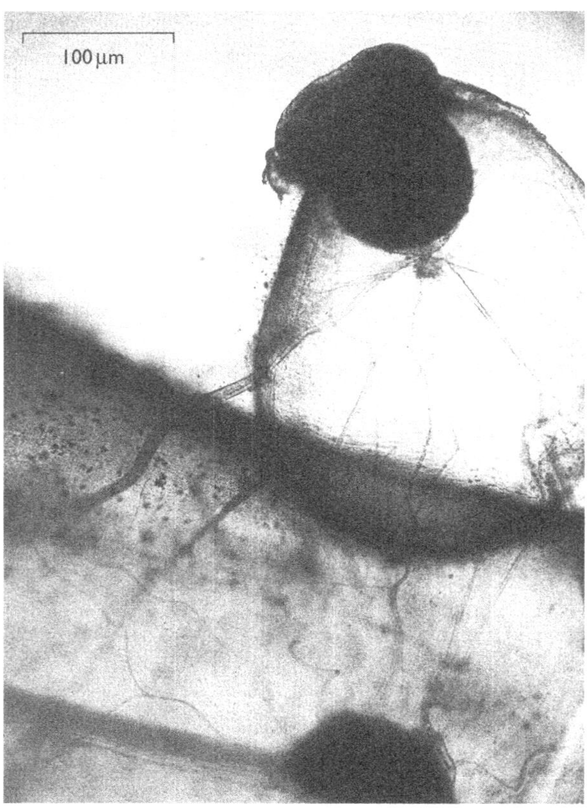

Fig. 30.3. *Vitreledonella richardi*. Part of an arm and a sucker showing a brachial ganglion with fibres running to the sucker ganglion.

30.4 Receptors and central nervous system

The eyes are set widely apart on either side of the head, at an angle of about 40° from the longitudinal axis of the body. Each eye is nearly twice as long dorsoventrally as anteroposteriorly, or laterally; the almost spherical lens is centrally placed (Fig. 30.5). The focal length gave values of about three lens radii, and did not reveal any differences in focal length for different axes. This shows that the shape of the eye is not to do with optical asymmetries of the lens. The axes of the eyes point obliquely forwards, making an angle of about 75° with each other. The retina itself fills the eye-cup from dorsal to ventral poles, but not from side to side. Here the retina covered only the gently curved back of the eye. No obvious differences were found in the retinal histology in different parts of the eye region. Each retina has a horizontal field of about 82° giving a small region of binocular vision of 7° in front of the animal, and a good coverage ahead. The retinal dimensions imply a vertical field of view of about 166°. Land (1992) suggested that the shape of the eye is dictated by the need for concealment from below.

The optic lobes are largely divided into two parts although the medulla is broadly united, and it is not clear whether there are distinct outputs to the central brain. The optic tracts are very long and the small peduncle lobe and optic gland lie close to the optic lobe (Figs 30.1, 30.6). The optic lobes form 215% of the total volume of the brain, and, as in other pelagic octopods, they are larger than in *Octopus* (Table 2.6). Their volume is presumably connected with the intensity of light encountered by the animal in its natural habitat rather than with a capacity for fine resolution. There are no long

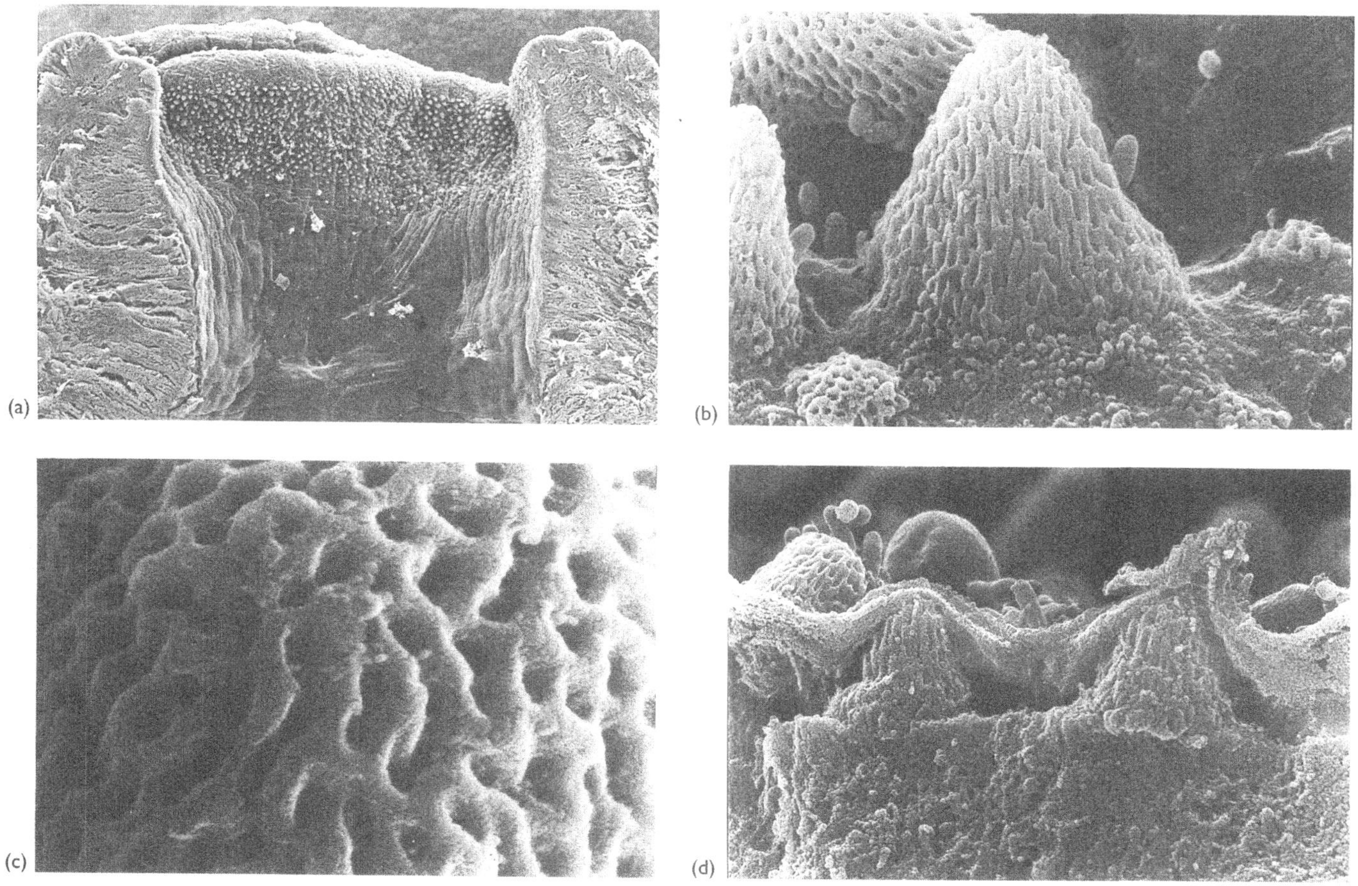

Fig. 30.4. *Vitreledonella richardi.* Scanning electron micrographs of a sucker: (a) A hemisection to show the pegs arranged in a restricted band (field width 625 μm). (b) A single peg to show its shape and form (field width 10 μm). (c) The surface of the peg at higher magnification (field width 2.5 μm). (d) A section to show the developing pegs underlying those on the surface (field width 13.8 μm). (Scanning electron micrographs kindly provided by K.P. Ryan.)

columns of fibres and neuropil in the optic lobes, such as are found in *Octopus* and *Loligo*.

The central brain is compact (Fig. 30.7). The anterior suboesophageal mass shows no division into pre- and postbrachial lobes. The wall of the anterior suboesophageal mass contains large and medium-sized cells. The brachiopalliovisceral connective is large (Fig. 30.8). The suprabrachial commissure is distinct. Chromatophore lobes are present, presumably for the control of the chromatophores when the animal is in the photic zone.

In the supraoesophageal mass the superior buccal lobe continues backwards as a large posterior buccal lobe (Fig. 30.7). The inferior frontal lobe is not apparent on the external surface and the sections show that it and the subfrontal lobe are reduced to a few scattered cells at the back of the posterior buccal lobe. The inferior frontal system forms 4.0% of the total brain volume (Table 2.6) (Maddock and Young 1987). The dorsal basal region contains many islands of cells, as in *Octopus*. There is no distinct subpedunculate nerve. The superior frontal lobe has no distinct divisions into lateral and medial parts. The more lateral part of the superior frontal lobe sends a large output to the subvertical lobe, with fibres perhaps reaching the precommissural lobe. The vertical lobe receives a small output from the more medial part of the superior frontal lobe; it has many small cells. The vertical lobe system is well developed and forms 11.9% of the total brain volume; the superior frontal lobe is larger than the vertical lobe.

The epistellar body is a single spherical vesicle (240 μm) lying on the posterior margin of the stellate ganglion (750 μm) in a specimen of 66 mm ML (R.E. Young 1978).

The statocysts are relatively large (Young 1989). Each has an outer and an inner sac with a typical octopod form. There is a single oval macula. The crista runs round the front, side, back and then vertically upwards, and it is divided into nine sections with alternate single and double rows of large cells. There is no anticrista. Kölliker's canal has a large inner section and funnel.

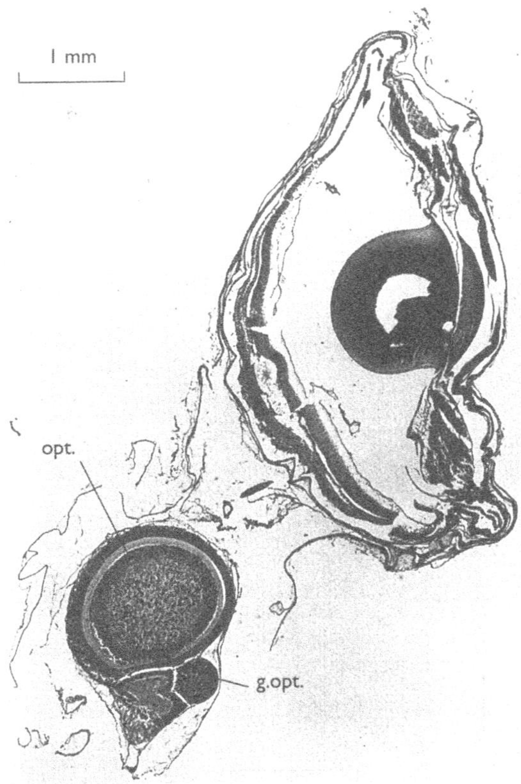

Fig. 30.5. *Vitreledonella richardi* (66 mm ML). Sagittal section of eye and optic lobe.

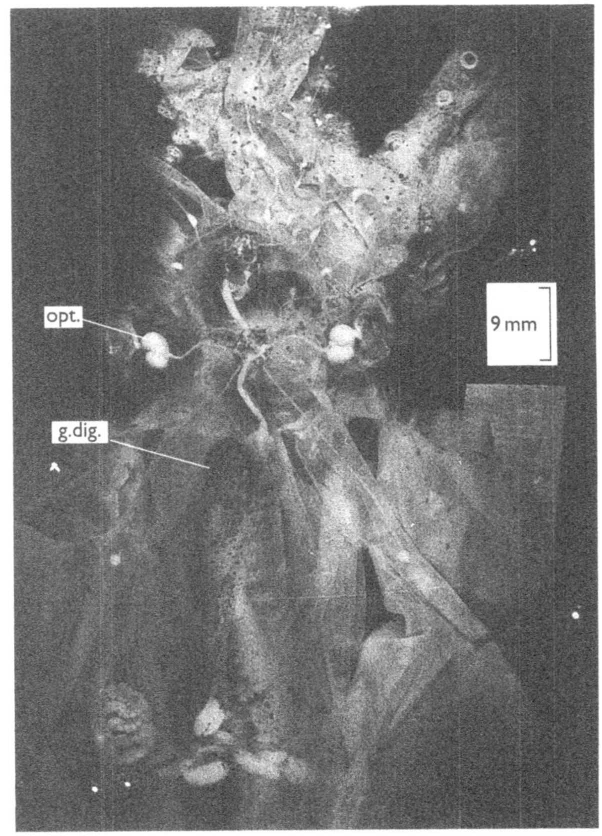

Fig. 30.6. *Vitreledonella richardi* (66 mm ML). Dissection to show the central nervous system and largely divided optic lobes, and the anterior position of the viscera.

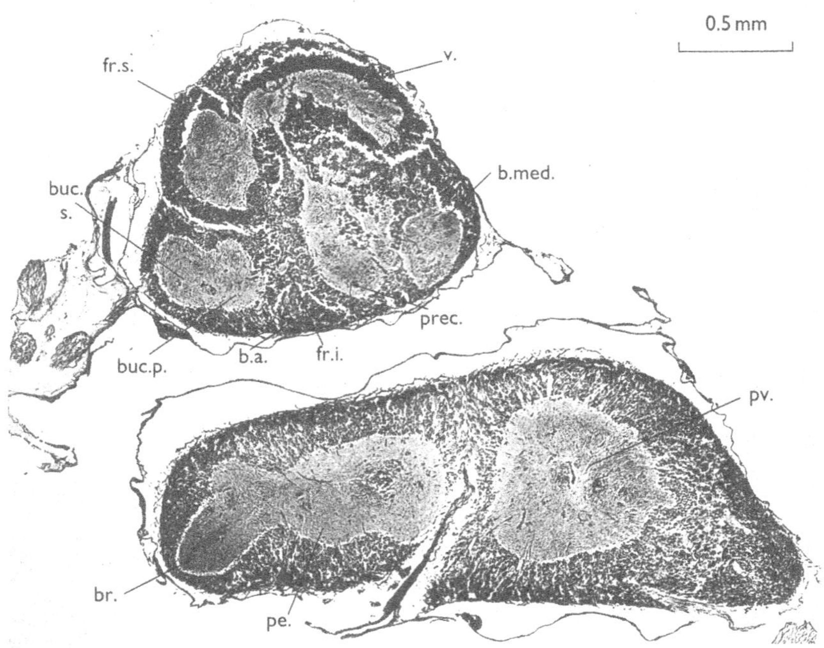

Fig. 30.7. *Vitreledonella richardi* (66 mm ML). Sagittal section near the midline showing the supra- and suboesophageal masses of the brain, and the reduced inferior frontal lobe.

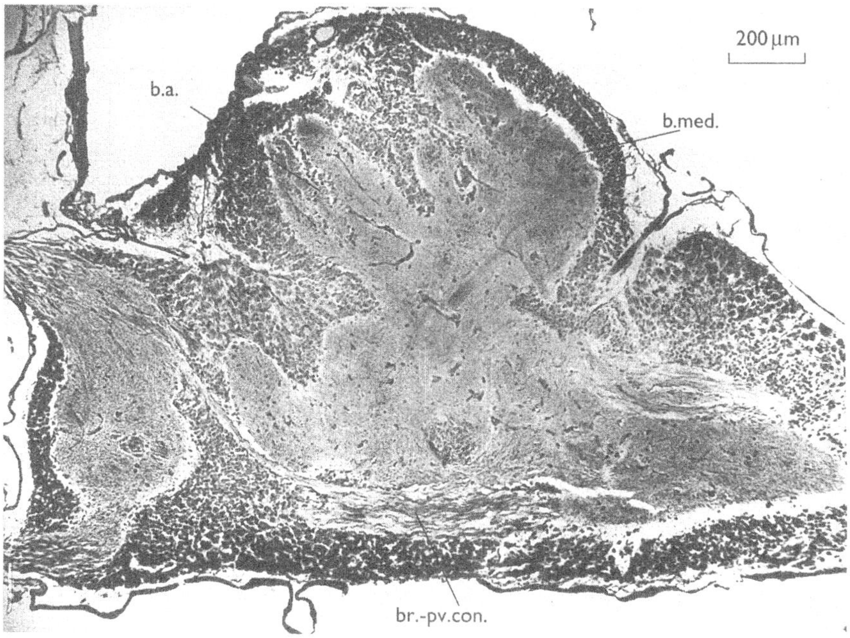

Fig. 30.8. *Vitreledonella richardi* (66 mm ML). Sagittal section further laterally than in previous figure.

31 Octopodidae

Members of this family live in coastal waters and are found around the world from the Arctic to the Antarctic. Their vertical distribution is from the littoral zone to abyssal depths. The adults are benthic in habit.

The family Octopodidae is the largest in the Class Cephalopoda. It includes four subfamilies, the Octopodinae, Eledoninae, Graneledoninae, and Bathypolypodinae (Hochberg 1998) with more than 20 genera. To these genera belong more than 200 species, 100 of which belong to the genus *Octopus*.

31.1 Octopodinae

The octopuses of this subfamily inhabit shallow waters to a depth of about 200 m. In size they range from tiny to very large. Each arm has two longitudinal rows of suckers and enlarged suckers may be present or absent. An ink sac is present. The copulatory organ is small to large and spermatophores are large and few in number.

31.1.1 Octopus

The members of this genus inhabit coastal regions in temperate and tropical waters (Mangold 1983a). There are notable differences in their horizontal distribution as the western edges of both the Atlantic and the Pacific Oceans are richer in the species of this genus than their eastern edges, and the coasts of the west Pacific and of Australia especially so (Nesis 1987, Norman 1997, Nixon in press b). The vertical distribution of *Octopus* extends from the intertidal zone to depths of 500 m (Mangold-Wirz 1963, Nixon in press a).

Fig. 31.1. *Octopus vulgaris.* Drawn from life (Jane de Vere).

Habitats and habits

Octopus vulgaris (Fig. 31.1) is found in the temperate littoral and sublittoral zones of the continental shelf of the Mediterranean, Atlantic, and west Pacific (Nesis 1987, Nixon in press b). It is solitary and inhabits a 'home' or 'den' among or under rocks, in crevices, in holes in rock facies, or in debris of human origin. A den can be recognized from the presence, near the entrance, of collections of arranged stones or shells, or middens containing empty mollusc shells and crustacean exoskeletons, although the latter are quickly lost to local currents.

Observations in the sea, at depths of 3–6 m, revealed this octopus to be highly active at night and early morning, and that some dens were 'permanent' but others only temporary (Altman 1967, Hanlon and Messenger 1996). The dens of adults are more highly modified than those of juveniles (Mather 1994), and are occupied for 1–34 days (Mather and O'Dor 1991). The animal remains in its den between 0800 and 1600 h, making only short excursions during this period, and spends the remaining hours in the open (Kayes 1974). During daylight octopuses use visual spatial information for navigation within their home ranges and to guide them during hunting trips (Fig. 31.2); laboratory trials show this octopus's ability to orient to visual landmarks although this environment is less complex than that in the sea (Mather 1991a).

The mobility of *O vulgaris* makes it difficult to estimate either their density on the sea floor or the area foraged by one animal.

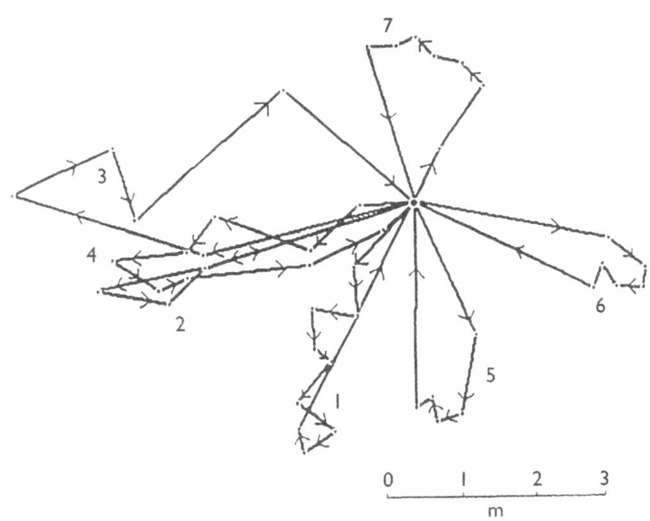

Fig. 31.2. *Octopus vulgaris.* Observations of the foraging trips of a juvenile octopus, in Bermuda, over a 10-day period in July. The drawing shows movements of the animal of more than 2 m from the den (Mather 1991a).

Trawls along the coast of the eastern Atlantic Ocean (23°N-26°N) over an estimated area of 138 000–150 000 m^2 resulted in the capture of 31–152 octopuses per hour (Guerra 1981). Juveniles (250 g) have been found to forage over areas of up to 200 m^2 and continuous hunting behaviour lasts for an average of 7.2 min. Their mode of hunting varies, depending on the microhabitat. They often crawl over the substratum and extend one or two arms into crevices, or envelope a rock within arms and web to detect any lurking prey present by chemotactile exploration (Mather 1991b).

Locomotion

After hatching the young *O vulgaris* enters the plankton where it remains for some weeks before settling to the adult habitat on the sea bed (Table 31.1). Once on the bottom the juvenile octopus can swim by jet propulsion, or walk using its muscular arms. In daylight, when foraging, the juvenile can cover distances of up to 10.5 m from its den, the forays lasting up to 75 min; the outward and return journeys may not necessarily follow the same paths (Fig. 31.2) (Mather 1991a). Little is known of the horizontal distance travelled but, off the coast of Japan, a small number of tagged octopuses were recovered up to 50 km from the site, and up to 40 days from the time of release (Mangold 1983a).

It is a vigorous swimmer and the mantle shows considerable changes in shape. The free part of the funnel is very flexible and controls the direction of the jet of water from the mantle. Besides the function of the mantle in jet propulsion it is also in continuous use for respiration. The muscles of the mantle are controlled by the central nervous system and the stellate ganglia and are occasionally used for maximal responses during escape (peak jet velocity 2.9 m/s for an octopus of 370 g) (Trueman and Packard 1968). The arms usually trail behind the mantle during jet propulsion.

The octopus has a highly developed mode of locomotion, using its arms and multiple suckers to walk over the substrate. The muscles of the arms are arranged three-dimensionally in closely packed blocks of constant volume with a connective tissue framework (Bone *et al.* 1981). The contraction of one particular set of muscles transmits forces to others in the block, an arrangement termed a muscular hydrostat (Kier 1985, 1988). There is repetition of nerve ganglia along the axis of each arm, one at the level of each sucker (Graziadei 1971). The nerves, muscles, and suckers are all involved in the propagation and integration of waves of contraction in a column of arm muscle, which is composed of serial blocks (Packard 1972). When walking the distal part of the arms are in contact with the substrate, and the animal appears to glide over the bottom. Such movement allows the huge numbers of chemotactile receptors present in the suckers and arms to detect and capture prey hidden below the web and arms (Mather 1991b).

Life history

The reproductive habits of several species of *Octopus* were reviewed in the first volume of *Cephalopod life cycles* (Boyle 1983b), and others have been described since that time (Rodaniche 1984, DeRusha *et al.* 1987, Iribarne 1991). The eggs range in length from 1.5 mm in *O. vulgaris* to 35.0 mm in *O. piscatorum*, and in number from 10 to

Table 31.1. *Octopus vulgaris.* Various aspects of the habits and habitats of newly hatched young until settlement (Nixon and Mangold 1996, 1998)

Mature adults	Migrate to inshore shallow water
Breeding period in the Mediterranean	Spring, summer, autumn
Spawning site	In crevices—inshore
Depth of spawning site	5–40 m
Bottom habitat	Rocks, sand, mud
Eggs	Laid in strings and attached to roof or wall of den
Number/individual	100 000–500 000
Capsule length	2 mm
Care	Brooded in den by female
Embryo development	47–54 days at 21 °C
Mantle length at hatching	2 mm
Total length at hatching	2.9 mm
Weight at hatching	1.0–1.4 mg
Time of hatching	No observation available
Hatchling	Translucent with very short arms. Young are expelled into water column by female
Type of paralarva	Dispersive
Denticles on oral surface of beak	Present
Buccal mass length at hatching	60% of mantle length
Number of suckers on each arm at hatching	3–4
Habitat in daylight and darkness	24 h in plankton
Response to light	Positive—moves towards light at hatching
Chromatophores	
At hatching	70–80/animal
In juvenile	Massive increase at settlement
Locomotion	Jet propulsion, and probably passive drifting
Buoyancy	May raft on surface film
Ink sac	Present, silver coated
Ink ejection	Probable
Feeding and prey in the sea	Predatory—planktotrophic on crustaceans
Prey capture	By the short arms, and perhaps by other as yet unknown modes
Paralysis of prey	Toxins secreted by posterior salivary glands
Habits	Entirely planktonic until close to time of settlement when it joins flotsam for short periods
Settlement on bottom	At 33–54 days after hatching, depending on body weight
Receptors	
Eyes	Present
Extraocular light organs	Present
Statocysts	Present
Olfactory organs	Present
Epidermal ciliated receptor cells	Present
Lateral line analogue	Present
Lifespan	15–24 months

700 000; the hatchlings are either benthic or pelagic in habit depending on the species (Hochberg *et al.* 1992). The early part of the life of *O. vulgaris*, from hatching to settlement, has been reviewed (Nixon and Mangold 1996) (Table 31.1).

Octopus vulgaris shows little external sexual dimorphism except that the male is slightly larger and develops two or three conspicuously enlarged suckers on arms II and III (Packard 1961). The size and age at maturation and spawning varies, mainly in relation to temperature. There is little courtship but sexual recognition is swift, presumably from chemotactile cues (Mangold 1983a). Mating is *à distance* in this species as the male can stretch right arm III to insert the hectocotylus into the female's mantle cavity and then pass spermatophores into the oviducts (Mangold 1987).

Reproduction in *Octopus vulgaris* is mainly under the control of the optic glands, two small rounded bodies (0.5–1.5 mm in diameter), situated on the dorsoposterior surface of the optic tracts (Young 1971). The glands develop from presumptive nerve cells, close to the olfactory lobe, when the posthatchling settles to its adult habitat on the sea bed (Bonichon 1967). At this stage the glands are pale yellow, but with maturity they become swollen and orange in colour. The optic glands, recognized to be endocrine organs, secrete a gonadotrophic hormone (Boycott and Young 1956, Wells 1978, Mangold 1989). Two cell types are present. Chief cells are unbranched stellate cells with relatively large round nuclei and prominent nucleoli; these cells secrete the optic gland hormone, and change in size and appearance as the gland matures (see Wells 1978). The other cells are probably fibrocytes. The optic gland is innervated by fibres from the subpedunculate lobe, and severance of the nerve supply to the gland results in its enlargement and subsequently that of the ovary or testis (Boycott and Young 1956, Wells and Wells 1959). A similar response is evoked when the optic nerves are cut, but the process takes longer. After removal of the optic glands the gonad remains small.

The hormone secreted by the optic gland of *Octopus vulgaris* (Mangold 1987) is necessary for development of the genital ducts, and it stimulates the production of proteinaceous yolk by the follicle cells of the ovary and of spermatogonia by the testis (Wells 1978). After removal of the optic gland, males were still able to perform the complete act of inserting the hectocotylus and the release of spermatophores (Wells and Wells 1972). Furthermore, there is no evidence of a gonadal hormone influencing behaviour. After castration and removal of the genital ducts males will undertake the act of passing non-existent spermatophores along the hectocotylized arm. The only operation known to interfere with this process is the removal of the tip of the hectocotylus: evidently this locates the oviduct and provides the signal to initiate the program of spermatophore transfer (Wells 1978).

Sex steroid hormones have been found in the gonad of male *O. vulgaris* (D'Aniello *et al.* 1996, Di Cosmo and Di Cristo 1996). Two neuropeptides, the gonadotropin-releasing factor (GnRH) and a neuromodulator (FMRF amide), have been localized, the former in the posterior olfactory lobe and the latter in the subpedunculate area of the brain, and may influence the secretory activity of the optic gland (Di Cosmo and Di Cristo 1998): these findings support those of Wells and Wells (1959). An investigation of the endocrinology of the female revealed receptors specific for progesterone in the reproductive tract (Di Cosmo *et al.* 1998); thus analogies and functional parallels with insects, crustaceans, and vertebrates are seen.

The eggs are elongated (2.2 × 0.9 mm) and the chorion is drawn out into a stalk. Secretions result in the stalks of the eggs adhering together, so forming strings which the female attaches to the roof of the den. The eggs are brooded, aerated, and cleaned by the female while she adds further egg strings over a period of 4–6 weeks. Females of 1–5 kg body weight produce between 100 000 and 500 000 eggs. Brooding can last as long as 5 months, during which time the female does not feed; she dies shortly after the last egg hatches (Mangold 1983a).

At hatching, young *Octopus vulgaris* are expelled, by a jet of water from the funnel of the female, into the surrounding water where they are carried by currents into the plankton. The paralarva (Fig. 31.3) is translucent and most of the chromatophores it possesses are on the ventral surface (Fioroni 1965, Packard and Sanders 1969, Packard 1985). Kölliker's tufts are present (Fig. 31.4) (Brocco *et al.* 1974) in paralarvae of up to 6 mm ML, which are still in the plankton (Rees 1950), and ciliated cells (Lenz 1997). Paralarvae remain in the plankton for up to 12 weeks, and then spend part of their time on flotsam until settlement (about 11 mm ML) (Mangold 1983a).

During the early period of paralarval life numerous morphological and behavioural changes take place (Table 31.1) (Nixon and

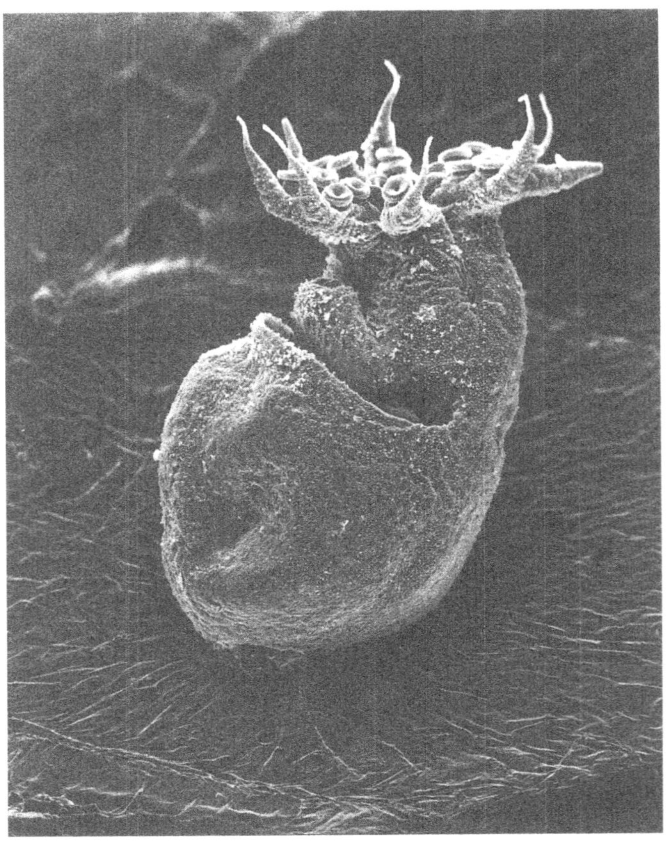

Fig. 31.3. *Octopus vulgaris.* Scanning electron micrograph of newly hatched paralarva to show the short arms, each with three suckers, and their long slender tips entirely lacking suckers (see Fig. 31.7). The mantle opening is wide and the funnel pronounced (kindly provided by P.N. Dilly).

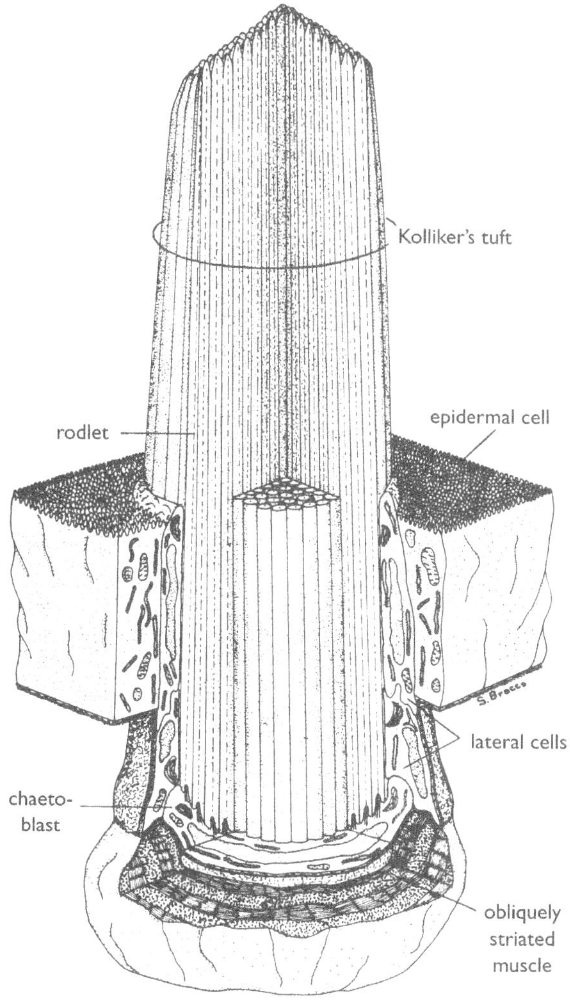

Fig. 31.4. *Octopus.* Kölliker's organ with an erupted tuft shown in a diagrammatic representation from scanning electron micrographs (Brocco *et al.* 1974).

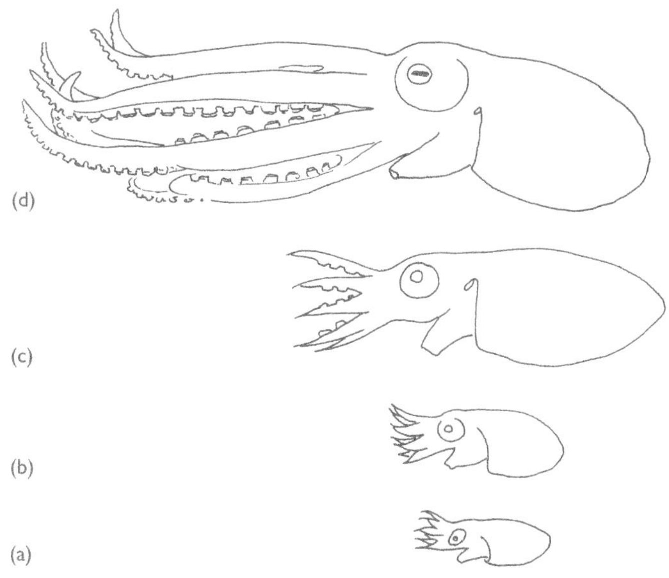

Fig. 31.5. *Octopus vulgaris.* Lateral view of young specimens, from the Bay of Naples: (a) newly hatched from egg (2.0 mm ML), (b) from the plankton (2.5 mm ML), (c) from the plankton (4.1 mm ML), and (d) a settled juvenile (8.6 mm ML) (Naef 1923, 1928).

Mangold 1996). There is a notable lengthening of the arms, and an increase in the number of suckers to more than 100 on each, as the animal approaches the time to settle on the bottom (Fig. 31.5). At hatching the buccal mass is very large in relation to the animal, and so too are the eyes. At 2 days, when each arm bears only three suckers, the paralarva can capture and hold live prey and feed on it (Boletzky 1974a). The upper and lower beaks bear denticles on their oral surfaces (Boletzky 1971). These are presumably of value in retaining prey aided by the closely encircling suckers (Fig. 31.6). Paralarvae are planktotrophic and can capture prey 50–100% of their own mantle length (Villanueva 1994). Most of their time is spent moving backwards, but they turn forwards to attack prey (Villanueva *et al.* 1996). The functions of the arms in the capture of prey has not been analysed but Naef (1923) noted that 'These stumpy larval arms, the whip-like ends of which are probably particularly sensitive, can be extended suddenly for a short distance' and that 'the finger-like ends of the arms at first and then the suckers fasten on to the prey'. The tips of the arms are slender and lack suckers, and their sensitivity is likely to be due to the receptors present (Fig. 31.7); the function of these tips has not been investigated.

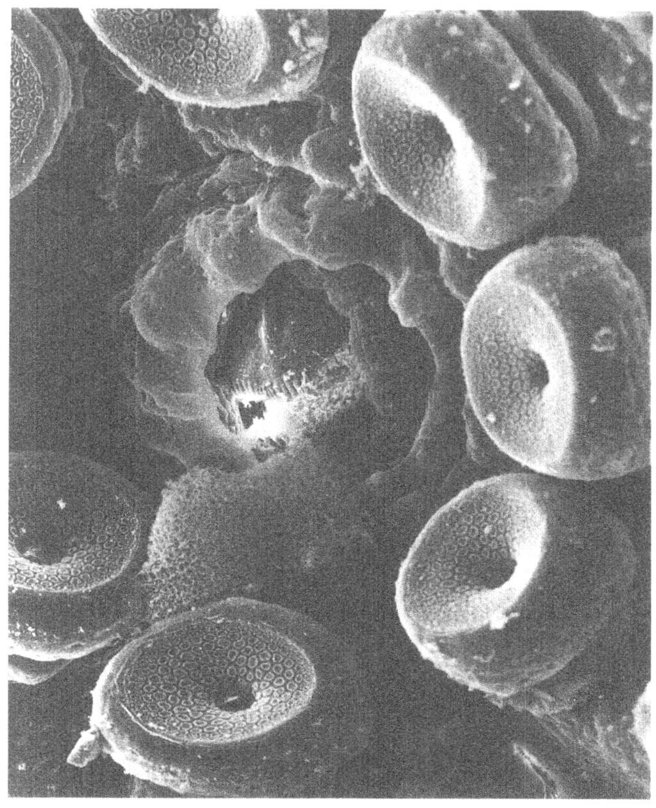

Fig. 31.6. *Octopus vulgaris.* Scanning electron micrograph showing the upper and lower beaks with denticles on their oral surfaces and the surrounding lips, and the bases of the arms and suckers (kindly provided by P.N. Dilly).

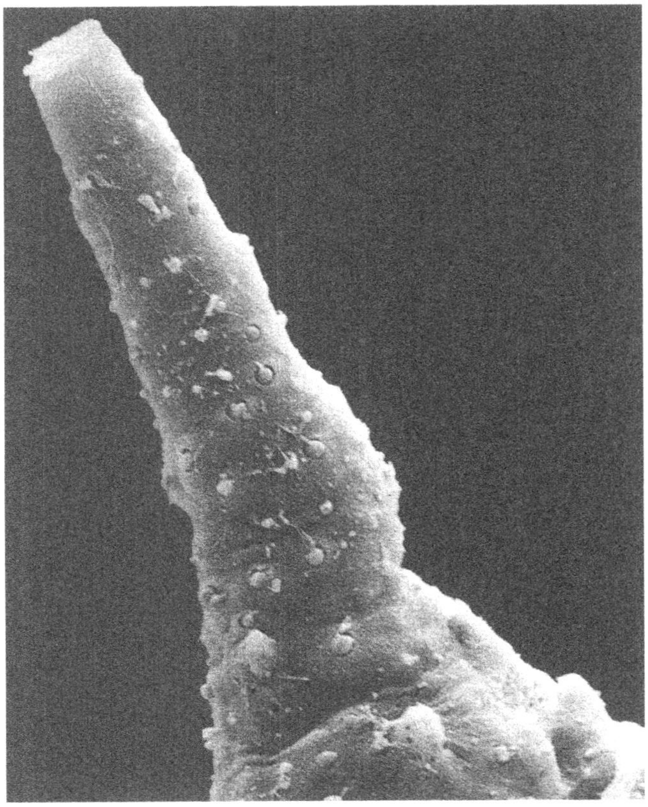

Fig. 31.7. *Octopus vulgaris.* Scanning electron micrograph of most of the terminal part of an arm showing the presence of numerous putative receptors (kindly provided by P.N. Dilly).

The brain is already prominent at hatching (Fig. 31.8) and its growth continues throughout the life of the animal. It weighs 0.2 mg at hatching, when it forms 20% of the total weight of the animal, and increases to an average of 20 mg at settlement, and at the same time there is a 25-fold increase in the number of cells (see Fig. 31.22) (Packard and Albergoni 1970). The lobes of the central nervous system are well defined at hatching, although some develop later than others (Frösch 1971); further changes continue until settlement (Nixon and Mangold 1996). The buccal and basal lobes are relatively large at hatching when compared with those in the newly settled animal. In contrast, the brachial lobes are small in the hatchling and are relatively much larger in settled and juvenile animals. The vertical lobe, subvertical lobe, and subfrontal lobes are very small initially when compared with those of the newly settled and juvenile animals (Figs 31.8, 31.23).

Food uptake is high and growth is rapid (Nixon 1966, Mangold 1983a). Off the African coast a female *Octopus vulgaris* can reach a weight of 4.5 kg in 8 months (Smale and Buchan 1981). The life span of the female ranges between 12 and 24 months, but the male may live longer (Mangold 1983a).

Arms and suckers

The arms are used in many activities including locomotion, jet propulsion, walking, exploration, chemoreception, capture of prey, in courtship display and copulation by the male, and in cleaning the eggs by the female. They are capable of considerable changes in size, and can be bent in any plane.

The musculature of the arms has been described by Cuvier (1817), Colasanti (1876), and Guèrin (1908). The muscles can be divided into the intrinsic muscles of the arms and of the suckers, and these are linked by a further set (Young 1971). There is an axial, or central, nerve cord, and this is surrounded by a mass of closely organized transverse muscles, thickest aborally, thinnest orally, and of intermediate thickness laterally. Bundles of these muscle fibres extend outwards in trabeculae, between the bundles of longitudinal fibres, to insert directly on to the fibrous connective sheath, which encircles most of the intrinsic muscles and is notably thick orally and aborally. As the trabeculae of the transverse muscle extend laterally they pass through three sets of oblique muscle layers and branch as they do so, before inserting on the connective tissue sheath (Kier 1988). The three-dimensional array of muscle fibres of the arms forms a solid mass of tissue, entirely devoid of skeletal tissue; this system is termed a muscular-hydrostat by Kier and Smith (1985). In such a system the volume remains essentially constant at physiological pressures during elongation, shortening, and torsion of the arms. Thus contraction of transverse muscles would decrease the cross-section and thereby increase the length of the arms, and conversely that of longitudinal muscles would shorten the arm. These latter muscles may be involved in bending movements and torsion probably results from contraction of oblique muscles. The complex system of muscles allows the many and diverse movements of which the arms are capable, including delicate and local movements. The action of a particular muscle fibre or group of muscle fibres is localized and suggests a highly subdivided and complex neural system (Kier 1988).

The complexity of the nervous system is confirmed by the presence of many more nerve cells in the arms than in the central nervous system of the octopus: the eight arms together contain nearly 350 million cells, compared with only 42 million cells in the brain and 92 million in the two optic lobes (Young 1963c). The axial nerve cord forms a prominent part of each arm and extends to the extreme tip (Fig. 31.9). It consists of a series of ganglia. Many neurons of each ganglion send axons to innervate the corresponding sucker. The axial nerve is linked by nerve bundles to peripheral nerve centres, consisting of four intramuscular nerve cords lying in the intrinsic musculature of the arm, and to sucker ganglia, one lying below the base of each sucker. The arm muscles are innervated by motoneurons of the axial nerve centres (Graziadei 1971). The arms receive efferent fibres directly from the posterior buccal and subvertical lobes. Afferent fibres pass from the arms to the lateral and median inferior frontal lobes, others to the lateral and median superior frontal lobes. These pathways are summarized in Fig. 31.30 (Budelmann and Young 1985). The efferent fibres are responsible for the control of movements of learned tactile discriminations (Young 1983).

Each sucker is attached to the arm by a short, muscular peduncle (Fig. 31.9). The sucker itself has a tightly packed, three-dimensional array of muscles, and a circular sphincter muscle (Guèrin 1908). This complex system, like that of the arms, functions as a muscular hydrostat. There are two main parts to each sucker: the acetabulum, a chamber which opens to the second part, the funnel-shaped

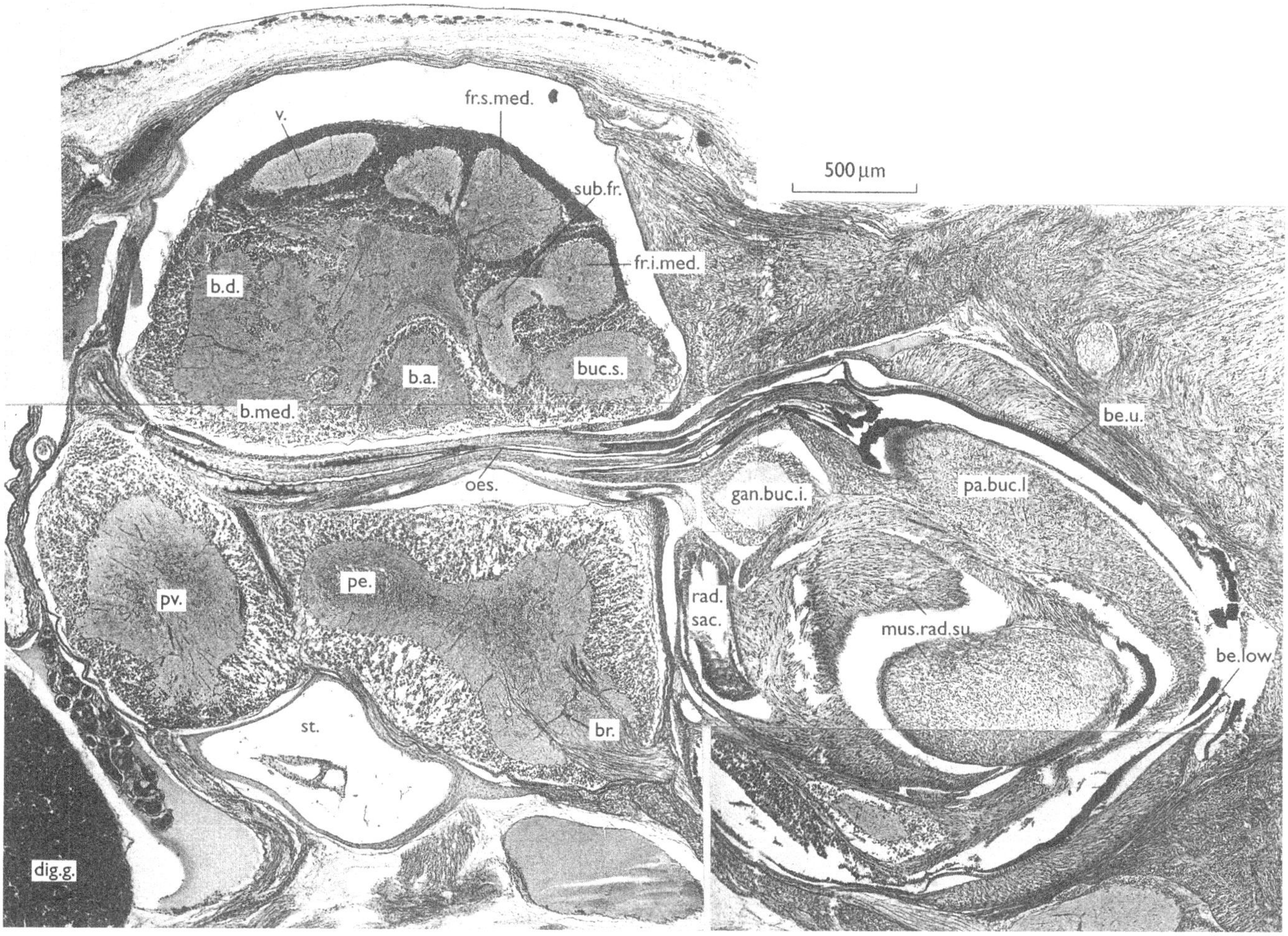

Fig. 31.8. *Octopus vulgaris*. Sagittal section of a recently settled specimen (9.2 mm ML) from the Bay of Naples, to show the major lobes of the supra- and suboesophageal masses of the brain, separated by the oesophagus, and the buccal mass (Nixon and Mangold 1996).

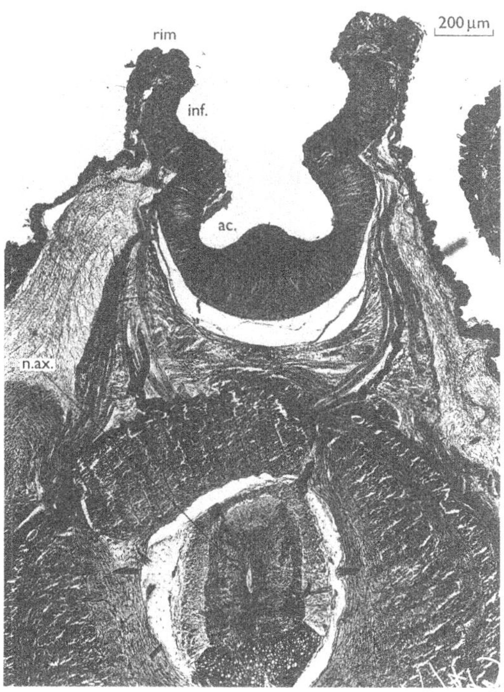

Fig. 31.9. *Octopus vulgaris* (25 mm ML). A transverse section through an arm to show its complex musculature, axial nerve cord, and the sucker with radial and sphincter muscles.

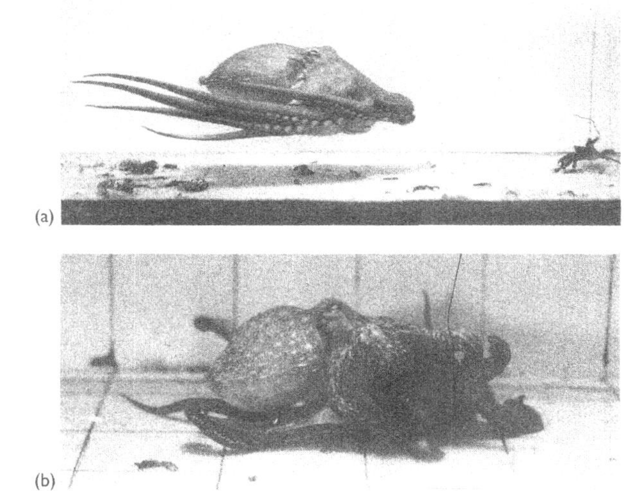

Fig. 31.10. *Octopus vulgaris* moving rapidly forward by jet propulsion (a) to seize and envelope the crab in its web (b) (photographs kindly provided by B.B. Boycott).

infundibulum. Internally a sphincter of circular muscle marks the boundary between the acetabulum and the infundibulum. The acetabular and infundibular walls consist of radial, circular, and meridional muscles, although the arrangement differs slightly in each (Kier and Smith 1990). The acetabulum and infundibulum have a cuticular covering of a chitin–protein complex (Hunt and Nixon 1981), secreted by the underlying epithelial cells. This cuticle is shed at intervals (Girod 1884, Packard 1988) and is continuously renewed (Naef 1923), presumably to maintain the condition of the sucker surface, which is subjected to continuous use.

The surface of the sucker has a rim, a region of complex radial folds (Graziadei 1971, Kier and Smith 1990). Within this is a narrow zone where the cuticle is smooth, a transition zone, only becoming sculptured where it covers the infundibulum and carries projecting pegs (Girod 1884). The infundibulum has a pattern of radial folds, in which the pegs become less numerous passing towards the centre of the suckers and are reduced in height or are absent. The cuticle continues into the acetabulum, where there are a few short pegs. The cells secreting the cuticle also exhibit differences, being short below the acetabular cuticle and tall beneath the infundibular cuticle. The infundibular cuticle has small pores, and at higher magnification each peg is seen to be formed by an aggregation of rods so as to leave fairly regular spaces (0.8 μm in diameter) (Nixon and Dilly 1977).

The suckers have numerous functions. They are used in chemotactile reactions, exploration of the substrate, in gathering or manipulating objects, the capture of prey (Wells 1978), in cleaning activities when shedding the sucker discs or brooding eggs, and as visual stimuli in defence and courtship (Packard 1988). For these activities the suckers are richly innervated, and have several types of morphologically distinct receptor cells present in the order of tens of thousands per sucker (Graziadei 1971). The receptor cells are scattered among the epithelial cells and are more numerous in the region abutting the junction between the smooth cuticle and that covered with pegs. They include three ciliated and one non-ciliated cell type, the latter lying under the peg-bearing cuticle. Axons from the receptors go either directly to the ganglion of the sucker, or terminate on interneurons. The receptors include two forms of ciliated cells that are considered to be chemoreceptors or mechanoreceptors; the function of the non-ciliated receptor is unknown (Graziadei and Gagne 1976a, b).

Many of the actions of the suckers are probably concerned with forming a seal followed by adhesion and suction. Girod (1884) estimated the maximum potential of suction force developed by the suckers along one arm, 300 mm in length, to be 15 kg. Parker (1921) measured the breaking force of isolated suckers of *Octopus bimaculatus*; the results were closely similar to those obtained by Dilly and his colleagues (1964) who measured the forces exerted by *O. vulgaris*; in animals weighing 900–1350 g the values obtained reached 18 kg. During attachment of a sucker to the substratum a seal is formed by the rim (Nixon and Dilly 1977, Kier and Smith 1990), followed by reduction of hydrostatic pressure in the sealed chamber by the action of the muscles (Smith 1991). Kier and Smith (1990) consider that only the musculature and connective tissue of the acetabulum are needed to generate the attachment force. The role of the infundibulum, with its numerous pegs and their many fine pores, in the actions of the sucker remains uncertain. It may be involved in the adhesion of the sucker to a foreign surface (Nixon and Dilly 1977).

The mechanism of release of a sucker from the substrate has been little investigated. It has a neural basis as lesions in the inferior frontal lobe of the brain result in the 'sticky sucker' syndrome; a condition in which the animal holds indefinitely any object which it touches.

Prey and its capture

The prey of *Octopus vulgaris* consists mainly of crustaceans, fishes, and molluscs, including cephalopods, as well as polychaetes, ophiuroids, and foraminiferans (Nigmatullin and Ostapenko 1976, Budelmann and Nixon 1984, Nixon 1987), and the spectrum of prey increases with growth (Smale and Buchan 1981). Regional and temporal differences in feeding habits have been observed (Guerra 1978, Hatanaka 1979), as well as preferences in the types of prey eaten (Taki 1941).

The shape, form, and morphology of the prey eaten varies widely, entailing different methods of capture and immobilization of the prey, and of ingestion. A benthic juvenile octopus will capture a crab within 3.8 s of entry into its visual field (Maldonado 1963). The prey is enveloped and sealed within the arms and web (Fig. 31.10). The saliva of *O. vulgaris* immobilizes a crab in less than 30 s (Ghiretti 1959, 1960, Cariello and Zanetti 1977). One route by which the saliva can enter a crab is via a puncture in the eye; this was observed in another octopod, *Eledone cirrhosa*, and the effect of the toxin was rapid (Grisley *et al.* 1996). *Octopus vulgaris* living in the Mediterranean does not usually penetrate the calcified exoskeleton of crustaceans to introduce saliva (Boyle and Knobloch 1981), but along the Atlantic coast of Spain it does sometimes drill the carapace of crabs (Guerra and Nixon 1987). The salivary enzymes act to breakdown the sites of attachment of the muscles to the exoskeleton of the crab. This is completed in about 20 min when the exoskeleton is expelled, neatly separated at the joints and entirely devoid of tissues but otherwise undamaged (Nixon 1984). The octopus usually ingests only soft tissues and these are still recognizable in the crop for up to 2 h after feeding (Altman and Nixon 1970), and may remain so for up to 6 h in the stomach (Andrews and Tansey 1983).

Many species of molluscs also are eaten, the shells of which may have holes drilled in them by the octopus. The cavity is usually small (2 mm diameter) at the surface of the shell, and slopes downwards to a small entry point (< 1 mm diameter) on the inner surface at the point of penetration of the shell. The breakdown of the more superficial part of the shell is due, presumably, to the combined actions of the salivary secretions and of the radular teeth, since many exhibit damage (Nixon 1969). The deepest part of the cavity in a mollusc shell is usually too small to be entered by the radula, but the tip of the posterior salivary gland duct is small and could enter (see Figs 31.16a, 31.18). The anterior part of the papilla of the duct of the posterior salivary glands is tooth-covered, as is the eversible tip of the single duct from these glands (Nixon 1979b, 1980). Once the shell is entered then saliva can be introduced on to, or into, the soft body of the prey. The saliva is presumably toxic to molluscs, though this has not been confirmed. It contains a substance (or substances) that acts at the sites of attachment of the muscle to the shell of the mollusc, since the soft body falls away to leave the shell devoid of adhering tissue. The saliva also acts on both the mineral and the organic matrix of the shell (Fig. 31.17b, c) (Nixon *et al.* 1980). The whole process of drilling and feeding may take less than 1 h but this depends both on the type of mineral present and on the structural organization of the shell (Nixon and Maconnachie 1988).

Buccal mass

The morphology and innervation of the buccal mass of *Octopus vulgaris* was described by Young (1965b, 1971). The buccal mass forms a considerable portion of the newly hatched young as its length is almost half that of the mantle; in a near adult it forms only about a tenth of the mantle length (Nixon 1985). The beaks at hatching have denticles on their oral surfaces (Fig. 31.6) (Boletzky 1971). These are presumably of value in retaining prey while the paralarval octopus lives and feeds in the plankton (Nixon and Mangold 1996).

The buccal mass is almost spherical and lies within the hollow formed by the bases of the arms. The buccal membrane is a pigmented web of folds of skin, attached to the arms, that can enclose the buccal mass. Within is the outer lip, a simple fold of tissue, and the prominent, much-folded, inner lip. The beaks are firm, and are composed of a chitin–protein complex (Hunt and Nixon 1981). They provide support and attachment for the mandibular muscles and form a buccal cavity enclosing the various structures within. The superior, inferior, and lateral mandibular muscles are responsible for the movements of the beaks in which the upper beak moves relative to the lower beak. Electrical stimulation of the interbuccal connective elicits a cycle of movements; the beaks open, then move together and close. When closing the upper beak retracts tightly into the lower beak before returning to the resting position, with the beaks slightly apart (Boyle *et al.* 1979). The mandibular muscles are separated from the lateral walls and crest of the beak by a single layer of tall columnar cells, the beccublast cells. Many beccublast cells have cell-long fibrils attached to complex trabeculae which extrude from the cell into the matrix of the beak, and at the opposite end are anchored to the muscle interface, so serving to attach the muscles to the beaks (Dilly and Nixon 1976c).

Within the buccal cavity lies the buccal complex, which comprises the radula, radular support system, lateral buccal palps, salivary papilla, and salivary glands (Fig. 31.11). The radular teeth are secreted within the radular sac (Raven 1958) by tall columnar cells (Figs 31.12, 31.13) (Nixon 1968), and the radula grows continuously forward to turn acutely over the bending plane. The radula has seven teeth across each transverse row. These comprise a central rhachidian tooth with, on either side, lateral teeth 1 and 2, marginal tooth 1, and a marginal plate (Nixon 1995). The newly developed teeth, as they emerge from the radular sac, are larger than their predecessors. The teeth in the region of the bending plane show signs of wear, the cusps being broken or fractured, and they lack the orderly arrangement and pristine condition of the newly formed teeth (Nixon 1969). Teeth are lost from the distal end of the radula, where it terminates in a fold, the walls of which are differentiated and have an area of putative

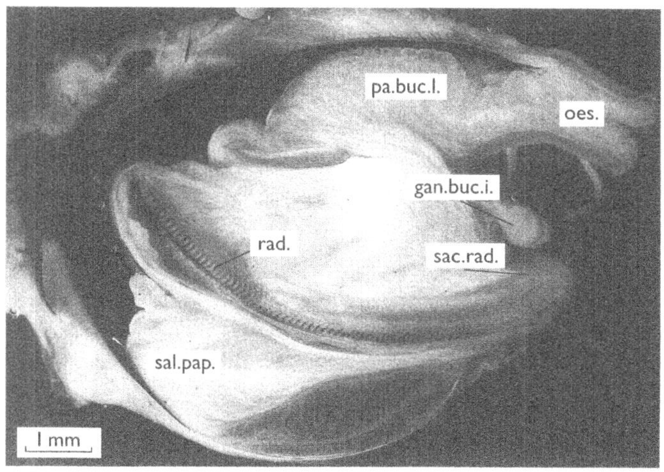

Fig. 31.11. *Octopus vulgaris* (110 mm ML). The features of the buccal cavity exposed after cutting the buccal mass in the medial sagittal plane (Nixon 1968).

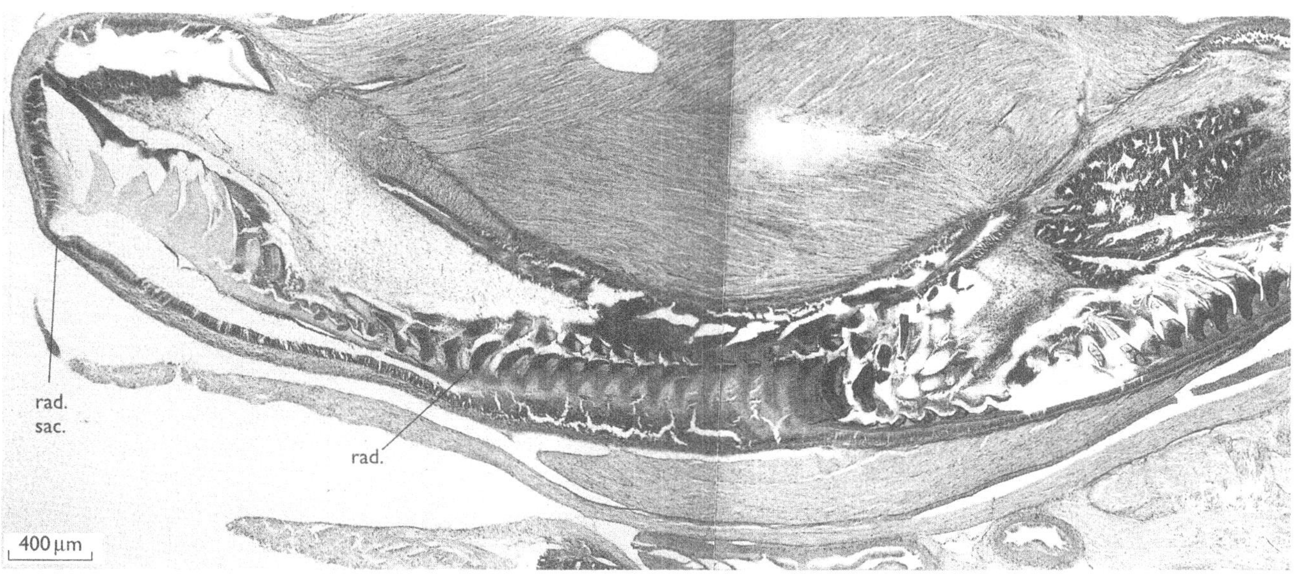

Fig. 31.12. *Octopus vulgaris.* A medial sagittal section of the radular sac and the radular teeth (Nixon 1968).

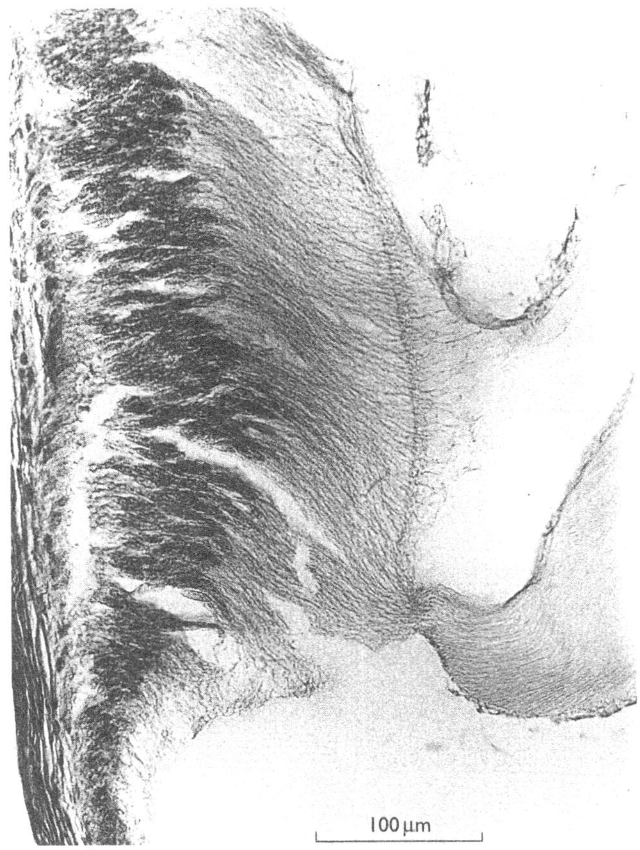

Fig. 31.13. *Octopus vulgaris.* The section seen in the previous figure, at higher magnification. It shows the postero-ventral part of the last-formed tooth and the odontoblasts behind, in the posterior wall of the radular sac (Nixon 1968).

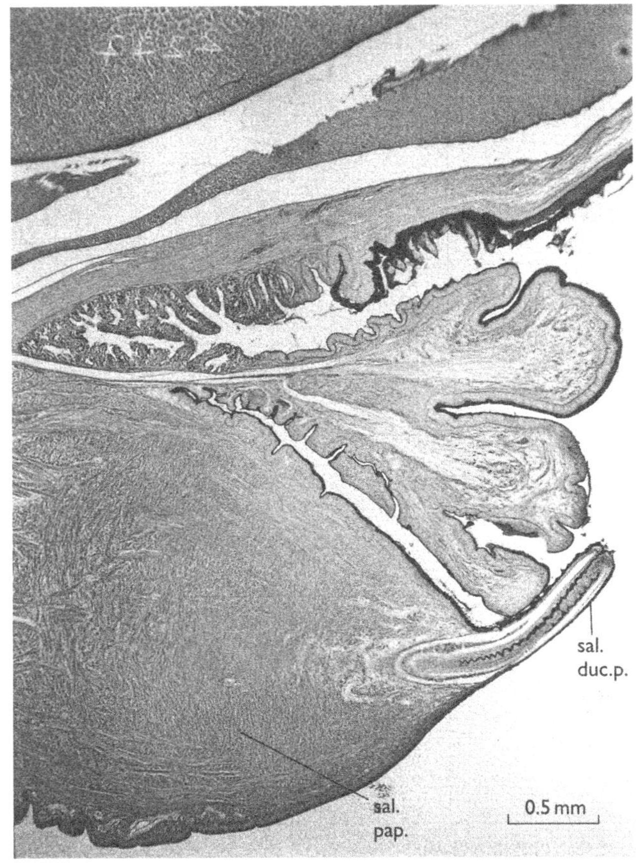

Fig. 31.14. *Octopus vulgaris.* Sagittal section of the posterior salivary gland duct and its papilla, and the terminal part of the radula, within a fold of tissue, beyond the bending plane (spirit blue).

glandular tissue; secretions from this may aid dissolution of the radula, so loosening the teeth from the ribbon. In addition there is a protuberance at the end of the radula, perhaps a buffer and an aid in the ejection of teeth from the ribbon (Fig. 31.14). Wear presumably results from the actions of the radula in drilling, as its teeth can act at the surface of the shell and enter the upper portion of the cavities drilled in both mollusc shells and the exoskeleton of crustaceans (Nixon 1979a). However, ablation of the radula at the bending plane results only in the neglect of the more delicate cleaning processes of crabs and fish, and a small reduction in the quantity of food ingested, suggesting other parts of the buccal apparatus are involved in feeding, especially the lower beak (Altman and Nixon 1970).

On either side of the radula, and within each radular support, there is a pair of crescent-shaped, laterally compressed masses, the bolsters. Each bolster consists of poorly staining tissue with loosely organized transverse muscle fibres (Fig. 31.15), and is innervated by large branches of the superior division of the radular nerve (Young 1965b). The bolsters are unattached except laterally at the hind end, and anteriorly below the radula where they are joined to each other via a band of muscle fibres. Within the enclosed space they occupy the two bolsters can change their shape and form but not their volume; they form a muscular-hydrostat (Kier and Smith 1985). As the bolsters elongate, so the radular teeth become erect as they pass over the bending plane and the lateral teeth are splayed laterally (Nixon 1968). As the bolsters shorten so food material is taken towards the oesophagus by the backward-pointing teeth (Messenger and Young 1999).

The lateral buccal palps are paired, one on either side of the radular support system. They have a cuticular covering and their anteromedial surfaces carry backward-pointing teeth. The palps are capable of considerable movement and can sweep forwards to almost enclose the radula before receding to the back of the buccal cavity, presumably aiding the passage of food particles towards the oesophagus (Nixon 1968).

Two pairs of salivary glands, anterior and posterior, lie outside the buccal cavity. The anterior salivary glands are paired and lie on either side of the oesophagus, behind the buccal mass; from each a duct opens on the anteromedial face of each lateral buccal palp, close to the radula. Several substances have been isolated from these glands, including hyaluronidase (Romanini 1952), and this may act as a spreading agent for the toxin secreted by the posterior salivary glands (Nixon and Maconnachie 1988). An unpaired submandibular gland lies close to the floor of the buccal cavity, behind the salivary papilla, and its secretions are expelled through a number of orifices on its ventral surface. The function of this gland and substances secreted by it are currently not known.

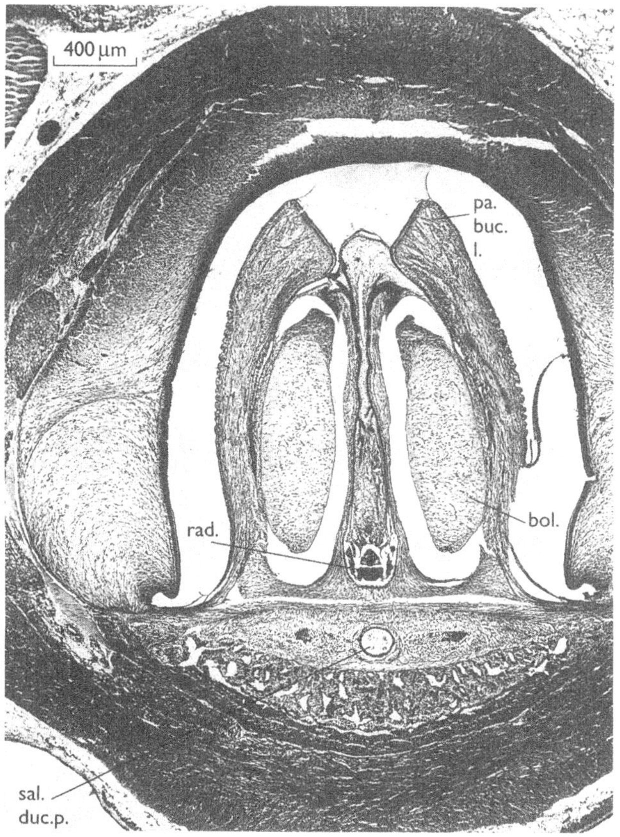

Fig. 31.15. *Octopus vulgaris* (70 mm ML). Horizontal section to show the radula with, on either side, the bolsters and the lateral buccal palps. It also shows the radula, the posterior salivary gland papilla and posterior salivary gland duct.

A pair of large posterior salivary glands lie behind the head and in front of the digestive gland. These glands secrete a number of substances (Russell 1965, Wells 1978) including a crab-paralysing toxin, cephalotoxin (Ghiretti 1960). The secretion is carried anteriorly from the ducts of the posterior salivary glands; the ducts unite and run forwards into the salivary papilla, which lies below the radula. The anterior face of the salivary papilla is covered with cuticle which bears small conical projections. The terminal part of the single duct from the posterior salivary glands can be everted to expose its cuticular covering, which bears tooth-like projections (Nixon 1979b, 1980) (Fig. 31.16). Eversion of the end of the duct is presumably due to the pressure of the secretion as it is expelled from the posterior salivary glands. The toothed tip of the duct may act to ensure that it is retained in position to introduce saliva into the prey. The salivary papilla itself is richly innervated by the two subradular ganglia, from which about ten trunks run forward to branch profusely in the muscles around the tip of the duct (Young 1965b). After the ablation of the anterior part of the salivary papilla in several octopuses they no longer drilled mollusc shells (Nixon 1979a).

Many species of crustaceans and molluscs fall prey to *Octopus vulgaris*, and their carapaces and shells may have small cavities. These are less than 2 mm in diameter at the surface, and slope downwards to a small opening, less than 1 mm in diameter, to penetrate the inner surface of the shell (Fig. 31.17). The size of this hole is usually insufficient for the radula to enter (Nixon 1979a), but the tip of the posterior salivary gland duct can do so (Nixon *et al.* 1980, Nixon and Boyle 1982) (Figs 31.16a, 31.18). Once the shell or carapace has been entered the saliva can be introduced on to, or into, the soft body of the prey. The saliva is toxic to crustaceans, and is presumably so to mollucs, although this has not been confirmed. It contains a substance or

(a) 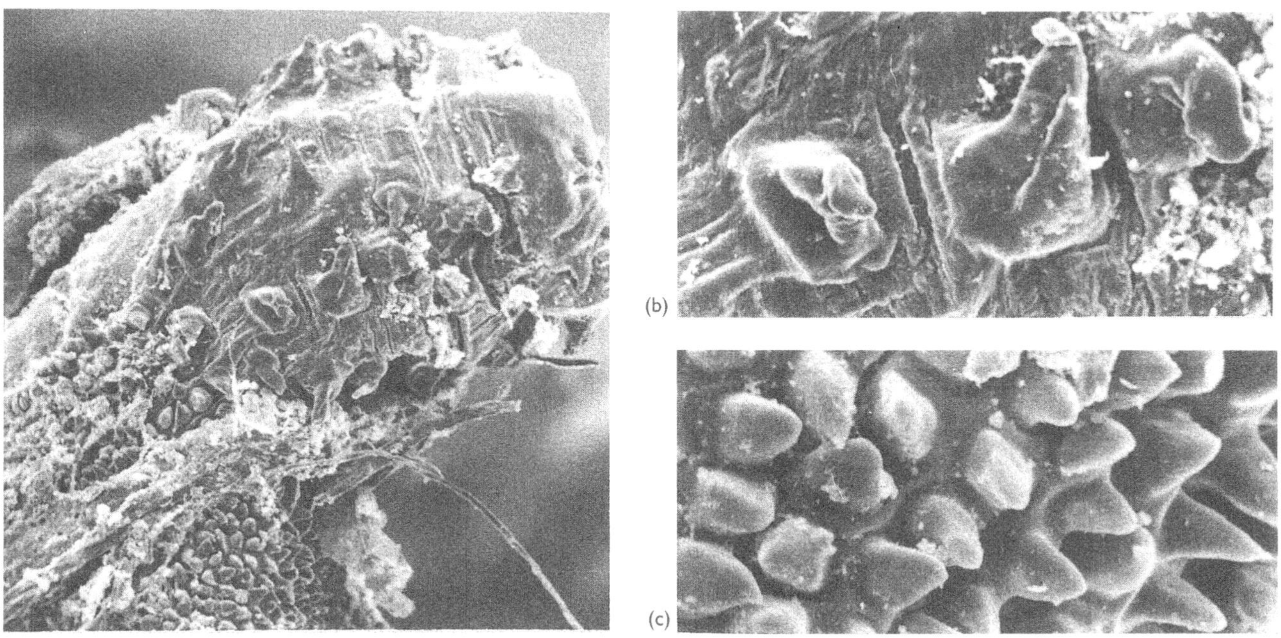
(b) (c)

Fig. 31.16. *Octopus vulgaris* (80 mm ML). Scanning electron micrograph of the salivary papilla and the posterior salivary gland duct: (a) The duct is everted so that it protrudes from the salivary papilla. (b) Part of the tip of the duct showing the teeth at higher magnification. (c) The small conical teeth on the face of the salivary papilla (Nixon 1979b). Field widths (a) 470 µm, (b) 190 µm, (c) 95 µm.

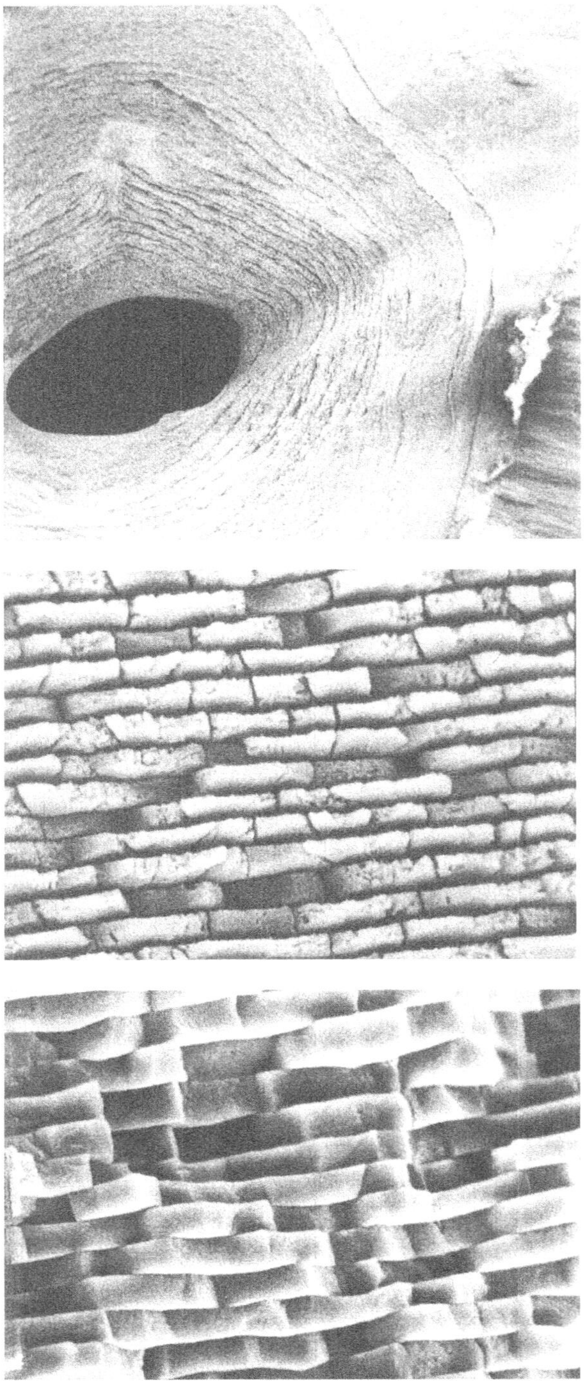

Fig. 31.17. *Octopus vulgaris*. Scanning electron micrographs of (a) a cavity drilled in *Mytilus provincialis* (field width 830 μm), and (b) the same cavity, at much higher magnification, to show the extent of the erosion resulting from drilling when compared with (c) showing a fracture surface from the same shell (both at the same magnification; field width 17 μm). The last micrograph shows the structural organization and the sharp edges of the lamellae, all in marked contrast with those of the surface after drilling (Nixon and Maconnachie 1988).

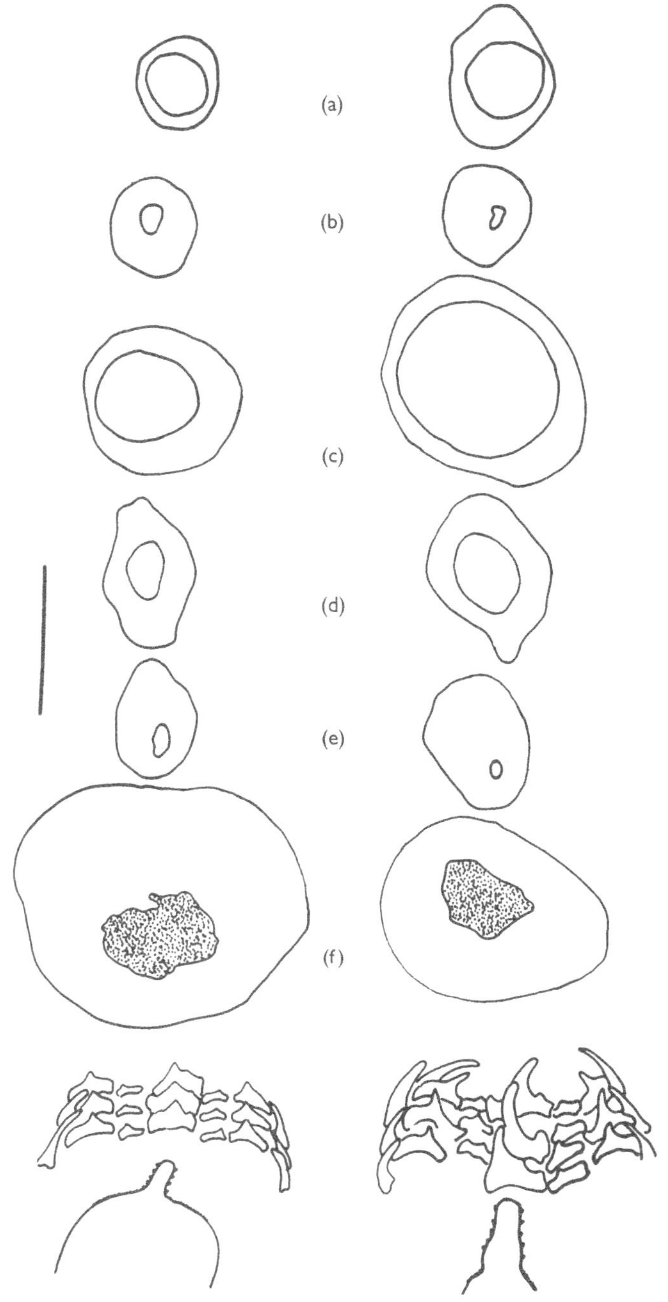

Fig. 31.18. *Octopus vulgaris*. Outlines of the upper and lower surfaces of holes drilled in six species of mollusc by two octopuses (left 200 g, and right 230 g), whose radulae and salivary papillae are shown below at the same magnification (scale bar 1 mm). The shape and form of the drill holes is dependent largely upon the structural features and mineral content of the shell: (a) *Aporrhais pes-pelecani* (A, XL), (b) *Euthria cornea* (A, XL), (c) *Phyllonotus trunculus* (A, XL), (d) *Gibbula magus* (A, P), (e) *Mytilus galloprovincialis* (A + C, P), (f) *Ostrea edulis* (C, P). (Nixon and Maconnachie 1988). Type of shell structure: A, aragonite; C, calcite; P, prismatic; XL, crossed lamellar.

substances which act at the sites of attachment of the muscle to the shell of the mollusc or crustacean exoskeleton: the soft body falls away to leave the shell devoid of adhering tissue. The saliva also acts on the mineral constituent and the organic matrix of the shell (Fig. 31.17b, c) (Nixon *et al.* 1980).

The shell and its properties determine the size, shape, and position of drill holes, and these differ quite markedly between species of gastropod and bivalve molluscs (Figs 31.18, 31.19). This is due at least in part to both the mineral and the organic components of the shell, and to its structural organization (Nixon *et al.* 1980, Nixon and

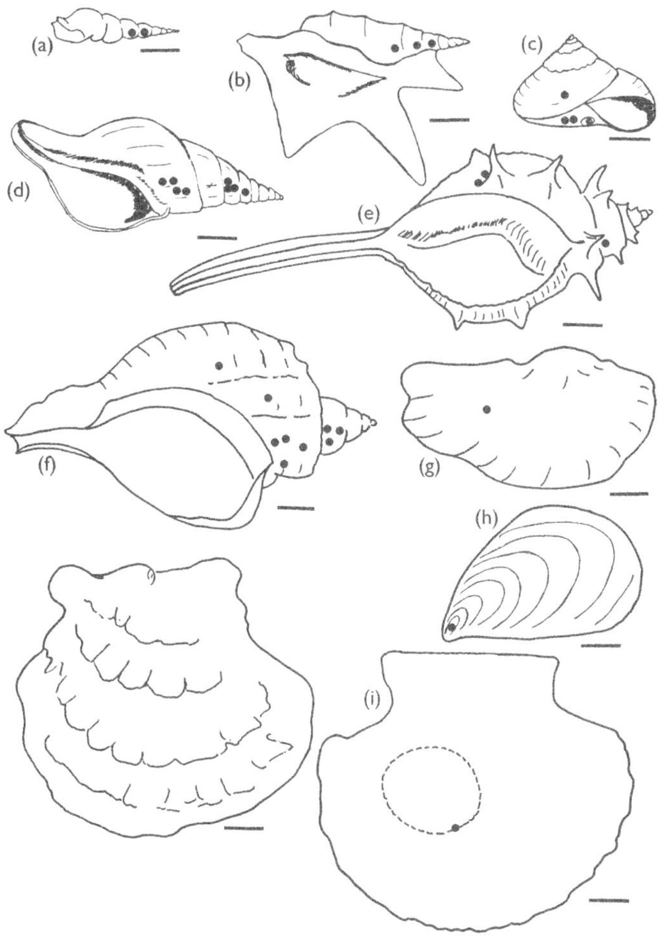

Fig. 31.19. *Octopus vulgaris.* The sites of holes drilled in molluscs: (a) *Bittium reticulatum*, (b) *Aporrhais pes-pelecani*, (c) *Gibbula magus*, (d) *Euthria cornea*, (e) *Bolinus brandaris*, (f) *Phyllonotus trunculus*, (g) *Arca barbata*, (h) *Mytilus galloprovincialis*, (i) *Pecten jacobeus*, (j) *Ostrea edulis* (scale bars 10 mm) (Nixon and Maconnachie 1988).

Maconnachie 1988). The breakdown of the more superficial part of the shell is due, presumably, to the combined actions of the salivary secretions and the teeth of the radula, since many exhibit damage (Nixon 1969). The toothed area of the salivary papilla, the teeth on the tip of the duct, the radula, and the secretion from the posterior salivary glands may be all involved in the mechanical and chemical breakdown of the organic matrix and the mineral part of mollusc shells or crustacean exoskeletons that takes place during drilling. The effects of the secretion are visible only at high magnification, and the face of the drilled surface can be seen to differ markedly from that of a surface after fracture in the scanning electron micrographs of Fig. 31.17 (Nixon and Maconnachie 1988).

In the laboratory the whole process of drilling and feeding on a mollusc may take less than 1 h but this depends both on the type of mineral present and the structural organization of the shell (Nixon and Maconnachie 1988). After ablation of the anterior part of the salivary papilla the octopus no longer drills cavities in shells (Nixon 1979b). Interruption of the nervous pathways by a cut through the middle of the superior buccal lobe renders an octopus incapable of inducing paralysis in crabs (Young 1965b).

Skin, and body patterns

Octopus vulgaris displays various body patterns and to the human eye many changes appear to take place almost instantaneously. The body pattern is the appearance of the whole animal and comprises components of particular behavioural sequences (Packard 1982, Hanlon and Messenger 1996, Messenger 2001). The displays are for concealment, defence, or courtship. While remaining in the same position *O. vulgaris* can change in three stages from a conspicuous warning pattern to one of camouflage (Packard 1995).

To the observer colour is the most conspicuous change to take place in an octopod, but alterations in posture, body form, and the skin are also of great importance in allowing an enormous array of patterns, including some that are fleeting, such as the startle or deimatic pattern, and others which are long lasting (Hanlon and Messenger 1996). The changes that take place in the body patterns are the result of actions in the chromatophores, the reflecting cells, iridophores (or iridocytes), leucophores, and the melanophores, all of which lie in the dermis. In addition the texture of the skin is also modified. Chromatophores contain pigments in a cytoelastic sac (see Fig. 8.17); their colours are straw-yellow, orange, red-brown, and black; most belong to a single biochemical series (Packard and Hochberg 1977, Packard 1988). The leucophores of *O. vulgaris* are much-branched cells, bearing small clubs all over the surface, that contain electron-dense material (Froesch and Messenger 1978). The leucophores and iridophores (Fig. 31.20) of *Enteroctopus (Octopus) dofleini* have been described by Brocco and Cloney (1980) and Cloney and Brocco (1983). Melanophores arise in young *O. vulgaris* as small clear, almost colourless spheres, and in a few days they darken progressively, passing through orange to become deep red, then muddy brown and finally black (Packard and Hochberg 1977).

The chromatic patterns exhibited by *O vulgaris* were described by Cowdry (1911), since when Packard and his colleagues have given detailed accounts of the repertoire and control of body patterns (Packard and Sanders 1969, 1971, Packard and Hochberg 1977, Packard 1988, 1995). They are built up hierarchically from morphological and physiological elements, units, components, and patterns (Hanlon and Messenger 1996) (Table 31.2). The elements comprise chromatophores, leucophores, iridophores, and melanophores and they are grouped into units, the skin patches. The patches, together with surrounding grooves, respond to signals from the nervous system to form dark and light components. Components, when combined in series with like, or unlike, ones, comprise whole patterns (Packard and Hochberg 1977).

The physiological units are largely concerned with the control of the radial muscles of the chromatophores. These muscles receive excitatory innervation, and excitation causes them to contract, allowing the pigment-containing cytoelastic sac to expand. In the absence of excitation the muscles of the chromatophore relax and the sac contracts. Expansion, or contraction, of the sac renders the octopus dark, or pale, respectively. The nerves to the radial muscles arise in the paired anterior and posterior chromatophore lobes of the brain, and nerves run, without synapse through the stellate ganglion (Boycott 1953) to the radial muscles of the chromatophores (see Fig. 8.17) (Froesch 1973a). The chromatophore lobes of the brain

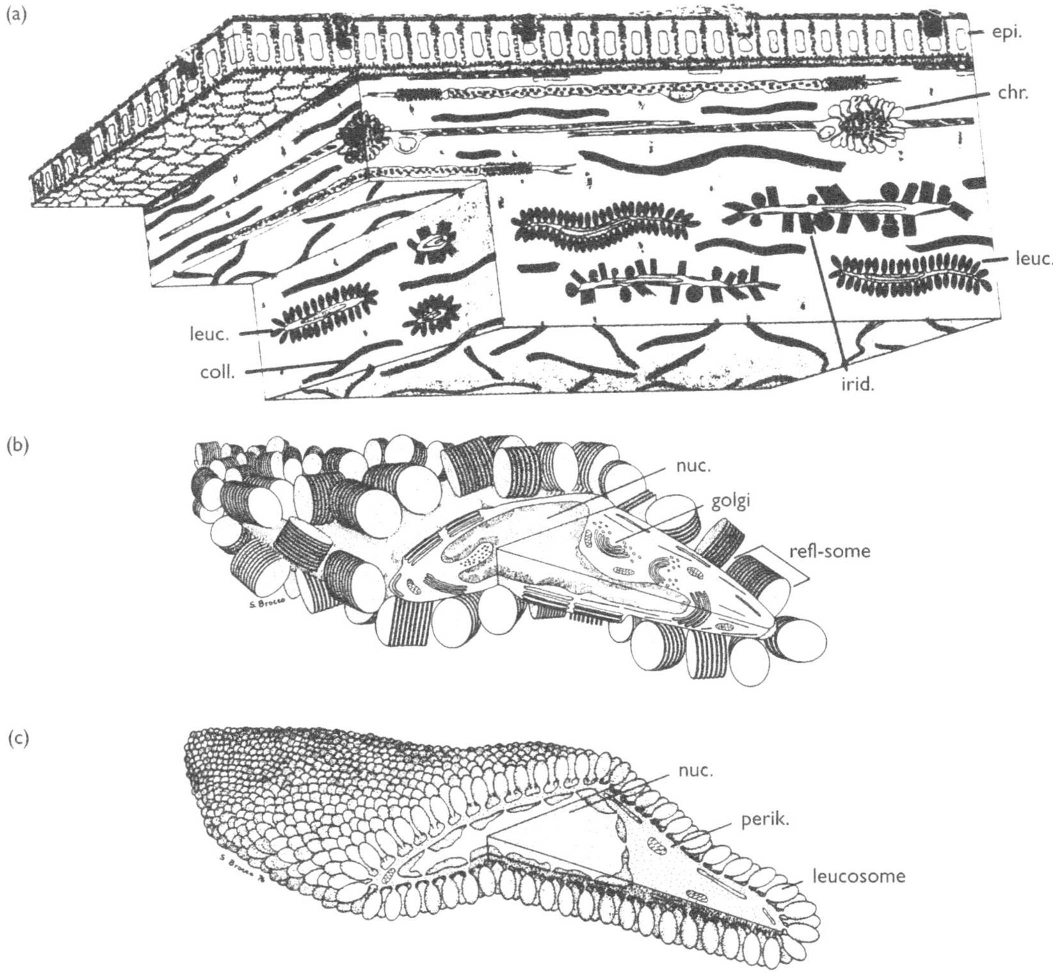

Fig. 31.20. *Octopus dofleini.* Diagrammatic representations from transmission electron micrographs of (a) the epidermis and dermis to show the distribution of chromatophores, reflector cells, and leucophores, (b) an iridophore, (c) a leucophore (b, Brocco and Cloney 1980, a and c, Cloney and Brocco 1983).

receive their principal input from the lateral basal lobes, which in turn receive input from the optic lobes as well as fibres from the median basal and peduncle lobes (Boycott 1953); these lobes form part of the hierarchical organization of the chromatophores (see Fig. 31.29) (Miyan and Messenger 1995). After interruption of the pallial nerve the specific chromatic patterns and skin papillae recover their precise positions within 2 months, showing that the nerve fibres re-make their original connections (Sanders and Young 1974).

The system in the brain concerned with the control of the chromatophores is complex, and comprises over half a million neurons in the chromatophore lobes (Young 1963c). Neurotransmitters, injected in small quantities into the blood supply of the central nervous system, induce almost instantaneous changes in the colour of the skin of the arms, head, and body of *O. vulgaris.* Dopamine, noradrenalin, octopamine, and γ-amino butyric acid all cause darkening of the octopus; acetylcholine and nicotine cause paling; and 5-hydroxytryptamine elicits mottle patterns (Andrews *et al.* 1981, 1983, Messenger 2001).

Besides the great array of body patterns to diminish the chances of a predator encountering the prey, the octopus can respond to detection by threat or unpredictable behaviour. When startled *O. vulgaris* pales, curves its arms, spreads the web, and develops dark edges to the suckers and a dark eye ring. The term 'deimatic' is used to describe this display. If a predator is not deterred by such a display, then the octopus ejects ink; this resembles a dark cloud from behind which it can escape. The ink may also act to block olfactory or taste receptors, as moray eels react to a large cloud of ink by hesitating (MacGinitie and MacGinitie 1968). There is now some biochemical evidence to support this suggestion (Gilly and Lucero 1992, Lucero *et al.* 1994).

Learning

Learning has been defined as 'that process which manifests itself by adaptive changes in individual behaviour as a result of experience' (Thorpe 1956). This capacity has been demonstrated in several cephalopods and is especially well documented from many studies of *O. vulgaris.* In the 1950s and 1960s three long series of experiments were undertaken at the Stazione Zoologica, Naples, with *O. vulgaris.* The first series was to establish a model for the basis of memory, investigating visual discrimination, and the second investigated tactile discrimination; both series also entailed lesions in the brain. The third series examined the visual capacities of this animal.

Table 31.2 The hierarchy of body patterns: levels and definitions (Hanlon and Messenger 1996)

Body patterns	
Chronic	Lasting for minutes or hours
Acute	Lasting seconds or minutes
Components	
Chromatic	Shapes of specific size and position in the skin
Dark	Expanded chromatophores or internal organs visible through the skin
Light	Pale features which may be enhanced by reflecting cells
Textural	Smoothness/papillation of skin
Postural	Various positions of body parts
Locomotion	Typical movements—walking, jetting
Units	
Morphological array of elements in the skin	Patch and groove
Elements	
Chromatophore organs	Pigmentary colours
Reflecting cells	Structural colours
Iridophores	
Reflector cells	
Leucophores	
Internal organs	Ink sac, testis, eyeball, digestive gland
Muscles	For skin texture, posture and locomotion

These studies resulted in many publications and also comprehensive reviews (Boycott and Young 1950, 1955, Young 1958, 1961, 1964, 1965c, 1966, 1983, Wells 1966, 1978, Sutherland 1957, 1958, 1969, Sanders 1975). Recently Robertson and his colleagues have examined the effects of nitric oxide and cytochalasin on visual and tactile discrimination in this octopus (Robertson and Lee 1990, Robertson 1994, Robertson *et al.* 1994, 1995, 1996).

For visual discrimination tests octopuses were maintained in isolation with an artificial 'home' at one end of the tank and objects presented at the opposite end. The animals were trained to take one shape, or object, and rewarded with food, and to reject another because small shocks were given. In this way the reactions to various shapes recorded the capacity of the octopus to learn. Such a training regime set up a memory which was found to last several days. The effects of brain lesions on learning was first investigated in *Sepia* by F.K. Sanders and Young in 1940, and the methods used were expanded for the study of *O. vulgaris*. A period of training was given, after which the octopuses were anaesthetized and a lesion or ablation made to a selected part of the brain. The effects were followed by determining changes in the ability of the animals to learn and retain their learning before and after ablation of different lobes of the brain. The brains of normal octopuses were sectioned histologically, as well as those from animals with lesions, and the pathways concerned in learning determined by correlating the octopuses' performance with intact and damaged parts of the brain. In addition to and in parallel with the discrimination experiments, a detailed description of the anatomy of the brain was compiled by Young (1971).

The appropriate response to a complex environment depends on the ability of an animal to identify objects, features, and situations in its immediate surroundings. The memory system of *O. vulgaris* comprises lobes forming paired centres with a lower long loop and an upper short loop (see Figs 2.7, 2.9). The paired-centres hypothesis is based largely on the anatomy of the brain and is largely compatible with the results obtained from brain lesion studies (Young 1963b). These centres have numerous parallel, interacting channels, and so the various lobes provide sequences of matrices of intersecting axes (see Fig. 2.8) (Young 1991b, 1995a,b). The function of the matrices is to provide for selection of paths that are used and inhibition of those that are not. The nervous system of the octopus provides for the categorization and setting up of memories for both vision and touch. It was found that all of the lobes of the paired centres are involved in tactile learning and constitute an example of distributed memory using a series of networks (Young 1991b), similar to the model suggested by Wells (1978).

Octopus vulgaris readily learns to make various visual discriminations (Boycott and Young 1955). This was further demonstrated in a series of studies by Sutherland (1969) who showed that octopuses have visual abilities comparable to those of vertebrates. Presenting visual figures to the same eye of an octopus and then testing with the opposite eye showed that interocular transfer of learned discrimination occurred, although it was rarely complete, and also that the memory trace is bilateral (Muntz 1961, 1963). However, an analysis of results from some of the early experiments on visual discrimination suggests that they were probably not free from inadvertent cues (Boal 1996).

To investigate the ability to make tactile discriminations, the octopus was first blinded (a simple operation to sever the optic nerves) to ensure that no visual cues of the test objects reached the animal. The animal was trained to accept one object (and rewarded) but not another with a different surface texture. Further sophistication of this procedure included splitting the supraoesophageal brain sagittally, so forming two symmetrical halves (Wells 1978, Young 1983), and later the sides were isolated by a thin barrier of mica (Robertson 1994). Each half-brain is known to control the four arms on its own side and is able to learn the opposite tactile paradigms (rough and smooth) mediated by these arms. After several days of training with a touch paradigm, structural changes were detected in cells in the subfrontal and posterior buccal lobes, and a significant increase in the number of filopodia extending directly from axons was found (Robertson and Lee 1990). It was then postulated that new synapses formed during learning might arise by a similar mechanism to that operating in embryogenesis. Filopodia extension is dependent on actin dynamics and this is profoundly affected by cytochalasin. It was found that cytochalsin injected into the subfrontal and posterior buccal lobes inhibited touch learning. It was then demonstrated that cytochalasin blocked touch learning completely when applied directly to the subfrontal lobe neuropil, but not when applied to the vertical lobe (Robertson 1994). Touch learning was also blocked by nitric oxide synthase (Robertson *et al.* 1994).

Field observations of *O. vulgaris* in Bermuda have demonstrated that its search for prey is mainly chemotactile, and that it has a spatial memory capable of using visual landmarks when foraging in various directions from its den (Mather 1991a, 1995). The use of the same den over a period of time also indicates a spatial memory, as the octopus returns after a foray which may last many hours (Altman 1967, Kayes 1974), probably using chemotactile and visual cues.

The memory mechanism of octopus is probably built-in multiple matrices with intersecting axes and cyclic re-excitation (Young 1995b) that find association between signals and their consequences (Young 1991b). The systems for visual and tactile learning are distinct but have some overlap. In the visual system the impulses from groups of retinal receptors are passed through the optic lobes and then a series of four further matrices of intersecting axons. With appropriate synaptic changes these form associations between conjunctions of signals of visual events and their consequences, producing attack or retreat by the octopus. The centres for touch memory consist primarily of four matrices, but the tactile impulses from the arms also have access to the visual centres. Injury to any one of the centres reduces the accuracy of learned tactile discriminations (Young 1995a).

Development of central nervous system in the late embryo, hatchling, and posthatchling

Marquis (1989) investigated gangliogenesis, and lobular and cellular differentiation, in *O. vulgaris*. During embryonic development the central nervous system arises as distinct groups of cells separating from the ectoderm. These groups of cells form the optic, cerebral, pedal, and visceral 'primordial ganglia' and the arm nerve cords (Fig. 31.21). Later the cerebral ganglia fuse and then differentiate into the various supra- and suboesophageal lobes, whereas the brachial lobe separates from the pedal lobe. The connectives and commissures are all secondarily formed by the ganglia. The origin of the pedal ganglia is of special interest in relation to the homology of the arms. A translation of Marquis's statement is that 'The front part of the paired suboesophageal ends of the brain complex has its origin in the region (*Bereich*) of the funnel rudiment behind the ventral arms'. Thus the pedal ganglia lie immediately behind the rudiments of the arms and above that of the funnel. The cells then proliferate and move upwards as a thin strand to join the brachial ganglia. This account does not support the suggestion that the anterior suboesophageal mass may be of cerebral origin (Young 1971). The nerve cords of the arms themselves develop early and independently. They are not at first related to the rudiment of the pedal ganglia and are not close to those of the funnel. So this careful study favours the classical view that the arms are related to the funnel, as has long been supposed.

The magnocellular lobe is considered by Marquis (1989) to be 'ein mit zell rinde eingekleidetes Konnektiv' and probably at least partially derived from the visceral ganglia. This author emphasizes that the nervous system of *O. vulgaris* clearly passes through a ganglionic stage of development, but admits that further embryological evidence is needed if the visceral ganglia of cephalopods are to be homologized with the pleural and abdominal ganglia of gastropod molluscs.

At hatching the major parts of the central nervous system are present although some only as Anlagen; the sub- and supraoesophageal centres, as well as the optic lobes, have already formed a central brain, although not all of the peripheral ganglia have reached the same stage of development (Marquis 1989). The suboesophageal mass consists mainly of the pedal and visceral lobes as the brachial lobes, lying anterior to the pedal lobe, are relatively small at this stage (Table 2.3) (Frösch 1971). The chromatophore lobes are undifferentiated and all of the basal lobes are poorly developed (Hanlon and Messenger 1988). The stellate ganglia are well developed at hatching, as are ganglia associated with the buccal complex. The visceral ganglia are poorly developed or even lacking at the time of hatching. The sucker ganglia are recognizable as cell masses but lack fibres and are not yet connected with the brachial ganglia (Marquis 1989).

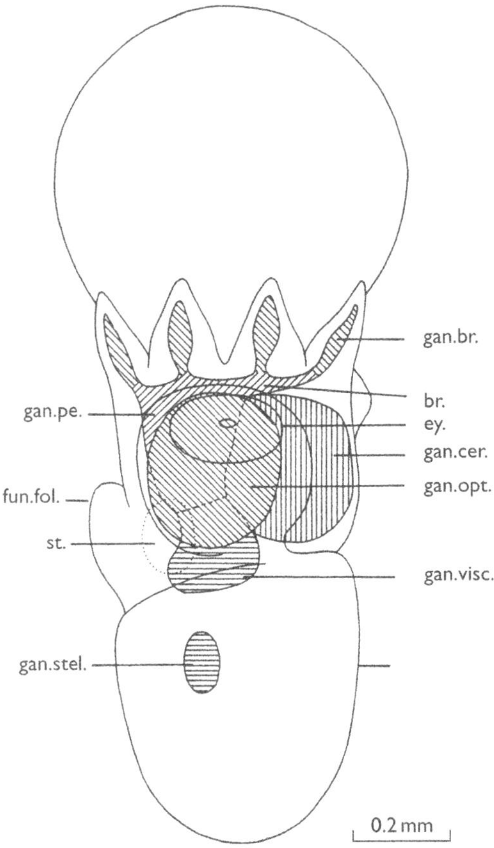

Fig. 31.21. *Octopus vulgaris.* Diagram showing the developing central nervous system in an embryo at stage XV of Naef (1928) (Marquis 1989).

The development of the optic lobes, the lens and the retina was followed until some days after hatching in two Australian octopuses, *Octopus australis* and *O. pallidus* (Wentworth and Muntz 1992). These two species both adopt benthic habits at hatching, and so are not strictly comparable with *O. vulgaris* since its posthatchling enters the plankton (Table 31.1). The inner part of the lens begins to form at about stage IX (Naef 1928); at this time the retina is still quite undifferentiated and the optic lobes have not begun to form. The outer part of the lens develops later. By hatching all of the relevant parts of the visual system are recognizable and essentially like the adult although growth continues. The relative length of the rhabdoms increases strikingly in the first 15 days after hatching. The optic lobes increase markedly in complexity in the adult, although

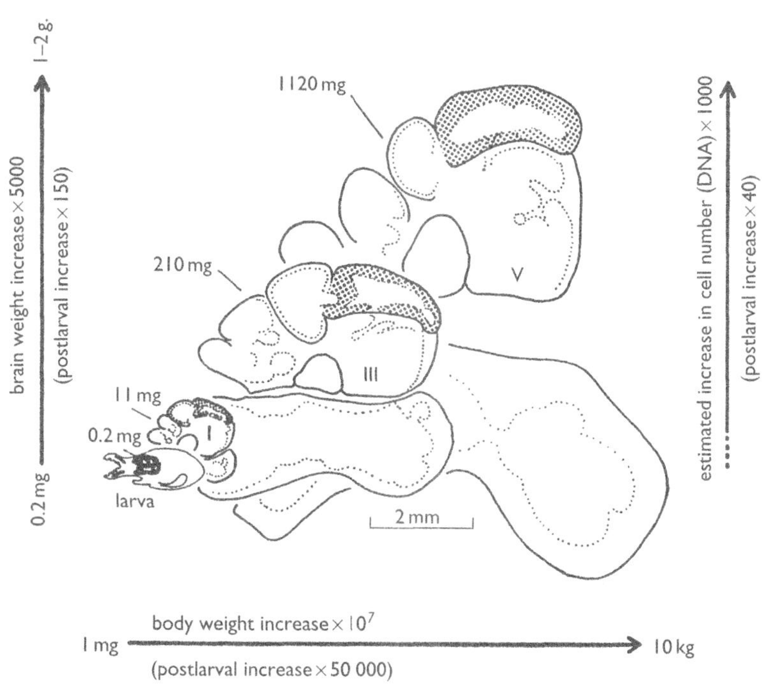

Fig. 31.22. *Octopus vulgaris*. Superimposed outlines of sagittal sections of brains from three body-weight groups, with brain weights shown in milligrams. At the left is a sagittal section of a specimen at hatching, with the brain shown in black. The approximate body weights are: I, 0.3 g; III, 40 g; V, 5 kg. The cleft running across each brain indicates the path of the oesophagus. The stippled areas indicate the cell body layer ('cortex') of the vertical lobe. The optic lobes are not shown (Packard and Albergoni 1970).

Wentworth and Muntz noted that their structure is much simpler in the plexiform zones than in *O. vulgaris* (Young 1962a).

The brain of *O. vulgaris* grows throughout its life (Fig. 31.22) and increases from 0.20 mg at hatching to 1120 mg in an adult (Packard and Albergoni 1970). Likewise the number of cells increase from 2×10^5 to 2×10^8 respectively (Giuditta *et al.* 1971). Quantitative studies of the volume of the various lobes of the central nervous system at hatching, at settlement and in the adult show that changes in proportions take place in some of the lobes (Wirz 1959, Frösch 1971, Nixon and Mangold 1996). At hatching the buccal lobes and the basal lobes are relatively large in comparison with those in newly settled and juvenile animals. The brachial lobes in contrast are small initially, and relatively much larger by the time of settlement. The vertical, subvertical, and subfrontal lobes are very small at hatching when compared with those in settled and juvenile specimens. The paralarval phase is spent in the plankton and the proportions of several lobes of the brain alter markedly by the time of settlement on the substrate for the juvenile and adult stages of life (Fig. 31.23). The relative increase in the proportion of the brachial lobe is reflected in the marked lengthening and development of the arms and increase in the number of suckers around the time of settlement. The basal lobes, concerned with the control of complex movements, are large at hatching, a time when the young octopus is living and feeding in the water column; these lobes become proportionally smaller after the animal has settled.

The changes in morphology and behaviour that take place between hatching and the time of settlement to the substrate are mirrored in the central nervous system of *O. vulgaris* (Nixon and Mangold 1996). Newly hatched octopuses move by jet propulsion. The life of the paralarva is active in pursuit of prey in order to fulfil its growth and metabolic requirements, and is one of almost continuous adjustment to maintain itself in the water column. This is reflected in the proportionally large anterior basal and peduncle lobes at this stage (Table 2.3). The former lobe is concerned with the control of posture and movement of the head and eyes in swimming (Young 1971), and the latter provide a system that controls locomotion based on visual cues (Messenger 1967b, 1971). These lobes receive visual and static inputs and constitute a higher motor centre, analogous to the vertebrate cerebellum (Hobbs and Young 1973). The buccal lobe is a motor centre that receives information from all of the organs concerned with the capture of prey and feeding (Young 1965b), and it is relatively much larger in the newly hatched than in the newly settled or juvenile octopus. This emphasizes the need for the paralarva to detect and capture prey to supply its metabolic requirements for locomotion and growth.

The brachial lobe, a lower motor centre for the control of the actions of the arms, is poorly developed at hatching. At this time the arms are short with three suckers on each and slender tips, which lack suckers (Fig. 31.3). The brachial lobe forms 8% of the total volume of the brain at hatching but in an early settled young it is 13%, and reaches almost 18% in an adult (Fig. 31.23) (Nixon and Mangold 1996). The subfrontal lobe, important in tactile learning (Wells and Young 1975), also increases in proportion to the rest of the brain between hatching and settlement, and so anticipates the development of the arms and suckers. The vertical and subvertical lobes, part of the tactile and visual learning system (Young 1971), also increase in volume at this time. By the early benthonic stage the learning centres are close to the condition found in juvenile animals. The chromatophore lobes, although not differentiated at hatching (Hanlon

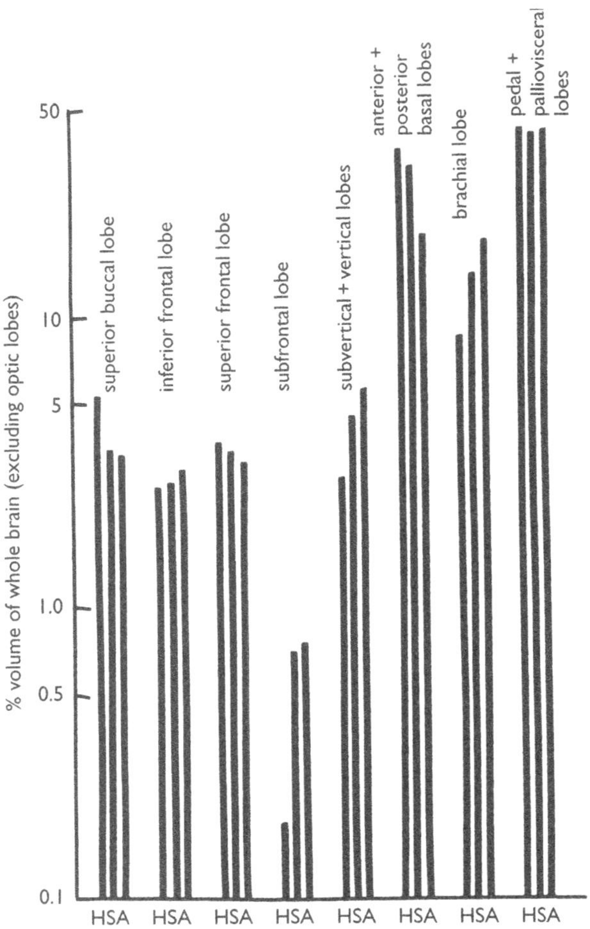

Fig. 31.23. *Octopus vulgaris.* Various lobes of the brain shown as percentages of the volume of the central nervous system, excluding the optic lobe, plotted on a log scale. H, newly hatched (Frösch 1971); S, settled (Nixon and Mangold 1996); A, 'adult' (Wirz 1959).

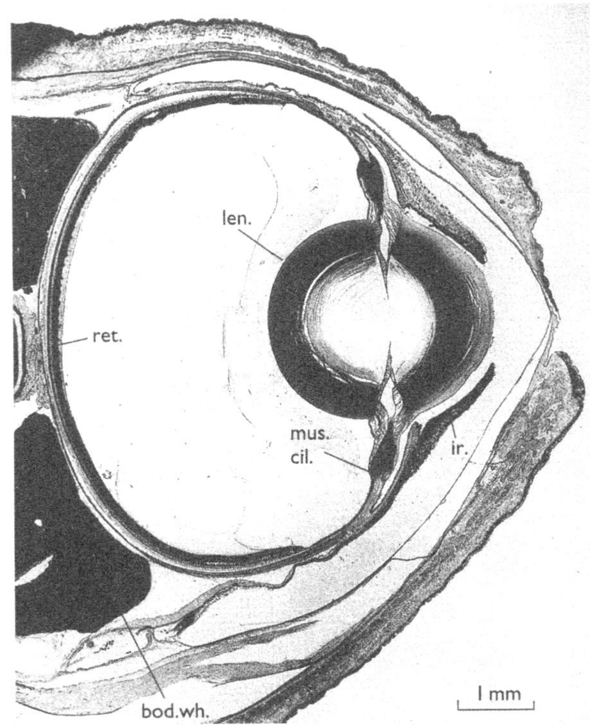

Fig. 31.24. *Octopus vulgaris.* Transverse section through the eye (Borax carmine stain).

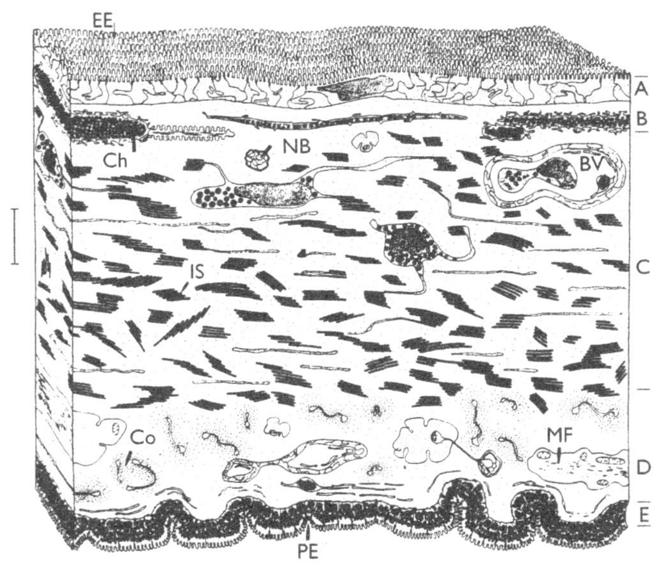

Fig. 31.25. *Octopus vulgaris.* The iris (500 μm in depth). A diagram drawn from transmission electron micrographs to show the five layers of the iris (A–E) (Froesch 1973a) (scale bar 100 μm) (see also Fig. 31.20). BV, blood vessel; Ch, chromatophore; Co, collagen; EE, external epithelium; IS, iridophores; MF, muscle fibres; NB, nerve bundle; PE, pigment epithelium.

and Messenger 1988) form 5% of the total brain volume in an octopus of 80 mm ML (Maddock and Young 1987). Packard (1985) commented that 'The dorsal spurt in chromatophore genesis at the end of the planktonic phase is so dramatic as to hint at something like metamorphosis', at which time the octopus displays a wide variety of patterns of concealment (Packard and Sanders 1969, Packard 1988).

Eyes

Cuvier (1817) described and illustrated the eye of *Octopus vulgaris.* The eyes are sited laterally and form a considerable portion of the head. They lie in orbits, partly protected by cartilage. Each is spherical with an iris, a lens, and a retina, behind which lies a choroid, a sclera, an external argentea, and a plexus of intrinisic eye muscles (Fig. 31.24). Vision is well developed in this species as various visual discriminations are readily learned (Boycott and Young 1955), and in this the abilities of the octopus are comparable with those of vertebrates (Sutherland 1957, 1969).

An iris covers the surface of the lens, and comprises five layers: an external epithelium, a chromatophore layer, an iridophore layer, a layer of muscles and collagen fibres, and a pigment epithelium (Fig. 31.25). The iris is a fast-moving diaphragm with light-reflecting and light-absorbing parts and a sphincter. The latter is innervated by a nerve entering the orbit together with the optic stalk (Froesch 1973b).

The lens is a relatively firm structure and is formed of α-, β-, γ-, and δ-crystallin (Bon *et al.* 1967). Structurally it takes the form of a stack of planoconvex and subspherical onion-like layers (Fig. 31.24).

The lens is suspended by ciliary muscles which allow for some accommodation for near vision. The lens has a focal length approximately 2.5 times the lens radius (Matthiessen's ratio) (Pumphrey 1961). In the absence of a pupil light passes through the lens and comes to focus on the retina and form a sharp image, whereas light passing through the periphery of the lens would strike the receptors at an angle and so cause blurring (Fig. 31.26a) (Muntz 1977). The receptors are shorter and pigment migration is greater in the peripheral zone, and so the sharpness of the image is preserved.

The retina (Young 1962b, 1971, Tonosaki 1965, Yamamoto *et al.* 1965) comprises the receptor cells (the retinal or retinula cells), the supporting cells, and the retinal glial cells. The retinal cells contain granules of screening pigment which move distally in the light and retract to the base of the outer segment in the dark (Young 1963a). The structure of the retina is simpler than that of the vertebrate eye as only two neural components are present, the receptor cells and efferent fibres. The density of visual cells in the retina is 70 000/mm^2, with a total number of 2.0×10^7 (Young 1962b). There is a horizontal strip of longer and thinner receptor cells that are presumed to form a kind of fovea or sensitive area (Young 1963a).

The receptor cells are primary sensory cells, and their outer segments face the lens (Fig. 31.26b). The receptor cells and rhabdoms are often arranged in a rectangular array lying in the horizontal and vertical planes. The rhabdoms contain subunits, the rhabdomeres, each of which consists of close-packed fine tubules (Messenger 1981). Each photoreceptor cell has an outer (distal) segment, which carries straight microvilli with a parallel arrangement, and an inner (proximal) segment, below the basal lamina, that contains the nucleus. The microvilli contain the photopigment rhodopsin (Brown and Brown 1958) and those of neighbouring cells have an orthogonal arrangement, either horizontally or vertically in a normally oriented eye, associated with the eye's sensitivity to the plane of polarized light (Moody and Parriss 1960, 1961). The presence of a single photopigment and behavioural findings indicate that *O. vulgaris* is colour blind (Messenger 1981, 1991, Budelmann 1994, Hanlon and Messenger 1996). The retinal plexiform layer comprises glial cells, fine collaterals of receptor cells, and tangentially running efferent fibres (Patterson and Silver 1983). Axons of the retinula cells leave the back of the eye and run to the optic lobe after making a complex dorsoventral chiasma (Fig. 2.13), which presumably corrects the inverted retinal image produced by the lens and projects a visual map that is in register with a gravitational map (Young 1962c).

The screening pigment in the retina reduces the amount of light reaching the rhodopsin in the rhabdoms: during light-adaptation the screening pigment migrates distally and vice versa for dark-adaptation (Young 1963a). The retraction of the screening pigment during dark-adaptation is mediated by dopamine released from efferent endings on the photoreceptor cell. The arrival of efferent signals in the retina that effect adaptational changes through the mediation of dopamine is a remarkable analogue to the vertebrate system (Gleadall *et al.* 1993).

Seven extraocular eye muscles move the eyeball of *O. vulgaris* (Budelmann and Young 1984). Three powerful recti muscles produce linear movement and four oblique muscles rotation of the eye (Fig. 31.27). Some of the latter muscles are very thin sheets passing halfway round the eyeball. Seven nerves control the eye muscles, and

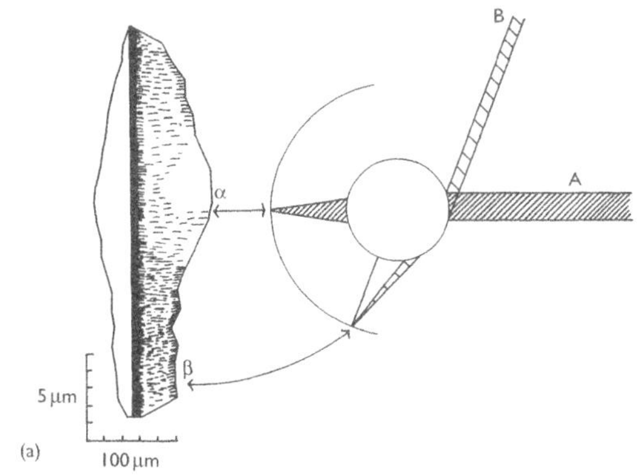

Fig. 31.26. (a) *Octopus vulgaris*. The relationship between the shape of the pupil and the retinal anatomy of the eye. The left half of the figure represents a straightened out vertical section through the centre of a light-adapted retina, and the length of the basal segments and the rhabdoms at different distances along the retina. The right half of the figure represents a vertical section through the eye with a constricted pupil to show how light reaches the retina (Muntz 1977).

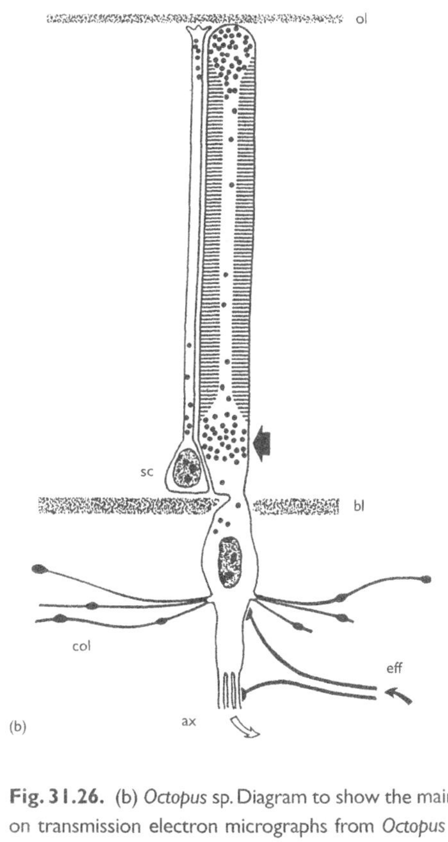

Fig. 31.26. (b) *Octopus* sp. Diagram to show the main parts of the retina based on transmission electron micrographs from *Octopus vulgaris* and *Amphioctopus (Octopus) fangsiao (ocellatus)* (Gleadall *et al.* 1993). ax, axon-like processes of the photoreceptor; bl, basement lamina; col, collaterals; eff, efferent fibres arising in the optic lobe end in synapses with the photoreceptor; ol, outer lamina; sc, supporting cell; filled curved arrow, path of efferents; large filled arrow, layer of proximal screening pigment which does not migrate; open curved arrow, path of afferents to the optic lobe.

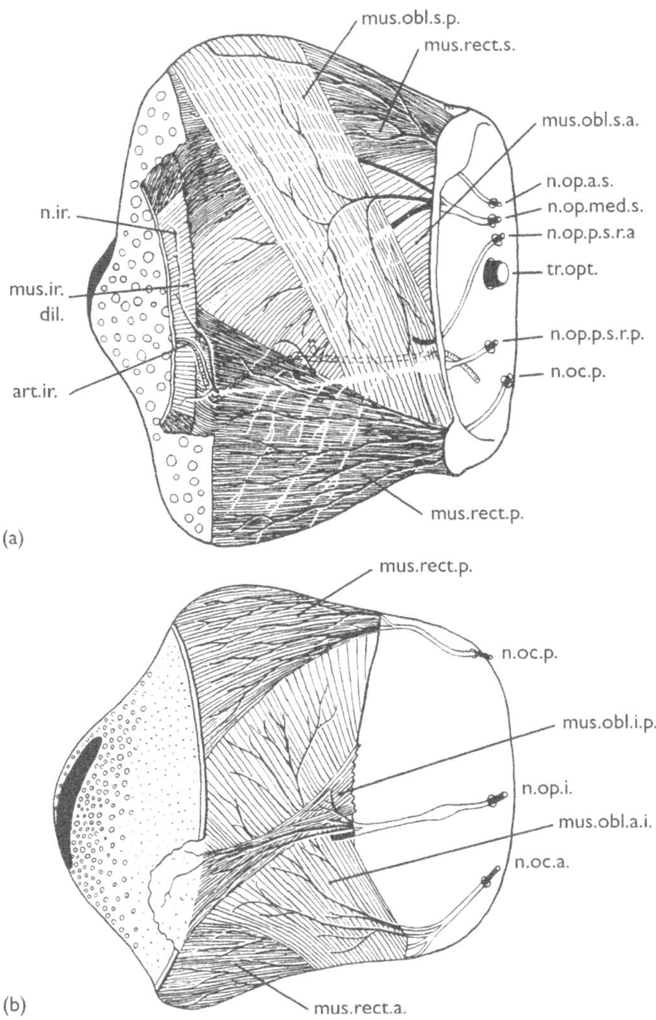

Fig. 31.27. *Octopus vulgaris.* The arrangement of the extraocular eye muscles of the dorsal (a) and the ventral (b) part of the left eye. The course of the nerves within the muscles are shown diagrammatically. The nerves are shown in white when passing underneath the outermost muscle layers (Budelmann and Young 1984).

several innervate more than one muscle. Stimulation of individual nerves results in the linear or rotatory movements, or both, to be expected from the morphological organization of the muscles they innervate (Table 31.3). Two of the nerves run only to extraocular eye muscles, the other five contain additional fibres for the iris, chromatophores, or skin. Cobalt filling of the central ends of the eye muscle nerves indicate that all have fibres originating in the anterior lateral pedal lobe, which is thus established as the oculomotor centre (see Fig. 31.33). Stimulation of two nerves resulted in expansion of the chromatophores of the iris; these nerves contain fibres with somata in the ipsilateral anterior chromatophore lobe. Two nerves gave constriction of the pupil and these have fibres with somata in an area between the posterior pedal and the magnocellular lobes, thus demonstrating a pupillary control centre. Stimulation of one nerve resulted in dilation of the pupil, but the origin of its cells remains unclear. The eye muscles produce voluntary and pronounced compensatory eye movements.

Table 31.3. *Octopus vulgaris.* The seven nerves that contain fibres for the extraocular eye muscles, the muscles they innervate, and the movements produced by stimulating the nerves (Budelmann and Young 1984)

Nerves	Muscles	Main movements of the eyes
Anterior oculomotor	Anterior rectus	Forwards–downwards
	Anterior inferior oblique	Rotation either way
Posterior oculomotor	Posterior rectus	Backwards–upwards, no rotation
Anterior superior ophthalmic	Superior rectus	Forwards–upwards
	Posterior superior oblique	Clockwise rotation
Medial superior ophthalmic	Superior rectus	Forwards–upwards
	Anterior superior oblique	Rotation either way
	Posterior superior oblique	
Anterior root of posterior superior ophthalmic	Anterior superior oblique	Forwards–upwards, anticlockwise rotation
Posterior root of posterior superior ophthalmic	Anterior superior oblique	Forwards–upwards, anticlockwise rotation
Inferior ophthalmic	Posterior inferior oblique	Forwards–downwards, anticlockwise rotation

Octopus vulgaris can discriminate between shapes on the basis of size, brightness, form, horizontal/vertical orientation, and the plane of polarization (Messenger 1979b). It can distinguish between targets on the basis of the presence or absence of a pattern produced by a polarization contrast pattern within a single visual target (Shashar and Cronin 1996). However, the octopus cannot learn to discriminate on the basis of hue, although a variety of training techniques have been employed to investigate colour vision in this animal (Messenger 1981, 1991, Hanlon and Messenger 1996).

Optic lobes

The visual information from the retina passes via the optic nerves to the optic lobes (see Fig. 2.13). The optic nerves consist of numerous bundles of fine fibres proceeding from the back of the eyeball to the surface of the optic lobe. They form chiasmata, the bundles interweaving so that fibres from the dorsal part of the retina enter the optic lobe ventrally and vice versa (Young 1962a,b). The central fibres projecting to the middle region of the optic lobe are shorter and the more peripheral ones longer.

The optic lobes of *O. vulgaris* have been described in detail at the light microscope and ultrastructural level (Young 1962a, 1971, Dilly *et al.* 1963) and their development in *O. australis* and *O. pallidus* by Wentworth and Muntz (1992) (see 'Development of central nervous system' in this chapter). The lobes lie on either side of the brain and are separated from it by optic tracts (see Fig. 2.1b). Each optic lobe lies behind the eye in an orbit bounded by cartilage and a tough fibrous membrane. The lobe is bean-shaped and is cushioned by the white body, a haemopoietic organ (Budelmann *et al.* 1997).

The optic lobes are the largest lobes of the central nervous system and together form 85% of the total brain volume (Maddock and Young

1987). Some 64 million cells, with nuclei of various diameters, are present in one optic lobe of an octopus of about 500 g (Young 1963c, 1971). The optic lobes, like other lobes in the brain, have an outer cortex and a central medulla, but their organization is the most elaborate of all (see Fig. 2.11). The cortex of the optic lobe has outer and inner granule cell layers, separated by a plexiform layer of neuropil (Young 1962a). The image from the eye is projected on to the plexiform layer, where the first-order neurons meet branches of at least four types of cells. From the plexiform layer second-order neurons run towards the centre of the lobe and form parallel bundles; these are surrounded by different types of nerve cells, including amacrine cells. The organization is columnar and the large cells with processes spreading tangentially presumably ensure interaction between the columns. Further centrally in the lobe there are fairly regularly arranged islands of nerve cells and many scattered small cells. The left and right optic lobes are directly connected via dorsal and ventral optic commissures (Young 1971).

The optic lobes receive afferent inputs from many regions as well as from the eyes. Fibres proceeding peripherally in the lobe can be seen to end in all parts including the plexiform zone (Young 1971). The connections of the optic lobes are shown in Table 31.4.

The optic lobes, with other lobes, form the site of the visual memory store (Young 1965c, 1991b), although the complex connections in the optic lobes have yet to be elucidated. The laying down of memories involves a series of lobes, notably the vertical and superior frontal lobes, which act in some way to balance the tendency to attack or retreat in the face of a given visual stimulus. Efferent pathways from the optic lobes control the action patterns of the octopus. Control of the body musculature is via the suboesophageal motor centres of the peduncle and basal lobes with their systems of fine parallel fibres. Control of the chromatophores is through a separate tract to the lateral basal lobes. The magnocellular lobe provides a direct pathway that probably initiates escape responses, and also provides the pathway for the regulation of the diameter of the pupil under direct control of the amount of light or the approach of a potential adversary.

Using a radio-autographic technique, physiological activity was examined in the optic lobes of unilaterally blinded octopuses (Novicki *et al.* 1992). The optic lobes so deprived of visual input took up less of the labelled tracer, presumably due to less metabolic activity than in the contralateral non-deprived lobes. The remaining neural activity in a deprived optic lobe could derive from non-visual inputs from various lobes in the central brain (Young 1971), or from the contralateral optic lobe as there is transfer of visual information between the two sides of the brain via the dorsal and ventral optic commissures (Muntz 1961, 1963).

Table 31.4 Classification of the connections of the optic lobes

Afferent pathways

A *From the periphery*
- 1 Retinal nerve fibres
- 2 Magnocellular to optic lobe tract (? fibres from pallial nerves)
- 3 Inferior frontal to optic lobe tract (? fibres from lips and arms)

B *From higher motor centres*
- 4 Peduncle to optic lobe tract
- 5 Anterior basal to optic lobe tract
- 6 Median basal to optic lobe tract
- 7 Lateral superior frontal to optic lobe tract
- 8 Dorsal basal to optic lobe tract

C *From the vertical lobe system*
- 9 Subvertical to optic lobe tract

D *Commissural fibres*
- 10 Ventral optic commissure
- 11 Dorsal optic commissure

Efferent fibres

A *To the retina*
- 1 Centrifugal fibres (originating in the outer plexiform layer)

B *To higher motor centres*
- 2 Optic to peduncle lobe tract
- 3 Optic to anterior basal lobe tract
- 4 Optic to median basal lobe tract
- 5 Optic to dorsal basal lobe tract

C *To the vertical lobe system*
- 6 Optic to superior frontal lobe tract

D *To other motor centres*
- 7 Optic to magnocellular lobe tract
- 8 Optic to lateral basal lobe tract

E *Commissural fibres*
- 9 Ventral optic commissure
- 10 Dorsal optic commissure

Central nervous system

The central nervous system was described by Young (1971) in his book *The anatomy of the nervous system of* Octopus vulgaris, the first book to be devoted solely to the brain of a cephalopod, which provides a detailed account of the subject. Later findings have been included here.

The central nervous system of *O. vulgaris* (Fig. 2.4) is protected and supported by a cartilaginous cranium. The space between the nervous tissue and the cranium is filled by a translucent, jelly-like matrix with embedded star-shaped cells (Fig. 31.28). The lateral walls of the cephalic cartilage are connected by four transverse bands of intracranial muscles, which are richly innervated. A tough, fibrous, cerebral membrane surrounds the central nervous system. The membrane is continuous with the epineurium of the nerves and outer walls of the blood vessels reaching the brain. It forms an attachment to the cartilage where the nerves and arteries leave the brain. On either side of the cranium the eyes and the optic lobes lie in orbits with membranous walls. The eye muscles are attached to the cranium (Budelmann and Young 1984).

The brain surrounds the oesophagus, which divides it into supra- and suboesophageal masses (Fig. 2.1). These are joined laterally by the cerebrobrachial and posterior cerebral connectives. Behind the anterior connective is a space through which the pharyngo-ophthalmic vein and a band of intracranial muscles pass (Young 1971).

In *O. vulgaris* and other octopods the various parts of the brain are brought close together: the superior buccal ganglion is united with the inferior frontal lobe system, and the brachial and pedal lobes are not widely separated as they are in *Loligo* (see Figs 2.2, 2.5). The neuropils of many of the lobes are continuous, making it difficult to define precise boundaries. Such condensation and continuity of the

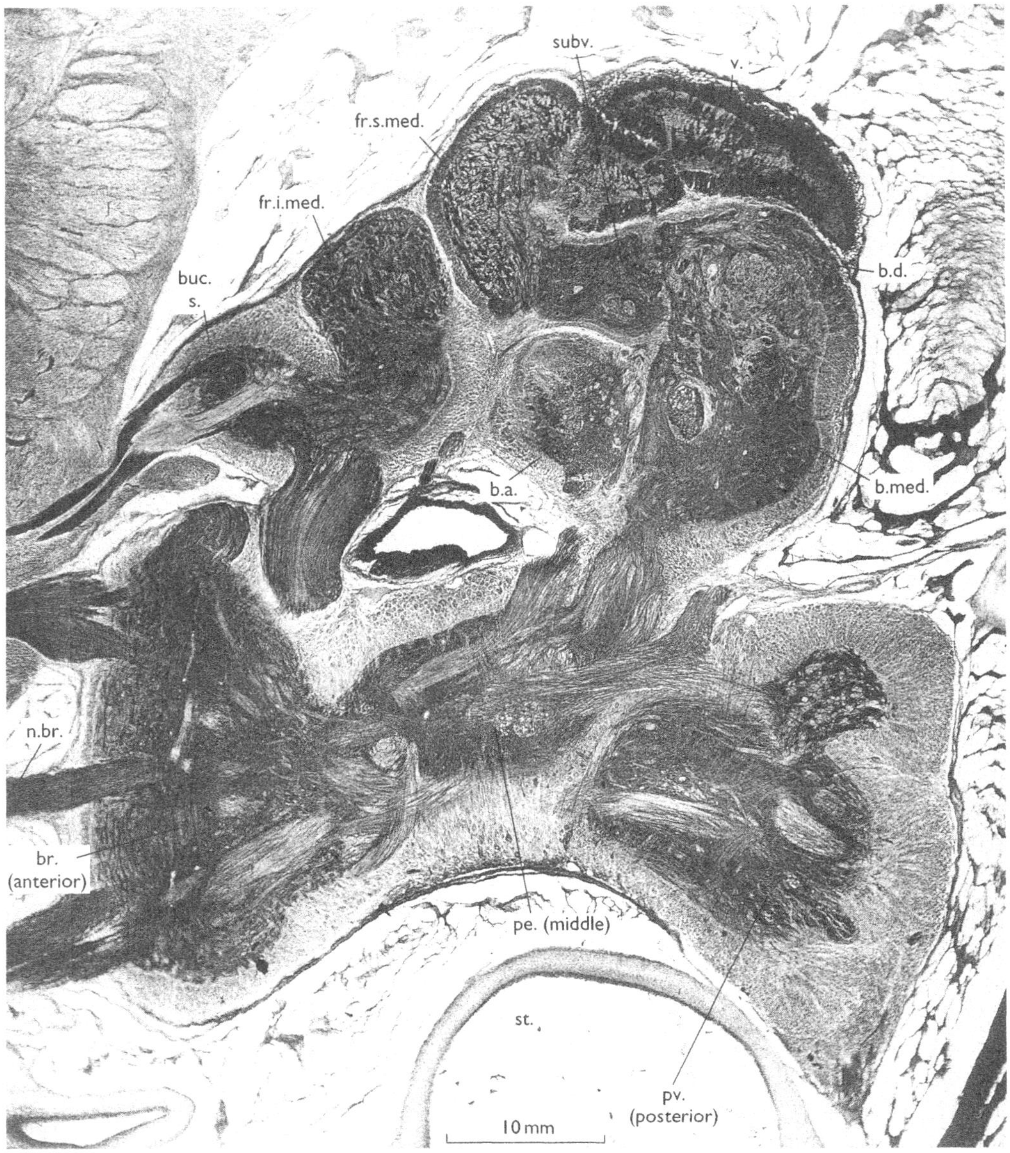

Fig. 31.28. *Octopus vulgaris*. A sagittal section to show the sub- and suboesophageal masses of the central nervous system, and the statocyst.

lobes allows for functional integration, and this unity provides the capacity for the characteristically complex behaviour patterns demonstrated in this cephalopod.

Suboesophageal mass

The suboesophageal mass (Fig. 31.28) is formed by anterior, middle, and posterior regions, each of which is further divided. The anterior region has two parts, and the middle and posterior regions each have five recognizable parts.

Anterior suboesophageal mass

The most anterior part of the brain is the anterior suboesophageal mass (Fig. 31.28), and from it arise the eight brachial nerves (although many of their fibres pass through it to and from other centres). Dorsally it is related to the perioesophageal sinus, the cerebrobrachial connectives enter laterally from above, and posteriorly it is continuous with the middle suboesophageal mass. On either side a brachiopalliovisceral connective carries fibres to and from the posterior suboesophageal mass. The anterior suboesophageal mass receives afferent fibres from the arms, other suboesophageal centres, as well as the inferior frontal lobe and the posterior buccal lobes. It sends fibres to the brachial nerves. A few other pathways to the supraoesophageal mass have been demonstrated (Plän 1987). The anterior suboesophageal mass consists of two parts, an anterior, prebrachial lobe, and a posterior, postbrachial lobe.

The prebrachial lobe is a ring of neural tissue encircling the oesophagus. Above the oesophagus is the suprabrachial commissure. The lateral parts of the prebrachial lobe comprise a neuropil with cells on the lateral, anterior, and medial surfaces. The neuropil of the prebrachial lobe is continuous with that of the postbrachial lobe, but has three transverse strands of cells on each side. These strands lie between the first and second, second and third, and third and fourth brachial nerves. Ventrally the prebrachial ring is completed by the infrabrachial commissure. The prebrachial lobe receives numerous fibres from the brachial nerves. The cell wall of the lobe consists of a characteristic mixture of large and medium-sized cells, with few very small ones. The neuropil contains two plexiform arrangements; an afferent plexus is formed by fine fibres entering from the medial roots of the brachial nerves, many of the fibres running in a dorsoventral direction. Other bundles run to the cerebrobrachial connective by which they reach the supraoesophageal lobes. The efferent plexus lies more laterally and consists of larger fibres coming from cells of all parts of the prebrachial lobe. These fibres join the lateral divisions of the brachial nerves. The plexuses contain mostly afferent or efferent fibres, but each contains some fibres of the other type. Between the two plexuses the neuropil is a dense network of fibres, containing masses of degenerating tissue in normal octopuses. This degenerating tissue perhaps represents the remains of afferent fibres that are damaged by injuries to the arms and suckers; injuries so common in the octopus that they can be regarded as 'normal'.

The anterior postbrachial lobe is attached at its face to the prebrachial lobe. It differs in having no wall of cells at its centre, and so is a median, unpaired structure. The largest cells in the wall of the postbrachial lobe are larger than the largest ones of the prebrachial lobe. The neuropil of the postbrachial lobe is an irregular tangle. Many fibres from the medial and lateral divisions of the brachial nerves enter the postbrachial lobe.

Middle suboesophageal mass

This is confluent in front and behind with the anterior and posterior suboesophageal masses respectively. In the midline the middle suboesophageal mass lies below the perioesophageal sinus, laterally it receives the anterior and median basal–suboesophageal connectives and smaller tracts from the optic and peduncle lobes. The cephalic arteries pass ventrally behind the middle suboesophageal mass.

The middle suboesophageal mass, or pedal lobe, includes five parts, definable by different structures and connections, but with their neuropils broadly continuous (Young 1971). They are:

- a median anterior pedal lobe
- a pair of anterior chromatophore lobes
- a pair of lateral pedal lobes, divisible into anterior and posterior parts
- a median posterior pedal lobe
- a pair of anterior funnel lobes.

The anterior pedal lobe is concerned predominantly with the innervation of the arms and the muscles of the front of the head. This lobe includes the cell wall and neuropil of the front part of the middle suboesophageal mass, and posteriorly its limit is marked by the conspicuous middle pedal commissure. The cell wall is thick, especially ventrally, and the cell composition of the ventral wall has a higher proportion of medium-sized cells (5–15 μm) than the dorsal wall which contains many large cells and few small ones. The neuropil of the lobe is a continuous dense tangle. The main input to the lobe is from the anterior basal to anterior pedal tract and fibres spread throughout the anterior pedal lobe. Other afferent connections come from the brachial, lateral pedal, posterior pedal, and precommissural lobes. Many fibres of the lateral pedal lobe, after crossing in the middle pedal commissure, end in the anterior pedal lobe. The efferent pathways from the anterior pedal lobe run to the lateral roots of the brachial nerves, the antorbital nerves. In addition there is the anterior pedal to posterior pedal lobe tract, and the anterior pedal to brachial tract.

The anterior chromatophore lobes form swellings on either side of the anterior part of the middle suboesophageal mass. The cell walls are thick compared with those of other lobes of the brain. The cell bodies have a regular arrangement, the large ones (with nuclei up to 20 μm) are at the periphery, and the smallest (with nuclei of less than 5 μm) at the centre. The walls have shallow pockets into which the neuropil extends, and a few fibres wander amongst the cell bodies. The neuropil is characterized by bundles of fibres forming a loose latticework with a rather regular pattern. Afferent fibres arrive from the lateral basal lobe. The efferent fibres probably all go to the muscles of the skin and chromatophores without synapse in a peripheral ganglion.

The lateral pedal lobes run from the back of the anterior chromatophore lobes along the remaining length of the middle suboesophageal mass. Each can be divided into anterior and posterior parts. The anterior part contains motoneurons for the control of extraocular muscles (Fig. 31.27). The posterior part has large cells whose axons run into the middle pedal commissure. The neuropil is continuous between the two parts and together they have many afferent connections. Besides the many peripheral nerve fibres which enter the lateral pedal lobe, this lobe receives fibres from many other parts of the brain. Afferent connections arise from the peduncle, brachial, anterior basal, median basal, precommissural, and anterior and posterior pedal, and magnocellular lobes. In addition there are the ophthalmic and statocyst nerves. The efferent pathways include fibres in the oculomotor nerves and ophthalmic nerves. The middle pedal commissure includes a complex system of fibres connecting the lateral pedal and anterior pedal lobes. This region has a wide range of afferent and efferent connections indicative of its importance in the control of eye movements (Budelmann and Young 1984).

The median posterior pedal lobe lies behind the middle pedal commissure, and its lateral limits are marked by the brachiopalliovisceral connectives and the magnocellular lobes. Dorsally the lobe receives a connective from the median basal lobe. The walls of the lobe consist mainly of large motor cells, and the neuropil is a uniform tangle. The afferent pathways enter the lobe in tracts from the anterior pedal, lateral pedal, median basal, optic, brachial, palliovisceral, and magnocellular lobes. The efferent connections include the anterior funnel nerves, median pallial retractor nerves, head retractor nerves, and tracts to the anterior pedal and palliovisceral lobes.

Each of the paired anterior funnel lobes has a small knot of tangled neuropil adjacent to the cell layers of the central side of the posterior pedal lobe.

Posterior suboesophageal mass

The posterior suboesophageal mass is the most posterior part of the brain (Fig. 31.28). In the past this has been regarded as a unit, the visceral or palliovisceral ganglion. It is impossible so to consider it because there is internal division into several lobes, and one, the magnocellular lobe, continues upwards as a perioesophageal structure. The six main regions recognized in the posterior suboesophageal mass are:

- a median palliovisceral lobe
- a pair of posterodorsal lobes
- a pair of posterior chromatophore lobes
- a pair of posterior funnel lobes
- the vasomotor lobes (dorsal and ventral median, and dorsal and ventral pairs)
- a pair of magnocellular lobes.

The median palliovisceral lobe occupies the centre of the posterior suboesophageal mass (Fig. 31.28). Its walls contain the cell bodies of the neurons of the four main pairs of nerves that leave the mass: the pallial nerves or mantle connectives (except for the fibres that arise in the posterior chromatophore lobes), the collar nerves, the posterior funnel nerves, and the visceral nerves (except for their vasomotor fibres). The lobe receives afferent fibres from most of these nerves, from the static and brachial nerves, and from many other parts of the brain including the median basal, pedal, and magnocellular lobes. The cells forming the walls of the palliovisceral lobe are of typical motor-centre type, with large cells towards the outside and a few small ones near the neuropil. The centre of the lobe is occupied by a large mass of neuropil containing bundles that are difficult to follow, many of which are large and constitute the roots of the nerves.

Each posterodorsal lobe is a small knot of neuropil surrounded by cell walls containing regular rows of small and medium cells, but no large ones, and the neuropil is a woven tangle. These characteristics differentiate this lobe from the rest of the palliovisceral lobe, and from the chromatophore and vasomotor lobes which lie nearby.

Each posterior chromatophore lobe lies on the dorsal side of the posterior suboesophageal mass (Fig. 2.4). Its cell wall is thick and cell bodies have a regular arrangement, the largest being at the periphery and the smallest at the centre. Using a whole-brain preparation, micropipettes were used to micro-inject dye into individual neurons of the posterior chromatophore lobe (Fig. 31.29) (Miyan and Messenger 1995). Nerves to the chromatophores were found to originate in the paired anterior and posterior chromatophore lobes, forming part of the hierarchically arranged chromatophore control system (Boycott 1953). This is driven mainly by visual cues, but also other influences (Budelmann and Young 1985, Ferguson and Messenger 1991). Intracellular recordings reveal three cell types in the posterior chromatophore lobes, suggesting that the chromatophore control system is complex.

The vasomotor lobe has subdivisions. They include the dorsal and ventral median vasomotor lobes, and paired dorsal and ventral lateral vasomotor lobes; together these lobes make up nearly the whole face of the posterior suboesophageal mass (Young 1971). Their neuropil is broadly continuous with that of the palliovisceral lobe.

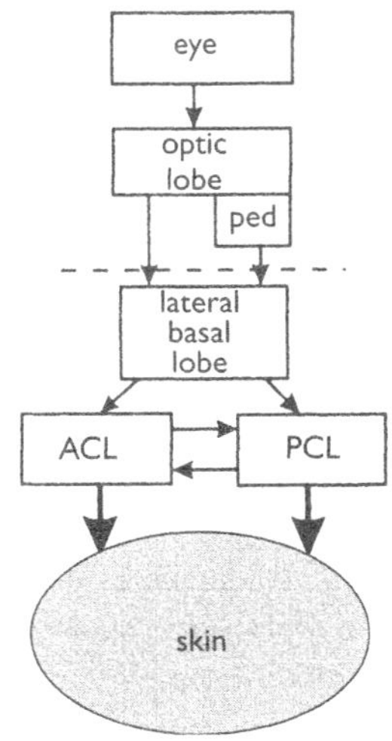

Fig. 31.29. *Octopus vulgaris*. A diagram to illustrate the hierarchical organization of the chromatophore system for one side of the brain. Commissures link the left and right anterior (ACL) and posterior (PCL) chromatophore lobes (Miyan and Messenger 1995).

Each posterior funnel lobe lies immediately in front of the lateral ventral vasomotor lobe, and the cell walls are similar in both lobes. The neuropil is a characteristic tangle of fibres.

Central pathways of the nerves of the arms

Further evidence of the organization of the tactile learning system was obtained with the use of centripetal cobalt filling of the brachial and pallial nerves (Budelmann and Young 1985). The afferent and efferent pathways are summarized in Fig. 31.30.

The inferior frontal system is wholly concerned with learned tactile reactions (Wells and Young 1972, Young 1983). The centripetal fillings of the brachial nerves have provided further evidence about its input and output pathways. The afferent fibres from the arms branch profusely in the lateral and median inferior frontal lobes, and many bundles cross the midline. There is thus widespread distribution of the signals from one arm to all parts of the lobe. The filled fibres in the posterior buccal lobe are all large and are assumed to be efferent fibres, but the possibility that a few are afferent fibres cannot be excluded. No afferent fibres pass directly to the subfrontal lobe, which receives its input from the median inferior frontal lobe. Large somata were filled in the posterior buccal lobe; these are of special interest as forming the final common pathways for the learned tactile reactions. They occur mostly close to the base of the subfrontal lobe and are presumably controlled by its output fibres. No somata were filled in the subfrontal lobe itself.

Degenerating fibres have been seen, after section of the cerebrobrachial tracts, in the prebrachial lobes and in the brachial nerves (Young 1971). Centripetal cobalt filling now provides evidence of the cells from which they arise. The absence of filled somata in the lateral and median inferior frontal lobes confirms that any descending fibres from these lobes end in the anterior suboesophageal mass of the brain and do not proceed to the arms.

Many somata and afferent fibres were filled in the magnocellular lobes after filling either the brachial or pallial nerves; this is probably a region from which escape reactions are initiated. The lower part of the median basal lobe also receives afferent fibres from both of these nerves and a few somata were filled at the lower edge of this lobe; this probably controls the magnocellular lobe lying below. After filling a brachial nerve or the pallial nerve, somata were filled in both the anterior and posterior chromatophore lobes, but few or no afferent fibres were filled in these lobes.

After filling a brachial nerve many somata and afferent fibres were filled in the prebrachial and brachial lobes and in the anterior pedal lobe, and many fine afferent fibres and a few somata were filled in the superior buccal lobe. After filling the pallial nerve some filled fibres run forwards to the brachial lobe. There was no evidence of fibres proceeding direct to the optic lobes from either the brachial or pallial nerves (Budelmann and Young 1985).

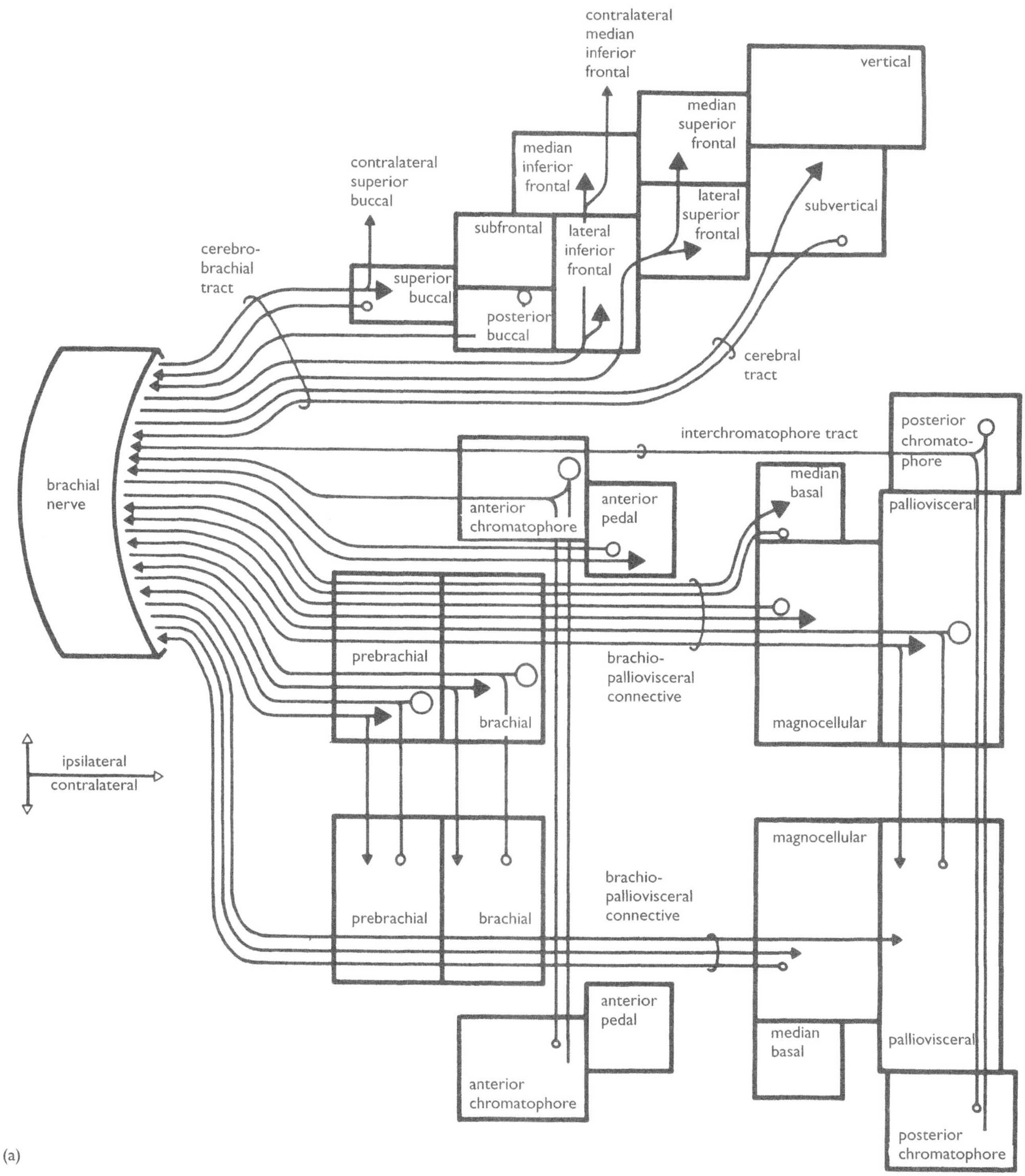

Fig. 31.30. *Continued*

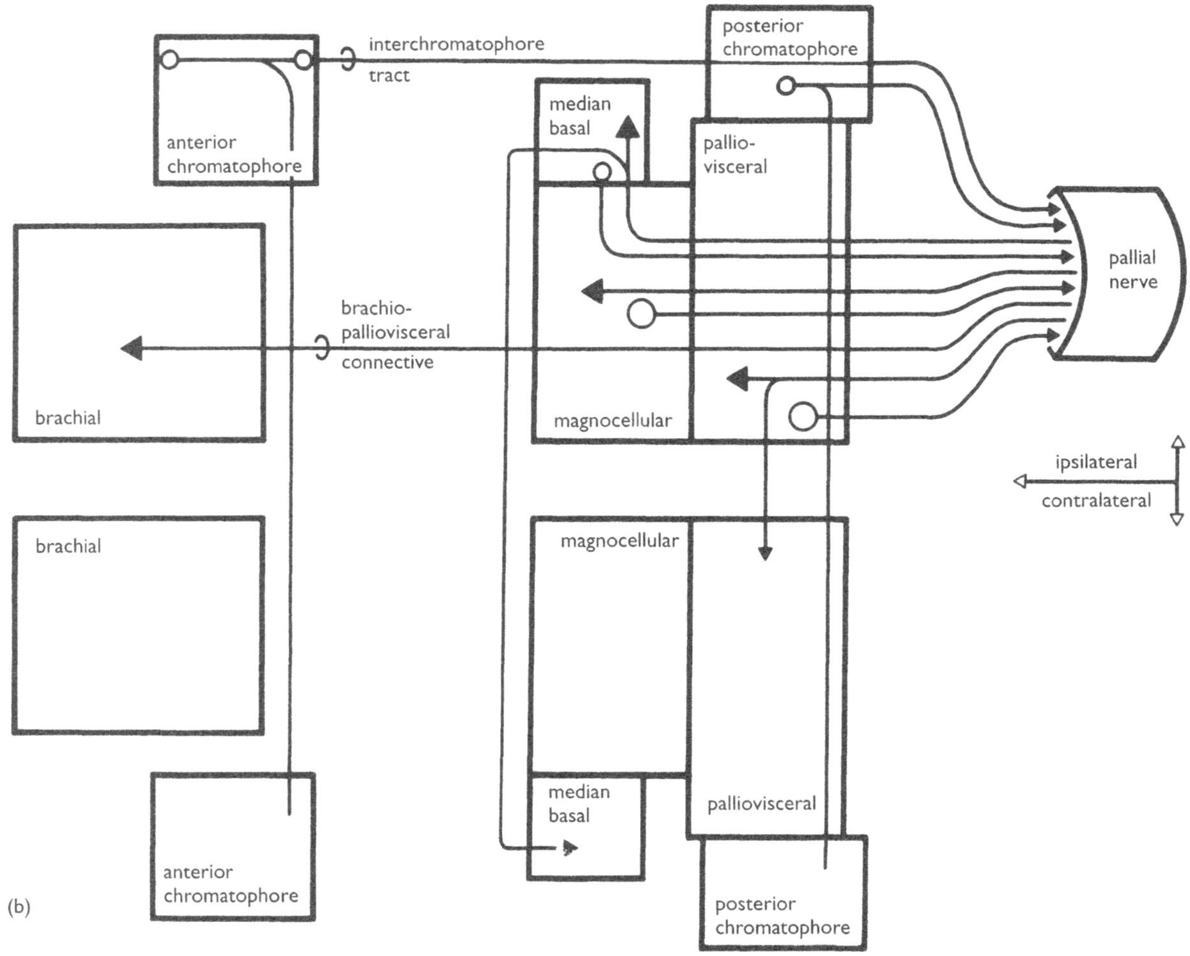

Fig. 31.30. *Octopus vulgaris.* A schematic presentation of the afferent and efferent brain pathways of the third brachial nerve (a) and the pallial nerve (b), demonstrated by centripetal cobalt filling. The rectangles represent lobes of the brain. The relevant contralateral lobes are shown only for the suboesophageal mass. The sizes of the open circles or arrowheads give an approximation of the number of cells or fibres filled. Open circles indicate the origin (somata) of the pathways and arrowheads their termination. Large open circles represent many filled somata, small open circles few. Large arrowheads indicate the termination of many filled afferents, small arrowheads few. Branched lines indicate continuation of the pathway, not necessarily of individual fibres. In the subvertical lobe some of the fibres are large and presumed to be of efferents, but their somata were not found (Budelmann and Young 1985).

Magnocellular lobes

The magnocellular lobes lie far laterally, around the sides of the brain, and are neither supra- nor suboesophageal. (Fig. 2.5). These lobes are best regarded as the part of the ancestral cephalopod nervous system at which the main cords meet. Each magnocellular lobe has a triangular form when seen from the side, with the upper corner connected to the dorsal basal lobe and the two lower ones with the hind end of the pedal and with the palliovisceral lobes.

In octopods the magnocellular lobes receive few afferent fibres direct from the periphery and send no efferent ones to it, and yet they have connections with almost every part of the brain. There are no giant nerve fibres; presumably the adoption of a mainly bottom-living habitat, and the use of the arms for walking has reduced the advantage obtained from rapid conduction in the nerve fibres that control swimming. Yet although the giant nerve cells have disappeared the magnocellular lobes remain as the centres responsible for initiating protective and defensive actions, such as the deimatic reaction. The magnocellular lobes thus receive most of their afferent fibres from the optic, subvertical, and median and lateral basal lobes (Young 1971).

Each of these lobes is divided according to the character of the cell walls with dorsal and ventral divisions. The dorsal region has mainly small cells forming a thin, closely set row. The ventral region has small cells mixed with some large ones. This forms the boundary and is defined by a well-marked groove, which anteriorly lies rather below the level of the oesophagus.

The afferent and efferent connections of the magnocellular lobes are listed in Table 31.5.

Supraoesophageal mass

The part of the brain above the oesophagus comprises numerous lobes. There are regions in front with relatively large cells, the superior

Table 31.5 Afferent and efferent connections of the magnocellular lobes

Afferent connections		Efferent connections	
1	Subvertical to magnocellular tract	1	Magnocellular to brachial tract
2	Dorsal basal to magnocellular tract	2	Magnocellular to lateral pedal tract
3	Optic to magnocellular tract	3	Magnocellular to posterior pedal tract
4	Peduncle to magnocellular tract	4	Magnocellular to palliovisceral tract
5	Median basal to magnocellular tract	5	Magnocellular to subvertical tract
6	Lateral basal to magnocellular tract		
7	Brachial to magnocellular tract		
8	Pallial to magnocellular tract		
9	Static to magnocellular tract		
10	Anterior infundibular to magnocellular fibres		

buccal and posterior buccal lobes, the higher motor lobes for the control of manipulation of prey by the arms and of feeding. Other higher motor centres lie behind, namely the anterior basal lobes, the median basal lobe, and lateral basal lobes, concerned with swimming, walking, colour change, and other actions of the animal. The upper part of the supraoesophageal mass consists of lobes that receive inputs from the eyes and from the arms. These lobes have distinctive patterns of organization and have been investigated experimentally. This has allowed an understanding of the principles by which they operate. Chemotactile and visual centres have been recognized. Each of these centres comprises four lobes, which differentiate from a single lobe during ontogenesis. The four lobes of each system are arranged in parallel, in two tiers (see Fig. 2.7). In the chemotactile system the lower tier is formed by the lateral inferior frontal and posterior buccal lobes, and the upper by the median inferior frontal and subfrontal lobes. In the visual system the lower pair are the lateral superior frontal and subvertical lobes, and the upper the median superior frontal and vertical lobes.

Superior buccal lobe and inferior buccal ganglia

The superior buccal lobe is a single lobe stretching across the midline and is continuous behind at the sides with the posterior buccal lobes (see Fig. 2.10). The place at which the single superior buccal lobe divides into two posterior buccal lobes makes a good formal division. There is only a slight change in the character of the cells and neuropil at this point; the superior and posterior buccal lobes are broadly continuous and must be related in function. A large buccal commissure crosses in the midline at the extreme hind end of the superior buccal lobe. Anteriorly the superior buccal lobe receives about 20 labial nerves. Below and laterally it receives on each side three large bundles: the interbuccal connectives, the bucco-subradular connectives, joining it to the inferior buccal lobe and subradular ganglia, and the nerves of the posterior salivary glands. Posteriorly the superior buccal lobe receives part of the cerebrobrachial connectives from below.

The cell wall is thick and contains large cells at the periphery and some smaller ones nearer to the neuropil. It resembles those of the suboesophageal lobes, as would be expected if the lobe contains motoneurons whose axons reach to the posterior salivary glands and the muscles of the lips. The neuropil is also typical of a motor centre, as it has a dense tangle of fibres.

The inferior buccal ganglia are joined in the midline (Fig. 2.10), and lie immediately behind the buccal mass in the buccal venous sinus. The oesophagus lies above the ganglia and the posterior salivary gland duct below. The ganglia receive interbuccal connectives at the sides, and give off several pairs of nerves to organs of the buccal mass and digestive tract.

Inferior frontal lobe system

The inferior frontal lobe system comprises the paired posterior buccal lobes, paired lateral inferior frontal lobes, a single median inferior frontal lobe, and paired subfrontal lobes. These lobes form a closely interrelated and functional unit concerned with chemotactile information (Wells and Young 1972). They constitute two loops, one above the other, in which signals from receptors on the suckers interact with those from the lips (Fig. 2.7). The efferent fibres to the arms are responsible for the control of the movements of learned tactile discrimination (Young 1983). However, little is known of the functional components of the fibre tracts described in the suboesophageal lobes (Young 1971). Some of the afferent pathways have been determined following degeneration after section of the nerves, but this technique does not allow identification of the somata of the efferent fibres. Centripetal cobalt filling of brachial nerves has given valuable evidence of the input and output pathways. Efferent fibres pass from the posterior buccal and subvertical lobes direct to the arms. Afferent fibres from the arms pass to the lateral and median inferior frontal lobes, others to the lateral and median superior frontal lobes, and a third set to the subvertical lobe. None reaches to the vertical or subfrontal lobes. These pathways are summarized in Fig. 31.30 (Budelmann and Young 1985).

The posterior buccal lobes are continuous in front with the superior buccal lobe, and posteriorly they pass above the anterior basal lobe. The two posterior buccal lobes are united by a commissure through the subfrontal lobes. The walls of the lobe anteriorly have a cell composition similar to that of the superior buccal lobe with a few large cells peripherally and many small ones adjacent to the neuropil. More posteriorly the proportion of small cells increases and further back still the proportion of large cells increases again. Inputs to the lobe come from the subfrontal, lateral inferior frontal, superior buccal, and subvertical lobes. The efferent connections include axons passing along tracts to the brachial, subvertical, and superior buccal lobes.

The paired lateral inferior frontal lobes and the single median inferior frontal lobe can be distinguished by their cell types and connections. Each lateral inferior frontal lobe has only a thin layer of cells, some of which have nuclei with a diameter of more than 10 μm, and these are larger than any of those in the thicker walls of the median inferior frontal lobe where the nuclei are less than 5 μm in diameter. Medially the neuropil of each lateral inferior frontal lobe is continuous with that of the posterior buccal lobe. The lateral inferior frontal lobes send fibres to the posterior buccal, the brachial and prebrachial, and to the superior frontal lobe. The input comes from the brachial, superior buccal, vertical, and subvertical lobes. The whole output of the median inferior frontal lobe consists of fine fibres to the subfrontal lobes.

The subfrontal lobes lie between the posterior buccal lobes and below the median inferior frontal lobe. The walls are thick and have densely packed minute cells together with a very small number of larger cells. The small cells are amacrine cells whose axons do not extend beyond the lobe. The subfrontal lobes have a very similar appearance to that of the vertical lobe and may be considered a precise analogue. The afferent pathways include a tract from the median inferior lobe in which numerous bundles of minute fibres run between the two lobes, and tracts from the posterior buccal lobe. Efferent fibres from the centre of the subfrontal lobe collect into bundles and run into the neuropil of the posterior buccal lobe.

Superior frontal–vertical lobe system

This system comprises highly differentiated lobes: the median superior frontal, the lateral superior frontal lobe, the vertical and subvertical lobes (see Fig. 31.28). These, the 'highest' lobes of the brain, are concerned with the storage and use of memory records. They have a distinct structure and complex interconnections; together they form two circuits, one above the other (see Fig. 2.7).

Superior frontal lobe

This lobe forms a band across the front of the supraoesophageal mass, between the inferior frontal and the vertical lobes. It is sharply divided into a median superior frontal lobe and a pair of lateral superior frontal lobes. The latter are not conspicuous externally, unlike the median superior frontal lobe, but they are important in the control of attack behaviour. In the lateral superior frontal lobes fibres from the inferior frontal and optic lobes interweave in a complicated but rather regular plexus, making synapses with the efferent fibres, which go to the vertical lobe. The lateral superior frontal lobes also receive tracts of fibres from the vertical lobe. The cell walls of the lateral superior frontal lobes cover only the lateral aspect of each lobe and contain a relatively small number of cells. The neuropil has two distinct zones, outer and inner. The former contains a dense plexus of fibres that come from the inferior frontal, vertical, and subvertical lobes. The inner zone contains optic fibres entering the lobe posteriorly and proceeding forward to the median superior frontal lobe.

The median superior frontal lobe occupies the front of the supraoesophageal mass, between the inferior frontal and vertical lobes. It is continuous with the lateral superior frontal lobes at the sides and receives its entire input through these lobes. Its main output passes to the vertical lobe but a few bundles may run to the subvertical lobe. The anterior and dorsal surfaces of the lobe are covered with many layers of small cells (all with nuclei of less than 5 μm diameter). The axons of these cells run an irregular course through the neuropil of the median superior frontal lobe, and give off collaterals. After crossing the neuropil the axons of this lobe pass through a plexiform arrangement to the five lobules of the vertical lobe. In this way fibres originating together may be widely distributed in the vertical lobe.

Vertical lobe

The vertical lobe occupies the top and sides of the supraoesophageal mass and forms a prominent feature of the brain (see Figs 2.4, 2.10, 31.28). It is divided into five gyri, or lobules, each approximately cylindrical, with their long axes running anteroposteriorly. The formation of lobules allows a large number of cell bodies to be concentrated in a relatively small lobular volume of the vertical lobe, and irregular subdivisions form additional tongues of cells projecting inwards towards the centre of the neuropil. The neuropil therefore forms a smaller proportion of the vertical lobe than it does in the median superior frontal lobe. Each lobule has a thick cell wall composed mainly of masses of minute cells (the smallest in the octopus brain) with a small number of larger cells. The great majority of the cells are minute amacrine cells whose fibres do not extend beyond the vertical lobe. The trunks of the amacrine cells are filled with mostly non-granular vesicles and make two types of contact: they synapse in the outer region of the lobe with incoming fibres and in the centre with synaptic spines on the efferent cells. The output of the vertical lobe comes from a small number of larger cells whose trunks turn and twist in the neuropil, and carry numerous dendritic side branches. These fibres leave the lobe and pass downwards in bundles to end in the subvertical lobe and the lateral superior frontal lobes.

Subvertical lobe

This lobe lies between the vertical, superior frontal, precommissural and optic lobes, and the cerebral tract (Fig. 31.28). It has functional relations with all of these parts and it is the main pathway by which the lateral superior frontal and vertical lobes send their output to the optic lobes, and probably also, through the precommissural lobe, to the lower motor centres. The subvertical lobe consists of three parts:

- a pair of large anterior lobes, with outer cell layers and a central mass of neuropil
- a posterior lobe projecting backwards beneath the vertical lobe with tongues of cells and neuropil
- small lateral lobes with very large cells.

The neuropil of the subvertical lobe consists of a dense network of fine fibres. The larger cells of the lobe have many wide-spreading and branching collaterals. These presumably receive impulses from the fibres of the vertical lobe that descend in a set of bundles with an elaborate system of crossing bundles. This arrangement has the effect of channelling the whole output of the 25 million cells of the vertical lobe into some 800 000 cells in the subvertical lobe that lead mainly back to the optic lobes.

The afferent connections come from the vertical and the superior frontal lobes, and fibres from the optic lobes reach only to the lateral part of the subvertical lobe, the only known output of which is to the anterior basal lobe. Other fibres run directly from the brachial lobes to the subvertical lobe (Budelmann and Young 1984, Plän 1987). Efferent fibres pass largely to the optic lobes where they spread out to all parts, some crossing in the dorsal optic commissure. Other fibres pass to the neuropil of the precommissural lobe, some to the lateral superior frontal lobes and others to the vertical lobe. Fibres also reach the dorsal and median basal lobes, and the magnocellular lobes.

Precommissural lobe

This lobe is of irregular shape and lies below the subvertical lobe at the centre of the supraoesophageal lobes (Young 1971). The precommissural lobe is broadly continuous with the subvertical lobe and may form a connection between the whole vertical lobe system and the anterior basal and suboesophageal ganglia.

The walls of the lobe are mostly thin and composed of a few layers of cells. The neuropil resembles that of the subvertical lobe with which it is continuous. The precommissural lobe contains many fibres running, in a vertical direction, to and from the suboesophageal centres.

This lobe differs sharply from the subvertical lobe in that it receives no fibres directly from the vertical lobe system. It receives many fibres from the optic lobe; other inputs come from the subvertical, the dorsal basal, anterior basal, and median basal lobes, the peduncle lobes, and the brachial and magnocellular lobes. The most anterior part of the precommissural lobe is continuous with the cerebral tract, which spreads out into it. The output from the lobe includes several large bundles to the pedal lobe. Other fibres go to the anterior basal and peduncle lobes, and some to the optic lobes.

Cerebral tracts

This name is given to the large bundles that run back from the cerebrobrachial connective to the subvertical and precommissural lobes. They contain fibres running in both directions. The main part of the cerebral tract proceeds upwards and backwards through the anterior limits of the precommissural lobe to spread out in the hinder part of the latter and in the subvertical lobe.

The ascending fibres come in part from the arms and run through the brachial lobe without interruption. These fibres spread out through the buccal, inferior frontal, subvertical, and precommissural lobes, some passing directly to the superior frontal lobe. Descending fibres arise in the precommissural and subvertical lobes and pass forward in the cerebral tract.

Basal lobe system

This system is formed by the anterior basal (which can be divided into anterior and posterior parts), interbasal, median basal, lateral basal, and dorsal basal lobes, together with the peduncle lobes. All show a rather similar organization and pattern of connections, as well as the presence of numerous very fine parallel fibres. This system influences the direction and speed of movement, and it has a direct connection with the memory system of the vertical and inferior frontal lobes only through the subvertical lobe. The lobes all exchange fibres with the optic lobes and with each other, and send their output to the motor centres of the suboesophageal lobes (Plän 1987).

Electrical stimulation of the anterior basal lobe elicits effective stepping movements by the arms, of the median basal lobe swimming, and of the lateral basal lobe changes in colour by the action of the chromatophores (Young 1971).

Anterior basal lobe

This lobe lies across the midline at the level of the anterior borders of the optic tracts, and immediately behind the posterior part of the superior buccal and subfrontal lobes. Dorsal to the anterior basal lobe and anteriorly lie the cerebral tracts, and more posteriorly the lower part of the subvertical lobe. The precommissural lobe lies directly behind. Laterally the anterior basal lobe is continuous with the optic tracts, with which it exchanges many fibres. In the ventral midline the lobe lies immediately above the perioesophageal venous sinus, while at the sides and below it is broadly continuous with the anterior pedal lobe.

The anterior basal lobe can be divided into anterior and posterior parts by an oblique row of bundles of fibres of the optic to anterior basal lobe tracts and by a narrow band of nerve cell bodies (Fig. 31.31). Each part of the anterior basal lobe contains some large diameter neurons, many small neurons, and a 'spine' region. The latter is a bank of small diameter parallel fibres arising from at least three sources—the optic lobe, the peduncle lobe, and the anterior basal

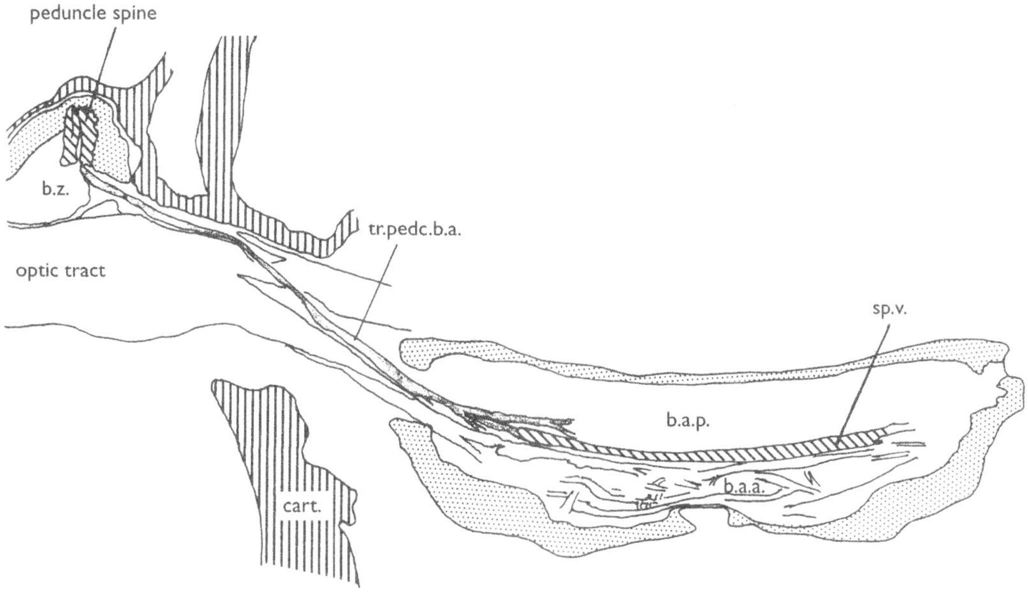

Fig. 31.31. *Octopus vulgaris.* A drawing of a transverse section showing the connections to the anterior basal lobe from the optic lobe and peduncle lobe of one side. The spine regions (diagonal hatching) of the peduncle and anterior basal lobes are connected by a peduncle to anterior basal lobe tract (Gleadall 1990). b.a.a., anterior anterior basal lobe; b.a.p., posterior anterior basal lobe; b.z., basal zone of peduncle lobe; cart. cranial cartilage; sp.v., ventral spine; tr.pedc.b.a., peduncle to anterior basal tract.

lobe itself. The most dorsal tracts contain fine fibres from the optic and peduncle lobes, and posteriorly the tracts originate in the lateral subvertical lobe. Their origins have been confirmed by degeneration after removal of an optic or peduncle lobe. The spine region of the anterior basal lobe, with its parallel fibres, shows some similarity with that of the peduncle lobe. It is suggested that the spine region of the anterior basal lobe, which is a higher motor centre, may correspond more closely in function with certain basal ganglia of the vertebrate brain (Gleadall 1990). The afferent fibres reaching to the anterior basal lobe pass from the optic, peduncle, median basal, precommissural, and probably from the statocyst. The efferent fibres pass to the optic and pedal lobes. Several other connections have been determined by cobalt filling techniques (Plän 1987).

The anterior anterior basal lobe lies more ventrally than the posterior anterior lobe. The former part has a ventral spine lying across the midline. Small cells at the side of the spine give rise to fine fibres (0.2–0.35 μm) (Gleadall 1982). The anterior anterior basal lobe includes two pairs of lobules situated anteroventrally. Each of the pair of anteromedial lobules is flanked by a smaller lateral lobe (Gleadall 1990). The upper extent of the anterior anterior basal lobe is marked by a thin strip of neuropil, the ventral spine. The remaining neuropil contains many bundles of fibres connecting with the optic lobe, some of which pass to the ventral parts of the contralateral anterior basal forming a conspicuous chiasm. The ventral spine is a bank of small-diameter fibres running along the axis of the anterior basal lobe across the width of the central brain. The fibres come from at least three sources: the peduncle lobes, the optic lobes and the intrinsic anterior basal lobe cells. The origin of these fibres has been confirmed by degeneration studies (Woodhams 1975, Gleadall 1990).

The posterior anterior basal lobe is without discrete lobules. Its neuropil consists mainly of fibres running transversely through the lobe rather like a large optic commissure, with many fibres running in from each optic tract, but there is no chiasm like that seen in the anterior anterior basal lobe. A region of small fibres, the dorsal spine, occupies the anterior part of the lobe, above the ventral spine. The dorsal spine receives fibres from the optic and peduncle lobes, and an input from the magnocellular to supraoesophageal tract. More posteriorly the spine is less well defined and larger fibres become mixed with the smaller ones. Some of the large fibres communicate with the pedal lobe, others with the optic lobe. The posterior anterior basal lobe receives small tracts of fibres from the precommissural and lateral subvertical lobes, and provides a connection between the vertical and basal lobe system. These tracts penetrate the thick connective tissue behind the anterior basal lobe with input from the vertical lobe system (Gleadall 1990). The lateral regions of the anterior basal lobe contain fibres entering and leaving in many directions, and this area may be termed the lateral anterior basal lobe. Large-diameter cells arising in the posterolateral wall send processes to the pedal lobes of both sides.

The anterior basal lobe exerts higher motor control over the arms, head and eyes, collating inputs from many areas of the brain, and produces coordination of locomotor activity (Gleadall 1990).

Interbasal lobe

This lobe is roughly spherical and is the most lateral and ventral part of the basal lobes. Its neuropil is continuous with those of the anterior basal, precommissural, and median basal lobes. The lateral wall of the interbasal lobe has characteristic sets of large and small cells. The neuropil is an irregular tangle of large and small fibres, some of which run in loosely organized bundles. The afferent pathways are tracts including the optic tract of the same side and from the median basal and peduncle lobes, and efferent pathways to the optic and peduncle lobes (Plän 1987).

Median basal lobe

This lobe forms the central part of the posterior basal part of the supraoesophageal mass (Fig. 31.28). Laterally and posteriorly the median basal lobe is continuous with the lateral basal lobes, above which it is in continuity at the sides with the magnocellular lobes. The median basal lobe has a voluminous neuropil continuous with that of several of its neighbours. Outside the neuropil there is a thin layer of nerve cells, made up of medium-sized and small cells. The cells lie mainly on the posterior face of the lobe, and so constitute the posterior wall of the whole supraoesophageal mass.

There are a number of input pathways and they come from the peduncle, magnocellular, anterior basal, interbasal, dorsal basal, precommissural, and olfactory lobes, and from the optic tracts (Young 1971), the statocysts (Budelmann and Young 1984), the brachial nerves, the pallial nerves (Budelmann and Young 1985), and the subvertical lobe (Plän 1987).

The axons of the cells of the median basal lobe reach many destinations including the anterior basal, lateral basal, interbasal, precommissural, posterior pedal, lateral pedal, magnocellular, peduncle, dorsal basal, subvertical, brachial, posterior buccal lobes and pallio-visceral lobes, and the optic tract.

Dorsal basal and subpedunculate lobes

The dorsal basal lobe occupies the posterior region of the supra-oesophageal lobe below the hind part of the vertical and subvertical lobes, and above the median basal lobe (Fig. 31.28). It is subdivided by a vertical wall of cells into anterior and posterior parts. The cells of this lobe are arranged partly as islands of neurons separated by neuropil and partly as islands of neuropil surrounded by cells. There is a thick dorsal wall of cells above each part of the lobe and vertical walls anteriorly and posteriorly, with a further wall of cells dividing the lobe into anterior and posterior parts. The cells of the lobe are small, but larger and less densely packed than those of the vertical lobe.

The connections of the dorsal basal lobe are numerous and complex, and appear to be concerned with the elaboration of patterns of defence and avoidance. Cobalt filling has revealed inputs from the inferior frontal, anterior basal, peduncle, and magnocellular lobes and outputs to the posterior buccal, pedal, anterior and lateral basal, precommissural, and peduncle lobes (Plän 1987). These add to, or confirm, those traced to the optic, median basal, subvertical, olfactory, magnocellular, and subpedunculate lobes by Young (1971).

The subpedunculate lobes are paired and are imperfectly marked off from the upper, hind end of the dorsal basal lobe. Each of the subpedunculate lobes thus lies medial to the posterior part of the subvertical lobe and below the hind end of the vertical lobe. The two subpedunculate lobes are the only part of the supraoesophageal mass (except the superior buccal lobe) from which efferent peripheral nerve fibres arise. The cell layers are not greatly different from those of the near-by dorsal basal and subvertical lobes. They contain a larger proportion of small cells and fewer very large ones. The neuropil is a dense tangle of exceedingly fine fibres. A large artery runs through the lobe and gives it a characteristic appearance.

The input of the lobe includes optic and olfactory fibres. The output of the lobe is presumed to consist entirely of fibres of the subpedunculate nerve and optic gland nerve. The subpedunculate nerve passes through the anterior basal and subvertical lobes to leave the cranium with the pharyngo-ophthalmic vein. The nerve and vein run to the floor of the orbit to join the suborbital sinus, where numerous small fibres of the nerve pass into an elaborate plexus of neurosecretory tissue in the walls of the veins. The optic gland nerve, in the subpedunculate area of the dorsal basal lobe, controls the onset of maturity and may be the sole connection of the optic gland.

Lateral basal lobes

These lobes lie on either side of the median basal lobe, and can just be discerned externally as slight swellings on the lower posterior face of the supraoesophageal mass. The neuropil of all of these lobes is broadly continuous. The cell walls of the lateral basal lobes are thinner than those of the median basal lobe, providing a means for distinguishing these lobes. Laterally the lateral basal lobes are in contact with the dorsal part of the magnocellular lobes. The posterior roots of the superior posterior ophthalmic nerves lie in the groove between the lateral basal and magnocellular lobes, whose neuropils are continuous. The lateral basal lobes have no cell walls anteriorly where their neuropils are continuous with the median basal and interbasal lobes.

The lateral basal lobes form a central part of the system for the control of the muscles of the chromatophores and skin. The optic lobes constitute a higher level of control and the anterior and posterior chromatophore lobes are lower ones.

Peduncle lobes—a visuo-motor centre

The peduncle lobes are paired. Each is elongated and lies on the optic tract. It receives a topographically regular series of fibres from the optic lobe. A commissure joins the peduncle lobe to that on the opposite optic tract. The structure and function of the peduncle lobe in *O. vulgaris* has been described by Messenger (1967b, c, 1971). Various experiments have shown that efferent fibres from the peduncle lobe project to motor centres and so it regulates motor control programs set up elsewhere. After ablation of the peduncle lobe the octopus shows aberrations in posture and locomotion (Messenger 1967b). A direct input of fibres from the statocyst to the peduncle lobe was demonstrated by Hobbs and Young (1973).

An ultrastructural study of the basal zone of the peduncle lobe showed large 'motor' cells (up to 20 μm in diameter), each with a rounded nucleus and large perikaryon, and also smaller cells (4–10 μm across) each with a dark irregularly-shaped nucleus. The latter receive synapses from afferent fibres to the lobe. Those from the principal source, the ipsilateral optic lobe, are characterized as large terminals with many small, clear vesicles. Other opticopeduncular afferents and the small-cell fibres themselves run directly to the second part of the lobe, the peduncle spine. These fibres of the spine cells form a conspicuous and characteristic array of parallel fibres in the neuropil, and they bear a striking resemblance to those of the vertebrate cerebellar granule cells. Serial synaptic relays are present along the length of these fibres (Fig. 31.32). The parallel fibres have a narrow diameter and hence a long conduction time, this is an essential feature of a fine motor control system. So their primary function is to provide cortical time delays in the multisecond range (Woodhams 1977).

The peduncle lobe receives a 'visual' input (from second- and third-order neurons in the optic lobe), a vestibular input (from first-order cells in the statocyst), and ascending pathways from higher motor centres (the basal lobes) (Fig. 31.33). The peduncle lobe also receives information from peripheral sense organs in the arms and mantle (Camm *et al.* 1985). Messenger (1979a, 1983) argued that the peduncle lobe, however exactly it regulates motor programs, has to match incoming sensory information of different modalities or from different sources. This was vindicated in the findings that the statocyst, arms, and mantle inputs all converge on the lateral bank of the spine of the peduncle lobe, while information from the contralateral optic and peduncle lobes converge in the median bank of the peduncle spine (Camm *et al.* 1985).

The peduncle lobe is thus analogous to the cerebellum of lower vertebrates. It is a centre concerned with regulating motor behaviour primarily on the basis of visual cues, although it also receives input from the statocyst and may be better termed a visuo-motor centre (Messenger 1967b).

Olfactory lobes

Each olfactory lobe (Boycott and Young 1956) lies medial to the peduncle lobe. Three areas are recognized, the posterior lobule, the middle lobule, and the anterior lobule (Messenger 1971), lying along an anteroposterior plane, the middle lobule being more medial than the other two. All three lobules are interconnected by their neuropil, and medially this becomes confluent with that of the basal zone of the peduncle lobe. The olfactory nerve arrives from the olfactory organ after passing into the orbit where it runs forwards and upwards to the edge of the optic tract.

Olfactory organs, mechano- and chemoreceptors

The olfactory organs are paired; each takes the form of a small pocket, or pit, in the skin. Each organ is found lying ventral and posterior to the eye and just ventral of the junction between the anterior edge of the mantle and the side of the head. The organ comprises an epithelium that appears pseudostratified, columnar and ciliated, and, below this, cells apparently vacuolated and ring-shaped. The ultrastructure of the superficial layer of the epithelium shows it to consist mainly of cells bearing microvilli. Below this, and often very

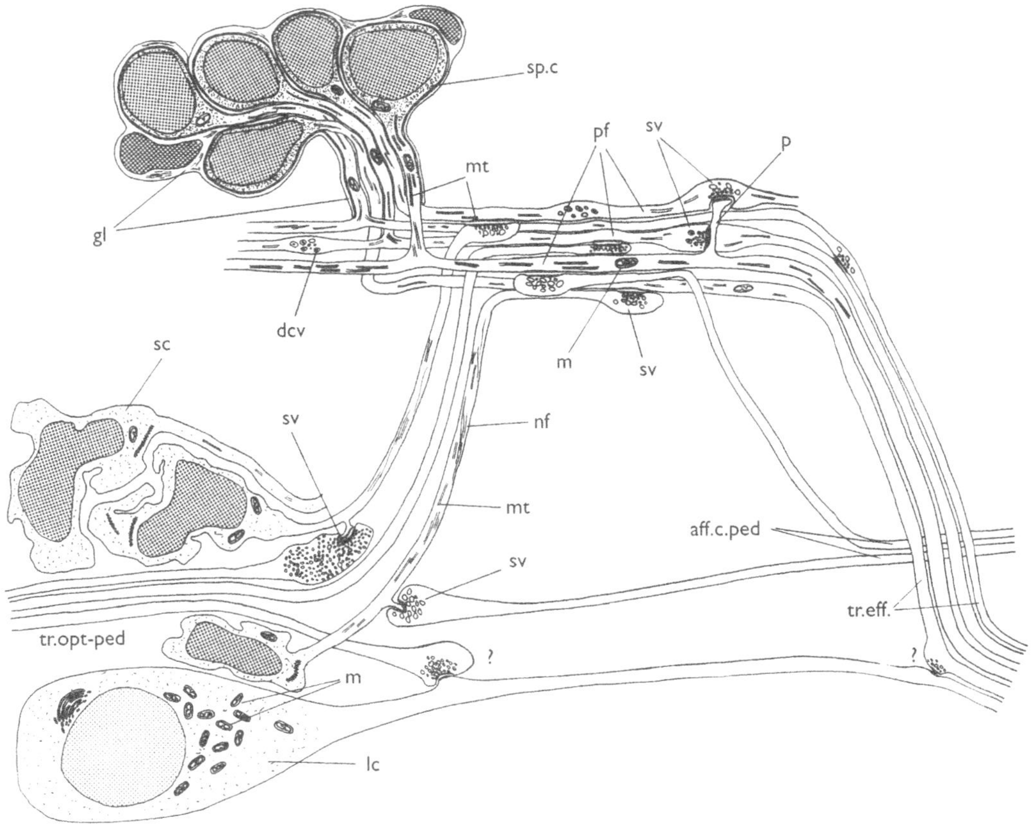

Fig. 31.32. *Octopus vulgaris.* Diagram to show some of the principal connections of the peduncle lobe seen in a transmission electron microscope study. aff.c.ped, contralateral peduncle lobe afferent fibres; dcv, dense-cored vesicles; gl, glia; lc, large cell of basal zone; m, mitochondria; mt, microtubules; nf, neurofilaments; p, spiny process; pf, parallel fibres; sc, small cell of basal zone; sp.c, cell of peduncle spine; sv, synaptic vesicles; tr.eff, efferent tract; tr.opt.-ped, optic-peduncular tract; ?, possible contacts to large cells (Woodhams 1977).

deep in this layer, lie many large cells that bear numerous cilia. These cells open to the surface via narrow processes running between the epithelial cells. The innervation of these cells has not been established, but the large cells are assumed to be receptors by comparison with olfactory organs of other animals, and it seems reasonable to suggest that the organ is chemosensory (Messenger 1967b, Woodhams and Messenger 1974). It has been suggested that this organ may be sensitive to sexual pheromones. In behavioural experiments *O. vulgaris* responds to various chemicals, many at low concentration (Wells *et al.* 1965, Chase and Wells 1986). Electrophysiological recordings from cells in the olfactory organ of *Loligo opalescens* provide further evidence for a chemosensory function (Gilly and Lucero 1992, Lucero and Gilly 1995).

Mechanoreceptors are apparently present, as an octopus will react to a light touch made anywhere on its body surface and can respond to food dropped on to the back of its mantle. There is physiological evidence for mechanoreceptors responding to stretching of the mantle; multipolar stellate cells with centripetal axons among the muscle layers of the mantle have been described (Bone *et al.* 1995).

Octopus is sensitive to chemical stimuli and is likely to have chemoreceptors on much of the body surface, including the lips and the suckers. Each sucker has about 10 000 primary receptors, and with some 200 suckers on each arm there must be a total of about 16 million of these cells in an adult octopus (Graziadei 1971). Training experiments have shown responses to dilute solutions of sugar, hydrochloric acid, and quinine in seawater, and between seawater with half, normal, and double-strength salinity (Wells 1978). Octopuses responded by moving upstream towards a solution of adenosine 5′-monophosphate (AMP), which attracts other species, and less strongly to glycine and glutamic acid (Chase and Wells 1986). Exposure of an isolated octopus to water in which crabs had lived resulted in an increase in ventilation rate (Boyle 1983c). Another species, *Octopus maya*, in a study of chemotaxis in a Y-maze, was strongly attracted to the compartment with crab extract, proline, or adenosine triphosphate, but movement was arrested by betaine (Lee 1992). These findings do suggest that distance chemoreception is probably of importance in the detection of prey.

Stellate ganglia

One stellate ganglion lies on each side of the animal, below the epithelium of the mantle, and anteromedial to each gill. Each ganglion (with approximately 120 000 nerve cells) forms one of the larger peripheral ganglia, and is divided into two distinct parts (Lund 1971, Young 1971). The differing structure of the two parts, one with small cells and one without them, was described in detail by Young (1972). A study of embryonic development until hatching showed

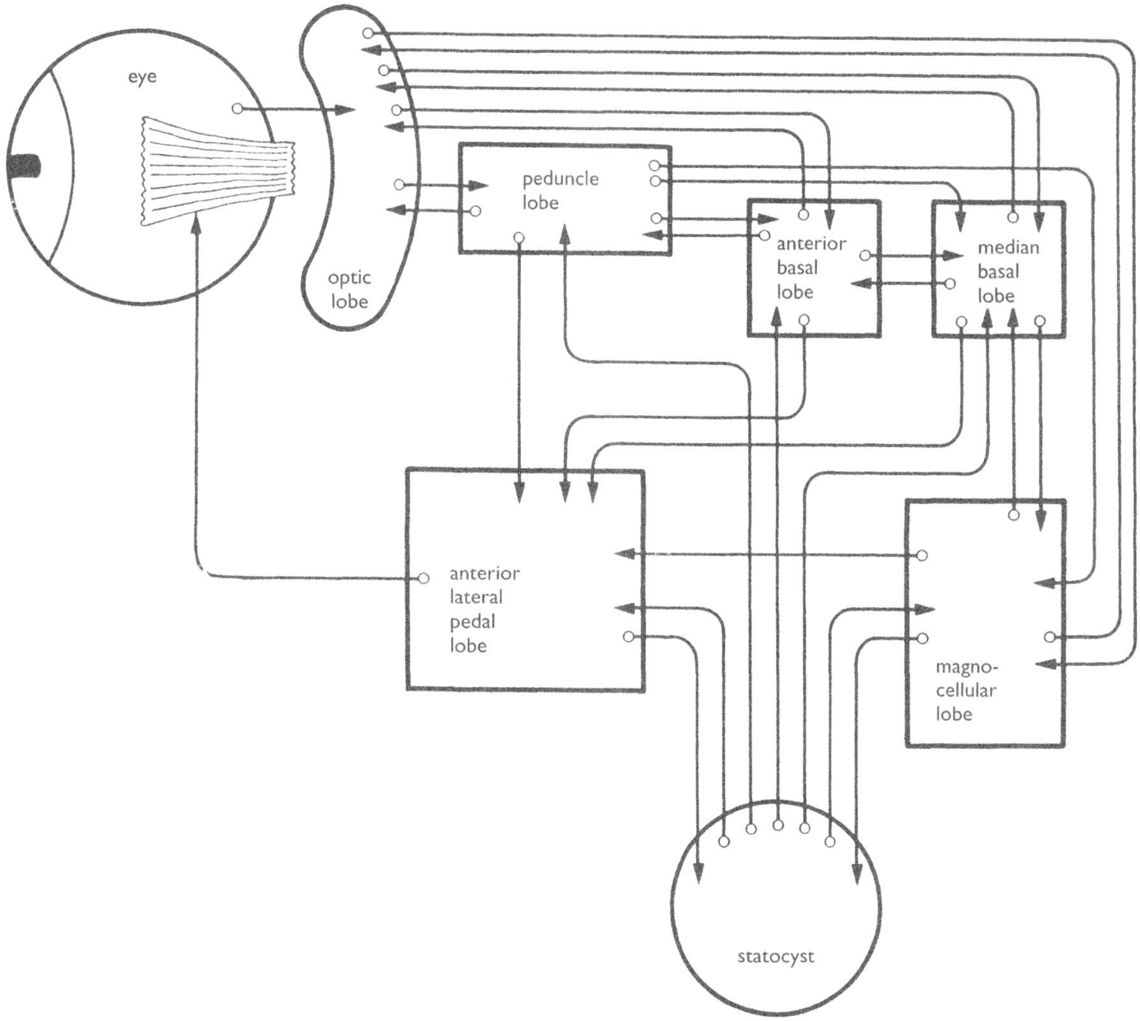

Fig. 31.33. *Octopus vulgaris*. The statocyst–oculomotor system and associated pathways in the central nervous system, shown diagrammatically. The figure includes pathways taken from Young (1971) and Saidel (1982) (Budelmann and Young 1984).

that the stellate ganglion has a direct connection with the ectodermal germinative layer (Marquis 1989).

The stellate ganglion is a way station on the control pathway from the central nervous system to the muscles of the mantle for respiration and locomotion. These muscles perform the functions both of respiration and locomotion. The ganglion is connected with the central nervous system by the pallial nerve, or mantle connective, and to the mantle muscles by the stellar nerves (Fig. 31.34). There are about twenty five stellar nerves.

The division of the ganglion into two parts is sharp, and is approximately in the horizontal plane. The ventromedial part faces the cavity of the mantle and so is exposed when the mantle is opened. This part comprises some 31 000 cells of 15–60 μm diameter, and the estimate for the number of fibres in the ventral roots of the stellar nerves gave a similar count. It seems likely that each cell of the ventral part

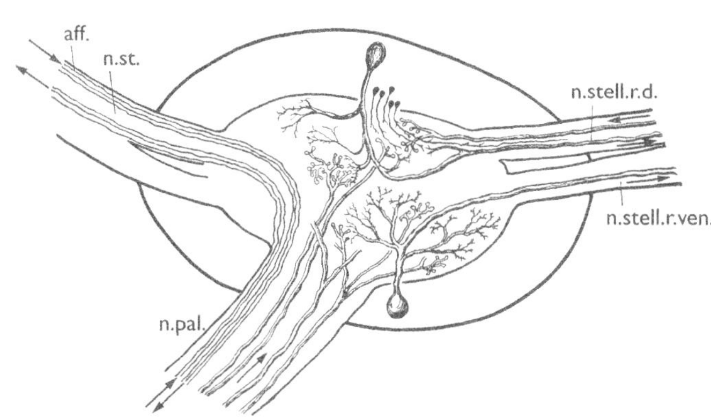

Fig. 31.34. *Octopus vulgaris*. Probable connections in the stellate ganglion; details of the finer branching remain speculative. The dorsal wall of the ganglion is at the top of the figure (Young 1972).

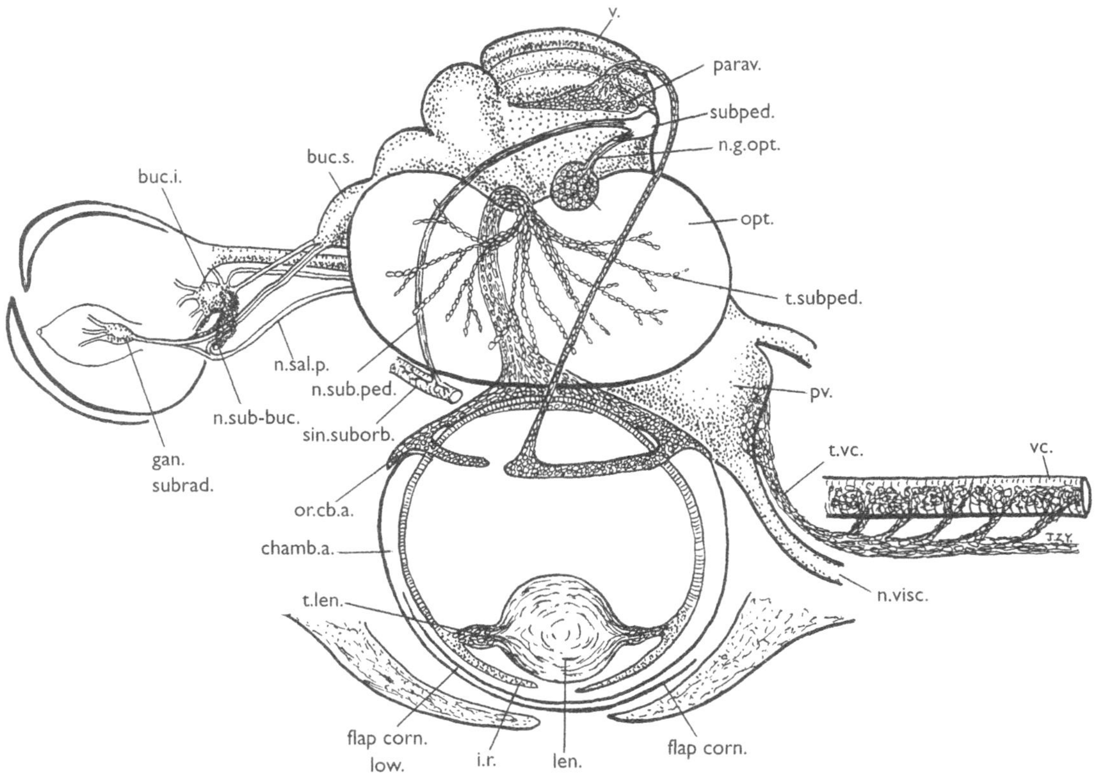

Fig. 31.35. *Octopus vulgaris*. Diagram to show the various neurovenous tissues (Young 1970b).

of the ganglion gives rise to one ventral root nerve fibre. The pallial nerve enters through the ventral, large-celled, region. The dorsal part of the ganglion is more differentiated with large cells and many small cells in its perikaryal layer. Its neuropil is more complex than that of the ventral part, and has tightly woven and irregularly arranged small fibres. This part of the ganglion is close to the muscles. The dorsal layer of the ganglion contains some 35 000 cells 15–60 μm in diameter, and 52 000 cells of 10 μm or less in diameter. Each stellar nerve has two roots, one from each part of the ganglion.

The stellate ganglia of *Sepia*, *Loligo*, and *Octopus* are built on the same plan, suggesting an ancient inheritance (Young 1972).

Epistellar bodies

An epistellar body lies on the ventral surface at the hind end of each stellate ganglion, and is a small colourless vesicle (Young 1971). In the 1960s the ultrastructure of this body revealed a striking resemblance to photosensitive structures in invertebrate animals. Investigations showed the presence of two cell types: large ovoid bipolar sensory cells with a cell body and a long apical process forming a rhabdomeric structure, and smaller, supporting or epithelial cells interspersed. No dioptric apparatus is associated with the vesicle (Mauro 1977). The innervation and electrophysiological responses of these vesicles have been recently investigated in *Eledone cirrhosa*, but their function remains to be determined (Cobb *et al.* 1995a,b, Cobb and Williamson 1998b).

Neurovenous tissues

There are a number of structures mostly found in relation to the central nervous system and the anterior vena cava (Fig. 31.35). These somewhat enigmatic tissues connect the nervous system with the venous system, and although there are differences these tissues appear to form a single set. They consist of strands of cells and nerve fibres forming columns. The strands arise from the outer surfaces of central or peripheral ganglia and end in relation to the lining of a cavity, such as a vein or the anterior chamber of the eye. The subbuccal tissue and the neurovenous tissue of the vena cava resemble each other and their fibres end in the walls of veins; they are well developed in *O. vulgaris* (Young 1970b). The physiology and possible functions of these neurosecretory tissues are as yet not known (Wells 1978).

Statocyst–oculomotor system

Like many other cephalopods the octopus shows pronounced compensatory eye movements. These are under the control of equilibrium receptor systems and visual reflexes. The eye movements are controlled by extraocular eye muscles, the arrangement of which was investigated by Budelmann and Young (1984) (Table 31.3).

Octopus vulgaris has seven extraocular eye muscles (Fig. 31.27). Powerful recti muscles produce linear movements and oblique muscles rotation. Some of the latter are very thin sheets and pass halfway round the eyeball. The eye muscles are controlled by seven nerves, three of which innervate more than one muscle. Two of the nerves run only to extraocular eye muscles, the other five contain additional fibres for the iris, chromatophores or skin.

After cobalt filling of the central ends of the eye muscle nerves it was found that all have fibres originating in the ipsilateral anterior lateral pedal lobe, the oculomotor centre. Stimulation of two nerves gave rise to the expansion of the chromatophores of the iris (Fig. 31.25) and were shown to contain fibres with somata in the ipsilateral anterior chromatophore lobe. Two nerves gave constriction of the pupil and proved to contain fibres with somata in an area between the posterior pedal and magnocellular lobes, thus demonstrating the position of the pupillary control centre. Stimulation of one nerve resulted in dilation of the pupil but the origin of the relevant cells is not clear.

Cobalt filling of the central ends of the macula and crista nerves showed the destinations of their afferent fibres in many parts of the brain, including the oculomotor centre and higher motor centres, the basal and peduncle lobes. Furthermore many somata of efferent fibres to the statocyst were filled in the anterior (oculmotor centre) and posterior lateral pedal lobe, and in the posterior pedal and magnocellular lobes.

The statocyst–oculomotor system includes two pathways from the statocyst equilibrium receptor organs to the motoneurons of the eyes (see Fig. 31.33). One is a direct pathway and another an indirect path via higher integrative centres where visual information about movement is combined with that coming from the statocysts. Thus there is a well-developed system for the control of eye movements based in an oculomotor centre.

Statocysts

The statocysts provide information to effect control of balance during swimming, jetting, and walking. There is a direct pathway from the equilibrium receptor system to the oculomotor neurons, as well as an indirect pathway. In the latter, visual information about movement initially interacts in various regions of the brain with that coming from the equilibrium receptor systems, before finally reaching the oculomotor centre (see Fig. 31.33) (Budelmann and Young 1984).

Young (1960b, 1971) gave the first detailed morphological description of the statocyst of *O. vulgaris*. This led to studies of their structure (Budelmann and Wolff 1973, Budelmann *et al.* 1987a, 1987b, 1997) and function (Wells 1978), and to morphometric analyses (Maddock and Young 1984, Young 1989), and to studies of the statocyst–oculomotor system (Budelmann and Young 1984, Budelmann 1994). In the statocyst of the newly hatched octopus the crista has three parts, and cupulae are absent, as are anticristae (Bülow and Fioroni 1989). Thus it is much less complex than in the adult.

The statocyst is approximately spherical in outline when seen from above and has several bulging regions. Each statocyst is situated just below the brain, and is associated with the cranial cartilage. The outer capsule of cartilage encloses an approximately oval cavity lying below the hind end of the suboesophageal mass. The spherical membranous sacs are smaller than the statocyst cavities and separate endolymph inside the sac from the surrounding perilymph (Amoore *et al.* 1959). The thin membranous wall contains three types of end-organ:

- an oval macula, on which hangs a statolith
- a continuous ridge of receptors, the crista, in three planes
- hair cells scattered over the inside of the wall, which has a rich nervous plexus (Fig. 2.14b).

The sac has two separate receptor systems. One, vertically oriented, is the macula/statolith system for the detection of gravity and other linear accelerations. The other is the crista/cupula system for the detection of angular accelerations, associated with this system is a

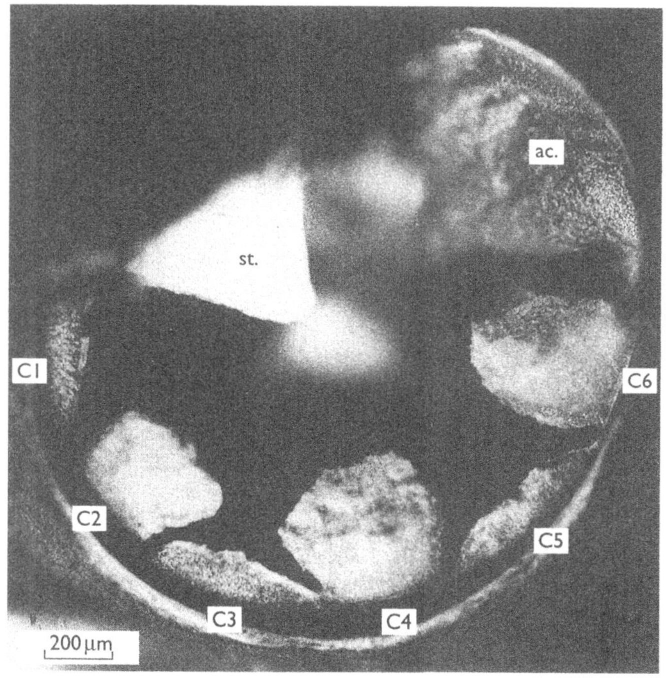

Fig. 31.36. *Octopus vulgaris*. The ventral view into an opened statocyst sac to show the cupulae of the six horizontally arranged crista segments (C1–C6) (scale bar 0.2 mm). (Budelmann *et al.* 1987a).

cartilaginous anticrista (Figs 2.14b, 31.36) (Budelmann *et al.* 1987a, 1987b). The gravity receptor system includes the oval macula which has 5000–6000 hair cells arranged in concentric rings (Budelmann *et al.* 1973). The hair cells are all secondary sensory cells that transmit impulses to the central nervous system via about 2000 first-order afferent neurons (Budelmann and Wolff 1976). The cellular, neuronal, and synaptic organization of the macula is shown in Fig. 31.37 (Colmers 1981). The principal macula statica is covered with a statolith. The functional and biological significance of the gravity receptor system was investigated by Wells (1960, 1978) who found that the octopus suffered some disorientation after ablation of the statocysts. Schöne and Budelmann (1970) demonstrated that the system is organized to eliminate the effect of changing magnitude of shear force while preserving information about change of direction.

The structural organization of the statocyst sensory epithelia, especially of the crista, is very sophisticated (Fig. 31.38) (Budelmann *et al.* 1987b). The crista forms a ridge that runs inside the statocyst in the three dimensions of space and is subdivided into nine segments (Young 1960b). Each segment of the crista carries its own cupula, a small sail-like flap which protrudes freely into the cavity. Two morphologically and physiologically different crista/cupula systems are present. One has a large cupula and the other a small cupula and they alternate along the length of the crista, thus the angular acceleration systems are separable into two receptor subsystems (Williamson and Budelmann 1985a, 1985b, Budelmann *et al.* 1987a). The first, with the large cupulae, has a threshold about 10 times lower than the second. It is suggested that the significance of these two subsystems may be related to slow walking and fast, jet-propulsive movements (see Budelmann 1994).

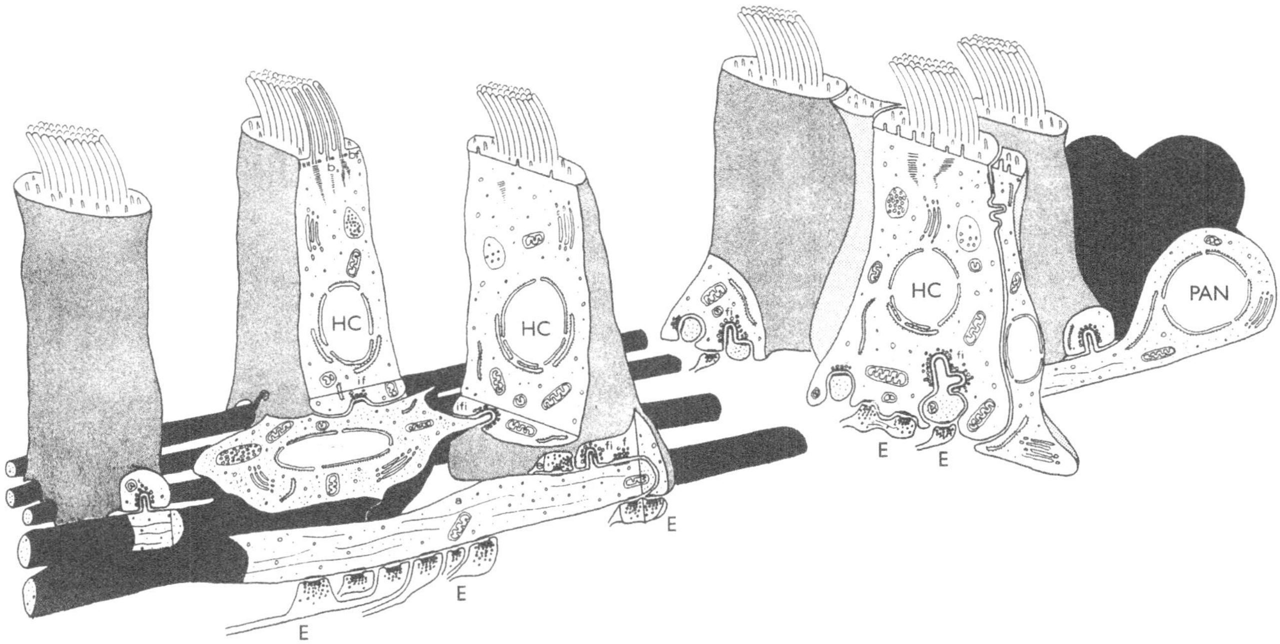

Fig. 31.37. *Octopus vulgaris.* A diagram to show the cellular, neuronal and synaptic organization of the macula of the statocyst (the synaptic contacts have been enlarged for emphasis). The secondary sensory hair cells (HC) are in afferent synaptic contact with two types of first-order afferent neurons; an intramacular afferent neuron (AN) and a perimacular afferent neuron (PAN). The hair cells and neurons receive many efferent endings (E) (Colmers 1981).

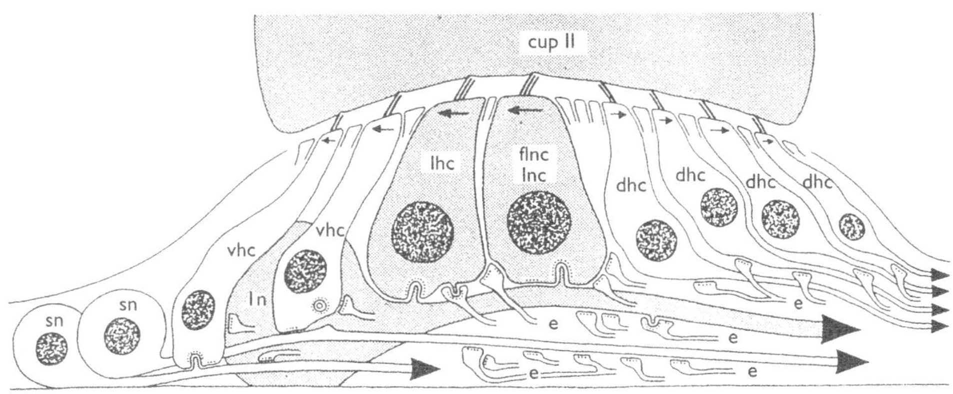

Fig. 31.38. *Octopus vulgaris.* Diagram of the crista. The crista comprises three types of receptor cells: primary sensory hair cells (dhc), large secondary sensory cells (flhc, lhc) in afferent synaptic contact with large first-order afferent neurons (ln), and small secondary sensory hair cells (vhc), which are in afferent synaptic contact with small first-order afferent neurons (sn). All of these elements receive multiple efferent endings (e). Only the basal part of the large cupola (cup.) (type II) is shown. Arrows in the distal ends of the hair cells indicate their direction of polarization (Budelmann *et al.* 1987a).

The synaptic organization of the sensory epithelia is complex (Figs 31.37, 31.38). Secondary sensory hair cells are in afferent synaptic contact with first-order afferent neurons, and two types of synapse are present (Budelmann *et al.* 1987a, 1997). The maculae and cristae segments are innervated by one macula nerve and three crista nerves, composed of afferent and efferent fibres, with fibres ranging from less than 1 μm to 18 μm.

31.1.2 Ameloctopus

Ameloctopus appears to be limited to the tropical waters of northern Australia from the southern Great Barrier Reef to north-western Australia (Norman 1992). Here it inhabits intertidal coral, rubble, sand, or mudflats in regions where the tidal range varies enormously and may be greater than 7 m.

Ameloctopus litoralis is the only species of the genus and is characterized by the absence of an ink sac, a feature it shares with the deep-living Bathypolypodinae. However, the very long arms, short webs, and well-developed crop indicate a closer alignment with the shallow-water Octopodinae.

This octopus forages at night, on open sand and on reef flats, in pools less than 100 mm in depth. It inhabits lairs excavated in the substrate and will extend one or more of its arms to capture prey. Examination of the stomach contents revealed fragments of small crustaceans. The salivary glands are notable as the anterior glands extend along approximately a quarter of the length of the buccal mass and the posterior salivary glands are up to twice the length of the buccal mass. The reason for their large size is not known.

The largest specimens recorded include a mature male of 30.4 mm ML and a submature female of 30.8 mm ML. The ovary of one female contained some 80 eggs (up to 10 mm in length) at several stages of development. Hatchlings have not yet been captured but are assumed to be benthic since the egg are large. The smallest specimens examined (6.7 and 8.1 mm ML) already possessed extremely long and fully developed arms (Norman 1992).

31.1.3 *Amphioctopus*

This genus is found mostly in tropical waters, extending into warm temperate waters. It is centred in the Pacific, in south-east Asia, where several species occur and here form perhaps the most fished group of octopuses. Two species apparently passed through the Straits of Panama before its closure, and one has penetrated the eastern end of the Mediterranean (Gleadall, in prep.). *Amphioctopus* inhabits shallow waters of the sublittoral zone, and is found mostly where the bottom is of sand, or mud, or where there are coral reefs. It is much smaller than *Octopus*, although similar in other respects.

Amphioctopus (Octopus) burryi spawns about 35 000 eggs (2.2–2.5 mm in length). In the laboratory one female laid eggs and then carried the entire brood in a 'sac' formed by the aboral surface of arms III and IV. She continuously ran her arm tips through the eggs and occasionally directed a jet of water into the brood. Egg laying continued for 2–3 weeks and the embryos formed three distinct groups, each differing in developmental stage by about 1 week. Embryonic development lasted for about 28 days. The hatchlings (1.5 mm ML) were planktonic and swam continuously with a 30–45° head-down orientation. Each arm bore four suckers, and each hatchling possessed over 200 chromatophores. At hatching *A. burryi* exhibits simple patterns, and when a mantle length of 16 mm is reached the young displays a wide repertoire of body patterns including longitudinal arm stripes (Forsythe and Hanlon 1985). The young apparently display diurnal planktonic/benthic behaviour before settling out of the plankton, possibly at 40–60 days of age. The male *A. burryi* reaches sexual maturity at about 38 mm ML, and the gravid female had a mantle length of 107 mm. The life cycle was estimated to be of 9–10 months duration (Forsythe and Hanlon 1985). Other species, such as *A. fangsiao*, produce relatively few lecithotrophic eggs. This species (160–200 mm total length) has a distinct spring and autumn reproductive season in Japanese waters, suggesting a life span of about 6 months; its eggs are around 8 mm in length (Gleadall and Naggs 1991).

When alive *Amphioctopus (Octopus) hummelincki* is rich reddish yellow-brown in colour with yellow mottling. It possesses an ocellus below each eye, between arms II and III, that is purplish blue within an irregular reddish-brown patch. During the deimatic response of *A. fangsiao* each ocellus is highlighted by a dark disc, while one or two dark stripes emphasize the outline of the arms; together this gives an illusion of a large face (Gleadall in prep.).

31.1.4 *Enteroctopus*

Enteroctopus is found in the coastal waters of the North Pacific Ocean, from the Okhotsk, Bering, and Japan Seas to San Diego, California. It inhabits depths from shallow subtidal waters to 1500 m and is benthic in habit. The genus includes *E. megalocyathus*, *E. magnificus*, and the giant Pacific octopus *Enteroctopus (Octopus) dofleini*, a new combination considered to represent a complex of species (Hochberg 1998).

The living animal has longitudinal ridges of skin on the mantle and arms, two large primary papillae over each eye, a pair of small, faint, white spots on the dorsal mantle, and a single small white spot on the front of the head (Anderson 1995). This octopus is generally nocturnal, and lives in a den which it occupies in daylight hours to gain protection from predators. It may use the same den for several months (Hartwick 1983).

There is considerable information on the biology, life history, and reproductive habits, of this octopod (Hochberg and Fields 1980, Hartwick 1983, Hochberg 1998). *Enteroctopus dofleini* is one of the largest octopod species known, and records include one with an arm spread of 9600 mm and weight of 272 kg (Hochberg and Fields 1980), a female of 350 mm ML, and a male of 445 mm ML (Hochberg 1998). Mature animals migrate offshore in the summer to deeper waters where mating takes place and this may last for 2–4 h. Each spermatophore is a metre or more in length and one or two are inserted into an oviduct; a portion is anchored while the rest hangs free (Mann *et al.* 1970). Spawning occurs after the octopus returns to relatively shallow water where there is rock and gravel and eggs are laid over a period of up to 15 days. The egg capsule is 6–8 mm in length, and the eggs are laid in strings or festoons, and suspended from the ceiling of the den. The number of eggs may reach 100 000, depending on the size of the female. Development takes 5–6 months, and is temperature related. Egg-laying and hatching both take place in darkness. After hatching the young (3–4 mm ML, 6 mm total length, and about 0.03 g in weight) joins the plankton. This phase has been estimated to last for up to 3 months, after which the paralarva becomes largely benthic, and the life span is probably of 2–3 years although a few individuals reach an age of 5 years (Hartwick 1983). Both hatchlings and paralarvae can feed on small euphausiid crustaceans at the surface (Marliave 1981).

31.1.5 *Hapalochlaena*

These octopods are found in shallow waters to depths of 10 m in temperate areas of Indo-Pacific waters (Norman and Sweeney 1997), and around the coasts of Australia they have become notorious for causing fatalities in humans (S.K. Sutherland 1983). They often inhabit crevices in the rocks, empty mollusc shells, or in close proximity to attached green algae. The horizontal distributions of three species differs. They have distinctive iridescent blue markings: the greater blue-ringed octopus occurs quite widely in Indo-Malay waters, the lesser blue-ringed octopus lives along the coasts of southern Australia and Tasmania, and the blue-lined octopus is restricted to two populations—eastern Australia (Norman and Sweeney 1997) and the South and East China Seas as far as southern Japan (Gleadall 1997).

Characteristically *Hapalochlaena* possesses conspicuous dark maculae with associated iridescent blue rings or lines on the mantle and small blue rings on the head, web, and arms (Roper and Hochberg 1988, Norman and Sweeney 1997). The greater blue-ringed octopus (*Hapalochlaena lunulata*) has rings 7–8 mm in diameter in regular longitudinal rows down the dorsal mantle; the lesser blue-ringed octopus (*H. maculosa*) has many small rings, only 1–2 mm in diameter, scattered across all dorsal surfaces; and the blue-lined octopus (*H. fasciata*) has lines and elongated dark maculae on the mantle and small rings elsewhere (Fig. 31.39).

Body patterns and behaviour

After observing the body patterns of *H. maculosa*, in the sea and in aquaria, Roper and Hochberg (1988) commented that this octopus 'looked like a piece of alga' as it clung to branches of a fixed alga (*Halmida*), which it closely matched in both colour and texture. One animal hovered motionless above the alga but on sighting the diver dropped instantly to the sea floor as if in 'free fall'. Once on the bottom it 'assumed the deimatic pattern with arms spread, web expanded, and pale in color'. On capture the octopus flashed a pattern of vivid blue rings surrounded by dark maculae, followed by a dark reddish-brown flush.

Life history

Hapalochlaena fasciata (identified as *H. maculosa* by Tranter and Augustine 1973), is unusual as the mother carries the eggs with her (Cropp 1972). One female, found in a partly opened oyster shell, was observed in an aquarium until her death. Six arms and the web formed a brood pouch, leaving only two arms for locomotion. Some 150 eggs (6.0–9.0 mm) in the pouch were cleaned and aerated at intervals, as the female made sweeps with the tips of the arms. During development reversal of the embryo took place accompanied by a brilliant chromatophore display, the whole process being completed in about 7 min. Hatching took place several days later and the young immediately settled on the substrate where they remained in the vicinity of the mother for 3–5 days. The hatchlings were capable of swimming vigorously and at 9–12 days the young would eject ink if disturbed, but rarely did so after 33 days of age. The characteristic blue rings were not apparent until 42 days of age. The mother survived for 8 weeks after the eggs hatched. The hatchlings were reared in the laboratory for 4 months, at the end of which time they were sexually mature (35–40 mm ML). During mating the male mounted the female from the dorsal side and wrapped six arms around her neck, leaving arms IV to trail over her body; the hectocotylized arm was inserted deeply into the mantle of the female (Tranter and Augustine 1973).

A gravid female *Hapalochlaena lunulata* (Philippines) spawned several clusters each of about 20 eggs (2.5–3.5 mm long) in an aquarium (Overath and Boletzky 1974). The clusters were attached to the wall of the tank, and brooded by the female. At hatching the young (2.3 mm ML) swam close to the surface of the water and remained in the plankton counteracting passive sinking by jet propulsion. The hatchlings would also attach themselves to the substrate by means of their suckers.

A newly hatched young *H. maculosa* (south and south-western Australia) is benthic in habit. At 5–6 mm ML the skin has scattered papillae, and a characteristic mottled pattern of dark brown chromatophores, but no iridescent markings at this stage. The arms are 1.25 times the mantle length, with four transverse bands of dark chromatophores. The ink sac is functional as young of this size discharge a thin, reddish-brown, cloud of ink, which dissipates rapidly. Larger specimens (6–20 mm ML) have the same characteristics of the skin except that small, iridescent, blue rings (0.1–1.0 mm in diameter) are present on the mantle, head, web, and arms; some 50–55 blue rings being on the mantle. The ink sac is no longer functional in animals of this size, as not one expelled ink even when repeatedly disturbed. The adult (up to 60 mm ML) exhibits the same skin patterns and the same number of blue rings (0.5–2.0 mm in diameter) remain on the mantle. Right arm III of the male is hectocotylized and is 20–30% shorter than left arm III. Mature eggs are 7–9 mm in length (Stranks and Lu 1991). Of the three species *H. lunulata* has the widest horizontal distribution and it also has a short planktonic phase in its life cycle, in contrast to the other two species.

Fig. 31.39. *Hapalochlaena fasciata* (40 mm ML). The pattern of rings and lines allows identification of the specimen.

Buccal mass and salivary glands

Histological sections of the buccal mass of *H. fasciata* show the anterior salivary glands to be large (Fig. 31.40). Ventrally the submandibular gland stretches across most of the width of the buccal mass. The posterior salivary glands are paired and lie in the mantle and their ducts join to form a single duct which passes forward through the papilla of the posterior salivary gland to terminate just below the radula (Fig. 31.41). The duct has a diameter of 140 μm and its tip is almost

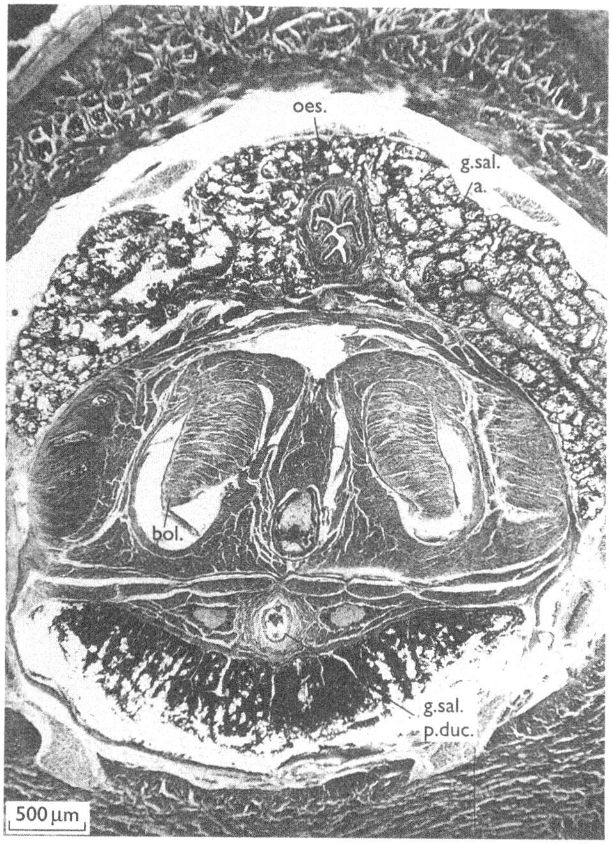

Fig. 31.40. *Hapalochlaena fasciata* (40 mm ML). Transverse section through the buccal mass showing the densely organized muscles and the well-developed glandular tissues.

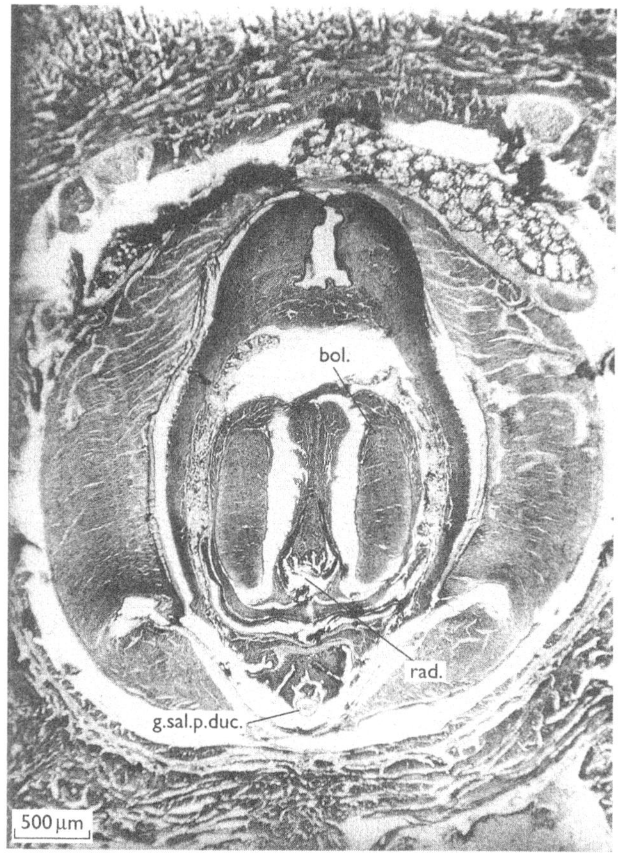

Fig. 31.41. *Hapalochlaena fasciata* (40 mm ML). A transverse section anterior to that in Fig. 31.40 to show the densely organized tissues of the lateral buccal palps and the posterior salivary gland papilla carrying the duct from the gland.

certainly eversible, as it is in *O. vulgaris* (Nixon 1979b, 1980). The tip of this duct has minute teeth which could penetrate prey and allow the toxic saliva to enter (Fig. 31.42). The papillary sheath has small chitinous teeth, approximately 9 μm in height, on its surface. The development of the posterior salivary glands of *Hapalochlaena maculosa* has been followed in 5, 15 and 42 day-old juveniles and adults; the glands have a compound tubulo-acinar exocrine structure and it is suggested that the movement of the saliva is by muscular contractions of the gland itself (Gibbs and Greenaway 1978).

The brilliant iridescent blue markings of these small, shallow-living octopuses make them attractive to swimmers, making a closer examination tempting. However, handling them may prove dangerous or even fatal (S.K. Sutherland 1983); several people have died after picking up and handling these animals. These fatalities prompted a series of studies of the posterior salivary glands of *Hapalochlaena*, which are the source of the toxin. Its effect is to block neuromuscular transmission, resulting in paralysis, respiratory impairment, and reduced blood pressure in humans (Trethewie 1965). A low-molecular-weight toxin was isolated from extracts of posterior salivary glands and named 'maculotoxin' by Freeman and Turner (1970); this substance has properties similar to tetrodotoxin and after purification was found to be the same neurotoxin (Sheumack *et al.* 1978). A further toxin has been isolated, hapalotoxin, which can also be lethal but takes longer to act than tetrodotoxin (Savage and Howden 1977). In addition three physiologically active amines (serotonin, tyramine, and histamine), are present (Howden and Williams 1974). When extracts from the posterior salivary glands or tetrodotoxin were injected into blue-ringed octopuses or into *in vitro* preparations of isolated nerves of this animal, there were no ill effects. However, both the extract and tetrodotoxin proved potent when tested on similar *in vitro* preparations from other species (Flachsenberger and Kerr 1985). Tetrodotoxin-producing bacteria have been recorded in one species (Hwang *et al.* 1989). Besides the toxin secreted by the posterior salivary glands of one blue-ringed octopus (species uncertain) it lays eggs containing a tetrodotoxin-like substance (Sheumack *et al.* 1984); this too proved active when tested on isolated nerve preparations and the effects were indistinguishable from those of tetrodotoxin. The function of this toxin is not known but it may act to poison an egg-eating predator quickly enough to save remaining eggs from ingestion.

31.1.6 *Danoctopus* and *Pteroctopus*

Confusion among previous identifications of octopuses occurring at mid-depth (100–1000 m) led Norman *et al.* (1997) to further investigate these animals. These authors placed *Danoctopus* and *Pteroctopus* in a genus-group complex, with two subgroups: species

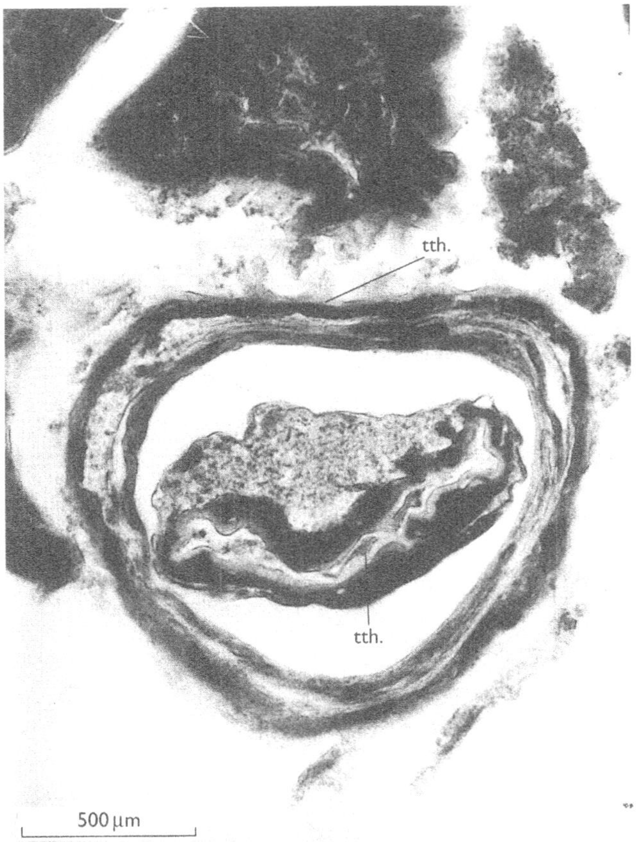

Fig. 31.42. *Hapalochlaena fasciata* (40 mm ML). The papilla of the posterior salivary gland the surface of which bears small, pale-staining teeth. Within the papilla is the duct of the posterior salivary gland; this part of the duct is eversible and has teeth which are exposed when the duct is everted (see Fig. 31.14).

of the genus *Danoctopus* Joubin, 1933, with the hectocotylus on right arm III of the male, and species of the genus *Pteroctopus* Fischer, 1882, with left arm III of the male hectocotylized.

Danoctopus is known from the tropical Atlantic and the Indo-West Pacific (Nesis 1987). The adults are usually in the upper bathyal zone in depths of 140–1200 m and are benthic in habit, but immature forms have been caught in midwater (Voss 1956). The maximum size recorded so far is a male of 38 mm ML (Cairns 1976).

Pteroctopus tetracirrhus lives on the outer part of the continental shelf and upper slope in depths of 100–800 m, but mostly between 250 and 500 m (Mangold-Wirz 1963). It is found in the Mediterranean Sea (Mangold and Boletzky 1988), in Madeiran waters, along the west coast of Africa to 4°N, the east coast of America from 34°N to 38°S (Roper *et al.* 1984) including the Gulf of Mexico (Voss 1956), and in Indonesian waters (Norman *et al.* 1997). In Madeiran waters one specimen was filmed from a submersible at 790 m depth on a bottom consisting of mud, coral, and rocks, and another at 517 m depth was moving slowly and searching among the rocky irregularities (Gonçalves and Martins 1992). The largest recorded specimens are a female of 132 mm ML and a male of 107 mm ML.

Our specimen of *Pteroctopus tetracirrhus* (Fig. 31.43) was alive when brought to the aquarium in Naples. It could swim, and the arms were seen to act together although they appeared to provide little aid to locomotion. Touching the funnel resulted in its retraction

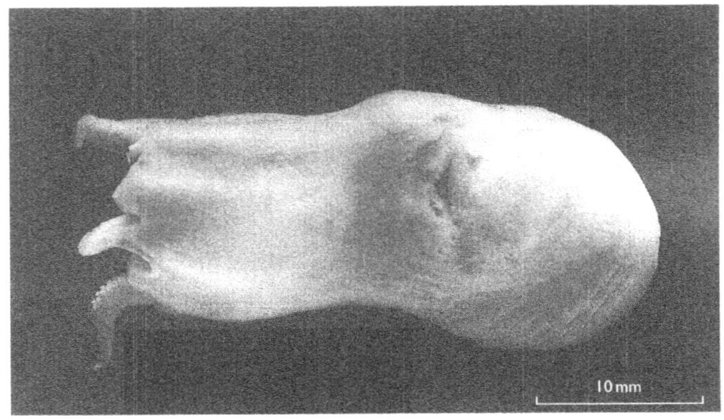

Fig. 31.43. *Pteroctopus tetracirrhus* (18 mm ML). Ventral view of specimen caught in the Bay of Naples, and preserved.

into the mantle. The animal was almost uniformly orange, although paler ventrally, and it displayed only slight changes, becoming paler or darker (B.B. Boycott, pers. comm.). The chromatophore lobe of this specimen forms 1.5% of the total brain volume (Maddock and Young 1987).

This octopus mates *à distance*, the only link being the outstretched hectocotylus of the male as this is introduced into the mantle cavity of the female (Mangold 1987). One female of 80 mm ML was alive after capture in the western Mediterranean, and subsequently laid 75 eggs in an aquarium. At its death many remained in the ovary. It laid eggs (7 × 2.5 mm) singly and these were attached directly to the substrate by a stalk (Boletzky 1981). On the basis of egg size it seems that the young probably join the plankton after hatching (Boletzky 1976). They may remain there for some time since our specimen of 18 mm ML still possessed Kölliker's bristles, which in other species of octopus are associated with early life in the plankton (Fig. 31.44). The bristles have a basal cell and a muscular 'cup' below which are elongated chambers, lined with cells but seemingly empty, and a thick layer of loosely organized tissue. Internal to these layers is the mantle muscle, only the inner and outer layers of which have a relatively dense organization.

The arms have a thick layer of tissue with loose organization, but there are no chambers such as those seen in the mantle. Some of the muscles of the arms are dense but these are largely restricted to the periphery. The suckers are partly embedded in the subcutaneous tissues of the arms in preserved material. The muscles of the suckers are dense, unlike those of the arms.

There is little cartilage associated with the central nervous system. The brain is typically octopodine. A subfrontal lobe can be recognized and there is a suprabrachial commissure (Young 1977a). The median basal lobe forms 12.8% of the total brain volume. Principal component analysis of the lobes of the brain shows that *Pteroctopus tetracirrhus* lies far from *O. vulgaris* and closer to *Scaeurgus unicirrhus* and *Bathypolypus sponsalis* (Maddock and Young 1987). The epistellar body is relatively large, and faces towards the viscera (Fig. 31.45).

31.1.7 *Scaeurgus*

This octopod lives in the tropical–subtropical regions of the Atlantic, Pacific, and Indian Oceans (Roper *et al.* 1984, Sánchez and Alvarez

(a)

(b)

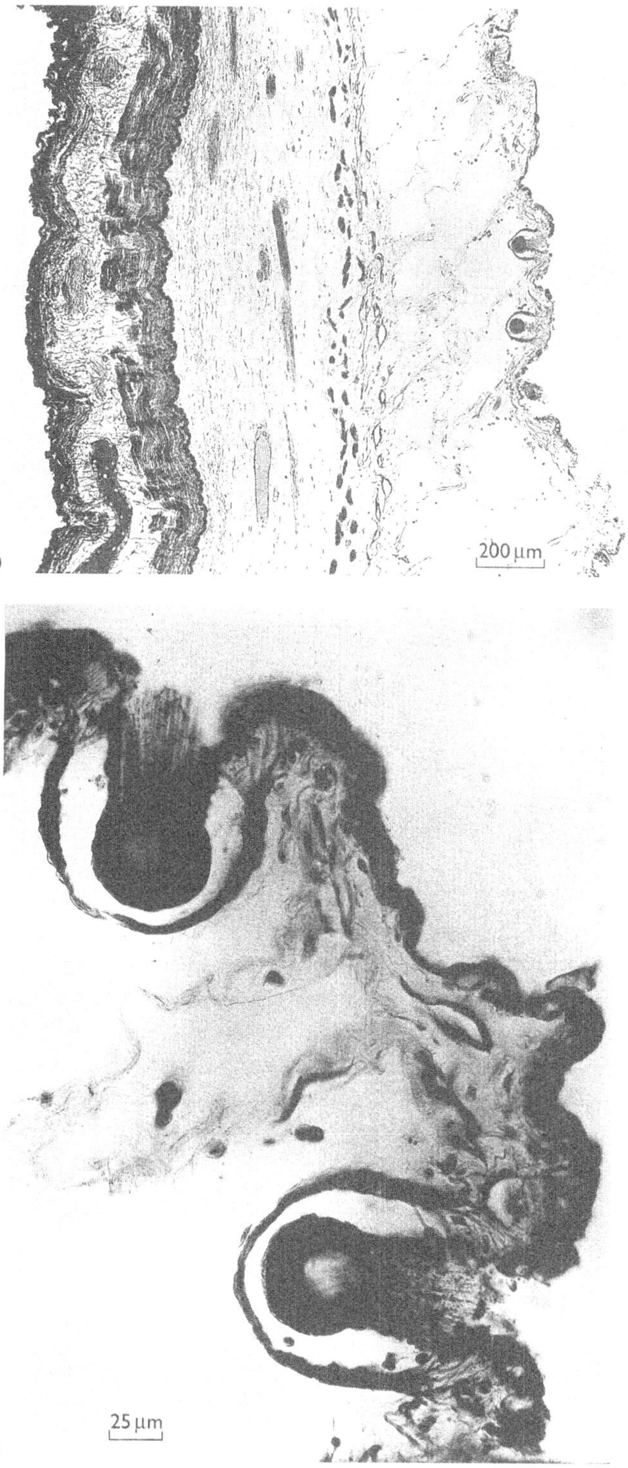

Fig. 31.44. *Pteroctopus tetracirrhus* (18 mm ML): (a) Section in sagittal plane through the mantle wall showing Kölliker's tufts, and the layer of fluid-filled chambers, lined with cells. (b) Kölliker's organs at higher magnification (see also Fig. 31.4).

1988), the Mediterranean Sea (Mangold-Wirz 1963), off the islands of the Azores, and in Madeiran waters (Gonçalves and Martins 1992). It is a characteristic inhabitant of the tops of seamounts (Nesis 1993a). Mostly benthic in habit in depths of 30–800 m, it is found where the bottom is of sand, madreporian coral, or barnacle beds (Mangold and Boletzky 1988, Ragonese *et al.* 1992a, 1992b, 1992c). A female of

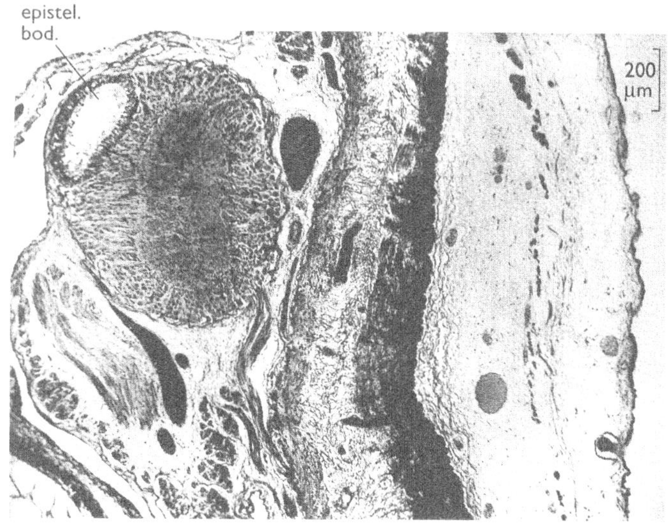

Fig. 31.45. *Pteroctopus tetracirrhus* (18 mm ML). The stellate ganglion with the epistellar body facing towards the viscera.

91 mm ML caught on the Valdivia Bank in the south-east Atlantic (Sánchez and Alvarez 1988) and a male of 90 mm ML caught on a seamount (Gonçalves 1991) are the largest specimens recorded so far.

A female *Scaeurgus unicirrhus* (50 mm ML) captured in the Straits of Florida was found with an egg mass in the ovary occupying half of the visceral mass; the eggs (1–1.5 mm long) looked like grains of rice (Cairns 1976). Females of 87 mm and 91 mm ML caught on a seamount of the Valdivia Bank were immature with small eggs and barely developed oviducal glands (Sánchez and Alvarez 1988). Ripe eggs reach a size of 2.5×1.0 mm and some 1000–2000 may be found in the ovary (Mangold 1989). Hatchlings (2.0–2.5 mm ML) have a planktonic phase immediately after hatching (Boletzky 1977, R.E. Young *et al.* 1989) and are distinctive with a short globose mantle, a wide funnel, a broad head with large eyes, short arms with filamentous tips, and skin densely packed with Kölliker's bristles (Boletzky 1984, Hochberg *et al.* 1992).

The suckers are numerous and closely set along the arms of the adult, and the arms and the suckers have dense musculature. The infundibulum of the sucker has closely packed, cone-shaped pegs arranged as narrow bands alternating with areas devoid of pegs, and a similar arrangement is found in *Octopus salutii* (Nixon and Dilly 1977). The proximal suckers of *Scaeurgus* are enlarged in the male and are displayed to the female at the time of mating (Jereb *et al.* 1992, Jereb and Ragonese 1993).

The central nervous system is octobrachian in form. The inferior frontal lobe system forms 6.4% and the vertical lobe system 15.2% of the total brain volume (Table 2.6) (Maddock and Young 1987).

31.2 Eledoninae

Eledonines inhabit intermediate depths. They possess one longitudinal row of suckers and an ink sac is present.

31.2.1 Eledone

Members of this genus live on the continental shelves of Europe, South America, Africa, and Australia (Boyle 1983a, Mangold 1983b,

Haimovici 1988, Lu and Stranks 1991). The juveniles and adults are benthic and inhabit the bottom where it is rather flat with sand, mud, or stones, and in areas of broken rocks and submarine cliffs, in depths of 10–800 m.

Life history

The eggs of eledonines range in size from 6 to 16 mm in length. The hatchlings of some species have a short planktonic phase, but others are benthic (Voss 1988a, Hochberg *et al.* 1992). The maximum size recorded is an animal of 950 mm ML (Roper *et al.* 1984).

Eledone cirrhosa and *E. moschata* (Fig. 31.46) both occur in the Mediterranean but the former is more widely spread and also lives along the coasts of western Europe, around the British Isles, and, as far north as the Lofoten Islands, Norway, and Iceland (Boyle 1986b). Where the two species are sympatric, *E. cirrhosa* occupies slightly deeper water. There is a marked difference in the size of the eggs and of the hatchlings of the two species, as well as the behaviour of their posthatchlings (Table 31.6). At hatching young *E. cirrhosa* may spend a very brief period in the plankton, probably remaining close to the bottom as small specimens have been captured in bottom trawls. This contrasts with posthatchling *E. moschata* which immediately adopts the benthic habits of the adult. The life cycles of both species have been reviewed (Boyle 1983a, Mangold 1983b).

The diet of the hatchling is not known, but the juveniles feed on crustaceans. The carapaces of crabs are often drilled by *E. cirrhosa*. The cavities made are small and the penetration hole is minute (Boyle and Knobloch 1981) but could be entered by the tip of the tall central cusp of the rhachidian tooth and by the apical tip of the eversible duct of the posterior salivary gland (Nixon and Boyle 1982). Entry of the tip of the duct would permit saliva which includes a toxin, cephalotoxin (Ghiretti 1959, 1960), to enter the prey. Another route of entry of the saliva is via the eye, which was found to be punctured by this octopus. Rapid immobilization of the crab ensued (Grisley *et al.* 1996). The exoskeletons of crabs are usually left totally devoid of tissue and separated at each joint. A proteolytic enzyme has been isolated and shown to selectively release the muscles from the exoskeleton of the crab (Grisley and Boyle 1987). Chitinase is also secreted by the posterior salivary gland, although its function is less obvious (Grisley and Boyle 1990), but during drilling may aid breakdown of the organic matrix of the crustacean carapace.

Fig. 31.46. *Eledone moschata* caught in the Bay of Naples and preserved.

Body patterns

The body patterns displayed by juvenile *E. cirrhosa* (Boyle and Dubas 1981) and *E. moschata* are more restricted than those of *O. vulgaris*. The characteristic components of the patterns are the same but lack the morphological separation by the grooves which are present in the latter (Packard and Hochberg 1977, Packard 1995). Chromatophores of the mantle and arms of *E. cirrhosa* are organized in circular patches, each centred above a leucophore, which forms a conspicuous component of the pattern when the former contract. The patches continuously vary in size and give a gradation of mottling. When the animal is disturbed there is an increase in contrast between light and dark areas of the skin. Extremes in the condition of the chromatophores are seen when the animal becomes either brick-red or entirely white with the exception of the eye ring; this last is the deimatic display (Boyle and Dubas 1981). These authors investigated the innervation of the chromatophore muscles by stimulating nerve bundles and monitoring movements of the chromatophores; the motor units were found to form precise parts of natural patterns. The horizontal keel around the mantle is another feature of the body pattern.

Receptors and central nervous system

The eye of *E. cirrhosa* has an ellipsoidal lens with a contractile, horizontal, slit-like diaphragm. This is effective in reducing optimal distortion, but the potential ability of the eye to produce images of relatively high resolution cannot be exploited because the grain of the retinal mosaic is not sufficiently fine (Sroczynski and Muntz 1985). These authors suggest that the pupil is used mainly for quick adaptation to changes in ambient illumination to increase the image contrast.

Table 31.6. *Eledone*. Various aspects of the early life of two European species. The ovarian eggs were counted gravimetrically (Mangold 1983b, Boyle 1983a)

	Eledone moschata	*Eledone cirrhosa*
Mantle length (mm)	140	>150
Weight (g)	600	750 male, 2000 female
Maximum size of eggs (mm)	16 × 5	7.5 × 2.5
Number of eggs	100–500	2000–54 000
Spawning period (days)	56–70	10–28
Egg development (days)	140–180	100–120
Mantle length at hatching (mm)	12	3
Habits of hatchling	Benthonic, crawl away	Planktonic
Life span (months)	15–24	15–20
Migratory pattern	Inshore–offshore	Inshore–offshore

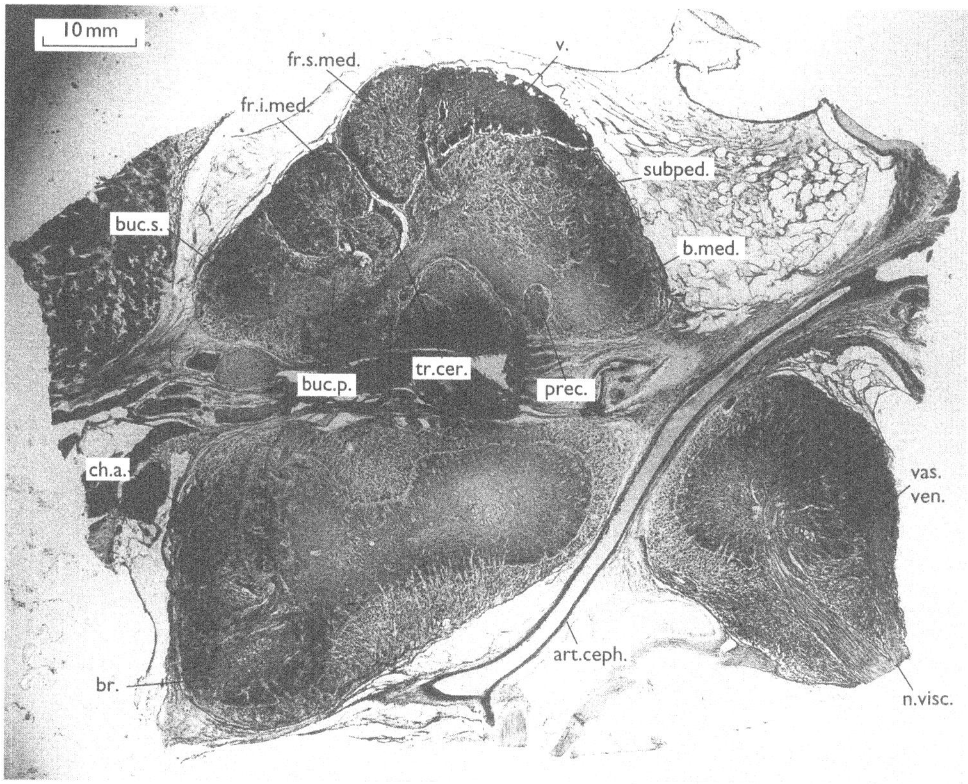

Fig. 31.47. *Eledone cirrhosa*. Sagittal section of the central nervous system close to the midline.

The central nervous system of *Eledone cirrhosa* (Fig. 31.47) is like that of *Octopus vulgaris*, the only clear distinction being the division of the cephalic aorta into two equal branches in the former. Measurements of the brain of *E. moschata* show the proportions are similar to those of *Octopus vulgaris* (Table 2.6) (Maddock and Young 1987). The inferior frontal lobe system forms 9.3%, the vertical lobe system 11.4%, the basal lobe system 5.2%, and the brachial lobes 20.9% of the total brain volume. The posterior chromatophore lobes of the former are 2.1%, and so occupy a much smaller proportion of the brain than those of *O. vulgaris*. Ten lobes of the brain of *E. cirrhosa* were measured in a newly hatched animal (Frösch 1971) and in an adult (Wirz 1959) (Fig. 31.48). The palliovisceral lobe and the higher motor centres of the hatchling form a relatively larger portion of the brain than in the adult. This is likely to be related to the locomotor activity during the planktonic phase of this species. The structure of the visual and chemotactile systems suggests that, like *O. vulgaris*, *E. cirrhosa* is capable of learning. The latter can respond to water-borne chemicals, being most sensitive to betaine (Boyle 1986c).

An epistellar body lies on each stellate ganglion (Fig. 31.49). This body was first described in *Eledone moschata* by Young in 1929. It is a small vesicle on the posteroventral surface of the stellate ganglion, and light microscopy revealed a reticular pattern thought at first to be glandular but later found to have a rhabdomeric organization (Young 1991d). At the ultrastructural level packed arrays of photoreceptors with microvilli, reminiscent of rhabdomeres, are present (Nishioka *et al.* 1966). Two cell types are present. The first type are large bipolar sensory cells with large apical processes, covered with rhabdomeric microvilli, and at the opposite extremity each tails off into an axon. The nerve fibres extend back as a single bundle (50 μm in diameter) containing some 1500 axons. The rhabdomeric processes are directed towards the centre of the epistellar body. The second type are supporting cells which lie between the sensory cells (Perrelet and Mauro 1972). Isolation of visual pigment with an absorption spectrum characteristic of rhodopsin supported the suggestion that the epistellar body is an extraocular photoreceptor, and this was confirmed by electrophysiological evidence (Mauro and Baumann 1968, Mauro 1977). After isolation epistellar bodies exhibit resting potentials, and are depolarized by flashes of white light but not of red light (Cobb and Williamson 1998a). Extracellular recordings of light-generated receptor potentials in the epistellar bodies of *E. cirrhosa* show their extraocular photoreceptors to be capable of responding to very low levels of light (0.1 $\mu W/cm^2$). The overlying mantle wall permits only some 1% of the ambient light to penetrate the epistellar bodies, yet this is sufficient for the receptors to be activated at the usual depths inhabited by this eledonine (Cobb *et al.* 1995a). *E. cirrhosa* has a nocturnal pattern of activity (Cobb *et al.* 1995b), but ablation of the epistellar bodies does not prevent these nocturnal activities, indicating that although the epistellar bodies could be involved in monitoring the onset of dusk and daylight they are not essential (Cobb *et al.* 1995a). The function of these bodies remains to be elucidated.

The statocyst of *E. cirrhosa* is relatively larger than that of *O. vulgaris* of the same body mass. The anticrista differs in that there is a ridge

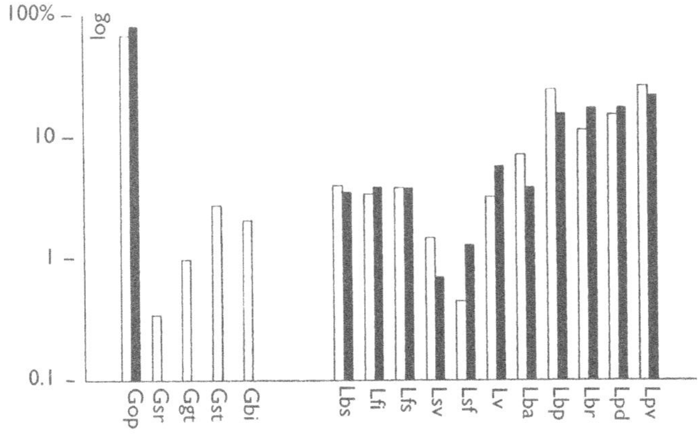

Fig. 31.48. *Eledone cirrhosa*. Histogram showing the volumes of each lobe as a percentage of the whole brain. Open bars, newly hatched (Frösch 1971); filled bars, adults (Wirz 1959). Gbi, inferior buccal ganglion; Ggt, gastric ganglion; Gop, optic ganglion; Gsr, subradular ganglion; Gst, stellate ganglion; Lba, anterior basal lobe; Lbp, posterior basal lobe; Lbr, brachial lobe; Lbs, superior buccal lobe; Lfi, inferior frontal lobe; Lfs, superior frontal lobe; Lpd, pedal lobe; Lpv, palliovisceral lobe; Lsf, subfrontal lobe; Lsv, subvertical lobe; Lv, vertical lobe.

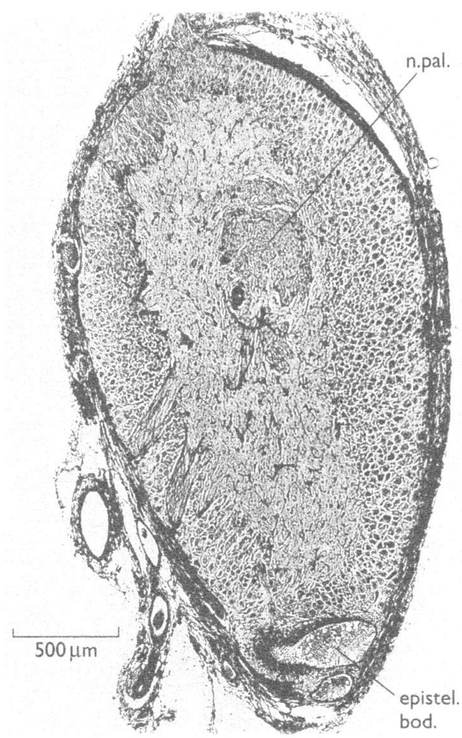

Fig. 31.49. *Eledone moschata*. Sagittal section of the stellate ganglion and epistellar body (Young 1936).

with an irregular outline and a slight peak at the dorsal end (Young 1989). The crista is sensitive to vibration and maximum sensitivity represented particle displacement of 0.12 μm. Additional units were recruited in response to increasing stimulus amplitude, implying that different units have different sensitivity thresholds. The sensitivity of the crista of *Eledone cirrhosa* is greater than that of some other aquatic invertebrates but is lower than in fish (Williamson 1988).

31.2.2 *Pareledone*

Members of this genus (Fig. 31.50) are restricted to the southern hemisphere where they are the dominant octopods of Antarctic waters (Roper *et al.* 1985). Benthic in habit, they are found from tide-pools to depths of 2000 m on the Antarctic shelf (Voss 1988b). Several valid species of *Pareledone* are recognized from the Southern Ocean and all are considered to be endemic, and others have been trawled off the Palmer Archipelago, south-eastern Argentina, south-eastern New Zealand, the Crozet Islands, and Showa Station, in depths of 100–500 m (Kubodera and Okutani 1994). The most abundant species caught off the South Shetland Islands, in bottom trawls at 50 and 500 m, were *P. charcoti*, *P. polymorpha*, and *P. turqueti* in that order (Kuehl 1988).

Pareledone adeliana was captured at 630 m where the bottom was soft and sponges were present. It was alive when brought to the surface and kept in an aquarium. The dorsal surface of the animal was a pinkish-purple-grey and white and it could darken (H.I. Daly and P.G. Rodhouse, pers. comm.). *P. nigra* was caught at 680 m together with sponges, bryozoans, and gorgonians, and its mantle had papillae, white spots, and bars; after fixation the animal became dark blue. Another species, *P. turqueti*, was a uniform purpulish-pink with white spots and bars, and it too could darken (Lu and Stranks 1994).

Histological sections show the epistellar body to be a partly divided cavity with no pigmentation (Fig. 31.51).

31.2.3 *Vosseledone*

Vosseledone charrua was first caught off the coast of Brazil and Uruguay in depths of 10–200 m. Its most extraordinary feature is the radula, with its simple rhachidian tooth, a faint trace of lateral tooth 1, a cuspless lateral tooth 2, and no marginal tooth or marginal plate (Palacio 1978, Nixon 1998).

The largest specimen recorded is a female of 76.5 mm ML. The ovaries of two gravid females (58 mm and 61 mm ML) contained closely clustered eggs differentially developed. In the smaller female

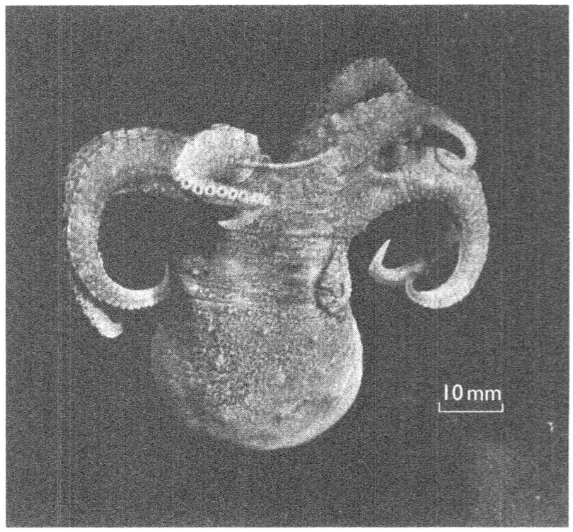

Fig. 31.50. *Pareledone*. Dorsal view.

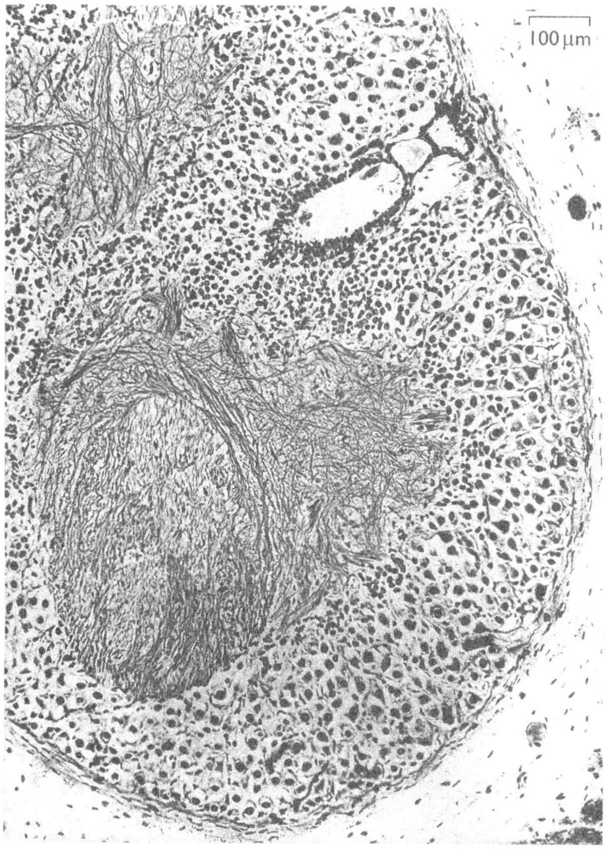

Fig. 31.51. *Pareledone.* The stellate ganglion and epistellar body seen in transverse section.

about 40 eggs were present, the apparently mature ones being 13 × 4 mm in size. Three spermatophores (56 mm long) were present in the Needham's sac of a male of 50 mm ML (Palacio 1978).

31.2.4 Velodona

Velodona togata (Fig. 31.52) is the only species of the genus and was described first from two specimens caught off Zanzibar (Chun 1915). A subspecies, found off the Cape Province at depths of 400–800 m, was recorded by Robson (1932).

The suckers of *Velodona togata* (Fig. 31.53) are set closely along the arms and are partially embedded in the tissues, at least after fixation. The infundibulum is deep and the cuticle has a radial arrangement. The cuticle is covered with fairly regularly organized pegs, each about 3 μm in height. The pegs have a somewhat blunted, conical form, with fine pores on the surface. Their features are closely similar to those of the suckers of *Octopus vulgaris* (Nixon and Dilly 1977).

Histological sections show the epistellar body to be a single sac, with pronounced rhabdoms, lying posteriorly on the stellate ganglion (Fig. 31.54).

31.3 Graneledoninae

These are deep-water octopuses of medium size. They have one longitudinal row of suckers but no enlarged suckers. There is no ink sac.

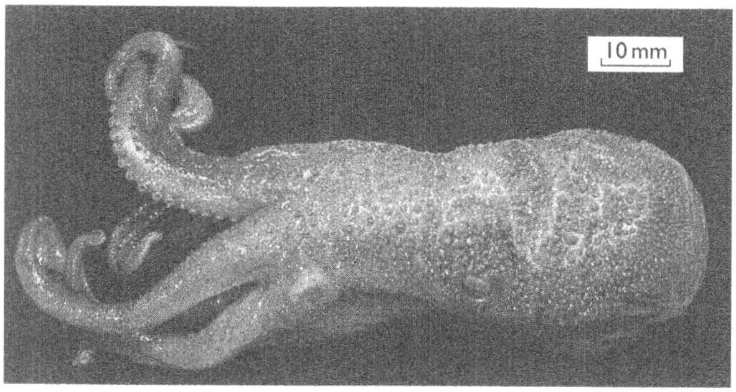

Fig. 31.52. *Velodona.* Dorsal view.

31.3.1 Graneledone

Graneledone is found in the north-west Atlantic, the north Pacific, the Gulf of Panama, off the north coast of New Zealand, and in the Southern Ocean, and it is deep-living between 700 and 2756 m (Voss 1988a,b, Voss and Pearcy 1990, Villanueva and Sánchez 1993, Kubodera and Okutani 1994). It has been observed at hydrothermal vent fields at a depth of 2480 m (Lutz and Voight 1997). The genus includes six species (Sweeney and Roper 1998).

Eggs, deposited in clusters, were collected in New Zealand waters in depths of 784 m and 1755 m. The egg capsules were teardrop-shaped (up to 26.2 × 3.6 mm) with short stalks, and individually cemented to the substratum. Young animals (9–13 mm ML) were caught and they possessed chromatophores, those on the ventral surface being more densely arranged than those on the dorsal surface (O'Shea and Kubodera 1996). A female *G. gonzalezi* (80 mm ML), was found to have oocytes of varying sizes (up to 10 mm in length) (Guerra *et al.* 2000); this suggests it is an intermittent spawner (Mangold *et al.* 1993). The maximum recorded size is a specimen with a mantle length of 147 mm (O'Shea and Kubodera 1996).

A specimen of *Graneledone* was isolated when captured at a hydrothermal vent. It had shell fragments and even whole shells of gastropod molluscs as well as bristles from polychaetes in its gut contents. The beaks of this specimen are large relative to the size of the animal which is clearly capable of fracturing and ingesting such hard material (Voight 2000). Polychaete bristles and amorphous material were recorded from the stomach of another specimen (Voss 1976).

31.3.2 Bentheledone

The first specimens to be described were caught in the southern part of the Pacific Ocean, at 3000–4000 m where the bottom was of mud or thick ooze, and named *Eledone rotunda* by Hoyle (1885a, 1886). This species has been found in the abyssal zone of the eastern Pacific and the Australia–Antarctic Basin (Nesis 1987).

Two species, *Bentheledone rotunda* and *B. albida*, are currently recognized (Sweeney and Roper 1998). Eggs 16 mm long were recorded by Hoyle (1886) and ones of 16 × 9.5 mm were present in a specimen of 55 mm ML (Robson 1932).

(a)
(b)
(c)
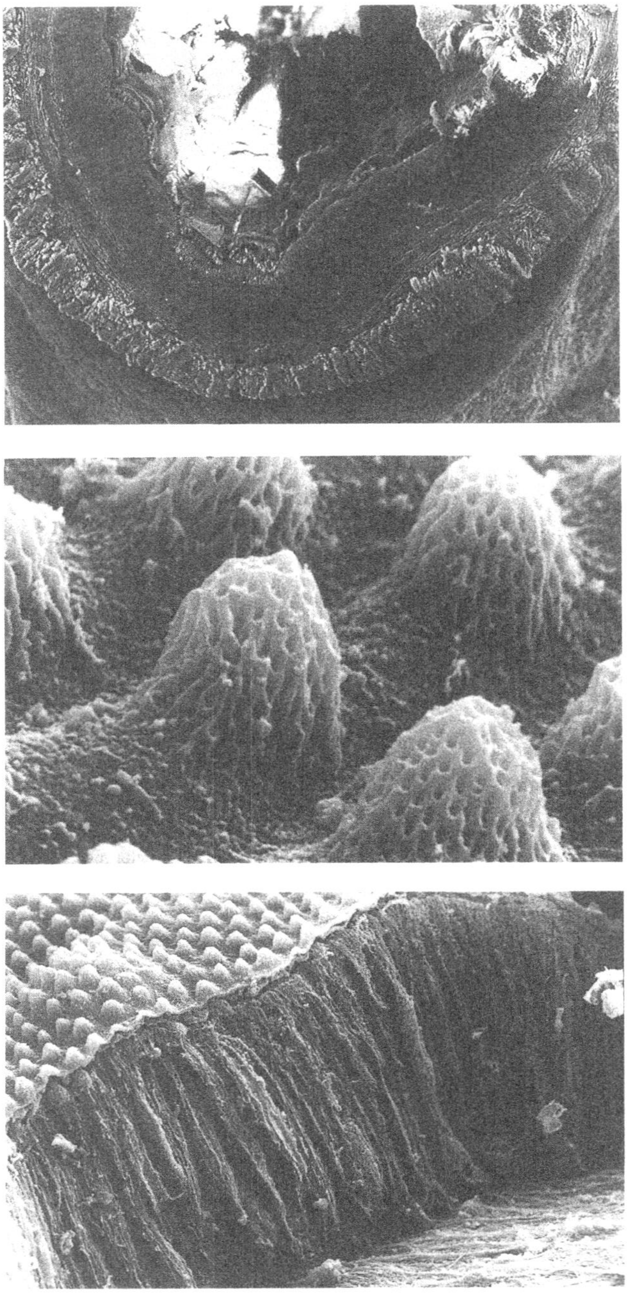

Fig. 31.53. *Velodona.* Scanning electron micrographs of a sucker: (a) The rim and the cushions of pegs (field width 1850 μm), (b) the pegs at high magnification (field width 10 μm), and (c) the pegs and the underlying tall columnar cells (field width 106 μm).

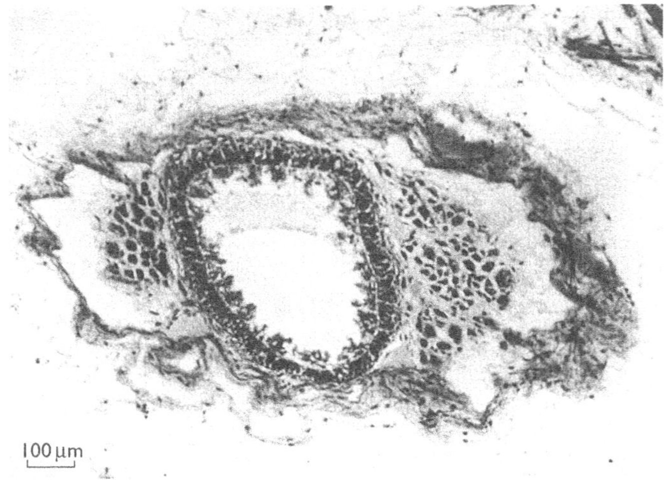

Fig. 31.54. *Velodona.* Transverse section of the epistellar body.

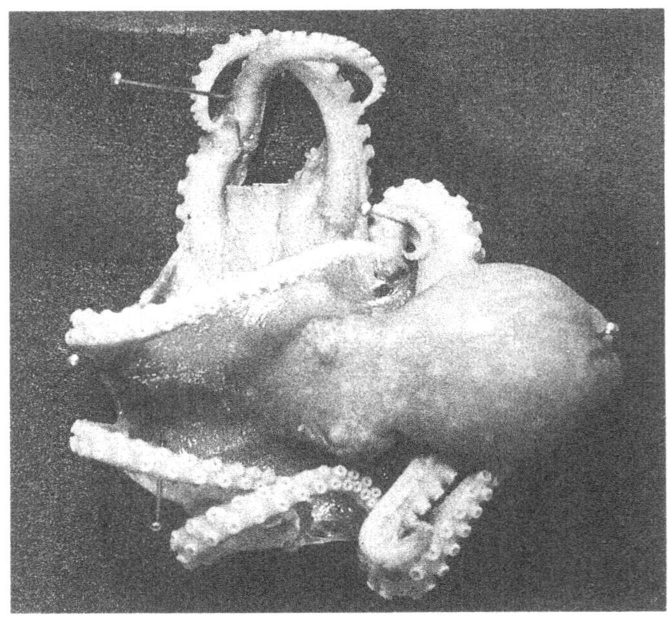

Fig. 31.55. *Bathypolypus arcticus.* (25 mm ML). Dorsal view of preserved specimen.

31.4 Bathypolypodinae

Bathypolypodines are mid-depth to deep-water octopuses. In size they are small to medium. The suckers are in two longitudinal rows, and enlarged suckers are occasionally present in males. There is no ink sac. Eggs are very large and the hatchling has no planktonic stage.

31.4.1 Bathypolypus

Members of this genus (Fig. 31.55) are widely distributed, being found in the North Atlantic, the Mediterranean, off South Africa and Namibia, and off Japan. They are caught to depths of 2500 m, mostly where the bottom is sandy or muddy (O'Dor and Macalaster 1983, Voss 1988a,b).

A maximum weight of 300 g and a mantle length of 53 mm has been recorded for *B. arcticus* (O'Dor and Macalaster 1983), and a mantle length of up to 450 mm for *B. valdiviae* (Nesis 1987).

Life history

Bathypolypus arcticus is perhaps the best-known species of the genus and is found in the North Atlantic from the Straits of Florida to Baffin Bay (Voss 1988b). It is small, squat, and muddy red in colour, and notable for the size of its reproductive organs. Arm III of the mature male has a ligula which may occupy more than a third of the length of the hectocotylus. The gonad of a mature female may reach 40% of the weight of the animal. The life history of *B. arcticus* has been followed in the laboratory for almost a year. Typically a female produces some 90–100 eggs (9 × 4 mm to 14 × 6 mm), attaches each by a stalk to an overhanging surface, and then broods them for 11 months until they hatch. At hatching the young resembles the adult and spends much of its time attached to a vertical surface. A female may need 3 or even 4 years to complete her life cycle: 1 year as an embryo, 1 year of growth, 1 year of gametogenesis, and 1 year of brooding (Macalaster 1981, O'Dor and Macalaster 1983). The female of another species, *B. sponsalis*, is mature at 30 mm ML, and the male at 27 mm ML, but growth continues to 50 mm and 40 mm respectively; gravid females have up to 60 pear-shaped ovarian eggs ranging from 6 to 13 mm in length (D'Onghia *et al.* 1993). The ovarian eggs were found to be at several stages of development in both species, indicating intermittent spawning over a long period of time (Mangold *et al.* 1993).

Feeding habits

In an aquarium *Bathypolypus arcticus* crawls over the substrate, and although it can move by jet propulsion it is not a strong swimmer. It detects objects primarily by touch and will capture any prey which brushes against it. The stomach contents of specimens of this species, collected from the sea, contained ophiuroids, crustaceans, polychaetes, bivalves, and gastropods for the most part, but foraminiferans, siphunculids, and cumaceans were occasionally also found; in some specimens only one food type was present (Macalaster 1981). *B. sponsalis*, in the Mediterranean, was found to have fed predominantly on benthic organisms (Würtz 1982).

Receptors and central nervous system

The eyes have rhabdoms longer at the centre than at the periphery, the maximum length being 200 μm. The optic lobes form 44.2% of the total brain volume (Maddock and Young 1987).

The central nervous system is of typical octopod form (Fig. 31.56). The inferior frontal lobe system occupies 9.6% and the vertical lobe system 11.2% of the total brain volume (Table 2.6). The subfrontal lobe is divided into lobules. The epistellar body is on the ventral side of the stellate ganglion (Fig. 31.57). The mantle wall has inner and outer layers of muscle separated by reticulate tissue.

The statocyst was seen only in histological sections, and has a single anticrista. The crista is divided into short sections.

31.4.2 Benthoctopus

Benthoctopus (Fig. 31.58) is found widely between the Arctic and the Southern Ocean but is poorly known. It has mostly been caught in the deep sea at depths to 3430 m (Voss and Pearcy 1990, Norman *et al.* 1997). One species, *Benthoctopus piscatorum*, is found from 40°N on the west side and 50°N on the east side of the Atlantic Ocean, although its complete range is not known (Voss 1988b). Members of perhaps three species have been photographed at hydrothermal vents, where they appear to be vagrant rather than resident (Lutz and Voight 1997).

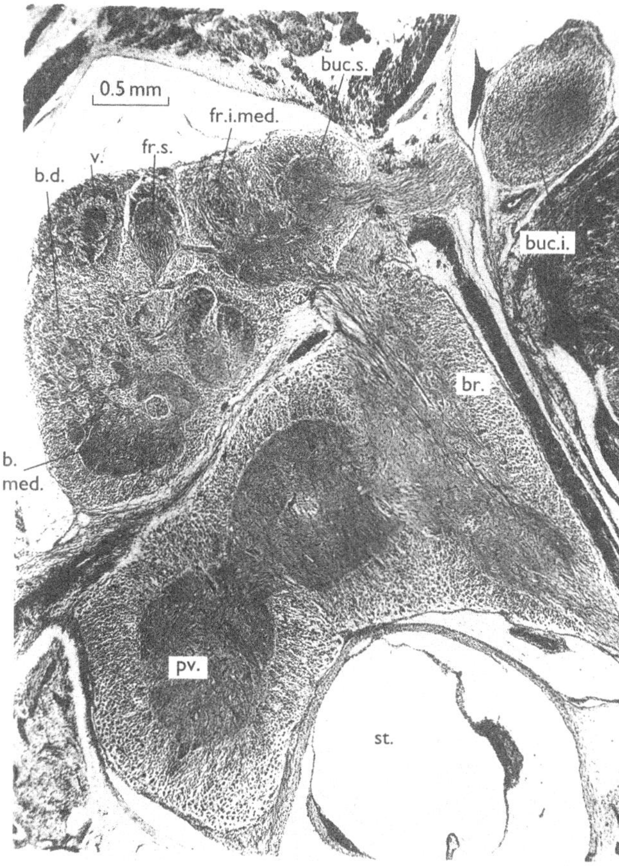

Fig. 31.56. *Bathypolypus arcticus* (25 mm ML). Sagittal section of the central nervous system close to the midline, also showing the statocyst. Figs 31.55–31.56 are of the same specimen of 25 mm ML.

Life history

A gravid female *Benthoctopus piscatorum* (89 mm ML) was caught at a depth of 120 m in Placentia Bay, Newfoundland (Aldrich and Lu 1968b). The optic glands of this gravid female were large, and so too was the ovary, which contained about 50 eggs ranging in size from 10 × 5 mm to 21 × 6 mm (Fig. 31.59) (Nixon 1991). The variation in the size of the eggs indicates that development and spawning takes place over a prolonged period and so this species can be considered to be an intermittent spawner (Mangold *et al.* 1993). A female *B. karubar*, had 150 submature eggs of up to 14 mm in length in the ovary (Norman *et al.* 1997). Specimens of up to 170 mm ML have been recorded (Nesis 1987).

The suckers of *Benthoctopus januarii*, although not well preserved, show a radial arrangement of the infundibulum, and a cuticle with regularly arranged pegs (Fig. 31.60). Each peg is somewhat conical in shape, about 3 μm in height, and covered with very small pores. The pores are closely similar in size to those of the pegs of *O. vulgaris* (Nixon and Dilly 1977).

Central nervous system

The central nervous system of the gravid female from Placentia Bay (Fig. 31.61) is similar to that of *O. vulgaris*. The optic lobes formed 49.2% of the total brain volume (Table 2.6). The inferior frontal system forms 10.7% and the vertical lobe system 7.6% of the total brain volume (Maddock and Young 1987). The subfrontal lobe lies behind the median inferior frontal lobe rather than below as in *O. vulgaris*. The anterior part of the subfrontal lobe is divided into islands, but more posteriorly the lobe forms a cylinder opening to the posterior buccal lobe. The suprabrachial commissure is relatively large. The epistellar body lies near the entrance of the mantle connective, close to the neuropil (Fig. 31.62).

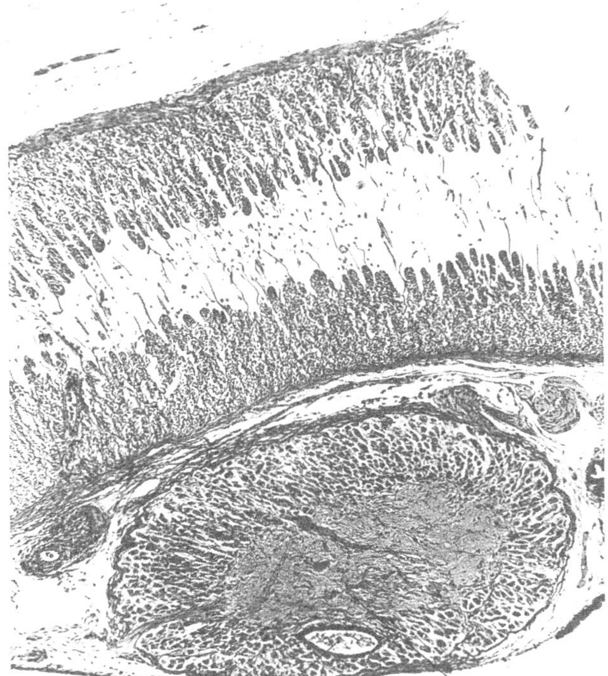

Fig. 31.57. *Bathypolypus articus* (25 mm ML). Section through the mantle and stellate ganglion, showing the epistellar body close to the entry of the pallial nerve.

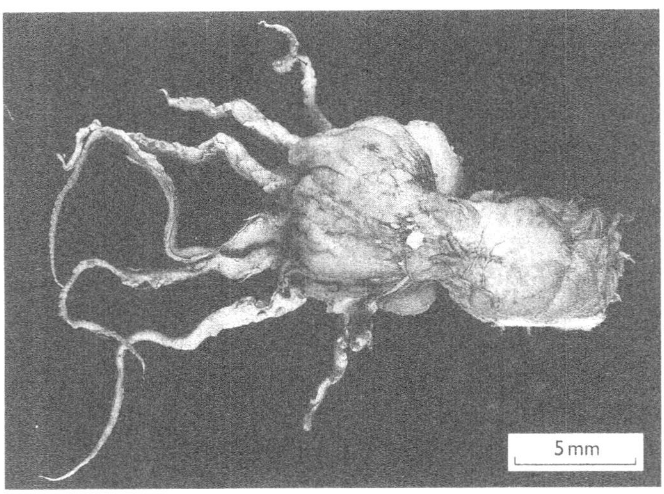

Fig. 31.58. *Benthoctopus piscatorum* (89 mm ML). Gravid female taken at a depth of 110 m in Placentia Bay, Newfoundland.

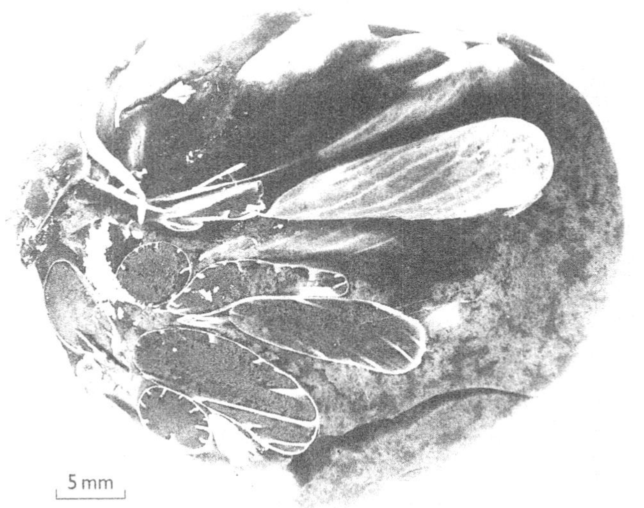

Fig. 31.59. *Benthoctopus piscatorum* (89 mm ML). Ovarian eggs found in the female (Nixon 1991).

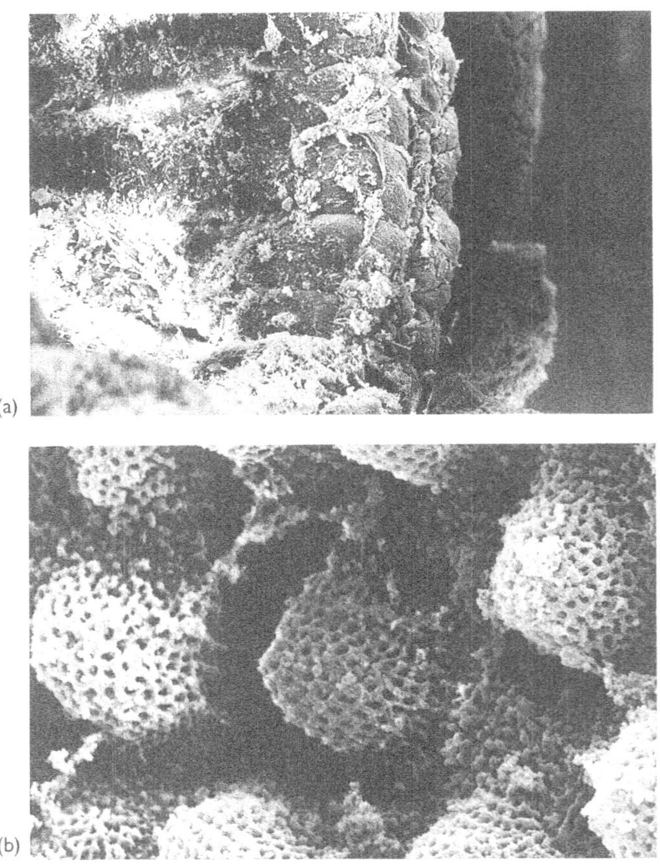

Fig. 31.60. *Benthoctopus januari*. Scanning electron micrographs of a sucker to show (a) the arrangement of the pegs (field width 1160 μm), and (b) the pegs at high magnification (field width 10 μm) (micrographs kindly provided by K.P. Ryan).

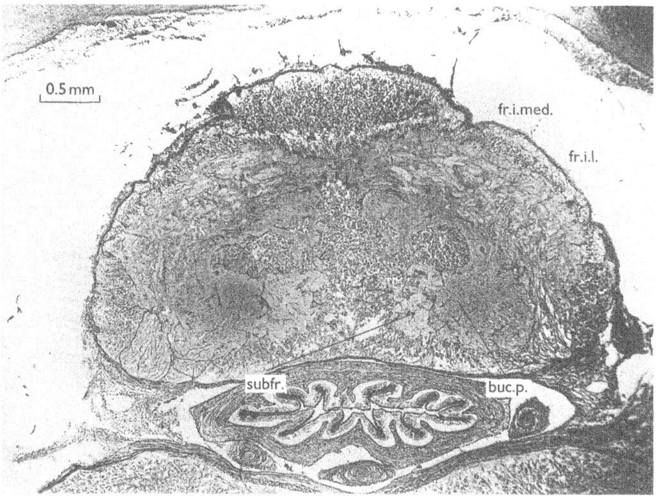

Fig. 31.61. *Benthoctopus piscatorum* (89 mm ML). Transverse section of the inferior frontal lobes of the brain.

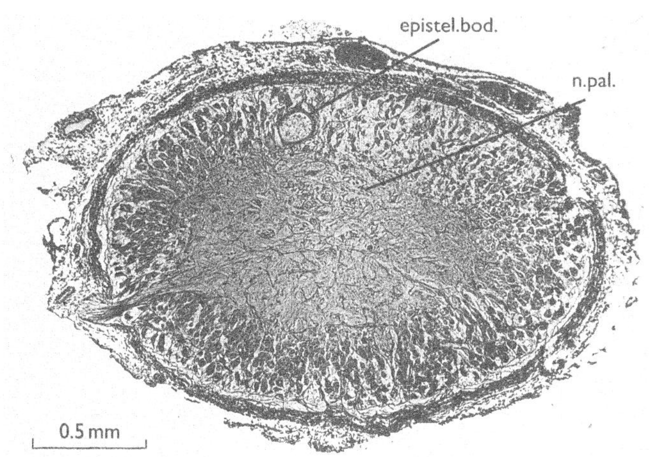

Fig. 31.62. *Benthoctopus piscatorum* (89 mm ML). Longitudinal section of the stellate ganglion and the epistellar body.

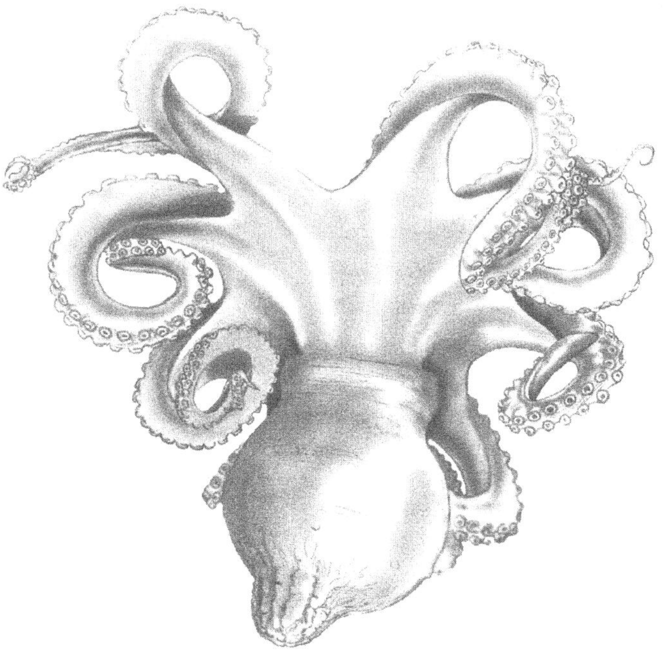

Fig. 31.63. *Vulcanoctopus hydrothermalis*. Dorsal view (scale bar 10 mm) (González *et al.* 1998).

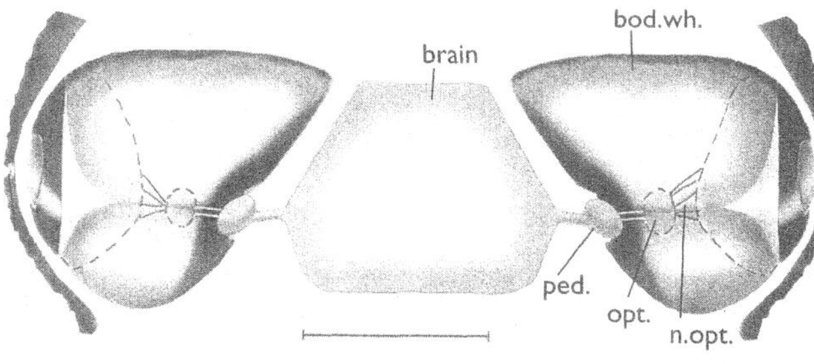

Fig. 31.64. *Vulcanoctopus hydrothermalis*. The central brain, peduncle lobe, and the eye, indicated with broken lines, the optic nerves, and optic lobe, shown diagrammatically (scale bar 3 mm) (González *et al.* 1998).

31.4.3 Vulcanoctopus

Two octopods were captured by a manned submersible in the east Pacific Ocean (12°48′N, 103°56′W), at a depth of 2647 m, and named *Vulcanoctopus hydrothermalis* (Fig. 31.63) (González *et al.* 1998). They were near chimneys of a hydrothermal vent, on a substrate of basaltic rocks covered by oxidative sulfides. One animal was 2 m from a black smoker, and the other 1 m from a small white smoker, and although found on separate occasions they were at sites just 10 m apart. White crustaceans, perhaps galatheids, were seen at the site of capture and the larger octopod contained the remains of similar animals in its stomach.

The body is pear-shaped and the mantle wall thin, pale, and semi-translucent in the two males (35 mm ML, mature, and 21 mm ML, immature). The head is narrower than the mantle and the neck region is slightly constricted. The long, slender arms have two rows of suckers, none of which is enlarged. Right arm III is hectocotylized. The funnel is short. The beaks are similar in shape and form to those of other octopods. The central rhachidian teeth of the radula have cusps, the first and second lateral teeth are wide, and marginal teeth and marginal plates are present.

The eyes are covered by skin except for a semitranslucent window. Each eye (6.8 mm diameter) has a lens but lacks an iris, and appears to lack a chiasm. Much of the eyeball is covered by white body, which also spreads over the optic lobe and optic nerves. In a stylized drawing the optic lobe is small in relation to the eye and the central nervous system (Fig. 31.64). Stellate ganglia are present.

The unusual features found in this octopod, include the absence of an iris, no optic chiasm, the presence of a multilobate digestive gland and of a dark swelling on the intestine, led González and his colleagues (1998) to suggest it may be necessary to erect a new subfamily.

32 Tremoctopodidae

32.1 *Tremoctopus*

Tremoctopus violaceus (Fig. 32.1) is the best known of the two species of the only genus of the family. It lives in the open sea and has a wide circum-tropical distribution between 36°S and 40°N, and in the Mediterranean Sea (Voss and Williamson 1971, Nesis 1987, Mangold and Boletzky 1988, Zeidler 1989). It is seen at the surface at night but may take small diel migrations (Clarke and Lu 1975). Juveniles are found from the surface to 250 m depth (Roper and Young 1975) *T. gelatus* is the other species. It is translucent and lives in deeper water (Thomas 1977).

32.2 Reproduction

Sexual dimorphism in *Tremoctopus violaceus* is extreme; a mature female can reach 320 mm ML but the male only 15 mm ML (Thomas 1977). The male resembles a small female except for right arm III, which is a huge hectocotylus that develops in a pouch between arms II and IV (Voss and Williamson 1971). Only one large spermatophore is produced and it is transferred to the tip of the hectocotylized arm. This arm is autotomized at copulation and several may be found in the mantle of one female (Thomas 1977). It is not known how long the spermatophores remain viable.

Off the coast of Japan one female was found with a large egg mass that weighed 500 g and contained more than 100 000 eggs (3.6 × 2.1 mm). The stalks of the eggs were cemented together to form a rod-like structure (Hamabe 1973); the rods were held by suckers of the dorsal arms where the eggs were brooded (Naef 1923). The egg mass was in five parts and in each the eggs were at the same stage of development, but at different stages from those in the other parts: the eggs ranged from early to late stages of development (Hamabe 1973).

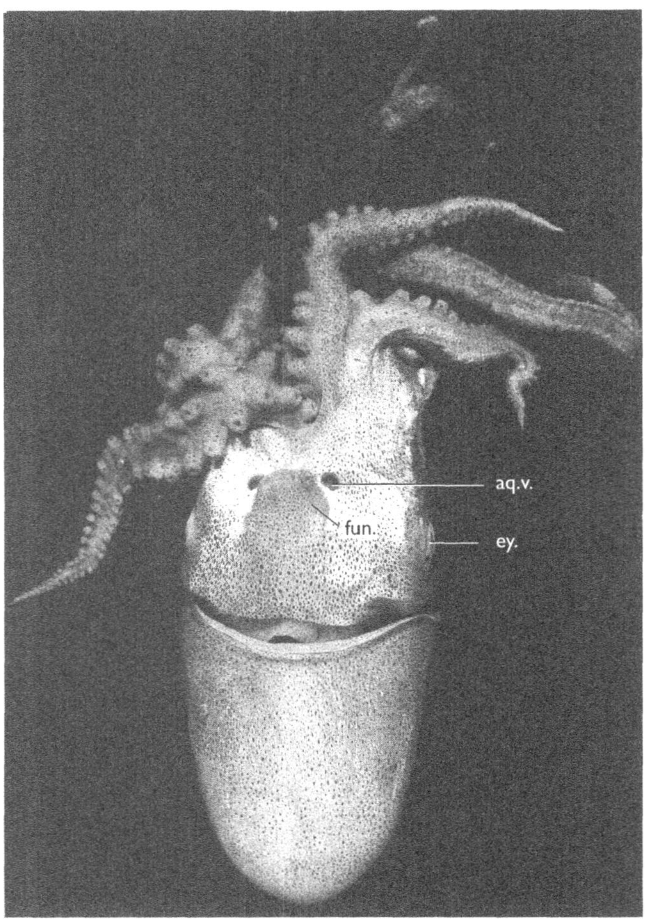

Fig. 32.1. *Tremoctopus violaceus* captured in Naples, in ventral view.

32.3 Behaviour

The appearance of the female is striking due to the wide, deep web attached to arms I and II (Portmann 1952), for which the name 'blanket' or 'handkerchief' octopus has been given (Voss and Williamson 1971). The web is most extensive between arms I and II and it can be expanded or contracted to a remarkable degree. Contractions of the web are due to transverse bands of muscle, each surrounded by elastic fibres which probably allow the web to expand after autotomy. Besides this the edge of the web, parallel with the arm, has a row of 'pouches' lined by glands; these may have a role in the attachment of the clusters of stalked eggs (Portmann 1952). Autotomy of the web occurs along visible fracture lines when the animal is distressed. The dorsal arms and web are held rolled back when the animal swims, but spread out if disturbed. The web undoubtedly has several functions, as it does in *Argonauta*.

One animal was still alive after capture, and was

> purple and gold... In the tank it suddenly disappeared in a cloud of ink. Then, like an apparition, a sheet of magnificent iridescent purple and white tissue with peacock-green eyespots billowed in the tank. The eye-like spots were clusters of green in patches of white tissue. After a few minutes the *Tremoctopus* contracted its arms until they appeared like small palps and the web regained its original shape.

The web when fully extended was 600 mm long and 300 mm wide, and the animal had a mantle length of 180 mm.

> When stroked... these webs flickered with iridescent sheens of purple, orange and blue-green. The contraction and expansion of the arm webs was repeated

several times at about 10 min intervals, until the animal died (Voss and Williamson 1971).

The arms may detect, or even repel, predators; the suckers of the four dorsal arms, of males and females, carry fragments of *Physalia* whose nematocysts remain functional. These may aid in the capture of plankton or fish, as the nematocysts are capable of paralysing small fish (Jones 1963). Thomas (1977) reports that when the dorsal surface of the mantle is stimulated the arms sweep forward over the mantle 'as though to remove the irritant'. This recalls similar actions of the arms of *Argonauta* when feeding (Young 1960a).

There are two observations of possible luminescence. Akimushkin (1965) reported the emission of a reddish light by *Tremoctopus*, and the females were found to bear a row of orange organs on each side of the distal part of the web; as the octopod swam placidly and stretched and flexed its arms the light emitted became brighter or dimmer. A diver snorkelling at night saw 'dozens of luminous eyes' and a photograph taken at a depth of 4 m showed the source to be *T. violaceus* (Goud 1989). Further observations are necessary, as well as microscopical examination of the organs.

32.4 Morphology

There are two pairs of aquiferous pores on the head (Figs 32.1, 32.2). Below the pores lie a series of large veins, some of which are swollen at intervals and prolonged between the tissues of the dorsal arms. The swellings are due to the presence of radially arranged muscle fibres. Each begins as a pear-shaped cell body peripherally continuing as a fibre around the lumen of the vessel. This tissue is well supplied with nerve fibres and presumably serves to pump blood into the veins, perhaps producing the rolling out of the dorsal arms and web when disturbed.

The arms are unequal in size and shape, and bear biserial suckers, each on a broad peduncle (Fig. 32.3). There is a deep, peripheral, layer of muscle surrounding a reticulate region, within which is a large brachial nerve. It is not known whether the reticulate tissues in the dorsal arms act as a buoyancy mechanism. The peduncle of the sucker is also reticulate but the sucker itself is muscular. A large sucker ganglion lies suspended in the reticulate tissue of the peduncle.

The buccal mass forms a relatively large proportion of the animal (Fig. 32.4). The beak is wide with an extremely broad hood but lacks a rostrum. It is only partially tanned, although quite thick. Thomas (1977) reported that the posterior salivary gland was small in an adult female and large in a male, but did not quantify his statement. The salivary papilla is a prominent part of the buccal complex and is almost entirely glandular. The mandibular muscles are large; these and the muscles of the buccal complex are densely organized (Fig. 32.2). The oesophagus is deeply folded, and has a thick cuticular lining. It leads to a large crop (Massy 1916), which extends forward above the brain and back to a muscular stomach. Thomas (1977) noted that adult females fed mainly on small fish. One of our specimens contained crustacean remains. Jaws and spicules from swarming polychaetes were found in the stomach of another (Zeidler 1989).

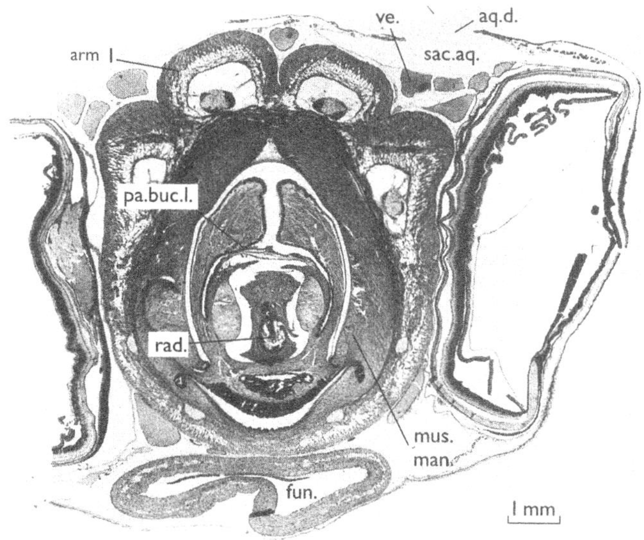

Fig. 32.2. *Tremoctopus violaceus.* Transverse section of the head showing the eye, buccal mass, dorsal arms, and the dorsal aquiferous pore and underlying sac and veins on one side.

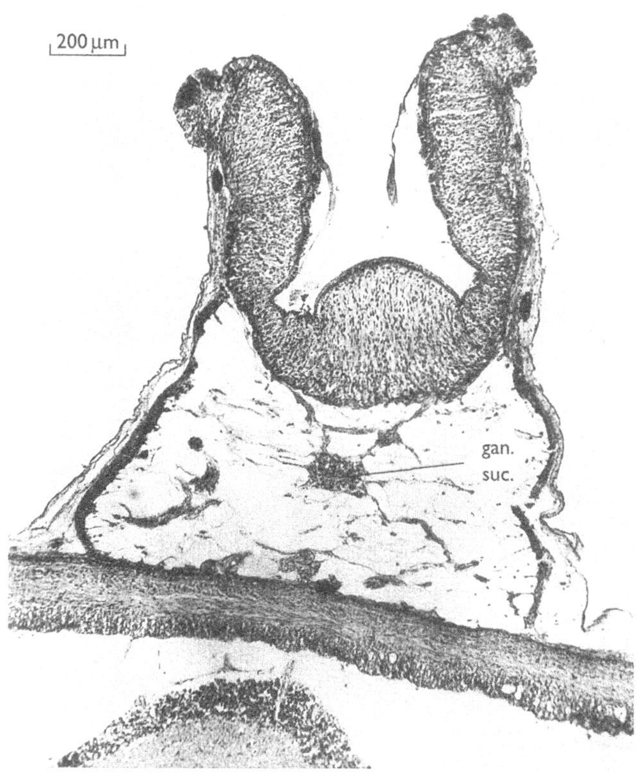

Fig. 32.3. *Tremoctopus violaceus.* Longitudinal section through a sucker with its dense radial muscles and underlying large sucker ganglion in amorphous tissue.

32.5 Receptors and central nervous system

The eyes are round and relatively large. In the retina the rhabdoms are of about equal length, 80μm. The optic tract is short. The peduncle spine runs vertically down the back of the optic tract, below

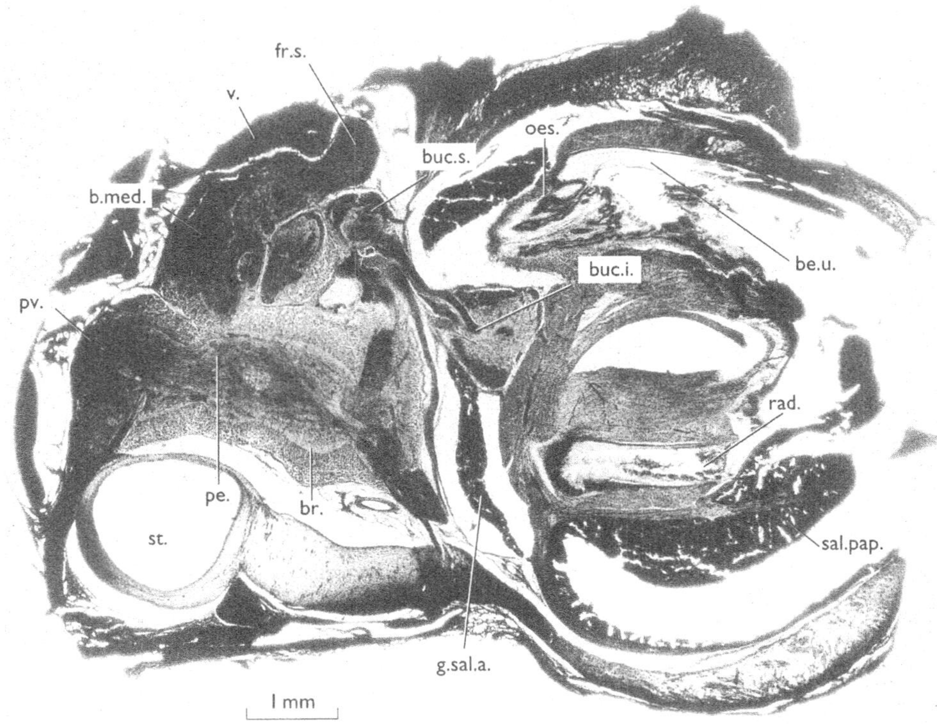

Fig. 32.4. *Tremoctopus violaceus.* Sagittal section through the head and buccal mass.

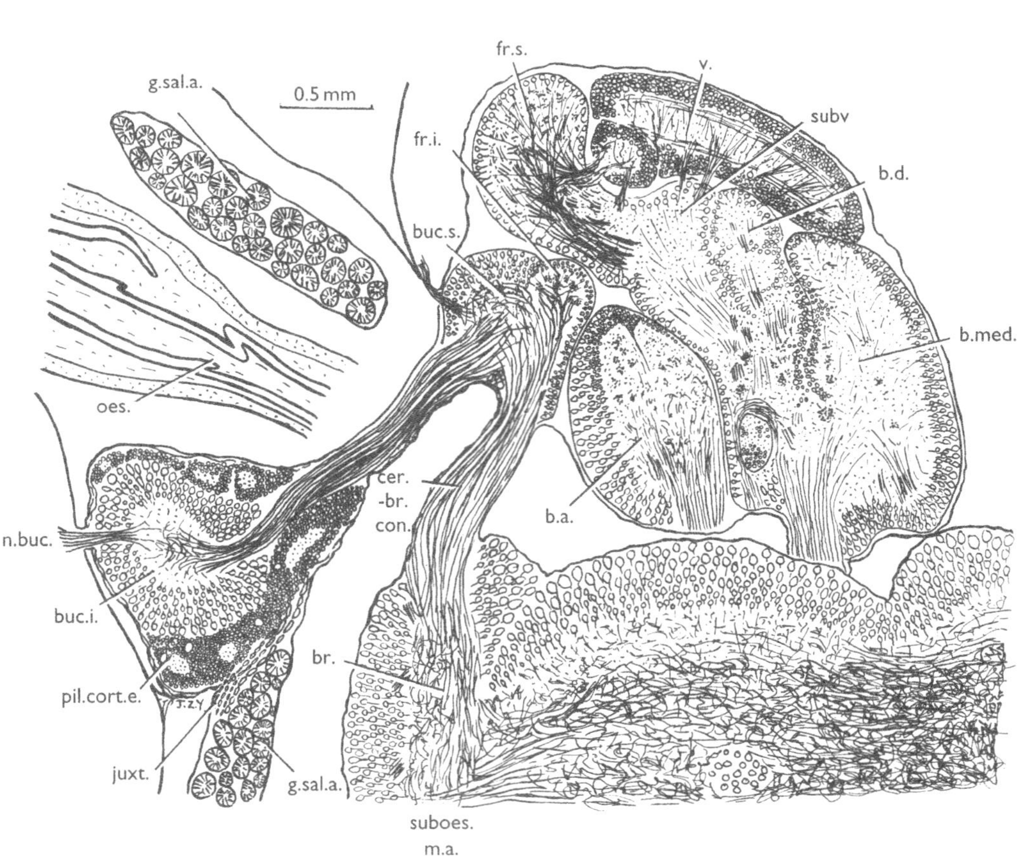

Fig. 32.5. *Tremoctopus violaceus.* Drawing of a sagittal section of the supraoesophageal mass and part of the suboesophageal mass of the nervous system.

which is the optic gland. The optic lobes form 259% of the total brain volume (Table 2.6).

The central nervous system is compact (Fig. 32.4). The large inferior frontal buccal ganglion is surrounded by much juxtaganglionic tissue (Fig. 32.5). This is unusual in that it takes the form of massive bundles of strands of cells, penetrating other tissues in various directions. The brachial lobe forms 8.7% of the total brain volume and shows little division into pre- and postbrachial lobes. The lateral basal and anterior and posterior chromatophore lobes are large, indicating the importance of the display of body patterns observed in this octopod. The inferior frontal lobe system forms 2.23% of the total brain volume (Table 2.6) (Maddock and Young 1987).

The median inferior frontal lobe consists only of a few bundles crossing the midline, between the posterior buccal lobes (Fig. 32.5). The subfrontal lobe is moderately well developed as cords of small cells in the midline reaching down to the posterior buccal lobe. This unequal development of the parts of the system remains to be explained.

The vertical lobe and median basal lobe systems form 13.9% and 14.2% respectively of the total brain volume (Table 2.6, Fig. 32.5).

The statocysts lie far laterally, behind the back of the orbits. The static nerves are long and enter the statocyst dorso-medially. The macula lies on the mediodorsal wall. There are nine sections in the crista, with hair-cell rows alternately single and double. The anticrista is a small mass terminating with a rounded knob (Young 1989).

33 Argonautidae

33.1 *Argonauta*

This is the only genus of the family (Fig. 33.1). *Argonauta* has a wide distribution in warm and tropical waters of the Pacific and Atlantic Oceans and the Mediterranean Sea (Voss and Williamson 1971, Okutani *et al.* 1987, Mangold and Boletzky 1988). It is found mostly at the surface with a few in deeper water at 50–155 m, and occasionally to 300–350 m (Roper and Young 1975, Clarke and Lu 1975, Lu and Clarke 1975b). Mass strandings have been recorded since 1758; more than 6000 specimens were captured in 1 week in Japan and large catches continued for months (Nishimura 1968, Kubota and Miyashita 1975, Okutani and Kawaguchi 1983).

33.2 The shell

The shell is found only in the female and forms the brood chamber. It is largest in *Argonauta argo* and can reach a diameter of 300 mm; the mantle length of the animal within being 120 mm (Dell 1952, Okutani *et al.* 1987). The shell is thin and fragile and is secreted by the shell glands of the two dorsal arms of the female (Naef 1923). It is calcareous and structurally the greater part of it is very finely prismatic, consisting of a thicker upper and a thinner under layer, between which is a thin layer of very fine, irregularly grained calcite (Bøggild 1930, Noda *et al.* 1986).

Fig. 33.1. *Argonauta argo.* A drawing by Jane de Vere from a film of a living, mature female with eggs in her brood shell (Young 1960a).

33.3 Life history

The male is a dwarf and is mature at 7 mm ML in *Argonauta boettgeri* and *A. hians*, and after autotomizing the hectocotylus it dies (Nesis 1977). Hectocotylized arms have been found in the brood chamber of female *A. argo*. Three were present in one shell, two of which 'were alive, the suckered and filament halves flexing and curling slowly like a Gecko's tail' (Voss and Williamson 1971). Females mature over a wide size range. At all sizes they have oocytes at different stages of maturity in the ovary, and varying stages of development are also found in the broods (Naef 1923, Nesis 1977, Okutani and Kawaguchi 1983). Female *A. hians* and *A. boettgeri* mate at 12–13 mm ML and begin to lay eggs at 15 mm ML. Meanwhile these animals continue to grow and reproduce for a long period; the maturation process being longer in *A. hians* than in *A. boettgeri* (Nesis 1977). Beebe (1926) found 13 000 eggs in the shell of one argonaut and embryos at various stages of development. Brood shells 88 mm and 59 mm in length contained 48 000 and 11 200 eggs respectively (Kubota and Miyashita 1973). At the end of the incubation period only a portion of paralarvae are released and Nesis (1977) concluded that female argonauts reproduce very early and continue to grow and reproduce for a long period of time.

Six argonauts, caught together in the waters of Hong Kong, all had clusters of eggs incubating inside the dorsal whorl of the shell. In each shell the eggs were at different stages of development. One female, placed in the shipboard aquarium, was resting on the bottom when she 'suddenly directed her siphon vertically upward and, with several powerful contractions of the mantle, puffed out a cloud of about 100 active free-swimming 1 mm long larvae' (Voss and Williamson 1971). Spawning may last almost a year (Nishimura 1968). Argonautids are clearly intermittent spawners (Mangold *et al.* 1993).

The embryos probably hatch and are released in darkness (Nesis 1977). At hatching the young have mantle lengths of 0.75 mm to 1.25 mm and will jet into the plankton immediately. The silvery eyes are pronounced. The digestive gland has a silvery sheen on its ventral surface and beneath it the ink sac is readily visible. Chromatophores are few and sparsely arranged. The funnel is relatively large and extends forwards beyond the anterior edge of the eye; it is muscular (Hochberg *et al.* 1992). Young *A. argo* are separable into males and females at a mantle length of 2 mm. In the male the pouch of the hectocotylus is visible at 2.5 mm ML. At this size the eye diameter is 40% of the mantle length. In females the shell sac is visible at 4 mm ML. Young with mantle lengths of 5–13 mm have been captured in opening/closing nets, from the surface to depths of 200 m, and at the surface at night (Roper and Young 1975).

33.4 Movement and locomotion

Argonauts have been seen at sea moving together as a 'sinuous row' swimming slowly along (Beebe 1926), and as a 'moving rod . . . swimming beneath the surface. It rose quickly, swam along and down out of sight . . . it was a string of six Argonauts holding on to each other swimming as a unit' (Voss and Williamson 1971). At night, at the surface, females have been seen to seize drifting objects, medusae, stomatopod larvae, empty pteropod shells, and other argonauts (Nesis 1977). Chains of salps were found with a single argonaut in one salp of each chain. Each argonaut was attached by its arms to the wall of the branchial cavity of the salp (Banas *et al.* 1982). Kramp (1956) observed a 'remarkable sight of a number of *Argonauta* ensconced on the top of jellyfish, *Crambionella orsini*, which drifted in large numbers round the ship' between Mombasa and the Seychelles. A female found holding a jellyfish tightly, with the suckers of the lateral and ventral arms, was watched by scuba divers for an hour (Heeger *et al.* 1992). Both animals were collected and when placed in an aquarium the argonaut ejected itself from the medusa. Large pieces of mesoglea were missing, and a few distinct holes in the centre of the bell of the medusa were interpreted as bites. Argonauts apparently feed, take shelter, and even obtain a free passage from other animals.

The movements of a female *A. argo* were followed in an aquarium. It swam gently, rocking backwards and forwards as if pivoted on a rod, and periodically moved faster when the buccal mass pointed almost directly upwards and the funnel was almost horizontal; after several strong contractions of the mantle it then glided some distance horizontally (B.B. Boycott pers comm.). The argonaut made vertical movements, swam forwards and backwards by directional movements of the funnel, and turned through a wide arc, again using the funnel. The female is neutrally buoyant, aided by a bubble of gas trapped in the upper part of the shell (Young 1960a, Voss and Williamson 1971, Roper 1972). Experimental removal of the air from the shell of a brooding female resulted in the animal sinking to the substratum; re-introduction of the air restored the initial neutrally buoyant condition (Boletzky 1983c). It is not known whether there is any other buoyancy mechanism, such as reduced salts (Clarke *et al.* 1979), but the tissues of the arms and sucker peduncle are reticulate and may also aid buoyancy (Nixon and Dilly 1977).

33.5 Colour

Argonauta argo floating close to the surface is a shining blue above and white below, the eyes an iridescent steel blue, and the dorsal arms blue (Voss and Williamson 1971). The web is a flattened expansion of the first pair of arms. It may spread over the shell or remain concealed, only to re-appear 'as a shining white mass near the apex of the shell. This then spreads ventrally as a silver stream . . . the movement proceeds in an uncannily steady manner so that the shell is covered before one realizes what is happening' (Young 1960a).

The colour of an adult female is striking but the range of patterns produced is limited. This is reflected in the small size of the chromatophore lobe (Table 2.6). Expansion of the red-purple chromatophores on the upper part of the web, dorsal arms, eyes, head, funnel, and mantle occurs when the animal is disturbed. The argonaut can become intensely red, the upper part colouring more readily than the lower; sudden movement results in a concentration of the chromatophores and blanching (B.B. Boycott pers comm.). The arms and the web show the same iridescence as the large web attached to arms I, blue dorsally and white ventrally. When seen from in front the animal has a striking appearance with dark patches above the eyes, a deep blue-black beak surrounded by the arms bearing white suckers standing out against their blue background (Young 1960a).

33.6 Morphology

The arms are slender and extremely mobile, and the suckers are borne on relatively long peduncles. The sucker has densely organized muscles and contrasts with the reticulate tissues of the sucker peduncle. The argonaut possesses stout cones around the periphery of the infundibulum (Niemec 1885). The surface of the large cones differs quite markedly from that of the pegs at the same magnification (Fig. 33.2). There is no evidence for the function of the cones but they may provide a better grip on the shell of the individual in front in a chain of argonauts, or for the attachment to other drifting animals. The infundibulum of the sucker is divided into sectors, each carrying closely organized pegs, which show variation in size and shape; the pegs have pores 0.08 μm in diameter (Nixon and Dilly 1977).

33.7 Feeding and the buccal mass

In the sea *A. boettgeri* and *A. hians* were found to be feeding on heteropods and pteropods (Nesis 1977). Two female *A. boettgeri* (shell lengths 13.2 and 14.6 mm) were captured each clutching a pteropod (*Cavolinia tridentata*) with a similar or greater shell length (Okutani 1960). Beebe (1926) observed that when given a small fish the argonaut 'bit its head off and ate the body'. When a piece of fish touched the web of an argonaut then a ventral arm of the same side swept the web until the flesh could be seized (Young 1960a). At sea argonauts have been observed feeding on the gonads or the centre of the bell of jellyfish (Clarke and Herring 1971, Heeger *et al.* 1992).

The buccal mass is relatively large with anteroposterior lengths of 48% and 21% of the mantle length in argonauts of 15 mm and 83 mm ML respectively (Morton and Nixon 1987). It has densely organized muscles. The beak is wide and flattened anteriorly, with a small and indistinct rostrum (Dell 1952, Clarke 1962a, 1986), and yet it is capable of reducing soft material to ingestible pieces. The radula forms relatively large part of the buccal complex and its teeth are relatively stout (Nixon 1998). Below the radula the large submandibular

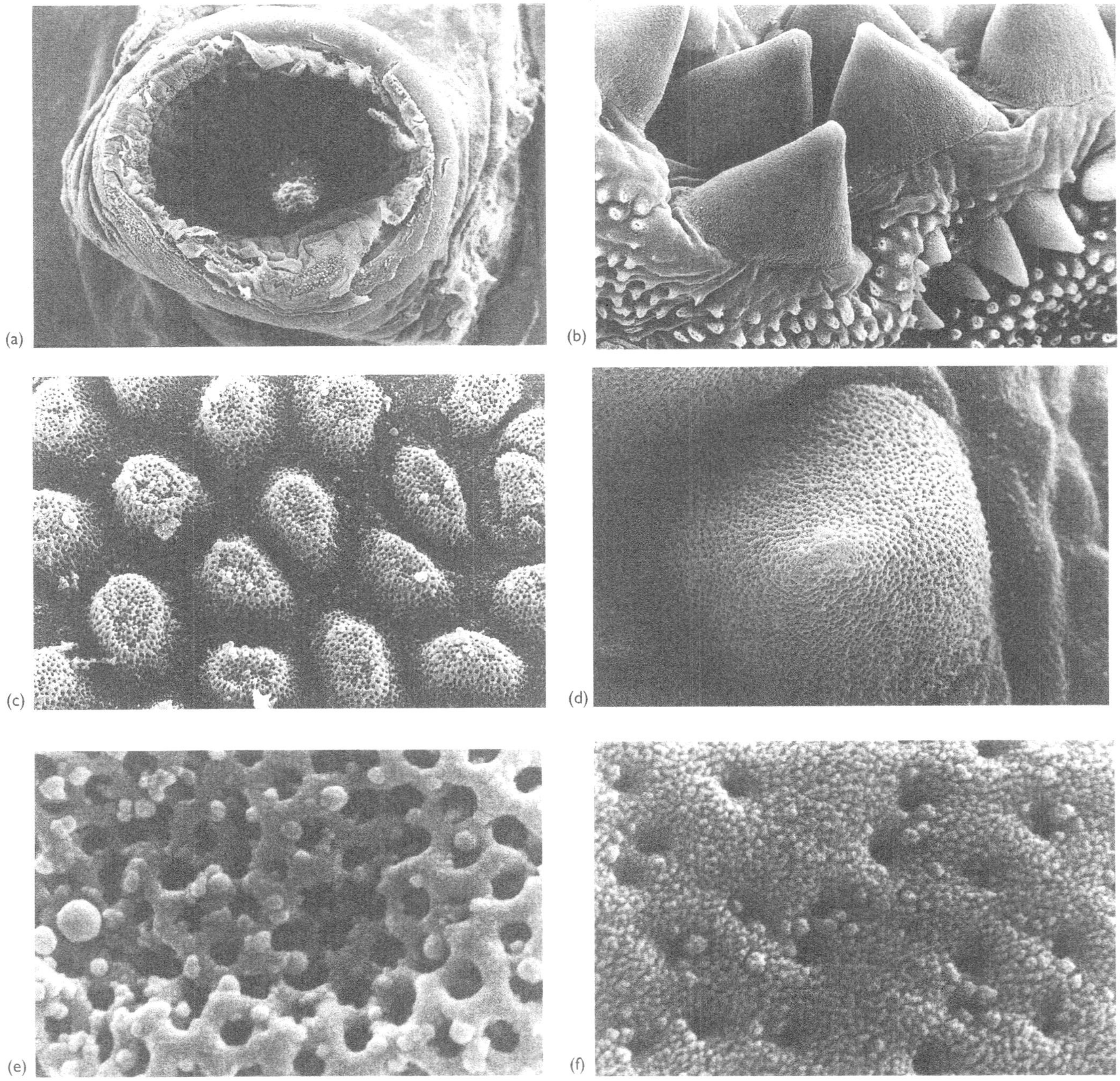

Fig. 33.2. *Argonauta argo.* Scanning electron micrographs of a sucker: (a) To show the arrangement of the pegs and the large cones around the periphery of the infundibulum (field width 2272 μm). (b) The size difference between the cones and the surrounding pegs is notable (field width 116 μm). (c) The structural features of a peg (c) and a cone (d) at the same magnification (field width 25 μm), can be compared, and a peg (e) and a cone (f) at the same higher magnification (field width 2.5 μm) (micrographs kindly provided by K.P. Ryan).

gland occupies much of the buccal cavity. The paired posterior salivary glands lie behind the central nervous system and are smaller than the anterior pair. The oesophagus is large and has a very thick cuticle (Fig. 33.3). This condition suggests that skeletal material, with hard, sharp, and rough edges, is ingested, and so the cuticle protects the encircling central nervous system (Fig. 33.4). The calcareous material is presumably important to the female for the production of the brood shell. Support for this suggestion comes from the feeding habits of *Sepia* (Guerra *et al.* 1988a) and *Nautilus* (Ward 1987), both of which ingest considerable quantities of skeletal material for building the calcareous cuttlebone and the shell respectively. After prolonged deprivation *Sepia* no longer builds a normal cuttlebone, instead this is shrunken and the chambers short (Boletzky 1974b).

33.8 Receptors and central nervous system

The eyes are large, round and silvery, and are laterally placed thus giving a wide visual field (Fig. 33.1). In a male of 2.5 mm ML the eye forms 40% of the mantle length; in a female with a shell diameter of

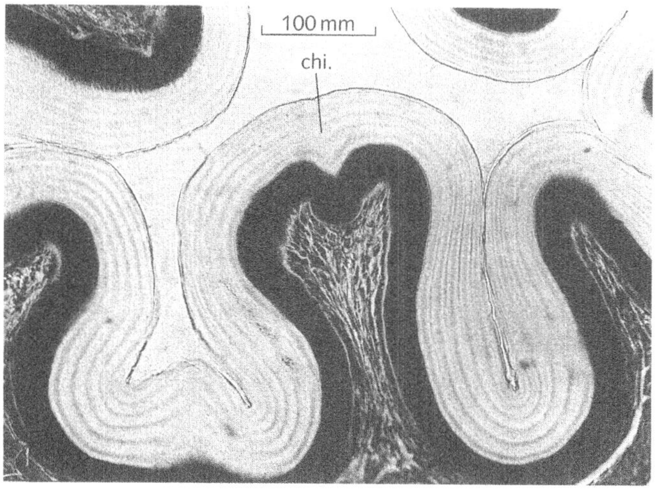

Fig. 33.3. *Argonauta argo.* Transverse section through the oesophagus showing the deep folds and extremely thick cuticular lining.

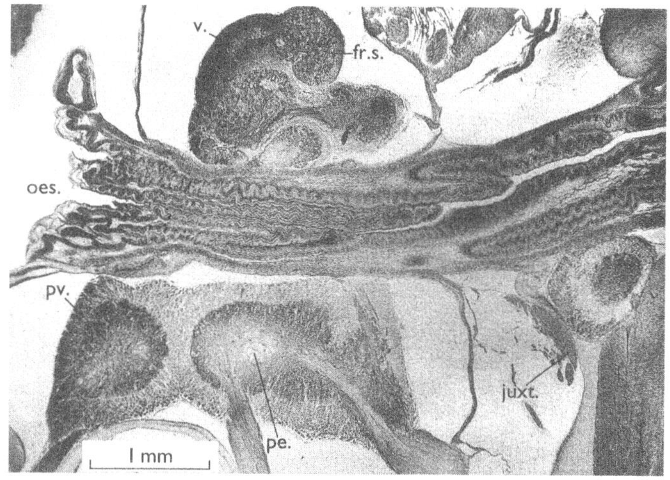

Fig. 33.5. *Argonauta argo.* Sagittal section of the central nervous system and the inferior buccal ganglion. The sub- and supraoesophageal masses are separated widely by the oesophagus.

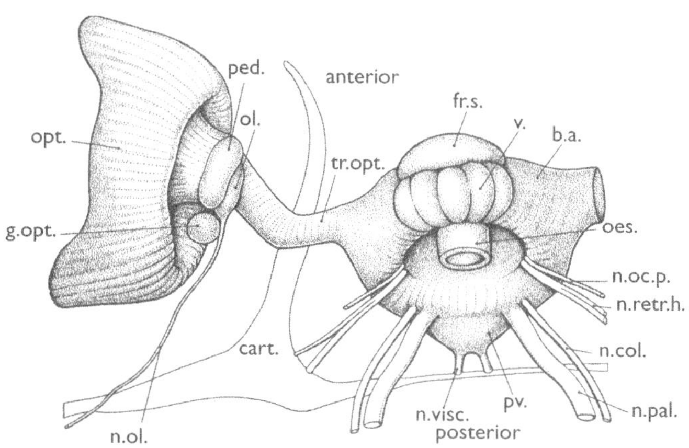

Fig. 33.4. *Argonauta argo.* Drawing of the central nervous system seen from above and slightly posteriorly (Boycott and Young 1956).

some 95 mm the eye diameter is 17% of the mantle length. The eyes are capable of rotation and will maintain their position by counter-rolling to compensate for the rocking movements which result from the expulsion of jets of water through the funnel (Boletzky 1983c). The muscles of the eyes are richly innervated, with nerve fibres carrying swellings *en passage*. In the retina the rhabdoms are of almost uniform length (140 μm).

The optic lobes form 227% of the total brain volume (Fig. 33.4) (Maddock and Young 1987). In the optic lobes the dendrites of the second-order nerve cells have been especially well stained in some preparations. Many proceed straight for long distances and the largest for at least 400 μm, with spreading branches that may extend twice as far. Argonauts are thus provided with the capacity for form discrimination, although the only evidence for this comes from reports of fast backwards movement when a large visual stimulus is placed at the side of the octopod (Beebe 1926, B.B. Boycott pers. comm.).

The optic tract is moderately long and the peduncle lobe lies close to the optic lobe (Fig. 33.4) (Boycott and Young 1956). The peduncle spine runs vertically downwards. The optic gland lies below the peduncle lobe. The sacs of the white body tissue are relatively small and do not fill the orbit; the remaining space is occupied by gelatinous tissue.

The central nervous system was described by Thore (1939). It is compact (Fig. 33.5), and in the suboesophageal mass the brachial lobes are short, and form only 7.0% of the total brain volume. The brachial lobes are the smallest recorded among the epipelagic octopods measured (Wirz 1959, Maddock and Young 1987). The pedal lobe provides much of the innervation of the arms, and many fibres run in the brachiopalliovisceral connective directly to and from the arms. The anterior chromatophore lobes are well developed, although the posterior chromatophore lobes are small; together they form 0.4% of the total brain volume (Table 2.6). The palliovisceral lobe shows prominent development of the circum-palliovisceral tract of fibres from the vasomotor lobes (Fig. 33.6). Many fibres leave the tract to join the anterior vena cava, innervating the plentiful juxtaganglionic tissue of its lining. Other fibres run to the cephalic arteries. This suggests an elaborate control of the circulation in *Argonauta*, by both nervous and hormonal mechanisms.

In the supraoesophageal mass the inferior buccal lobe is relatively large and is surrounded by a layer of juxtaganglionic tissue (Young 1977a). The inferior frontal lobe system forms 3.7%, and the vertical lobe system 12% of the total brain volume (Fig. 33.7). The volume of the vertical lobe is larger than that of the superior frontal lobe (Maddock and Young 1987).

In two studies, both using the same method and species, the volumes of 10 lobes of the brain were measured in newly hatched and in adult specimens (Wirz 1959, Frösch 1971). The palliovisceral lobe and the higher motor centres are all relatively larger at hatching than in the adult (Fig. 33.8). The inferior frontal lobe, the main centre for tactile discrimination, is very large at hatching but relatively smaller in the adult. These findings are reflected in the observations of newly hatched argonauts described earlier in the chapter.

The epistellar bodies are small sacs at the hind end of the stellate ganglion. The cell bodies of their rhabdoms are similar to those of the nearby neurons of the ganglion.

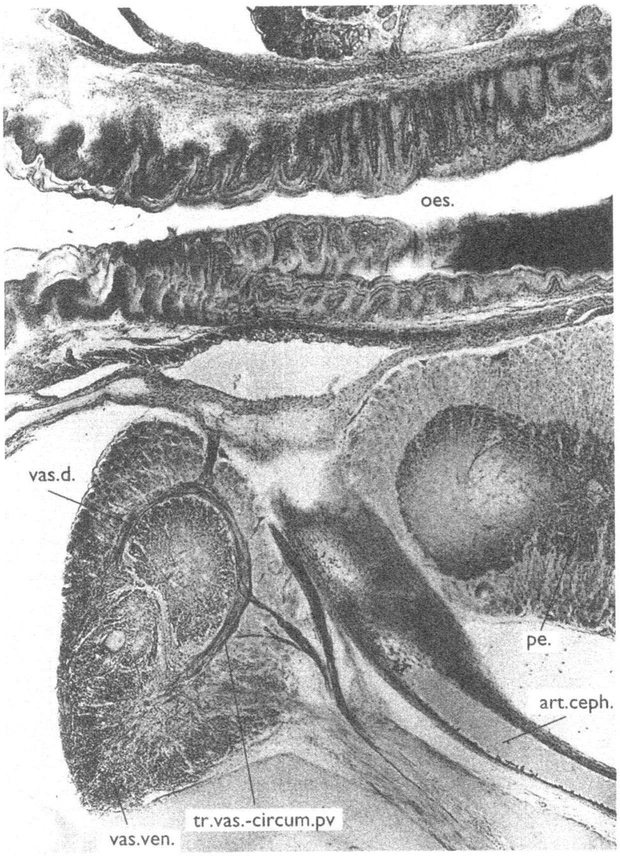

Fig. 33.6. *Argonauta argo.* Sagittal section of the palliovisceral lobe to show the vasomotor lobes and the tracts leading from them to the cephalic artery and vena cava.

The statocyst is almost spherical (Young 1989). The crista has nine sections, the last being short and indistinct; the sections show alternately single and double rows of large hair cells. The anticrista is an elongated somewhat flat plate, with six lobes. It extends forward almost reaching the ovoid macula to provide a channel to confine and restrain the flow during turning in the yawing plane. Kölliker's canal has a large ciliated funnel, with a broad proximal section and a long thin distal tube ending, apparently blindly, close to the base of the anticrista.

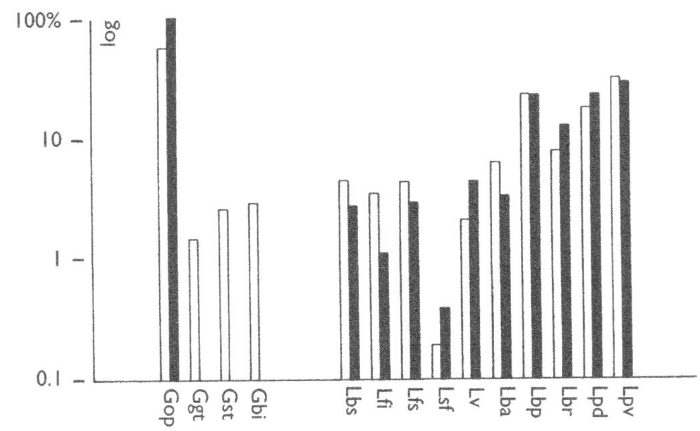

Fig. 33.8. *Argonauta argo.* Histogram showing the volumes of each lobe as a percentage of the whole brain. Open bars, newly hatched (Frösch 1971); filled bars, adults (Wirz 1959). Gbi, inferior buccal ganglion; Ggt, gastric ganglion; Gop, optic ganglion; Gst, stellate ganglion; Lba, anterior basal lobe; Lbp, posterior basal lobe; Lbr, brachial lobe; Lbs, superior buccal lobe; Lfi, inferior frontal lobe; Lfs, superior frontal lobe; Lpd, pedal lobe; Lpv, palliovisceral lobe; Lsf, subfrontal lobe; Lsv, subvertical lobe; Lv, vertical lobe.

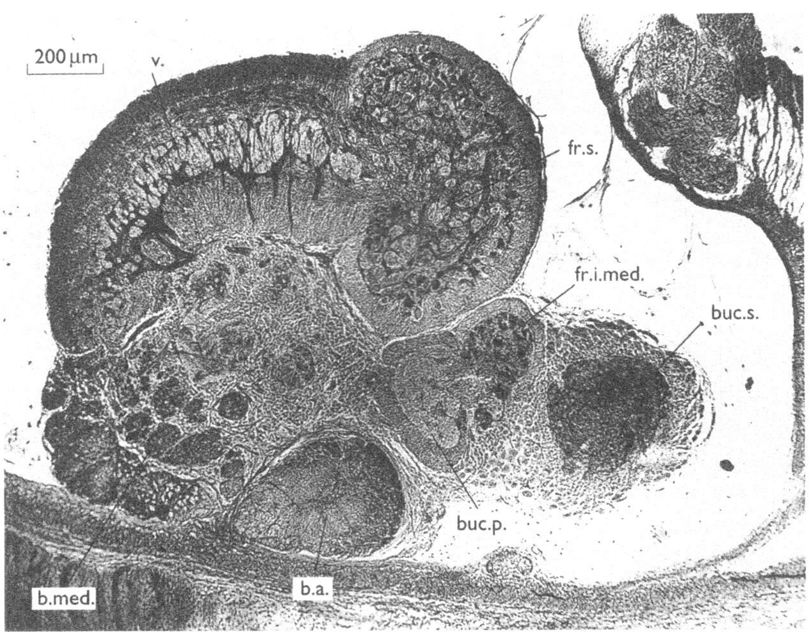

Fig. 33.7. *Argonauta argo.* The supraoesophageal mass to show the prominent superior frontal lobe.

34 Alloposidae

34.1 *Haliphron*

Haliphron atlanticus (*Alloposus mollis*) (Fig. 34.1) is the only genus and species of the family. It is oceanic and has a circumglobal distribution in the Atlantic, Indian, and Pacific Oceans between 43°N and 40°S (Thore 1949, Alvariño and Hunter 1981). The vertical distribution is poorly known but captures with opening/closing nets indicate a possible diel migration as some animals have been caught from the surface to 100 m at night and at 600–700 m during the day (Lu and Clarke 1975a, b, Lu and Roper 1979). A single otter trawl, fished at 682 m, caught 26 large specimens, suggesting that this octopod may not be so rare as captures suggest (Roper and Young 1975).

34.2 Life history

Little is known of the life cycle of this octopod but an increase in size from winter through to the autumn was found from a small number of captures, suggesting that spawning may occur in winter (Lu and Roper 1979). One brooding female was seen from a submersible to be drifting near the ocean floor (R.E. Young 1995). Immature eggs, varying in size from 0.3 to 1.5 mm, were present in the ovary of one female (100 mm ML) (Thore 1949), and an oviducal egg 6 mm in length was found in another female (450 mm ML) (Willassen 1986). The size difference seen in ovarian eggs suggests that *Haliphron* is an intermittent spawner, like a number of other cephalopods (Mangold *et al.* 1993). A male of 103 mm ML was considered to be an adult, and it cannot be considered a dwarf in comparison with the female (Lu and Roper 1979).

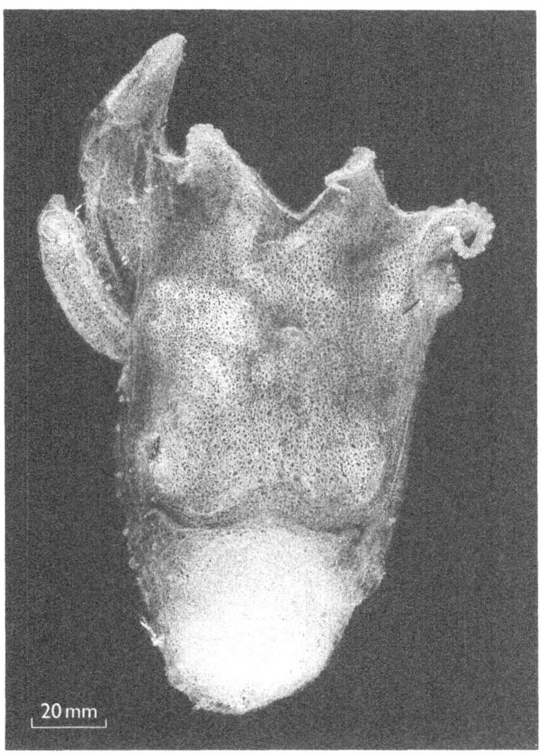

Fig. 34.1. *Haliphron atlanticus.* A male in ventral view showing the wide head, laterally placed eyes, wide mantle opening, and the funnel. This specimen of 58 mm ML is also seen in Figs 34.2–34.6.

34.3 Morphology

The mantle has quite deep musculature. Stellate ganglia are present (Willasen 1986). The mantle and funnel are connected by strap-like muscular septa of which Thore (1949) said 'these septa give the impression that the power of expulsion of water from the mantle cavity must be great, thus making the animal a rather fast backward swimmer'. This observation has been confirmed, as biologists watching from the deck of a boat saw one young *Haliphron* efficiently avoiding capture when the net was towed in direct pursuit; the eventual capture confirmed identification (Alvariño and Hunter 1981).

The arms are short and the suckers small. Apart from the sucker itself there are few muscles in the arms and sucker peduncle (Fig. 34.2). The brachial nerve is made notable by the surrounding pale-staining tissues which appear to have little structure. However, this octopod is capable of capturing prey since the crop of our specimen was filled with shrimps whose heads had been neatly severed, and the stomach contained the remains of a small but almost whole *Haliphron*. The buccal mass forms quite a large proportion of the animal; the length of the crest of the upper beak measured 53 mm in a specimen of 450 mm ML (Willasen 1986). Thore (1949) illustrated the beak and radula of an animal with a total length of 400 mm. The paired anterior salivary glands lie on the dorsoposterior surface of the buccal mass, and the paired posterior salivary glands lie behind the brain. The oesophagus is of note because of its considerable size, being 2 mm in diameter in our specimen of 58 mm ML. The oesophagus is deeply folded and has a thick lining of chitin (Fig. 34.3). This lining would provide protection when prey with a hard and often sharp exoskeleton, such as shrimps, is ingested. Immediately behind the brain the oesophagus opens into a large, thin-walled crop. The stomach has a thick wall. The digestive gland is relatively large and the ink sac is buried within it.

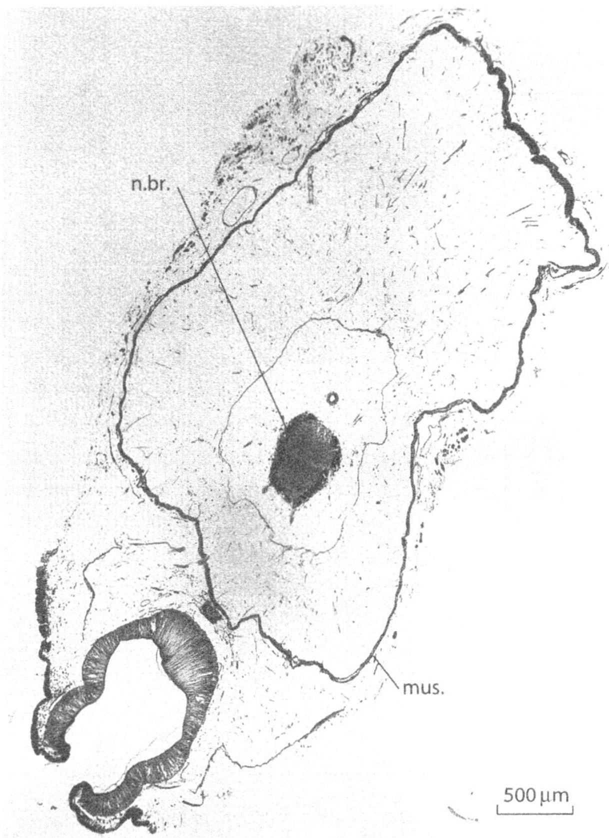

Fig. 34.2. *Haliphron atlanticus.* Transverse section through an arm and a sucker.

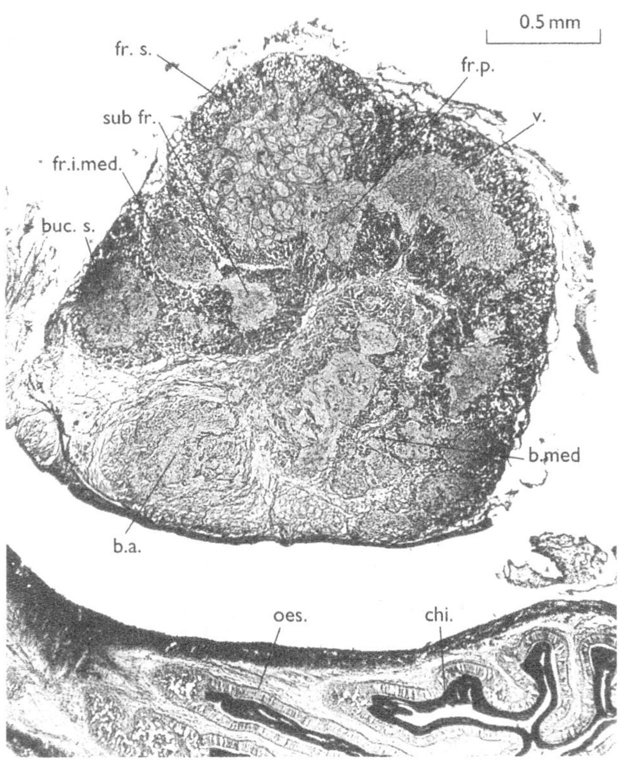

Fig. 34.3. *Haliphron atlanticus.* Oblique sagittal section of supraoesophageal lobes. The posterior buccal lobe is not included because of the angle of the section.

34.4 Receptors and central nervous system

The eyes are set on either side of the broad head and are relatively large (Fig. 34.1). The diameter of the eye can be as much as 52% of the mantle length (Table 34.1). The eye and the pupil are round and in living animals the small central opening of the lid has a sphincter that can expand to expose the whole eye (Alvariño and Hunter 1981). The retina is almost uniform, with rhabdoms 300 µm in length at the centre and 200 µm peripherally.

The optic lobes are widely separated and lie close to the eyes, but are far from the central nervous system, at the end of long optic tracts (Fig. 34.4). The optic lobes have a uniform structure, and form 373% of the total brain volume (Table 2.6). The optic nerves form the familiar dorsoventral chiasma.

The peduncle lobe and optic gland form distinct protuberances on the optic tract, and they lie approximately halfway between the optic lobes and the central nervous system. The peduncle spine runs obliquely anteroposteriorly and turns down behind the optic tract. The nerve fibres to the peduncle lobe run as a series of separate bundles, distinct from the rest of the optic tract. They maintain their order in the horizontal plane as far as their entry to the peduncle lobe, but immediately after entering they diverge to form a plexus. The optic gland lies nearby and has large cells and amoebocytes containing pigment.

The central nervous system encircles the relatively huge oesophagus (Fig. 34.4). The lobes are thus widely separated but are, nevertheless, compact anteroposteriorly (Fig. 34.3). The vertical lobe system is well developed and forms 11.6% of the total brain volume (Table 2.6). The vertical lobe comprises five irregular cylinders and forms 4.23% of the total brain volume, and the superior frontal lobe comprises 4.27% of the total brain volume (Maddock and Young 1987).

The brachiopalliovisceral connective is large (Fig. 34.5). Each brachial nerve has two fibres of up to 60 µm in diameter. The third right brachial nerve of this male is especially large with a tangle of nerve fibres at its origin; this may be concerned with autotomy of the hectocotylized arm.

The posterior chromatophore lobes form 1.7% of the total brain volume (Fig. 34.5) (Table 2.6). Behind the palliovisceral lobe is a mass of extracortical neuropil. The cerebrobrachial connective contains a bundle of fibres that breaks up in the lateral inferior frontal lobe. From this lobe a relatively large interfrontal tract runs to the

Table 34.1. *Haliphron atlanticus.* The diameter of the eye and its relationship with the mantle length

Mantle length (mm)	Eye diameter (mm)	Eye diameter/ mantle length (%)	Source
17	8	47	Alvariño and Hunter (1981)
50	26	52	Alvariño and Hunter (1981)
58	30	52	Our specimen (Fig. 34.1)
115	38	34	R.E. Young (1972b)
178	64	35	Verrill (1881)

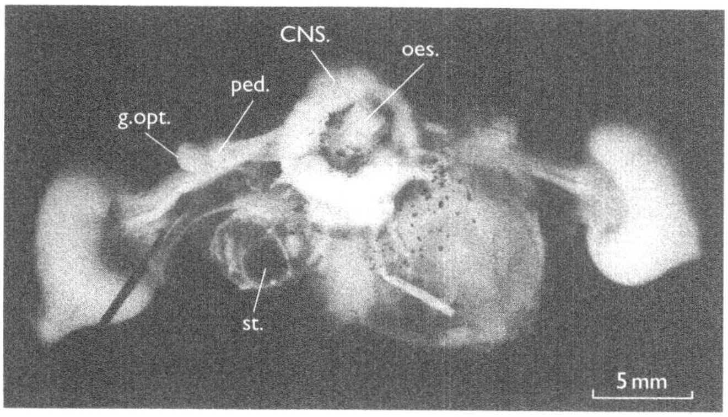

Fig. 34.4. *Haliphron atlanticus*. Dissection to show the optic lobes joined to the central nervous system by a long optic tract, and the peduncle lobe and optic gland lying halfway along the optic tract. The statocyst is also visible.

superior frontal lobe. Fibres from the median inferior frontal lobe pass into the subfrontal lobe (Fig. 34.3).

The arteries of the central nervous system are of notable size (Fig. 34.5) but the need for such large arteries is not understood.

The mantle and funnel are connected by lateral septa which lie close to the stellate ganglion (Willasen 1986). Each epistellar body is near the entry of the mantle connective. There are about seven distinct vesicles and their nerves run into the neuropil where they appear to spread outwards (Fig. 34.6).

The statocysts lie behind and below the brain and are separated by the oesophagus (Fig. 34.4) (Young 1989). The macula is round. The crista is divided into nine sections. The anticrista is a small plate set in the transverse plane close to the last turn of the crista. Multiple discriminant analysis of measurements of the statocyst place *Haliphron* amongst buoyant cephalopods (Maddock and Young 1984).

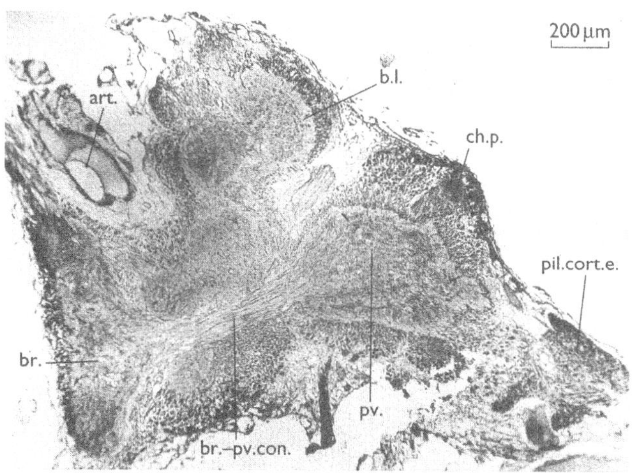

Fig. 34.5. *Haliphron atlanticus*. Oblique sagittal section of the central nervous system; note the large brachiopalliovisceral connective.

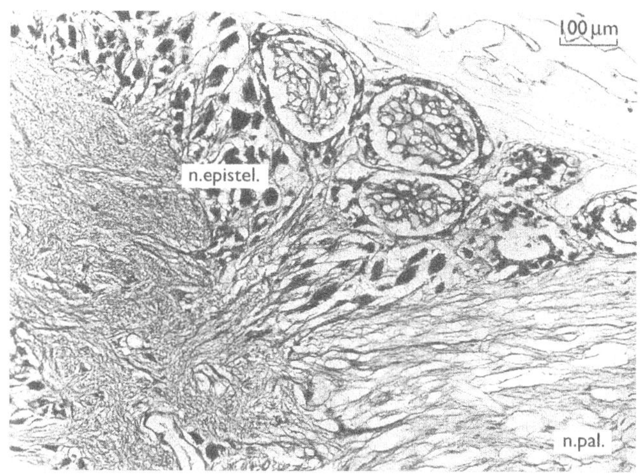

Fig. 34.6. *Haliphron atlanticus*. Sagittal section of epistellar bodies showing the nerve fibres entering the neuropil of the stellate ganglion.

35 Ocythoidae

35.1 *Ocythoe*

Ocythoe tuberculata (Fig. 35.1) is the only genus and species of the family. It is cosmopolitan in tropical to temperate seas and has been recorded from widely scattered localities between latitudes 50°N and 40°S, including the Mediterranean Sea (Berry 1955b, 1970, Roper and Sweeney 1976, Lu and Phillips 1985, Cardoso 1991). The majority have been caught between 40°N and 20°N in the Atlantic and Pacific Oceans, mostly in the vicinity of land masses and islands; not one, however, has been captured near Hawaii in spite of intensive sampling efforts over many years. *Ocythoe* has been caught in closing nets at depths of 200 m and 100 m, and at the surface (Roper and Sweeney 1976). Others have been caught in pelagic trawls at 10–40 m and 180 m and at 30–40 m in a purse seine, and divers have observed males in shallow waters (Berry 1970, Hardwick 1970, Okutani and Osuga 1986).

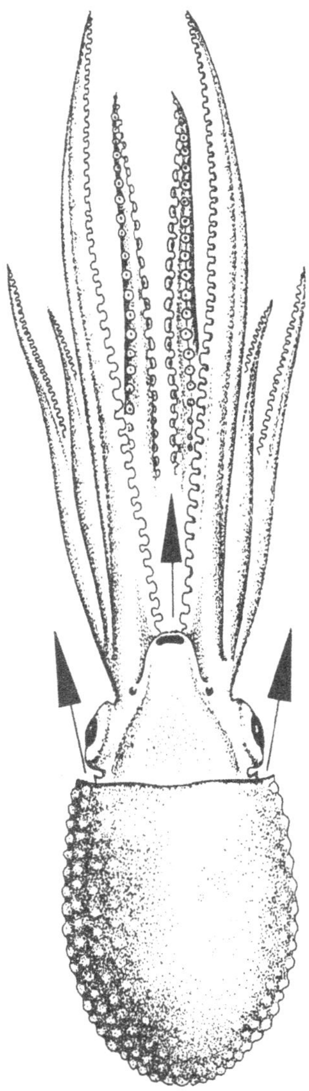

Fig. 35.1. *Ocythoe tuberculata.* Drawing of an animal, seen from below, to show the use of triple jets (arrows) during an escape manoeuvre (Packard and Wurtz 1994).

The colour after preservation is purplish-brown dorsally and paler ventrally. In living animals the flesh is said to be

> translucent, milky-white, ... the lateral and ventral sides of the mantle have a bright silvery sheen ... but the predominant tones are blue and violet, mainly on the dorsal side of the mantle, head and dorsal arms (Naef 1923).

The colour of the dorsal surface of the animal can change from sky blue to deep violet, and even almost to black. One specimen swimming at the surface of a tank was orange-brown at the sides and tips of arms, and blue changing to violet dorsally (Packard and Wurtz 1994).

35.2 Locomotion and buoyancy

An *Ocythoe* (150 mm ML) lived in an aquarium for 2 days, during which time it was buoyant and swam strongly. When swimming the mantle aperture was found to be converted into transient accessory funnels, through which water was ejected from the upper chamber of the mantle cavity. The accessory funnels could be deployed to jet water from the mantle upwards, sideways and backwards. The mantle cavity is divided horizontally into an upper compression chamber for the paired lateral accessory funnels, and a lower chamber for the relatively huge funnel. The animal thus possesses effectively three jets for the expulsion of water during locomotion (Fig. 35.1) (Packard and Wurtz 1994).

The adult female has a unique and complex system for buoyancy. There is a large organ in the midline, on top of the visceral mass, immediately above the digestive gland, and in the most dorsal part of the mantle, the dorsal sac (Fig. 35.2). It is an elongated, tongue-like structure lying horizontally above the midline, and directly above the crop. In large specimens the sac can reach 40% of the mantle length. In juveniles the sac is relatively small and is distended with 'jelly'. In mature females at the egg-bearing stage the sac is large and empty, and at this time it occupies 5% of the volume of the whole animal (Packard and Wurtz 1994).

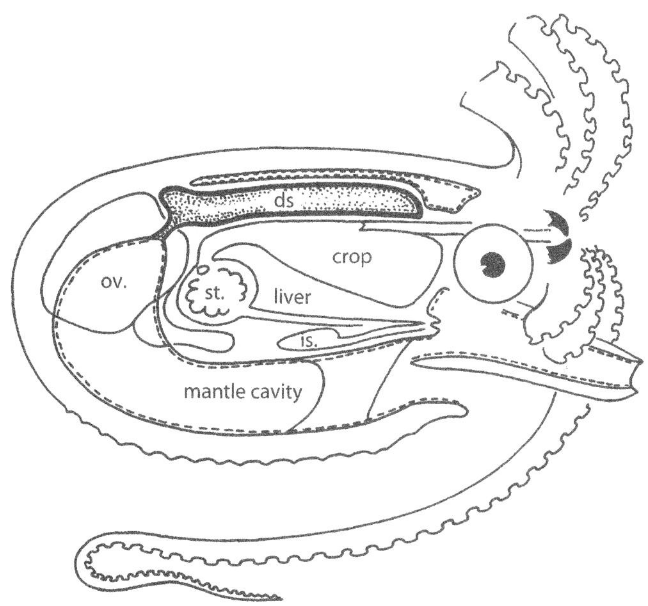

Fig. 35.2. *Ocythoe tuberculata.* Drawing of a hemisection of a female to show the dorsal sac, which lies above the crop. ds, dorsal sac; is, ink sac; liver is the digestive gland; ov, ovary; st, stomach.

The dorsal sac, or swimbladder, is supplied with a nerve, and blood vessels. Notably this sac is absent from the dwarf male. The sac has thick walls of a form of connective tissue, with a matrix and ovoid nuclei. At the periphery there are scattered connective tissue fibres and some muscle fibres. There is a transition at the periphery from connective tissue to the central region of isolated nuclei and matrix. The inner lining is a much folded epithelium and the ventral wall of the dorsal sac has an outer layer of circular muscles. The thick dorsal wall of the sac consists mostly of connective tissue with a few scattered muscle fibres. The thickened base of the dorsal sac contains much loose connective tissue and blood vessels, and there are also large bundles of nerve fibres. The jelly-like substance in the dorsal sac is probably part of the flotation system. In ripe females, the dorsal sac is a distended bladder full of a gas, but no evidence of such a gland was found in sections of an immature female of 80 mm ML. The dorsal sac opens into the mantle cavity, its opening being hidden beneath and behind the base of the right gill where it connects with the mantle wall. The nature and origin of the gas are not known but it may come from the jelly-like substance. It is possible that air is trapped in the mantle as the animal swims at the surface, and so enters the dorsal sac. The specimen observed by Packard and Wurtz (1994) emitted several bubbles of gas before descending to the bottom of the tank. This occurred after the animal had been inspected to free any trapped air bubbles. Female *Argonauta* make use of air, trapped in the apex of the shell for buoyancy. The pelagic habit of the animal, the anatomical position, the size and structure of the dorsal sac, and the posture and behaviour of the animal indicate that it serves a hydrostatic function, and forms a flotation chamber, when the animal is heavy with young (Packard 1972). Another possible buoyancy mechanism lies below the skin of the mantle where, between the tubercles, there is loosely organized tissue with cells similar to those in the jelly of the dorsal sac. It is not known whether there is any replacement of heavier ions by lighter ions.

35.3 Reproduction

The smallest specimen captured so far is a posthatchling of 2.3 mm ML. Arms I and IV are already greatly elongated and the funnel extends far forward in front of the bases of the arms. The dorsal surface of the mantle of males and females is covered with a uniform pattern of tiny chromatophores; the ventral surface in both sexes is devoid of colour cells. The ventral water pores are evident at all stages (Naef 1923, Hochberg *et al.* 1992).

Ocythoe is ovoviviparous and the female has extremely long and convoluted oviducts in which embryonic development takes place as the eggs move along the duct (Steenstrup 1880, Jatta 1896, Naef 1923). As many as 104 000 ovarian eggs were found in a large female of 310 mm ML: the average size of the eggs was 1.75×1.0 mm (Roper and Sweeney 1976). Jatta (1896) observed that some of the eggs in the oviduct of a female (100 mm ML) were ready to hatch, indicating that the eggs develop and hatch in batches over some considerable period of time, during which the female continues to grow. The whole genital organ has been found to occupy one-tenth of the body weight in a mature female. It is thus conceivable that the function of the dorsal sac is indeed connected with the habit of bearing live young, especially as there is no trace of such a sac in the dwarf male (Packard and Wurtz 1994).

Dwarf males are often recorded as inhabiting the body cavity of salps. One male of about 30 mm ML was found sheltering in *Salpa tilesi*, its hectocotylus still within the sac (Jatta 1896). Divers in waters 0–10 m deep have observed this phenomenon, the salp being *Thetys vagina*; in one instance the diver grasped the salp and the male *Ocythoe* abandoned its 'home' and swam some 20 m before it could be recaptured (Berry 1970, Hardwick 1970, Okutani and Osuga 1986). Arm III is the longest and the hectocotylus is enclosed in a sac (Naef 1923). This arm is autotomized during mating and remains in the mantle cavity of the female during the period of fertilization (Roper and Sweeney 1976). The hectocotylus was described by Naef (1923). The spermatophores must presumably remain viable during this time, although the arm is detached from the male. In one female a hectocotylus was found in the region of the oviducts. A thread consisting of spermatophores had become extruded from the base of two of the suckers; this seems unlikely to be the normal means of exit. The distal sac contains the coiled penis, which protrudes slightly. In transverse section this arm appears flattened, but it has a deep band of muscles and a relatively large axial nerve cord. The sucker is muscular and has a wide, ovoid infundibulum, and the chamber opening is eccentric.

35.4 Morphology

There is a pair of ventral water pores on the head but no dorsal water pores are present. The lateral pores lead to a space below the eyes and

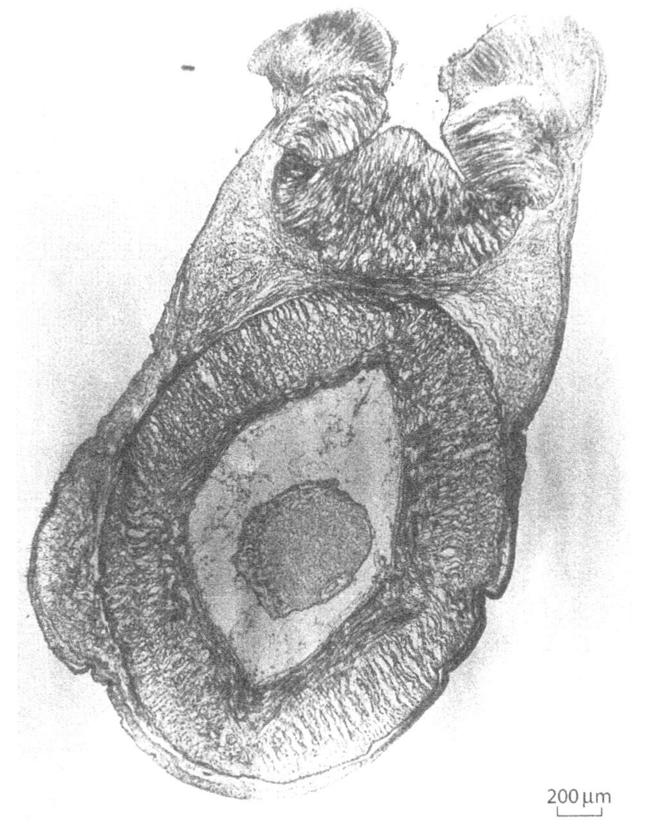

Fig. 35.3. *Ocythoe tuberculata* (62 mm ML). Transverse section through the arm and sucker of a female. It shows the broad band of muscle, and the relatively large axial nerve cord surrounded by loosely organized tissue (haematoxylin and eosin).

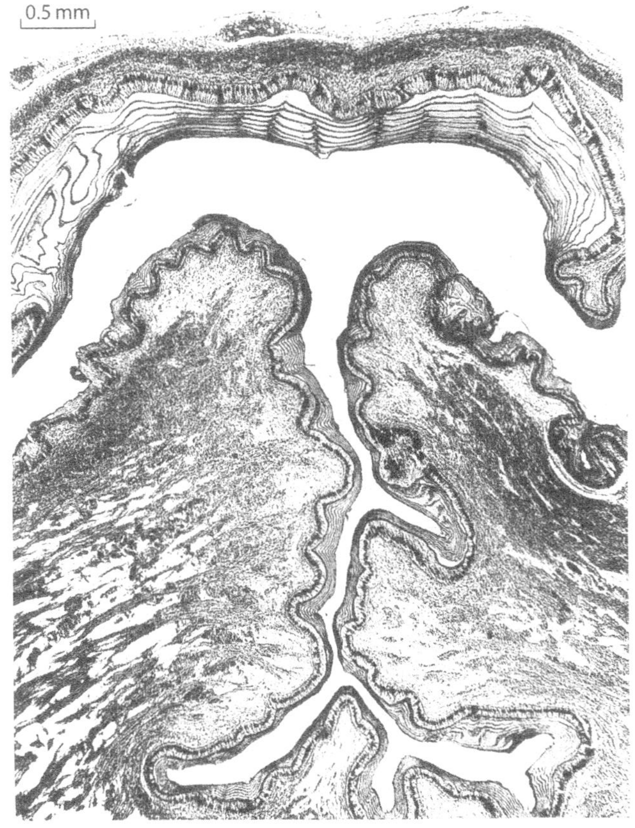

Fig. 35.4. *Ocythoe tuberculata*. Transverse section of the buccal mass showing part of the upper beak and the lateral buccal palps with laminated cuticle. Figs 35.4–35.8 are of an animal of 700 g body weight.

extend laterally and upwards, between the bases of the arms, to the dorsal side. Here the cavity is crossed by large blood vessels which form a ring. From the ring smaller vessels proceed to the arms and branches pass to the skin and other tissues at the base of the arms. There are no swellings such as those seen in *Tremoctopus*. The significance of this system remains enigmatic.

The arms are quite slender especially when compared with the suckers (Fig. 35.3). There is a broad peripheral band of muscle tissue encircling a region which appears 'empty' except for sparsely distributed delicate strands of tissue. This surrounds the relatively large axial nerve cord. The muscles of the sucker have, like those of the arms, a somewhat loose organization.

The beak is broad and has a small 'pinched in' rostrum (Clarke 1986). It protrudes from the base of the reflected arms (Fig. 35.1). The buccal mass forms a relatively large proportion of the head. In the buccal cavity the lateral buccal palps have tall, columnar cells which secrete a thick, laminated, cuticular covering (Fig. 35.4). A similarly thick, layered, cuticle lines the oesophagus and continues as a thick lining into the large crop. Below the radula the salivary papilla is surrounded by deeply invaginated glandular tissue. Sardines and crustaceans were found in the crops of three specimens (A. Packard, pers. comm.). A live animal showed stereotyped reactions to an anchovy dropped straight ahead of it; the dorsal arms were thrown out simultaneously and symmetrically forward to capture the fish from which large bites were taken (Packard and Wurtz 1994). It has also been observed to feed on pteropod and heteropod molluscs (Robson 1932).

35.5 Receptors and central nervous system

The eye is relatively large, with a diameter of 17 mm in a female of 62 mm ML. It is round, as is the pupil. In the retina the rhabdoms are of nearly equal length but anteriorly, dorsally, and posteriorly (150 µm) they are slightly longer than those at the centre (125 µm).

The optic lobes are laterally flattened (Fig. 35.5). They form 172.4% of the total brain volume, and so are relatively smaller than those of other pelagic octopods measured (Table 2.6). The plexiform zone of the optic lobe is well differentiated, with dendrites of the several layers oriented in various directions. The optic lobe lies at a distance from the peduncle lobe and optic gland. The opticopeduncular fibres enter the lobe over its full extent and derive from a wide area within the cell islands of the medulla of the optic lobe. Afferent fibres entering the peduncle lobe maintain the same linear order as their points of origin in the optic lobe. Once they have traversed the cell layer the constituent fibres of the incoming bundles disperse in an apparently regular fashion.

The central nervous system was described briefly by Pfefferkorn (1915) who commented on the wide separation of the supra- and suboesophageal masses to accommodate the oesophagus. Thore (1939) added further to the description of the brain. The suboesophageal mass forms 72.3% of the total brain volume, and is amongst the largest among the cephalopods measured, and the pallioviscerall lobe occupies most of this volume (Table 2.6). The brachial lobes (8.1%), are rather small compared with those of most of the benthic octopods. The anterior and posterior chromatophore lobes form 2.8%, and the lateral basal lobes 1.1% of the brain volume, indicating the importance of the control of colour for this epipelagic octopod. The palliovisceral lobe continues backwards as a large mass of 'extracortical neuropil' as in *Argonauta* and *Tremoctopus* (Young 1971).

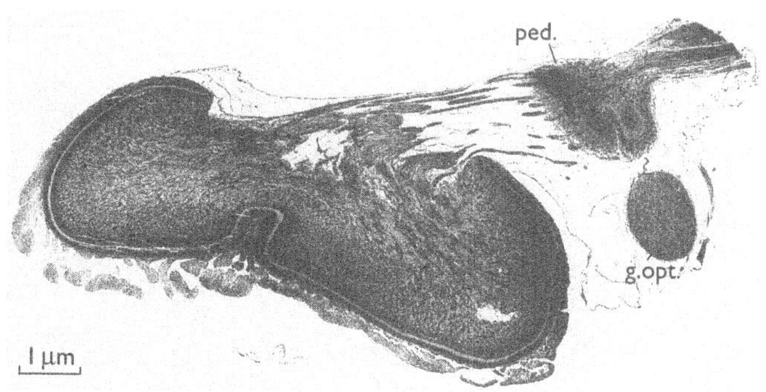

Fig. 35.5. *Ocythoe tuberculata.* Transverse section through the head to show the laterally flattened optic lobe, and the optic tract, the peduncle lobe, and optic gland.

The inferior buccal ganglion is relatively large, with abundant masses of juxtaganglionic tissue around it (Young 1971), as in *Argonauta* (Young 1970b). Strands proceed to the wall of the buccal sinus and among tissues of the buccal mass. The juxtaganglionic tissue springs from characteristic 'extracortical neuropil' consisting of groups of small neurons around a centre of fibrous neuropil.

The supraoesophageal mass is compact (27.7%) (Fig. 35.6). The inferior frontal system occupies 3.7% of the brain volume, and is relatively small as it is in the other pelagic octopods measured (Maddock and Young 1987). The lateral and median inferior frontal lobes receive an input from the arms and the fibres interweave. There is a conspicuous tract from the inferior frontal to the lateral superior frontal lobe. The median inferior frontal lobe is notable for its small size (Fig. 35.7). The subfrontal lobe is a backward prolongation of the posterior buccal lobe, with whose inner cell layers it is clearly continuous. The subfrontal lobe lies far back and its tissue reaches up behind the median inferior frontal lobe almost reaching the anterior ventral end of the vertical lobe. This indicates how the whole inferior frontal system and the vertical lobe system is a backward extension of the superior buccal lobe. All of the parts of this system in *Octopus vulgaris* are concerned with learned reactions to touch.

The vertical lobe system forms 7.8% of the total brain volume (Fig. 35.6). The lateral and median superior frontal lobes are

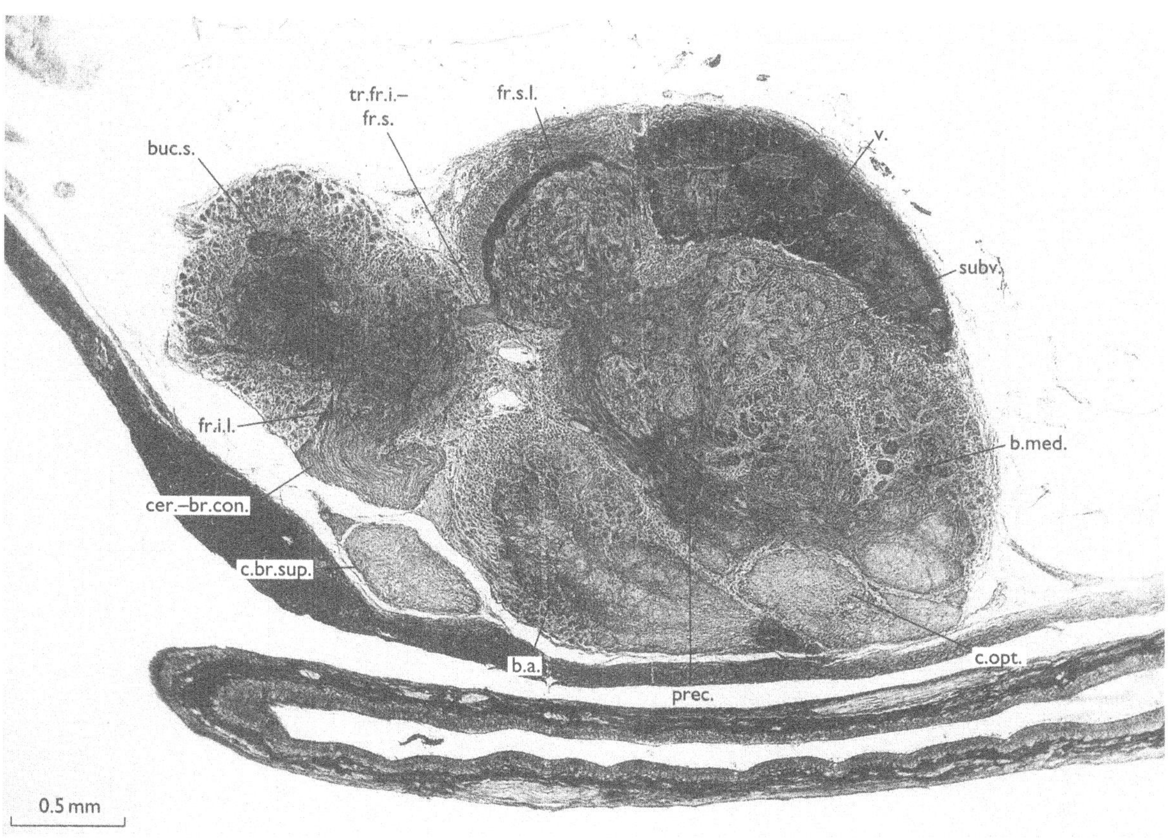

Fig. 35.6. *Ocythoe tuberculata.* Sagittal section to show the supraoesophageal lobes of the central nervous system.

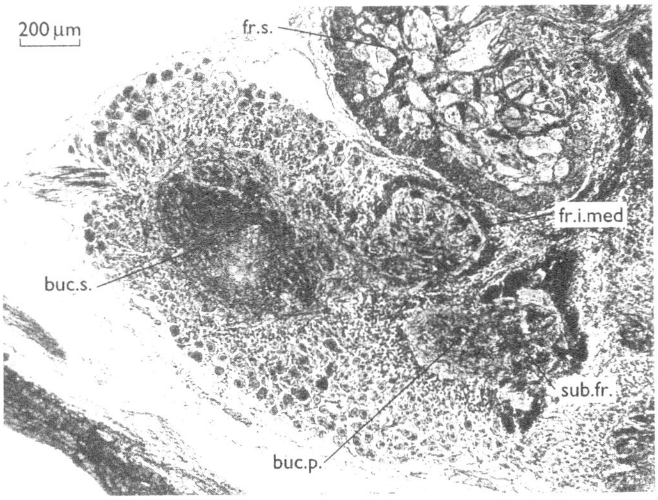

Fig. 35.7. *Ocythoe tuberculata.* Sagittal section showing the small median inferior frontal lobe.

moderately large. The lateral inferior frontal lobe has a distinct layer of fibres. The subpedunculate lobes are associated with masses of paravertical tissue. The subpedunculate nerves proceed to the veins at the base of the orbit, perhaps to juxtaganglionic tissues there.

The epistellar body is a small vesicle situated at the extreme hind end of the stellate ganglion (Fig. 35.8) (Young 1936). It is a vesicle of irregular shape and diameter, and is lined by cubical cells with processes extending into the central lumen. A drawing was made in 1936, before the epistellar body was found to be an extraocular photoreceptor organ, and what is now recognized to be a rhabdom was labelled a 'process of a neurosecretory cell'. The subsequent discoveries and the evidence gathered show that these cells are closely similar to the receptor cells in the retina (Young 1971, 1991d).

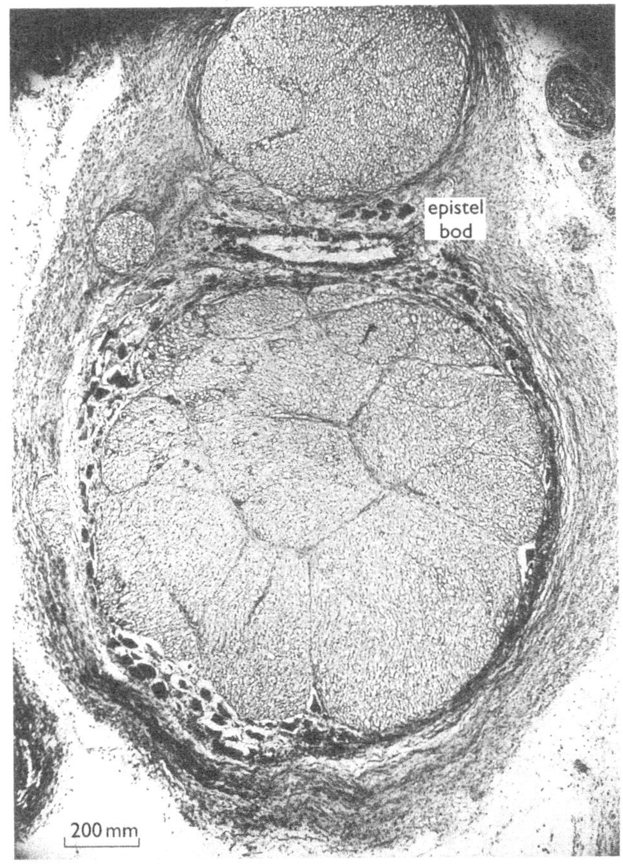

Fig. 35.8. *Ocythoe tuberculata.* Transverse section of the stellar nerves and epistellar body.

References

Abbott, N.J., Williamson, R., and Maddock, L. (ed.) (1995). *Cephalopod neurobiology: Neuroscience studies in squid, octopus and cuttlefish.* Oxford University Press, Oxford.

Adam, W. (1950). Un céphalopode nouveau: *Pholidoteuthis boschmai* gen. et sp. nov. *Koninklijke Nederlandse Akademie van Wetenschappen*, 53, (10), 1592–8.

Adam, W. (1954). Cephalopoda. IV. Céphalopodes à l'exclusion des genres *Sepia, Sepiella* et *Sepioteuthis. Sibogá-Expeditie*, **144**, 121–98.

Adam, W. (1979). The Sepiidae (Cephalopoda, Decapoda) in the collections of the Western Australian Museum. *Records of the Western Australian Museum*, 7, (2), 113–212.

Adam, W. (1986). La radule et les mandibules de quelques espèces d'*Idiosepius* Steenstrup, 1881 (Mollusca, Cephalopoda, Decapoda). *Bulletin de l'Institut Royal des Sciences Naturelles de Belgique*, **56**, 149–54.

Adam, W. and Rees, W.J. (1966). A review of the cephalopod family Sepiidae. *The John Murray Expedition 1933–1934, Scientific Reports*, **11**, No. 1, 1–165.

Adamo, S.A. and Hanlon, R.T. (1996). Do cuttlefish (Cephalopoda) signal their intentions to conspecifics during agonistic encounters? *Animal Behaviour*, **52**, 73–81.

Akimushkin, I.I. (1965). *Cephalopods of the seas of the U.S.S.R.* (trans. A. Mercado). Israel Program for Scientific Translations, Jerusalem.

Aldred, R.G. (1974). Structure, growth and distribution of the squid *Bathothauma lyromma* Chun. *Journal of the Marine Biological Association, U.K.*, **54**, 995–1006.

Aldred, R.G., Nixon, M., and Young, J.Z. (1978). The blind octopus, *Cirrothauma. Nature, London*, **275**, 547–49.

Aldred, R.G., Nixon, M., and Young, J.Z. (1983). *Cirrothauma murrayi* Chun, a finned octopod. *Philosophical Transactions of the Royal Society, London*, **B 301**, 1–54.

Aldred, R.G., Nixon, M., and Young, J.Z. (1984). Ganglia not light organs in the suckers of octopods. *Journal of Molluscan Studies*, **50**, 67–9.

Aldrich, F.A. (1991). Some aspects of the systematics and biology of the genus *Architeuthis* based on a study of specimens from Newfoundland waters. *Bulletin of Marine Science*, **49**, 457–81.

Aldrich, F.A. and. Lu, C.C. (1968a). Report on the larvae, eggs, and egg mass of *Rossia* sp. (Decapoda, Cephalopoda) from Bonavista Bay, Newfoundland. *Canadian Journal of Zoology*, **46**, 369–71.

Aldrich, F.A. and Lu, C.C. (1968b). On an octopod from Placentia Bay, Newfoundland. *The Veliger*, **11**, 70–1.

Alexandrowicz, J.-S. (1927). Contribution à l'étude des muscles, des nerfs et du mécanisme de l'accommodation de l'oeil des céphalopodes. *Archives de Zoologie Expérimentale et Génerale*, **66**, 71–134.

Alexandrowicz, J.-S. (1964). The neurosecretory system of the vena cava in cephalopods. I. *Eledone cirrosa. Journal of the Marine Biological Association, U.K.*, **44**, 111–32.

Alexandrowicz, J.-S. (1965). The neurosecretory system of the vena cava in Cephalopoda II. *Sepia officinalis* and *Octopus vulgaris. Journal of the Marine Biological Association, U.K.*, **45**, 209–28.

Allen, A., Michels, J., and Young, J.Z. (1985). Memory and visual discrimination by squids. *Marine Behaviour and Physiology*, **11**, 271–82.

Altman, J.S. (1967). The behaviour of *Octopus vulgaris* Lam. in its natural habitat: a pilot study. *Underwater Association Reports*, 1966–67, 77–83.

Altman, J.S., and Nixon, M. (1970). Use of the beaks and radula by *Octopus vulgaris* in feeding. *Journal of Zoology, London*, **161**, 25–38.

Alvariño, A. and Hunter, J.R. (1981). New records of *Alloposus mollis* Verrill (Cephalopoda, Octopoda) from the Pacific Ocean. *The Nautilus*, **95**, 26–32.

Amoore, J.E., Rodgers, K., and Young, J.Z. (1959). Sodium and potassium in the endolymph and perilymph of the statocyst and eye of *Octopus. Journal of Experimental Biology*, **36**, 709–14.

Anderson, E.A. (1978). Notes on the cephalopods of Monterey Bay, California, with new records for the area. *The Veliger*, **21**, 255–62.

Anderson, F.E. (2000). Phylogenetic relationships among loliginid squids (Cephalopoda: Myopsida) based on analyses of multiple data sets. *Zoological Journal of the Linnaean Society, London*, **130**, 603–33.

Anderson, R.C. (1995). West coast octopuses. *Diver*, **21**, 28–9.

Anderson, R.C. and Mather, J.A. (1996). Escape responses of *Euprymna scolopes* Berry, 1911 (Cephalopoda: Sepiolidae). *Journal of Molluscan Studies*, **62**, 543–5.

Andrews, P.L.R., and Tansey, E.M. (1983). The digestive tract of *Octopus vulgaris*: the anatomy, physiology and pharmacology of the upper tract. *Journal of the Marine Biological Association, U.K.*, **63**, 109–34.

Andrews, P.L.R., Messenger, J.B., and Tansey, E.M. (1981). Colour changes in cephalopods after neurotransmitter injection into the cephalic aorta. *Proceedings of the Royal Society, London*, **B 213**, 93–9.

Andrews, P.L.R., Messenger, J.B., and Tansey, E.M. (1983). The chromatic and motor effects of neurotransmitter injection in intact and brain-lesioned *Octopus. Journal of the Marine Biological Association, U.K.*, **63**, 355–70.

Appellöf, A. (1893). Die Schalen von *Sepia, Spirula* und *Nautilus.* Studien über den Bau und das Wachsthum. *Kongliga Svenska Vetenskaps Akademiens Handlingar*, **25**, 1–106.

Arata, G.F. (1954). A note on the flying behaviour of certain squids. *The Nautilus*, **68**, 1–3.

Aristotle (384–322 BCE) see D'A.W. Thompson (1910).

Arkhipkin, A.I. (1996). Age and growth of planktonic squids *Cranchia scabra* and *Liocranchia reinhardti* (Cephalopoda, Cranchiidae) in the epipelagic waters of the central-east Atlantic. *Journal of Plankton Research*, **18**, 1675–83.

Arkhipkin, A.I. (1997). Age of the micronektonic squid *Pterygioteuthis gemmata* (Cephalopoda: Pyroteuthidae) from the central-east Atlantic based on statolith growth increments. *Journal of Molluscan Studies*, **63**, 287–90.

Arkipkhin, A. and Bizikov, V.A. (2000). Role of the statolith in functioning of the acceleration receptor system in squids and sepioids. *Journal of Zoology*, **250**, 31–55.

Arkhipkin, A. and Nekludova, N. (1993). Age, growth and maturation of the loliginid squids *Alloteuthis africana* and *A. subulata* on the West African shelf. *Journal of the Marine Biological Association, U.K.*, 73, 949–61.

Arkhipkin, A.I. and Nigmatullin, C.M. (1997). Ecology of the oceanic squid *Onychoteuthis banksi* and the relationship between the genera *Onychoteuthis* and *Chaunoteuthis* (Cephalopoda: Onychoteuthidae). *Journal of the Marine Biological Association, U.K.*, 77, 839–69.

Arkhipkin, A.I. and Schetinnikov, A.S. (1989). Fauna and distribution of the young and adult pelagic cephalopods of the Gulf of Guinea. *Zoologicheskii Zhurnal*, **68**, 26–32.

Arkhipkin, A., Zuev, G.V., Nesis, K.N., Nigmatullin, C.M., Pinchukov, M.A., Tsimbal, M.A. *et al.* (1983a). Distribution and main peculiarities of life cycle of the squid *Thysanoteuthis rhombus*. In *Taxonomy and ecology of cephalopods: Scientific papers*, pp. 130–1. U.S.S.R. Academy of Sciences, Zoological Institute, Leningrad.

Arkhipkin, A., Nigmatullin, C.M., Sabirov, R.M., and Shilin, K.D. (1983b). Morphology and functional peculiarities of the reproductive system of the squid *Thysanoteuthis rhombus*. In *Taxonomy and ecology of cephalopods: Scientific papers*, pp. 59–61. U.S.S.R. Academy of Sciences, Zoological Institute, Leningrad.

Arnold, J.M. (1965). Observations on the mating behavior of the squid. *Bulletin of Marine Science*, **15**, 216–22.

Arnold, J.M. (1987). Reproduction and embryology of *Nautilus*. In Nautilus: *The biology and paleobiology of a living fossil* (ed. W.B. Saunders and N. Landman), pp. 353–72. Plenum Press, New York.

Arnold, J.M., and Carlson, B.A. (1991). Living *Nautilus* embryos: *in ova* movements. *Journal of Cephalopod Biology*, **2**, 1–11.

Arnold, J.M., and Williams-Arnold, L.D. (1980). Development of the ciliature pattern on the embryo of the squid *Loligo pealei*. A scanning electron microscope study. *Biological Bulletin, Woods Hole*, **159**, 102–16.

Arnold, J.M., and Young, R.E. (1974). Ultrastructure of a cephalopod photophore. I. Structure of the photogenic tissue. *Biological Bulletin, Woods Hole*, **147**, 507–21.

Banas, P.T., Smith, D.E., and Biggs, D.C. (1982). An association between a pelagic octopod, *Argonauta* sp. Linnaeus, 1758, and aggregate salps. *Fishery Bulletin, U.S.A.*, **80**, 648–50.

Bandel, K., and Boletzky, S. v. (1979). A comparative study of the structure, development and morphological relationships of chambered cephalopod shells. *The Veliger*, **21**, 313–54.

Bandel, K. and Leich, H. (1986). Jurassic Vampyromorpha (dibranchiate cephalopods). *Neues Jahrbuch für Geologie und Paläontologie, Monatschrift*, **H3**, 129–48.

Barber, V.C., and Wright, D.E. (1969). The fine structure of the sense organs of the cephalopod mollusc *Nautilus*. *Zeitschrift für Zellforschung*, **102**, 293–312.

Barnard, K.H. (1934). Cephalopods from the Cape of Good Hope. *Journal of Conchology, London*, **20**, 44–5.

Bather, F.A. (1888). Professor Blake and shell-growth in Cephalopoda. *Geological Magazine*, (6), **1**, 421–7.

Bather, F.A. (1895). The habits of the young Sepia. *Journal of Malacology*, **4**, 33–4.

Beebe, W. (1926). The Arcturus adventure: an account of the New York Zoological Society's first oceanic expedition. Putnam, New York.

Belcari, P., Biagi, F., and Sartor, P. (1989). Sepiolinae (Mollusca, Cephalopoda) del mar Tirreno settentrionale. *Atti della Società Toscana di Scienze Naturali, Memorie* **B96**, 207–18.

Bello, G. (1995). A key to the identification of the Mediterranean sepiolods, (Mollusca: Cephalopoda). *Bulletin de l'Institut océnographique, Monaco*, No. spécial **16**, 41–55.

Bello, G., and Biagi, V. (1995). How benthic are sepiolids? *Bulletin de l'Institut océnographique, Monaco*, No. spécial **16**, 57–61.

Bergström, B.I. (1985). Aspects of natural foraging by *Sepietta oweniana* (Mollusca, Cephalopoda). *Ophelia*, **24**, 65–74.

Bergstöm, B., and Summers, W.C. (1983). *Sepietta oweniana*. In *Cephalopod life cycles* Vol. I. *Species accounts* (ed. P.R. Boyle), pp. 75–91. Academic Press, London.

Berry, S.S. (1909). Diagnoses of new cephalopods from the Hawaiian Islands. *Proceedings of the United States National Museum*, **37**, 407–19.

Berry, S.S. (1911). A new sepiolid from Japan. *Zoologischer Anzeiger*, **37**, 39–41.

Berry, S.S. (1913). *Nematolampas*, a remarkable new cephalopod from the South Pacific. *Biological Bulletin, Woods Hole*, **25**, 208–12.

Berry, S.S. (1916). Cephalopoda of the Kermadec Islands. *Proceedings of the Academy of Natural Sciences, Philadelphia*, **68**, 45–66.

Berry, S.S. (1920). Light production in cephalopods I. An introductory survey. *Biological Bulletin, Woods Hole*, **38**, 141–95.

Berry, S.S. (1921). A review of the cephalopod genera *Sepioloidea*, *Sepiadarium*, and *Idiosepius*. *Records of the South Australian Museum*, **1**, 347–64.

Berry, S.S. (1949). A new *Opisthoteuthis* from the eastern Pacific. *Leaflets in Malacology*, **1**, 23–6.

Berry, S.S. (1952). The flapjack devilfish, *Opisthoteuthis*, in California. *California Fish and Game*, **38**, 183–8.

Berry, S.S. (1955 a). The male flapjack devilfish. *California Fish and Game*, **41**, 219–24.

Berry, S.S. (1955b). On recent Californian occurrences of the rare octopod *Ocythoe*. *California Fish and Game*, **41**, 177–81.

Berry, S.S. (1970). A note on the behavior of the octopod *Ocythoe tuberculata*. *California Fish and Game*, **56**, 68–70.

Bidder, A.M. (1950). The digestive mechanism of the European squids *Loligo vulgaris*, *Loligo forbesi*, *Alloteuthis media*, and *Alloteuthis subulata*. *Quarterly Journal of Microscopical Science*, **91**, 1–43.

Bidder, A.M. (1962). Use of the tentacles, swimming and buoyancy control in the Pearly Nautilus. *Nature, London*, **196**, 451–4.

Birchall, J.D. and Thomas, N.L. (1983). On the architecture and function of the cuttlefish bone. *Journal of Materials Science*, **18**, 2081–86.

Bizikov, V.A. (1995). Growth of *Sthenoteuthis oualaniensis*, using a new method based on gladius structure. *International Council for the Exploration of the Sea, Marine Science Symposia*, **199**, 445–8.

Bizikov, V.A. and Arkhipkin, A.I. (1997). Morphology and microstructure of the gladius and statolith from the boreal Pacific giant squid *Moroteuthis robusta* (Oegopsida, Onychotetuthidae). *Journal of Zoology, London*, **241**, 475–92.

Blackburn, S., Sauer, W.H.H., and Lipinski, M.R. (1998). The embryonic development of the chokka squid *Loligo vulgaris reynaudii* d'Orbigny, 1845. *The Veliger*, **41**, 249–58.

Blackith, R.E. and Reyment, R.A. (1971). *Multivariate morphometrics*. Academic Press, London.

Bleckmann, H., Budelmann, B.U., and Bullock, T.H. (1991). Peripheral and central nervous responses evoked by small water movements in a cephalopod. *Journal of Comparative Physiology*, **A168**, 247–57.

Boal, J. (1996). A review of simultaneous visual discrimination as a method of training octopus. *Biological Reviews, Cambridge Philosophical Society*, **71**, 157–89.

Boettcher, K.J. and Ruby, E.G. (1990). Depressed light emission by symbiotic *Vibro fischeri* of the sepiolid squid *Euprymna scolopes*. *Journal of Bacteriology*, **172**, 3701–6.

Bøggild, O.B. (1930). The shell structure of the mollusks. *Det Kongelige Danske Videnskabernes Selskabs Skrifter, Naturvidenskabelig og Mathematisk Afdeling*, **2**, 232–325.

Boletzky, S. v. (1970). On the presence of light organs in *Semirossia* Steenstrup, 1887 (Mollusca: Cephalopoda). *Bulletin of Marine Science*, **20**, 374–88.

Boletzky, S. v. (1971). Mandibules denticulées chez les larves des Teuthoidés et des Octopodes (Mollusca, Cephalopoda). *Compte rendu de l'Académie des sciences, Paris*, **272**, 2904–6.

Boletzky, S. v. (1974a). The 'larvae' of Cephalopoda: a review. *Thalassia Jugoslavica*, **10**, 45–76.

Boletzky, S. v. (1974b). Effets de la sous-nutrition prolongé e sur le développement de la coquille de *Sepia officinalis* L. (Mollusca, Cephalopoda). *Bulletin de la Societé Zoologique de France*, **99**, 667–3.

Boletzky, S. v. (1976). Quelques observations sur *Pteroctopus tetracirrhus* (Mollusca, Cephalopoda). *Rapport Commission Internationale pour la Mer Méditerranée*, **23**, 95–102.

Boletzky, S. v. (1977). Post-hatching behaviour and mode of life in cephalopods. *Symposia Zoological Society, London*, No. 38, 557–67.

Boletzky, S. v. (1978). Nos connaissances actuelles sur le développment embryonnaire du Sepiolidae pelagique *Heteroteuthis* (Mollusca: Cephalopoda). *Haliotis*, **9**, 81–4.

Boletzky, S. v. (1979). Observations on the early post-embryonic development of *Loligo vulgaris* (Mollusca, Cephalopoda). *Rapport Commission Internationale pour la Mer Mediterranee*, **25/26**, 159–68.

Boletzky, S. v. (1981). Morphologie de l'oeuf et mode de ponte chez *Pteroctoopus tetracirrhus* (Mollusca, Cephalopoda). *Vie et Milieu*, **31**, 255–9.

Boletzky, S. v. (1982). On eggs and embryos of cirromorph octopods. *Malacologia*, **22**, 197–204.

Boletzky, S. v. (1983a). *Sepia officinalis* . In *Cephalopod life cycles* Vol. I. *Species accounts* (ed. P.R. Boyle), pp. 31–52. Academic Press, London.

Boletzky, S. v. (1983b). *Sepiola robusta*. In: *Cephalopod life cycles* Vol. I. *Species accounts* (ed. P.R. Boyle), pp. 53–67. Academic Press, London.

Boletzky, S. v. (1983c). Laboratory observations on a female Argonauta argo (Mollusca: Cephalopoda). Rapports de la Commission Internationales pour l'Exploration Scientifique de la Mer Méditerrané, **28**, 289–91.

Boletzky, S. v. (1984). The embryonic development of the octopus *Scaeurgus unicirrhus* (Mollusca, Cephalopoda), additional data and discussion. *Vie et Milieu*, **34**, 87–93.

Boletzky, S. v. (1987). Fecundity variation in relation to intermittent or chronic spawning in the cuttlefish, *Sepia officinalis* L. (Mollusca, Cephalopoda). *Bulletin of Marine Science*, **40**, 382–7.

Boletzky, S. v. (1995). The systematic position of the Sepiolidae (Mollusca: Cephalopoda). *Bulletin de l'Institut Océanographique, Monaco*, No. spécial **16**, 99–104.

Boletzky, S. v. and Boletzky, M.V. v. (1973). Observations on the embryonic and early post-embryonic development of *Rossia macrosoma* (Mollusca, Cephalopoda). *Helgoländer wissenschaftliche Meeresuntersuchungen*, **25**, 135–61.

Boletzky, S. v. and Roeleveld, M.A.C. (2000). 'Ventral adhesion' to hard substrates: a thigmotactic response in sepiid cuttlefish (Mollusca, Cephalopoda). *Vie et Milieu*, **50**, 59–54.

Boletzky, S. v. and Wiedmann, J. (1978). Schulp-Wachstum bei Sepia officinalis. Abhängigkeit von ökologischen Parametern. Neues Jahrbuch für Geologis und Paläontologie Abhandlungen, **157**, 103–7.

Boletzky, S. v. , Rio, M., and Roux, M. (1992). Octopod 'ballooning' response. *Nature, London*, **356**, 199.

Bon, W.F., Dohrn, A., and Batink, H. (1967). The lens protein of a marine invertebrate, *Octopus vulgaris*. *Biochimica et Biophysica Acta*, **140**, 312–8.

Bone, Q. (1972). Buoyancy and hydrodynamic functions of integument in the castor oil fish, *Ruvettus pretiosus* (Pisces: Gempylidae). *Copeia*, 1972(1), 78–87.

Bone, Q. and Ryan, K.P. (1976). Cupular sense organs in *Ciona* (Tunicata: Ascidiacea). *Journal of Zoology, London*, **186**, 417–429.

Bone, Q., Pulsford, A.L., and Chubb, A.D. (1981). Squid mantle muscle. *Journal of the Marine Biological Association, U.K.*, **61**, 327–42.

Bone, Q., Brown, E.R., Nicholson, D., and Travers, G. (1994). On the respiratory flow in the cuttlefish, *Sepia*. *Journal of Experimental Biology*, **194**, 153–65.

Bone, Q., Brown, E.R., and Usher, M. (1995). The structure and physiology of cephalopod muscle fibres. In *Cephalopod neurobiology: Neuroscience studies in squid, octopus and cuttlefish* (ed. N.J. Abbott, R. Williamson and L. Maddock), pp. 301–29. Oxford University Press.

Bonichon, A. (1967). Contribution à l'étude de la neurosécrétion et de l'endocrinologie chez les Céphalopodes. I. *Octopus vulgaris*. *Vie et Milieu*, **18**, 227–63.

Bott, R. (1938). Kopula und Eiablage von Sepia officinalis L. Zeitschrift für Morphologie und Oekologie der Tiere, **34**, 150–60.

Boucaud-Camou, E. (1968). Étude histologique et histochimique de l'appareil digestif de *Sepiola atlantica* D'Orbigny et *Sepia officinalis* L. *Bulletin de la Société Linnéenne de Normandie* (10) **9**, 220–43.

Boucaud-Camou, E. and Boismery, J. (1991). The migrations of the cuttlefish (*Sepia officinalis* L.) in the English Channel. In *La seiche: The cuttlefish. Ist International Symposium on the cuttlefish* Sepia *Caen, June 1–3, 1989*, (ed. E. Boucaud-Camou), pp. 179–89. University of Caen.

Boucaud-Camou, E. and Boucher-Rodoni, R. (1983). Feeding and digestion in cephalopods. In: *The Mollusca* (ed. K.M. Wilbur), Vol. 5, *Physiology* Part 2 (ed. A.S.M. Saleuddin and K.M. Wilbur), pp. 149–187. Academic Press, San Diego.

Boucher, L.M. (1983). Aspects of reproduction in some enoploteuthid squids from Hawaiian waters. *The Veliger*, **26**, 106–12.

Bower, J., Seki, M.P., Young, R.E., Bigelow, K.A., Hirota, J., and Flament, P. (1999). Cephalopod paralarvae assemblages in Hawaiian waters. *Marine Ecology Progress Series*, **185**, 203–12.

Boycott, B.B. (1953). The chromatophore system of cephalopods. *Proceedings of the Linnean Society, London*, **164**, 235–40.

Boycott, B.B. (1960). The functioning of the statocysts of *Octopus vulgaris*. *Proceedings of the Royal Society, London*, **B152**, 78–87.

Boycott, B.B. (1961). The functional organization of the brain of the cuttlefish *Sepia officinalis*. *Proceedings of the Royal Society*, **B153**, 503–34.

Boycott, B.B. (1965). A comparison of living *Sepioteuthis sepioidea* and *Doryteuthis plei* with other squids, and with *Sepia officinalis*. *Journal of Zoology, London*, **147**, 344–51.

Boycott, B.B. (1998). John Zachary Young 18 March 1907–4 July 1997. *Biographical Memoirs of Fellows of The Royal Society of London*, **44**, 485–509.

Boycott, B.B. and Young, J.Z. (1950). The comparative study of learning. *Symposia of the Society of Experimental Biology*, **4**, 432–53.

Boycott, B.B. and Young, J.Z. (1955). A memory system in *Octopus vulgaris* Lamarck. *Proceedings of the Royal Society, London*, **B143**, 449–80.

Boycott, B.B. and Young, J.Z. (1956). The subpedunculate body and nerve and other organs associated with the optic tract of cephalopods. *Bertil Hanström. Zoological papers in honour of his sixty-fifth birthday* (ed. K.G. Wingstrand), pp. 76–105. Zoological Institute, Lund.

Boyde, A. (1974). Histological and cytological methods for the SEM in biology and medicine. In *Scanning electron microscopy* (ed. O.C. Wells), pp. 257–64. McGraw-Hill, New York.

Boyle, P.R. (1983a). *Eledone cirrhosa*. In *Cephalopod life cycles* Vol. I. *Species accounts* (ed. P.R. Boyle), pp. 365–86. Academic Press, London.

Boyle, P.R. (ed.) (1983b). *Cephalopod life cycles*, Vol. I. *Species accounts*. Academic Press, London.

Boyle, P.R. (1983c). Ventilation rate and arousal in the octopus. *Journal of Experimental Marine Biology and Ecology*, **69**, 129–36.

Boyle, P.R. (1986a). Report on a specimen of *Architeuthis* stranded near Aberdeen, Scotland. *Journal of Molluscan Studies*, **52**, 81–2.

Boyle, P.R. (1986b). A descriptive ecology of *Eledone cirrhosa* (Mollusca: Cephalopoda) in Scottish waters. *Journal of the Marine Biological Association, U.K.*, **66**, 855–65.

Boyle, P.R. (1986c). Responses to water-borne chemicals by the octopus *Eledone cirrhosa* (Lamarck, 1795). *Journal of Experimental Marine Biology and Ecology*, **104**, 23–9.

Boyle, P.R. (ed.). (1987) *Cephalopod life cycles*, Vol. II. *Comparative reviews*. Academic Press, New York.

Boyle, P.R. and Dubas, F. (1981). Components of body pattern displays in the octopus *Eledone cirrhosa*. *Marine Behaviour and Physiology*, **8**, 135–48.

Boyle, P.R. and Knobloch, D. (1981). Hole boring of crustacean prey by the octopus *Eledone cirrhosa* (Mollusca: Cephalopoda). *Journal of Zoology, London*, **193**, 1–10.

Boyle, P.R., Mangold, K., and Froesch, D. (1979). The mandibular movements of *Octopus vulgaris*. *Journal of Zoology, London*, **188**, 53–67.

Breneman, J.W., Robles, L.J., and Bok, D. (1986). Light-activated retinoid transport in cephalopod photoreceptors. *Experimental Eye Research*, **42**, 645–58.

Brocco, S.L. and Cloney, R.A. (1980). Reflector cells in the skin of *Octopus dofleini*. *Cell and Tissue Research*, **205**, 167–86.

Brocco, S.L., O'Clair, R.M., and Cloney, R.A. (1974). Cephalopod integument: the ultrastructure of Kölliker's organs and their relationship to setae. *Cell and Tissue Research*, **151**, 293–308.

Brown, P.K. and Brown, P.S. (1958). Visual pigments of the octopus and cuttlefish. *Nature, London*, **182**, 1288–90.

Bruggen, A.C. van (1962). *Thysanoteuthis rhombus* in Algoa Bay. *Annals of the Natal Museum*, **15**, 267–72.

Brunnmüller, E. (1929). Zur Genus- und Artcharakteristik von *Vampyroteuthis infernalis* Chun. *Zeitschrift für wissenschaftliche Zoologie*, **135**, 302–22.

Bruun, A.F. (1943). The biology of *Spirula spirula* (L.). *DANA-Report*, No. **24**, 1–44.

Bruun, A.F. (1955). New light on the biology of *Spirula*, a mesopelagic cephalopod. In *Essays in the natural sciences in honour of Captain Allan Hancock: on the occasion of his birthday July 26th 1955*, pp. 61–72. Los Angeles, University of Southern California Press.

Budelmann, B.U. (1977). Structure and function of the angular acceleration receptor systems in the statocysts of cephalopods. *Symposia of the Zoological Society, London*, No. **38**, 309–24.

Budelmann, B.U. (1994). Cephalopod sense organs, nerves and the brain: adaptations for high performance and life style. *Marine and Freshwater Behaviour and Physiology*, **25**, 13–33.

Budelmann, B.U. (1995). The cephalopod nervous system: what evolution has made of the molluscan design. In *The nervous systems of invertebrates: an evolutionary and comparative approach* (ed. O. Breidbach and W. Kutsch), pp. 115–38. Birkhauser Verlag, Basel.

Budelmann, B.U. (1996). Active marine predators: the sensory world of cephalopods. *Marine and Freshwater Behavioural Physiology*, **27**, 59–75.

Budelmann, B.U. and Bleckmann, H. (1988). A lateral line analogue in cephalopods: water waves generate microphonic potentials in the epidermal head lines of *Sepia* and *Lolliguncula*. *Journal of Comparative Physiology* **A164**, 1–5.

Budelmann, B.U. and Nixon, M. (1984). Scale worms—occasional food of *Octopus*. *Journal of Molluscan Studies*, **50**, 39–42.

Budelmann, B.U. and Tu, Y. (1997). The statocyst-oculomotor reflex of cephalopods and the vestibulo-motor reflex of verteberates: a tabular comparison. *Vie et Milieu*, **47**, 95–9.

Budelmann, B.U. and Wolff, H.G. (1973). Gravity response from angular acceleration receptors in *Octopus vulgaris*. *Journal of Comparative Physiology*, **85**, 283–90.

Budelmann, B.U. and Wolff, H.G. (1976). Mapping of neurones in the gravity receptor system of the *Octopus* statocyst by iontophoretic cobalt staining. *Cell and Tissue Research*, **171**, 403–6.

Budelmann, B.U. and Young, J.Z. (1984). The statocyst-oculomotor system of *Octopus vulgaris*: extraocular eye muscles, eye muscle nerves, statocyst nerves and the oculomotor centre in the central nervous system. *Philosophical Transactions of the Royal Society, London*, **B306**, 159–189.

Budelmann, B.U. and Young, J.Z. (1985). Central pathways of the nerves of the arms and mantle of *Octopus*. *Philosophical Transactions of the Royal Society, London*, **B310**, 109–22.

Budelmann, B.U. and Young, J.Z. (1987). Brain pathways of the brachial nerves of *Sepia* and *Loligo*. *Philosophical Transactions of the Royal Society, London*, **B315**, 345–52.

Budelmann, B.U. and Young, J.Z. (1993). The oculomotor system of decapod cephalopods: eye muscles, eye muscle nerves, and the oculomotor neurons in the central nervous system. *Philosophical Transactions of the Royal Society, London*, **B340**, 93–125.

Budelmann, B.U., Barber, V.C., and West, S. (1973). Scanning electron microscopical studies of the arrangements and numbers of hair cells in the statocysts of *Octopus vulgaris, Sepia officinalis* and *Loligo vulgaris*. *Brain Research*, **56**, 25–41.

Budelmann, B.U., Sachse, M., and Staudigl, M. (1987a). The angular acceleration receptor system of *Octopus vulgaris*: morphometry, ultrastructure, and neuronal and synaptic organization. *Philosophical Transactions of the Royal Society, London*, **B315**, 305–43.

Budelmann, B.U., Williamson, R., and Auerbach, B. (1987b). Structure and function of the angular acceleration system of *Octopus* with special reference to its efferent innervation. In *The vestibular system: neurophysiological and clinical research* (ed. M.D. Graham and J.L. Kemink), pp. 165–8. Raven Press, New York.

Budelmann, B.U., Reise, U., and Bleckmann, H. (1991). Structure, function, biological significance of the cuttlefish 'lateral lines'. *La seiche – The cuttlefish, 1st International Symposium on the Cuttlefish Sepia, Caen, June 1–3, 1989* (ed. E. Boucaud-Camou), pp. 201–209. University of Caen.

Budelmann, B.U., Bullock, T.H., and Williamson, R. (1995). Cephalopod brains: promising preparations for brain physiology. In *Cephalopod neurobiology: Neuroscience studies in squid, octopus and cuttlefish* (ed. N.J. Abbott, R. Williamson and L. Maddock), pp. 399–413. Oxford University Press.

Budelmann, B.U., Schipp, R., and Boletzky, S. v. (1997). Cephalopoda. In *Microscopic anatomy of invertebrates, Vol. 6A: Mollusca II* (ed. F.W. Harrison and A.J. Kohn), pp. 119–414. Wiley-Liss, New York.

Bullock, T.H. (1965). Mollusca: Cephalopoda. In: *Structure and function in the nervous system of invertebrates* (ed. T.H. Bullock and G.A. Horridge), pp.1433–515. Freeman, San Francisco.

Bullock, T.H. and Budelmann, B.U. (1991). Sensory evoked potentials in unanaesthetized unrestrained cuttlefish: a new preparation for brain physiology in cephalopods. *Journal of Comparative Physiology*, **A168**, 141–50.

Bülow, B. v. and Fioroni, P. (1989). Der Bau der Statocyste in Schlüpfstadium der Cephalopoden *Loligo vulgaris* Lam. und *Octopus vulgaris* Lam. *Zoologische Jahrbücher (Anatomie)*, **119**, 89–105.

Burgess, L.A. (1982). Four new species of squid (Oegopsida: *Enoploteuthis*) from the central Pacific and a description of adult *Enoploteuthis reticulata*. *Fishery Bulletin, U.S.A.*, **80**, 703–34.

Burrows, T., Campbell, I.A., and Howe, E. (1965). Conduction velocity and diameter of nerve fibres of cephalopods. *Journal of Physiology, London*, **179**, 39–40*P*.

Butcher, S., Dilly, P.N., and Herring, P.J. (1982). The comparative morphology of the photophores of the squid *Pyroteuthis margaritifera* (Cephalopoda: Enoploteuthidae). *Journal of Zoology, London*, **196**, 133–50.

Cairns, S.D. (1976). Cephalopods collected in the Straits of Florida by the R/V Gerda. *Bulletin of Marine Science*, **26**, 233–72.

Cajal, S.R. y (1917). Contribución al conocimiento de la retina y centros ópticos de los cefalópodos. *Trabajos del Laboratorio de Investigaciones Biológicas de la Universidad de Madrid*, **15**, 1–82.

Camm, J-P., Messenger, J.B., and Tansey, E.M. (1985). New pathways to the 'cerebellum' in *Octopus*. Studies using a modified Fink-Heimer technique. *Cell and Tissue Research*, **242**, 649–56.

Cardoso, F. (1991). First record of net collected *Ocythöe tuberculata* (Cephalopoda: Octopoda) from Peruvian waters. *American Malacological Bulletin*, **8**, 143–44.

Cariello, L. and Zanetti, L. (1977). α- and β-cephalotoxin: two paralysing proteins from posterior salivary glands. *Comparative Biochemistry and Physiology*, **57C**, 169–73.

Carlson, B.A. (1991). *Nautilus* hatches at Waikiki Aquarium. *Chambered Nautilus Newsletter*, **63**, 2–3.

Chase, R. and Wells, M.J. (1986). Chemotactic behaviour in *Octopus*. *Journal of Comparative Physiology*, **A158**, 375–81.

Chen, D.S., van Dykhuizen, G., Hodge, J., and Gilly, W.F. (1996). Ontogeny of copepod predation in juvenile squid (*Loligo opalescens*). *Biological Bulletin, Woods Hole*, **190**, 69–81.

Chichery, R. and Chanelet, J. (1976). Motor and behavioural responses obtained by stimulation with chronic electrodes of the optic lobes of *Sepia officinalis*. *Brain Research*, **105**, 525–32.

Chichery, R. and Chanelet, J. (1978). Motor responses obtained by stimulation of the peduncle lobe of *Sepia officinalis* in chronic experiments. *Brain Research*, **150**, 188–93.

Chichery, M.P. and Chichery, R. (1987). The anterior basal lobe and control of prey-capture in the cuttlefish (*Sepia officinalis*). *Physiology and Behavior*, **40**, 329–36.

Chichery, M.P. and Chichery, R. (1991a). The predatory behaviour of *Sepia officinalis*. Ethological and neurophysiological studies. In *La seiche – The cuttlefish, 1st International Symposium on the Cuttlefish* Sepia, *Caen, June 1–3, 1989* (ed. E. Boucaud-Camou), pp. 281–288. University of Caen.

Chichery, M.P. and Chichery, R. (1991b). The role of the inferior frontal lobe of the cuttlefish brain in the control of prey manipulation and ingestion motot activity. *La seiche – The cuttlefish, 1st International Symposium on the Cuttlefish* Sepia, *Caen, June 1–3, 1989* (ed. E. Boucaud-Camou), p. 295. University of Caen.

Chichery, M.P. and Chichery, R. (1992a). Behavioural and neurohistological changes in aging *Sepia*. *Brain Research*, **574**, 77–84.

Chichery, R. and Chichery, M.P. (1992b). Learning performances and aging in cuttlefish (*Sepia officinalis*). *Experimental Gerontology*, **27**, 233–9.

Choe, S. (1963). Daily age markings in the shell of cuttlefishes. *Nature, London*, **197**, 306–7.

Choe, S. (1966). On the eggs, rearing, habits of the fry, and growth of some Cephalopoda. *Bulletin of Marine Science of the Gulf and Caribbean*, **16**, 330–48.

Chun, C. (1903). *Aus den Tiefen des Weltmeeres* (2nd edn). Gustav Fischer, Jena.

Chun, C. (1910a). Die Cephalopoden, I: Oegopsida. *Wissenschaftliche Ergebnisse der Deutschen Tiefsee-Expedition auf dem dampfer 'Valdivia' 1898–1899*, **18**, 1–401. [trans. A. Mercado (1975). *The Cephalopoda*. Israel Program for Scientific Translations, Jerusalem.]

Chun, C. (1910b). *Spirula australis* Lam. *Berichter der Mathematisch-physischen Klasse der Königlich Sächsischen Gesellschaft der Wissenschaften zu Leipzig*, **62**, 171–88.

Chun, C. (1911). *Cirrothauma*, ein blinder Cephalopod. Doctoral Dissertation, Leipzig.

Chun, C. (1913). Cephalopoda from the 'Michael Sars' North Atlantic Deep-sea Expedition 1910. *Report on the Scientific Results of the 'Michael Sars' Expedition, 1910*, **3**, Part 1, 1–21.

Chun, C. (1915). Die Cephalopoden, II: Myopsida. Octopoda. *Wissenschaftliche Ergebnisse der Deutschen Tiefsee-Expedition auf dem dampfer 'Valdivia' 1898–1899*, **18**, 405–552. [trans. A. Mercado (1975). *The Cephalopoda*. Israel Program for Scientific Translations, Jerusalem.]

Church, R. (1971). Deepstar explores the ocean floor. *National Geographic Magazine*, **139**, 110–29.

Clarke, M.R. (1962a). The identification of cephalopod 'beaks' and the relationship between beak size and total body weight. *Bulletin of the British Museum (Natural History) Zoology*, **8**, 421–80.

Clarke, M.R. (1962b). Respiratory and swimming movements in the cephalopod *Cranchia scabra*. *Nature, London*, **196**, 351–2.

Clarke, M.R. (1964). Young stages of *Lepidoteuthis grimaldi* (Cephalopoda, Decapoda). *Proceedings of the Malacological Society, London*, **36**, 69–78.

Clarke, M.R. (1965). Large light organs on the dorsal surfaces of the squids *Ommastrephes pteropus*, *Symplectoteuthis oualensis* and *Dosidicus gigas*. *Proceedings of the Malacological Society, London*, **36**, 319–21.

Clarke, M.R. (1966). A review of the systematics and ecology of oceanic squids. *Advances in Marine Biology*, **4**, 91–300.

Clarke, M.R. (1967). A deep-sea squid, *Taningia danae* Joubin, 1931. *Symposia of the Zoological Society, London*, No. **19**, 127–43.

Clarke, M.R. (1969). Cephalopoda collected on the SOND cruise. *Journal of the Marine Biological Association, U.K.*, **49**, 961–76.

Clarke, M.R. (1970). Growth and development of *Spirula spirula*. *Journal of the Marine Biological Association, U.K.*, **50**, 53–64.

Clarke, M.R. (1980). Cephalopoda in the diet of sperm whales of the southern hemisphere and their bearing on sperm whale biology. *Discovery Reports*, **37**, 1–324.

Clarke, M.R. (1983). Cephalopod biomass – estimation from predation. *Memoirs of the National Museum of Victoria*, No. **44**, 95–107.

Clarke, M.R. (ed.) (1986). *A handbook for the identification of cephalopod beaks*. Clarendon Press, Oxford.

Clarke, M.R. (1988). Evolution of Recent cephalopods – a brief review. In *The Mollusca* (ed. K.M. Wilbur), Vol. 12. *Paleontology and neontology* (ed. M.R. Clarke and E.R. Trueman), pp. 331–40. Academic Press, San Diego.

Clarke, M.R. (1992a). Family Lepidoteuthidae Pfeffer, 1912. *Smithsonian Contributions to Zoology*, No. **513**, 167–8.

Clarke, M.R. (1992b). Family Pholidoteuthidae Adam, 1950. *Smithsonian Contributions to Zoology*, No. **513**, 168–70.

Clarke, M.R. (1992c). Family Onychoteuthidae Gray, 1849. *Smithsonian Contributions to Zoology*, No. **513**, 127–37.

Clarke, M.R. (ed.) (1996a). Cephalopods in the world's oceans. Philosophical Transactions of the Royal Society, London, **B351**, 979–1112.

Clarke, M.R. (1996b). Cephalopods as prey. III. Cetaceans. *Philosophical Transactions of the Royal Society, London*, **B351**, 1053–65.

Clarke, M.R. (1998). The value of statolith shape for systematics, taxonomy, and identification. *Smithsonian Contributions to Zoology*, No. **586**, 69–76.

Clarke, M.R. and Herring, P.J. (1971). Animal miscellany: A survey of oceanic families. In *Deep oceans* (ed. P.J. Herring and M.R. Clarke), pp. 164–204. Barker, London.

Clarke, M.R. and Lu, C.C. (1974). Vertical distribution of cephalopods at 30°N 23°W in the North Atlantic. *Journal of the Marine Biological Association, U.K.*, **54**, 969–84.

Clarke, M.R. and Lu C.C. (1975). Vertical distribution of cephalopods at 18°N 25°W in the North Atlantic. *Journal of the Marine Biological Association, U.K.*, **55**, 165–82.

Clarke, M.R. and Maul, G.E. (1962). A description of the 'scaled' squid *Lepidoteuthis grimaldi* Joubin, 1895. *Proceedings of the Zoological Society, London*, **139**, 97–118.

Clarke, M.R., Denton E.J., and Gilpin-Brown, J.B. (1979). On the use of ammonium for buoyancy in squids. *Journal of the Marine Biological Association, U.K.*, **59**, 259–76.

Cloney, R.A. and Brocco, S.L. (1983). Chromatophore organs, reflector cells, iridocytes and leucophores in cephalopods. *American Zoologist*, **23**, 581–92.

Cloney, R.A. and Florey, E. (1968). Ultrastructure of cephalopod chromatophore organs. *Zeitschrift für Zellforschung*, **89**, 250–80.

Cobb, C.S. and Williamson, R. (1998a). Electrophysiology of extraocular photoreceptors in the squid *Loligo forbesi* (Cephalopoda: Loliginidae). *Journal of Molluscan Studies*, **64**, 111–17.

Cobb, C.S. and Williamson, R. (1998b). Electrophysiology and innervation of the photosensitive epistellar body in the lesser octopus *Eledone cirrhosa*. *Biological Bulletin, Woods Hole*, **195**, 78–87.

Cobb, C.S., Pope, S.K., and Williamson, R. (1995a). Circadian rhythms to light-dark cycles in the lesser octopus, *Eledone cirrhosa*. *Marine and Freshwater Behaviour and Physiology*, **26**, 47–57.

Cobb, C.S., Williamson, R., and Pope, S.K. (1995b). The responses of the epistellar photoreceptors to light and their effect on circadian rhythms in the lesser octopus, *Eledone cirrhosa*. *Marine and Freshwater Behaviour and Physiology*, **26**, 59–69.

Cohen, A.I. (1973a). An ultrastructural analysis of the photoreceptors of the squid and their synaptic connections. I. Photoreceptive and non-synaptic regions of the retina. *Journal of Comparative Neurology*, **147**, 351–78.

Cohen A.I. (1973b). An ultrastructural analysis of the photoreceptors of the squid and their synaptic connections. II. Intraretinal synapses and plexus. *Journal of Comparative Neurology*, **147**, 379–98.

Colasanti, G. (1876). Anatomische und physiologische Untersuchungen über den Arm der Kephalopoden. *Archiv für Anatomie und Physiologie*, 480–500.

Cole, K.S. and Gilbert, D.L. (1970). Jet propulsion of squid. *Biological Bulletin, Woods Hole*, **138**, 245–6.

Colmers, W.F. (1981). Afferent synaptic connections between hair cells and the somata of intramuscular neurons in the gravity receptor system of the statocyst of *Octopus vulgaris*. *Journal of Comparative Neurology*, **197**, 385–94.

Cooper, K.M. and Hanlon, R.T. (1986). Correlation of iridescence with changes in iridophore platelet ultrastructure in the squid *Lolliguncula brevis*. *Journal of Experimental Biology*, **121**, 451–5.

Cooper, K.M., Hanlon, R.T., and Budelmann, B.U. (1990). Physiological color change in squid iridophores II. Ultrastructural mechanisms in *Lolliguncula brevis*. *Cell and Tissue Research*, **259**, 15–24.

Corner, B.D. and Moore, H.T. (1980). Field observations on the reproductive behavior of *Sepia latimanus*. *Micronesia*, **16**, 235–60.

Cornwell, C.J., Messenger, J.B., and Hanlon, R.T. (1997). Chromatophores and body patterning in the squid *Alloteuthis subulata*. *Journal of the Marine Biological Association, U.K.*, 77, 1243–6.

Cowdry, E.V. (1911). The colour changes of *Octopus vulgaris*. *University of Toronto Studies in Biology Series*, No. **10**, 1–53.

Cropp, B. (1972). The mini-killers. *Oceans*, **5**, 40–1.

Cuvier, G.L.C.F.D. (1795). Second mémoire sur l'organisation et les rapports des animaux á sang blanc, dans lequel on traite de la structure des Mollusques et leur division en ordre. *Magazin Encyclopédique, ou Journal des Sciences, des Lettres et des Arts* (Anné 1), **2**, 433–49.

Cuvier, G.L.C.F.D. (1802). *Lectures on comparative anatomy*. Translated from the French of G. Cuvier... by William Ross; under the inspection of James Macartney. *1. On the organs of motion*. Longman and Rees, London.

Cuvier, G.L.C.F.D. (1817). Mémoire sur les Céphalopodes et sur leur anatomie. In *Mémoires pour servir à l'histoire et à l'anatomie des Mollusques*, pp. 1–54. Deterville, Paris.

Dall, W.H. (1871). Descriptions of sixty new forms of mollusks from the west coast of North America and the North Pacific Ocean, with notes on others already described. *American Journal of Conchology*, 7, 93–169.

D'Aniello, A., Di Cosmo, A., Di Cristo, C., Assissi, C., Botte, V., and Di Fiore, M.M. (1996). Occurrence of sex steroid hormones and their binding proteins in *Octopus vulgaris* Lam. *Biophysical Biochmemical Research Communications*, **227**, 782–8.

Davis, R.A. (1987). *Nautilus* studies – the first twenty-two centuries. In Nautilus: *The biology and paleobiology of a living fossil* (ed. W.B. Saunders and N. Landman), pp. 3–21. Plenum, New York.

Davis, R.A. and Mohorter, W. (1973). Juvenile *Nautilus* from the Fiji Islands. *Journal of Paleontology*, **47**, 925–8.

Dawe, E.G. and Stephen, S.J. (1988). Cephalopod assemblage of the Gulf Stream system east of 60°W. *Malacologia*, **29**, 235–45.

Dell, R.K. (1952). The Recent Cephalopoda of New Zealand. *Dominion Museum Bulletin*, No. **16**, 1–157.

Dell, R.K. (1959). Some additional New Zealand cephalopods from Cook Strait. *Zoology Publications from Victoria University of Wellington*, No. **25**, 1–12.

Dell, R.K. (1970). A specimen of the giant squid *Architeuthis* from New Zealand. *Records of the Dominion Museum*, **7**, 25–36.

Denton, E.J. (1961). The buoyancy of fish and cephalopods. *Progress in Biophysics and Biophysical Chemistry*, **11**, 178–234.

Denton, E.J. (1971). Examples of the use of active transport of salts and water to give buoyancy in the sea. *Philosophical Transactions of the Royal Society, London*, **B262**, 277–87.

Denton, E.J. (1974). On buoyancy and the lives of modern and fossil cephalopods. *Proceedings of the Royal Society, London*, **B185**, 273–99.

Denton, E.J. (1990). Light and vision at depths greater than 200 metres. In *Light and life in the sea* (ed. P.J. Herring, A.K. Campbell, M. Whitfield and L. Maddock), pp. 127–46. Cambridge University Press.

Denton, E.J. and Gilpin-Brown, J.B. (1961a). The buoyancy of the cuttlefish *Sepia officinalis* (L.). *Journal of the Marine Biological Association, U.K.*, **41**, 319–42.

Denton, E.J. and Gilpin-Brown, J.B. (1961b). The effect of light on the buoyancy of the cuttlefish. *Journal of the Marine Biological Laboratory, U.K.*, **41**, 343–50.

Denton, E.J. and Gilpin-Brown, J.B. (1966). On the buoyancy of the pearly *Nautilus*. *Journal of the Marine Biological Laboratory, U.K.*, **46**, 723–59.

Denton, E.J. and Gilpin-Brown, J.B. (1971). Further observations on the buoyancy of *Spirula*. *Journal of the Marine Biological Association, U.K.*, **51**, 363–73.

Denton, E.J. and Gilpin-Brown, J.B. (1973). Flotation mechanisms in modern and fossil cephalopods. *Advances in Marine Biology*, **11**, 197–268.

Denton, E.J. and Land, M.F. (1971). Mechanism of reflexion in silvery layers of fish and cephalopods. *Proceedings of the Royal Society, London*, **A178**, 43–61.

Denton, E.J. and Locket, N.A. (1989). Possible wavelength discrimination by multi-bank retinae in the deep-sea fishes. *Journal of the Marine Biological Association, U.K.*, **69**, 409–35.

Denton, E.J. and Taylor, D.W. (1964). The composition of gas in the chambers of the cuttlebone of *Sepia officinalis*. *Journal of the Marine Biological Association, U.K.*, **44**, 203–7.

Denton, E.J. and Warren, F.J. (1957). The photosensitive pigments in the retinae of deep-sea fish. *Journal of the Marine Biological Association, U.K.*, **36**, 651–62.

Denton, E.J. and Warren, F.J. (1968). Eyes of Histioteuthidae. *Nature, London*, **219**, 400–1.

Denton, E.J., Gilpin-Brown, J.B., and Howarth, J.V. (1961). The osmotic mechanism of the cuttlebone. *Journal of the Marine Biological Association, U.K.*, **41**, 351–64.

Denton, E.J., Gilpin-Brown, J.B., and Howarth, J.V. (1967). On the buoyancy of *Spirula spirula*. *Journal of the Marine Biological Association, U.K.*, **47**, 181–91.

Denton, E.J., Gilpin-Brown, J.B., and Shaw, T.I. (1969). A buoyancy mechanism found in cranchid squid. *Proceedings of the Royal Society, London*, **B174**, 271–79.

Denucé, J.M. (1985). How embryos escape from their envelopes: a new look at the (phylogenetically) old problem of hatching. *Meddelingen van de Kononklijke Academie voor Wetenschapped, Letteren en Schone Kunsten van Belgie, Klasse der Wetenschappen*, Jahrgang 46, No. **3**, 1–30.

Denucé, J.M. and Formisano, A. (1982). Circumstantial evidence for an active contribution of Hoyle's gland to enzymatic hatching of cephalopod embryos. *Archives internationales de physiologie et de biochimie*, **90B**, 185–6.

DeRusha, R.H., Forsythe, J.W., and Hanlon, R.T. (1987). Laboratory growth, reproduction and life span of the Pacific pygmy octopus, *Octopus digueti*. *Pacific Science*, **41**, 104–21.

Di Cosmo, A. and Di Cristo, C. (1996). Endocrine control of the reproduction of *Octopus vulgaris*. *Proceedings of the 57th National Meeting of Italian Zoological Union*, Abstract, p. 104.

Di Cosmo, A. and Di Cristo, C. (1998). Neuropeptidergic control of the optic gland of *Octopus vulgaris*; FMRF-amide and GnRH immunoreactivity. *Journal of Comparative Neurology*, **398**, 1–12.

Di Cosmo, A., Paolucci, M., Di Cristo, C., Botte, V., and Ciarcia, G. (1998). Progesterone receptor in the reproductive system of the female of *Octopus vulgaris*: characterization and immunolocalization. *Molecular Reproduction and Development*, **50**, 451–60.

Dietl, J. (1878). Untersuchungen über die Organisation des Gehirns wirbelloser Tiere (Cephalopoden, Tethys). *Sitzungberichte der Akademie der Wissenschaften in Wien*, part 1, 77, 481–533.

Dilly, P.N. (1972). *Taonius megalops*, a squid that rolls up into a ball. *Nature, London*, **237**, 403–4.

Dilly, P.N. (1973). The enigma of colouration and light emission in deep-sea animals. *Endeavour*, **32**, 25–9.

Dilly, P.N. and Herring, P.J. (1974). The ocular light organ of *Bathothauma lyromma* (Mollusca: Cephalopoda). *Journal of Zoology, London*, **172**, 81–100.

Dilly, P.N. and Herring, P.J. (1978). The light organ and ink sac of *Heteroteuthis dispar* (Mollusca: Cephalopoda). *Journal of Zoology, London*, **186**, 47–59.

Dilly, P.N. and Herring, P.J. (1981). Ultrastructure and features of the light organs of *Histioteuthis macrohista* (Mollusca: Cephalopoda). *Journal of Zoology, London*, **195**, 255–66.

Dilly, P.N. and Nixon, M. (1976a). Growth and development of *Taonius megalops* (Mollusca: Cephalopoda), and some phases of its life cycle. *Journal of Zoology, London*, **179**, 19–83.

Dilly, P.N. and Nixon, M. (1976b). The dermal tubercles of *Cranchia scabra* (Mollusca, Cephalopoda); surface structure and development. *Journal of Zoology, London*, **179**, 291–5.

Dilly, P.N. and Nixon, M. (1976c). The cells that secrete the beaks in octopods and squids (Mollusca, Cephalopoda). *Cell and Tissue Research*, **167**, 229–41.

Dilly, P.N., Gray, E.G. and Young, J.Z. (1963). Electron microscopy of optic nerves and optic lobes of *Octopus* and *Eledone*. *Proceedings of the Royal Society, London*, **B158**, 446–56.

Dilly, P.N., Nixon, M. and Packard, A. (1964). Forces exerted by *Octopus vulgaris*. *Pubblicazioni della Stazione Zoologica di Napoli*, **34**, 86–97.

Dilly, P.N., Stephens, P.R. and Young, J.Z. (1975). Receptors in the statocyst of squids. *Journal of Physiology, London*, **249**, 59–61*P*.

Dilly, P.N., Nixon, M., and Young, J.Z. (1977). *Mastigoteuthis*- the whip-lash squid. *Journal of Zoology, London*, **181**, 527–59.

Doguzhaeva, L.A. (1996). Two early Cretaceous spirulid coleoids of the north-western Caucasus: their shell ultrastructure and evolutionary implications. *Palaeontology*, **39**, 681–707.

Doino, J.A., and McFall-Ngai, M.J. (1995). A transient exposure to symbiosis competent bacteria induces light organ morphogenesis in the host squid. *Biological Bulletin, Woods Hole*, **189**, 347–55.

D'Onghia, G., Tursi, A., Matarrese, A., and Panetta, P. (1993). Some aspects of the biology of *Bathypolypus sponsalis* (Mollusca, Cephalopoda) in the north Aegean Sea (Eastern Mediterranean Sea). *Vie et Milieu*, **43**, 161–4.

Donovan, D.T. (1977). Evolution of the dibranchiate Cephalopoda. *Symposia of the Zoological Society, London*, No. **38**, 15–48.

Donovan, D.T. (1996). Origins of the terms Cephalopod, Cephalopoda and Gastropoda, and early subdivisions of the Mollusca. *Bulletin of Zoological Nomenclature*, **53**, 247–52.

Donovan, D.T. (in press a). Suprafamilial classification. In *Treatise on Invertebrate Paleontology, Part M, Mollusca 5, Cephalopoda Coleoidea* (ed. D.T. Donovan). The Geological Society of America and The University of Kansas Press, Boulder, CO.

Donovan, D.T. (in press b). History of classification of Coleoidea. *Treatise on Invertebrate Paleontology, Part M, Mollusca 5, Cephalopoda Coleoidea* (ed. D.T. Donovan). Geological Society of America and The University of Kansas Press, Boulder, CO.

Donovan, D.T. (ed.) (in press c) *Treatise on Invertebrate Paleontology, Part M. Mollusca 5, Cephalopoda Coleoidea* Geological Society of America and The University of Kansas Press, Boulder, CO.

d'Orbigny, A. (1838–1848) Livraison 12. In *Histoire naturelle générale et particulière des Céphalopodes acétabulifères vivants et fossiles* (ed. A.E. de Férussac and A. d'Orbigny). J.-B. Ballière, Paris. (See A. Tillier and R. Boucher-Rodoni (1993) for publication dates of sections.)

d'Orbigny, A. (1845) *Mollusques vivant et fossiles, ou description de toutes les espèces de coquilles et de mollusques*. Adolphe Delahays, Paris.

Doyle, P., Donovan, D.T., and Nixon, M. (1994). Phylogeny and systematics of the Coleoidea. *The University of Kansas Paleontological Contributions*, N.S. No. **5**, 1–15.

Dubas, F., Hanlon, R. T., Ferguson, G.P., and Pinsker, H.M. (1986). Localization and stimulation of chromatophore motoneurones in the brain of the squid, *Lolliguncula brevis*. *Journal of Experimental Biology*, **121**, 1–25.

Dunning, M.C. (1998). A review of the systematics, distribution, and biology of the arrow squid genera *Ommastrephes* Orbigny, 1835, *Sthenoteuthis* Verrill, 1880, and *Ornithoteuthis* Okada, 1927 (Cephalopoda: Ommastrephidae). *Smithsonian Contributions to Zoology*, No. **586**, 425–33.

Dunning, M. and Brandt S.B. (1985). Distribution and life history of deep-water squid of commercial interest from Australia. *Australian Journal of Marine and Freshwater Research*, **36**, 343–59.

Dunning, M.C. and Förch, E.C. (1998). A review of the systematics, distribution, and biology of arrow squids of the genus *Nototodarus* Pfeffer, 1912 (Cephalopoda: Ommastrephidae). *Smithsonian Contributions to Zoology*, No. **586**, 393–404.

Dunning, M.C. and Wormuth, J.H. (1998). The ommastrephid squid genus *Todarodes*: a review of systematics, distribution and biology (Cephalopoda: Teuthoidea). *Smithsonian Conntributions to Zoology*, No. **586**, 385–91.

Durchon, M. and Richard, A. (1967). Étude, en culture organotypique, du rôle endocrine de la glande optique dans la maturation ovarienne chez *Sepia officinalis* L. (Mollusque, Céphalopode). *Compte Rendu de l'Académie des Sciences, Paris*, **D264**, 1492–500.

Duval, P. Chichery, M.-P., and Chichery, R. (1984). Prey capture by the cuttle-fish (*Sepia officinalis* L.): an experimental study of two strategies. *Behavioural Processes*, **9**, 13–21.

Ebersbach, A. (1915). Zur Anatomie von *Cirroteuthis umbellata* Fischer und *Stauroteuthis* sp. *Zeitschrift für Wissenschaftliche Zoologie*, **113**, 361–483.

Edwards, J. (1984). *The Roman cookery of Apicius. Translated and adapted for the modern kitchen*. Century, London.

Elder, H.Y. (1980). Peristaltic mechanisms. *Society for Experimental Biology, Seminar Series*, **5**, 71–92.

Elder, H.Y. and Owen, G. (1967). Occurence of 'elastic' fibres in the invertebrates. *Journal of Zoology, London*, **163**, 277–84.

Emery, D.G. (1975). The histology and fine structure of the olfactory organ of the squid *Lolliguncula brevis* Blainville. *Tissue and Cell*, **7**, 357–67.

Erhardt, N.M. (1991). Potential impact of a seasonal migratory jumbo sqiud (*Dosidicus gigas)* stock on a Gulf of California sardine (*Sardinia caerulea*) population. *Bulletin of Marine Science*, **49**, 325–32.

Eschricht, D.F. (1836). Cirroteuthis mülleri, eine neue Gattung der Cephalopoden Bildend. *Nova Acta Physico-Medica Academiae Caesarea Leopoldino-Carolinae Naturae Curiosorum*, **18**, 627–34.

Ferguson, G.P. and Messenger, J.B. (1991). A countershading reflex in cephalopods. *Proceedings of the Royal Society, London*, **B243**, 63–7.

Ferguson, G.P., Martini, F.M., and Pinsker, H.M. (1988). Chromatophore motor fields in the squid, *Lolliguncula brevis*. *Journal of Experimental Biology*, **134**, 281–95.

Ferguson, G.P., Messenger, J.B., and Budelmann, B.U. (1994). Gravity and light influence the countershading reflexes of the cuttlefish *Sepia officinalis*. *Journal of Experimental Biology*, **191**, 247–56.

Filippova, J.A. (1970). New data on the Cephalopoda of the Indian Ocean. *Proceedings of Symposium on Mollusca* (India), Part I, 257–64.

Filippova, J.A. (1972). New data on the squids (Cephalopoda: Oegopsida) from the Scotia Sea (Antarctic). *Malacologia*, **11**, 391–406.

Filippova, J.A., and Yukhov, V.L. (1982). New data on the genus *Alluroteuthis* Odhner, 1923. *Antarktika, Moscow*, **18**, 175–85.

Finke, E., Pörtner, H.O., Lee, P.G., and Webber, D.M. (1996). Squid (*Lolliguncula brevis*) life in shallow waters: oxygen limitation of metabolism and swimming performance. *Journal of Experimental Biology*, **199**, 911–921.

Fioroni, P. (1965). Die embryonale Musterentwicklung bei einigen mediterranean Tintenfischarten. *Vie et Milieu*, **16** (2A), 655–756.

Fioroni, P. (1981). Die Sonderstellung der Sepioliden, ein Vergleich der Ordnungender rezenten Cephalopoden. *Zoologische Jährbücher, Abteilung für Systematik Ökologie und Geographie*, **108**, 178–228.

Fioroni, P. (1991). Our recent knowledge of the development of the cuttlefish (*Sepia officinalis*). *Zoologischer Anzeiger*, **224**, 1/2, S. 1–25.

Flachsenberger, W. and Kerr, D.I.B. (1985). Lack of effect of tetrodotoxin and of an extract from the posterior salivary gland of the blue-ringed octopus following injection into the octopus and following application to its brachial nerve. *Toxicon*, **23**, 997–9.

Flood, P.R., Deibel, D., and Morris, C.C. (1990). Visualization of the transparent, gelatinous house of the pelagic tunicate *Oikopleura vanhoeffini* using *Sepia* ink. *Biological Bulletin, Woods Hole*, **178**, 118–25.

Flores, E.C. (1983a). Laboratory observations on the visual attack of the squid, *Todarodes pacificus*. *Memoirs of the National Museum, Victoria*, No. **44**, 205–12.

Flores, E.C. (1983b). Visual discrimination testing in the squid *Todarodes pacificus*: experimental evidence for lack of color vision. *Memoirs of the National Museum, Victoria*, No. **44**, 213–27.

Florey, E. (1966). Nervous control and spontaneous activityof the chromatophores of a cephalopod, *Loligo opalescens*. *Comparative Biochemistry and Physiology*, **18**, 305–24.

Florey, E. (1969). Ultrastructure and function of cephalopod chromatophores. *American Zoologist*, **9**, 429–42.

Florey, E. and Kriebel, M.E. (1969). Electrical and mechanical responses of chromatophore muscle fibers of the squid, *Loligo opalescens*, to nerve stimulation and drugs. *Zeitschrift für vergleichende Physiologie*, **65**, 98–130.

Förch, E.C. (1986). Rhynchoteuthion larvae from New Zealand waters (Cephalopoda: Ommstrephidae). *Vie et Milieu*, **36**, 177–83.

Förch, E.C. and Uozumi, Y. (1990). Discovery of a specimen of *Lycoteuthis lorigera* (Steenstrup, 1875) (Cephalopoda:Teuthoidea) from New Zealand and additional notes on its morphology. *New Zealand Journal of Marine and Freshwater Research*, **24**, 251–58.

Forsythe, J.W. and Hanlon, J.T. (1985). Aspects of egg development, posthatching behavior, growth and reproductive biology of *Octopus burryi* Voss, 1950 (Mollusca: Cephalopoda). *Vie et Milieu*, **35**, 273–82.

Forsythe, J.W. and Hanlon, J.T. (1989). Growth of the eastern Atlantic squid, *Loligo forbesi* Steenstrup (Mollusca: Cephalopoda). *Aquaculture Fishery Management*, **20**, 14.

Foyle, T.P. and O'Dor, R.K. (1987). Predatory strategies of squid (*Illex illecebrosus*) attacking small and large fish. *Marine Behaviour and Physiology*, **13**, 155–68.

Freeman, S.E. and Turner, R.J. (1970). Maculotoxin, a potent toxin secreted by *Octopus maculosus* Hoyle. *Toxicology and Applied Pharmacology*, **16**, 681.

Froesch, D. (1973a). Projection of chromatophore nerves on the body surface of *Octopus vulgaris*. *Marine Biology*, **19**, 153–5.

Froesch, D. (1973b). On fine structure of the *Octopus* iris. *Zeitschrift für Zellforschung*, **145**, 119–29.

Froesch, D. and Messenger, J.B. (1978). On leucophores and the chromatic unit of *Octopus vulgaris*. *Journal of Zoology, London*, **186**, 163–73.

Frösch, D. (1971). Quantitative Untersuchungen am Zentralnervensystem der Schlüpfstadien von zehn mediterranean Cephalopodenarten. *Revue Suisse de Zoologie*, **78**, 1069–122.

Gabe, M. and Prenant, M. (1957). Recherches sur la gaine radulaire des mollusques. IV. L'apparei redulaire de quelques Céphalopodes. *Annales des Sciences naturelles, Zoologie*, **19**, 587–602.

Gauldie, R.W., West, I.F., and Förch, E.C. (1994). Statocyst, statolith, and age estimation of the giant squid, *Architeuthis kirki*. *The Veliger*, **37**, 93–109.

Ghiretti, F. (1959). Cephalotoxin: the crab-paralysing agent of the posterior salivary glands of cephalopods. *Nature, London*, **183**, 1192–3.

Ghiretti, F. (1960). Toxicity of octopus saliva against crustacea. *Annals of the New York Academy of Sciences*, **90**, 726–41.

Gibbs, P.J. and Greenaway, P. (1978). Histological structure of the posterior salivary glands in the blue-ringed octopus *Hapalochlaena maculosa* Hoyle. *Toxicon*, **16**, 59–70.

Gilly, W.F. and Lucero, M.T. (1992). Behavioral responses to chemical stimulation of the olfactory organ in the squid *Loligo opalescens*. *Journal of Experimental Biology*, **162**, 209–22.

Gilly, W.F., Hopkins, B., and Mackie, G.O. (1991). Development of giant motor axons and neural control of escape responses in squid embryos and hatchlings. *Biological Bulletin, Woods Hole*, **180**, 209–20.

Gilly, W.F., Preuss, T. and McFarlane, M.B. (1996). All-or-none contraction and sodium channels in a subset of circular muscle fibers of squid mantle. *Biological Bulletin, Woods Hole*, **191**, 337–40.

Girod, P. (1884). Recherches sur la peau des Céphalopodes. La ventouse. *Archives zoologique expérimentale génerale*, (2) **2**, 379–401.

Girsch, S.J., Herring, P.J., and McCapra, F. (1976). Structure and preliminary biochemical characterization of the bioluminescent system of *Ommastrephes pteropus* (Steenstrup) (Mollusca; Cephalopoda). *Journal of the Marine Biological Associoation, U.K.*, **56**, 707–22.

Giuditta, A., Libonati, M., Packard, A., and Prozzo, N. (1971). Nuclear counts in the brain lobes of *Octopus vulgaris* as a function of body size. *Brain Research*, **25**, 55–62.

Gleadall, I.G. (1982). The anatomy of the anterior basal lobe of octopus. Thesis, Doctor of Philosophy, University of Sheffield.

Gleadall, I.G. (1990). Higher motor function in the brain of *Octopus*: the anterior basal lobe and its analogies with the vertebrate basal ganglia. *Annals of Applied Information Sciences*, **16**, 1–30.

Gleadall, I.G. (1994). A model for enhancing the visual information available under low-level light conditions: multiple contrast channels created by stepwise changes in detector parameters in the banked ventral retina of the firefly squid. *Inter disciplinary Information Sciences*, **1**, 67–75.

Gleadall, I.G. (1997). Hong Kong Cephalopoda: a brief review of current knowledge and identification of specimens collected in 1995. In *The marine flora and fauna of Hong Kong* (ed. B. Morton). Hong Kong University Press, Hong Kong.

Gleadall, I.G. (in prep.).

Gleadall, I.G. and Naggs, F.C. (1991). The Asian ocellate octopuses. II. The validity of *Octopus fangsiao* d'Orbigny. *Annals of Applied Information Sciences*, **16**, 173–180.

Gleadall, I.G., Ohtsu, K., Gleadall, E., and Tsukahara, Y. (1993). Screening-pigment migration in the octopus retina includes control by dopaminergic efferents. *Journal of Experimental Biology*, **185**, 1–16.

Glockauer, A. (1915). Zur Anatomie und Histologie des Cephalopodenauges. *Zeitschrift für wissenschaftliche Zoologie*, **113**, 325–60.

Gonçalves, J.M. (1991). The Octopoda (Mollusca: Cephalopoda) of the Azores. *Arquipélago, Ciências da Natureza*, **9**, 75–81.

Gonçalves, J.M. and Martins, H.R. (1992). Additions to the Octopoda (Mollusca: Cephalopoda) fauna of Madeira. *Bocagiana, Museu Municipal do Funchal*, **10**, No. 157, 3pp.

González, A.F., and Rodhouse, P.G. (1998). Fishery biology of the seven star flying squid *Martialia hyadesi* at South Georgia during winter. *Polar Biology*, **19**, 231–6.

González, A.F., Trathan, P.N., Yau, C., and Rodhouse, P.G. (1997). Interactions between oceanography, ecology and fishery biology of the ommastrephid squid *Martialia hyadesi* in the South Atlantic. *Marine Ecology Progress Series*, **152**, 205–15.

González, A.F., Guerra, A., Pascual, S., and Briand, P. (1998). *Vulcanoctopus hydrothermalis* gen. et sp. nov. (Mollusca, Cephalopoda): an octopod from a deep-sea hydrothermal vent site. *Cahiers de Biologie Marine*, **39**, 169–84.

Gosline, J.M. and DeMont, M.E. (1985). Jet-propelled swimming in squids. *Scientific American*, **252** (January), 74–80.

Gosline, J.M. and Shadwick, R.E. (1983a). Molluscan collagen and its mechanical organization in squid mantle. In *The Mollusca* Vol. 1 *Metabolic biochemistry and molecular biomechanics* (ed. P.W. Hochachka), pp. 371–98. Academic Press, New York.

Gosline, J.M. and Shadwick, R.E. (1983b). The role of elastic energy storage mechanisms in swimming: an analysis of mantle elasticity in escape jetting in the squid, *Loligo opalescens*. *Canadian Journal of Zoology*, **61**, 1421–31.

Gosline, J.M., Sleeves , J.D., Harman, A.D., and DeMont, M.E. (1983). Patterns of circular and radial mantle muscle activity in respiration and jetting of the squid *Loligo opalescens*. *Journal of Experimental Biology*, **104**, 97–109.

Goud, J. (1989). Een bijzondere waarneming tijdens de Rode Zee-reis. *Vita Marina*, **39**, 75–78.

Graziadei, P. (1959). Sulla presenza di elementi nervosi negli epiteli di rivestimento della ventosa di *Sepia officinalis*. *Zeitschrift für Anatomie und Entwicklungsgeschichte*, **121**, 103–15.

Graziadei, P. (1964). Receptors in the sucker of the cuttlefish. *Nature, London*, **203**, 384–6.

Graziadei, P. (1965). Sensory receptor cells and related neurons in cephalopods. *Cold Spring Harbour Symposia Quantitative Biology*, **30**, 45–57.

Graziadei, P. (1971). The nervous system of the arms. In *The anatomy of the nervous system of* Octopus vulgaris (by J.Z. Young), pp. 45–61. Clarendon Press, Oxford.

Graziadei, P.P.C. and Gagne, H.T. (1976a). Sensory innervation on the rim of the octopus sucker. *Journal of Morphology*, **150**, 639–76.

Graziadei, P.P.C., and Gagne, H.T. (1976b). An unusual receptor in octopus. *Tissue and Cell*, **8**, 229–40.

Greenwald, L. and Ward, P.D. (1987). Buoyancy in *Nautilus*. In Nautilus: *The biology and paleobiology of a living fossil* (ed. W.B. Saunders and N.H. Landman), pp. 547–60. Plenum, New York.

Grieg, J.A. (1934). Cephalopods from the west coast of Norway. *Bergens Museum Årbok 1933 Naturvidenskapeling*, No. 4, 1–25.

Griffin, L.E. (1900). The anatomy of *Nautilus pompilius*. *Memoir of the National Academy of Sciences*, **8**, 101–97.

Grimpe, G. (1916). *Chunioteuthis*: Eine neue Cephalopoden-gattung. *Zoologischer Anzeiger*, **46**, 349–59.

Grimpe, G. (1921). Systematische Übersicht der Nordseecephalooden. *Zoologischer Anzeiger*, **52**, 296–304.

Grisley, M.S. and Boyle, P.R. (1987). Bioassay and proeolytic enzyme from octopus saliva. *Comparative Biochemistry and Physiology*, **88B**, 1117–23.

Grisley, M.S. and Boyle, P.R. (1990). Chitinase, a new enzyme in octopus saliva. *Comparative Biochemistry and Physiology*, **95B**, 311–6.

Grisley, M.S., Boyle, P.R., and Key, L.N. (1996). Eye puncture as a route of entry for saliva during predation on crabs by the octopus *Eledone cirrhosa*. *Journal of Experimental Marine Biology and Ecology*, **202**, 225–37.

Guèrin, J. (1908). Contribution à l'étude des systèmes cutanés musculaire et nerveux de l'appareil tentaculaire de Céphalopodes. *Archives de zoologie expérimentale et générale*, (4) **8**, 1–178.

Guerra, A. (1978). Sobre la alimentación y comportamiento alimentario de *Octopus vulgaris*. *Investigación Pesquera*, **42**, 351–64.

Guerra, A. (1981). Spatial distribution pattern of *Octopus vulgaris*. *Journal of Zoology, London*, **195**, 133–146.

Guerra, A. (1982a). Cefalópodos capturados en la campaña 'Golfo de Cádiz-81'. *Resultados Expediciones Cientificas Supplement Investigaciones Pesqueras*, No. **10**, 17–49.

Guerra, A. (1982b). *Rondeletiola minor* (Naef, 1912) (Cephalopoda: Sepioidea): New record for the central-east Atlantic. *The Veliger*, **24**, 300–2.

Guerra, A. (1985a). An extension of the known depth range for *Sepia elegans* Blainville, 1827 (Cephalopoda: Sepioidea). *The Veliger*, **28**, 217–8.

Guerra, A. (1985b). Food of the cuttlefish *Sepia officinalis* and *S. elegans* in the Ria de Vigo (NW Spain) (Mollusca: Cephalopoda). *Journal of Zoology, London*, **207**, 511–9.

Guerra, A. (1992). *Fauna Iberica*, Vol. 1, *Mollusca Cephalopoda*. Museo Nacional de Ciencias Naturales Consejo Superior de Investigaciones Cientifica, Madrid.

Guerra, A. and Nixon, M. (1987). Crab and mollusc shell drilling by *Octopus vulgaris* (Mollusca: Cephalopoda) in the Ria de Vigo (north-west Spain). *Journal of Zoology, London*, **211**, 515–23.

Guerra, A., Nixon, M., and Castro, B. (1988a). Initial stages of food ingestion by *Sepia officinalis* (Mollusca: Cephalopoda). *Journal of Zoology, London*, **214**, 189–97.

Guerra, A., Villanueva, R., Nesis K.N., and Bedoya, J. (1998). Redescription of the deep-sea cirrate octopod *Cirroteuthis magna* Hoyle, 1885, and considerations on the genus *Cirroteuthis* (Mollusca: Cephalopoda). *Bulletin of Marine Science*, **63**, 51–81.

Guerra, A., González, A.F., and Cherel, Y. (2000). *Graneledone gonzalezi* sp. nov. (Mollusca: Cephalopoda): a new octopod from the Îles Kerguelen. *Antarctic Science*, **12**, 33–40.

Haimovici, M. (1988). *Eledone gaucha*, a new species of eledonid octopod (Cephalopoda: Octopodidae) from Southern Brazil. *The Nautilus*, **102**, 82–7.

Haimovici, M., Perez, J.A.A., and Costa, P.A.S. (1989). A review of cephalopods occurring in the waters of Rio de Janiero State, Brazil with first record of four species. *Revista Brasileira Biologia*, **49**, 503–10.

Hamabe, M. (1961). Experimental studies on breeding habits and development of the squid, *Ommastrephes sloani pacificus* Steenstrup 1. Copulation. *Zoological Magazine, Tokyo*, **70**, 378–84.

Hamabe, M. (1973). Egg mass and newborns of *Tremoctopus violaceus* Delle Chiaje, caught in the harbour of Kasumi, Hyogo Prefecture. *Bulletin of Tokai Regional Fisheries Research Laboratory, Tokyo*, No. **75**, 1–5.

Hamada, T., Obata, I., and Okutani, T. (1980). Nautilus macromphalus *in captivity*. Tokai University Press, Tokyo.

Hamada, T., Mikami, S., and Okutani, T. (1987). Experience with aquarium rearing of *Nautilus* in Japan. In Nautilus: *The biology and paleobiology of a living fossil* (ed. W.B. Saunders and N.H. Landman), pp. 579–83. Plenum, New York.

Haneda, Y. (1955). Luminous organisms of Japan and the Far East. In *The luminescence of biological systems* (ed. F.H. Johnson), pp. 335–85. American Association for the Advancement of Science, Washington, D.C.

Haneda, Y. (1963). Observations on the luminescence of the shallow-water squid *Uroteuthis bartschi*. *Science Report Yokosuka City Museum*, **8**, 10–16.

Hanlon, R.T. (1982). The functional organization of chromatophores and iridescent cells in the body patterning of *Loligo plei* (Cephalopoda: Myopsida). *Malacologia*, **23**, 89–119.

Hanlon, R.T. (1988). Behavioral and body patterning characters useful in taxonomy and field identification of cephalopods. *Malacologia*, **29**, 247–64.

Hanlon, R.T. and Messenger, J.B. (1988). Adaptive coloration in young cuttlefish (*Sepia officinalis* L.): The morphology and development of body patterns and their relation to behaviour. *Philosophical Transactions of the Royal Society, London*, **B320**, 437–87.

Hanlon, R.T. and Messenger, J.B. (1996). *Cephalopod behaviour*. Cambridge University Press.

Hanlon, R.T., Hixon, R.F., and Hulet, W.H. (1983). Survival, growth, and behavior of the loliginid squids *Loligo plei, Loligo pealei*, and *Lolliguncula brevis* (Mollusca: Cephalopoda) in closed sea water systems. *Biological Bulletin, Woods Hole*, **165**, 637–85.

Hanlon, R.T., Cooper, K.M., Budelmann, B.U., and Pappas. T. (1990). Physiological color change in squid iridophores I. Behavior, morphology and pharmacology in *Lolliguncula brevis*. *Cell and Tissue Reseach*, **259**, 3–14.

Hanlon, R.T., Boletzky, S. v., Okutani, T., Perez-Gandras, G., Sanchez, P., Sousa-Reis, C. *et al.* (1992). Suborder Myopsida Orbigny, 1845. *Smithsonian Contributions to Zoology*, No. **513**, 37–53.

Hanlon, R.T., Smale, M., and Sauer, W.H.H. (1994). An ethogram of body patterning behavior in the squid *Loligo vulgaris reynaudii* on spawning grounds in South Africa. *Biological Bulletin, Woods Hole*, **187**, 363–72.

Hanlon, R.T., Claes, M.F., Ashcraft, S.E., and Dunlap, P.V. (1997). Laboratory culture of the sepiolid squid *Euprymna scolopes*: A model system for bacteria-animal symbiosis. *Biological Bulletin, Woods Hole*, **192**, 164–74.

Hanlon, R.T., Maxwell, M.R., Shashar, N., Loew, E.R., and Boyle, K.-L. (1999). An ethogram of body patterning behaviour in the biomedically and commercially valuable squid *Loligo pealei* off Cape Cod, Massachusetts. *Biological Bulletin, Woods Hole*, **197**, 49–62.

Hardwick, J.E. (1970). A note on the behavior of the octopod *Ocythöe tuberculata*. *California Fish and Game*, **56**, 68–70.

Harman, R.F. and Seki, M.P. (1990). *Iridoteuthis iris* (Cephalopoda: Sepiolidae): New records from the central north Pacific and first description of the adults. *Pacific Science*, **44**, 171–9.

Harman, R.F. and Young, R.E. (1985). The larvae of ommastrephid squids (Cephalopoda, Teuthoidea) from Hawaiian waters. *Vie et Milieu*, **35**, 211–22.

Harman, R.F., Young, R.E., Reid, S.B., Mangold, K., Suzuki, T., and Hixon, R.F. (1989). Evidence for multiple spawning in the tropical oceanic squid *Sthenoteuthis oualaniensis* (Teuthoidea: Ommastrephidae). *Marine Biology*, **101**, 513–9.

Hartline, P.H. and Lange, G.D. (1974). Optic nerve responses to visual stimuli in squid. *Journal of Comparative Physiology*, **93**, 37–54.

Hartline, P.H. and Lange, G.D. (1984). Visual systems of cephalopods. In *Comparative physiology of sensory systems* (ed. L. Bolis, R.D. Keynes and S.H.P. Maddrell), pp. 335–55. Cambridge University Press.

Hartline, P.H., Hurley, A.C. and Lange, G.D. (1979). Eye stabilization by statocyst mediated oculomotor reflex in *Nautilus*. *Journal of Comparative Physiology*, **A 132**, 117–26.

Hartwick, B. (1983). *Octopus dofleini*. In *Cephalopod life cycles* Vol. I. *Species accounts* (ed. P.R. Boyle), pp. 277–91. Academic Press, London.

Hastie, L.C., Joy, J.B., Pierce, G.J., and Yau, C. (1994). Reproductive biology of *Todaropsis eblanae* (Cephalopoda: Ommastrephidae) in Scottish waters. *Journal of the Marine Biological Association, U.K.*, **74**, 367–82.

Hastings, J.W. and Nealson, K.H. (1981). The symbiotic luminous bacteria. In *The prokaryotes* (ed. M.P. Starr, H. Stolop, H.G. Truper, A. Balows, and H.G. Schlegel), pp. 1332–45. Springer-Verlag, New York.

Hatanaka, H. (1979). Studies on the fisheries biology of common octopus off the northwest coast of Africa. *Bulletin far Seas fisheries Research Laboratory, Shimizu*, **17**, 13–124.

Haven, N. (1972). The ecology and behavior of *Nautilus pompilius* in the Philippines. *The Veliger*, **15**, 75–81.

Haven, N. (1977a). The reproductive biology of *Nautilus pompilius* in the Philippines. *Marine Biology*, **42**, 177–84.

Haven, N. (1977b). Cephalopoda: Nautiloidea. In: *Reproduction of marine invertebrates* Vol. 4 (ed. A.C. Giese and J.S. Pearse), pp. 227–41. Academic Press, New York.

Hayashi, S. (1993). Growth and maturation of the firefly squid, *Watasenia scintillans* (Berry), in Toyama Bay and adjacent waters of the Sea of Japan. In *Recent advances in fisheries biology* (ed. T. Okutani, R.K. O'Dor and T. Kubodera), pp. 173–8. Tokai University Press, Tokyo.

Hayashi, K., Kishimura, H., and Sakurai Y. (1990). Level and composition of diacyl glycerol ethers in the different tissues and stomach contents of giant squid *Moroteuthis robusta*. *Nippon Suisan Gakkaishi*, **56**, 1635–39.

Healy, J.M. (1989). Spermatozoa of the deep-sea cephalopod *Vampyroteuthis infernalis* Chun: ultrastructure and possible phylogenetic significance. *Philosophical Transactions of the Royal Society, London*, **B323**, 589–608.

Healy, J.M. (1990a). Ultrastructure of spermatozoa and spermiogenesis in *Spirula spirula* (L.): systematic importance and comparison with other cephalopods. *Helgoländer Meeresunters*, **44**, 109–23.

Healy, J.M. (1990b). Ultrastructure of spermiogenesis in *Vampyroteuthis infernalis* Chun – a relict cephalopod mollusc. *Helgoländer Meeresunters*, **44**, 95–107.

Healy, J.M. (1995). Molluscan sperm ultrastructure: correlation with taxonomic units within the Gastropoda, Cephalopoda and Bivalvia. In *Origin and evolutionary radiation of the Mollusca* (ed. J. Taylor), pp. 99–113. The Malacological Society of London, Oxford University Press, Oxford.

Heeger, T., Piatkowski, U., and Möller, H. (1992). Predation on jellyfish by the cephalopod *Argonauta argo*. *Marine Ecology Progress Series*, **88**, 293–6.

Hendrix, J.P., Hulet, W.H., and Greenberg, M.J. (1981). Salinity tolerance and the responses to hypo-osmotic stress of the bay squid *Lolliguncula brevis*, a euryhaline cephalopod mollusc. *Comparative Biochemistry and Physiology*, **69A**, 641–8.

Henry, J. and Boucaud-Camou, E. (1991). Diet of very young *Sepia officinalis* L. in Bay of Seine. In *La seiche – The cuttlefish, 1st International Symposium on the cuttlefish*, Sepia, *Caen June 1–3, 1989* (ed. E. Boucaud-Camou), p. 195. University of Caen.

Herring, P.J. (1977). Luminescence in cephalopods and fish. *Symposia of the Zoological Society, London*, No. **38**, 127–59.

Herring, P.J. (1983). The spectral characteristics of luminous marine organisms. *Proceedings of the Royal Society, London*, **B220**, 183–217.

Herring, P.J. (1988). Luminescent organs. In: *The Mollusca* Vol. 11. *Form and function* (ed. E.R. Trueman and M.R. Clarke), pp.449–89. Academic Press, San Diego.

Herring, P.J., Clarke, M.R., Boletzky, S. v., and Ryan, K.P. (1981). The light organs of *Sepiola affinis* and *Spirula spirula* (Mollusca: Cephalopoda): bacterial and intrinsic systems in the order Sepioidea. *Journal of the Marine Biological Association, U.K.*, **61**, 901–16.

Herring, P.J., Dilly, P.N., and Cope, C. (1985). The photophore morphology of *Selenoteuthis scintillans* Voss and other lycoteuthids (Cephalopoda: Lycoteuthidae). *Journal of Zoology, London*, **A206**, 567–89.

Herring, P.J., Dilly, P.N., and Cope, C. (1987). The morphology of the bioluminescent tissue of the cephalopod *Japetella diaphana* (Octopoda: Bolitaenidae). *Journal of Zoology, London*, **12**, 245–54.

Herring, P.J., Dilly, P.N., and Cope, C. (1992a). Different types of photophore in the oceanic squids *Octopoteuthis* and *Taningia* (Cephalopoda: Octopoteuthidae). *Journal of Zoology, London*, **227**, 479–91.

Herring, P.J., Widder, E.A., and Haddock, S.H.D. (1992b). Correlation of bioluminescence emissions with ventral photophores in the mesopelagic squid *Abralia veranyi* (Cephalopoda: Enoploteuthidae). *Marine Biology*, **112**, 293–8.

Herring, P.J., Dilly, P.N., and Cope, C. (1994). The bioluminescent organs of the deep-sea cephalopod *Vampyroteuthis infernalis* (Cephalopoda: Vampyromorpha). *Journal of Zoology, London*, **233**, 45–55.

Herring, P.J., Dilly, P.N., and Cope, C. (2002). The photophores of the squid family Cranchiidae (Cephalopoda: Oegopsida). *Journal of Zoology, London*, **258**, 73–90.

Hewitt, R.A. (1975). Analysis of aragonite from the cuttlebone of *Sepia officinalis* L. *Marine Geology*, **18**, M1–5.

Hillig, R. (1912). Das Nervensystem von *Sepia officinalis* L. *Zeitschrift für. Wissenschaftliche Zoologie*, **101**, 736–800.

Hixon, R.F. (1983). *Loligo opalescens*. In *Cephalopod life cycles* Vol. I. *Species accounts* (ed. P.R. Boyle), pp. 95–114. Academic Press, London.

Hixon, R.F., Hanlon, R.T., Gillespie, S.M., and Griffin, W. l. (1980). Squid fishery in Texas: Biological, economic, and market considerations. *Marine Fisheries Review*, **42**, 44–50.

Hobbs, M.J. and Young, J.Z. (1973). A cephalopod cerebellum. *Brain Research*, **55**, 424–30.

Hochachka, P., Moon, T.W., Mustafa, T., and Storey, K.B. (1975). Metabolic sources of power for mantle muscle of fast swimming squid. *Comparative Biochemistry and Physiology*, **52B**, 151–8.

Hochachka, P.W., French, C.J., and Meredith, J. (1978). Metabolic and ultrastructural organization in *Nautilus* muscles. *Journal of Experimental Zoology*, **205**, 51–62.

Hochberg, F.G. (1998). Class Cephalopoda. In *Taxonomic atlas of the benthic fauna of the Santa Maria Basin and western Santa Barbara Channel. Vol. 8—The Mollusca Part 1, The Aplacophora, Polyplacophora, Scaphopoda, Bivalvia and Cephalopoda* (ed. P.V. Scott and J. Blake), pp. 175–236. Santa Barbara Museum of Natural History, California.

Hochberg, F.G. and Couch, J.A. (1971). Biology of cephalopods. In *Tektite II, Scientists-in-the-sea, Mission 8–50* (ed. J.W. Miller, J.G. Vanderwalker, and R. Waller), pp. V1 221–8. U.S. Department of the Interior, Washington.

Hochberg, F.G. and Fields, W.G. (1980). Cephalopoda: the squids and octopuses. In: *Intertidal invertebrates of California* (ed. R.H. Morris, D.P. Abbott and E.C. Haderlie), pp. 428–45. Stanford University Press, California.

Hochberg, F.G., Nixon, M., and Toll, R.B. (1992). Order Octopoda Leach, 1818. *Smithsonian Contributions to Zoology*, No. **513**, 213–79.

Holland, C.H. (1987). The nautiloid cephalopods: A strange success. *Journal of the Geological Society, London*, **144**, 1–15.

Holmes, W. (1940). The colour changes and colour patterns of *Sepia officinalis. Proceedings of the Zoological Society, London*, **A110**, 17–35.

Holmes, W. (1955). The colour changes of cephalopods. *Endeavour*, **14**, 78–82.

House, M.R. (1987). Geographic distribution of *Nautilus* shells. In Nautilus: *The biology and paleobiology of a living fossil* (ed. W.B. Saunders and N.H. Landman), pp. 53–64. Plenum, New York.

House, M.R. (1988). Major features of cephalopod evolution. In *Cephalopods – present and past* (ed. J. Wiedmann and J. Kullmann), pp. 1–16. Schweizerbart'sche Verlagsbuchhandlung, Stuttgart.

Howden, M.E.H. and Williams, P.A. (1974). Occurrence of amines in the posterior salivary glands of the octopus *Hapalochlaena maculosa* (Cephalopoda). *Toxicon*, **12**, 317–20.

Hoyle, W.E. (1885a). Diagnoses of new species of Cephalopoda, collected during the cruise of H.M.S. *Challenger* – Part I. The Octopoda. *Annals and Magazine of Natural History*, (5), **15**, 222–36.

Hoyle, W.E. (1885b). Narrative of the Challenger Expedition, Report on the Scientific Results of the voyage of the H.M.S. *Challenger* during the years 1873–1876, Narrative, **1** (1), 269–74.

Hoyle, W.E. (1886). Report on the Cephalopoda collected by H.M.S. *Challenger* during the years 1873–1876. *Report on the Scientific Results of the Voyage of H.M.S.* Challenger *1873–1876, Zoology*, **16**, 1–245.

Hoyle, W.E. (1902). The luminous organs of *Pterygioteuthis margaritifera*, a Mediterranean cephalopod. *Manchester Memoirs*, **46**, 1–14.

Hoyle, W.E. (1904). Reports on the dredging operations off the west coast of Central America to the Galapagos, to the west coast of Mexico, and in the Gulf of California, in charge of Alexander Agassiz, carried on by the U.S. Fish Commission Steamer 'Albatross' during 1891, Lieut. Commander Z.L. Tanner, U.S.N., Commanding. XXIX. Reports on the Cephalopoda. *Bulletin of the Museum of Comparative Zoology, Harvard*, **43**, 1–72.

Hulet, W.H. (1982). Commentary on the international symposium on functional morphology of cephalopods. *Malacologia*, **23**, 203–8.

Hunt, S. and Nixon, M. (1981). A comparative study of protein composition in the chtin-protein complexes of the beak, pen, sucker disc, radula and oesophageal cuticle in cephalopods. *Comparative Biochemical Physiology*, **B68**, 535–46.

Hunt, S. and Sherief, A. El (1990). A periodic structure in the 'pen' chitin of the squid *Loligo vulgaris. Tissue and Cell*, **22**, 191–7.

Hurley, A.C. (1978). School structure of the squid *Loligo opalescens. Fishery Bulletin, U.S.A.*, **76**, 433–42.

Hurley, A.C., Lange, G.D., and Hartline, P.H. (1978). The adjustable 'pinhole camera' eye of *Nautilus. Journal of Experimental Zoology*, **205**, 37–44.

Hwang, D.F., Arakawa, O., Saito, T., Noguchi, T., Simidu, U., Tsukamoto, K., Shida, Y., and Hashimoto, K. (1989). Tetrodotoxin-producing bacteria from the blue ringed-octopus. *Marine Biology*, **100**, 327–32.

Ijema, I. and Ikeda, S. (1895). Description of *Opisthoteuthis depressa* n. sp. *Journal of the College of Science, Imperial University, Japan*, **8**, 323–35.

Ijema, I. and Ikeda, S. (1902). Notes on a specimen of *Amphitretus* obtained in the Sagami Sea. *Annotationes Zoologicæ Japonenses*, 4, 85–101.

Iribarne, O.O. (1991). Life history and distribution of the small south-western Atlantic octopus, *Octopus tehuelchus. Journal of Zoology, London*, **223**, 549–65.

Ishikawa, C. (1913). Einige Bemerkungen über den leuchtenden Tintenfisch, *Watasea* nov. gen. (*Abraliopsis* der Autoren) *scintillans*, Berry, aus Japan. *Zoologische Anzeiger*, **43**, 162–72.

Jackson, G.D. (1989). The use of statolith microstructure to analyze life-history events in the small tropical cephalopod *Idiosepius pygmaeus. Fishery Bulletin, U.S.A.*, **87**, 265–72.

Jackson, G.D. (1993). Seasonal variation in reproductive investment in the triopical loliginid squid *Loligo chinensis* and the small tropical sepioid *Idiosepius pygmaeus. Fishery Bulletin, U.S.A.*, **91**, 260–70.

Jackson, G.D. and Mladenov, P.V. (1994). Terminal spawning in the deep-water squid *Moroteuthis ingens* (Cephalopoda: Onychoteuthidae). *Journal of Zoology, London*, **234**, 189–201.

Jackson, G.D. and O'Dor, R.K. (2001). Time, space and the ecophysiology of squid growth, life in the fast lane. *Vie et Milieu*, **51**, 205–15.

Jackson, G.D., Lu, C.C., and Dunning, M. (1991). Growth rings within the statolith microstructure of the giant squid *Architeuthis. The Veliger*, **34**, 331–4.

Jacob, T.J.C. and Duncan, G. (1981). Electrical coupling between fibre cells in amphibian and cephalopod lenses. *Nature, London*, **290**, 704–6.

Jander, R., Daumer, K., and Waterman, T.H. (1963). Polarized light orientation by two Hawaiian decapod cephalopods. *Zeitschrift für vergleichende Physiologie*, **46**, 383–94.

Jarre, A., Clarke, M.R., and Pauly, D. (1991). Re-examination of growth estimates in oceanic squids: The case of *Kondakovia longimana* (Onychoteuthidae). *International Council for the Exploration of the Sea, Journal of Marine Science*, **48**, 195–200.

Jatta, G. (1896). I Cefalopodi viventi nel Golfo di Napoli (Sistemica). *Fauna und Flora des Golfes von Neapel Monographie*, **23**, 1–268.

Jefferts, K. and Roper, C.F.E. (1992). Family Cycloteuthidae Naef, 1923. *Smithsonian Contributions to Zoology*, No. **513**, 161–3.

Jereb, P. and Ragonese, S. (1993). On the photographic documentation on the unihorn octopus (*Scaeurgus unicirrhus* Delle Chiaje in Ferrusac & Orbigny, 1840) of the Sicilian Channel. *N.T.R. – I.T.P.P.* No. **42**, 2–7. Pubblicazione dell'Istituto di Tecnologia della pesca e del pescato, Mazara del Vallo.

Jereb, P., Ragonese, S. and Di Stefano, L. (1992). Occurrences of unihorn octopus *Scaeurgus unicirrhus* in the Sicilian Channel III: Description of features. *Rapport Commission Internationale pour la Mer Méditerranée*, **33**, 296.

Johnsen, S. and Kier, W.M. (1993). Intramuscular crossed connective tissue fibres: skeletal support in the lateral fins of squid and cuttlefish (Mollusca: Cephalopoda). *Journal of Zoology, London*, **231**, 311–38.

Jones, E.C. (1963). *Tremoctopus violaceus* uses *Physalia* tentacles as weapons. *Science, N.Y.*, **139**, 764–6.

Joubin, L. (1919). Études préliminaires sur les Céphalopodes recueillis au cours des croisières de S.A.S. le Prince de Monaco, 7e note: *Cycloteuthis sirventi* nov. gen. et sp. *Bulletin de l'Institut Océanographique, Monaco*, No. **351**, 1–7.

Joubin, L. (1920). Céphalopodes provenant de Campagnes de 1a 'Princesse Alice' (1898–1910). *Résultats des Campagnens Scientifiques accomplies par le Prince Albert I, Monaco*, **54**, 1–95.

Joubin, L. (1937). Les Octopodes de la croisière du Dana (1921–1922). *Dana-Report*, no. **11**, 1–49.

Kälin, I. (1967). Ein wunderwurk der Statik: der Schlup des Tintenfisches. *Mikrokosmos* (Stuttgart), **8**, 230–8.

Kawaguti, S. and Ohgishi, S. (1962). Electron microscope study on iridophores of a cuttlefish, *Sepia esculenta. Biological Journal of Okayama University*, **8**, 115–29.

Kawahara, M., Gleadall, I.G., and Tsukahara, Y. (1998). A note on the fibre-optic light-guides in the eye photophores of *Watasenia scintillans. South African Journal of Marine Science*, **20**, 123–7.

Kayes, R.J. (1974). The daily activity pattern of *Octopus vulgaris* in a natural habitat. *Marine Behaviour and Physiology*, **2**, 337–43.

Kerr, J.G. (1895). On some points in the anatomy of *Nautilus pompilius*. *Proceedings of the Zoological Society, London* (1895), 664–86.

Kerr, J.G. (1931). DANA specimens of *Spirula* and upon certain problems of cephalpod morphology. *DANA-Report*, No. **8**, 1–34.

Khromov, D.N. (1998). Distribution patterns of Sepiidae. *Smithsonian Contributions to Zoology*, No. **586**, 191–206.

Khromov, D.N., Lu, C.C., Guerra, A., Dong, Zh., and Boletzky, S. v. (1998). A synopsis of Sepiidae outside Australian waters (Cephalopoda: Sepioidea). *Smithsonian Contributions to Zoology*, No. **586**, 77–157.

Kier, W.M. (1982). The functional morphology of the musculature of squid (Loliginidae) arms and tentacles. *Journal of Morphology*, **172**, 179–82.

Kier, W.M. (1985). The musculature of squid arms and tentacles: ultrastructural evidence for functional differences. *Journal of Morphology*, **185**, 223–39.

Kier, W.M. (1987). The functional morphology of the tentacle musculature of *Nautilus pompilius*. In Nautilus: *The biology and paleobiology of a living fossil* (ed. W.B. Saunders and N.H. Landman), pp. 257–69. Plenum, New York.

Kier, W.M. (1988). The arrangement and function of molluscan muscle. In *The Mollusca* (ed. K.M. Wilbur), Vol. 11. *Form and function* (ed. M.R. Clarke and E.R. Trueman), pp. 211–52. Academic Press, San Diego.

Kier, W.M. (1989). The fin musculature of cuttlefish and squid (Mollusca, Cephalopoda): morphology and mechanics. *Journal of Zoology, London*, **217**, 23–38.

Kier, W.M. (1991). Squid cross-striated muscle: the evolution of a specialized muscle fiber type. *Bulletin of Marine Science*, **49**, 389–403.

Kier, W.M. (1996). Muscle development in squid: ultrastructural differentiation of a specialized muscle fiber type. *Journal of Morphology*, **229**, 271–88.

Kier, W.M. and Schachat, F.H. (1992). Biochemical comparison of fast and slow contracting squid muscle. *Journal of Experimental Biology*, **168**, 41–53.

Kier, W.M. and Smith, K.K. (1985). Tongues, tentacles and trunks: the biomechanics of movement in muscular-hydrostats. *Zoological Journal Linnean Society, London*, **83**, 307–24.

Kier, W.M. and Smith, A.M. (1990). The morphology and mechanics of octopus suckers. *Biological Bulletin, Woods Hole*, **178**, 126–36.

Kier, W.M. and van Leeuwen, J.L. (1997). A kinematic analysis of tentacle extension in the squid *Loligo pealei*. *Journal of Experimental Biology*, **200**, 41–53.

Kier, W.M., Messenger, J.B., and Miyan, J.A. (1985). Mechanoreceptors in the fins of the cuttlefish, *Sepia officinalis*. *Journal of Experimental Biology*, **119**, 369–73.

Kier, W.M., Smith, K.K., and Miyan, J.A. (1989). Electromyography of the fin musculature of the cuttlefish *Sepia officinalis*. *Journal of Experimental Biology*, **143**, 17–31.

Kishimoto, H. and Kohno H. (1992). Development of the luminous organ in the purpleback flying squid, *Sthenoteuthis oualaniensis*, as shown by Alcian Blue stain technique. *Bulletin Institute Oceanic Research and Development, Tokai University*, **13**, 71–83.

Knudsen, J. (1957). Some observations on a mature male specimen of *Architeuthis* from Danish waters. *Proceedings of the Malacological Society, London*, **32**, 189–98.

Knudsen, J. (1981). Three Sepioidae new to the eastern Mediterranean (Mollusca: Cephalopoda). *Argamon*, **7**, 45–50.

Koueta, N. and Boucaud-Camou, E. (1986). Comparative study of secretion of the posterior sallivary glands of decapod cephalopods – I. *Sepia officinalis* L. *Cellular and Molecular Biology*, **32**, 333–41.

Kramp, P.L. (1956). Pelagic fauna. In *The Galathea deep sea expedition* (ed. A.F. Bruun, Sv. Greve, H. Mielche and R. Spärek), pp. 65–86. Allen and Unwin, London.

Kristensen, T.K. (1977a). Scanning electron microscopy of hook development in *Gonatus fabricii* (Lichtenstein, 1818) (Mollusca: Cephalopoda). *Videnskabelige Meddelelser fra Dansk naturhistorik Forening i Kjøbenhavn*, **140**, 111–16.

Kristensen, T.K. (1977b). Hatching, growth, and distribution of juvenile *Gonatus fabricii* (Mollusca: Cephalopoda) in Greenland waters. *Astarte*, **10**, 21–8.

Kristensen, T.K. (1980). Large mature female of *Histioteuthis bonnellii* (Férrusac, 1835) (Mollusca: Cephalopoda) recorded from the Davis Strait, West Greenland. *Steenstrupia*, **6**, 73–9.

Kristensen, T.K. (1981a). The genus *Gonatus* Gray, 1849 (Mollusca: Cephalopoda) in the North Atlantic. A revision of the North Atlantic species and description of *Gonatus steenstrupi* n. sp. *Steenstrupia*, **7**, 61–99.

Kristensen, T.K. (1981b). First record of a mature female of the squid *Gonatus fabrici* (Lichtestein, 1818) (Cephalopoda: Teuthoidea). *Steenstrupia*, **7**, 101–8.

Kristensen, T.K. (1982). Multivariate analysis of geographic variation in the squid *Gonatus fabricii* (Lichtenstein, 1818) (Mollusca: Cephalopoda). *Malacologia*, **22**, 581–6.

Kristensen, T.K. (1983). *Gonatus fabricii*. In: *Cephalopod life cycles* Vol. I. *Systematic accounts* (ed. P.R. Boyle), pp. 159–73. Academic Press, New York.

Kristensen, T.K. (1984). Biology of the squid *Gonatus fabricii* (Lichtenstein, 1818) from West Greenland waters. *Meddelelser om Grønland, Bioscience*, **13**, 1–17.

Kubodera, T. (1989). Young squids collected with 10-foot IKPT net during the JARE-28 Cruise, 1987. *Proceedings National Institute of Polar Research, Symposium on Polar Biology*, No. **2**, 71–7.

Kubodera, T. and Jefferts, K. (1984). Distribution and abundance of the early life stages of squid, primarily Gonatida (Cephalopoda, Oegopsida), in the northern North Pacific (part 2). *Bulletin of the National Science Museum, Series A (Zoology)*, **10**, 165–93.

Kubodera, T. and Okutani, T. (1981). The systematics and identification of larval cephalopods from the northern North Pacific. *Research Institute of North Pacific Fisheries, Hokkaido University, Special Volume*, pp. 131–59.

Kubodera, T. and Okutani, T. (1986). New and rare cephalopods from the Antarctic waters. *Memoirs of National Institute of Polar Research, Special Issue*, No. **44**, 129–43.

Kubodera, T. and Okutani, T. (1994). Eledonine octopods from the Southern Ocean: systematics and distribution. *Antarctic Science*, **6**, 205–14.

Kubodera, T., Pearcy, W.G., Murakami K., Kobayashi, T., Nakata, J., and Mishima, S. (1983). Distribution and abundance of squids caught in surface gillnets in the subarctic Pacific, 1977–1981. *Memoirs of the Faculty of Fisheries, Hokkaido University*, **30**, 1–49.

Kubodera, T., Piatkowski, U., Okutani, T., and Clarke, M.R. (1998). Taxonomy and zoogeography of the Family Onychoteuthidae (Cephalopoda: Oegopsida). *Smithsonian Contributions to Zoology*, No. **586**, 277–91.

Kubota, T. and Miyashita, A. (1973). Two species of *Argonauta* collected from Suruga Bay. *Collecting and Breeding*, **35**, 128–30.

Kubota, T. and Miyashita, A. (1975). Abundant appearance of paper-nautilus, *Argonauta hians* in Suruga Bay, Japan. *Collecting and Breeding*, **37**, 72–4.

Kubota, T., Koshiga, M., and Okutani, T. (1981). Rare and interesting squid from Japan-VII. Some biological data on *Chiroteuthis imperator* from Suruga Bay, Japan. *Venus, Japanese Journal of Malacology*, **40**, 150–9.

Kubota, T., Iizuka, K., and Okutani, T. (1982). Some biological aspects of *Abralia andamanica* from Suruga Bay, Japan (Cephalopoda: Enoploteuthidae). *Journal Faculty Marine Science and Technology, Tokai University*, **15**, 333–43.

Kuehl, S. (1988). A contribution to the reproductive biology and geographical distribution of Antarctic Octopodidae (Cephalopoda). *Malacologia*, **29**, 89–100.

Lam, D.M.K., Wiesel, T.N., and Kaneko, A. (1974). Neurotransmitter synthesis in cephalopod retina. *Brain Research*, **82**, 365–8.

Land, M.F. (1992). A note on the elongated eye of the octopus *Vitreledonella richardi*. *Journal of the Marine Biological Association, U.K.*, **72**, 89–92.

Landman, N.H. and Cochran, J.K. (1987). Growth and longevity of *Nautilus*. In Nautilus: *The biology and paleobiology of a living fossil* (ed. W.B. Saunders and N. Landman), pp. 401–20. Plenum, New York.

Laptikhovsky, V.V. and Murzov, S.A. (1990). Epipelagic egg mass of the squid *Sthenoteuthis pteropus* collected in the tropical eastern Atlantic. *Biologiya Morya*, No. **3**, 62–3.

Laptikhovsky, V.V. and Nigmatullin, C.M. (1993). Egg size, fecundity, and spawning in females of the genus *Illex* (Cephalopoda: Ommastrephidae). *International Council for the Exploration of the Sea, Journal of Marine Science*, **50**, 393–403.

Laughlin, R.A. and Livingston, R.J. (1982). Environmental and trophic determinants of the spatial/temporal distribution of the brief squid (*Lolliguncula brevis*) in the Apalachicola estuary (North Florida, U.S.A.). *Bulletin of Marine Science Gulf and Caribbean*, **32**, 489–97.

Leach, W.E. (1817). Synopsis of the orders, families and genera of the class Cephalopoda. In *The zoological miscellany; being descriptions of new, or interesting animals* 3 (30), 137–41 (W.E. Leach and R.P. Nodder).

Lee, H. (1875). *The Octopus; or the 'devil-fish' of fiction and of fact*. Chapman and Hall, London.

Lee, P.G. (1992). Chemotaxis by *Octopus maya* Voss et Solis in a Y-maze. *Journal of Experimental Marine Biology and Ecology*, **153**, 53–67.

Le Goff, R., Gauvrit, E., Pinczon du Sel, G., and Daguzan, J. (1998). Age group determination by analysis of the cuttlebone of the cuttlefish *Sepia officinalis* L. in reproduction in the Bay of Biscay. *Journal of Molluscan Studies*, **64**, 183–93.

Lehmann, U. (1981). *The ammonites: their life and their world* (trans. J. Lettau). Cambridge University Press.

Leisman, G., Cohn, D.H., and Nealson, K.H. (1980). Bacterial origin of luminescence in marine animals. *Science*, **208**, 1271–3.

Lenz, S. (1997). Cilia in the epidermis of late embryonic stages and paralarvae of *Octopus vulgaris* (Mollusca: Cephalopoda). *Vie et Milieu*, **47**, 143–7.

Lewis, A.R. and Choat, J.H. (1993). Spawning mode and reproductive output of the tropical cephalopod *Idiosepius pygmaeus*. *Canadian Journal of Fisheries and Aquatic Sciences*, **50**, 20–8.

Linnaeus, C. (1758). *Systema naturae per regna tria naturae, secundum Classes, Ordines, Genera, Species, cum characteribus, differentiis, synonymis, locis. Tomus I. Regnum animale. Editio decima, reformata*. Holmiae Laurentius Salvius, Stockholm.

Lipinski, M.R. (1993). Description of mature males of the histioteuthid cephalopod *Histioteuthis atlantica* (Hoyle, 1885) (Cephalopoda: Oegopsida) from the South Atlantic Ocean. *South African Journal of Marine Science*, **18**, 299–303.

Lipinski, M.R. (1997). Morphology of giant squid *Architeuthis* statoliths. *South African Journal of Marine Science*, **18**, 299–303.

Lipinski, M. and Jackson, S. (1989). Surface-feeding on cephalopods by procellariform seabirds in the southern Benguela region, South Africa. *Journal of Zoology, London*, **218**, 549–63.

Lipinski, M.R. and Linkowski, T.B. (1988). Food of the squid *Ommastrephes bartrami* (Lesueur, 1821) from the south-west Atlantic Ocean. *South African Journal of Marine Science*, **6**, 43–6.

Lipinski, M. and Turoboyski, K. (1983). The ammonium content in the tissues of selected species of squid (Cephalopoda: Teuthoidea). *Journal of Experimental Marine Biology and Ecology*, **69**, 145–50.

Llewellyn, J. (1984). The biology of *Isancistrum subulatae* n. sp., a monogenean parasitic on the squid, *Alloteuthis subulata*, at Plymouth. *Journal of the Marine Biological Association, U.K.*, **64**, 285–302.

Lo Bianco, S. (1908). Notizie biologiche riguardanti specialmennte il periodi di maturità sessulae degli animali del golfo di Napoli. *Mitteilungen aus der Zoologischen Station zu Neapel*, **19**, 512–761.

Long, T.M., Hanlon, R.T., Maat, A.T., and Pinsker, H.M. (1989). Non-associative learning in the squid *Lolliguncula brevis* (Mollusca, Cephalopoda). *Marine Behaviour and Physiology*, **16**, 1–9.

Lordan, C., Collins, M.A., and Perales-Raya, C. (1998). Observations on morphology, age and diet of three *Architeuthis* caught off the west coast of Ireland. *Journal of the Marine Biological Association, U.K.*, **78**, 903–17.

Lowenstam, H.A., Traub, W., and Weiner, S. (1984). *Nautilus* hard partrs: a study of the mineral and organic constituents. *Paleobiology*, **10**, 268–79.

Lu, C.C. (1977). A new species of squid, *Chiroteuthis acanthoderma*, from the southwest Pacific (Cephalopoda, Chiroteuthidae). *Steenstrupia*, **4**, 179–88.

Lu, C.C. (1982). First record of *Todaropsis eblanae* (Ball, 1841) (Cephalopoda: Oegopsida) in the Pacific Ocean. *Venus, Japanese Journal of Malacology*, **41**, 67–70.

Lu, C.C. (1998). A synopsis of Sepiidae in Australian waters (Cephalopoda: Sepioidea). *Smithsonian Contributions to Zoology*, No. **586**, 159–90.

Lu, C.C. and Clarke, M.R. (1975a). Vertical distribution of cephalopods at 40°N, 53°N and 60°N at 20°W in the North Atlantic. *Journal of the Marine Biological Association, U.K.*, **55**, 143–63.

Lu. C.C. and Clarke, M.R. (1975b). Vertical distribution of cephalopods at 11°N, 20°W in the North Atlantic. *Journal of the Marine Biological Association, U.K.*, **55**, 369–89.

Lu, C.C. and Phillips, J.U. (1985). An annotated checklist of the Cephalopoda from Australian waters. *Occasional Papers from the Museum of Victoria*, **2**, 21–36.

Lu, C.C. and Roper, C.F.E. (1979). Cephalopods from deepwater Dumpsite 106 (Western Atlantic): vertical distribution and seasonal abundance. *Smithsonian Contributions to Zoology*, No. **288**, 1–36.

Lu, C.C. and Stranks, T.N. (1991). *Eledone palari*, a new species of octopus (Cephalopoda: Octopodidae) from Australia. *Bulletin of Marine Science*, **49**, 73–87.

Lu, C.C. and Stranks, T.N. (1994). Synopsis of *Pareledone* and *Megaleldone* species, with descriptions of two new species from East Antarctica (Cephalopoda: Octopodidae). *Memoirs of the Museum of Victoria*, **54**, 221–42.

Lu, C.C. and Tait, R.W. (1983). Taxonomic studies on *Sepioteuthis* Blainville (Cephalopoda: Loliginidae) from the Australian region. *Proceedings Royal Society Victoria*, **95**, 181–204.

Lu, C.C. and Williams, R. (1994a). *Kondakovia longimana* Filippova, 1972 (Cephalopoda: Onychoteuthidae) from the Indian Ocean sector of the Southern Ocean. *Antarctic Science*, **6**, 231–4.

Lu, C.C. and Williams, R. (1994b). Contribution to the biology of squid in the Prydz Bay Region, Antarctica. *Antarctic Science*, **6**, 223–9.

Lu, C.C., Roper, C.F.E., and Tait, R.W. (1985). A revision of *Loliolus* (Cephalopoda; Loliginidae), including *L. noctiluca*, a new species of squid from Australian waters. *Proceedings of the Royal Society, Victoria*, **97**, 59–85.

Lucero, M.T. and Gilly, W.T. (1995). Physiology of squid olfaction. In *Cephalopod neurobiology: Neuroscience studies in squid, octopus and cuttlefish* (ed. N.J. Abbott, R. Williamson and L. Maddock), pp. 521–34. Oxford University Press, Oxford.

Lucero, M.T., Horrigan, F.T., and Gilly, W.F. (1992). Electrical responses to chemical stimulation of squid olfactory receptor cells. *Journal of Experimental Biology*, **162**, 231–49.

Lucero, M.T., Farrington, H., and Gilly, W.F. (1994). Quantification of L-dopa and dopamine in squid ink: implications for chemoreception. *Biological Bulletin, Woods Hole*, **187**, 55–63.

Lund, R.D. (1971). Stellate ganglion. In *The anatomy of the nervous system of* Octopus vulgaris (by J.Z. Young), pp. 621–40. Clarendon Press, Oxford.

Lutz, R.A. and Voight, J.R. (1997). *Graneledone* spp. *Benthoctopus* spp. In *Handbook of deep-sea hydrothermal vent fauna* (ed. D. Desbruyères and M. Segonzac), pp. 158–9. IFREMER, Brest.

Macalaster, E.G. (1981). Dark-water octopus. *Sea Frontiers*, **27**, 78–81.

MacGinitie, G.E. and MacGinitie, N. (1968). *Natural history of marine animals*. McGraw-Hill, New York.

Maddock, L. and Young, J.Z. (1984). Some dimensions of the angular acceleration receptor systems of cephalopods. *Journal of the Marine Biological Association, U.K.*, **64**, 55–79.

Maddock, L. and Young, J.Z. (1987). Quantitative differences among the brains of cephalopods. *Journal of Zoology, London*, **212**, 739–67.

Magnier, Y. and Laboute, P. (1978). *Guides sous-marin de Nouvelle Caledonia.* Les Editions du Pacifique, Paris.

Maldonado, H. (1963). The positive learning process in *Octopus vulgaris. Zeitschrift für vergleichende Physiologie*, **47**, 191–214.

Mangold, K. (1966). *Sepia officinalis* de la Mer Catalane. *Vie et Milieu*, **A17**, 961–1012.

Mangold, K. (1983a). *Octopus vulgaris.* In *Cephalopod life cycles* Vol. I. *Species accounts* (ed. P.R. Boyle), pp. 335–64. Academic Press, London.

Mangold, K. (1983b). *Eledone moschata.* In *Cephalopod life cycles* Vol. I. *Systematic accounts* (ed. P.R. Boyle), pp. 387–400. Academic Press, New York.

Mangold, K. (1987). Reproduction. In *Cephalopod life cycles* Vol. II. *Species accounts* (ed. P.R. Boyle), pp. 157–200. Academic Press, London.

Mangold, K. (1989). Reproduction, croissance et durée de vie. In *Traité de zoologie, anatomie systématique, biologie* (ed. P.-P. Grassé) *Tome V, Céphalopodes, Fascicule 4* (ed. K. Mangold), pp. 493–552. Masson, Paris.

Mangold, K. and Boletzky, S. v. (1988). Mediterranean cephalopod fauna. In *The Mollusca* Vol.12 *Paleontology and neontology of cephalopods* (ed. M.R. Clarke and E.R. Trueman), pp. 315–30. Academic Press, San Diego.

Mangold, K. and Boletzky, M. (1989). Distribution géographique. In *Traité de zoologie: anatomie, systématique, biologie* (P.-P. Grassé), *Tome V, Céphalopodes, fascicule 4* (ed. K. Mangold), pp. 609–18. Masson, Paris.

Mangold, K., Young, R.E., and Nixon, M. (1993). Growth versus maturation in cephalopods. In *Advances in Cephalopod fisheries biology* (ed. T. Okutani, R.K. O'Dor, and T. Kubodera), pp. 697–704. Tokai University Press, Tokyo.

Mangold-Wirz, K. (1963). Biologie des Céphalopodes benthiques et nectoniques de la Mer Catalane. *Vie et Milieu*, Supplemént No. **13**, 1–285.

Mangum, C. (1991). Salt sensitivity of the haemocyanin of eury- and stenohaline squids. *Comparative Biochemistry and Physiology*, **99A**, 159–61.

Mann, T., Martin, A.W., and Thiersch, J.B. (1970). Male reproductive tract, spermatophores and spermatophoric reaction in the giant octopus of the North Pacific, *Octopus dofleini martini. Proceedings of the Royal Society of London*, **B175**, 31–61.

Marliave, J.B. (1981). Neustonic feeding in early larvae of *Octopus dofleini* (Wulker). *Veliger*, **23**, 350–1.

Marquis, F. (1989). Die Embryonalentwicklung des Nervensystems von *Octopus vulgaris* Lam. (Cephalopoda, Octopoda), eine histologische Analyse. *Verhandlungen der Naturforschenden Gesellschaft in Basel*, **99**, 23–76.

Marshall, N.B. (1979). *Developments in deep-sea biology.* Blandford Press, Poole, Dorset.

Marshall, N.J. and Messenger, J.B. (1996). Colour-blind camouflage. *Nature, London*, **382**, 408–9.

Martin, R. (1965). On the structure and development of the giant fibre system of the squid *Loligo vulgaris. Zeitschrift für Zellforschung und mikroskopische Anatomie*, **67**, 77–85.

Martin, R. (1969). The structural organization of the intracerebral giant fibre system of cephalopods. The chiasma of the first order giant fibre axons. *Zeitschrift Zellforschung und mikroskopische Anatomie*, **97**, 50–68.

Martin, R. (1977). The giant nerve fibre system of cephalopods. Recent structural findings. *Symposia of the Zoological Society, London*, No. **38**, 261–75.

Martin, R. and Miledi, R. (1986). The form and dimensions of the giant fibre system of squids. *Philosophical Transactions of The Royal Society, London*, **B312**, 355–77.

Massy, A.L. (1916). The Cephalopoda of the Indian Museum. *Records of the Indian Museum*, **12**, 185–247.

Mather, J. (1991a). Navigation by spatial memory and use of landmarks by octopuses. *Journal of Comparative Physiology*, **168A**, 491–7.

Mather, J. (1991b). Foraging, feeding and prey remains in middens of juvenile *Octopus vulgaris* (Mollusca: Cephalopoda). *Journal of Zoology, London*, **224**, 27–39.

Mather, J.A. (1994). 'Home' choice and modification by juvenile *Octopus vulgaris*: specialized intelligence and tool use? *Journal of Zoology, London*, **233**, 359–68.

Mather, J.A. (1995). Cognition in cephalopods. *Advances in the study of behavior*, **24**, 317–53.

Mather, J.A. and O'Dor, R.K. (1991). Foraging strategies and predation risk shape the natural history of juvenile *Octopus vulgaris. Bulletin of Marine Science*, **49**, 256–69.

Mäthger, L. and Denton, E.J. (2001). Reflective properties of iridophores and fluorescent 'eyespots' in the loliginid squids *Alloteuthis subulata* and *Loligo vulgaris. Journal of Experimental Biology*, **204**, 2103–18.

Matsui, S., Seidou, M., Horiuchi, S., Uchiyama, I., and Kito, Y. (1988). Adaptation of a deep-sea cephalopod to the photic environment. *Journal of General Physiology*, **92**, 55–66.

Mattacola, A.D., Maddock, L., and Denton, E.J. (1984). Weights and lengths of *Sepia officinalis* trawled by the laboratory's boats, 1978–1983. *Journal of the Marine Biological Association, U.K.*, **64**, 735–7.

Mauro, A. (1977). Extra-ocular photoreceptors in cephalopods. *Symposia of the Zoological Society, London*, No. **38**, 287–308.

Mauro, A. and Baumann, F. (1968). Electrophysiological evidence of photoreceptors in the epistellar body of *Eledone moschata. Nature, London*, **220**, 1332–4.

McConathy, D.A., Hanlon, R.T., and Hixon, R.F. (1980). Chromatophore arrangements of hatchling loliginid squids (Cephalopoda, Myopsida). *Malacologia*, **19**, 279–88.

McFall-Ngai, M. (1998). The development of cooperative associations between animals and bacteria: Establishing détente among domains. *American Zoologist*, **38**, 593–608.

McFall-Ngai, M. and Montgomery, M.K. (1990). The anatomy and morphology of the adult bacterial light organ of *Euprymna scolopes* Berry (Cephalopoda: Sepiolidae). *Biological Bulletin, Woods Hole*, **179**, 332–9.

McSweeny, E.S. (1970). Description of the juvenile form of the Antarctic squid *Mesonychoteuthis hamiltoni* Robson. *Malacologia*, **10**, 323–32.

McSweeny, E.S. (1978). Systematics and morphology of the Antarctic cranchiid squid *Galiteuthis glacialis* (Chun). *Antarctic Research Series*, **27**, 1–39.

Messenger, J.B. (1967a). Parolfactory vesicles as photoreceptors in a deep-sea squid. *Nature, London*, **213**, 836–8.

Messenger, J.B. (1967b). The peduncle lobe: A visuo-motor centre in *Octopus. Proceedings of the Royal Society, London*, **B167**, 225–51.

Messenger, J.B. (1967c). The effects on locomotion of lesions to the visuo-motor centre in *Octopus. Proceedings of the Royal Society, London*, **B167**, 252–81.

Messenger, J.B. (1968). The visual attack of the cuttlefish, *Sepia officinalis. Animal Behaviour*, **16**, 342–57.

Messenger, J.B. (1971). The optic tract lobes. In *The anatomy of the nervous system of* Octopus vulgaris (by J.Z. Young), pp. 481–506. Clarendon Press, Oxford.

Messenger, J.B. (1973). Learning performance and brain structure: a study in development. *Brain Research*, **58**, 519–23.

Messenger, J.B. (1974). Reflecting elements in cephalopod skin and their importance for camouflage. *Journal of Zoology, London*, **174**, 387–95.

Messenger, J.B. (1977). Prey-capture and learning in the cuttlefish, *Sepia. Symposia of the Zoological Society, London*, No. **38**, 347–76.

Messenger, J.B. (1979a). The nervous system of *Loligo* IV. The peduncle and olfactory lobes. *Philosophical Transactions of the Royal Society, London*, **B285**, 275–309.

Messenger, J.B. (1979b). The eyes and skin of *Octopus*: Compensating for sensory deficiencies. *Endeavour*, **3**, 92–8.

Messenger, J.B. (1981). Comparative physiology of vision in molluscs. In *Handbook of sensory physiology* Vol. VII/6C. *Comparative physiology and evolution of vision in invertebrates* (ed. H. Autrum), pp. 93–200. Springer-Verlag, Berlin.

Messenger, J.B. (1983). Multimodal convergence and the regulation of motor programs in cephalopods. *Fortschritte der Zoologie*, **28**, 77–98.

Messenger, J.B. (1991). Photoreception and vision in molluscs. In *Evolution of the eye and the visual system* (ed. J.R. Cronly-Dillon and R.L. Gregory), pp. 364–97. MacMillan, Basingstoke.

Messenger, J.B. (1996). Neurotransmitters of cephalopods. *Invertebrate Neuroscience*, **2**, 95–114.

Messenger, J.B. (2001). Cephalopod chromatophores: neurobiology and natural history. *Biological Reviews*, **76**, 473–528.

Messenger, J.B. and Miyan, J.A. (1986). Neural correlates of colour change in cuttlefish. *Journal of Experimental Biology*, **125**, 395–400.

Messenger, J.B. and Woodhams, P.L. (1976). A molluscan 'cerebellum': parallel fibres and visuo-motor control. *Journal of Physiology, London*, **259**, 20–1*P*.

Messenger, J.B. and Young, J.Z. (1999). The radular apparatus of cephalopods. *Philosophical Transactions of the Royal Society, London*, **B354**, 161–82.

Messenger, J.B., DeSantis, A., and Ogden, D.C. (1995). Chemical transmission at hte squid giant synapse. In *Cephalopod neurobiology: neuroscience studies in squid, octopus and cuttlefish* (ed. N.J. Abbott, R. Williamson and L. Maddock), pp. 283–97. Oxford University Press.

Messenger, J.B., Cornwell, C.J., and Reed, C.M. (1997). L-Glutamate and serotonin are endogenous in squid chromatophore nerves. *Journal of Experimental Biology*, **200**, 3043–54.

Meyer, W.T. (1906). Die Anatomie von *Opisthoteuthis depressa* (Ijema und Ikeda). *Zeitschrift für Wissenschaftliche Zoologie*, **85**, 183–269.

Michinomae, M., Masuda, H., Seido, M., and Kito, Y. (1994). Structural basis for wavelength discrimination in the banked retina of the firefly squid *Watasenia cintillans. Journal of Experimental Biology*, **193**, 1–12.

Mirow, S. (1972a). Skin color in the squids *Loligo pealii* and *Loligo opalescens* I. Chromatophores. *Zeitschrift für Zellforschung*, **125**, 143–75.

Mirow, S. (1972b). Skin color in the squids *Loligo pealii* and *Loligo opalescens* II. Iridophores. *Zeitschrift für Zellforschung*, **125**, 176–90.

Misaki, H. and Okutani, T. (1976). Studies on early life history of decapodan Mollusca-VI. An evidence of spawning of an oceanic squid, *Thysanoteuthis rhombus* Troschel, in the Japanese waters. *Venus, Japanese Journal of Malacology*, **35**, 211–3.

Miyan, J.A. and Messenger, J.B. (1995). Intracellular recordings from the chromatophore lobes of *Octopus*. In *Cephalopod neurobiology: Neuroscience studies in squid, octopus and cuttlefish* (ed. N.J. Abbott, R. Williamson, and L. Maddock), pp. 416–29. Oxford University Press.

Moiseev, S.L. (1991). Observation of the vertical distribution and behaviour of nektonic squids using manned submersible. *Bulletin of Marine Science*, **49**, 446–56.

Mommsen, T.P., Ballantyne, D., MacDonald, D., Gosline, J., and Hochachka, P.W. (1981). Analogues of red and white muscle in squid mantle. *Proceedings National Academy Science, U.S.A.*, **78**, 3274–8.

Montgomery, M.K. and McFall-Ngai, M.J. (1994). Bacterial symbiosis induce host organ morphogenesis during early postembryonic development of the squid *Euprymna scolopes. Development*, **120**, 1719–29.

Moody, M.F. and Parriss, J.R. (1960). The visual system of *Octopus*: Discrimination of polarized light by *Octopus. Nature, London*, **186**, 839–40.

Moody, M.F. and Parriss, J.R. (1961). The discrimination of polarized light by *Octopus*: A behavioural and morphological study. *Zeitschrift für vergleichende Physiologie*, **44**, 268–91.

Moolenbeek, R.G. (1984). *Pickfordiateuthis pulchella* (Cephalopoda: Myopsida) range extension to Bonaire (N.A.). *Studies on West Indian Marine Molluscs*, **4**, 104–6.

Moon, T.W. and Hulbert, W.C. (1975). The ultrastructure of the mantle musculature of the squid *Symplectoteuthis oualaniensis. Comparative Biochemical Physiology*, **52B**, 145–9.

Morales, E. (1962). Cefalopodos de Cataluna II. *Investigación Pesquera*, **21**, 97–111.

Morales, E. (1981). Presencia de *Thysanoteuthis rhombus* Troschel, en el puerto de Mahón (Menorca). *Investigación Pesquera*, **45**, 17–20.

Morton, N. and Nixon, M. (1987). Size and function of ammonite aptychi in comparison with buccal masses of modern cephalopods. *Lethaia*, **20**, 231–8.

Moynihan, M. (1983a). Notes on the behavior of *Idiosepius pygmaeus* (Cephalopoda: Idiosepiidae). *Behaviour*, **85**, 42–57.

Moynihan, M. (1983b). Notes on the behaviour of *Euprymna scolopes* (Cephalopoda: Sepioidae). *Behaviour*, **85**, 25–41.

Moynihan, M. (1985). *Communication and non-communication by cephalopods.* Indiana University Press, Bloomington.

Moynihan, M. and Rodaniche, A.F. (1982). The behavior and natural history of the Caribbean reef squid *Sepioteuthis sepioidea. Fortschritte der Verhaltensforschung, Beihefte zur Zeitschrift für Tierphysiologie*, **25**, 1–150.

Mugglin, F. (1939). Beiträge zur Kenntnis der Anatomie von *Nautilus macrompahlus* G.B. Sow. *Vierteljahrsschrift Naturforschung Gesellschaft, Zürich, Jahrgang*, **84**, 25–118.

Munk, O. and Frederiksen, R.D. (1974). On the function of aphakic aperture in teleosts. *Videnskabelige Meddelelser fra Dansk naturhistorik Forening i Kjøbenhavn*, **137**, 65–94.

Muntz, W.R.A. (1961). Interocular transfer in *Octopus vulgaris*; Bilaterality of the engram. *Journal of Comparative Physiological Psychology*, **54**, 186–91.

Muntz, W.R.A. (1963). Interocular transfer and the function of the optic lobes in *Octopus. Quarterly Journal of Experimental Psychology*, **15**, 116–24.

Muntz, W.R.A. (1976). On yellow lenses in mesopelagic animals. *Journal of the Marine Biological Association, U.K.*, **56**, 963–76.

Muntz, W.R.A. (1977). Pupillary response of cephalopods. *Symposia of the Zoological Society, London*, No. **38**, 277–85.

Muntz, W.R.A. (1987a). A possible function of the iris groove of *Nautilus*. In Nautilus *the biology and paleobiolgy of a living fossil* (ed. W.B. Saunders and N.H. Landman), pp. 231–44. Plenum, New York.

Muntz, W.R.A. (1987b). Visual behavior and visual sensitivity of *Nautilus pompilius*. In Nautilus *the biology and paleobiolgy of a living fossil* (ed. W.B. Saunders and N.H. Landman), pp. 245–7. Plenum, New York.

Muntz, W.R.A. (1994a). Effects of light on the efficacy of traps for *Nautilus pompilius. Marine Behaviour and Physiology*, **24**, 189–93.

Muntz, W.R.A. (1994b). Spatial summation in the photactic behaviour of *Nautilus pompilius. Marine Behaviour and Physiology*, **24**, 183–7.

Muntz, W.R.A. and Johnson, M.S. (1978). Rhodopsins of oceanic decapods. *Vision Research*, **18**, 601–2.

Muntz, W.R.A. and Raj, U. (1984). On the visual system of *Nautilus pompilius. Journal of Experimental Biology*, **109**, 253–63.

Muntz, W.R.A. and Wentworth, S.L. (1987). An anatomical study of the retina of *Nautilus pompilius. Biological Bulletin, Woods Hole*, **173**, 387–97.

Mutvei, H. (1964). On the shells of *Nautilus* and *Spirula* with some notes on the shell secretion in non-cephalopod molluscs. *Arkiv för Zoologi*, **16**, 221–78.

Mutvei, H. (1970). Ultrastructure of the mineral and organic componenets of molluscan nacreous layers. *Biomineralisation*, **2**, 48–61.

Mutvei, H. and Doguzhaeva. L. (1997). Shell ultrastructure and ontogenetic growth in *Nautilus pompilius* L. (Mollusca: Cephalopoda). *Palaeontographica Beiträge zur Naturgeschichte der Vorzeit*, **A246**, 33–52.

Mutvei, H., Arnold, J.M., and Landman, N.H. (1993). Muscles and attachment of the body to the shell in embryos and adults of *Nautilus belauensis* (Cephalopoda). *American Museum Novitates*, No. **3059**, 15p.

Naef, A. (1916). Systematische uebersicht der Mediterranen Cephalopoden. *Pubblicazioni della Stazione Zoologica di Napoli*, **1**, 11–19.

Naef, A. (1921, 1923, 1928). Die Cephalopoden. *Fauna e Flora del Golfo di Napoli*, Monographie **35**, Vol. I, Parts I and II, Systematik, pp. 1–863; Vol. II, Embryologie, pp. 1–364. (The first part of the first volume was published in 1921 and the second part in 1923; volume II in 1928.) (trans. A. Mercado, 1972, Israel Program for Scientific Translations, Jerusalem.)

Natsukari, Y. (1970). Egg-laying behavior, embryonic development and hatched larva of the pygmy cuttlefish, *Idiosepius pygmeus paradoxus* Ortmann. *Bulletin of the Faculty of Fisheries, Nagasaki University*, No. **30**, 15–29.

Natsukari, Y. and Tashiro, M. (1991). Neritic squid resources and cuttlefish resources in Japan. *Marine Behavioural Physiology*, **18**, 149–226.

Natsukari, Y., Hirata, S., and Washizaki, M. (1991). Growth and seasonal change of cuttlebone characters of *Sepia esculenta*. In *La seiche, The cuttlefish, 1st International Symposium on the cuttlefish* Sepia, *Caen: June 1–3, 1989* (ed. E. Boucaud-Camou), pp. 49–67. University of Caen.

Nazumi, T. (1975). Notes on the fishery and the ecology of the squid, *Thysanoteuthis rhombus*, Troschel, in the east San'in waters. *Bulletin Hyogo Prefecture Fisheries Experimental Station*, **15**, 15–34.

Nazumi T., Kasahara, S., and Hamabe, M. (1979). Contribution of biological information useful for development of inshore squid fishery in the Japan Sea. III. Supplements and amendments to the previous paper on reproduction and distribution of *Berryteuthis magister* (Berry). *Bulletin of the Japan Sea Regional Fisheries Research Laboratory*, No. **30**, 1–14.

Nemoto, T., Okiyama, M., and Takahashi, M. (1985). Aspects of the roles of squid in food chains of marine Antarctic ecosystems. In: *Antarctic nutrient cycles and food webs* (ed. W.R. Siegfied, P.R. Condy and R.M. Laws). Springer-Verlag, Berlin.

Nesis, K.N. (1965). The distribution and nutrition of young squiuds *Gonatus fabricii* (Licht.) in the Labrador Sea and the Norwegian sea. *Oceanology*, **5**, 102–8.

Nesis, K.N. (1972). Two new species of gonatid squid from the North Pacific. *Zoologicheskii Zhurnal*, **51**, 1300–7.

Nesis, K.N. (1977). The biology of paper nautiluses, *Argonauta boettgeri* and *A. hians* (Cephalopoda, Octopoda), in the Western Pacific and the seas of the East Indian Archipelago. *Zoologicheskii Zhurnal*, **56**, 1004–14.

Nesis, K.N. (1983). *Dosidicus gigas*. In: *Cephalopod life cycles* Vol. I. *Species accounts* (ed. P.R. Boyle), pp. 215–31. Academic Press, London.

Nesis, K.N. (1985). *Oceanic cephalopods: Distribution, life forms, evolution*. Nauka, Moscow.

Nesis, K.N. (1986). Cephalopods from the seamounts of the western Indian Ocean. *Okeanologia*, **26**, 123–30.

Nesis, K.N. (1987). *Cephalopods of the world. Squids, cuttlefishes, octopuses and allies* (trans. B.S. Levitov). T.F.H. Publications, Neptune City.

Nesis, K.N. (1989). Teuthofauna of the Okhotsk Sea. Biology of gonatid squids *Berryteuthis magister* and *Gonatopsis borealis*. *Zoologicheskii Zhurnal*, **68**, 45–56.

Nesis, K.N. (1993a). Cephalopods of seamounts and submarine ridges. In *Recent advances in cephalopod fisheries biology* (ed T. Okutani, R.K. O'Dor, and T. Kubodera), pp. 365–73. Tokai University Press, Tokyo.

Nesis K.N. (1993b). Okhotsk Sea – the home of giant cephalopods. In *Recent advances in cephalopod fisheries biology* (ed. T. Okutani, R.K. O'Dor, and T. Kubodera), pp. 745–6. Tokai University Press, Tokyo.

Nesis, K.N. (1993c). Spent females of deep-water squid *Gonatopsis octopedatus* Sasaki, 1920 (Gonatidae) in the epipelagic layer of the Okhotsk and Japan seas. *Ruthenica*, **3**, 153–8.

Nesis, K.N. (1994). Teuthofauna of Walters Shoals, a seamount in the south-western Indian Ocean. *Ruthenica*, **4**, 67–77.

Nesis, K.N. (1997). Gonatid squids in the subarctic North Pacific: Ecology, biogeography, niche diversity and role in the ecosystem. *Adavnccs in Marine. Biology*, **32**, 243–324.

Nesis, K.N. and Nikitina, I.V. (1986). New genus and species of squid of the family Neoteuthidae (Cephalopoda, Oegopsida) from the south-eastern part of the Pacific Ocean. *Zoologicheskii Zhurnal*, **65**, 290–4.

Nesis, K.N and Nikitina, I.V. (1992). New records of oceanic squids *Walvisteuthis virilis* Nesis et Nikitina, 1986 and *Nototeuthis dimegacotyle* Nesis et Nikitina, 1986 (Cephalopoda, Oegopsida) from the South Atlantic and the South Pacific. *Ruthenica*, **2**, 55–8.

Nesis, K.N., Nigmatullin, C.M., and Nikitina, I.V. (1998). Spent females of deepwater squid *Galiteuthis glacialis* under the ice at the surface of the Weddell Sea (Antarctica). *Journal of Zoology, London*, **244**, 185–200.

Neumeister, H. and Budelmann, B.U. (1997). Structure and function of the *Nautilus* statocyst. *Philosophical Transactions of the Royal Society, London*, **B352**, 1565–88.

Nicol, S. and O'Dor, R.K. (1985). Predatory behaviour of squid (*Illex illecebrosus*) feeding on surface swarms of euphausiids. *Canadian Journal of Zoology*, **63**, 15–17.

Niemec, J. (1885). Recherches morphologiques sur les ventouses dans le règne animal. *Recueil Zoologique Suisse*, (1) **2**, 1–147.

Nigmatullin, C.M. and Ostapenko, A.A. (1976). Feeding of *Octopus vulgaris* Lam. from the northwest coast African coast. *International Council for the Exploration of the Sea, CM 1976/K: Shellfish and Benthos Committee*, **6**, 1–15.

Nigmatullin, C., Arkhipkin, A.I., and Sabirov, R.M. (1991). Structure of the reproductive system of the squid *Thysanoteuthis rhombus* (Cephalopoda: Oegopsida). *Journal of Zoology, London*, **224**, 271–83.

Nishimura, S. (1966). Notes on the occurrence and biology of the oceanic squid, *Thysanoteuthis rhombus* Troschel, in Japan. *Publications Seto Marine Biological Laboratory*, **14**, 327–49.

Nishimura, S. (1968). Glimpse of the biology of *Argonauta argo* Linnaeus (Cephalopoda: Octopodida) in the Japanese waters. *Publications of the Seto Marine Biology Laboratories*, **16**, 61–70.

Nishioka, R.S., Yasumasu, I., Packard, A., Bern, H.A. and Young, J.Z. (1966). Nature of vesicles associated with the nervous system of cephalopods. *Zeitschrift für Zellforschung und Mikroskopische Anatomie*, **75**, 301–16.

Nixon, M. (1966). Changes in body weight and intake of food by *Octopus vulgaris*. *Journalof Zoology, London*, **150**, 1–9.

Nixon, M. (1968). Feeding mechanisms and growth in *Octopus vulgaris*. Thesis, Doctor of Philosophy, University of London.

Nixon, M. (1969). Growth of the beak and radula of *Octopus vulgaris*. *Journal of Zoology, London*, **159**, 363–79.

Nixon, M. (1971). Some parameters of growth of *Octopus vulgaris*. *Journal of Zoology*, **163**, 277–84.

Nixon, M. (1979a). Hole-boring in shells by *Octopus* Cuvier in the Mediterranean. *Malacologia*, **18**, 431–43.

Nixon, M. (1979b). Has *Octopus vulgaris* a second radula? *Journal of Zoology, London*, **187**, 291–6.

Nixon, M. (1980). The salivary papilla of *Octopus* as an accessory radula for drilling shells. *Journal of Zoology, London*, **190**, 53–7.

Nixon, M. (1983). *Teuthowenia megalops*. In *Cephalopod life cycles* Vol. I. *Systematic accounts* (ed. P.R. Boyle), pp. 233–47. Academic Press, London.

Nixon, M. (1984). Is there external digestion by *Octopus*? *Journal of Zoology, London*, **202**, 441–7.

Nixon, M. (1985). Capture of prey, diet and feeding of *Sepia officinalis* and *Octopus vulgaris* (Mollusca: Cephalopoda) from hatchling to adult. *Vie et Milieu*, **35**, 255–61.

Nixon, M. (1987). Cephalopod diets. In *Cephalopod life cycles* Vol II. *Comparative reviews* (ed. P.R. Boyle), pp. 201–19. Academic Press, London.

Nixon, M. (1988). The buccal mass of fossil and Recent Cephalopoda. In *The Mollusca*, Vol. 12. *Paleontology and Neontology of cephalopods* (ed. M.R. Clarke and E.R. Trueman), pp. 103–22. Academic Press, San Diego.

Nixon, M. (1991). A note on the eggs of *Benthoctopus piscatorum* (Cephalopoda: Octopoda). *Journal of Zoology, London*, **223**, 499–500.

Nixon, M. (1995). A nomenclature for the radula of the Cephalopoda (Mollusca)- living and fossil. *Journal of Zoology, London*, **236**, 73–81.

Nixon, M. (1996). Morphology of the jaws and radula in ammonoids. In *Ammonoid biology* (ed. N.H. Landman, K. Tanabe and R.A. Davis), pp. 23–42. Plenum, New York.

Nixon, M. (1998). The radulae of Cephalopoda. *Smithsonian Contributions to Zoology*, No. **586**, 39–53.

Nixon, M. (in press (a)). Ecology. In *Treatise on Invertebrate Paleontology: Part M, Mollusca 5: Coleoidea Cephalopoda* (ed. D.T. Donovan). University of Kansas Paleontological Institute, Kansas Boulder, CO.

Nixon, M. (in press (b)). Biogeography. In *Treatise on Invertebrate Paleontology: Part M, Mollusca 5: Coleoidea Cephalopoda* (ed. D.T. Donovan). University of Kansas Paleontological Institute, Kansas Boulder, CO..

Nixon, M. and Boyle, P.R. (1982). Hole-drilling in crustaceans by *Eledone cirrhosa* (Mollusca: Cephalopoda). *Journal of Zoology, London*, **196**, 439–44.

Nixon, M. and Dilly, P.N. (1977). Sucker surfaces and prey capture. *Symposia of the Zoological Society, London*, No. **38**, 447–511.

Nixon, M. and Maconnachie, E. (1988). Drilling by *Octopus vulgaris* (Mollusca: Cephalopoda) in the Mediterranean. *Journal of Zoology, London*, **216**, 687–716.

Nixon, M. and Mangold, K. (1996). The early life of *Octopus vulgaris* (Cephalopoda: Octopodidae) in the plankton and at settlement: A change in lifestyle. *Journal of Zoology, London*, **239**, 301–27.

Nixon, M. and Mangold, K. (1998). The early life of *Sepia officinalis*, and the contrast with that of *Octopus vulgaris* (Cephalopoda). *Journal of Zoology, London*, **245**, 407–21.

Nixon, M. and Messenger, J.B. (eds) (1977). *The biology of cephalopods. Symposia of the Zoological Society of London*, **No. 38**, pp. 615.

Nixon, M., Maconnachie, E., and Howell, P.G.T. (1980). The effects on shells of drilling by *Octopus. Journal of Zoology, London*, **191**, 75–88.

Noda, H., Ogasawara, K., and Nomura, R. (1986). Systematic and paleobiogeographic studies on the Japanese Miocene argonautid *'Nautilus' izomoensis. Science Reports of the Institute of Geoscience, University of Tsukuba, Section B, Geological Sciences*, 7, 15–42.

Norman, M. (1992). *Ameloctopus litoralis* n. gen., n. sp. (Cephalopoda: Octopodidae), a new shallow water octopus from tropical Australian waters. *Invertebrate Taxonomy*, **6**, 567–82.

Norman, M.D. (1997). Australian octopuses. *Nature Australia*, Summer 1997, pp. 41–9.

Norman, M.D. and Hochberg, F.G. (1994). Shallow-water octopuses (Cephalopoda: Octopodidae: Octopodidae) from Hong Kong waters. In *The Malacofauna of Hong Kong and Southern China III* (ed. B. Morton), pp. 141–60. Hong Kong University Press, Hong Kong.

Norman, M.D. and Lu, C.C. (1997a). Redescription of the southern dumpling squid *Euprymna tasmanica* and a revision of the genus *Euprymna* (Cephalopolpoda: Sepiolidae). *Journal of the Marine Biological Association, U.K.*, **77**, 1109–37.

Norman, M.D. and Lu, C.C. (1997b). Sex in giant squid. *Nature, London*, **389**, 683–4.

Norman, M.D. and Sweeney, M.J. (1997). The shallow-water octopuses (Cephalopoda: Octopodidae) of the Philippines. *Invertebrate Taxonomy*, **11**, 89–140.

Norman, M.D., Hochberg, F.G., and Lu, C.C. (1997). Mollusca Cephalopoda: Mid-depth octopuses (200–1000 m) of the Banda and Arafura Seas (Octopodidae and Alloposidae). *Mémoires Muséum national d'histoire naturelle*, **172**, 357–83.

Novicki, A., Budelmann, B.U., and Hanlon, R.T. (1990). Brain pathways of the chromatophore system in the squid *Lolliguncula brevis. Brain Research*, **519**, 315–23.

Novicki, A., Messenger, J.B., Budelmann, B.U., Terrell, M.L., and Kadekaro, M. (1992). [^{14}C]-deoxyglucose labelling of functional activity in the cephalopod nervous system. *Proceedings of the Royal Society, London*, **B249**, 77–82.

O'Dor, R.K. (1982). Respiratory metabolism and swimming performance of the squid *Loligo opalescens. Canadian Journal of Fisheries and Aquatic Sciences*, **39**, 580–7.

O'Dor, R.K. (1983). *Illex illecebrosus*. In *Cephalopod life cycles* Vol. 1. *Species accounts* (ed. P.R. Boyle), pp. 175–99. Academic Press, London.

O'Dor, R.K. (1988a). Limitations on locomotor performance in squid. *Journal of Applied Physiology*, **64**, 128–34.

O'Dor, R.K. (1988b). The forces acting on swimming squid. *Journal of Experimental Biology*, **137**, 421–42.

O'Dor, R.K. and Coelho, M.L. (1993). Big squid, big currents and big fisheries. In *Recent advances in cephalopod fisheries biology* (ed. T. Okutani, R.K. O'Dor, and T. Kubodera), pp. 385–96. Tokai University Press, Tokyo.

O'Dor, R.K. and Macalaster, E.G. (1983). *Bathypolypus arcticus*. In *Cephalopod life cycles* Vol. I. *Species accounts* (ed. P.R. Boyle), pp. 401–10. Academic Press, London.

O'Dor, R.K. and Webber, D.M. (1986). The constraints on cephalopods: Why squid aren't fish. *Canadian Journal of Zoology*, **64**, 1591–627.

O'Dor, R.K., Durward, R.D., and Balch, N. (1977). Maintenance and maturation of squid (*Illex illecebrosus*) in a 15 meter circular pool. *Biological Bulletin, Woods Hole*, **153**, 322–35.

O'Dor, R.K., Helm, P., and Balch, N. (1985). Can rhynchoteuthions suspension feed? (Mollusca, Cephalopoda). *Vie et Milieu*, **35**, 267–71.

O'Dor, R.K., Wells, J., and Wells, M.J. (1990). Speed, jet pressure and oxygen consumption relationships in free-swimming *Nautilus. Journal of Experimental Biology*, **154**, 383–96.

O'Dor, R.K., Forsythe, J., Webber, D.M., Wells, J., and Wells, M.J. (1993). Acitivity levels of *Nautilus* in the wild. *Nature, London*, **362**, 626–8.

Ohshima, Y. and Choe, S. (1961). On the rearing of young cuttlefish and squid. *Bulletin of the Japanese Society of Scientific Fisheries*, **27**, 979–86.

Okada, Y.K. (1966). Observations on rod-like contents in the photogenic tissues of *Watasenia scintillans* through the electron microscope. In *Bioluminescence in progress* (ed. F.H. Johnson and Y. Haneda), pp. 611–25. Princeton University Press.

Okada, Y.K. (1968). Studies of luminous Cephalopoda I. Luminous squids belonging to the Ommastrephidae. *Science Report of the Yokosuka City Museum*, No. **14**, 88–94.

Okiyama, M. (1993). Kinds, abundance and distribution of the oceanic squids in the Sea of Japan. In *Recent advances in cephalopod fisheries biology* (ed. T. Okutani, R. O'Dor and T. Kubodera), pp. 403–15. Tokai University Press, Tokyo.

Okubo, S., Tsujii, T., Watabe, N. and Williams, D.F. (1995). Hatching of *Nautilus belauensis* Saunders, 1989, in captivity: culture, growth and stable isotope composition of shells, and histology and immunochemistry of the mantle epithelium of the juveniles. *The Veliger*, **38**, 192–202.

Okutani, T. (1960). *Argonauta boettgeri* preys on *Cavolinia tridentata. Venus, Japanese Journal of Malacology*, **21**, 39–41.

Okutani, T. (1966). Studies on early life history of decapodan Mollusca-II. Planktonic larvae of decapodan cephalopods from the northern North Pacific in summer seasons during 1952–1959. *Bulletin of the Tokai Regional Fisheries Research Laboratory*, No. **45**, 61–79.

Okutani, T. (1967). Preliminary catalogue of Decapodan Mollusca from Japanese waters. *Bulletin of the Tokai Regional Fisheries Research Laboratory*, No. **50**, 1–16.

Okutani, T. (1968). Studies on early life history of decapodan Mollusca-III. Systematics and distribution of decapod cephalopods collected from the sea surface on the Pacific coast of Japan, 1960–1965. *Bulletin of the Tokai Regional Fisheries Research Laboratory*, No. **55**, 9–57.

Okutani, T. (1974). Epipelagic decapod cephalopods collected by micronekton tows during the EASTROPAC Expeditions, 1967–1968 (Systematic part). *Bulletin of the Tokai Regional Fisheries Research Laboratory*, No. **80**, 29–118.

Okutani, T. (1975). Supplements and amendments for 'Guide and keys to squid in Japan'. *Bulletin of the Tokai Regional Fisheries Research Laboratory*, No. **83**, 41–4.

Okutani, T. (1976). Rare and interesting squid from Japan – V. A gravid female of *Ancistrocheirus lesueuri* (d'Orbigny, 1839) collected in the Kuroshio Area

(Oegopsida: Enoploteuthidae). *Venus, Japanese Journal of Malacology*, **35**, 73–81.

Okutani, T. (1978). Studies on early life history of decapodan Mollusca-VII. Eggs and newly hatched larvae of *Sepia latimanus* Quoy and Gaimard. *Venus, Japanese Journal of Malacology*, **37**, 245–8.

Okutani, T. (1981). Two new species of the squid genus *Onykia* from the tropical Indian Ocean (Cephalopoda, Onychoteuthidae). *Bulletin of the National Science Museum*, series A, 7, 155–63.

Okutani, T. (1983a). *Todarodes pacificus. Cephalopod life cycles* Vol. I. *Species accounts* (ed. P.R. Boyle), pp. 201–14. Academic Press, London.

Okutani, T. (1983b). A new species of an oceanic squid, *Moroteuthis pacifica* from the North Pacific (Cephalopoda: Onychoteuthidae). *Bulletin of the National Science museum*, series A, **9**, 105–112.

Okutani, T. (1983c). Rare and interesting squid from Japan – VIII. Rediscovery of *Promachoteuthis megaptera* Hoyle, 1885 (Oegopsida: Promachoteuthidae). *Venus, Japanese Journal of Malacology*, **42**, 241–7.

Okutani, T. (1983d). *Abralia marisarabica*, a new enoploteuthid squid from the Arabian Sea (Cephalopoda: Oegopsida). *Bulletin National Science Museum, Tokyo*, **9**, 161–8.

Okutani, T. (1987). Juvenile morphology. In: *Cephalopod life cycles* Vol. II. *Comparative Reviews* (ed. P.R. Boyle), pp. 33–44. Academic Press, London.

Okutani, T. (1988). Evidence of spawning of *Berryteuthis magister* in the northeastern Pacific (Cephalopoda: Gonatidae). *Bulletin of the Ocean Research Institute, University of Tokyo*, **26–I**, 193–200.

Okutani, T. (1995). *Cuttlefish and squids of the world in colour*. National Cooperative Association of Squid Processors, Tokyo.

Okutani, T. and Clarke, M.R. (1992). Family Gonatidae Hoyle, 1886. *Smithsonian Contributions to Zoology*, No. **513**, 139–56.

Okutani, T. and Ida, H. (1986). Rare and interesting squid in Japan-IX. A mass occurrence of *Chaunoteuthis mollis* Appellöf, 1891 (Oegopsida: Onychoteuthidae) from off Japan. *Venus, Japanese Journal of Malacology*, **45**, 53–60.

Okutani, T. and Kawaguchi, T. (1983). A mass occurrence of *Argonauta argo* (Cephalopoda: Octopoda) along the coast of Shimane Prefecture, western Japan Sea. *Venus, Japanese Journal of Malacology*, **41**, 281–90.

Okutani, T. and Kubota, T. (1972). Rare and interesting squid from Japan-I. *Joubiniteuthis portieri* (Joubin, 1912), the first occurrence from the Pacific (Cephalopoda: Oegopsida). *Venus, Japanese Journal of Malacology*, **31**, 35–40.

Okutani, T. and Kubota, T. (1976). Cephalopods eaten by lancet-fish, *Alepisaurus ferox* Lowe, in Suruga Bay. *Japanese Bulletin of the Tokai Regional Fisheries Research Laboratory*, No. **84**, 1–9.

Okutani, T. and McGowan, J.A. (1969). Systematics, distribution, and abundance of the epiplanktonic squid (Cephalopoda, Decapoda) larvae of the California Current April, 1954-March, 1957. *Bulletin Scripps Institution Oceanography University of California, San Diego, La Jolla, California*, **14**, 1–90.

Okutani, T. and Murata, M. (1983). A review of the biology of the oceanic squid *Onychoteuthis borealijaponicus*. *Memoirs of the National Museum, Victoria*, No. **44**, 189–195.

Okutani, T. and Osuga, K. (1986). A peculiar nesting behavior of *Ocythoe tuberculata* in the test of a gigantic salp, *Tethys vagina*. *Venus, Japanese Journal of Malacology*, **45**, 67–9.

Okutani, T. and Tsukada, S. (1988). Squids eaten by lancetfish and tunas in the tropical Indo-Pacific Oceans. *Journal of the Tokyo University of Fisheries*, **75**, 1–44.

Okutani, T. and Tung, I-H. (1978). Reviews of biology of commercially important squids in Japanese and adjacent waters I. *Symplectoteuthis oualaniensis* (Lesson). *The Veliger*, **21**, 87–94.

Okutani, T., Satake, Y., Ohsumi, S., and Kawakami, T. (1976). Squids eaten by sperm whales caught off Joban District, Japan, during January–February, 1976. *Bulletin of the Tokai Regional Fisheries Research Laboratory*, No. **87**, 67–113.

Okutani, T., Tagawa, M., and Horikawa, H. (1987). *Cephalopods from continental shelf and slope around Japan*. Japan Fisheries Resouce Conservation Association, Tokyo.

Okutani, T., Kubodera, T., and Jefferts, K. (1988). Diversity, distribution and ecology of gonatid squids in the subarctic Pacific: A review. *Bulletin of the Ocean Research Institute University, Tokyo*, **26**, 159–92.

O'Shea, S. and Kubodera, T. (1996). Eggs and larvae of *Graneledone* sp. (Mollusca, Octopoda) from New Zealand. *Bulletin of the National Science Museum*, series A (*Zoology*), **22**, 153–64.

O'Sullivan, D. and Cullen, J.M. (1983). Food of the squid *Nototodarus gouldi* in Bass Strait. *Australian Journal of Marine and Freshwater Research*, **34**, 261–85.

O'Sullivan, D.B., Johnstone, G.W., Kerry, K.R., and Imber, M.J. (1983). A mass stranding of squid *Martialia hyadesi* Rochebrunne and Mabille (Teuthoidea: Ommastrephidae) at Macquarie Island. *Papers and Procedings of the Royal Society of Tasmania*, **117**, 161–3.

Otis, T. and Gilly, W.F. (1990). Jet propelled escape in the squid *Loligo opalescens*: concerted control by giant and non-giant motor axon pathways. *Proceedings National Academy Science, U.S.A.*, **87**, 2911–5.

Overath, H. and Boletzky, S. v. (1974). Laboratory observations on spawning and embryonic development of a blue-ringed octopus. *Marine Biology*, **27**, 333–7.

Owen R. (1832). *Memoir on the pearly* Nautilus (Nautilus Pompilius, *Linn.*) *with illustrations of its external form and internal structure*. Council of the Royal College of Surgeons, London.

Owen, R. (1835). Mollusca-Cephalopoda. Nov. Gen.-Rossia (Owen). In *Appendix to the narrative of a second voyage in search of a north-west passage and of a residence in the Arctic regions during the years 1829, 1830, 1831, 1832, 1833, including reports of James Clark Ross*. Webster, London.

Owen, R. (1885). Description of some new and rare Cephalopoda, part II. *Transactions of the Zoological Society, London*, **11**, 131–70.

Packard, A. (1961). Sucker display of *Octopus*. *Nature, London*, **190**, 736–7.

Packard, A. (1969). Jet propulsion and the giant fibre response of *Loligo*. *Nature, London*, **221**, 875–7.

Packard, A. (1972). Cephalopods and fish: The limits of convergence. *Biological Reviews*, **47**, 241–307.

Packard, A. (1982). Morphogenesis of chromatophore patterns in cephalopods: Are morphological and physiological 'units' the same? *Malacologia*, **23**, 193–201.

Packard, A. (1985). Sizes and distribution of chromatophores during post-embryonic development in cephalopods. *Vie et Milieu*, **35**, 285–98.

Packard, A. (1988). The skin of cephalopods (Coleoids): general and species adaptations. In *Mollusca* (ed. K.M. Wilbur), Vol. **11** *Form and function*, (ed. E.R. Trueman and M.R. Clarke), pp. 37–67. Academic Press, San Diego.

Packard, A. (1995). Organization of cephalopod chromatophore systems: A neuromuscular image-generator. In *Cephalopod neurobiology: Neuroscience studies in squid, octopus and cuttlefish* (ed. N.J. Abbott, R. Williamson and L. Maddock), pp. 331–67. Oxford University Press.

Packard, A. and Albergoni, V. (1970). Relative cell growth, nucleic acid content and cell numbers of the brain in *Octopus vulgaris* (Lamarck). *Journal of Experimental Biology*, **52**, 539–52.

Packard, A. and Hochberg, F.G. (1977). Skin patterning in *Octopus* and other genera. *Symposia of the Zoological Society, London*, No. **38**, 191–231.

Packard, A. and Sanders, G.D. (1969). What the octopus shows to the world. *Endeavour*, **28**, 992–9.

Packard, A. and Sanders, G.D. (1971). Body patterns of *Octopus vulgaris* and maturation of the response to disturbance. *Animal Behaviour*, **19**, 780–90.

Packard, A. and Trueman, E.R. (1974). Muscular activity of the mantle of *Sepia* and *Loligo* (Cephalopoda) during respiratory movements and jetting, and its physiological interpretation. *Journal of Experimental Biology*, **61**, 411–9.

Packard, A. and Würtz, M. (1994). An octopus, *Ocythoe*, with a swimbladder and triple jets. *Philosophical Transactions of the Royal Society, London*, **B344**, 261–75.

Packard, A., Bone, Q., and Highnette, M. (1980). Breathing and swimming movements in a captive *Nautilus*. *Journal of Experimental Biology*, **60**, 313–27.

Packard, A., Karlsen, H.E., and Sand, O. (1990). Low frequency hearing in cephalopods. *Journal of Comparative Physiology*, **A166**, 501–5.

Palacio, F.J. (1978). *Vosseledone charrua*: a new Patagonian cephalopod (Octopodidae) with notes on related genera. *Bulletin of Marine Science*, **28**, 282–96.

Palumbo, A., Di Cosmo, A., Poli, A., Di Cristo, C., and d'Ischia, M. (1999). A calcium/calmodulin-dependent nitiric oxide synthase, NMDA R2/3 receptor subunits, and glutamate in the CNS of the cuttlefish *Sepia officinalis*: localization in specific neural pathways controlling the inking system. *Journal of Neurochemistry*, **73**, 1254–63.

Parker, G.H. (1921). The power of adhesion in the suckers of *Octopus bimaculatus* Verrill. *Journal of Experimental Zoology*, **33**, 391–4.

Passarella, K.C. and Hopkins, T.L. (1991). Species composition and food habits of the micronektonic cephalopod assemblage in the eastern Gulf of Mexico. *Bulletin of Marine Science*, **49**, 638–59.

Patterson, J.A. and Silver, S.C. (1983). Afferent and efferent components of *Octopus* retina. *Journal of Comparative Physiology*, **151**, 381–7.

Paulij, W.P., Herman, P.M.J., Hannen, E.J. van, and Denucé, J.M. (1990). The impact of photoperiodicity on hatching of *Loligo vulgaris* and *Loligo forbesi*. *Journal of the Marine Biological Association, U.K.*, **70**, 597–610.

Paulij, W.P., Herman, P.M.J., Roozen, M.E.F., and Denucé, J.M. (1991). The influence of photoperiodicity on hatching of *Sepia officinalis*. *Journal of the Marine Biological Association, U.K.*, **71**, 665–78.

Pearcy, W.G. and Beal, A. (1973). Deep-sea cirromorphs (Cephalopoda) photographed in the Arctic Ocean. *Deep-Sea Research*, **20**, 107–8.

Pecl, G.T. and Moltschaniwskyj, N.A. (1997). Changes to muscle structure associated with somatic growth in *Idiosepius pygmaeus*, a small tropical cephalopod. *Journal of Zoology, London*, **242**, 751–64.

Perrelet, A. and Mauro, A. (1972). Ultrastructure of the nerves associated with the epistellar body of the octopod *Eledone moschata* and the parolfactory vesicles of the squid *Todarodes sagittatus*. *Brain*, **37**, 161–71.

Pereyra, W.T. (1965). New records and observations in the flapjack devilfish, *Opisthoteuthis californiana* Berry. *Pacific Science*, **19**, 427–41.

Peréz-Gándaras, G. and Guerra, A. (1978). Nueva cita de *Architeuthis* (Cephalopoda: Teuthoidea): Descripción y alimentación. *Investigación Pesquera*, **42**, 401–14.

Person, P. (1969). Cartilaginous dermal scales in cephalopods. *Science*, **164**, 1404–5.

Pfeffer, G. (1912). Die Cephalopoden der Plankton-Expedition. Zugleich eine monographische Ubersicht der oegopsiden Cephalpoden. *Ergebenisse der Plankton-Expedition Humboldt-Stiftung*, **2**, pp. 1–815. (1993) *The Cephalopoda of the plankton expedieion* (trans. I. Kohli), Smithsonian Institution Libraaries and National Science Foundation, Washington, D.C.)

Pfefferkorn, A. (1915). Das Nervensystem der Octopoden. *Zeitschrift Wissenschaftliche Zoologie*, **114**, 425–531.

Piatkowski, U. and Hagen, W. (1994). Distribution and lipid composition of early life stages of the cranchiid squid *Galiteuthis glacialis* (Chun) in the Weddell Sea, Antarctica. *Antarctic Science*, **6**, 235–9.

Piatkowski U. and Pütz, K. (1994). Squid diet of emperor penguins (*Aptenodytes forsteri*) in the eastrern Weddell Sea, Antarctica during late summer. *Antarctic Science*, **6**, 241–7.

Pickford, G.E. (1936). A new order of dibranchiate cephalopods. *Anatomical Record*, **67** (suppl. 1), 77–8.

Pickford, G.E. (1938). The Vampyromorha – a new order of dibranchiate cephalopods. *Report of the British Association for the Advancement of Science*, p. 427.

Pickford, G.E. (1940). The Vampyromorpha, living-fossil Cephalopoda. *Transactions of the New York Academy of Sciences*, (II), 2, 169–81.

Pickford, G.E. (1946). *Vampyroteuthis infernalis* Chun, an archaic dibranchiate cephalopod. I. Natural history and distribution. *DANA-Report*, No. **29**, 1–40.

Pickford, G.E. (1949a). *Vampyroteuthis infernalis* Chun, an archaic dibranchiate cephalopod. II. External anatomy. *Dana-Report*, No. **32**, 1–132.

Pickford, G.E. (1949b). The distribution of the eggs of *Vampyroteuthis infernalis* Chun. *Sears Foundation, Journal of Marine Research*, **8**, 73–83.

Pickford, G.E. (1952). The Vampyromorpha of the Discovery Expeditions. *Discovery Reports*, **26**, 197–210.

Pickford, G.E. (1957). Vampyromorpha. *Galathea Report*, **1**, 243–53.

Plän, T. (1987). Functional neuroanatomy of sensory-motor lobes of the brain of *Octopus vulgaris*. Dissertation, University of Regensburg, Regensburg, Germany.

Portmann, A. (1952). Les bras dorsaux de *Tremoctopous violaceus* Delle Chiaje. *Revue suisse de zoologie*, **59**, 288–93.

Posselt, H.J. (1891). *Todarodes sagittatus* (Lmk.) Stp. En anatomisk Studie. Med Bemaerkninger om Slaegtskabsforholdet mellem Ommatostrephfamiliens Genera. *Videnskabelige Meddelelser fra den Dansk naturhistorisk Forening i Kjøbenhavn*, (1890), pp. 301–59. (trans. J. Knudsen and M. Roeleveld (1991). Henrik J. Posselt: *Todarodes sagittatus* (Lmk.) Stp. An anatomical study. With remarks on the relationship between the genera of the ommatostrephid family. *Steenstrupia*, **17**, 161–96.)

Preuss, T and Budelmann, B.U. (1995a). Proprioceptive hair cells on the neck of the squid *Lolliguncula brevis*: A sense organ in cephalopods for the control of head-to-body position. *Philosophical Transactions of the Royal Society, London*, **B349**, 153–78.

Preuss, T and Budelmann, B.U. (1995b). A dorsal light reflex in squid. *Journal of Experimental Biology*, **198**, 1157–9.

Preuss, T., Lebaric, Z.N., and Gilly, W.F. (1997). Post-hatching development of circular mantle muscles in rthe squid *Loligo opaslescens*. *Biological Bulletin, Woods Hole*, **192**, 375–87.

Pringgenies, D. and Jørgensen, J.M. (1994). Morphology of the luminous organ of the squid *Loligo duvauceli* d'Orbigny, 1839. *Acta Zoologica*, **75**, 305–9.

Prosser, C.L. and Young, J.Z. (1937). Responses of muscles of the squid to repetitive stimulation of giant nerve fibres. *Biological Bulletin, Woods Hole*, **73**, 237–41.

Pumphrey, R.J. (1961). Concerning vision. In *The cell and the organism. Essays presented to Sir James Gray* (ed. J.A. Ramsay and V.B. Wigglesworth). Cambridge University Press.

Pumphrey, R.J., and Young, J.Z. (1938). The rates of conduction of nerve fibres of various diameters in cephalopods. *Journal of Experimental Biology*, **15**, 453–66.

Rae, B.B. (1950). Description of a giant squid stranded near Aberdeen. *Proceedings of the Malacological Society of London*, **28**, 163–7.

Ragonese, S., Jereb, P., and Stefano, L. Di. (1992a). Occurrence of the unihorn octopus *Scaeurgus unicirrhus* in the Sicilian Channel. I. Spatial distribution and abundance. *Rapport Commission Internationale pour la Mer Méditerranée*, **33**, 307.

Ragonese, S., Jereb, P., and Stefano, L. Di. (1992b). Occurrence of the unihorn octopus *Scaeurgus unicirrhus* in the Sicilian Channel. II. Bathymaetric distribution and abundance. *Rapport Commission Internationale pour la Mer Méditerranée*, **33**, 308.

Ragonese, S., Jereb, P., and Stefano, L. Di. (1992c). On the association of the unihorn octopus *Scaeurgus unicirrhus* with white corals (Anthozoa-Madreporaria) in the Sicilian Channel. *Rapport Commission Internationale pour la Mer Méditerranée*, **33**, 306.

Ramon y Cajal *see* Cajal

Rancurel, P. (1970). Les contenus stomacaux d'*Alepisaurus ferox* dans le sud-oest Pacifique (Céphalopodes). *Cahiers O.R.S.T.O.M. séries Océanographique*, **8**, 3–87.

Rancurel, P. (1971). *Mastigoteuthis grimaldi* (Joubin, 1895). Chiroteuthidae peu connu de l'Atlantique tropical (Cephalopoda-Oegopsida). *Cahiers O.R.S.T.O.M. séries Océanographique*, **9**, 125–45.

Rasero, M., González, A.F., and Guerra, A. (1995). Spawning pattern and fecundity of the ommastrephid squid *Todaropsis eblanae* in northeastern Atlantic waters. *International Council for the Exploration of the Sea, CM, Shellfish Committee*, K. **9**, 1–14.

Raven, C.P. (1958). *Morphogenesis: The analysis of molluscan development.* Pergamon Press, Oxford.

Rees, W.J. (1949). Note of the hooked squid, *Onychoteuthis banksi. Proceedings of the Malacological Society, London*, **28**, 43–5.

Rees, W.J. (1950). The distribution of *Octopus vulgaris* Lamarck in British waters. *Journal of the Marine Biological Association, U.K.*, **29**, 361–78.

Rees, W.J. and Maul, G.E. (1956). The Cephalopoda of Madeira, records and distribution. *Bulletin of the British Museum (Natural History) Zoology*, 3, 259–81.

Rehder, H.A. (1945). A new genus and species of squids from the Philippines. *Proceedings of the Biological Society, Washington*, **58**, 21–6.

Reid, A.L. (1991). Taxonomic review of the Australian Rossiinae (Cephalopoda: Sepiolidae), with a description of a new species, *Neorossia leptodons*, and redescription of *N. caroli* (Joubin, 1902). *Bulletin of Marine Science*, **49**, 748–831.

Reinhardt, J.T. and Prosch, V. (1846). *Om Sciadephorus Mülleri (Eschr.).* Bianco Lunos Bogtrykkeri, Kjöbenhavn.

Relini, L.O. (1995). Notes on midwater collections of *Heteroteuthis dispar* (Cephalopoda, Sepiolidae). *Bulletin de l'Institut océanographique, Monaco*, No. spécial, **16**, 63–72.

Relini, L.O. and Bertuletti, M. (1989). Sepiolinae (Mollusca, Cephalopoda) from the Ligurian Sea. *Vie et Milieu*, **39**, 183–90.

Relini, L.O., and Massi, D. (1988). Feeding of *Sepietta oweniana* (d'Orbigny 1839) along the slope of the Ligurian Sea: A preliminary note. *Rapports de la Commission internationale pour l'exploration scientifique de la Mer Méditeranée*, **31**, 255.

Relini, L.O. and Massi, D. (1991). The butterfly squid *Stoloteuthis leucoptera* in the Mediterranean. *Journal of the Marine Biological Association, U.K.*, 71, 47–51.

Richard, A. (1969). The part played by temperature in the rhythm of formation of markings on the shell of cuttlefish (*Sepia officinalis* L.) (Cephalopoda, Mollusca). *Experientia*, **25**, 1051.

Richter, K. (1913). Das nervensystem der Oegopsiden. *Zeitschrift für Wissenschaftliche Zoologie*, **106**, 289–408.

Riddell, D.J. (1985). The Enoploteuthidae (Cephalopoda: Oegopsida) of the New Zealand region. *Fisheries Research Bulletin*, No. **27**, 1–52.

Robertson, J.D. (1994). Cytochalasin D blocks touch learning in *Octopus vulgaris. Proceedings of the Royal Society, London*, **B258**, 61–6.

Robertson, J.D. and Lee, P. (1990). An electron microscopic and behavioral study of tactile learning in *Octopus vulgaris. Progress in Cell Research*, 1, 287–306.

Robertson, J.D., Bonaventura, J., and Kohn, A.P. (1994). Nitric oxide is required for tactile learning in *Octopus vulgaris. Proceedings of the Royal Society, London*, **B256**, 269–73.

Robertson, J.D., Bonaventura, J., and Kohn, A.P. (1995). Nitric oxide synthase inhibition blocks *Octopus* touch learning without producing sensory or motor dysfunction . *Proceedings of the Royal Society, London*, **B261**, 167–72.

Robertson, J.D., Bonaventura, J., Kohm, A., and Hiscat, M. (1996). Nitric oxide is necessary for visual learning in *Octopus vulgaris. Proceedings of the Royal Society, London*, **B263**, 1739–43.

Robison, B.H. (1989). Depth of occurrence and partial chemical composition of a giant squid, *Architeuthis*, off Southern California. *The Veliger*, **32**, 39–42.

Robison, B.H. and Young, R.E. (1981). Bioluminescence in pelagic octopods. *Pacific Science*, **35**, 39–44.

Robson, G.C. (1929, 1932). *A monograph of the Recent Cephalopoda, Part I, Octopodinae and Part II, Octopoda.* The Trustees of the British Museum, London.

Robson, G.C. (1933). On *Architeuhthis clarkei*, a new species of giant squid, with observations on the genus. *Proceedings of the Zoological Society, London*, 1933, 681–97.

Rodaniche, A.F. (1984). Iteroparity in the lesser Pacific striped octopus *Octopus chierchiae. Bulletin Marine Science*, **35**, 99–104.

Rodhouse, P.G. (1988). Distribution of the neotetuhid squid *Alluroteuthis antarcticus* Odhner in the Atlantic sector of the Southern Ocean. *Malacologia*, **29**, 267–74.

Rodhouse, P.G. and Clarke, M.R. (1985). Growth and distribution of young *Mesonychoteuthis hamiltoni* Robson (Mollusca: Cephalopoda): An Antarctic squid. *Vie et Milieu*, **35**, 222–30.

Rodhouse, P.G. and Clarke, M.R. (1986). Distribution of the early-life phase of the Antarctic squid *Galiteuthis glacialis* in relation to the hydrology of the Southern Ocean in the sector 15° to 30°E. *Marine Biology*, **91**, 353–7.

Rodhouse, P.G. and Yeatman, J. (1990). Redescription of *Martialia hyadesi* Rochebrune and Mabille, 1889 (Mollusca: Cephalopoda) from the Southern Ocean. *Bulletin British Museum Natural History (Zoology)*, **56**, 135–143.

Rodhouse, P.G., Swinfen, R.C., and Murray, A.W.A. (1988). Life cycle, demography and reproductive investment in the myopsid squid *Alloteuthis subulata. Marine Ecology-Progress Series*, **45**, 245–53.

Rodhouse, P.G., Symon, C., and Hatfield, E.M.C. (1992). Early life cycle of cephalopods in relation to the major oceanographic features of the southwest Atlantic. *Marine Ecology-Progress Series*, **89**, 183–95.

Roeleveld, M.A. (1972). A review of the Sepiidae (Cephalopoda) of Southern Africa. *Annals of the South African Museum*, **59**, 193–313.

Roeleveld, M.A. (1988). Generic interrelationships within the Ommastrephidae (Cephalopoda). In *The Mollusca* (ed. K.M. Wilbur), Vol. 12 *Paleontology and neontology of cephalopods* (ed. M.R. Clarke and E.R. Trueman), pp. 277–91. Academic Press, San Diego.

Roeleveld, M.A.C. and Lipinski, M.R. (1991). The giant squid *Architeuthis* in southern African waters. *Journal of Zoology, London*, **224**, 431–77.

Roeleveld, M.A. and Pheiffer, F. (1987). The diamond squid: a large squid that regularly attracts attention when stranded on South African beaches. *Sagittarius*, **2**, 20–2.

Romanini, M.G. (1952). Osservazioni sulla ialuronidasi delle ghiandole salivari anteiori e posteriori degli Octopodi. *Pubblicazioni della Stazione Zoologica di Napoli*, **23**, 251–70.

Roper, C.F.E. (1963). Observations on bioluminescence in *Ommastrephes pteropus* (Steenstrup, 1855), with notes on its occurrence in the family Ommastrephidae (Mollusca: Cephalopoda). *Bulletin of Marine Science of the Gulf and Caribbean*, **13**, 343–53.

Roper, C.F.E. (1966). A study of the genus *Enoploteuthis* (Cephalopoda: Oegopsida) in the Atlantic Ocean with a redescription of the type species *E. leptura* (Leach, 1817). *DaANA-Report* No. **66**, 1–46.

Roper, C.F.E. (1968). Preliminary descriptions of two new species of the bathypelagic squid *Bathyteuthis* (Cephalopoda: Oegopsida). *Proceedings of the Biological Society, Washington*, **81**, 161–72.

Roper, C.F.E. (1969). Systematics of the worldwide bathypelagic squid *Bathyteuthis* (Cephalopoda: Oegopsida). *United States National Museum Bulletin*, No. **291**, 1–210.

Roper, C.F.E. (1972). Ecology and vertical distribution of Mediterranean pelagic cephalopods. In: *Mediterranean biological studies—final report*, Vol. I (ed. R.H. Gibbs, Jr., *et al.*), pp. 282–386. Smithsonian Institution, Washington, D.C.

Roper, C.F.E. and Boss, K.J. (1982). The giant squid. *Scientific American*, **246**, 96–105.

Roper, C.F.E. and Brundage, W.L. (1972). Cirrate octopods with associated deep-sea organisms: new biological data based on deep benthic photographs (Cephalopoda). *Smithsonian Contributions to Zoology*, No. **121**, 1–146.

Roper, C.F.E. and Clarke, M.R. (1992). Family Joubiniteuthidae Naef, 1922. *Smithsonian Contributions to Zoology*, No. **513**, 179.

Roper, C.F.E. and Hochberg, F.G. (1987). Cephalopods of Lizard island, Great Barrier Reef, Australia. *Occasional Papers from the Museum of Victoria*, **3**, 15–20.

Roper, C.F.E. and Hochberg, F.G. (1988). Behavior and systematics of cepehalopods from Lizard Island, Australia, based on color and body patterns. *Malacologia*, **29**, 153–93.

Roper, C.F.E. and Lu, C.C. (1979). Rhynchoteuthion larvae of ommastrephid squids of the western North Atlantic, with the first description of larvae and juveniles of *Illex illecebrosus*. *Proceedings of the Biological Society, Washington*, **91**, 1039–59.

Roper, C.F.E. and Lu, C.C. (1990). Comparative morphology and function of dermal structures in oceanic squids (Cephalopoda). *Smithsonian Contributions to Zoology*, No. **493**, 1–40.

Roper, C.F.E. and Sweeney, M.J. (1976). The pelagic octopod *Ocythoe tuberculata* Rafinesque, 1814. *Bulletin of the American Malacological Union for 1975*, 21–28.

Roper, C.F.E. and Sweeney, M.J. (1992). Family Brachioteuthidae Pfeffer, 1908. *Smithsonian Contributions to Zoology*, No. **513**, 157–9.

Roper, C.F.E. and Vecchione, M. (1993). A geographic and taxonomic review of *Taningia danae* Joubin, 1931 (Cephalopoda: Octopoteuthidae), with new records and observations on bioluminescence. In *Recent advances in cephalopod fisheries biology* (ed. T. Okutani, R. O'Dor, and T. Kubodera), pp. 441–56. Tokai University Press, Tokyo.

Roper, C.F.E. and Vecchione, M. (1997). *In situ* observations test hypotheses of functional morphology in *Mastigoteuthis* (Cephalopoda, Oegopsida). *Vie et Milieu*, **47**, 87–93.

Roper, C.F.E. and Voss, G.L. (1983). Guidelines for taxonomic descriptions of cephalopod species. *Memoirs of the National Museum of Victoria*, **44**, 49–63.

Roper, C.F.E. and Young, R.E. (1967). A review of the Valbyteuthidae and an evaluation of its relationship with the Chiroteuthidae (Cephalopoda: Oegopsida). *Proceedings of the United States National Museum*, **123** (No. 3612), 1–9.

Roper, C.F.E. and Young, R.E. (1968). The family Promachoteuthidae (Cephalopoda: Oegopsida). I. A re-evaluation of its systematic position based on new material from Antarctic and adjacent waters. *Antarctic Research Series*, **II**, 203–14.

Roper, C.F.E. and Young, R.E. (1972). First records of juvenile giant squid, *Architeuthis* (Cephalpoda: Oegopsida). *Proceedings of the Biological Society, Washington*, **85**, 205–22.

Roper, C.F.E. and Young, R.E. (1975). Vertical distribution of pelagic cephalopods. *Smithsonian Contributions to Zoology*, No. **209**, 1–51.

Roper, C.F.E., Young, R.E., and Voss, G.L. (1969). An illustrated key to the families of the order Teuthoidea (Cephlaopoda). *Smithsonian Contributions to Zoology*, No. **13**, 1–32.

Roper, C.F.E., Sweeney, M.J., and Nauen, C.E. (1984). *Cephalopods of the World. An annotated and illustrated catalogue of species of interest to fisheries. FAO Fisheries synopsis* No. **125**, Vol. 3, 1–277. Food and Agricultural Organization of the United Nations, Rome.

Roper, C.F.E., Sweeney, M.J., and Clarke, M.R. (1985). *FAO Species identification sheets for fishery purposes. Southern ocean*, Vol. 1 (ed. W. Fischer and J.C. Hureau), pp. 117–205. Food and Agricultural Organization of the United Nations, Rome.

Ruby, E.G. (1996). Lessons from a cooperative, bacterial-animal association: the *Vibrio fischeri—Euprymna scolopes* light organ symbiosis. *Annual Review of Microbiology*, **50**, 591–624.

Ruby, E.G. and McFall-Ngai, M.J. (1992). A squid that glows in the light: development of an animal-bacterium mutualism. *Journal of Bacteriology*, **174**, 4865–70.

Russell, E.S. (1965). Marine toxins and venomous and poisonous marine animals. *Advances in Marine Biology*, **3**, 255–384.

Sabirov, R.M., Arkhipkin, A.I., Tsygankov, V. Yu., and Shchetinnikov, A.S. (1987). Egg-laying and embryonal development of diamond-shaped squid, *Thysanoteuthis rhombus* (Oegopsida, Thysanoteuthidae). *Zoologicheskii Zhurnal*, **66**, 1155–63.

Saibil, H.R., Langmack, K.A., Venien-Bryan, C., and Wilkinson, J.R. (1995). Squid rhodopsin. In *Cephalopod neurobiology: neuroscience studies in squid, octopus and cuttlefish* (ed N.J. Abbott, R. Williamson and L. Maddock), pp. 479–89. Oxford University Press.

Saidel, W.M. (1982). Connections of the *Octopus* optic lobe: an HRP study. *Journal of Comparative Neurology*, **206**, 346–58.

Saidel, W.M., Lettvin, J.Y., and MacNichol, E.F. (1983). Processing of polarized light by squid photoreceptors. *Nature, London*, **304**, 534–6.

Sánchez, P. (1985). Sobre la presencia de *Teuthowenia megalops* (Prosch, 1849) (Cephalopoda, Cranchiidae) en el Mediterraneo. *Investigación Pesquera*, **49**, 315–8.

Sánchez, P. and Alvarez, J.A. (1988). *Scaeurgus unicirrhus* (Orbigny, 1840) (Cephalopoda, Octopodidae): First record from the south-east Atlantic. *South African Journal of Marine Science*, **7**, 69–74.

Sánchez, P. and Guerra, A. (1989). A new species of cirrate octopod *Opisthoteuthis vossi* from the southeast Atlantic (Cephalopoda: Octopoda). *Bulletin of Marine Science*, **44**, 1159–65.

Sanders, F.K. and Young J.Z. (1940). Learning and other functions of the higher nervous centres of *Sepia*. *Journal of Neurophysiology*, **3**, 501–26.

Sanders, G.D. (1975). The cephalopods. In *Invertebrate learning* Vol. 3. *Cephalopods and echinoderms* (ed. W.C. Corning, J.A. Dyal, and A.O.D. Willows), pp. 1–101. Plenum Press, New York.

Sanders, G.D. and Young, J.Z. (1974). Reappearance of specific colour patterns after nerve regeneration in *Octopus*. *Proceedings of the Royal Society, London*, **B186**, 1–11.

Sangiovanni, G. (1819). Descrizione d'un particolare sistema di organi cromophoro espansivo-dermoideo e dei fenomei ch'esso produce, scoperto nei molluschi cefalopodi. *Giorn. Enciclopedico di Napoli*, **13**, No. 9.

Sartor, P. and Belcari, P. (1995). Sepiolidae (Mollusca, Cephalopoda) of the Northern Tyrrhenian Sea. *Bulletin de l'Institut océanographique, Monaco*, No. spécial, **16**, 15–17.

Sasaki, M. (1914). Observations on Hotaru-ika *Watasenia scintillans*. *Journal of the College of Agriculture, Tohoku Imperial University*, **6**, 75–105.

Sasaki, M. (1915). On three interesting new oegopsids from the Bay of Sagami. *Journal of the College of Agriculture Tohoku Imperial, University of Sapporo, Japan*, **6**, 131–50.

Sasaki, M. (1920). Report on the cephalopods collected during 1906 by the United States Bureau of Fisheries steamer *Albatross* in the northwestern Pacific. *Proceedings of the United States National Museum*, **57**, 163–203.

Sasaki, M. (1923). On an adhering habit of a pygmy cuttlefish, *Idiosepius pygmaeus* Steenstrup. *Annotationes Zoologicae Japonenses*, **10**, 209–13.

Sasaki, M. (1925). The anatomy of the 'surume' squid, *Ommastrephes sloani pacificus* (*Todarodes pacificus*) Steenstrup. *Dobutsugaku Zasshi (Zoological Magazine), Tokyo*, **37**, 240–61.

Sasaki, M. (1929). A monograph of the dibranchiate cephalopods of the Japanese and adjacent waters. *Journal of the Faculty of Fisheries and Animal Husbandry, Hiroshima University*, **5**, 277–330.

Sato, T. (1985). Reports on the squid tagging in New Zealand waters. *Far Seas Fisheries Research Laboratory*, S Series, **14**, 1–73.

Sauer, W.H.H. and Smale, M.J. (1993). Spawning behaviour of *Loligo vulgaris reynaudii* in shallow coastal waters of the South-Eastern Cape, South

Africa. In *Recent advances in cephalopod fisheries biology* (ed. T. Okutani, R.K. O'Dor, and T. Kubodera), pp. 489–98. Tokai University Press, Tokyo.

Sauer, W.H.H., Smale, M.J., and Lipinski, M.R. (1992). The location of spawning grounds, spawning and schooling behaviour of the squid *Loligo vulgaris reynaudii* (Cephalopoda: Myopsida) off the Eastern Cape coast, South Africa. *Marine Biology*, **114**, 97–107.

Sauer, W.H.H., Roberts, M.J., Lipinski, M.R., Smale, M.J., Hanlon, R.T., Webber, D.M. *et al.* (1997). Choreography of the squid's 'nuptial dance'. *Biological Bulletin, Woods Hole*, **192**, 203–7.

Saunders, W.B. (1983). Natural rates of growth and longevity of *Nautilus belauensis*. *Paleobiology*, **9**, 280–8.

Saunders, W.B. (1984). *Nautilus* growth and longevity: evidence from marked and recaptured animals. *Science*, **224**, 990–2.

Saunders, W.B. (1987). The specie of *Nautilus*. In Nautilus: *The biology and paleobiology of a living fossil* (ed. W.B. Saunders and N.H. Landman), pp. 35–52. Plenum, New York.

Saunders, W.B. and Landman, N.H. (1987) (ed.). Nautilus: *The biology and paleobiology of a living fossil*. Plenum, New York.

Saunders, W.B. and Ward, P.D. (1987). Ecology, distribution, and population characteristics of *Nautilus*. In Nautilus: *The biology and paleobiology of a living fossil* (ed. W.B. Saunders and N.H. Landman), pp. 137–62. Plenum, New York.

Savage, I.V.E. and Howden, M.E.H. (1977). Hapalotoxin, a second lethal toxin from the octopus *Hapalochlaena maculosa*. *Toxicon*, **15**, 463–6.

Schkaff, B. (1914). Zur Kenntnis des Nervensystems der Myopsiden. *Zeitschrift für wissenschaftliche Zoologie*, **109**, 591–630.

Schmidt, J. (1922). Live specimens of *Spirula*. *Nature, London*, **110**, 788–90.

Schöne, H. and Budelmann, B.U. (1970). Function of the gravity receptor of *Octopus vulgaris*. *Nature, London*, **226**, 864–5.

Schraermeyer, U. (1994). Fine structure of melanogenesis in the ink sac of *Sepia officinalis*. *Pigment Cell Research*, **7**, 52–60.

Seapy, R.R. and Young, R.E. (1986). Concealment in epipelagic pterotracheid heteropods (Gastropoda) and cranchiid squids (Cephalopoda). *Journal of Zoology, London* (A), **210**, 137–47.

Segawa, S., Yang, W.T., Marthy, H.-J., and Hanlon, R.T. (1988). Illustrated embryonic stages of the eastern Atlantic squid *Loligo forbesi*. *The Veliger*, **30**, 230–43.

Seidou, M., Sugahara, M., Uchiyama, M., Hiraki K., Hamanaka, T., Michinomae, M. *et al.* (1990). On the three visual pigments in the retina of the firefly squid, *Watasenia scintillans*. *Journal of Comparative Physiology*, **A166**, 769–73.

Shashar, N., and Cronin, T.W. (1996). Polarization contrast vision in *Octopus*. *Journal of Experimental Biology*, **199**, 999–1004.

Shashar, N., Rutledge, P.S., and Cronin, T.W. (1996). Polarization vision in cuttlefish – a concealed communication channel? *Journal of Experimental Biology*, **199**, 2077–84.

Shears, J. (1988). The use of a sand-coat in relation to feeding and diel activity in the sepiolid squid *Euprymna scolopes*. *Malacologia*, **29**, 121–33.

Sheumack, D.D., Howden, M.E.H., Spence, I., and Quinn, R.J. (1978). Maculotoxin: a neurotoxin from the venom glands of the octopus *Hapalochlaena maculose*, identified as tetrodotoxin. *Science*, **199**, 188–9.

Sheumack, D.D., Howden, M.E.H., and Spence, I. (1984). Occurrence of a tetrodotoxin-like compound in the eggs of the venomous blue-ringed octopus (*Hapalochlaena maculosa*). *Toxicon*, **22**, 811–2.

Shimek, R. (1984). Escape behavior of *Rossia pacifica* Berry, 1911. *American Malacological Bulletin*, **2**, 91–2.

Shimomura, O., Inoue, S., Johnson, S.F.H., and Haneda, Y. (1980). Widespread occurrence of coelenterazine in marine bioluminescence. *Comparative Biochemistry and Physiology*, **65B**, 435–7.

Singley, C.T. (1980). Ultrastructure of the bacterial photophore of the sepiolid squid *Euprymna scolopes*. *Abstracts of the 46th Annual Meeting of the American Malacological Union*, p. 20. Louisville, Kentucky.

Singley, C.T. (1982). Histochemistry and fine structure of the ectodermal epithelium of the sepiolid squid *Euprymna scolopes*. *Malacologia*, **23**, 177–92.

Singley, C.T. (1983). *Euprymna scolopes*. In *Cephalopod life cycles* Vol. I. *Systematic accounts* (ed. P.R. Boyle), pp. 69–74. Academic Press, London.

Smale, M.J. and Buchan, P.R. (1981). The biology of *Octopus vulgaris* off the east coast of South Africa. *Marine Biology*, **65**, 1–12.

Smith, A.M. (1991). Negative pressure generated by *Octopus* suckers: A study of tensile strength of water in nature. *Journal of Experimental Biology*, **157**, 257–71.

Sroczynski, S. and Muntz, W.R.A. (1985). Image structure in *Eledone cirrhosa*, an octopus. *Zoologische Jahrbücher Physiologie*, **89**, 157–68.

Steenstrup, J. (1857). Oplysninger om Atlanterhavets colossale Blaeksprutter. *Forhandlinger ved de Skandinaviske Naturforskeres*, **7**, 182–5 (see A. Volsøe, J. Knudsen and W.J. Rees (trans.) (1962)).

Steenstrup, J. (1880). Orientering i de Ommatostrephagtige Blaekspruttters indbyrdes Forhold. *Oversigt over det Kongelige Danske Videnskabernes Selskabs Forhandlinger, 1880–1881*, 73–110 (see A. Volsøe, J. Knudsen and W.J. Rees (trans.), 1962).

Steenstrup, J. (1881). *Sepiadarium* og *Idiosepius* to nye Slaegter af Sepiernes Familie. Med Bemaerkninger om de to beslaegtede Former *Sepioloidea* D'Orb. og *Spirula* Lmk., (6), **1**, 213–42. *Kongelige Danske Videnskabernes Selskabs Skrifter*, (6), 1, 213–42 (see A. Volsøe, J. Knudsen and W.J. Rees (trans.), 1962).

Steenstrup, J. (1882). Prof. A.E. Verrill's to nye Cephalopodslaegter: *Sthenoteuhis* og *Lesoteuthis*. Bemaerkninger og Berigtigeleser. *Oversigt over det Kongelige Danske Videnskabernes Selskabs Forhandlinger, 1881–1882*, 1–27 (see A. Volsøe, J. Knudsen and W.J. Rees (trans.), 1962).

Stenzel, H.B. (1964). Living *Nautilus*. In *Treatise on invertebrate paleontology, Part K, Mollusca 3, Cephalopoda – general features. Endoceratoidea – Actinoceratoidea – Nautiloidea – Bactritoidea* (ed. R.C. Moore), pp. K59–K93. The Geological Society of America and the University of Kansas, Kansas.

Stephen, A.C. (1962). The species of *Architeuthis* inhabiting the North Atlantic. *Proceedings of the Royal Society, Edinburgh*, **68**, 147–61.

Stephen, S.J. (1985). The distribution of larvae of the genus *Octopoteuthis* Rüppell, 1844 (Cephalpoda, Teuthoidea). *Vie et Milieu*, **35**, 175–9.

Stephen, S.J. (1992). Family Thysanoteuthidae Keferstein, 1866. *Smithsonian Contributions to Zoology*, No. **513**, 121.

Stephen, S.J. and Jefferts, K. (1992). Family Octopoteuthida Berry, 1912. *Smithsonian Contributions to Zoology*, No. **513**, 165–6.

Stephens, P.R. (1971). Appendix II. Histological methods. In *The anatomy of the nervous system of* Octopus vulgaris (by J.Z. Young), pp. 646–9. Clarendon Press, Oxford.

Stephens, P.R. and Young, J.Z. (1976). The statocyst of *Vampyroteuthis infernalis* (Mollusca: Cephalopoda). *Journal of Zoology, London*, **180**, 565–88.

Stephens, P.R. and Young, J.Z. (1982). The statocyst of the squid *Loligo*. *Journal of Zoology, London*, **197**, 241–66.

Stevens, G.R. (1988). Giant ammonites: a review. In *Cephalopods- present and past* (ed. J. Wiedmann and J. Kullmann), pp. 140–66. Schweizerbart'sche Verlagsbuchhandlung, Stuttgart.

Stranks, T.N. and Lu, C.C. (1991). Post-embryonic development of the blue-ringed octopus *Hapalochlaena maculosa*. In *The marine flora and fauna of Albany, Western Australia* Vol. **2** (ed. F.E. Wells, D.I. Walker, H. Kirkman, and R. Lethbridge), pp. 713–22. Western Australian Museum, Perth.

Summers, W.C. (1983). *Loligo pealei*. In *Cephalopod life cycles Vol. I. Species accounts* (ed. P.R. Boyle), pp. 115–42. Academic Press, London.

Summers, W.C. (1985). Ecological implications of life stage timing determined from the cultivation of *Rossia pacifica* (Mollusca: Cephalopoda). *Vie et Milieu*, **35**, 249–54.

Sundermann, G. (1983). The fine structure of epidermal lines on arms and head of post-embryonic *Sepia officinalis* and *Loligo vulgaris* (Mollusca, Cephalopoda). *Cell and Tissue Research*, **232**, 669–77.

Sundermann, G. (1990). Development and hatching state of ectodermal vesicle-organs in the head of *Sepia officinalis, Lolilgo vulgaris* and *Lolilgo forbesi* (Cephalopoda, Decabrachia). *Zoomorphology*, **109**, 343–352.

Sundermann-Meister, G. (1978). Ein neuer Typ von Cilienzellen in der Haut von spätembryonalen und juvenilen *Loligo vulgaris* (Mollusca, Cephalopoda). *Zoolologische Jahrbücher Anatomie*, **99**, 493–99.

Sutherland, N.S. (1957). Visual discrimination of orientation and shape by the octopus. *Nature, London*, **179**, 11–3.

Sutherland, N.S. (1958). Visual discrimination of shape by *Octopus*. Squares and triangles. *Quarterly Journal of Experimental Psychology*, **10**, 40–7.

Sutherland, N.S. (1969). Shape discrimination in rat, octopus and goldfish: a comparative study. *Journal of Comparative and Physiological Psychology*, **67**, 160–76.

Sutherland, S.K. (1983). *Australian animal toxins: the creatures, their toxins and care of the poisoned patient.* Oxford University Press, Melbourne.

Suzuki, S., Misaki ,H., and Okutani ,T. (1979). Studies on early life history of decapodan Mollusca – VIII. A supplementary note on floating egg mass of *Thysanoteuthis rhombus* Troschel in Japan – the first underwater photography. *Venus, Japanese Journal of Malacology*, **38**, 153–5.

Sweeney, M.J. and Roper, C.F.E. (1998). Classification, type localities, and type repositories of Recent Cephalopoda. *Smithsonian Contributions to Zoology*, No. **586**, 561–99.

Sweeney, M.J. and Vecchione, M. (1998). Generic and specific names introduced in the squid family Loliginidae (Cephalopoda: Myopsida). *Smithsonian Contributions to Zoology*, No. **586**, 223–37.

Sweeney, M.J., Roper, C.F.E., Mangold, K.M., Clarke, M.R., and Boletzky, S. v. (ed.) (1992). 'Larval' and juvenile cephalopods: a manual for their identification. *Smithsonian Contributions to Zoology*, No. **513**, 1–282.

Takahama, H., Kinoshita, T., Sato, M., and Sasaki, F. (1991). Fine structure of the spermatophores and their ejaculated forms, sperm reservoirs, of the Japanese common squid, *Todarodes pacificus. Journal of Morphology*, **207**, 241–51.

Taki, I. (1941). On keeping octopods in an aquarium for physiological experiments, with remarks on some operative techniques. *Venus, Japanese Journal of Malacology*, **10**, 140–56.

Tang, I. and Khoo, H.-W. (1974). The food and feeding habits of the cuttlefish *Sepiella inermis* (Férussac and d'Orbigny). *The Veliger*, **16**, 405–10.

Teichert, C. (1988). Main features of cephalopod evolution. In *The Mollusca* (ed. K.M. Wilbur), Vol. 12. *Paleontolgy and neontology of cephalopods* (ed. M.R. Clarke and E.R. Trueman), pp. 11–79. Academic Press, San Diego.

Teichert, C. and Matsumoto, T. (1987). The ancestry of the genus *Nautilus*. In Nautilus: *The biology and paleobiology of a living fossil* (ed. W.B. Saunders and N.H. Landman), pp. 25–32. Plenum, New York.

Thiele, J. (1915). Die Cephalopoden. II Teil: Myopsida Octopoda. *Wissenschaftliche Ergebnisse der Deutschen Tiefsee-Expedition 1901–1903*, **18**, 403–552.

Thomas, R.F. (1977). Systematics, distribution, and biology of cephalopods of the genus *Tremoctopus* (Octopoda: Tremoctopodidae). *Bulletin of Marine Science*, **27**, 353–92.

Thompson, D'A.W. (trans.) (1910). *Historia animalium.* In *The works of Aristotle*, volume IV (ed. J.A Smith and W.D. Ross), pp. 486–633. Clarendon Press, Oxford.

Thore, S. (1939). Beiträge zur Kenntnis der vergleichenden Anatomie des zentralen Nervensystems der dibranchiaten Cephalopoden. *Pubblicazioni della Stazione zoologica di Napoli*, **176**, 313–506.

Thore, S. (1942). Ein beitrag zur Kenntnis der Seh-zentren im Gehirn von *Sepia. Acta Universitatis Lundensis*, **38**, 18pp.

Thore, S. (1949). Investigations on the 'DANA' Octopoda Part I. Bolitaenidae, Amphitretidae, Vitreledonellidae, and Alloposidae. *Dana-Report* No. **33**, 1–85.

Thorpe, W.H. (1956). *Learning and instinct in animals.* Methuen, London.

Tillier, A. and Boucher-Rodoni R. (1993). Férrusac and d'Orbigny's '*Histoire naturelle générale et partiulière des Céphalopodes acétabulifères*': Dates of publication of plates and text. *The Nautilus*, **107**, 97–103.

Tinbergen, L. (1939). Zur Fortpflanzungsetholgie von *Sepia officinalis* L. *Archives Néerlandaises de Zoologie*, **3**, 323–64.

Toll, R.B. (1982). The comparative morphology of the gladius in the order Teuthoidea (Mollusca: Cephalopoda) in relation to systematics and phylogeny. Thesis, Doctor of Philosophy, University of Miami.

Toll, R.B. (1983). The lycoteuthid genus *Oregoniateuthis* Voss, 1956, a synonym of *Lycoteuthis* Pfeffer, 1900 (Cephalopoda: Teuthoidea). *Proceedings of the Biological Society of Washington*, **96**, 365–69.

Toll, R.B. and Hess, S.C. (1981). A small, mature male *Architeuthis* (Cephalopoda: Oegopsida) with remarks on maturation in the family. *Proceedings of the Biological Society, Washington*, **94**, 753–60.

Tompsett, D.H. (1939). *Sepia. Liverpool Marine Biology Committee Memoirs on Typical British Marine Plants and Animals*, **32**, 1–184.

Tonosaki, A. (1965). The fine structure of the retinal plexus in *Octopus vulgaris. Zeitschrift für Zellforschung*, **67**, 521–32.

Tranter, D.J. and Augustine, O. (1973). Observations on the life history of the blue-ringed octopus, *Hapalochlaena maculosa. Marine Biology*, **18**, 115–28.

Trethewie, E.R. (1965). Pharmacological effects of the venom of the common octopus *Hapalochlaena maculosa. Toxicon*, **3**, 55–9.

Trueman, E.R. (1980). Swimming by jet propulsion. *Society for Experimental Biology, Seminar Series*, **5**, 93–105.

Trueman, E.R. and Packard, A. (1968). Motor performances of some cephalopods. *Journal of Experimental Biology*, **49**, 495–507.

Tsuchiya, K. and Okutani, T. (1991). Growth stages of *Moroteuthis robusta* (Verrill, 1881) with the re-evaluation of the genus. *Bulletin of Marine Science*, **49**, 137–47.

Tsuji, F.I. (1985). ATP-dependent bioluminescence in the firefly squid, *Watasenia scintillans. Proceedings of the National Academy of Sciences, U.S.A.*, **82**, 4629–32.

Tsuji, F.I. and Leisman, G.B. (1981). K^+/N^+-triggered bioluminescence in the oceanic squid *Symplectoteuthis oualaniensis. Proceedings of the National Academy of Science, U.S.A.*, **78**, 6719–23.

Tunnicliffe, V., McArthur, A.G., and McHugh, D. (1998). A biogeographical perspective of the deep-sea hydrothermal vent fauna. *Advances in Marine Biology*, **34**, 353–442.

Uozumi, Y., Förch, E.C., and Okazaki, T. (1991). Distribution and morphological characters of immature *Martialia hyadesi* (Cephalopoda: Oegopsida) in New Zealand waters. *New Zealand Journal of Marine and Freshwater Research*, **25**, 275–82.

Vacchi, M., Greco, S., and La Mesa, M. (1994). *Kondakovia longimana* Filippova, 1972 (Onychoteuthidae) from Terra Nova Bay, Ross Sea. *Antarctic Science*, **6**, 283.

van Leeuwen, J.L. and Kier, W.M. (1997). Functional design of tentacles in squid: linking sarcomere ultrastructure to gross morphological dynamics. *Philosophical Transactions of the Royal Society, London*, **B352**, 551–71.

Vecchione, M. (1982). Morphology and development of planktonic *Lolliguncula brevis* (Cephalopoda: Myopsida). *Proceedings of the Biological Society of Washington*, **95**, 602–9.

Vecchione, M. (1987). Juvenile ecology. In *Cephalopod Life cycles* Vol. II. *Comparative Reviews* (ed. P.R. Boyle), pp. 61–84. Academic Press, London.

Vecchione, M. (1991). Dissolved oxygen and the distribution of the euryhaline squid *Lolliguncula brevis. Bulletin of Marine Science* (Abstract), **49**, 668–9.

Vecchione, M. and Grant, G.C. (1983). A multivariate analysis of planktonic molluscan distribution in the Middle Atlantic Bight. *Continental Shelf Research*, **1**, 405–44.

Vecchione, M. and Roper, C.F.E. (1986). Occurrence of larval *Illex illecebrosus* and other young cephalopods in the slope water/Gulf Stream interface. *Proceedings of the Biological Society, Washington*, **99**, 703–8.

Vecchione, M. and Roper, C.F.E. (1991). Cephalopods observed from submersibles in the western north Atlantic. *Bulletin of Marine Science*, **49**, 433–45.

Vecchione, M. and Roper, C.F.E. (1992). *Taningia danae*, 1931 (Mollusca, Cephalpoda): proposed precedence over *Taningia persica* (Naef, 1923). *Bulletin of Zoological Nomenclature*, **49**, 261–3.

Vecchione, M. and Young, R.E. (1997) Aspects of the functional morphology of cirrate octopods: Locomotion and feeding. *Vie et Milieu*, **47**, 101–10.

Vecchione, M. and Young, R.E. (1998). The Magnapinnidae, a newly discovered family of oceanic squid (Cephalopoda: Oegopsida). *South African Journal of Marine Science*, **20**, 429–37.

Vecchione, M., Robison, B.H., and Roper, C.F.E. (1992). A tale of two species: Tail morphology in paralarval *Chiroteuthis* (Cephalopoda: Chiroteuthidae). *Proceedings of the Biological .Society, Washington*, **105**, 683–92.

Vecchione, M., Brakioniecki, T.F., Natsukari, Y., and Hanlon, R.T. (1998). A provisional generic classification of the family Loliginidae. *Smithsonian Contributions to Zoology*, No. **586**, 215–22.

Verrill, A.E. (1878). Notice of recent additions to the marine fauna of the eastern coast of North America, No. 2. *American Journal of Science and Arts*, **16**, 371–8.

Verrill, A.E. (1880). The cephalopods of the north-eastern coast of America. Part I. The gigantic squids (*Architeuthis*) and their allies; with observations on similar large species from foreign localities. *Transactions of the Connecticut Academy of Sciences*, **5**, 177–257.

Verrill, A.E. (1881). The cephalopods of the northeastern coast of America, Part II: The smaller cephalopods, including the 'squids' and octopi, with other allied forms. *Transaction of the Connecticut Academy of Arts and Sciences*, **5**, 259–446.

Verrill, A.E. (1882). Report on the cephalopods of the northeastern coast of America. *Annals and Reports of the United States Commissioner of Fish and Fisheries for 1879*: 211–455.

Verrill, A.E. (1883). Supplementary report on the 'Blake' cephalopods. *Bulletin of the Museum of Comparative Zoology, Harvard*, **11**, 105–15.

Verrill, A.E. (1885). Third catalogue of Mollusca recently added to the fauna of the New England coast and the adjacent parts of the Atlantic, consisting mostly of deep-sea species, with notes in others previously recorded. *Transactions of the Connecticut Academy Arts and Sciences*, **6**, 395–452.

Vidal, E.A.G. (1994). Relative growth of paralarvae and juveniles of *Illex argentinus* (Castellanos, 1960) in southern Brazil. *Antarctic Science*, **6**, 275–82.

Villanueva, R. (1992a). Continuous spawning in the cirrate octopods *Opisthoteuthis agassizii* and *O. vossi*: Features of sexual maturation defining a reproductive strategy in cephalopods. *Marine Biology*, **114**, 265–75.

Villanueva, R. (1992b). Deep-sea cephalopods of the north-western Mediterranean: Indications of up-slope ontogenetic migrations in two bathybenthic species. *Journal of Zoology, London*, **227**, 267–76.

Villanueva, R. (1994). Decapod crab zoeae as food for rearing cephalopod paralarvae. *Aquaculture*, **128**, 143–52.

Villanueva, R. (1995). Distribution and abundance of bathyal sepiolids (Mollusca: Cephalopoda) in the northwestern Mediterranean. *Bulletin de l'Institut océanographique, Monaco*, spécial No. **16**, 19–26.

Villanueva, R. and Guerra, A. (1991). Food and prey detection in two deep-sea cephalopods: *Opisthoteuthis agassizii* and *O. vossi* (Octopoda: Cirrata). *Bulletin of Marine Science*, **49**, 288–99.

Villanueva, R. and Sánchez, P. (1993). Cephalopods in the Benguela Current off Namibia: New additions and considerations on genus *Lycoteuthis*. *Journal of Natural History*, **27**, 15–46.

Villanueva, R., Nozals, C., and Boletzky, S. v. (1996). Swimming behaviour and food searching in planktonic *Octopus vulgaris* Cuvier from hatching to settlement. *Journal of Experimental Marine Biology and Ecology*, **208**, 168–84.

Villanueva, R. Segonzac, M., and Guerra, A. (1997). Locomotion modes of deep-sea cirrate octopods (Cephalooda) based on observations from video-recordings on the mid-Atlantic Ridge. *Marine Biology*, **129**, 113–22.

Villegas, G.M. (1969). Electron microscope study of the giant nerve fiber of the giant squid *Dosidicus gigas*. *Journal of Ultrastructural Research*, **26**, 501–14.

Villegas, G.M. and Villegas, R. (1984). Squid axon ultrastructure. *Current Topics in Membranes and Transport*, **22**, 3–37 (ed. P. Baker). Academic Press, New York.

Voight, J.R. (1995). Sexual dimorphism and niche divergence in a mid-water octopod (Cephalopoda: Bolitaenidae). *Biological Bulletin, Woods Hole*, **189**, 113–9.

Voight, J. (2000). A deep-sea octopus (*Graneledone* cf. *boreopacifica*) as a shell-crushing hydrothermal vent predator. *Journal of Zoology, London*, **252**, 335–41.

Volsøe, A., Knudsen, J., and Rees, W.J. (1962) (trans.). *The cephalopod papers of Japetus Steenstrup*. Danish Science Press, Copenhagen.

Voss, G.L. (1953). A new family, genus and species of myopsid squid from the Florida Keys. *Bulletin of Marine Science of the Gulf and Caribbean*, **2**, 602–9.

Voss, G.L. (1956). A review of the cephalopods of the Gulf of Mexico. *Bulletin of Marine Science of the Gulf and Caribbean*, **6**, 85–178.

Voss, G.L. (1957). *Ornithoteuthis antillarum* Adam, 1957 an ommastrephid squid from the West Indies. *Bulletin of Marine Science of the Gulf and Caribbean*, **7**, 370–8.

Voss, G.L. (1958). The cephalopods collected by the R/V Atlantis during the West Indian cruise of 1954. *Bulletin of Marine Science of the Gulf and Caribbean*, **8**, 369–89.

Voss, G.L. (1962a). A monograph of the Cephalopoda of the North Atlantic I. The family Lycoteuthidae. *Bulletin of Marine Science of the Gulf and Caribbean*, **12**, 264–305.

Voss, G.L. (1962b). South African cephalopods. *Transactions of the Royal Society of South Africa*, **36**, 245–72.

Voss, G.L. (1963). Cephalopods of the Philippine Islands. *Bulletin of the U.S. National Museum*, **234**, 1–180.

Voss, G.L. (1967). The biology and bathymetric distribution of deep-sea cephalopods. *Studies in Tropical Oceanography, Miami*, **5**, 511–35.

Voss, G.L. (1974). On the absence of cuttlefish in the western Atlantic. *The Veliger*, **16**, 367–9.

Voss, G.L. (1976). Two new species of octopods of the genus *Graneledone* (Mollusca: Cephalopoda) from the Southern Ocean. *Proceedings of the Biological Society of Washington*, **88**, 447–58.

Voss, G.L. (1977). Appendix II. Classification of Recent cephalopods. *Symposia of the Zoological Society, London*, No. **38**, 575–9.

Voss, G.L. (1988a). Evolution and phylogenetic relationship of deep-sea cephalopods (Cirrata and Incirrata). In: *The Mollusca* (ed. K.M. Wilbur), Vol. 12. *Paleontology and neontology of cephalopods* (ed. M.R. Clarke and E.R. Trueman), pp. 253–91. Academic Press, San Diego.

Voss, G.L. (1988b). The biogeography of the deep-sea Octopoda. *Malacologia*, **29**, 295–307.

Voss, G.L. and Erdman, D.S. (1959). *Thysanoteuthis rhombus*, large cephalopod new to the western Atlantic. *The Nautilus*, **73**, 23–5.

Voss, G.L. and Pearcy, W.G. (1990). Deep-water octopods (Mollusca; Cephalopoda) of the northeastern Pacific. *Proceedings of the California Academy of Sciences*, **47**, 47–94.

Voss, G.L. and Stephen, S.J. (1992). Family Lycoteuthidae Pfeffer, 1908. *Smithsonian Contributions to Zoology*, No. **513**, 67–71.

Voss, G.L. and Williamson, G. (1971). *Cephalopods of Hong Kong*. Government Press, Hong Kong.

Voss, N.A. (1969). A monograph of the Cephalopoda of the North Atlantic. The family Histioteuthidae. *Bulletin of Marine Science*, **19**, 713–867.

Voss N.A. (1974). Studies on the cephalopod family Cranchiidae. A redescription of *Egea inermis* Joubin, 1933. *Bulletin of Marine Science*, **24**, 939–56.

Voss, N.A. (1980). A generic revision of the Cranchiidae (Cephalopoda; Oegopsida). *Bulletin of Marine Science*, **30**, 365–412.

Voss, N.A. (1985). Systematics, biology and biogeography of the cranchiid cephalopod genus *Teuthowenia* (Oegopsida). *Bulletin of Marine Science*, **36**, 1–85.

Voss, N.A. (1988a). Systematics and zoogeography of cephalopods. *Malacologia*, **29**, 209–14.

Voss, N.A. (1988b). Evolution of the cephalopod family Cranchiidae (Oegopsida). In *The Mollusca* (ed. K.M. Wilbur), Vol. 12. *Paleontology and neontology* (ed. M.R. Clarke and E.R. Trueman), pp. 293–314. Academic Press, San Diego.

Voss, N.A. (1992). Family Promachoteuthidae Naef, 1912. *Smithsonian Contributions to Zoology*, No. **513**, 183–5.

Voss, N.A., Stephen, S.J., and Dong, Zh. (1992a). Family Histioteuthidae Verrill, 1881. *Smithsonian Contributions to Zoology*, No. **513**, 73–91.

Voss, N.A., Stephen, S.J., and Dong, Zh. (1992b). Family Cranchiidae Prosch, 1849. *Smithsonian Contributions to Zoology*, No. **513**, 187–210.

Voss, N.A., Nesis, K.N., and Rodhouse, P.G. (1998). The cephalopod family Histioteuthidae (Oegopsida): systematics, biology, and biogeography. *Smithsonian Contributions to Zoology*, No. **586**, 293–372.

Vovk, A.N. (1983). Food relations of long-finned squid *Loligo pealei* Lesueur, in the northwest Atlantic and its position in the ecosystem. *North Atlantic Fisheries Organisation SCR Scientific Council Studies*, **1**, 65–9.

Ward, D.V. (1972). Locomotory function of the squid mantle. *Journal of Zoology, London*, **167**, 487–99.

Ward, D.V. and Wainwright, S.A. (1972). Locomotory aspects of squid mantle structure. *Journal of Zoology, London*, **167**, 437–49.

Ward, P.D. (1987). *The natural history of* Nautilus. Allen and Unwin, London.

Ward, P.D. (1988). *In search of Nautilus: three centuries of of scientific adventures in the deep Pacific to capture a prehistoric-living-fossil.* Simon and Schuster, New York.

Ward, P.D., and Boletzky, S.v. (1984). Shell implosion depth and implosion morphologies in three species of *Sepia* (Cephalopoda) from the Mediterraean. *Journal of the Marine Biological Association, U.K.*, **64**, 955–66.

Ward, P.D. and Martin, A.W. (1978). On buoyancy of the pearly *Nautilus. Journal of Experimental Zoology*, **205**, 5–12.

Ward, P.D. and Wicksten, M.K. (1980). Food sources and feeding behavior of *Nautilus macromphalus. The Veliger*, **23**, 119–24.

Ward, P.D., Stone, R., Westermann, G., and Martin, A. (1977). Notes on animal weight, cameral fluids, swimming speed, and color polymorphism of the cephalopod *Nautilus pompilius* in the Fiji Islands. *Paleobiology*, **3**, 377–88.

Ward, P.D., Carlson, B., Weekly, M., and Brumbaugh, B. (1984). Remote telemetry of daily vertical and horizontal movement of *Nautilus* in Palau. *Nature, London*, **309**, 248–50.

Watase, S. (1905). Luminous organs of *Abraliopsis*, a new phosphorescent cephalopod from the Japan Sea. *Zoological Magazine, Tokyo*, **17**, 119–22.

Watkinson, G.B. (1909). Untersuchungen über die sogenannten Geruchsorgane der Cephalopoden. *Jenaische Zeitschrift für Naturwissenschaft*, **44**, 353–414.

Weber, W. (1968). Multiple Innervation der Chromatophorenmuskelzellen von *Loligo vulgaris. Zeitschrift für Zellforschung*, **92**, 367–76.

Weber, W. (1970). Zur Ultrastruktur der Chromatophorenmuskelzellen von *Loligo vulgaris. Zeitschrift für Zellforschung*, **108**, 446–56.

Weber, W. (1973). Peculiarities of innervation in chromatophore muscle cells of *Loligo vulgaris. Marine Biology*, **19**, 224–6.

Wei, S.L. and Young, R.E. (1989). Development of symbiotic bacterial bioluminescence in a nearshore cephalopod *Euprymna scolopes. Marine Biology*, **103**, 541–6.

Wells, M.J. (1958). Factors affecting reactions to *Mysis* by newly-hatched *Sepia. Behaviour*, **13**, 96–111.

Wells, M.J. (1960). Proprioception and visual discrimination of orientation in *Octopus. Journal of Experimental Biology*, **37**, 489–99.

Wells, M.J. (1962). Early learning in *Sepia. Symposia of the Zoological Society, London*, No. **8**, 149–69.

Wells, M.J. (1966). Learning in the *Octopus. Symposia of the Society for Experimental Biology*, **20**, 477–507.

Wells, M.J. (1978). *Octopus: Physiology and behaviour of an advanced invertebrate.* Chapman and Hall, London.

Wells, M.J. (1987). Oxygen uptake and the effect of feeding in *Nautilus. The Veliger*, **30**, 69–75.

Wells, M.J. and Wells, J. (1957). The function of the brain of *Octopus* in tactile discrimination. *Journal of Experimental Biology*, **34**, 131–42.

Wells, M.J. and Wells, J. (1959). Hormonal control of sexual maturity in *Octopus. Journal of Experimental Biology*, **36**, 1–33.

Wells, M.J. and Wells, J. (1972). Sexual displays and mating of *Octopus vulgaris* Cuvier and *O. cyanea* Gray and attempts to alter the performance by manipulating the glandular condition of the animals. *Animal Behaviour*, **20**, 293–308.

Wells, M.J. and Young, J.Z. (1965). Split-brain preparations and touch learning in the *Octopus. Journal of Experimental Biology*, **43**, 565–79.

Wells, M.J. and Young, J.Z. (1972). The median inferior frontal lobe and touch learning in *Octopus. Journal of Experimental Biology*, **56**, 381–402.

Wells, M.J. and Young, J.Z. (1975). The subfrontal lobe and touch learning in the octopus. *Brain Research*, **92**, 103–21.

Wells, M.J., Freeman, N.H., and Ashburner, M. (1965). Some experiments on the chemotactile sense of octopuses. *Journal of Experimental Biology*, **43**, 553–63.

Wells, M.J., Hanlon, R.T., Lee, P.G., and Dimarco, F.P. (1988). Respiratory and cardiac performance in *Lolliguncula brevis* (Cephalopoda, Myopsida): The effects of activity, temperature and hypoxia. *Journal of Experimental Biology*, **138**, 17–36.

Wells, M.J., Wells, J., and O'Dor, R.K. (1992). Life at low oxygen tensions: The behaviour and physiology of *Nautilus pompilius* and the biology of extinct forms. *Journal of the Marine Biological Association, U.K.*, **72**, 313–28.

Wentworth, S.L. and Muntz, W.R.A. (1989). Asymmetries in the sense organs and central nervous system of the squid *Histioteuthis. Journal of Zoology, London*, **219**, 607–19.

Wentworth, S.L. and Muntz, W.R.A. (1992). Development of the eye and optic lobe of *Octopus. Journal of Zoology, London*, **227**, 673–84.

Widder, E.A., Latz, M.F., and Case, J.F. (1983). Marine bioluminescence spectra measured with an optical multichannel detection system. *Biological Bulletin, Woods Hole*, **165**, 791–810.

Wildenburg, G. (1991). Ultrastructure of the olfactory organ of *Sepia officinalis* at the end of embryonic deveiopment. In *La seiche – The cuttlefish, 1st International Svmposium on the cuttlefish* Sepia, Caen, 1–3 June 1989 (ed. E. Boucaud-Camou), pp. 294–295. University of Caen.

Wildenburg, G. and Fioroni, P. (1989). Ultrastructure of the olfactory organ during embryonic development and at the hatching stage of *Loligo vulgaris. Journal of Cephalopod Biology*, **1**, 56–70.

Willassen, E. (1986). *Haliphron atlanticus* Steenstrup (Cephalopoda, Octopoda) from the coast of Norway. *Sarsia*, **71**, 35–40.

Willekens, B., Vrensen, G., Jacob, T., and Duncan, G. (1984). The ultrastructure of the lens of the cephalopod *Sepiola*: a scanning electron microscope study. *Tissue and Cell*, **16**, 941–50.

Willey, A. (1902). Contribution to the natural history of the pearly nautilus. In *Zoological results based on material from New Britain, New Guinea, Loyalty Islands and elsewhere, collected during the years 1895, 1896 and 1897*, part IV, pp. 691–830. Cambridge University Press, Cambridge.

Williams, L.W. (1909). *The anatomy of the common squid* Loligo pealii, *Lesueur*. Brill, Leiden.

Williamson, R. (1988). Vibration sensitivity in the statocyst of the northern octopus, *Eledone cirrosa. Journal of Experimental Biology*, **134**, 451–4.

Williamson, R. (1991). The responses of the sensory hair cells in the statocyst of *Sepia. La seiche: The cuttlefish, 1st International Symposium on the cuttlefish* Sepia, *Caen, June 1–3, 1989* (ed. E. Boucaud-Camou), pp. 211–21. University of Caen.

Williamson, R. (1995). The statocysts of cephalopods. *Cephalopod neurobiology: Neuroscience studies in squid, octopus and cuttlefish* (ed. N.J. Abbott, R. Williamson and L. Maddock), pp. 503–20. Oxford University Press.

Williamson, R., and Budelmann, B.U. (1985a).The response of the *Octopus* angular acceleration receptor system to sinusoidal stimulation. *Journal of Comparative Physiology*, **A156**, 403–12.

Williamson, R. and Budelmann, B.U. (1985b). An angular acceleration receptor system of dual sensitivity in the statocyst of *Octopus*. *Experientia*, **41**, 1321–2.

Williamson, R. and Budelmann, B.U. (1991). Convergent inputs to octopus oculomotor neurones demonstrated in a brain slice preparation. *Neuroscience Letters*, **121**, 215–8.

Wilson, D.M. (1960). Nervous control of movement in cephalopods. *Journal of Experimental Biology*, **37**, 57–72.

Wilson, D.P. (1946). A note on the capture of prey by *Sepia officinalis* L. *Journal of the Marine Biological Association, U.K.*, **26**, 421–5.

Wirz, K. (1954). Étude quantitatives sur le système nerveux des Céphalopodes. *Compte rendu hebdomadaire des séances de l'Académie des sciences, Paris*, **238**, 1353–5.

Wirz, K. (1959). Étude biométrique du système nerveux des Céphalopodes. *Bulletin Biologique de la France et de la Belgique*, **93**, 78–117.

Woodhams, P.L. (1975). The ultrastructure of a visuomotor centre in *Octopus*. Thesis, Doctor of Philosophy, University of Sheffield.

Woodhams, P.L. (1977). The ultrastructure of a cerebellar analogue in *Octopus*. *Journal of Comparative Neurology*, **174**, 329–45.

Woodhams, P.L. and Messenger, J.B. (1974). A note on the ultrastructure of the *Octopus* olfactory organ. *Cell and Tissue Research*, **152**, 253–8.

Worms, J. (1983). *Loligo vulgaris*. In *Cephalopod life cycles* Vol. I. *Species accounts* (ed. P.R. Boyle), pp. 143–57. Academic Press, London.

Wormuth, J.H. (1976). The biogeography and numerical taxonomy of the oegopsid squid family Ommastrephidae in the Pacific Ocean. *Bulletin of the Scripps Institution of Oceanography*, **23**, 1–90.

Wormuth, J.H. (1998). Workshop deliberations on the Ommastrephidae: A brief history of their systematics and a review of the systematics, distribution, and biology of the genera *Martialia* Rochebrune and Mabille, 1889, *Todaropsis* Girard, 1890, *Dosidicus* Steenstrup, 1857, *Hyaloteuthis* Gray, 1849, and *Eucleoteuthis* Berry, 1916. *Smithsonian Contributions to Zoology*, No. **586**, 373–83.

Wormuth, J.H., O'Dor, R.K., Balch, N., Dunning, M.C., Forch, E.C., Harman, R.F. *et al.* (1992). Family Ommastrephidae Steenstrup, 1857. *Smithsonian Contributions to Zoology*, No. **513**, 105–19.

Würtz, M. (1982). Osservazioni sull' alimèntazione di *Bathypolypus sponsalis* (P.H. Fischer) (Cephalopoda, Octopoda). *Naturalista Sicilia*, S VI suppl., **1**, 124–6.

Würtz, M., Matricardi, G., and Repetto, N. (1995). Sepiolidae (Mollusca, Cephalopoda) from the lower Tyrrhenian Sea, central Mediterranean. *Bulletin l'Institut océanographique, Monaco*, No spécial, **16**, 35–9.

Yamamoto, M. (1982). Normal stages in the development of the cuttlefish, *Sepiella japonica* Sasaki. *Zoological Magazine, Tokyo (Dobutsugaku Zasshi)*, **91**, 147–57.

Yamamoto, M. (1985). Ontogeny of the visual system in the cuttlefish, *Sepiella japonica*. I. Morphological differentiation of the visual cell. *Journal of Comparative Neurology*, **232**, 347–61.

Yamamoto, M. (1988). Normal embryonic stages of the pygmy cuttlefish, *Idiosepius pygmaeus paradoxus* Ortmann. *Zoological Science*, **5**, 989–98.

Yamamoto, M. and Okutani, T. (1975). Studies on early life history of decapodan Mollusca—V. Systematics and distribution of epipelagic larvae of decapod cephalopods in the southwestern waters of Japan during the summer in 1970. *Bulletin Tokai Regional Fisheries Research Laboratory*, No. **83**, 45–96.

Yamamoto, M., Tasaki, K., Sugawara, Y., and Tonosaki, A. (1965). Fine structure of the octopus retina. *Journal of Cell Biology*, **25**, 345–59.

Yamamoto, M., Takasu, N., and Uragami, I. (1985). Ontogeny of the visual system in the cuttlefish, *Sepiella japonica*. II. Intramembrane particles, histofluorescence, and electrical responses in the developing retina. *Journal of Comparative Neurology*, **232**, 362–71.

Yasumoto, K. (1938). Sex-correlated pattern on the dorsal side of a cuttlefish, *Sepiella japonica* Sasaki. *Venus, Japanese Journal of Malacology*, **8**, 45–7.

Young, J.Z. (1929). Sopra un nuovo organo dei cefalopodi. *Bolletino della Società italiana di biologia sperimentale*, **4**, 1022–4.

Young, J.Z. (1934). Structure of nerve fibres in *Sepia*. *Journal of Physiology, London*, **83**, 27–28*P*.

Young, J.Z. (1936). The giant nerve fibres and epistellar body of cephalopods. *Quarterly Journal of Microscopical Science*, **78**, 367–86.

Young, J.Z. (1938). The functioning of the giant nerve fibre of the squid. *Journal of Experimental Biology*, **15**, 170–85.

Young, J.Z. (1939). Fused neurones and synaptic contacts in the giant nerve fibres of cephlaopods. *Philosophical Transactions of the Royal Society, London*, **B229**, 465–503.

Young, J.Z. (1958). Effect of removal of various amounts of the vertical lobes on visual discrimination by *Octopus*. *Proceedings of the Royal Society, London*, **B149**, 441–62.

Young, J.Z. (1960a). Observations on *Argonauta* and especially its method of feeding. *Proceedings of the Zoological Society, London*, **133**, 471–9.

Young, J.Z. (1960b). The statocysts of *Octopus vulgaris*. *Proceedings of the Royal Society, London*, **B152**, 3–29.

Young, J.Z. (1961). Learning and discrimination in the octopus. *Biological Reviews*, **36**, 32–96.

Young, J.Z. (1962a). The optic lobes of *Octopus vulgaris*. *Philosophical Transactions of the Royal Society, London*, **B245**, 19–58.

Young, J.Z. (1962b). The retina of cephalopods and its degeneration after optic nerve section. *Philosophical Transactions of the Royal Society, London*, **B245**, 1–18.

Young, J.Z. (1962c). Why do we have two brains? In *Interhemispheric relations and cerebral dominance* (ed. V.B. Mountcastle), pp. 7–24. Johns Hopkins Press, Baltimore.

Young, J.Z. (1963a). Light- and dark-adaption in the eyes of some cephalopods. *Proceedings of the Zoological Society, London*, **140**, 255–72.

Young, J.Z. (1963b). Some essentials of neural memory systems. Paired centres that regulate and address the signals of results of action. *Nature, London*, **198**, 626–30.

Young, J.Z. (1963c). The number and sizes of nerve cells in *Octopus vulgaris*. *Proceedings of the Zoological Society, London*, **140**, 229–54.

Young, J.Z. (1964). *A model of the brain*. Clarendon Press, Oxford.

Young, J.Z. (1965a). The central nervous system of *Nautilus*. *Philosophical Transactions of the Royal Society, London*, **B249**, 1–25.

Young, J.Z. (1965b). The buccal nervous system of *Octopus*. *Philosophical Transactions of the Royal Society, London*, **B249**, 27–44.

Young, J.Z. (1965c). The organization of a memory system. *Proceedings of the Royal Society, London*, **B163**, 285–320.

Young, J.Z. (1966). *The memory system of the brain*. Oxford University Press.

Young, J.Z. (1967). The visceral nerves of *Octopus*. *Philosophical Transactions of the Royal Society, London*, **B253**, 1–22.

Young, J.Z. (1970a). The stalked eyes of *Bathothauma* (Mollusca, Cephalopoda). *Journal of Zoology, London*, **162**, 437–47.

Young, J.Z. (1970b). Neurovenous tissues in cephalopods. *Philosophical Transactions of the Royal Society, London*, **B257**, 309–21.

Young, J.Z. (1971). *The anatomy of the nervous system of Octopus vulgaris*. Clarendon Press, Oxford.

Young, J.Z. (1972). The organization of a cephalopod ganglion. *Philosophical Transactions of the Royal Society, London*, **B263**, 409–29.

Young, J.Z. (1973). The giant fibre synapse of *Loligo*. *Brain Research*, 57, 457–60.

Young, J.Z. (1974). The central nervous system of *Loligo* I. The optic lobe. *Philosophical Transactions of the Royal Society, London*, **B267**, 263–302.

Young, J.Z. (1976). The nervous system of *Loligo* II. Suboesophageal centres. *Philosophical Transactions of the Royal Society, Lonodon*, **B274**, 101–67.

Young, J.Z. (1977a). Brain, behaviour and evolution of cephalopods. *Symposia of the Zoological Society, London*, No. **38**, 377–434.

Young, J.Z. (1977b). The nervous system of *Loligo* III. Higher motor centres: the basal supraoesophageal lobes. *Philosophical Transactions of the Royal Society, London*, **B276**, 351–98.

Young, J.Z. (1979). The nervous system of *Loligo* V. The vertical lobe complex. *Philosophical Transactions of the Royal Society, London*, **B285**, 311–54.

Young, J.Z. (1983). The distributed tactile memory system of *Octopus*. *Proceedings of the Royal Society, London*, **B218**, 135–76.

Young, J.Z. (1984). The statocysts of cranchiid squids (Cephalopoda). *Journal of Zoology, London*, **203**, 1–21.

Young, J.Z. (1987). The central nervous system. In Nautilus: *The biology and paleobiology of a living fossil* (ed. W.B. Saunders and N.H. Landman), pp. 215–21. Plenum Press, New York.

Young, J.Z. (1988a). Evolution of the cephalopod brain. In *The Mollusca* (ed. K.M. Wilbur), Vol. 12. *Paleontology and neontology of cephalopods* (ed. M.R. Clarke and E.R. Trueman), pp. 215–28. Academic Press, San Diego.

Young, J.Z. (1988b). Evolution of the cephalopod statocyst. In *The Mollusca* (ed. K.M.Wilbur), Vol. 12, *Paleontology and neontology of cephalopods* (ed. M.R. Clarke and E.R. Trueman), pp. 229–39. Academic Press, San Diego.

Young, J.Z. (1989). The angular acceleration receptor system of diverse cephalopods. *Philosophical Transactions of the Royal Society, London*, **B325**, 198–238.

Young, J.Z. (1991a). Light has many meanings for cephalopods. *Visual Neuroscience*, 7, 1–12.

Young, J.Z. (1991b). Computation in the learning system of cephalopods. *Biological Bulletin, Woods Hole*, **180**, 200–8.

Young, J.Z. (1991c). *Ctenopteryx* the comb-fin squid is related to *Loligo*. *Bulletin of Marine Science*, **49**, 148–61.

Young, J.Z. (1991d). The epistellar body and what followed from its discovery. *Bollettino della Società Italiana di Biologia Sperimentale*, **66**, 607–13.

Young, J.Z. (1995a). Multiple matrices in the memory system of *Octopus*. In *Cephalopod neurobiology: Neuroscience studies in squid, octopus, and cuttlefish* (ed. N.J. Abbott, R. Williamson, and L. Maddock), pp. 431–43. Oxford University Press.

Young, J.Z. (1995b). Emerging ideas on memory in multiple matrices and with cyclic re-excitation. In *Learning and memory* (ed. N. Elsner and R. Menzel), *Proceedings of the 23rd Göttingen Neurobiology Conference 1995*, pp. 61–82. Georg Thieme Verlag, Stuttgart.

Young, R.E. (1964a). The anatomy of the vampire squid. Thesis, Master of Science, University of Southern California.

Young, R.E. (1964b). A note on three specimens of the squid *Lampadioteuthis megaleia* Berry, 1916 (Cephalopoda: Oegopsida) from the Atlantic Ocean, with a description of the male. *Bulletin of Marine Science of the Gulf and Caribbean*, **14**, 444–52.

Young, R.E. (1967). Homology of retractile filaments of vampire squid. *Science*, **156**, 1633–4.

Young, R.E. (1972a). Function of extra-ocular photoreceptors in bathypelagic cephalopods. *Deep-Sea Research*, **19**, 651–60.

Young, R.E. (1972b). The systematics and areal distribution of pelagic cephalopods from the seas off Southern California. *Smithsonian Contributions to Zoology*, No. **97**, 1–159.

Young, R.E. (1972c). Brooding in a bathypelagic octopod. *Pacific Science*, **26**, 400–4.

Young, R.E. (1973a). Evidence for spawning by *Gonatus* sp. (Cephalopoda: Teuthoidea) in the high Arctic Ocean. *The Nautilus*, **87**, 53–8.

Young, R.E. (1973b). Information feedback from photophores and ventral countershading in midwater squid. *Pacific Science*, **27**, 1–7.

Young, R.E. (1975a). Function of the dimorphic eyes in the midwater squid *Histioteuthis dofleini*. *Pacific Science*, **29**, 211–8.

Young, R.E. (1975b). A brief review of the biology of the oceanic squid, *Symplectoteuthis oualaniensis* (Lesson). *Comparative Biochemistry and Physiology*, **52B**, 141–3.

Young, R.E. (1975c). *Leachia pacifica* (Cephalopoda, Teuthoidea): spawning habitat and function of the brachial photophores. *Pacific Science*, **29**, 19–25.

Young, R.E. (1975d). Transitory eye shapes and the vertical distribution of two midwater squids. *Pacific Science*, **29**, 243–55.

Young, R.E. (1977). Ventral bioluminescent countershading in midwater cephalopods. *Symposia of the Zoological Society, London*, No. **38**, 161–190.

Young, R.E. (1978). Vertical distribution and photosensitive vesicles of pelagic cephalopods from Hawaiian waters. *Fishery Bulletin, U.S.A.*, **76**, 583–615.

Young, R.E. (1981). Color of bioluminescence in pelagic organisms. In *Bioluminescence: Current perspectives* (ed. K.H. Nealson), pp. 72–81. Burgess, Minneapolis, Minnesota.

Young, R.E. (1983). Oceanic bioluminescence: An overview of general functions. *Bulletin of Marine Science*, **33**, 829–45.

Young, R.E. (1991). Chiroteuthid and related paralarvae from Hawaiian waters. *Bulletin of Marine Science*, **49**, 162–85.

Young, R.E. (1995). Aspects of natural history of pelagic cephalopods of the Hawaiian mesopelagic-boundary region. *Pacific Science*, **49**, 143–55.

Young, R.E. and Arnold J.M. (1982). The functional morphology of a ventral photophore from the mesopelagic squd, *Abralia trigonura*. *Malacologia*, **23**, 135–63.

Young, R.E. and Bennett, T. (1988). Photophore structure and evolution within the Enoploteuthinae (Cephalopoda). In: *The Mollusca* (ed. K.M. Wilbur), Vol. 12. *Paleontology and neontology of cephalopods* (ed. M.R. Clarke and E.R. Trueman), pp. 241–51. Academic Press, San Diego.

Young, R.E. and Harman, R.F. (1985a). Descriptions of the larvae of three species of the *Onychoteuthis banksii* complex from Hawaiian waters. *The Veliger*, **29**, 313–21.

Young, R.E. and Harman, R.F. (1985b). Early life history stages of enoploteuthin squids (Cephalopoda: Teuthoidea: Enoploteuthidae) from Hawaiian waters. *Vie et Milieu*, **35**, 181–201.

Young, R.E. and Harman, R.F. (1988). 'Larva'. 'paralarva' and 'subadult' in cephalopod terminology. *Malacologia*, **29**, 201–7.

Young, R.E. and Mencher, F.M. (1980). Bioluminescence in mesopelagic squid: Diel color change during counterillumination. *Science*, **208**, 1286–8.

Young, R.E. and Roper, C.F.E. (1968). The Batoteuthidae, a new family of squid (Cephalopoda; Oegopsida) from Antarctic waters. *Antarctic Research Series*, **II**, 185–202.

Young, R.E. and Roper, C.F.E. (1969a). A monograph of the Cephalopoda of the North Atlantic: the family Cycloteuthidae. *Smithsonian Contributions to Zoology*, No. **5**, 1–24.

Young, R.E. and Roper, C.F.E. (1969b). A monograph of the Cephalopoda of the North Atlantic: The family Joubiniteuthidae. *Smithsonian Contributions to Zoology*, No. **15**, 1–10.

Young, R.E. and Roper, C.F.E. (1976). Bioluminescent countershading in midwater animals: evidence from living squid. *Science*, **191**, 1046–8.

Young, R.E. and Roper, C.F.E. (1977). Intensity regulation of bioluminescence during countershading in living midwater animals. *Fishery Bulletin, U.S.A.*, 75, 239–52.

Young, R.E. and Vecchione, M. (1999). Morphological observations on a hatching and a paralarva of the vampire squid, *Vampyroteuthis infernalis* Chun (Mollusca: Cephalopoda). *Proceedings of the Biological Society, Washington*, **112**, 661–6.

Young, R.E., Roper, C.F.E., and Walters, J.F. (1979a). Eyes and extraocular photoreceptors in midwater cephalopods and fishes: Their roles in detecting downwelling light for counterillumination. *Marine Biology*, **51**, 371–80.

Young, R.E., Roper, C.F.E., Mangold, K., Leisman, G., and Hochberg, F.G. (1979b). Luminescence from non-luminescent tissue in oceanic cepahlopods. *Marine Biology*, **53**, 69–77.

Young, R.E., Kampa, E.M., Maynard, S.D., Mencher, F.M., and Roper, C.F.E. (1980). Counterillumination and the upper depth limits of midwater animals. *Deep-Sea Research*, **27A**, 671–91.

Young, R.E., Seapy, R.R., Mangold, K.M., and Hochberg, F.G. (1982). Luminescent flashing in the midwater squids *Pterygioteuthis microlampas* and *P. giardi*. *Marine Biology*, **69**, 299–308.

Young, R.E., Harman, R.F., and Mangold, K.M. (1985). The eggs and larvae of *Brachioteuthis* sp. (Cephalopoda: Teuthoidea) from Hawaiian waters. *Vie et Milieu*, **35**, 203–9.

Young, R.E., Harman, R.F., and Hochberg, F.G. (1989). Octopodid paralarvae from Hawaiian waters. *The Veliger*, **32**, 152–65.

Young, R.E., Mangold, K.M., and Vecchione, M. (1992). The enoploteuthid group of families, Family Enoploteuthidae Pfeffer, 1900, Family Ancistrocheiridae Pfeffer, 1912, Family Pyroteuthidae Pfeffer, 1912. *Smithsonian Contributions to Zoology*, No. **513**, 55–66.

Young, R.E., Burgess, L.A., Roper, C.F.E., Sweeney, M.J., and Stephen, S.J. (1998a). Classification of the Enoploteuthidae, Pyroteuthidae, and Ancistrocheiridae. *Smithsonian Contributions to Zoology*, No. **586**, 239–55.

Young, R.E., Vecchione, M., and Donovan, D.T. (1998b). The evolution of coleoid cephalopods and their present diversity and ecology. *South African Journal of Marine Science*, **20**, 393–420.

Yuki, Y. (1985). Spawning and growth of *Watasenia scintillans* in the south-western Japan Sea. *Report Shimane Prefecture Fishery Experimental Station*, **49**, 1–6.

Zahl, P.A. (1971). Nature's night lights. *National Geographic*, **140**, 45–69.

Zeidler, W. (1987). The scaled-squid *Lepidoteuthis grimaldi* Joubin from southern Australian waters. *Records of the South Australian Museum*, **21**, 157–59.

Zeidler, W. (1989). The pelagic octopus *Tremoctopus violaceus* Delle Chiaje, 1830, from southern Australian waters. *The Veliger*, **32**, 166–70.

Zuev, G.V. (1975). Growth and size of brain during development of oceanic squids. *Soviet Journal of Ecology*, **6**, 347–50.

References for Table 1.1

Adam, W. (1938). *Robsonella* nom. nov. für *Joubinia* Robson, 1929 (Cephalopoda, Octopoda). *Zoologischer Anzeiger*, **121**, 223–224.

Adam, W. (1950). Un Céphalopode nouveau: *Pholidoteuthis boschmai* gen. et sp. nov. *Koninklijke Nederlandse Academie van Wetenschappen*, **53**, 1592–1598.

Adam, W. (1960). Notes sur les céphalopodes, XXIV: contribution à la connaissance de l'Hectocotyle chez les Ommastrephidae. *Bulletin de l'Institut Royal des Sciences Naturelles de Belgique*, **36**, 1–10.

Agassiz, L. (1848). *Nomenclatoris Zoologici Index Universalis, continens nomina systematica classium, ordinum, familiarum et generum Animalium... homonymiis plantarum*. Jent and Gassman, Soloduri.

Appellöf, A. (1890). Teuthologische Beiträge I. *Chtenopteryx* n. g., *Veranya sicula* Krohn, *Calliteuthis* Verrill. *Bergens Museums Aarsberetning*, 1889, 1–34.

Appellöf, A. (1898). Cephalopoden von Ternate, 2: Untersuchungen über *Idiosepius*, *Sepiadarium* und verwandte formen, ein beitrag zur beleuchtung der heklokotylisation und ihrer systematischen bedeutung. *Abhandlungen hrsg. von der Senckenbergischen Naturforchenden Gesellschaft*, **24**, 570–637.

Bather, F.A. (1888). Professor Blake and shell-growth in Cephalopoda. *Geological Magazine*, series 6, **1**, 421–427.

Berry, S.S. (1912). A catalogue of Japanese Cephalopoda. *Proceedings of the Academy of Natural Sciences of Philadelphia*, **64**, 380–444.

Berry, S.S. (1913). *Nematolampas*, a remarkable new cephalopod from the South Pacific. *Biological Bulletin* (Woods Hole), **25**, 208–212.

Berry, S.S. (1916). Cephalopoda of the Kermadec Islands. *Proceedings of the Academy of Natural Sciences of Philadelphia*, **68**, 45–66.

Berry, S.S. (1920). Light production in cephalopods, I. An introductory survey. *Biological Bulletin (Woods Hole)*, **38**, 141–169.

Blainville, H.M.D. de (1824). Mollusques, Mollusca (Malacoz.). In F. Cuvier (ed.) *Dictionnaires des Sciences Naturelles*, **32**, 1–392. Paris and Strasbourg.

Blainville, H.M.D. de (1825). *Manuel de Malacologie et de Conchyliologie*. Paris.

Boletzky, S. v. (1971). *Neorossia* n. g. pro *Rossia (Allorossia) caroli* Joubin, 1902, with remarks on the generic status of *Semirossia* Steenstrup, 1887 (Mollusca: Cephalopoda). *Bulletin of Marine Science*, **21**, 964–969.

Cantraine, F. (1841). Malacologie méditerranéenne. *Nouvelles Mémoire de l'Académie Belgique*, **13**, 1–21.

Chiaie, S. d. (1823–1831). *Memorie sulla storia e notomia degli Animali senze vertebre del regno di Napoli*. Stamperia della Società Tipografica, Napoli.

Chun, C. (1903). *Aus den Tiefen des Weltmeeres*, (2nd edn). Gustav Fischer, Jena.

Chun, C. (1906). System der Cranchien. *Zoologischer Anzeiger*, **31**, 82–86.

Chun, C. (1910). Die Cephalopoden, I. Oegopsida. In *Wissenschaftliche Ergebnisse der Deutschen Tiefsee-Expedition auf dem Dampfer Valdivia 1898–1899*, **18**, 1–401.

Chun, C. (1911). *Cirrothauma*, ein Blinder Cephalopod. Doctoral dissertation, Leipzig.

Chun, C. (1915). Die Cephalopoden, II. Myopsida. In *Wissenschaftliche Ergebnisse der Deutschen Tiefsee-Expedition auf dem Dampfer Valdivia 1898–1899*, **18**, 405–552.

Clarke, M.R. (1980). Cephalopoda in the diet of sperm whales of the southern hemisphere and their bearing on sperm whale biology. *Discovery Reports*, **37**, 1–324.

Cuvier, G.L.C.F.D. (1795). Second mémoire sur l'organisation et les rapports des animaux á sang blanc, dans lequel on traite de la structure des Mollusques et leur division en ordre. *Magazin Encyclopédique, ou Journal des Sciences, des Lettres et des Arts* (Année 1), **2**, 433–449.

Cuvier, G.L.C.F.D. (1797). *Tableau élémentaire de l'histoire naturelle des animaux*. Paris.

Eschricht, D.F. (1836). Cirroteuthis mülleri, eine neue Gattung der Cephalopoden Bildend. *Nova Acta Physico-Medica Academiae Caesarea Leopoldino-Carolinae Naturae Curiosorum*, **18**, 627–634.

Férussac, A. de and Orbigny, A. d' (1834–1848). In *Histoire naturelle générale et particulière des Céphalopodes acétabulifères vivants et fossiles*. J.-B. Baillière, Paris. [See Tillier and Boucher-Rodoni (1993) for dates of publication of sections.]

Filippova, Y.A. (1972). New data on the squids (Cephalopoda: Oegopsida) from the Scotia Sea. *Malacologia*, **11**, 391–406.

Fischer, H. (1896). Note préliminaire sur le *Pterygioteuthis Giardi*, céphalopode nouveau recuelli dans le Cours de l'Expédition scientifique du Talisman (1883). *Journal de Conchyliologie, Paris*, **43**, 205–211.

Fischer, P. (1880–1887). *Manuel de conchyliologie et de paléontologie conchyliologique ou histoire naturelle des mollusques vivants et fossiles*. F. Savy, Paris.

Girard, A.A. (1890). Nota sobre os Céphalopodes de Portugal. *Journal de Sciencias Mathematicas, Physicas e Naturaes*, series 2, **1**, 200–205.

Gonzalez, A.F., Guerra, A., Pascual, S. and Briand, P. (1998). *Vulcanoctopus hydrothermalis* gen. et sp. nov. (Mollusca, Cephalopoda): an octopod from a deep-sea hydrothermal vent site. *Cahiers de Biologie Marine*, **39**, 169–184.

Gray, J.E. (1849). *Catalogue of the Mollusca in the collection of the British Museum, I: Cephalopoda Antepedia*. British Museum (Natural History), London.

Grimpe, G. (1916). *Chunioteuthis*: Eine neue Cephalopoden-gattung. *Zoologischer Anzeiger*, **46**, 349–359.

Grimpe, G. (1921). Teuthologische Mitteilungen, VII: Systematische Übersicht der Nordsee-cephalopoden. *Zoologischer Anzeiger*, **52**, 297–305.

Grimpe, G. (1922). Systematische Übersicht der europäischen Cephalopoden. *Sitzungsberichte der Naturforschenden Gesellschaft zu Leipzig*, **45**, 36–52.

Haeckel, E.H.P.A. (1866). *Generelle Morphologie der Organismen*. Band 2, *Allgemeine Entwicklungsgeschichte der Organismen*. Reiner, Berlin.

Haeckel, E.H.P.A. (1896). *Systemischer Phylogenie. Entwurf eines Naturlichen Systems der Organismen auf Grund ihrer Stammesgeschichte. Zweiter Theil; Systemische Phylogenie der Wirbellosen Thiere (Invertebrata)*. Georg Reimer, Berlin.

Hoyle, W.E. (1885a). Diagnoses of new species of Cephalopoda collected during the cruise of H.M.S. *Challenger*, I: The Octopoda. *Annals and Magazine of Natural History*, series 5, **15**, 222–236.

Hoyle, W.E. (1885b). Diagnoses of new species of Cephalopoda collected during the cruise of H.M.S. *Challenger*, II: The Decapoda. *Annals and Magazine of Natural History*, series 5, **16**, 181–203.

Hoyle, W.E. (1885c). Narrative of the Challenger Expedition. *Report on the Scientific Results of the Voyage of H.M.S.* Challenger. *1873–76, Narrative*, **1**, 269–274.

Hoyle, W.E. (1886). Report on the Cephalopoda collected by H.M.S. *Challenger* during the years 1873–1876. *Report on the Scientific Results of the Voyage of H.M.S.* Challenger. *1873–1876, Zoology*, **16**, 1–245.

Hoyle, W.E. (1904). Reports on the Cephalopoda. *Bulletin of the Museum of Comparative Zoology, Harvard*, **43**, 1–72.

Ishikawa, C. (1913). Einige Bermerkungen über den Leuchtenden Tintenfisch, *Watasea* (*Abraliopsis* der Autoren) *scintillans* Berry, aus Japan. *Zoologischer Anzeiger*, **43**(4), 162–172.

Issel, R. (1908). Diagnosi preliminari di nuovo genre e di due nuove specie di Cefalopodi allartenenti all fam. Cranchiidae raccolti della R. Nave 'Liguria'. *Monitore Zoologico Italiano*, **19**, 102–4.

Joubin, L. (1895). Céphalopodes recuellis dans l'estomac d'un cachalot, capture aux Iles Açores. *Comptes Rendu des Séances de l'Académie des Sciences*, **121**, 1172–1174.

Joubin, L. (1896). Observations sur divers Céphalopodes, premièrre note: *Abraliopsis pfefferi* (nov. gen. et spec.). *Bulletin de la Société Scientifique et Médicale de l'Oest*, **5**, 19–35.

Joubin, L. (1898a). Observations sur divers Céphalopodes, quatrième note: *Grimalditeuthis richardi. Bulletin de la Société Zoologique de France*, **20**, 21–28.

Joubin, L. (1898b). Note sur une nouvelle familie de Céphalopodes. *Annales des Sciences Naturelles (Zoologie)*, series 8, **6**, 279–292.

Joubin, L. (1918). Étude préliminaires sur les Céphalopodes recueillis au cours des croisières de S.A.S. le Prince de Monaco, 6e note: *Vitreledonella richardi* Joubin. *Bulletin de l'Institut Océanographique, Monaco*, No. 340, 1–40.

Joubin, L. (1919). Études préliminaire sur les Céphalopodes recueillis au cours des croisières de S.A.S. le Prince de Monaco, 7e note: *Cycloteuthis sirventi* nov. gen. et sp. *Bulletin de l'Institut Océanographique*, Monaco, No. 351, 1–7.

Joubin, L. (1931). Notes préliminaires sur les Céphalopodes des croisières du DANA (1921–1922), 3e partie. *Annales de l'Institut Océanographique, Monaco*, **10**, 169–211.

Joubin, L. (1933). Notes préliminaires sur les Céphalopodes des croisières du DANA (1921–1922), 4e partie. *Annales de l'Institut Océanographique, Monaco*, **13**, 1–49.

Keferstein, W. (1862–1866). Malacozoa Cephalophora. In H. G. Bronn, *Die Klassen und Ordnungen des Thier-Reichs wissenschaftlich dargestellt in Wort und Bild*, **3** (2), 519–1500. C. F. Winter, Leipzig and Heidelberg.

Lamarck, J.B.P.A. de M. de (1798). Extrait d'un mémoire sur le genre de la Séche, du Calmar et du Poulpe, vulgairement nommés, Polypes de mer. *Bulletin des Sciences, par la Société Philomatique de Paris*, **17**, 129–131.

Lamarck, J.B.P.A. de M. de (1799). Prodrome d'une nouvelle classification des coquilles. *Mémoires de la Société d'Historie Naturelle de Paris*, **1**, 63–91.

Leach, W.E. (1817). Synopsis of the Orders, Families and Genera of the Class Cephalopoda. In W.E. Leach and R.P. Nodder, *The zoological miscellany; being descriptions of new, or interesting animals*, **3** (30), 137–41.

Lesueur, C.A. (1821). Descriptions of several new species of cuttle-fish. *Journal of the Academy of Natural Sciences of Philadelphia*, **2**, 86–101.

Lichtenstein, H.C. (1818). Von den Sepien mit Krallen. *Isis, oder Encyclopädische Zeitung*, 1818 (9), 1591–1592.

Linnaeus, C. (1758). *Systema Naturae per regna tria naturae, secundum classes, ordines, genera, species cum characteribus, diferentiis, synonymis, locis* (10th edn). Holmiae (Stockholm).

Massy, A.L. (1907). Preliminary notice of new and remarkable cephalopods from the south-west coast of Ireland. *Annals and Magazine of Natural History*, series 7, **20**, 377–384.

Naef, A. (1912a). Teuthologische Notizen, 2: Die Gattungen der Sepioliden. *Zoologischer Anzeiger*, **39**, 244–248.

Naef, A. (1912b). Teuthologische Notizen, 1: Die Familien der Myopsiden. *Zoologischer Anzeiger*, **39**, 241–244.

Naef, A. (1916). Systematische uebersicht der Mediterranen Cephalopoden. *Pubblicazioni della Stazione Zoologica di Napoli*, **1**, 11–19.

Naef, A. (1921). Das System der dibranchiaten Cephlopoden und die mediterranen Arten dersellben. *Mitteilungen aus der Zoologischen Station zu Neapel*, **22**, 527–542.

Naef, A. (1922). *Die Fossilen Tintenfische*. Gustav Fischer, Jena.

Naef, A. (1923). Die Cephalopoden. *Fauna e Flora de Golfo di Napoli, Monograph* **35**, 149–863.

Nesis, K.N. (1972). Two new species of gonatid squid from the North Pacific. *Zoologicheskii Zhurnal*, **51**, 1300–1307.

Nesis, K.N. (1980). Taxonomic position of *Chiroteuthis famelica* Berry (Cephalopoda, Oegopsida). *Biuletyn Moskovskoe Obshchestvo Ispytatelei Prirody, Otdel Biologicheskii*, **85**, 59–66.

Nesis, K.N. and Nikitina, I.V. (1986a). New genus and species of squid of the family Neoteuthidae (Cephalopoda, Oegopsida) from the south-eastern part of the Pacific Ocean. *Zooloicheskii Zhurnal*, **65**, 290–294.

Nesis, K.N. and Nikitina, I.V. (1986b). A new family of abyssal squids (Cephalopoda, Oegopsida) from the south-eastern Atlantic. *Zooloicheskii Zhurnal*, **65**, 47–54.

Norman, M. (1992). *Ameloctopus litoralis* n. gen., n. sp. (Cephalopoda: Octopodidae), a new shallow water octopus from tropical Australian waters. *Invertebrate Taxonomy*, **6**, 567–582.

Odhner, N.H. (1923). Die Cephalopoden. *Further Zoological Results of the Swedish Antarctic Expedition 1901–03*, **1**, 1–7.

Okada, Y.K. (1927). Contribution à l'étude des céphalopodes lumineaux (notes préliminaires), IV: *Ommastrephes volatilis* Sasaki est une forme lumineuse; établissement d'un nouveau genre: *Ornithoteuthis. Bulletin de l'Institut Océanographique de Monaco*, **494**, 13–16.

Orbigny, A. d' (1834) (1834–1847). Mollusques. In *Voyages dans l'Amérique Méridionale*, 5(3), 1-758. Paris. (See Sherborn and Griffin, 1934, for publication dates of sections).

Orbigny, A. d' (1841). In A. de Ferrusac and A. d'Orbigny 1834–1848.

Orbigny, A. d' (1845). *Mollusques vivants et fossiles ou description de toutes les espèces de coquilles et de mollusqus classées suivant leur distribution géologique et géographique*. Gide, Paris.

Owen, R. (1835). Mollusca-Cephalopoda. Nov. gen.-Rossia. (Owen.). In *John Ross, Appendix to the narrative of a second voyage in search of a north-west passage and of a residence in the Arctic regions during the years 1829–1833 including the reports of James Clark Ross*, pp. xcii-xcix. Webster, London.

Palacio, F.J. (1978). *Vosseledone charrua*: a new patagonian cephalopod (Octopodidae) with notes on related genera. *Bulletin of Marine Science*, **28**, 282–296.

Pfeffer, G. (1884). Die Cephalopoden des Hamburger Naturhistorischen Museums. *Abhandlungen aus dem Gebiete der Naturwissenschaften, Hamburg*, **8**, 1–30.

Pfeffer, G. (1900). Synopsis der oegopsiden Cephalopoden. *Mitteilungen aus dem Naturhistorischen Museum in Hamburg*, **17**, 147–198.

Pfeffer, G. (1908a). Teuthologische Bemerkungen. *Mitteilungen aus dem Naturhistorischen Museum in Hamburg*, **25**, 289–295.

Pfeffer, G. (1908b). Die Cephalopoden. *Nordisches Plankton*, **2**, 9–116.

Pfeffer, G. (1912). Die Cephalopoden der Plankton-Expedition: Zugleich eine Monographische Übersicht der Oegopsiden Cephalopoden. In V.A.C. Hensen, *Ergebnisse der Plankton-Expedition der Humboldt-Stiftung*, **2F**(a), 1–815.

Posselt, H.J. (1891). *Todarodes sagittatus* (Lmk.) Stp., En anatomisk Studie med Bemaerkninger om Slaegtskabsforholdet mellem Ommatostrephfamiliens Genera. *Videnskabelige Meddelelser danske naturhistoriske Forening i Kjøbenhavn*, 1890, 301–359.

Prosch, V. (1847). Nogle nye Cephalopoder, beskrevne og anatomisk undersogte. *Konglige Danske Videnskabernes Selskabs Skrifter*, series 5, **1**, 53–72.

Rafinesque, C.S. (1814). Classe Malacosia – Le Mollusques. In *Précis des découvertes et travaux somologiques de C.S. Rafinesque-Schmaltz... zoologie et en botanique entre 1800 et 1814*, pp. 28–30. Palermo.

Rafinesque, C.S. (1815). *Analyse de la nature ou tableau de l'univers et des corps organisés*. Palermo.

Rehder, H.A. (1945). A new genus and species of squids from the Philippines. *Proceedings of the Biological Society of Washington*, **58**, 21–26.

Robson, G.C. (1925). On *Mesonychoteuthis*, a new genus of oegopsid Cephalopoda. *Annals and Magazine of Natural History*, series 9, **16**, 272–277.

Robson, G.C. (1928). Notes on the Cephalopoda, VI: On *Grimpella*, a new genus of Octopoda, with remarks on the classification of the Octopodidae. *Annals and Magazine of Natural History*, series 10, **2**, 108–114.

Robson, G.C. (1929). *A monograph of the Recent Cephalopoda, I: The Octopodinae.* British Museum (Natural History), London.

Robson, G.C. (1930). Cephalopoda, I. Octopoda. *Discovery Reports*, **2**, 371–402.

Robson, G.C. (1932). *A monograph of the Recent Cephalopoda, II: The Octopoda.* British Museum (Natural History), London.

Rochebrune, A.T. d' and Mabille, J. (1889). Mollusques. In *Mission Scientifique du Cap Horn*, 1882–1858, **6** (2), H1–H143.

Rüppell, E. (1844). Intorno ad alcuni cefalopodi del mare di Messina: Lettera del Dr Eduardo Rüppell di Frankfort sul Meno al Prof. Anastasio Cocco. *Giornale del Gabinetto Letteràrio di Messina*, 5, 129–135.

Sasaki, M. (1920). Report on Cephalopoda collected during 1906 by the United States Bureau of Fisheries Steamer 'Albatross' in the Northwestern Pacific. *Proceedings of the United States National Museum*, **57**, 163–203.

Sherborn, C.D. and Griffin, F.J. (1934). On the dates of publication of the natural history portions of Alcide d'Orbigny's "Voyage Amérique Méridionale". *Annals and Magazine of Natural History*, series 10, **13**, 130–4.

Steenstrup, J. (1856). Hectocotyldannelsen hos Octopodslaegterne *Argonaute* og *Tremoctopus* oplyst ved lagttagelse af lilgnende Dannelser hos Blaeksrutterne i Almindelighed. *Kongelige Danske Videnskabernes Selskabs Skrifter*, series 5, **4**, 185–216.

Steenstrup, J. (1857a). Prof. Steenstrup foreviste og beskrev nogle nye Blaekspruter, *Dosidicus Eschrictii* Stp. og *Onychoteuthis* (?). *Videnskabelige Meddelelser danske Naturhistoriske Forening i Kjøbenhavn*, 1856–1857, 120–121. (English translation in Volsøe *et al.* 1962).

Steenstrup, J. (1857b). Oplysninger om Atlanterhavets Colossale Blaekspruttter. *Forhandlinger ved de Skandinaviske Naturforskeres*, 7, 182–185. (English translation in Volsøe *et al.* 1962).

Steenstrup, J. (1857c). Oplysning om en ny Art Blaekspruter, Dosidicus Eschrichtii. *Oversigt over det Kongelige Danske Videnskabernes Selskabs Forhandlinger*, 1857, 11–14. (English translation in Volsøe *et al.* 1962).

Steenstrup, J. (1859). Prof. Stp. gav en meddelelse om to nye Cephalopodstaegten *Bolitaena* og *Haliphron*. *Videnskabellige Meddelelser Danske Naturhistoriske Forening i Kjøbenhavn*, 1858, 180–183. (English translation in Volsøe *et al.* 1962).

Steenstrup, J. (1861). Overblik over de i Kjøbenhavns Museer opbevarede Blaekspruter fra det aabne Hav (1860–61). *Oversigt over det Konglelige Danske Videnskabernes Selskabs Forhandlinger*, 1861, 69–86. (English translation in Volsøe *et al.* 1962).

Steenstrup, J. (1880). De Ommatostrephagtige Blaekspruter indbyrdes Forhold, en Oreintering. *Oversigt over det Kongelige Danske Videnskabernes Selskabs Forhandlinger*, 1880, 73–110. (English translation in Volsøe *et al.* 1962).

Steenstrup, J. (1881). *Sepiadarium* og *Idiosepius* to nye Slaegter af Sepiernes Familie; Med Bemaerkninger om de to beslaegtede former *Sepioloidea* D'Orb. og *Spirula* Lmk. *Kongelige Danske Videnskabernes Selskabs Skrifter*, series 6, **1**, 213–242. (English translation in Volsøe *et al.* 1962).

Steenstrup, J. (1887a). Notae Teuthologicae, 6: Species Generis Sepiolae Maris Mediterranei. *Oversigt over det Kongelige Danske Videnskabernes Selskabs Forhandlinger*, 1887, 47–66. (English translation in Volsøe *et al.* 1962.)

Steenstrup, J. (1887b). Notae Teuthologicae, 7. *Oversigt over det Kongelige Danske Videnskabernes Selskabs Forhandlinger*, 1887, 67–126. (English translation in Volsøe *et al.* 1962.)

Taki, I. (1961). On two new eledonid octopods from the Antarctic Sea. *Journal of the Faculty of Fisheries and Animal Husbandry, Hiroshima University*, 3(2), 297–316.

Thiele, J. (1915). In Chun (1915).

Thiele, J. (1920). Die Cephalopoden der deutschen Südpolar-Expedition 1901–1903. *Deutsche Südpolar-Expedition 1901–1903*, **16**, 433–465.

Tillier, A. and Boucher-Rodoni, R. (1993). Férussac and d'Orbigny's 'Histoire naturelle générale et particulière des Céphalopodes acétabulifères': Dates of publication of plates and text. *The Nautilus*, **107** (3), 97–103.

Troschel, F.H. (1857). Bermerkkungen über die Cephalopoden von Messina. *Archiv für Naturgeschichte*, **23**, 40–76.

Tryon, G. (1879). *Manual of conchology; structural and systematic, I. Cephalopoda.* Tryon, Philadelphia.

Vecchione, M. and Young, R.E. (1998). The Magnapinnidae, a newly discovered family of oceanic squid (Cephalopoda: Oegopsida). *South African Journal of Marine Science*, **20**, 429–437.

Verrill, A.E. (1879). Notice of recent additions to the marine fauna of the eastern coast of North America, No. 7. *American Journal of Science and Arts*, **18**, 468–470.

Verrill, A.E. (1880). The cephalopods of the north-eastern coast of America, Part I: The gigantic squids (*Architeuthis*) and their allies; with observations on similar large species from foreign localities. *Transactions of the Connecticut Academy of Sciences*, **5**, 177–257.

Verrill, A.E. (1881). The cephalopods of the north-eastern coast of America, Part II: The smaller cephalopods including the 'squids' and the octopi, with other allied forms. *Transactions of the Connecticut Academy of Sciences*, **5**, 259–446.

Verrill, A.E. (1881a). Notice of the remarkable marine fauna occupying the outer banks of the southern coast of New England, No. 2. *American Journal of Science*, series 3, **22**, 292–303.

Verrill, A.E. (1881b). Report on the cephalopods, and on some additional species dredged by the U.S. Commission Steamer 'Fish Hawk', during the season of 1880. *Bulletin of the Museum of Comparative Zoology, Harvard*, **8**, 99–116.

Verrill, A.E. (1883). Supplementary Report on the 'Blake' cephalopods. *Bulletin of the Museum of Caomparative Zoology, Harvard*, **11**, 105–115.

Verrill, A.E. (1884). Second catalogue of the Mollusca, recently added to the fauna of the New England coast and the adjacent parts of the Atlantic, consisting mostly of deepsea species, with notes on others previously recorded. *Transactions of the Connecticut Academy of Sciences*, **6**, 139–294.

Verrill, A.E. (1896). The Opisthoteuthidae: A remarkable new family of deepsea Cephalopoda, with remarks on some points in molluscan morphology. *American Journal of Science*, series 4, **2**, 74–80.

Volsøe, A., Knudsen, J., and Rees, W.J. (trans.) (1962). *The cephalopod papers of Japetus Steenstrup.* Danish Science Press, Copenhagen.

Voss, G.L. (1953). A new family, genus and species of myopsid squid from the Florida Keys. *Bulletin of Marine Science of the Gulf and Caribbean*, **2**, 602–609.

Voss, G.L. (1955). The Cephalopoda obtained by the Harvard-Havana Expedition off the coast of Cuba in 1938–39. *Bulletin of Marine Science of the Gulf and Caribbean*, **5**, 81–115.

Voss, G.L. (1959). The cephalopods collected by the R/V Atlantis during the West Indian cruise of 1954. *Bulletin of Marine Science of the Gulf and Caribbean*, **8**, 369–89. (Publication was in 1959, but the title page gives 1958.)

Voss, G.L. (1971). Cephalopods collected by the R/V *John Elliott Pillsbury* in the Gulf of Panama in 1967. *Bulletin of Marine Science*, **21**, 1–34.

Voss, G.L. (1988). Evolution and phylogenetic relationships of deep-sea octopods (Cirrata and Incirrata). In *The Mollusca*, Vol. 12: *Paleontology and neontology of cephalopods* (ed. M.R. Clarke and E.R. Trueman) pp. 253–276. Academic Press, San Diego.

Young, J.Z. (1989). The angular acceleration receptor system of diverse cephalopods. *Philosophical Transactions of the Royal Society, London*, **B325**, 198–238.

Young, R.E. and Roper, C.F.E. (1968). The Batoteuthidae, a new family of squid (Cephalopoda; Oegopsida) from Antarctic waters. *Antarctic Research Series*, **2**, 185–202.

Young, R.E. and Roper, C.F.E. (1969). A monograph of the Cephalopoda of the North Atlantic: The family Cycloteuthidae. *Smithsonian Contributions to Zoology*, No. 5, 1–24.

Abbreviations

a.	anterior
ab.	aboral
ac.	acetabulum of sucker
acc.	accessory lobe
aff.	afferent
am.	amacrine cell
amoeb.	amoebocyte
an.	anus
anticr.	anticrista (1–7)
aor.a.	anterior aorta
aper.	aperture
aq.	aquiferous pore
aq.d.	dorsal aquiferous pore
aq.v.	ventral aquiferous pore
ar.	arm [arms I–IV]
art.	artery
art.ceph.	cephalic artery
art.gon.	gonadal artery
art.ir.	artery of iris
art.pal.	pallial artery
art.pe.	pedal artery
art.pe.le.	left pedal artery
au.	point of autotomy in arm
au.li.	autotomy line
aur.	auricle
ax.	axon
ax.sh.	axon sheath
ax.syn.f.gi.3	axon to synapse with third order giant fibre
ax.ter.	axon terminal
b.	basal lobe
b.a.	anterior basal lobe
b.a.a.	anterior portion of the anterior basal lobe
b.a.a.sp.	spine of anterior anterior basal lobe
b.a.l.	lateral anterior basal lobe
b.a.med.a.	anteromedial lobule of anterior basal lobe
b.a.p.	posterior portion of the anterior basal lobe
b.a.p.sp.	spine of posterior anterior basal lobe
b.d.	dorsal basal lobe
b.d.a.	anterior dorsal basal lobe
b.d.p.	posterior dorsal basal lobe
b.inf.	infrabasal lobe
b.int.	interbasal lobe
b.l.	lateral basal lobe
b.med.	median basal lobe
be.	beak
be.low.	lower beak
be.u.	upper beak
bl.	blood vessel
blad.	bladder
bod.cil.	ciliary body
bod.w.	body wall
bod.wh.	white body
bol.	bolster
br.	brachial lobe
br.a.	anterior brachial lobe
br.intra.-pe.con.	intrabrachial–pedal connective
br.-mag.con.	brachiomagnocellular connective
br.-pe.con.	brachiopedal connective
br.po.	postbrachial lobe
br.-pv.con.	brachiopalliovisceral connective
br.-pv.con.div.d.	dorsal division, brachiopalliovisceral connective
br.-pv.con.div.ven.	ventral divison, brachiopalliovisceral connective
bran.ht.	branchial heart
bran.ht.le.	left branchial heart
bridge ax.int.	interaxonic bridge
bu.	bundle(s)
buc.	buccal
buc.-br.con.	buccobrachial connective
buc.cav.	buccal cavity
buc.fun.	buccal funnel
buc.i.	inferior buccal lobe
buc.i.gan.	inferior buccal ganglion
buc.int.con.	interbuccal connective
buc.l.	lateral buccal lobe
buc.m.	buccal mass
buc.mem.	buccal membrane
buc.p.	posterior buccal lobe
buc.s.	superior buccal lobe
c.br.int.	interbrachial commissure
c.br.sup.	suprabrachial commissure
c.bu.	commissural bundles
c.ch.p.	posterior chromatophore commissure
c.mag.a.	anterior magnocellular commissure
c.mag.p.	posterior magnocellular commissure
c.mag.ven.	ventral magnocellular commissure
c.opt.	optic commissure
c.pe.med.	medial pedal commissure
c.pe.p.	posterior pedal commissure
c.pe.sup.	suprapedal commissure
c.ped.	peduncle lobe commissure
c.suboes.a.	anterior suboesophageal commissure
caec.	caecum
caec.spi.	spiral caecum
can.wat.le.	left water canal
can.wat.rt.	right water canal
cap.	capillary
caps.	capsule
cart.	cartilage

cart.ceph.	cephalic cartilage
cart.ceph.vac.	vacuolated (reticulate) cephalic cartilage
cart.fin	fin cartilage
cart.retic.	reticulate cartilage
cart.troc.	trochlear cartilage
cart.vac.	vacuolated (reticulate) cartilage
cav.	cavity
cav. coel.	coelemic cavity
cb. a.	anterior chamber
ce.am.	amacrine cell
ce.am.in.	inner amacrine cell
ce.am.lar.	large amacrine cell
ce.bi.	bipolar cell
ce.cent.	cell with centrifugal nerve fibres
ce.gi.	giant cell
ce.gi.1–3	giant cell, first-, second- or third-order
ce.gi.1, ax	axon of giant cell 1
ce.gi.1, core cen.	central core of giant cell 1
ce.gi.acc.	accessory giant cell
ce.gi.den.a.	anterior dendrite, giant cell
ce.gi.1, den.l.	lateral dendrite, giant cell 1
ce.gi.1, den.ven.	ventral dendrite, giant cell 1
ce.gi.2a–3	giant cell(s) 2a–3
ce.gl.	glial cell
ce.gr.lay.out.	outer granule cell layer
ce.gr.lay.in.	inner granule cell layer
ce.lar.	large cell
ce.mult.	multipolar cell
ce.mult.lar.	large multipolar cell
ce.mult.sm.	small multipolar cell
ce.n.	nerve cell
ce.palis.	cell of palisade layer
ce.sec.	secretory cell
ce.vis.2	second-order visual cell
ce.vis.2,lar.	large second-order visual cell
ce.vis.2,out	second-order visual cell
ce.vis.3	third-order visual cell
cer.-br.con.	cerebrobrachial connective
cer.-buc.con.	cerebrobuccal connective
cer.-buc.con.l.	lateral cerebrobuccal connective
cer.-buc.con.med.	median cerebrobuccal connective
cer.cor.	cerebral cord
cer.cor.cen.	central zone of cerebral cord
cer.cor.lam.	laminated zone of cerebral cord
cer.cor.plex.	plexiform zone of cerebral cord
cer.l.	lateral cerebral lobe
cer.-subrad.con.	cerebrosubradular connective
cer.-subrad.con. (+n.sal.p.)	cerebrosubradular connective (plus posterior salivary gland nerve)
ch.	chromatophore
ch.a.	anterior chromatophore lobe
ch.p.	posterior chromatophore lobe
chamb.	chamber
chamb.a.	anterior chamber
chamb.br.	brachial chamber
chamb.d.	dorsal chamber
chamb.elong.	elongated chamber
chamb.epi.sub.	subepidermal chamber
chamb.in.	inner chamber
chamb.out.	outer chamber
chamb.sh.	chamber of shell
chi.	chitin
chr.(chrom.)	chromatophore
cil.	cilia
cir.	cirrus (cirri)
co.	collateral
co.ax.f.gi.2a	axonic collaterals of second order giant fibre
co.(f.gi.3)ter.m.	terminal mass collateral of third order giant fibre
co.tip	tip of collateral
coll.	collagen
core dist.	distal core
core out.	outer core
corn.	cornea
cort.	cortical
cr.	crista
cr.long.	longitudinal crista of statocyst
cr.trans.	transverse crista of statocyst
cr.trans.a.	anterior transverse crista of statocyst
cr.trans.p.	posterior transverse crista of statocyst
cr.v.	vertical crista of statocyst
cran.	cranium
cro.	crop
cryst.	crystalloid
cup.	cupula
cut.	cuticle
d.	dorsal
d.l.	dorsolateral lobe
d.w.	dorsal wall
den.	dendrite
der.	dermis
duc.dig.	digestive ducts
duc.kid.le.	left kidney duct
ec.	ectoderm
egg case out.	outer egg case
end.	endolymph of statocyst
ep.	epithelium
ep.bod.	epithelial body
epi.	epidermis
epistel.bod.	epistellar body
eros.	erosion
esoph	oesophagus
ey.	eye
ey.chamb.a.	anterior chamber of eye
ey.le.	left eye
ey.ri.	right eye
f.	fibre
f.crossing pv.	fibre crossing palliovisceral lobe
f.gi.	giant fibre

f.gi.1–3	giant fibres, first, second or third order fibres
f.gi.2a–e	giant fibres, second order a–e
f.gi.2a (branch)	branch, second order giant fibre synapse with third order giant fibre
f.gi. 2a, co.ax.	axonic collateral, giant fibre 2a
f.gi.a. (2, 3)	anterior giant fibre, second, third order
f.gi.a 3	anterior giant fibre, third order
f.syn.post.	postsynaptic fibre
f.syn.pre.	presynaptic fibre
fas.	fascia
fas.d.	dorsal fascia
fas.med.	media fascia
fas.ven.	ventral fascia
fil.	filament
filt.	filter
fin rud.	fin rudiment
fl.	fin lobe (of central nervous system)
for.	foramen
fov.	fovea
fr.	frontal
fr.i.	inferior frontal lobe
fr.i.l.	lateral inferior frontal lobe
fr.i.med.	median inferior frontal lobe
fr.p.	posterior frontal lobe
fr.s.	superior frontal lobe
fr.s.l.	lateral superior frontal lobe
fr.s.med.	median superior frontal lobe
fun.	funnel
fun. fol.	funnel fold
fr.s.p.	posterior superior frontal lobe
fun.ap.	funnel aperture
funl.	funnel lobe
g.acc.	accessory gland
g.dig.	digestive gland
g.opt.	optic gland
g.ovid.	oviducal gland
g.sal.	salivary gland
g.sal.a.	anterior salivary gland
g.sal.p.	posterior salivary gland
g.sal.p.duc.	duct of posterior salivary gland
g.subman.	submandibular gland
gan.br.	brachial ganglion
gan.br.intra.	intrabrachial ganglion
gan.buc.i.	inferior buccal ganglion
gan.buc.mem.	ganglion of buccal membrane
gan.buc.s.	superior buccal ganglion
gan.cer.	cerebral ganglion
gan.gas.	gastric ganglion
gan.ol.	olfactory ganglion
gan.opt.	optic ganglion
gan.pe.	pedal ganglion
gan.stel.	stellate ganglion
gan.subrad.	subradular ganglion
gan.suc.	ganglion of sucker
gan.visc.	visceral ganglion
gfl.	giant fibre lobe (of stellate ganglion)
gill le.	left gill
gl.fib.	fibrous glia
gl.nuc.	nucleus of glial cell
glad.	gladius
gol.she.	golden sheen region
gr.in.	inner granule cell layer
gr.out.	outer granule cell layer
gui.li.	light guide
h.(h.1–5)	hamulus (hamuli 1–5)
hear.	heart
hec.	hectocotylus
ho.	hood
hy.	hyaline shield
i.	inferior
in.ring	inner ring of sucker
inf.	infundibulum of sucker
intest.	intestine
intra.	intra
ir.	iris
irid.	iridophore
irid.ven.	ventral iridophores (of brachial chamber)
je.	'jelly'
je.orb.	orbital 'jelly'
juxt.	juxtaganglionic tissue
ke.	keel
kid.	kidney
Kol.	Kölliker's canal
Kol.ap.	interal aperture of Kölliker's canal
l.	lateral
lab.i.	inferior labial lobe
le.	left
len.	lens
leuc.	leucophore
li.	light
li. gui.	light guides
lip in.	inner lip
low.	lower
m.buc.	buccal mass
m.suboes.	suboesophageal mass
m.suboes.a.	anterior suboesophageal mass
m.suboes.p.	posterior suboesophageal mass
m.supr.	supraoesophageal mass
mac.	macula of statocyst
mac.neg.	macula neglecta
mac.neg.i.	inferior macula neglecta
mac.neg.s.	superior macula neglecta
mac.prin.	macula princeps
mag.	magnocellular lobe
mag.d.	dorsal magnocellular lobe
mag.p.	posterior magnocellular lobe
mag.ven.	ventral magnocellular lobe
mant.	mantle

mant.ap.	mantle aperture
mant.con.	mantle connective
med.	median, medial, middle
mem.buc.	buccal membrane
mem.prot.	protective membrane
mem.she.	membrane sheath
mem.troc.	trochlear membrane
mo.	mouth
mus.	muscle
mus.a.(I, II)	anterior muscle (I–II)
mus.ac.-br.	acetabulobrachial muscle (peduncular muscle)
mus.arm	arm muscle
mus.bol.med.	median muscle linking bolsters
mus.buc.	muscles of buccal mass
mus.cil.	ciliary muscle
mus.circ.	circular muscle
mus.conj.	conjunctive muscle
mus.conj.a.	anterior conjunctive muscle
mus.conj.a.(I, II)	anterior conjunctive muscles (I, II)
mus.conj.s.	superior conjunctive muscle
mus.i.(I, II)	inferior muscle (I, II)
mus.d.ven.	dorsoventral muscle
mus.d.ven.d.	dorsal dorsoventral muscle
mus.d.ven.ven.	ventral dorsoventral muscle
mus.dep.	depressor muscle
mus.ey.	eye muscle
mus.f.ra.	radial muscle fibres
mus.fin	fin muscle
mus.hel.	helical muscle
mus.i. (I, II)	inferior muscle (I, II)
mus.int.rad.su.	interradular suport muscle
mus.ir.dil.	dilator muscle of iris
mus.lev.	levator muscle
mus.long.	longitudinal muscle
mus.long.ke.sw.	longitudinal muscle, swimming keel
mus.long.sf.	superficial longitudinal muscle
mus.man.	mandibular muscles
mus.mant.	mantle muscle
mus.obl.	oblique muscle
mus.obl.a.i.	inferior anterior oblique muscle
mus.obl.i.p.	posterior inferior oblique muscle
mus.obl.s.a.	anterior superior oblique muscle
mus.obl.s.p.	posterior superior oblique muscle
mus. oc.	oculomotor muscle
mus.oc.a. I	anterior oculomotor muscle I
mus.oc.conj.a. I–II	anterior conjunctive oculomotor muscles I–II
mus.oc.i. II	inferior oculomotor muscle II
mus.oc.p. I–II	posterior oculomotor muscles I–II
mus.oc.s. I–II	superior oculomotor muscles I–III
mus.oc.troc. I–III	trochlear oculomotor muscles I–III
mus.p. I–II	posterior muscle I–II
mus.rad.protr.	radular protractor muscle
mus.rad.retr.	retractor muscle of radula
mus.rad.-sal.pap.	radula–salivary papilla muscle
mus.rad.su.	radula support muscle
mus.rect.a.	anterior rectus muscle
mus.rect.p.	posterior rectus muscle
mus.rect.s.	superior rectus muscle
mus.retr.	retractor muscle
mus.retr.h.	head retractor muscle
mus.retr.h.a.	anterior head retractor muscle
mus.retr.i.	inferior retractor muscle
mus.retr.rad.d.	dorsal radular retractor muscle
mus.retr.rad.ven.	ventral radular retractor muscle
mus.s. I–III	superior muscles I–III
mus.sal.pap.	salivary papilla muscle
mus.sheet con.	connecting sheet muscle
mus.sph.	sphincter muscle
mus.st.	muscle fibres of statocyst
mus.ta.	tail muscle
mus.ton.a.-p.	anteroposterior tongue muscle
mus.ton.med.	median tongue muscle
mus.trans.	transverse muscle
mus.trans.d.	dorsal transverse muscle
mus.trans.rad.su.	transverse muscle of radular supports
mus.trans.trab.	trabeculae of transverse muscle
mus.trans.ven.	ventral transverse muscle
mus.troc. I–II	trochlear muscle I–II
n.	nerve
n.antorb.i.	inferior antorbital nerve
n.antorb.s.	superior antorbital nerve
n.ax.	axial nerve
n.br.	brachial nerve
n.br.1–4	brachial nerves to arms I–IV
n.br.-mag.	brachiomagnocellular nerve
n.buc.	buccal nerve
n.col.	collar nerve
n.cord.mus.intra.	intramuscular nerve cord
n.cr.	crista nerve
n.cr.a.	anterior crista nerve
n.cr.med.	median crista nerve
n.cr.p.	posterior crista nerve
n.cr.-st.	static–crista nerve
n.epistel.	epistellar nerve
n.f.	nerve fibre(s)
n.f.cent.	centrifugal nerve fibre
n.fil.	filament nerve
n.fin	fin nerve
n.fun.	funnel nerve
n.fun.a.	anterior funnel nerve
n.fun.med.	median funnel nerve
n.fun.p.	posterior funnel nerve
n.g.opt.	optic gland nerve
n.ho.	hood nerve
n.ir.	iris nerve
n.lab.	labial nerve
n.mac	macula nerve
n.mac.st.d.	dorsal statica macula nerve

n.mac.st.ven.	ventral statica macula nerve
n.max.	maxillary nerve
n.oc.a.	anterior oculomotor nerve
n.oc.a.i.	inferior anterior oculomotor nerve
n.oc.a.s.	superior anterior oculomotor nerve
n.oc.p.	posterior oculomotor nerve
n.oc.p.i.	inferior posterior oculomotor nerve
n.oc.p.s.	superior posterior oculomotor nerve
n.oc.+n.op.p.s.	superior posterior ophthalmic nerve + oculomotor nerve
n.ol.	olfactory nerve
n.op.i.	inferior ophthalmic nerve
n.op.a.s.	superior anterior ophthalmic nerve
n.op.med.s.	superior medial ophthalmic nerve
n.op.p.i.	inferior posterior ophthalmic nerve
n.op.p.s.	superior posterior ophthalmic nerve
n.op.p.s.r.a.	anterior root of the superior posterior ophthalmic nerve
n.op.p.s.r.p.	posterior root of the superior posterior ophthalmic nerve
n.opt.	optic nerve
n.orb.po.	postorbital nerve
n.pal.	pallial nerve
n.pal.-mag.	pallial to magnocellular nerve
n.post-oc.tent.	postocular tentacle nerve
n.pre-oc.tent.	preocular tntacle nerve
n.prel.	prelingual nerve
n.ret.1–3	retinal nerve fibre (three fibre types 1–3)
n.retr.h.	head retractor nerve
n.retr.h.a.	anterior head retractor nerve
n.retr.h.p.	posterior head retractor nerve
n.retr.i.a.	anterior inferior retractor nerve
n.retr.i.med.	median inferior retractor nerve
n.retr.i.p.	posterior inferior retractor nerve
n.sal.duc.p.	posterior salivary gland duct nerve
n.sal.p.	posterior salivary nerve
n.st.	static nerve
n.st.cr.	crista statica nerve
n.st.-b.a.	anterior basal to static nerve
n.st.-b.med.	fibres from static nerve to median basal lobe
n.st.-ped.	static nerve fibres to peduncle lobe
n.stell.	stellar nerve
n.stell.r.d.	dorsal root stellar nerve
n.stell.r.ven.	ventral root stellar nerve
n.subped.	subpedunculate nerve
n.symp.	sympathetic nerve
n.tent.	tentacle nerve
n.tent.p.-ocul.	postocular tentacle nerve
n.tent.pre-ocul.	preocular tentacle nerve
n.vc.a.	anterior vena cava nerve
n.ves.pho.	nerve of photosensitive vesicles
n.ves.pho.d.	nerve of dorsal photosensitive vesicles
n.ves.pho.ven.	nerve of ventral photosensitive vesicles
n.visc.	visceral nerve
n.visc.d.r.	dorsal root of visceral nerve
n.visc.ven.r.	ventral root of visceral nerve
Nissl	Nissl substance
nuc.	nucleus
nuc.ph.	photocyte nucleus
ocul.	ocular
oes.	oesophagus
ol.	olfactory lobe
ol.pap.	olfactory papilla
ol.pit	olfactory pit
opt.	optic lobe
optic st.	optic stalk
or.	organ
or.cb.a.	anterior chamber organ
or.comp.	composite organ
or.ol.	olfactory organ
orb. po. and n.co.	postorbital and collar nerves to magnocellular lobe
orif.	orifice
out.	outer
ov.	ovary
ovi.	oviduct
p.	posterior
p.suboes.m.	posterior suboesophageal mass
pa.buc.l.	lateral buccal palp
pa.buc.med.	median buccal palp
palis.	palisade layer
pap.	papilla
parav.	paravertical tissue
pe.	pedal lobe
pe.a.	anterior pedal lobe
pe.l.	lateral pedal lobe
pe.l.a.	anterior lateral pedal lobe
pe.l.p.	posterior lateral pedal lobe
pe.p.	posterior pedal lobe
pe.-pv.con.	pedopalliovisceral connective
ped.	peduncle lobe
ped.sp.	peduncle spine
ped.sub.	subpedunculate lobe
perik.	perikaryon
peril.	perilymph
phot.(ph.)	photophore
phot.anal	anal photophore
phot.fin	fin photophore
phot.ocul.	ocular photophore
phot.ocul.ven.	ventral ocular photophore
phot.ves.	photosensitive vesicles
photog.	photogenic core
pig.	pigment
pig.s.	pigment screen
pil.cort.e.	extracortical neuropil
plex.	plexiform zone
plex.in.	inner plexiform zone
prec.	precommissural lobe

prel.pro.	prelingual process
pup.	pupil
PV	photosensitive vesicle
pv.	palliovisceral lobe
pv.a.	anterior palliovisceral lobe
pv.a.ven.	anteroventral palliovisceral lobe
pv.cen.	central palliovisceral lobe
pv.l.p.	posterolateral palliovisceral lobe
rad.('rad.')	radula (vestigial radula)
rad.can.	radular canal
rad.grv.	radular groove
rad.rib.	radular ribbon
rad.sac	radular sac
rad.su.	radular support
rad.tth.	radular teeth
rect.	rectum
ref.b.	basal reflector
ref.l.	lateral reflector
refl.	reflector
refl.b.	basal reflector
refl.cup.p.	posterior cup reflector
refl.l.	lateral reflector
refl.-some	reflectosome
ret.	retina
ret.acc.	accessory retina
ret.cen.	central part of retina
ret.dist.	distal part of retina
ret.prox.	proximal part of retina
retic.	reticulate
rh.	rhabdom
rhi.	rhinophore
ri.(rt.)	right
rl.1–4	radial layers of plexiform zone
ros.	rostrum
sac 1–3	subradular sac 1–3
sac aq.	sac of aquiferous pore
sac d.	dorsal sac
sac fil.	filament sac
sac gon.	gonadial sac
sac hy.	hyaline sac
sac in.	inner sac
sac out.	outer sac
sac p.	posterior sac of statocyst
sac rad.	radular sac
sac tent.	tentacular sac
sal.duc.a.	anterior salivary gland duct
sal.duc.p.	posterior salivary duct
sal.pap.	papilla of the duct from the posterior salivary glands
sal.pap.shd tth.	toothed shield of salivary papilla
sep.	septum
sept.trans.	transverse septum
sh.	shell
silv.	silvery
sin.bl.	blood sinus
sin.opt.	optic sinus
sin.suborb.	suborbital sinus
siph.	siphuncle
spe.	spermatophore
spe.out.tunic	outer tunic of spermatophore
spermpkt.	sperm packet
sph.	sphincter
st.	statocyst
sth.(stl.)	statolith
stom.	stomach
subfr.	subfrontal lobe
suboes.	suboesophageal lobes
suboes.m.a.	anterior suboesophageal mass
subped.	subpedunculate lobe
subv.	subvertical lobe
suc.	sucker
suc.disc	disc of cuticle shed from sucker
suc.inf.	infundibulum of sucker
suc.ped.	peduncle of sucker
superf.	superficial
supraoes.m.	supraoesophageal mass
sw.	swimming
syn.	synapse
t.con.	connective tissue
t.con.ab.	aboral connective tissue
t.con.der.	dermal connective tissue
t.con.f.	connective tissue fibre
t.con.oral	oral connective tissue
t.fo.len.	lens-forming tissue
t.gel.	gelatinous tissue
t.len.	lens-forming tissue
t.phg.	photogenic tissue
t.retic.	reticulate tissue (vacuolated)
t.sub.-buc.	subbuccal tissue
t.subped.	subpedunculate tissue
t.vc.	vena cava neurosecretory tissue
tap.	tapetum
tend.	tendon
tent.	tentacle
tent.dig.	digital tentacle
tent.in.	inner tentacles
tent.oc.p.	postocular tentacle
tent.oc.pr.	preocular tentacle
test.	testis
tis.vac.	vacuolated tissue
tl.1–4	tangential layers of the plexiform zone
ton.	tongue
tr.b.a.-b.med.	anterior basal–median basal lobe tract
tr.b.a.-pe.	anterior basal–pedal lobe tract
tr.b.a.-pe.a.	anterior basal–anterior pedal lobe tract
tr.b.a.-pe.l.	anterior basal–lateral pedal lobe tract

tr.b.a.-ped.	anterior basal—peduncle lobe tract
tr.b.a.a.-opt.	anterior anterior basal–optic lobe tract
tr.b.a.a.-pe.l.	anterior anterior basal–lateral pedal lobe tract
tr.b.a.p.-opt.	posterior anterior basal–optic lobe tract
tr.b.a.p.-pe.l.	posteior anterior basal–lateral pedal lobe tract
tr.b.l.-ch.a.	lateral basal–anterior chromatophore lobe tract.
tr.b.l.-ch.p.	lateral basal–posterior chromatophore lobe tract
tr.b.med.-b.a.	median basal–anterior vasal tract
tr.b.med.-pe.	median basal–pedal lobe tract
tr.b.med.-pe.a.	median basal–anterior pedal lobe tract
tr.b.med.-pe.l.	median basal–lateral pedal lobe tract
tr.b.med.-pe.p.	median basal–posterior pedal lobe tract
tr.b.med.-pv.	median basal–pallioviscera lobe tract
tr.b.med.-pv.cen.	median basal–central pallioviscera lobe tract
tr.b.med.-pv.(ext.)	median basal–pallioviscera lobe tract (external)
tr.b.med.-pv. +mag.	median basal–pallioviscera and magnocellular lobe tract
tr.b.med.-visc.	median basal–visceral lobe tract
tr.br.-mag.	brachiomagnocellular tract (part of brachio magnocellular connective)
tr.br.-opt.	brachial–optic lobe tract
tr.br.-pe.	brachial–pedal lobe tract
tr.br.-pv.	brachial–pallioviscera lobe tract
tr.cer.	cerebral tract
tr.cer.-br.	cerebrobrachial tract
tr.ch.int.	interchromatophore tract
tr.fr.i.-b.med.	inferior frontal–median basal lobe tract
tr.fr.i.-fr.s.	inferior frontal–superior frontal lobe tract
tr.fr.i.-pv.	inferior frontal–pallioviscera lobe tract
tr.fr.int.	interfrontal tract
tr.fr.s.-subv.	superior frontal–subvertical lobe tract
tr.fr.s.-v.	superior frontal–vertical lobe tract
tr.mag.-pv.	magnocellular–pallioviscera lobe tract
tr.mag.d.-pe.	dorsal magnocellular–pedal lobe tract
tr.mag.d.-pv.	dorsal magnocellular–pallioviscera lobe tract
tr.mag.ven.-mag.d.	ventral magnocellular–dorsal magnocellular lobe tract
tr.mag.ven.-pe.	ventral magnocellular–pedal lobe tract
tr.mag.ven.-pv.	ventral magnocellular–pallioviscera lobe tract
tr.opt.	optic tract
tr.opt.-b.a.	optic–anterior basal lobe tract
tr.opt.-b.a.(cruc.)	optic–anterior basal lobe tract (crossed)
tr.opt.-b.a.(rect.)	optic–anterior basal lobe tract (uncrossed)
tr.opt.-b.a.a.d.	optic–dorsal anterior anterior basal lobe tract
tr.opt.-b.a.l.	optic–lateral anterior basal lobe tract
tr.opt.-b.a.p.	optic–posterior anterior basal lobe tract
tr.opt.-b.a.ven.	optic–ventral anterior basal lobe tract
tr.opt.-fr.s.	optic–superior frontal lobe tract
tr.opt.l.-mag.	lateral optic-magnocellular lobe tract
tr.opt.-mag.	optic–magnocellular lobe tract
tr.opt.-mag.l.	optic–lateral magnocellular lobe tract
tr.opt.-mag.med.	optic–median magnocellular lobe tract
tr.opt.-mag.ven.	optic–ventral magnocellular lobe tract
tr.opt.-ped.	optic–peduncle lobe tract
tr.opt.-prec.	optic–precommissural lobe tract
tr.opt.-prec.ven.	optic–ventral precommissural lobe
tr.opt.l.-mag.	lateral optic–magnocellular lobe tract
tr.opt.med.-mag.	median optic–magnocellular lobe tract
tr.pe.-pv.	pedal–pallioviscera lobe tract
tr.pe.l.p.-pe.a.	posterior lateral pedal–anterior pedal lobe tract
tr.pe.p.-fl.	posterior pedal–fin lobe tract
tr.pe.p.-pv.	posterior pedal–pallioviscera lobe tract
tr.pe.p.-pv.a.	posterior pedal–pallioviscera lobe tract
tr.ped.-b.a.	peduncle–anterior basal lobe tract
tr.ped.-pe.l.	peduncle–lateral pedal lobe tract
tr.prec.-b.med.	precommissural–median basal lobe tract
tr.prec.-br.	precommissural–brachial lobe tract
tr.prec.-mag.	precommissural–magnocellular lobe tract
tr.prec.-pe.	precommissural–pedal lobe tract
tr.prec.-pv.	precommissural–pallioviscera lobe tract
tr.pv.-fr.i.	pallioviscera–inferior frontal lobe tract
tr.pv.a.-fl.	anterior pallioviscera–fin lobe tract
tr.suboes.-supraoes.	suboesophageal–supraoesophageal tract
tr.subv.-fr.s.	subvertical–superior frontal lobe tract
tr.subv.-opt.	subvertical–optic lobe tract
tr.v.-fr.s.	vertical–superior frontal lobe tract
tr.v.-subv.	vertical–subvertical lobe tract
tr.vas.-circum.pv.	circumpallioviscera–vasomotor tract
trab.	trabeculae
trans.	transverse
tth	tooth (or teeth)
u.	upper
v.	vertical lobe
v.cen.	central region of vertical lobe
v.pe.	peripheral region of vertical lobe
vac.	vacuolated
vas.	vasomotor lobe
vas.d.	dorsal vasomotor lobe
vas.ven.	ventral vasomotor lobe
vasdef.	vas deferens
vc.	vena cava
vc.neur.sec.	neurosecretory tissue
ve.	vein
ve.br.sf.	suuperficial brachial nerve
ve.tent.	tentacular vein
ve.tent. sf.	superficial tentacular veins
ven.	ventral
ventr.	ventricle
ves.	vesicle(s)
ves.pho.	photosensitive vesicles
ves.pho.a.	anterior photosensitive vesicles
ves.pho.cen.	central photosensitive vesicles
ves.pho.d.	dorsal photosensitive vesicles
ves.pho.med.	middle photosensitive vesicles
ves.pho.p.	posterior photosensitive vesicles

ves.pho.scre.	screen for photosensitive vesicles
ves.pho.ven.	ventral photosensitive vesicles
ves.pho.ven.a.	anteroventral photosensitive vesicles
ves.pho.ven.mid.	mid-ventral photosensitive vesicles
ves.sem.	seminal vesicle
visc.	visceral lobe
web	interbrachial web
web d.	dorsal web
web ven.	ventral web
win.	window
yo.	yolk
z.fro.	frontier zone of optic lobe
z.ra.	zone of radial columns of medulla
z.tan.	zone of tangential bundles
z.tr.opt.	zone of optic tract bundles in optic lobe

INDEX

The manufacturer's authorised representative in the EU for product safety is Oxford University Press España S.A. of El Parque Empresarial San Fernando de Henares, Avenida de Castilla, 2 - 28830 Madrid (www.oup.es/en or product.safety@oup.com). OUP España S.A. also acts as importer into Spain of products made by the manufacturer.
Printed and bound by CPI Group (UK) Ltd, Croydon, CR0 4YY
14/09/2026
02216298-0003